INTRODUCTION TO FINITE STRAIN THEORY FOR CONTINUUM ELASTO-PLASTICITY

WILEY SERIES IN COMPUTATIONAL MECHANICS

Series Advisors:

René de Borst
Perumal Nithiarasu
Tayfun E. Tezduyar
Genki Yagawa
Tarek Zohdi

Introduction to Finite Strain Theory for Continuum Elasto-Plasticity	Hashiguchi and Yamakawa	October 2012
Nonlinear Finite Element Analysis of Solids and Structures: Second edition	De Borst, Cris eld, Remmers and Verhoosel	August 2012
An Introduction to Mathematical Modeling: A Course in Mechanics	Oden	November 2011
Computational Mechanics of Discontinua	Munjiza, Knight and Rougier	November 2011
Introduction to Finite Element Analysis: Formulation, Veri cation and Validation	Szabó and Babuška	March 2011

INTRODUCTION TO FINITE STRAIN THEORY FOR CONTINUUM ELASTO-PLASTICITY

Koichi Hashiguchi
Kyushu University, Japan
Yuki Yamakawa
Tohoku University, Japan

WILEY
A John Wiley & Sons, Ltd., Publication

Reprinted with corrections May 2013

This edition rst published 2013

Registered office
John Wiley & Sons Ltd, The Atrium, Southern Gate, Chichester, West Sussex, PO19 8SQ, United Kingdom

For details of our global editorial of ces, for customer services and for information about how to apply for permission to reuse the copyright material in this book please see our website at www.wiley.com.

Library of Congress Cataloging-in-Publication Data

Hashiguchi, Koichi.
Introduction to nite strain theory for continuum elasto-plasticity / Koichi Hashiguchi, Yuki Yamakawa.
p. cm.
Includes bibliographical references and index.
ISBN 978-1-119-95185-8 (cloth)
1. Elastoplasticity. 2. Strains and stresses. I. Yamakawa, Yuki. II. Title.
TA418.14.H37 2013
620.1′1233–dc23

2012011797

A catalogue record for this book is available from the British Library.

Print ISBN: 9781119951858

Typeset in 10/12pt Times by Aptara Inc., New Delhi, India

Contents

Preface

The first author of this book recently published the book *Elastoplasticity Theory* (2009) which addresses the fundamentals of elasto-plasticity and various plasticity models. It is mainly concerned with the elasto-plastic deformation theory within the framework of the hypoelastic-based plastic constitutive equation. It has been widely adopted and has contributed to the prediction of the elasto-plastic deformation behavior of engineering materials and structures composed of solids such as metals, geomaterials and concretes. However, the hypoelastic-based plastic constitutive equation is premised on the additive decomposition of the strain rate (symmetric part of velocity gradient) into the elastic and the plastic strain rates and the linear relation between the elastic strain rate and stress rate. There is no one-to-one correspondence between the time-integrations of the elastic strain rate and stress rare and the energy may be produced or dissipated during a loading cycle in hypoelastic equation. Therefore, the elastic strain rate does not possess the elastic property in the strict sense so that an error may be induced in large-deformation analysis and accumulated in the cyclic loading analysis. An exact formulation without these defects in the infinitesimal elasto-plasticity theory has been sought in order to respond to recent developments in engineering in the relevant fields, such as mechanical, aeronautic, civil, and architectural engineering.

There has been a great deal of work during the last half century on the finite strain elasto-plasticity theory enabling exact deformation analysis up to large deformation, as represented by the epoch-making works of Oldroyd (1950), Kroner (1959), Lee (1969), Kratochvil (1971), Mandel (1972b), Hill (1978), Dafalias (1985), and Simo (1998). In this body of work the multiplicative decomposition which is the decomposition of the deformation gradient into the elastic and plastic parts, introducing the intermediate configuration obtained by unloading to the stress free-state, was proposed and the elastic part is formulated as a hyperelastic relation based on the elastic strain energy function. Further, the Mandel stress, the work-conjugate plastic velocity gradient with the Mandel stress, the plastic spin, and various physical quantities defined in the intermediate configuration have been introduced. The physical and mathematical foundations for the exact finite elasto-plasticity theory were established by 2000 and the constitutive equation based on these foundations – the hyperelastic-based plastic constitutive equation – has been formulated after 2005. However, a textbook on the hyperelastic-based finite elasto-plasticity theory has not been published to date.

Against this backdrop, the aim of this book is to give a comprehensive explanation of the finite elasto-plasticity theory. First, the classification of elastoplasticity theories from the view-point of the relevant range of deformation will be given and the prominence of the hyperelastic-based finite elastoplasticity theory will be explained in Introduction. Exact knowledge of the basic mechanical ingredients – finite strain (rate) tensors, the Lagrangian and Eulerian tensors, the objectivity of the tensor and the systematic definitions of pull-back and push-forward operations, the Lie derivative and the corotational rate – is required in the formulation of finite strain theory. To this end, descriptions of the physical quantities and relations in the embedded (convected) coordinate system, which turns into the curvilinear coordinate system under the deformation of material, are required, since their physical meanings can be captured clearly by observing them in a coordinate system which not only moves but also deforms and rotates with material itself. In other words, the essentials of continuum mechanics cannot be captured without the incorporation of the general curvilinear coordinate system, although numerous books with 'continuum mechanics' in their title and confined to the rectangular coordinate system have been published to date. On the other hand, knowledge of the curvilinear coordinate system is not required for users of finite strain theory. It is sufficient for finite element analyzers using finite strain theory to master the descriptions in the rectangular coordinate system and the relations between Lagrangian and Eulerian tensors through the deformation gradient, for instance. This book, which aims to impart the exact finite strain theory, gives a comprehensive explanation of the mathematical and physical fundamentals required for continuum solid mechanics, and provides a description in the general coordinate system before moving on to an explanation of finite strain theory.

In addition to the above-mentioned issues on the formulation of the constitutive equation, the formulation and implementation of a numerical algorithm for the state-updating calculation are of utmost importance. The state-updating procedure in the computational analysis of elasto-plasticity problems usually requires a proper algorithm for numerical integration of the rate forms of the constitutive laws and the evolution equations. The return-mapping scheme has been developed to a degree of common acceptance in the field of computational plasticity as an effective state-updating procedure for elasto-plastic models. In the numerical analysis of boundary-value problems, a consistent linearization of the weak form of the equilibrium equation and use of the so-called consistent (algorithmic) elasto-plastic tangent modulus tensor are necessary to ensure effectiveness and robustness of the iterative solution procedure. A Fortran program for the return-mapping and the consistent tangent modulus tensor which can readily be implemented in finite element codes is provided along with detailed explanation and user instructions so that readers will be able to carry out deformation analysis using the finite elasto-plasticity theory by themselves.

Chapters 1–7 were written by the first author, Chapter 8 by both authors, and Chapters 9 and 10 by the second author in close collaboration, and the computer programming and calculations were performed by the second author based on the theory formulated by both authors in Chapter 8. The authors hope that readers of this book will capture the fundamentals of the finite elasto-plasticity theory and will contribute to the development of mechanical designs of machinery and structures in the field of engineering practice by applying the theories addressed in this book. A reader is apt to give up reading a book if he encounters matter which is difficult to understand. For this reason, explanations of physical concepts in elasto-plasticity are given, and formulations and derivations/transformations for all equations are given without abbreviation for Chapters 1–8 for the basic formulations of finite strain theory. This is not a complete book on the finite elasto-plasticity theory, but the authors will be quite satisfied if it provides a foundation for further development of the theory by stimulating the curiosity of young researchers and it is applied widely to the analyses of engineering problems in practice. In addition, the authors hope that it will be followed by a variety of books on the finite elasto-plasticity.

The first author is deeply indebted to Professor B. Raniecki of the Institute of Fundamental Technological Research, Poland, for valuable suggestions and comments on solid mechanics, who has visited several times Kyushu University. His lecture notes on solid mechanics and regular private communications and advice have been valuable in writing some parts of this book. He wishes also to express his gratitude to Professor O.T. Bruhns of the Ruhr University, Bochum, Germany, and Professor H. Petryk of the Institute of Fundamental Technological Research, Poland, for valuable comments and for notes on their lectures on continuum mechanics delivered at Kyushu University. The second author would like to express sincere gratitude to Professor Kiyohiro Ikeda of Tohoku University for valuable suggestions and comments on nonlinear mechanics. He is also grateful to Professor Kenjiro Terada of Tohoku University for providing enlightening advices on numerical methods for finite strain elasto-plasticity. He also thanks Dr. Ikumu Watanabe of National Institute for Materials Science, Japan, for helpful advices on numerical methods for finite strain elasto-plasticity. The enthusiastic support of Dr. Keisuke Sato of Terrabyte Inc., Japan, and Shoya Nakaichi, Toshimitsu Fujisawa, Yosuke Yamaguchi and Yutaka Chida at Tohoku University in development and implementation of the numerical code is most appreciated. The authors thank Professor S. Reese of RWTH Aachen University, Professor J. Ihlemann and Professor A.V. Shutov of Chemnitz University, Germany, and Professor M. Wallin of Lund University, Sweden for valuable suggestions and for imparting relevant articles on finite strain theory to the authors.

The authors express their deep gratitude to Professor Genki Yagawa of Toyo University (Emeritus Professor, The University of Tokyo) for the appreciation and recommendation of this book to the Wiley Series in Computational Mechanics. Finally, they would like to state that the enthusiastic support of Eric Willner, Executive Commissioning Editor at Wiley, was indispensable for the publication of this book.

Koichi Hashiguchi
Yuki Yamakawa
February 2012

Series Preface

The series on *Computational Mechanics* is a conveniently identifiable set of books covering interrelated subjects that have been receiving much attention in recent years and need to have a place in senior undergraduate and graduate school curricula, and in engineering practice. The subjects will cover applications and methods categories. They will range from biomechanics to fluid-structure interactions to multiscale mechanics and from computational geometry to meshfree techniques to parallel and iterative computing methods. Application areas will be across the board in a wide range of industries, including civil, mechanical, aerospace, automotive, environmental and biomedical engineering. Practicing engineers, researchers and software developers at universities, industry and government laboratories, and graduate students will find this book series to be an indispensible source for new engineering approaches, interdisciplinary research, and a comprehensive learning experience in computational mechanics.

Over the last three decades, the finite strain theory for elasto-plasticity has been extensively developed to provide very precise descriptions of large elasto-plastic deformations. These theoretical developments have been augmented with robust computational methods, which provide accurate solutions to the corresponding boundary-value problems. Deep mathematical and physical knowledge of related continuum mechanics is required to learn the theory, which is difficult for students, engineers and even researchers in the field of applied mechanics, to capture in depth.

This book is one of the first introductory books to address finite strain elasto-plasticity theory, where the mathematical and physical foundations are comprehensively described. In particular, the representation of physical quantities in convected (curvilinear) coordinate systems is employed, which is required for the substantial interpretation of basic concepts, such as pull-back and push-forward operations and convected (Lie) derivatives, that is, the general objective rate in continuum mechanics. Furthermore, all the mathematical derivations and transformations of equations are shown without any abbreviation, explaining the numerical method for the finite elasto-plasticity in detail.

In addition, a Fortran program for the stress-update algorithm based on the return-mapping scheme and the consistent (algorithmic) tangent modulus tensor, which can readily be implemented in finite element codes, is appended with a detailed explanation and user instructions, so that the readers will be able to implement the numerical analysis on the basis of the finite elasto-plasticity theory without difficulty.

It is our hope that the readers of this book will contribute to the improvement of mechanical design of machinery and structures in the field of engineering by adopting and widely using the basic ideas of the finite elasto-plasticity theory and the corresponding numerical methods introduced in this text.

Introduction

Prominence of the finite strain elasto-plasticity theory

This book addresses the *finite strain elasto-plasticity theory*, abbreviated as the *finite elasto-plasticity*. Then, the prominence of finite strain elasto-plasticity theory and the necessity of its incorporation to deformation analysis is first reviewed by comparing with the *infinitesimal elasto-plasticity theory*, abbreviated as the *infinitesimal elastoplasticity*, prior to the detailed explanation of the finite elastoplasticity in the subsequent chapters.

Elastoplasticity theory is classified from the view point of relevant range of deformation as follows:

A) *Infinitesimal elastoplasticity*
The infinitesimal strain tensor defined by the symmetric part of displacement gradient tensor is additively decomposed into an elastic and a plastic parts, while there does not exist the distinction between the reference and the current configurations in which the infinitesimal strain tensor is based. The spin of material is described by the skew-symmetric part of the rate of displacement gradient tensor. This theory is limited to the description of infinitesimal deformation and rotation. Meanwhile, for the elastic part, the hyperelastic constitutive equation with a stored energy function can be formulated, which provides the one-to-one correspondence between the stress tensor and the infinitesimal elastic strain tensor. Therefore, the return-mapping and the consistent (algorithmic) tangent modulus tensor can be employed in numerical calculations under infinitesimal deformation and rotation.

B) *Finite elastoplasticity*
The frameworks of the theories describing the finite elastoplastic deformation and rotation are classified further as follows:

B-1) *Hypoelastic-based plasticity*
It is premised on the following assumptions, which would be called the *Hill–Rice approach*.

i) Deformation and rotation are described by the strain rate and the spin tensors, that is, the symmetric and the skew-symmetric parts, respectively, of velocity gradient tensor.
ii) The strain rate and the spin tensors are additively decomposed into elastic and plastic parts. Here, it should be noticed that the additive decomposition is derived from the multiplicative decomposition of the deformation gradient tensor

only when an elastic deformation is infinitesimal, whereas the multiplicative decomposition is the rigorous approach for the exact partition of the deformation into the elastic and the plastic parts.

iii) The elastic part of the strain rate tensor, that is, elastic strain rate tensor is related linearly to an appropriate corotational rate of the Cauchy stress tensor. It falls within the framework of the so-called *hypoelasticity* (Truesdell, 1955).

iv) The elastic part of the spin tensor, that is, elastic spin tensor, which is given by the subtraction of the plastic part of the spin tensor, that is, *plastic spin tensor* (Dafalias, 1985) from the continuum spin tensor, is regarded as the spin of substructure, that is, the *substructure* (*corotational*) *spin tensor*. Then, it is adopted in corotational rates of stress tensor and tensor-valued internal variables. The plastic spin tensor is formulated as the skew-symmetric part of the multiplication of the stress tensor by the plastic strain rate tensor (Zbib and Aifantis, 1988).

v) All variables adopted in the constitutive relation are Eulrian tensors based in the current configuration.

Constitutive behavior under finite deformation and rotation is properly described by incorporating corotational rate tensors based on an appropriate substructure spin tensor under the limitation of infinitesimal elastic deformation

However, the hypoelastic-based plasticity would possess following problems:

(1) Hypoelastic equation does not fulfill the complete integrability condition. Therefore, one-to-one correspondence between integrations of stress rate tensor and elastic strain rate tensor, that is, stress vs. strain relation cannot be obtained in general. Then, the hypoelastic deformation remains even after a closed stress cycle. In addition, work done during a closed loading cycle may not be zero as an energy is produced or dissipated during a closed loading cycle even when a plastic deformation is not induced. Therefore, the elastic strain rate tensor does not possess the reversibility in the strict sense, so that the hypoelastic constitutive equation is merely an elastic-like constitutive equation. This approach possesses the pertinence on the premise that the elastic deformation is infinitesimal.

(2) Stress cannot be calculated directly from strain variable but it has to be calculated by the time-integration of stress rate. Also, tensor-valued internal variables cannot be calculated directly from the elastic strain-like variable but they have to be calculated by the time-integration of their rates. Here, note that rates of stress and tensor-valued internal variables in the current configuration are influenced by the rigid-body rotation of material. Consequently, their pertinent objective corotational rate tensors must be adopted and their pertinent time-integration procedures must be incorporated in order that unrealistic and/or impertinent calculation such as oscillatory stress and tensor-valued internal variables are not resulted. Proper time-integrations would be given by the time-integration reflecting the convected derivative process (Simo and Hughes, 1998).

(3) The return-mapping and the consistent (algorithmic) tangent modulus tensor, which enable drastically accurate and efficient numerical calculations, cannot be employed since they are premised on the exact evaluation of stress by the hyperelastic constitutive equation, while the plastic strain increment in the plastic corrector step is calculated based on the overstress from the yield surface.

Nevertheless, the hypoelasticity with constant elastic moduli leads to the one-to-one correspondence between the time-integrations of objective stress rate and elastic strain rate tensors by performing proper time-integrations reflecting the convected derivative process, so that the return-mapping and the consistent tangent modulus tensor can be employed under the finite deformation and rotation. On the other hand, it should be noticed that the automatic controlling function to attract the stress to the yield surface in a plastic deformation process is furnished in the subloading surface model (Hashiguchi, 1989; Hashiguchi et al. 2012) as will be described in Chapter 8, so that numerical analyses of materials, in which elastic parameters in a hypoelastic equation are not constant, can be enforced drastically by this function in the Euler forward-integration method.

B-2) *Hyperelastic-based plasticity*

The hyperelastic-based plasticity has been formulated to overcome the above-mentioned problems in the hypoelastic-based plasticity by incorporating the following notions.

1. Deformation itself (not rate) is decomposed into the elastic and the plastic parts to provide the one-to-one correspondence between elastic deformation and stress, although the strain rate is decomposed into them in the hypoelastic-based plasticity. It has been materialized by the decomposition of the deformation gradient tensor into the elastic and the plastic parts in the multiplicative form, that is, the *multiplicative decomposition* by Kroner (1959), Lee and Liu (1967), Lee (1969), etc., while the elasto-plasticity based on this decomposition would be called the *Mandel–Lee approach*. Therein, the *intermediate configuration* is incorporated, which is attained by unloading to the stress free state along the hyperelastic deformation. Then, the deformation gradient is multiplicatively decomposed into the plastic deformation gradient tensor induced in the process from the initial to the intermediate configuration and the elastic deformation gradient tensor induced in the process from the intermediate to the current configuration.

   Formulations by principal values based on the Hencky strain, that is, principal logarithmic strain have been studied by a lot of workers (cf. Simo and Meschke, 1993; Borja and Tamagnini, 1998; Tamagnini et al., 2002; Borja et al., 2001; Rosati and Valoroso, 2004, Raniecki and Nguyen, 2005; Yamakawa et al., 2010). However, they have been limited to the description of isotropic materials in which the principal directions of stress, elastic strain and plastic strain rate coincide with each other leading to the *co-axiality*.

   Then, the formulation based on the following notions has been developed, by which general elastoplastic constitutive behavior can be described accurately over the finite deformation and rotation.
2. Both of the elastic and the plastic deformation gradient tensors are the two-point tensors, the one of the two base vectors of which lives in the intermediate configuration which is independent of the superposition of rigid-body rotation. Therefore, constitutive relation is formulated originally by tensor variables in the intermediate configuration.
3. Elastic deformation is described by an elastic strain tensor based on the elastic deformation gradient tensor and it is related to the stress tensor by the

hyperelastic constitutive equation possessing the elastic strain energy function. Then, the one-to-one correspondence between stress tensor and elastic strain tensor holds and the work done during a closed stress cycle is zero exactly when the plastic strain rate is not induced.

4. In addition, the plastic deformation gradient tensor is further decomposed into the energy-storage part causing the variation of substructure and the energy-dissipative part causing the slip between substructures (Lion, 2000). Then, the kinematic hardening variable, that is, the back stress is also formulated in relation to the elastic strain-like variable induced by the energy-storage part as a hyperelastic-like equation possessing a potential energy function of the variable.
5. Stress and back stress are formulated in the intermediate configuration and calculated directly from the elastic strain and the elastic strain-like variable of the kinematic hardening without performing the time-integrations of stress rate and back stress rate. Therefore, material rotation is independent of a rigid-body rotation and it would not be influenced severely be a plastic deformation on the calculation of stress and back stress.
6. Plastic flow rule is given by the relationship of the plastic velocity gradient tensor to the *Mandel stress tensor* (Mandel, 1972b) in the intermediate configuration, where these tensors fulfill the *plastic work-conjugacy*. Concurrently, the yield surface is described by the Mandel stress tensor which is obtained by the pull-back of the Kirchhoff stress tensor or the push-forward of the second Piola–Kirchhoff stress tensor.
7. In addition to the above-mentioned issues on the formulation of constitutive equation, the formulation and implementation of a numerical algorithm for the state-updating calculation are of utmost importance. The state-updating procedure in the computation of elastoplasticity problems usually requires a proper algorithm for numerical integration of the rate forms of the constitutive laws and the evolution equations. Responding to the formulation of hyperelastic-based plastic constitutive equation, the return-mapping scheme has achieved a degree of common acceptance in the field of computational plasticity as an effective state-updating procedure for elastoplastic models. In the numerical analysis of boundary-value problems, a consistent linearization of the weak form of equilibrium equation and a use of the so-called consistent elasto-plastic tangent modulus tensor are necessary to ensure effectiveness and robustness of the iterative solution procedure.

The hyperelastic-based plasticity may be called the *finite strain elastoplasticity*. On the other hand, the hypoelastic-based plasticity should be called merely the *finite deformation elastoplasticity*, since it does not use the elastic strain itself but it is based on the velocity gradient.

The hyperelastic-based plasticity will be explained exhaustively in this book. A certain amount of advanced mathematical knowledge is required to capture the essentials of continuum mechanics and to formulate constitutive equations in the framework of the finite strain theory. In order to capture the meanings of physical quantities and relations exactly, it is indispensable to describe them in the embedded coordinate system which not only moves but also deforms and rotates with material itself, which belongs to the general curvilinear coordinate system. Then, the incorporation of the curvilinear coordinate system is one of the distinctions of the mathematical methodology in the finite strain theory from that in the infinitesimal strain theory. The concise explanation of vector and tensor analysis in the curvilinear coordinate system will be given in chapter **2** in addition to the analysis in the Cartesian coordinate system in chapter **1** as the preliminary to the study of finite strain theory.

1

Mathematical Preliminaries

Advanced mathematical knowledge is required to learn the finite elasto-plasticity theory. The basics of vector and tensor analyses are described in preparation for the explanations of advanced theory in later chapters. Various representations of tensors, for example, the eigenvalues and principal directions, the Cayley–Hamilton theorem, the polar and spectral decompositions, the isotropic function and various differential equations, the time-derivatives, and the integration theorems are described. Component descriptions of vectors and tensors in this chapter are limited to the *normalized rectangular coordinate system*, that is, the rectangular coordinate system with unit base vectors, while the terms *orthogonal*, *orthonormal* and *Cartesian* are often used instead of *rectangular*. However, the derived tensor relations hold even in the general curvilinear coordinate system of the Euclidian space described in later chapters.

1.1 Basic Symbols and Conventions

An index appearing twice in a term is summed over the specified range of the index. For instance, we may write

$$u_r v_r = \sum_{r=1}^{n} u_r v_r, \quad T_{ir} v_r = \sum_{r=1}^{n} T_{ir} v_r, \quad T_{rr} = \sum_{r=1}^{n} T_{rr}, \tag{1.1}$$

where the range of index is taken to be $1, 2, \ldots, n$. The indices used repeatedly are arbitrary, and thus they are called *dummy indices*: note that $u_r v_r = u_s v_s$ and $T_{ir} v_r = T_{is} v_s$, for example. This is termed *Einstein's summation convention*. Henceforth, repeated indices refer to this convention unless specified by the additional remark '(no sum)'.

The symbol $\delta_{ij}(i, j = 1, 2, 3)$ defined in the following equation is termed the *Kronecker delta*:

$$\delta_{ij} = 1 \quad \text{for } i = j, \quad \delta_{ij} = 0 \quad \text{for } i \neq j \tag{1.2}$$

from which it follows that

$$\delta_{ir}\delta_{rj} = \delta_{ij}, \quad \delta_{ii} = 3. \tag{1.3}$$

Introduction to Finite Strain Theory for Continuum Elasto-Plasticity, First Edition. Koichi Hashiguchi and Yuki Yamakawa.

Furthermore, the symbol ε_{ijk} defined by the following equation is called the *alternating* (or *permutation*) *symbol* or *Eddington's epsilon* or *Levi-Civita 'e' tensor*:

$$\varepsilon_{ijk} = \begin{cases} 1 & \text{for cyclic permutation of } ijk \text{ from 123,} \\ -1 & \text{for anticyclic permutation of } ijk \text{ from 123,} \\ 0 & \text{for others.} \end{cases} \tag{1.4}$$

1.2 Definition of Tensor

In this section the definition of an objective tensor is given and, based on it, the criteria for a given physical quantity to be a tensor and its order are provided.

1.2.1 Objective Tensor

Let the set of n^m functions be described as $T(p_1, p_2, \ldots, p_m)$ in the coordinate system $\{\mathrm{O}, x_i\}$ with the origin O and the axes x_i in n-dimensional space, where each of the indices $p_1, p_2, \ldots, p_m$ takes a numberical value $1, 2, \ldots, n$. This set of functions is defined as the mth-order tensor in n-dimensional space, if the set of functions is observed in the other coordinate system $\{\mathrm{O}, x_i^*\}$ with the origin O and the axes x_i^* as follows:

$$T^*(p_1, p_2, \ldots, p_m) = Q_{p_1q_1}Q_{p_2q_2}\cdots Q_{p_mq_m}T(q_1, q_2, \ldots, q_m) \tag{1.5}$$

or

$$T^*(p_1, p_2, \ldots, p_m) = \frac{\partial x^*_{p_1}}{\partial x_{q_1}}\frac{\partial x^*_{p_2}}{\partial x_{q_2}}\cdots\frac{\partial x^*_{p_m}}{\partial x_{q_m}}T(q_1, q_2, \ldots, q_m), \tag{1.6}$$

provided that only the directions of axes are different but the origin is common and relative motion does not exist. Here, Q_{ij} in equation (1.5) is defined by

$$Q_{ij} = \frac{\partial x_i^*}{\partial x_j} \tag{1.7}$$

which fulfills

$$Q_{ir}Q_{jr} = \delta_{ij} \tag{1.8}$$

because of

$$Q_{ir}Q_{jr} = \frac{\partial x_i^*}{\partial x_r}\frac{\partial x_r}{\partial x_j^*}.$$

Denoting $T(p_1, p_2, \ldots, p_m)$ by the symbol $T_{p_1p_2\cdots p_m}$, equation (1.6) is expressed as

$$\boxed{T^*_{p_1p_2\cdots p_m} = Q_{p_1q_1}Q_{p_2q_2}\cdots Q_{p_mq_m}T_{q_1q_2\cdots q_m}} \tag{1.9}$$

that is,

$$T^*_{p_1p_2\cdots p_m} = \frac{\partial x^*_{p_1}}{\partial x_{q_1}}\frac{\partial x^*_{p_2}}{\partial x_{q_2}}\cdots\frac{\partial x^*_{p_m}}{\partial x_{q_m}}T_{q_1q_2\cdots q_m}. \tag{1.10}$$

Noting that

$$\begin{aligned}Q_{p_1r_1}Q_{p_2r_2}\cdots Q_{p_mr_m}T^*_{p_1p_2\ldots p_m} &= Q_{p_1r_1}Q_{p_2r_2}\cdots Q_{p_mr_m}Q_{p_1q_1}Q_{p_2q_2}\cdots Q_{p_mq_m}T_{q_1q_2\ldots q_m}\\ &= (Q_{p_1r_1}Q_{p_1q_1})(Q_{p_2r_2}Q_{p_2q_2})\cdots(Q_{p_mr_m}Q_{p_mq_m})T_{q_1q_2\ldots q_m}\\ &= \delta_{r_1q_1}\delta_{r_2q_2}\cdots\delta_{r_mq_m}T_{q_1q_2\ldots q_m}\end{aligned}$$

with equation (1.8), the inverse relation of equation (1.9) is given by

$$T_{r_1r_2\cdots r_m} = Q_{p_1r_1}Q_{p_2r_2}\cdots Q_{p_mr_m}T^*_{p_1p_2\cdots p_m}. \tag{1.11}$$

Indices put in a tensor take the dimension of the space in which the tensor exists. The number of indices, which is equal to the number of operators Q_{ij}, is called the order of tensor. For instance, the transformation rule of the first-order tensor, that is, the vector v_i, and the second-order tensor T_{ij} are given by

$$\left.\begin{aligned} v_i^* = Q_{ir}v_r, \quad v_i = Q_{ri}v^*_{rs}\\ T^*_{ij} = Q_{ir}Q_{js}T_{rs}, \quad T_{ij} = Q_{ri}Q_{sj}T^*_{rs}\end{aligned}\right\}. \tag{1.12}$$

Consequently, in order to prove that a certain quantity is a tensor, one needs only to show that it obeys the tensor transformation rule (1.9) or that multiplying the quantity by a tensor leads to a tensor by virtue of the quotient rule described in the next section.

The coordinate transformation rule in the form of equation (1.9) or (1.11) is called the *objective transformation*. A tensor obeying the objective transformation rule even between coordinate systems with a relative rate of motion, that is, relative parallel and rotational velocities, is called an *objective tensor*. Vectors and tensors without the time-dimension, for example, force, displacement, rotational angle, stress and strain, are objective vectors and tensors. On the other hand, time-rate quantities, for example, rate of force, velocity, spin and the material-time derivatives of physical quantities, for example, stress and strain, are not objective vectors and tensors in general; they are influenced by the relative rate of motion between coordinate systems. Constitutive equations of materials have to be formulated in terms of objective tensors, since material properties are not influenced by the rigid-body rotation of material and therefore must be described in a form independent of the coordinate systems, as will be explained in Section 5.3.

Tensors obviously fulfill the linearity

$$\left.\begin{aligned} T_{p_1p_2\cdots p_m}(G_{p_1p_2\cdots p_l} + H_{p_1p_2\cdots p_l}) = T_{p_1p_2\cdots p_m}G_{p_1p_2\cdots p_l} + T_{p_1p_2\cdots p_m}H_{p_1p_2\cdots p_l}\\ T_{p_1p_2\cdots p_m}(aA_{p_1p_2\cdots p_l}) = aT_{p_1p_2\cdots p_m}A_{p_1p_2\cdots p_l}\end{aligned}\right\}, \tag{1.13}$$

where a is an arbitrary scalar variable. Therefore, the tensor plays the role of linearly transforming one tensor into another and thus it is also called a *linear transformation*. The operation that lowers the order of a tensor by multiplying it by another tensor is called *contraction*.

Denoting by $\mathbf{e}_1^*, \mathbf{e}_2^*, \ldots, \mathbf{e}_m^*$ the unit base vectors of the coordinate axes $x_1^*, x_2^*, \ldots, x_m^*$, the quantity Q_{ij} in equation (1.7) is represented in terms of the base vectors as follows:

$$Q_{ij} = \mathbf{e}_i^* \cdot \mathbf{e}_j, \tag{1.14}$$

noting that

$$Q_{ij} = \frac{\partial x_i^*}{\partial x_j} = \mathbf{e}_i^* \cdot \frac{\partial x_s^*}{\partial x_j}\mathbf{e}_s^* = \mathbf{e}_i^* \cdot \frac{\partial x_s}{\partial x_j}\mathbf{e}_s = \mathbf{e}_i^* \cdot \delta_{js}\mathbf{e}_s,$$

where the coordinate transformation operator Q_{ij} is interpreted as

$$Q_{ij} \equiv \cos(\text{angle between } \mathbf{e}_i^* \text{ and } \mathbf{e}_j) \tag{1.15}$$

which fulfills equations (1.8), that is,

$$Q_{ir}Q_{jr} = Q_{ri}Q_{rj} = \delta_{ij} \tag{1.16}$$

which can be also verified by

$$Q_{ir}Q_{jr} = (\mathbf{e}_i^* \cdot \mathbf{e}_r)(\mathbf{e}_j^* \cdot \mathbf{e}_r) = \mathbf{e}_i^* \cdot (\mathbf{e}_j^* \cdot \mathbf{e}_r)\mathbf{e}_r = \delta_{ij}.$$

The transformation rule for base vectors is given by

$$\mathbf{e}_i = Q_{ri}\mathbf{e}_r^*, \quad \mathbf{e}_i^* = Q_{ir}\mathbf{e}_r, \tag{1.17}$$

noting that

$$\mathbf{e}_i = (\mathbf{e}_i \cdot \mathbf{e}_r^*)\mathbf{e}_r^*, \quad \mathbf{e}_i^* = (\mathbf{e}_i^* \cdot \mathbf{e}_r)\mathbf{e}_r.$$

1.2.2 Quotient Law

There is a convenient law, referred to as the *quotient law*, which enables us to judge whether or not a given quantity is tensor and to find its tensorial order as follows: If a set of functions $T(p_1, p_2, \ldots, p_m)$ becomes $B_{p_{l+1}p_{l+2}\cdots p_m}$ ($(m-l)$th-order tensor lacking the indices $p_1 \sim p_l$) through multiplying it by $A_{p_1p_2\cdots p_l}$ (lth-order tensor ($l \le m$)), the set is an mth-order tensor.

(*Proof*) This convention is proved by showing that the quantity $T(p_1, p_2, \ldots, p_m)$ is an mth-order tensor when the relation

$$T(p_1, p_2, \ldots, p_m)A_{p_1p_2\cdots p_l} = B_{p_{l+1}p_{l+2}\cdots p_m} \tag{1.18}$$

holds, which is described in the coordinate system $\{\text{O-}x_i^*\}$ as follows:

$$T^*(p_1, p_2, \ldots, p_m)A^*_{p_1p_2\cdots p_l} = B^*_{p_{l+1}p_{l+2}\cdots p_m}. \tag{1.19}$$

Here, $A_{p_1p_2\cdots p_l}$ is the lth-order tensor and $B_{p_{l+1}p_{l+2}\cdots p_m}$ is the $(m-l)$th-order tensor. Therefore, the following relation holds:

$$\begin{aligned} B^*_{p_{l+1}p_{l+2}\cdots p_m} &= Q_{p_{l+1}r_{l+1}}Q_{p_{l+2}r_{l+2}}\cdots Q_{p_mr_m}B_{r_{l+1}r_{l+2}\cdots r_m} \\ &= Q_{p_{l+1}r_{l+1}}Q_{p_{l+2}r_{l+2}}\cdots Q_{p_mr_m}T(r_1, r_2, \ldots, r_m)A_{r_1r_2\cdots r_l} \\ &= \underbrace{Q_{p_{l+1}r_{l+1}}Q_{p_{l+2}r_{l+2}}\cdots Q_{p_mr_m}}_{(l+1\sim m)}T(r_1, r_2, \ldots, r_m)\underbrace{Q_{p_1r_1}Q_{p_2r_2}\cdots Q_{p_lr_l}}_{(1\sim l)}A^*_{p_1p_2\cdots p_l}. \end{aligned} \tag{1.20}$$

Substituting equation (1.19) into the left-hand side of equation (1.20) yields

$$\{T^*(p_1, p_2, \ldots, p_m) - Q_{p_1r_1}Q_{p_2r_2}\cdots Q_{p_mr_m}T(r_1, r_2, \ldots, r_m)\}A^*_{p_1p_2\cdots p_l} = 0,$$

from which it follows that

$$T^*(p_1, p_2, \ldots, p_m) = Q_{p_1r_1}Q_{p_2r_2}\cdots Q_{p_mr_m}T(r_1, r_2, \ldots, r_m). \quad (1.21)$$

Equation (1.21) satisfies the definition of tensor in equation (1.5). Therefore, the quantity $T(p_1, p_2, \ldots, p_m)$ is an mth-order tensor.

According to the proof presented above, equation (1.18) can be written as

$$T_{p_1p_2\cdots p_m}A_{p_1p_2\cdots p_l} = B_{p_{l+1}p_{l+2}\cdots p_m}. \quad (1.22)$$

For instance, if the quantity $T(i, j)$ transforms the first-order tensor, that is, vector v_i, to the vector u_i by the operation $T(i, j)v_j = u_i$, one can regard $T(i, j)$ as the second-order tensor T_{ij}.

1.3 Vector Analysis

In this section some basic rules for vectors are given which are required to understand the representation of tensors in the general coordinate system described in the next chapter.

1.3.1 Scalar Product

The *scalar* (or *inner*) *product* of the vectors **a** and **b** is defined by

$$\mathbf{a}\cdot\mathbf{b} = ||\mathbf{a}||\,||\mathbf{b}||\cos\theta = a_ib_i, \quad (1.23)$$

where θ is the angle between the vectors **a** and **b**, and || || means the magnitude, that is,

$$||\mathbf{v}|| = \sqrt{v_iv_i} = \sqrt{\mathbf{v}\cdot\mathbf{v}}. \quad (1.24)$$

Here, the following relations hold for the scalar product:

$$\mathbf{a}\cdot\mathbf{b} = \mathbf{b}\cdot\mathbf{a} \text{ (commutative law)}, \quad (1.25)$$

$$\mathbf{a}\cdot(\mathbf{b}+\mathbf{c}) = \mathbf{a}\cdot\mathbf{b} + \mathbf{a}\cdot\mathbf{c} \text{ (distributive law)}, \quad (1.26)$$

$$s(\mathbf{a}\cdot\mathbf{b}) = (s\mathbf{a}\cdot\mathbf{b}) = \mathbf{a}\cdot(s\mathbf{b}) = (\mathbf{a}\cdot\mathbf{b})s, \quad (1.27)$$

$$(a\mathbf{a} + b\mathbf{b})\cdot\mathbf{c} = a\mathbf{a}\cdot\mathbf{c} + b\mathbf{b}\cdot\mathbf{c} \quad (1.28)$$

for arbitrary scalars s, a, b.

The vector is represented in terms of components with base vectors as follows:

$$\mathbf{v} = v_i\mathbf{e}_i, \quad (1.29)$$

where the components v_i are given by the projection of **v** onto the base vector $\mathbf{e}_i$, that is, their scalar product and thus it follows that

$$v_i = \mathbf{v}\cdot\mathbf{e}_i,\ \ \mathbf{v} = (\mathbf{v}\cdot\mathbf{e}_i)\mathbf{e}_i. \quad (1.30)$$

1.3.2 Vector Product

The *vector* (or *outer* or *cross*) *product* of vectors is defined by

$$\mathbf{a}\times\mathbf{b} = ||\mathbf{a}||\,||\mathbf{b}||\sin\theta\mathbf{n} = a_i\mathbf{e}_i\times b_j\mathbf{e}_j = \varepsilon_{ijk}a_ib_j\mathbf{e}_k$$
$$= (a_2b_3 - a_3b_2)\mathbf{e}_1 + (a_3b_1 - a_1b_3)\mathbf{e}_2 + (a_1b_2 - a_2b_1)\mathbf{e}_3, \tag{1.31}$$

where $\mathbf{n}$ is the unit vector which forms the right-handed bases $(\mathbf{a}, \mathbf{b}, \mathbf{n})$ in this order. It follows for base vectors from equation (1.31) that

$$\mathbf{e}_i\times\mathbf{e}_j = \varepsilon_{ijk}\mathbf{e}_k. \tag{1.32}$$

Here, the following equations hold for the vector product:

$$\mathbf{a}\times\mathbf{a} = \mathbf{0}, \tag{1.33}$$

$$\mathbf{a}\times\mathbf{b} = -\mathbf{b}\times\mathbf{a}, \tag{1.34}$$

$$\mathbf{a}\times(\mathbf{b}+\mathbf{c}) = \mathbf{a}\times\mathbf{b}+\mathbf{a}\times\mathbf{c}\ \text{(distributive law)}, \tag{1.35}$$

$$s(\mathbf{a}\times\mathbf{b}) = (s\mathbf{a}\times\mathbf{b}) = \mathbf{a}\times(s\mathbf{b}) = (\mathbf{a}\times\mathbf{b})s, \tag{1.36}$$

$$(a\mathbf{a}+b\mathbf{b})\times\mathbf{c} = a\mathbf{a}\times\mathbf{c}+b\mathbf{b}\times\mathbf{c}, \tag{1.37}$$

$$\|\mathbf{a}\times\mathbf{b}\|^2 + (\mathbf{a}\cdot\mathbf{b})^2 = (\|\mathbf{a}\|\,\|\mathbf{b}\|)^2. \tag{1.38}$$

1.3.3 Scalar Triple Product

The *scalar triple product* of vectors is defined by

$$[\mathbf{abc}] \equiv \mathbf{a}\cdot(\mathbf{b}\times\mathbf{c}) = \varepsilon_{ijk}a_ib_jc_k, \tag{1.39}$$

fulfilling

$$[\mathbf{abc}] = [\mathbf{bca}] = [\mathbf{cab}] = -[\mathbf{bac}] = -[\mathbf{cba}] = -[\mathbf{acb}]. \tag{1.40}$$

Denoting the vectors $\mathbf{a}, \mathbf{b}, \mathbf{c}$ by $\mathbf{v}_1, \mathbf{v}_2, \mathbf{v}_3$, it follows from equation (1.39) that

$$[\mathbf{v}_i\mathbf{v}_j\mathbf{v}_k] = e_{ijk}[\mathbf{v}_1\mathbf{v}_2\mathbf{v}_3], \tag{1.41}$$

noting the fact that the term on the right-hand side of this equation is $+[\mathbf{v}_1, \mathbf{v}_2, \mathbf{v}_3]$, $-[\mathbf{v}_1, \mathbf{v}_2, \mathbf{v}_3]$ and 0 when indices i, j, k are even and odd permutations and two of indices coincide with each other, respectively.

Here, the following equations hold for the scalar triple product.

$$[\mathbf{e}_i\mathbf{e}_j\mathbf{e}_k] = \varepsilon_{ijk}, \tag{1.42}$$

$$[s\mathbf{a}, \mathbf{b}, \mathbf{c}] = [\mathbf{a}, s\mathbf{b}, \mathbf{c}] = [\mathbf{a}, \mathbf{b}, s\mathbf{c}] = s[\mathbf{abc}], \tag{1.43}$$

$$[a\mathbf{a}+b\mathbf{b}, \mathbf{c}, \mathbf{d}] = a[\mathbf{acd}] + b[\mathbf{bcd}], \tag{1.44}$$

$$[\mathbf{a}\times\mathbf{b}, \mathbf{b}\times\mathbf{c}, \mathbf{c}\times\mathbf{a}] = [\mathbf{abc}]^2, \tag{1.45}$$

$$[\mathbf{abc}]\mathbf{d} = [\mathbf{bcd}]\mathbf{a} + [\mathbf{adc}]\mathbf{b} + [\mathbf{abd}]\mathbf{c}, \tag{1.46}$$

$$[\mathbf{a}\times\mathbf{b}, \mathbf{c}\times\mathbf{d}, \mathbf{e}\times\mathbf{f}] = [\mathbf{abd}][\mathbf{cef}] - [\mathbf{abc}][\mathbf{def}]. \tag{1.47}$$

1.3.4 Vector Triple Product

The *vector triple product* is defined by $\mathbf{a} \times (\mathbf{b} \times \mathbf{c})$. Setting $\mathbf{v} = \mathbf{a} \times (\mathbf{b} \times \mathbf{c})$, it follows from equation (1.31) that

$$\begin{aligned} v_1 &= a_2(\mathbf{b} \times \mathbf{c})_3 - a_3(\mathbf{b} \times \mathbf{c})_2 = a_2(b_1c_2 - b_2c_1) - a_3(b_3c_1 - b_1c_3) \\ &= (a_1c_1 + a_2c_2 + a_3c_3)b_1 - (a_1b_1 + a_2b_2 + a_3b_3)c_1 \\ &= (\mathbf{a} \cdot \mathbf{c})b_1 - (\mathbf{a} \cdot \mathbf{b})c_1. \end{aligned}$$

v_2 and v_3 can be represented analogously. Collecting these equations yields $\mathbf{v} = (\mathbf{a} \cdot \mathbf{c})\mathbf{b} - (\mathbf{a} \cdot \mathbf{b})\mathbf{c}$, and then one has the formulae of the *vector product*:

$$\left.\begin{aligned} \mathbf{a} \times (\mathbf{b} \times \mathbf{c}) &= (\mathbf{a} \cdot \mathbf{c})\mathbf{b} - (\mathbf{a} \cdot \mathbf{b})\mathbf{c} \\ (\mathbf{a} \times \mathbf{b}) \times \mathbf{c} &= -(\mathbf{b} \cdot \mathbf{c})\mathbf{a} + (\mathbf{a} \cdot \mathbf{c})\mathbf{b} \; (\neq \mathbf{a} \times (\mathbf{b} \times \mathbf{c})) \end{aligned}\right\}, \tag{1.48}$$

noting

$$(\mathbf{a} \times \mathbf{b}) \times \mathbf{c} = -\mathbf{c} \times (\mathbf{a} \times \mathbf{b}) = -\{(\mathbf{b} \cdot \mathbf{c})\mathbf{a} - (\mathbf{a} \cdot \mathbf{c})\mathbf{b}\}$$

to derive the second equation in equation (1.48), while the first equation in equation (1.48) is exploited to the second term in this equation. Besides, equation (1.48) can be derived easily using equation (1.56) below:

$$\begin{aligned} \mathbf{a} \times (\mathbf{b} \times \mathbf{c}) &= a_i\mathbf{e}_i \times \varepsilon_{pqr}b_qc_r\mathbf{e}_p = \varepsilon_{pqr}a_ib_qc_r\mathbf{e}_i \times \mathbf{e}_p = \varepsilon_{pqr}a_ib_qc_r\varepsilon_{ipk}\mathbf{e}_k \\ &= \varepsilon_{pqr}\varepsilon_{pki}a_ib_qc_r\mathbf{e}_k = (\delta_{qk}\delta_{ri} - \delta_{qi}\delta_{rk})a_ib_qc_r\mathbf{e}_k = a_ic_ib_k\mathbf{e}_k - a_ib_ic_k\mathbf{e}_k. \end{aligned} \tag{1.49}$$

By virtue of equation (1.48), we have

$$\mathbf{a} \times (\mathbf{b} \times \mathbf{c}) + \mathbf{b} \times (\mathbf{c} \times \mathbf{a}) + \mathbf{c} \times (\mathbf{a} \times \mathbf{b}) = \mathbf{0} \tag{1.50}$$

because of

$$\left.\begin{aligned} \mathbf{a} \times (\mathbf{b} \times \mathbf{c}) &= (\mathbf{a} \cdot \mathbf{c})\mathbf{b} - (\mathbf{a} \cdot \mathbf{b})\mathbf{c} \\ \mathbf{b} \times (\mathbf{c} \times \mathbf{a}) &= (\mathbf{b} \cdot \mathbf{a})\mathbf{c} - (\mathbf{b} \cdot \mathbf{c})\mathbf{a} \\ \mathbf{c} \times (\mathbf{a} \times \mathbf{b}) &= (\mathbf{c} \cdot \mathbf{b})\mathbf{a} - (\mathbf{c} \cdot \mathbf{a})\mathbf{b} \end{aligned}\right\}.$$

Furthermore, it follows, noting equation (1.48), that

$$(\mathbf{a} \times \mathbf{b}) \cdot (\mathbf{c} \times \mathbf{d}) = \mathbf{a} \cdot \{\mathbf{b} \times (\mathbf{c} \times \mathbf{d})\} = (\mathbf{a} \cdot \mathbf{c})(\mathbf{b} \cdot \mathbf{d}) - (\mathbf{a} \cdot \mathbf{d})(\mathbf{b} \cdot \mathbf{c}) \tag{1.51}$$

$$(\mathbf{a} \times \mathbf{b}) \cdot (\mathbf{a} \times \mathbf{b}) = (\|\mathbf{a}\|\|\mathbf{b}\|)^2 - (\mathbf{a} \cdot \mathbf{b})^2. \tag{1.52}$$

noting the following equation obtained by $\mathbf{x} \cdot (\mathbf{c} \times \mathbf{d}) = (\mathbf{x} \times \mathbf{c}) \cdot \mathbf{d}$ with $\mathbf{x} = \mathbf{a} \times \mathbf{b}$.

$$(\mathbf{a} \times \mathbf{b}) \cdot (\mathbf{c} \times \mathbf{d}) = \{(\mathbf{a} \times \mathbf{b}) \times \mathbf{c}\} \cdot \mathbf{d} = \{\mathbf{b}(\mathbf{a} \cdot \mathbf{c})\} - \mathbf{a}(\mathbf{b} \cdot \mathbf{c})\} \cdot \mathbf{d}.$$

Moreover, we can derive

$$(\mathbf{a} \times \mathbf{b}) \times (\mathbf{c} \times \mathbf{d}) = \begin{cases} ((\mathbf{a} \times \mathbf{b}) \cdot \mathbf{d})\mathbf{c} - ((\mathbf{a} \times \mathbf{b}) \cdot \mathbf{c})\mathbf{d} = [\mathbf{abd}]\mathbf{c} - [\mathbf{abc}]\mathbf{d} \\ ((\mathbf{c} \times \mathbf{d}) \cdot \mathbf{a})\mathbf{b} - ((\mathbf{c} \times \mathbf{d}) \cdot \mathbf{b})\mathbf{a} = [\mathbf{cda}]\mathbf{b} - [\mathbf{cdb}]\mathbf{a} \end{cases},$$

from equation (1.48), leading to

$$[\mathbf{abd}]\mathbf{c} - [\mathbf{abc}]\mathbf{d} = [\mathbf{cda}]\mathbf{b} - [\mathbf{cdb}]\mathbf{a}, \tag{1.53}$$

which in the special case of $\mathbf{c} = \mathbf{b}$ and $\mathbf{d} = \mathbf{c}$ reduces to

$$(\mathbf{a} \times \mathbf{b}) \times (\mathbf{b} \times \mathbf{c}) = [\mathbf{abc}]\mathbf{b} \tag{1.54}$$

An arbitrary vector $\mathbf{v}$ is represented by

$$\boxed{\mathbf{v} = \frac{[\mathbf{bcv}]\mathbf{a} + [\mathbf{cav}]\mathbf{b} + [\mathbf{abv}]\mathbf{c}}{[\mathbf{abc}]}} \tag{1.55}$$

by solving the following equation for $\mathbf{v}$:

$$[\mathbf{abv}]\mathbf{c} - [\mathbf{abc}]\mathbf{v} = [\mathbf{cva}]\mathbf{b} - [\mathbf{cvb}]\mathbf{a},$$

which is obtained from equation (1.53) with the replacement of $\mathbf{d}$ by $\mathbf{v}$.

The left- and right-hand sides of equation (1.51) are described in component form as follows:

$$(\mathbf{a} \times \mathbf{b}) \cdot (\mathbf{c} \times \mathbf{d}) = (\varepsilon_{ijk} a_j b_k \mathbf{e}_i) \cdot (\varepsilon_{prs} c_r d_s \mathbf{e}_p) = \varepsilon_{ijk}\varepsilon_{irs} a_j b_k c_r d_s,$$

$$(\mathbf{a} \cdot \mathbf{c})(\mathbf{b} \cdot \mathbf{d}) - (\mathbf{a} \cdot \mathbf{d})(\mathbf{b} \cdot \mathbf{c}) = \delta_{jr} a_j c_r \delta_{ks} b_k d_s - \delta_{js} a_j d_s \delta_{kr} b_k c_r = (\delta_{jr}\delta_{ks} - \delta_{js}\delta_{kr}) a_j b_k c_r d_s.$$

Equating these equations, we obtain

$$\varepsilon_{ijk}\varepsilon_{irs} = \varepsilon_{jki}\varepsilon_{rsi} = \delta_{jr}\delta_{ks} - \delta_{js}\delta_{kr}. \tag{1.56}$$

Setting $r = j$, equation (1.56) reduces to

$$\varepsilon_{ijk}\varepsilon_{ijs} = 2\delta_{ks}. \tag{1.57}$$

Furthermore, setting $s = k$, equation (1.57) reduces to

$$\varepsilon_{ijk}\varepsilon_{ijk} = 2\delta_{kk} = 3! = 6. \tag{1.58}$$

1.3.5 Reciprocal Vectors

The three vectors $\mathbf{a}'$, $\mathbf{b}'$, $\mathbf{c}'$ defined by the following equations, are called the *reciprocal vectors* of the *primary vectors* $\mathbf{a}$, $\mathbf{b}$, $\mathbf{c}$:

$$\boxed{\mathbf{a}' \equiv \frac{\mathbf{b} \times \mathbf{c}}{[\mathbf{abc}]}, \quad \mathbf{b}' \equiv \frac{\mathbf{c} \times \mathbf{a}}{[\mathbf{abc}]}, \quad \mathbf{c}' \equiv \frac{\mathbf{a} \times \mathbf{b}}{[\mathbf{abc}]}} \tag{1.59}$$

The inverse expression of equation (1.59) is given by

$$\mathbf{a} \equiv \frac{\mathbf{b}' \times \mathbf{c}'}{[\mathbf{a}'\mathbf{b}'\mathbf{c}']}, \quad \mathbf{b} \equiv \frac{\mathbf{c}' \times \mathbf{a}'}{[\mathbf{a}'\mathbf{b}'\mathbf{c}']}, \quad \mathbf{c} \equiv \frac{\mathbf{a}' \times \mathbf{b}'}{[\mathbf{a}'\mathbf{b}'\mathbf{c}']}, \tag{1.60}$$

noting equation (1.54). These vectors fulfill the relations:

$$\mathbf{a} \cdot \mathbf{a}' = \mathbf{b} \cdot \mathbf{b}' = \mathbf{c} \cdot \mathbf{c}' = 1, \tag{1.61}$$

$$\mathbf{a}' \cdot \mathbf{b} = \mathbf{a}' \cdot \mathbf{c} = \mathbf{b}' \cdot \mathbf{c} = \mathbf{b}' \cdot \mathbf{a} = \mathbf{c}' \cdot \mathbf{a} = \mathbf{c}' \cdot \mathbf{b} = 0, \tag{1.62}$$

$$[\mathbf{abc}][\mathbf{a}'\mathbf{b}'\mathbf{c}'] = 1, \tag{1.63}$$

so that the normality relations $\mathbf{a}' \perp \mathbf{b}, \mathbf{a}' \perp \mathbf{c}; \mathbf{b}' \perp \mathbf{c}, \mathbf{b}' \perp \mathbf{a}; \mathbf{c}' \perp \mathbf{a}, \mathbf{c}' \perp \mathbf{b}$ hold. In the particular case of the primary triad $(a\mathbf{i}, b\mathbf{j}, c\mathbf{k})$ where $\mathbf{i}, \mathbf{j}, \mathbf{k}$ are the orthonormal unit vectors, the reciprocal triad is given by $((1/a)\mathbf{i}, (1/b)\mathbf{j}, (1/c)\mathbf{k})$.

Using the reciprocal vectors, we can write an arbitrary vector $\mathbf{v}$ from equations (1.55) and (1.59) as

$$\mathbf{v} = (\mathbf{v} \cdot \mathbf{a}')\mathbf{a} + (\mathbf{v} \cdot \mathbf{b}')\mathbf{b} + (\mathbf{v} \cdot \mathbf{c}')\mathbf{c}. \tag{1.64}$$

It is known that the vector is represented by the assembly of primary vectors with the projections onto directions of the reciprocal vectors.

Further, it follows from equations (1.50) and (1.59) that

$$\mathbf{a} \times \mathbf{a}' + \mathbf{b} \times \mathbf{b}' + \mathbf{c} \times \mathbf{c}' = \mathbf{0}. \tag{1.65}$$

1.3.6 Tensor Product

A second-order tensor is constructed by the *dyadic* (or *tensor*) *product* of two vectors, designated as $\mathbf{a} \otimes \mathbf{b}$ and possessing the following property for another vector $\mathbf{c}$:

$$\boxed{\mathbf{a} \otimes \mathbf{b}\mathbf{c} = \mathbf{a}(\mathbf{b} \cdot \mathbf{c})} \tag{1.66}$$

Therefore, $\mathbf{a} \otimes \mathbf{b}\mathbf{c}$ is the vector possessing the direction of $\mathbf{a}$, while $\mathbf{b}$ is projected in the direction of $\mathbf{c}$ resulting in a scalar. Here, we have

$$(\mathbf{a} \otimes \mathbf{b})^T = \mathbf{b} \otimes \mathbf{a}, \tag{1.67}$$

$$\mathbf{a} \otimes (\mathbf{b} + \mathbf{c}) = \mathbf{a} \otimes \mathbf{b} + \mathbf{a} \otimes \mathbf{c}, \tag{1.68}$$

$$(\mathbf{b} \otimes \mathbf{c} - \mathbf{c} \otimes \mathbf{b})\mathbf{a} = \mathbf{a} \times (\mathbf{b} \times \mathbf{c}) = (\mathbf{a} \cdot \mathbf{c})\mathbf{b} - (\mathbf{a} \cdot \mathbf{b})\mathbf{c}. \tag{1.69}$$

The vector product is represented in the direct notation of the alternating tensor as follows:

$$\mathbf{a} \times \mathbf{b} = \boldsymbol{\varepsilon} : (\mathbf{a} \otimes \mathbf{b}), \quad \mathbf{e}_i \times \mathbf{e}_j = \boldsymbol{\varepsilon} : (\mathbf{e}_i \otimes \mathbf{e}_j) \tag{1.70}$$

because of $\mathbf{a} \times \mathbf{b} = \varepsilon_{ijk} a_j b_k \mathbf{e}_i = \varepsilon_{ijk} \mathbf{e}_i a_r (\mathbf{e}_j \cdot \mathbf{e}_r) b_s (\mathbf{e}_k \cdot \mathbf{e}_s) = (\varepsilon_{ijk} \mathbf{e}_i \otimes \mathbf{e}_j \otimes \mathbf{e}_k) : (a_r \mathbf{e}_r \otimes b_s \mathbf{e}_s)$, where $\boldsymbol{\varepsilon} = \varepsilon_{ijk} \mathbf{e}_i \otimes \mathbf{e}_j \otimes \mathbf{e}_k$ and $\boldsymbol{\varepsilon} : \mathbf{T}$ for $\varepsilon_{ijk} T_{jk}$.

1.4 Tensor Analysis

Various tensors and their algebra are addressed in this section. They are used often throughout this book.

1.4.1 Properties of Second-Order Tensor

The general definition of a tensor was given in Section 1.2. Here, based on that definition, basic properties of second-order tensors are described below.

Two tensors $\mathbf{A}$ and $\mathbf{B}$ are same when they yield the same transformation of an arbitrary vector $\mathbf{v}$, that is,

$$\mathbf{A} = \mathbf{B} \text{ when } \mathbf{A}\mathbf{v} = \mathbf{B}\mathbf{v}. \tag{1.71}$$

The following equations hold by virtue of the linear transformation in equation (1.13):

$$\mathbf{T}(a\mathbf{a} + b\mathbf{b}) = a(\mathbf{Ta}) + b(\mathbf{Tb}), \tag{1.72}$$

$$\mathbf{A} + \mathbf{B} = \mathbf{B} + \mathbf{A}, \tag{1.73}$$

$$s(\mathbf{AB}) = (s\mathbf{A})\mathbf{B} = \mathbf{A}(s\mathbf{B}), \tag{1.74}$$

$$(\mathbf{A} + \mathbf{B})\mathbf{v} = \mathbf{Av} + \mathbf{Bv}, \tag{1.75}$$

$$(\mathbf{AB})\mathbf{v} = \mathbf{A}(\mathbf{Bv}) \tag{1.76}$$

$$\mathbf{A}(\mathbf{B} + \mathbf{C}) = \mathbf{AB} + \mathbf{AC}, \quad (\mathbf{A} + \mathbf{B})\mathbf{C} = \mathbf{AC} + \mathbf{BC}, \tag{1.77}$$

$$\mathbf{A}(\mathbf{BC}) = (\mathbf{AB})\mathbf{C}, \tag{1.78}$$

where a, b and s are arbitrary scalar variables.

The magnitude of a tensor is given by

$$||\mathbf{T}|| = \sqrt{T_{rs}T_{rs}} = \sqrt{\mathbf{T} : \mathbf{T}} = \sqrt{\mathrm{tr}(\mathbf{T}\mathbf{T}^T)}. \tag{1.79}$$

Here, $\mathrm{tr}(\cdot)$ is called the *trace* and is defined by the scalar product with the identity tensor, that is,

$$\mathrm{tr}\mathbf{T} \equiv \mathbf{T} : \mathbf{I} = T_{ii}, \tag{1.80}$$

which fulfills

$$\mathrm{tr}\mathbf{T}^T = \mathrm{tr}\mathbf{T}, \quad \mathrm{tr}(\mathbf{AB}) = \mathrm{tr}(\mathbf{BA}), \quad \mathrm{tr}(\mathbf{AB}) = \mathbf{A} : \mathbf{B}^T, \quad \mathrm{tr}(\mathbf{a} \otimes \mathbf{b}) = \mathbf{a} \cdot \mathbf{b}, \tag{1.81}$$

where $\mathbf{I}$ is the identity tensor, that is, $\mathbf{I} = \delta_{ij}\mathbf{e}_i \otimes \mathbf{e}_j$.

Powers of tensors are defined as follows:

$$\mathbf{T}^0 = \mathbf{I}, \quad \mathbf{T}^n = \overbrace{\mathbf{T} \cdots \mathbf{T}}^{n}. \tag{1.82}$$

These obey the rules of *exponentiation*:

$$\mathbf{T}^m\mathbf{T}^n = \mathbf{T}^{m+n} = \mathbf{T}^n\mathbf{T}^m, \quad (a\mathbf{T})^n = a^n\mathbf{T}^n, \quad (\mathbf{T}^m)^n = \mathbf{T}^{mn} \tag{1.83}$$

for arbitrary integers m and n.

The *zero tensor* $\mathbf{O}$ and the *identity tensor* $\mathbf{I}$ transform an arbitrary vector $\mathbf{v}$ to the *zero vector* $\mathbf{0}$ and the vector $\mathbf{v}$ to itself, respectively, that is,

$$\mathbf{Ov} = \mathbf{0}, \quad \mathbf{Iv} = \mathbf{v}. \tag{1.84}$$

1.4.2 Tensor Components

Equation (1.29) describes a vector in terms of base vectors. We also require the expression for a tensor referring to base vectors. We obtain the following expression of the second-order tensor in terms of the components on the base vectors.

$$\mathbf{T} = T_{ij}\mathbf{e}_i \otimes \mathbf{e}_j, \quad T_{ij} = \mathbf{e}_i \cdot \mathbf{T}\mathbf{e}_j \tag{1.85}$$

based on

$$\mathbf{T} = \mathbf{TI} = \mathbf{T}(\delta_{ij}\mathbf{e}_i \otimes \mathbf{e}_j) = \mathbf{Te}_j \otimes \mathbf{e}_j = (\mathbf{Te}_j \cdot \mathbf{e}_i)\mathbf{e}_i \otimes \mathbf{e}_j \tag{1.86}$$

noting the expression $\mathbf{v} = (\mathbf{v} \cdot \mathbf{e}_i)\mathbf{e}_i$ for arbitrary vector $\mathbf{v}$ which is chosen as $\mathbf{v} = \mathbf{Te}_j$ in equation (1.86) which is the expression for a second-order tensor in terms of base vectors.

1.4.3 Transposed Tensor

The tensor $\mathbf{T}^T$ satisfying the following equation for arbitrary vectors **a** and **b** is defined as the transposed tensor of a tensor **T**:

$$\boxed{\mathbf{a} \cdot \mathbf{Tb} = \mathbf{b} \cdot \mathbf{T}^T\mathbf{a}} \tag{1.87}$$

The relations $\mathbf{T}^T = \mathbf{T}$ and $\mathbf{T}^T = -\mathbf{T}$ hold for a symmetric tensor and a skew-symmetric tensor, respectively (see Section 1.4.6).

In the rectangular Cartesian coordinate system, the particular selection of $\mathbf{a} = \mathbf{e}_i$, $\mathbf{b} = \mathbf{e}_j$ in equation (1.87) reads:

$$\left.\begin{aligned} \mathbf{e}_i \cdot \mathbf{Te}_j &= \mathbf{e}_i \cdot T_{rs}\mathbf{e}_r \otimes \mathbf{e}_s\mathbf{e}_j = T_{ij} \\ \mathbf{e}_j \cdot \mathbf{T}^T\mathbf{e}_i &= \mathbf{e}_j \cdot (\mathbf{T}^T)_{rs}\mathbf{e}_r \otimes \mathbf{e}_s\mathbf{e}_i = T_{ji}^T \end{aligned}\right\},$$

designating $T_{ij}^T \equiv (\mathbf{T}^T)_{ij}$ for the sake of simplicity. Equating these components, we have

$$T_{ij}^T = T_{ji}. \tag{1.88}$$

It follows that

$$\mathbf{T}^T = (T_{ij}\mathbf{e}_i \otimes \mathbf{e}_j)^T = T_{ij}\mathbf{e}_j \otimes \mathbf{e}_i. \tag{1.89}$$

The transposed tensor of $\mathbf{T} = T_{ij}\mathbf{e}_i \otimes \mathbf{e}_j$ can also be written $\mathbf{T}^T = T_{ji}\mathbf{e}_i \otimes \mathbf{e}_j$ in the rectangular Cartesian coordinate system. However, the transposed tensor of the mix of covariant and contravariant bases and components in the general curvilinear coordinate system is given by a tensor which possesses the same components but exchanged base vectors, as will be addressed in Chapter 2.

The following equations hold for a transposed tensor:

$$||\mathbf{T}^T|| = ||\mathbf{T}||, \tag{1.90}$$

$$\mathrm{tr}\mathbf{T}^T = \mathrm{tr}\mathbf{T}, \quad \mathrm{tr}(\mathbf{AB})^T = \mathrm{tr}(\mathbf{B}^T\mathbf{A}^T), \tag{1.91}$$

$$\mathbf{vT} = \mathbf{T}^T\mathbf{v}\ (v_jT_{ji} = T_{ji}v_j), \tag{1.92}$$

$$\left.\begin{aligned} \mathbf{a} \cdot \mathbf{Tb} &= \mathbf{T}^T\mathbf{a} \cdot \mathbf{b}\ (a_i(T_{ir}b_r) = (T_{ir}a_i)b_r) \\ \mathbf{a} \cdot \mathbf{bT} &= \mathbf{Ta} \cdot \mathbf{b}\ (a_i(b_rT_{ri}) = (T_{ri}a_i)b_r) \end{aligned}\right\}, \tag{1.93}$$

$$\left.\begin{aligned} \mathbf{a} \otimes \mathbf{Tb} &= \mathbf{a} \otimes \mathbf{bT}^T\ (a_i(T_{jr}b_r) = a_i(b_rT_{jr})) \\ \mathbf{Ta} \otimes \mathbf{b} &= \mathbf{aT}^T \otimes \mathbf{b}\ ((T_{ir}a_r)b_i = (a_rT_{ir})b_i)) \end{aligned}\right\}, \tag{1.94}$$

$$(\mathbf{A}\otimes\mathbf{B})^T = \mathbf{B}\otimes\mathbf{A}\ (A_{kl}B_{ij} = B_{ij}A_{kl}), \tag{1.95}$$

$$(\mathbf{AB})^T = \mathbf{B}^T\mathbf{A}^T\ (A_{jr}B_{ri} = B_{ri}A_{jr}). \tag{1.96}$$

1.4.4 Inverse Tensor

A tensor $\mathbf{T}$ fulfilling $\det\mathbf{T} \neq 0$ is called a *non-singular tensor*, for which there exists a tensor, called the *inverse tensor* and designated by $\mathbf{T}^{-1}$, satisfying the relation

$$\mathbf{T}\mathbf{T}^{-1} = \mathbf{T}^{-1}\mathbf{T} = \mathbf{I}, \quad T_{ir}T_{rj}^{-1} = T_{ir}^{-1}T_{rj} = \delta_{ij}, \tag{1.97}$$

where the components are denoted as $T_{ij}^{-1} \equiv (\mathbf{T}^{-1})_{ij}$ for the sake of simplicity. It follows that:

$$(\mathbf{AB})^{-1} = \mathbf{B}^{-1}\mathbf{A}^{-1} \tag{1.98}$$

because of

$$(\mathbf{AB})(\mathbf{B}^{-1}\mathbf{A}^{-1})(= \mathbf{ABB}^{-1}\mathbf{A}^{-1} = \mathbf{AIA}^{-1} = \mathbf{AA}^{-1}) = \mathbf{I}.$$

It follows also that

$$\mathbf{T}^{-n} \equiv (\mathbf{T}^{-1})^n = (\mathbf{T}^n)^{-1}, \tag{1.99}$$

and thus the rule of exponentiation extends to negative powers.

1.4.5 Orthogonal Tensor

An *orthogonal tensor* $\boldsymbol{Q}$ is defined as a tensor which keeps a scalar product of vectors constant, so that it fulfills

$$\boxed{(\boldsymbol{Q}\mathbf{a})\cdot(\boldsymbol{Q}\mathbf{b}) = \mathbf{a}\cdot\mathbf{b}} \tag{1.100}$$

By virtue of equation (1.87), the left-hand side in equation (1.100) becomes

$$(\boldsymbol{Q}\mathbf{a})\cdot(\boldsymbol{Q}\mathbf{b}) = \mathbf{a}\cdot(\boldsymbol{Q}^T\boldsymbol{Q}\mathbf{b}). \tag{1.101}$$

Comparing equation (1.101) with equation (1.100), an orthogonal tensor must fulfill

$$\boldsymbol{Q}\boldsymbol{Q}^T = \boldsymbol{Q}^T\boldsymbol{Q} = \mathbf{I}, \tag{1.102}$$

leading to

$$\boldsymbol{Q}^T = \boldsymbol{Q}^{-1}. \tag{1.103}$$

The component description of equation (1.102) is given by equation (1.16).

Choosing $\mathbf{b} = \mathbf{a} = \mathbf{v}$ in equation (1.100) gives

$$||\boldsymbol{Q}\mathbf{v}|| = ||\mathbf{v}||. \tag{1.104}$$

Therefore, an orthogonal transformation leaves the magnitude of a vector unchanged. This fact, along with equation (1.100), means an orthogonal transformation also leaves the angle formed by vectors unchanged.

The magnitude of the orthogonal tensor is given from equation (1.102) as

$$||\boldsymbol{Q}|| = \sqrt{3}, \tag{1.105}$$

noting that $||\boldsymbol{Q}|| = \sqrt{\mathrm{tr}(\boldsymbol{Q}\boldsymbol{Q}^T)} = \sqrt{\mathrm{tr}\mathbf{I}}$.

As well as leaving the scalar product of two vectors unchanged (see equation (1.100)), an orthogonal transformation also leaves the trace of two tensors unchanged, that is,

$$\mathrm{tr}\{(\boldsymbol{Q}\mathbf{A}\boldsymbol{Q}^T)(\boldsymbol{Q}\mathbf{B}\boldsymbol{Q}^T)\} = \mathrm{tr}(\mathbf{AB}). \tag{1.106}$$

Furthermore, in addition to leaving the magnitude of a vector unchanged (see equation (1.104)), an orthogonal transformation also leaves the magnitude of a tensor unchanged, that is,

$$||\boldsymbol{Q}\mathbf{T}\boldsymbol{Q}^T|| = ||\mathbf{T}||, \tag{1.107}$$

noting equations (1.81), (1.102) and

$$\sqrt{\mathrm{tr}\{\boldsymbol{Q}\mathbf{T}\boldsymbol{Q}^T(\boldsymbol{Q}\mathbf{T}\boldsymbol{Q}^T)^T\}} = \sqrt{\mathrm{tr}(\boldsymbol{Q}\mathbf{T}\boldsymbol{Q}^T\boldsymbol{Q}\mathbf{T}^T\boldsymbol{Q}^T)} = \sqrt{\mathrm{tr}(\mathbf{T}\mathbf{T}^T)}.$$

Then, the following tensor possessing the components Q_{ij} in equation (1.14) satisfies equation (1.16), the direct notation of which is identical to equation (1.102) and thus it is an orthogonal tensor:

$$\mathbf{Q} = Q_{ij}\mathbf{e}_i \otimes \mathbf{e}_j = Q_{ij}\mathbf{e}_i^* \otimes \mathbf{e}_j^* = \mathbf{e}_i \otimes \mathbf{e}_i^* \tag{1.108}$$

noting that

$$Q_{ij}\mathbf{e}_i \otimes \mathbf{e}_j = \mathbf{e}_i \otimes (\mathbf{e}_i^* \cdot \mathbf{e}_j)\mathbf{e}_j = \mathbf{e}_i \otimes \mathbf{e}_i^* = (\mathbf{e}_i \cdot \mathbf{e}_r^*)\mathbf{e}_r^* \otimes \mathbf{e}_i^* = Q_{ri}\mathbf{e}_r^* \otimes \mathbf{e}_i^*$$

by use of equation (1.17). Furthermore, because of

$$\mathbf{e}_i = \mathbf{e}_r\delta_{ir} = \mathbf{e}_r \otimes \mathbf{e}_r^*\mathbf{e}_i^*, \quad \mathbf{e}_i^* = \mathbf{e}_r^*\delta_{ir} = \mathbf{e}_r^* \otimes \mathbf{e}_r\mathbf{e}_i,$$

one has the expressions

$$\mathbf{e}_i = \mathbf{Q}\mathbf{e}_i^*, \quad \mathbf{e}_i^* = \mathbf{Q}^T\mathbf{e}_i. \tag{1.109}$$

An orthogonal tensor is expressed in matrix form as follows:

$$[Q_{ij}] = \begin{bmatrix} \cos\varphi\cos\theta - \cos\phi\cos\theta\sin\varphi & \cos\varphi\sin\theta + \cos\phi\cos\theta\sin\varphi & \sin\varphi\sin\phi \\ -\sin\varphi\cos\theta - \cos\phi\cos\theta\cos\varphi & -\sin\varphi\sin\theta + \cos\phi\cos\theta\cos\varphi & \cos\varphi\sin\phi \\ \sin\phi\sin\theta & -\sin\phi\cos\theta & \cos\phi \end{bmatrix}, \tag{1.110}$$

where θ, ϕ, φ are the Euler angles shown in Figure 1.1.

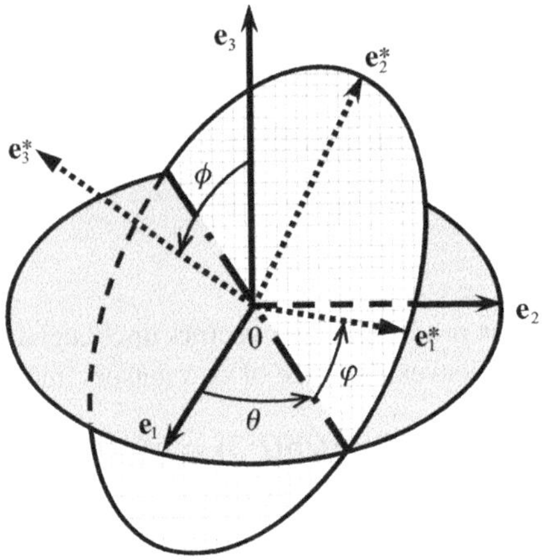

Figure 1.1 Rotation of coordinate system shown by Euler angles

The rotational transformation tensor $\mathbf{Q}^{(i)}$ around the base vector $\mathbf{e}_i$ is given from equation (1.110) as follows:

$$\mathbf{Q}^{(1)} = \begin{bmatrix} 1 & 0 & 0 \\ 0 & \cos\phi & \sin\phi \\ 0 & -\sin\phi & \cos\phi \end{bmatrix}, \quad \mathbf{Q}^{(2)} = \begin{bmatrix} \cos\varphi & 0 & -\sin\varphi \\ 0 & 1 & 0 \\ \sin\varphi & 0 & \cos\varphi \end{bmatrix},$$

$$\mathbf{Q}^{(3)} = \begin{bmatrix} \cos\theta & \sin\theta & 0 \\ -\sin\theta & \cos\theta & 0 \\ 0 & 0 & 1 \end{bmatrix}. \tag{1.111}$$

Hereinafter, consider only $\mathbf{Q}^{(3)}$ and denote it simply by $\mathbf{Q}$, that is,

$$\begin{aligned} \mathbf{Q} &= Q_{11}\mathbf{e}_1 \otimes \mathbf{e}_1 + Q_{12}\mathbf{e}_1 \otimes \mathbf{e}_2 + Q_{21}\mathbf{e}_2 \otimes \mathbf{e}_1 + Q_{22}\mathbf{e}_2 \otimes \mathbf{e}_2 + \mathbf{e}_3 \otimes \mathbf{e}_3 \\ &= \cos\theta\mathbf{e}_1 \otimes \mathbf{e}_1 + \sin\theta\mathbf{e}_1 \otimes \mathbf{e}_2 - \sin\theta\mathbf{e}_2 \otimes \mathbf{e}_1 + \cos\theta\mathbf{e}_2 \otimes \mathbf{e}_2 + \mathbf{e}_3 \otimes \mathbf{e}_3 \\ &= (\mathbf{e}_1 \otimes \mathbf{e}_1 + \mathbf{e}_2 \otimes \mathbf{e}_2)\cos\theta + (\mathbf{e}_1 \otimes \mathbf{e}_2 - \mathbf{e}_2 \otimes \mathbf{e}_1)\sin\theta + \mathbf{e}_3 \otimes \mathbf{e}_3 \end{aligned} \tag{1.112}$$

or

$$\mathbf{Q} = \begin{bmatrix} Q_{11} & Q_{12} & Q_{13} \\ Q_{21} & Q_{22} & Q_{23} \\ Q_{31} & Q_{32} & Q_{33} \end{bmatrix} = \begin{bmatrix} \cos\theta & \sin\theta & 0 \\ -\sin\theta & \cos\theta & 0 \\ 0 & 0 & 1 \end{bmatrix}, \tag{1.113}$$

for which we can write

$$\begin{Bmatrix} \mathbf{e}_1^* \\ \mathbf{e}_2^* \\ \mathbf{e}_3^* \end{Bmatrix} = \begin{Bmatrix} Q_{1r}\mathbf{e}_r \\ Q_{2r}\mathbf{e}_r \\ Q_{3r}\mathbf{e}_r \end{Bmatrix} = \begin{bmatrix} Q_{11} & Q_{12} & Q_{13} \\ Q_{21} & Q_{22} & Q_{23} \\ Q_{31} & Q_{32} & Q_{33} \end{bmatrix} \begin{Bmatrix} \mathbf{e}_1 \\ \mathbf{e}_2 \\ \mathbf{e}_3 \end{Bmatrix} = \begin{bmatrix} \cos\theta & \sin\theta & 0 \\ -\sin\theta & \cos\theta & 0 \\ 0 & 0 & 1 \end{bmatrix} \begin{Bmatrix} \mathbf{e}_1 \\ \mathbf{e}_2 \\ \mathbf{e}_3 \end{Bmatrix} \tag{1.114}$$

(see Figure 1.2). Equation (1.112) is referred to as the *canonical expression for orthogonal tensors*.

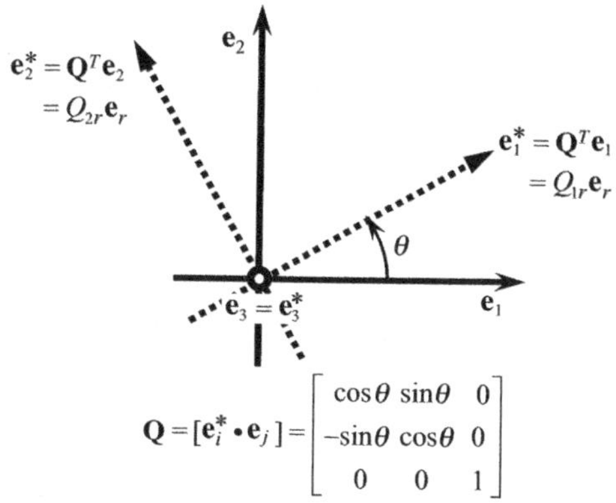

Figure 1.2 Coordinate system with base $\{\mathbf{e}_i^*\}$ rotated around the axis $\mathbf{e}_3$ (illustrated in two-dimensional state $\mathbf{e}_3 = \mathbf{e}_3^*$)

The orthogonal tensor causing a π*-rotation* is given by

$$\mathbf{Q}_i^{\pi} = 2\mathbf{e}_i \otimes \mathbf{e}_i - \mathbf{I} \text{ (no sum)}, \tag{1.115}$$

that is,

$$\mathbf{Q}_1^{\pi} = \begin{bmatrix} 1 & 0 & 0 \\ 0 & -1 & 0 \\ 0 & 0 & -1 \end{bmatrix}, \quad \mathbf{Q}_2^{\pi} = \begin{bmatrix} -1 & 0 & 0 \\ 0 & 1 & 0 \\ 0 & 0 & -1 \end{bmatrix}, \quad \mathbf{Q}_3^{\pi} = \begin{bmatrix} -1 & 0 & 0 \\ 0 & 1 & 0 \\ 0 & 0 & -1 \end{bmatrix} \tag{1.116}$$

by which an arbitrary vector $\mathbf{v}$ is π-rotated around the axis $\mathbf{e}_i$.

The orthogonal tensor causing a *reversal rotation*, that is, *reflection*, is given by

$$\mathbf{Q}_i^{R} = \mathbf{I} - 2\mathbf{e}_i \otimes \mathbf{e}_i \text{ (no sum)}, \tag{1.117}$$

that is,

$$\mathbf{Q}_1^{R} = \begin{bmatrix} -1 & 0 & 0 \\ 0 & 1 & 0 \\ 0 & 0 & 1 \end{bmatrix}, \quad \mathbf{Q}_2^{R} = \begin{bmatrix} 1 & 0 & 0 \\ 0 & -1 & 0 \\ 0 & 0 & 1 \end{bmatrix}, \quad \mathbf{Q}_3^{R} = \begin{bmatrix} 1 & 0 & 0 \\ 0 & 1 & 0 \\ 0 & 0 & -1 \end{bmatrix} \tag{1.118}$$

by which an arbitrary vector $\mathbf{v}$ is reflected in the plane perpendicular to the base vector $\mathbf{e}_i$.

1.4.6 Tensor Decompositions

Several types of tensor decomposition are often used for convenience, as will be described below.

Symmetric and skew-symmetric tensors

The tensor $\mathbf{T}$ is additively decomposed into the *symmetric tensor* $\boldsymbol{S}$ and the *skew-symmetric* (or *anti-symmetric*) *tensor* $\boldsymbol{\Omega}$ as follows:

$$\mathbf{T} = \boldsymbol{S} + \boldsymbol{\Omega}, \tag{1.119}$$

where

$$\boldsymbol{S} \equiv \mathrm{sym}[\mathbf{T}] \equiv \frac{1}{2}(\mathbf{T} + \mathbf{T}^T), \quad \boldsymbol{\Omega} \equiv \mathrm{ant}[\mathbf{T}] \equiv \frac{1}{2}(\mathbf{T} - \mathbf{T}^T) \tag{1.120}$$

which satisfy

$$\boldsymbol{S}^T = \boldsymbol{S}, \quad \boldsymbol{\Omega}^T = -\boldsymbol{\Omega}, \tag{1.121}$$

$$\left.\begin{array}{l} \mathbf{a} \cdot \boldsymbol{S}\mathbf{b} = \mathbf{b} \cdot \boldsymbol{S}\mathbf{a} \\ \mathbf{a} \cdot (\boldsymbol{\Omega}\mathbf{b}) = -\mathbf{b} \cdot (\boldsymbol{\Omega}\mathbf{a}) \end{array}\right\}, \tag{1.122}$$

noting equation (1.87). Equation (1.119) is called the *Cartesian decomposition*. Here, it follows that

$$\boldsymbol{S}\boldsymbol{S}^T = \boldsymbol{S}^2, \tag{1.123}$$

$$\mathrm{tr}(\boldsymbol{S}\boldsymbol{\Omega}) = \mathrm{tr}(\boldsymbol{S}\boldsymbol{\Omega}^T) = 0 \tag{1.124}$$

because of $\mathrm{tr}(\boldsymbol{S}\boldsymbol{\Omega}) = \mathrm{tr}((\boldsymbol{S}\boldsymbol{\Omega})^T) = \mathrm{tr}(\boldsymbol{\Omega}^T\boldsymbol{S}) = -\mathrm{tr}(\boldsymbol{\Omega}\boldsymbol{S}) = -\mathrm{tr}(\boldsymbol{S}\boldsymbol{\Omega})$, noting equation (1.91). The skew-symmetric tensor fulfills

$$\mathrm{tr}\boldsymbol{\Omega} = 0, \quad \mathbf{a} \cdot (\boldsymbol{\Omega}\mathbf{a}) = 0, \tag{1.125}$$

exploiting equations $(1.81)_1$, (1.87), and $(1.121)_2$.

By virtue of equation (1.124), it follows that

$$\mathrm{tr}(\mathbf{AB}) = \mathrm{tr}(\mathrm{sym}[\mathbf{A}]\mathrm{sym}[\mathbf{B}]) + \mathrm{tr}(\mathrm{ant}[\mathbf{A}]\mathrm{ant}[\mathbf{B}]). \tag{1.126}$$

Spherical and deviatoric tensors

The tensor $\mathbf{T}$ can decomposed as follows:

$$\mathbf{T} = \mathbf{T}_m + \mathbf{T}', \tag{1.127}$$

$$\mathbf{T}_m \equiv T_m\mathbf{I}, \quad T_m \equiv \frac{1}{3}\mathrm{tr}\mathbf{T}, \quad \mathbf{T}' \equiv \mathbf{T} - T_m\mathbf{I} \ (\mathrm{tr}\mathbf{T}' = 0), \tag{1.128}$$

where $\mathbf{T}_m$ and $\mathbf{T}'$ are called the *spherical* (or *mean*) *part* and the *deviatoric part*, respectively, of the tensor $\mathbf{T}$. A prime $(\,\bullet\,)'$ is used for the deviatoric part throughout this book.

Normal and tangential tensors: Projection tensors

The vector $\mathbf{v}$ can be decomposed as follows:

$$\mathbf{v} = \mathbf{v}_n + \mathbf{v}_t, \tag{1.129}$$

where

$$\mathbf{v}_n \equiv \mathbf{P}_n\mathbf{v}, \quad \mathbf{v}_t \equiv \mathbf{P}_t\mathbf{v} \tag{1.130}$$

with

$$\left.\begin{aligned} \mathbf{P}_n &\equiv \mathbf{n}\otimes\mathbf{n} \\ \mathbf{P}_t &\equiv \mathbf{I}-\mathbf{P}_n = \mathbf{I}-\mathbf{n}\otimes\mathbf{n} \end{aligned}\right\}, \tag{1.131}$$

$\mathbf{n}$ being an arbitrary unit vector. $\mathbf{P}_n$ and $\mathbf{P}_t$ are called the second-order *normal projection tensor* and *tangential projection tensor*, respectively, and fulfill

$$\mathbf{P}_n^2 = \mathbf{P}_n, \quad \mathbf{P}_n\mathbf{P}_t = \mathbf{0}. \tag{1.132}$$

$\mathbf{v}_n$ and $\mathbf{v}_t$ are the projection of the vector $\mathbf{v}$ onto the vector $\mathbf{n}$ and the orthogonal projection of $\mathbf{v}$ onto the plane normal to the vector $\mathbf{n}$, respectively. Here, $\mathbf{v}_n$ and $\mathbf{v}_t$ are also interpreted as the vector decompositions into the normal part and the tangential part of the plane normal to the vector $\mathbf{n}$.

Similarly, the second-order tensor $\mathbf{T}$ can be decomposed as follows:

$$\mathbf{T} = \mathbf{T}_n + \mathbf{T}_t, \tag{1.133}$$

where

$$\mathbf{T}_n \equiv \mathcal{P}_n\mathbf{T}, \quad \mathbf{T}_t \equiv \mathcal{P}_t\mathbf{T} \tag{1.134}$$

with

$$\left.\begin{aligned} \mathcal{P}_n &\equiv \mathbf{N}\otimes\mathbf{N} \\ \mathcal{P}_t &\equiv \overline{\mathcal{I}}-\mathcal{P}_n = \overline{\mathcal{I}}-\mathbf{N}\otimes\mathbf{N} \end{aligned}\right\}, \tag{1.135}$$

$\mathbf{N}$ being an arbitrary second-order unit tensor and $\overline{\mathcal{I}} = \delta_{ik}\delta_{jl}\mathbf{e}_i\otimes\mathbf{e}_j\otimes\mathbf{e}_k\otimes\mathbf{e}_l$ being the fourth-order identity tensor which will be described in Section 1.5.4. The fourth-order tensors $\mathcal{P}_n$ and $\mathcal{P}_t$ are called also the *normal-projection tensor* and the *tangential-projection tensor*, respectively, and fulfill

$$\mathcal{P}_n^2 = \mathcal{P}_n, \quad \mathcal{P}_n\mathcal{P}_t = \mathbf{O}. \tag{1.136}$$

1.4.7 Axial Vector

The anti-symmetric tensor fulfills the following properties as known by comparing equations (1.85) and $(1.122)_2$ regarding $\mathbf{a}$ and $\mathbf{b}$ as the base vectors.

1. Corresponding non-diagonal components have the same absolute value but opposite sign.
2. Diagonal components are zero and thus eigenvalues are zero. Therefore, the general anti-symmetric tensor $\boldsymbol{\Omega}$ is given by

$$[\Omega_{ij}] = \begin{bmatrix} 0 & \Omega_{12} & \Omega_{13} \\ & 0 & \Omega_{23} \\ \text{ant.} & & 0 \end{bmatrix}. \tag{1.137}$$

The anti-symmetric tensor thus possesses only three components, and we can infer that it can be related to a vector uniquely. We examine it below.

The *axial vector* $\boldsymbol{\omega}$ is defined as the vector which fulfills the following equation for the skew-symmetric tensor $\boldsymbol{\Omega}$ and an arbitrary vector $\mathbf{a}$:

$$\boldsymbol{\Omega}\mathbf{a} = \boldsymbol{\omega} \times \mathbf{a}. \tag{1.138}$$

Choosing $\mathbf{a}$ as the base vector $\mathbf{e}_j$ and making the scalar product with $\mathbf{e}_i$, we find

$$\Omega_{ij} = \mathbf{e}_i \cdot \boldsymbol{\Omega}\mathbf{e}_j = \mathbf{e}_i \cdot \boldsymbol{\omega} \times \mathbf{e}_j = \varepsilon_{pqr}(\mathbf{e}_i)_p \omega_q (\mathbf{e}_j)_r = \varepsilon_{pqr}\delta_{ip}\omega_q\delta_{jr} = \varepsilon_{iqj}\omega_q = -\varepsilon_{ijq}\omega_q,$$

that is,

$$\Omega_{12} = -\omega_3, \quad \Omega_{23} = -\omega_1, \quad \Omega_{13} = \omega_2. \tag{1.139}$$

Then $\boldsymbol{\Omega}$ is expressed as follows:

$$\boxed{\boldsymbol{\Omega} = -\boldsymbol{\varepsilon}\boldsymbol{\omega} = -\varepsilon_{ijk}\omega_k \mathbf{e}_i \otimes \mathbf{e}_j} \tag{1.140}$$

that is,

$$[\Omega_{ij}] = \begin{bmatrix} 0 & -\omega_3 & \omega_2 \\ & 0 & -\omega_1 \\ & \text{ant.} & 0 \end{bmatrix}, \tag{1.141}$$

noting that

$$\boldsymbol{\Omega} = -\boldsymbol{\varepsilon}\boldsymbol{\omega} = \varepsilon_{ijk}\mathbf{e}_i \otimes \mathbf{e}_j \otimes \mathbf{e}_k \omega_r \mathbf{e}_r. \tag{1.142}$$

The inverse relation of equation (1.140) is given by

$$\boldsymbol{\omega} = -\frac{1}{2}\boldsymbol{\varepsilon}\boldsymbol{\Omega} = -\frac{1}{2}\Omega_{ij}\mathbf{e}_i \times \mathbf{e}_j = -\frac{1}{2}\varepsilon_{irs}\Omega_{rs}\mathbf{e}_i \tag{1.143}$$

because of $\omega_p \mathbf{e}_p = \delta_{pq}\omega_q\mathbf{e}_p = \varepsilon_{ijp}\varepsilon_{ijq}\omega_q\mathbf{e}_p/2 = -\varepsilon_{ijp}\Omega_{ij}\mathbf{e}_p/2 = -\Omega_{ij}\mathbf{e}_i \times \mathbf{e}_j/2$ by virtue of equation (1.57).

It is follows from equation (1.140) that

$$\boldsymbol{\Omega} = \mathbf{I} \times \boldsymbol{\omega} = \boldsymbol{\omega} \times \mathbf{I} \tag{1.144}$$

because of

$$-\varepsilon_{ijk}\omega_k \mathbf{e}_i \otimes \mathbf{e}_j = \begin{cases} \mathbf{e}_i \otimes \varepsilon_{jik}\omega_k \mathbf{e}_j = \mathbf{e}_i \otimes \mathbf{e}_i \times \omega_k \mathbf{e}_k = \mathbf{e}_i \otimes \mathbf{e}_i \times \boldsymbol{\omega} \\ -\mathbf{e}_j \times \mathbf{e}_k \omega_k \otimes \mathbf{e}_j = \mathbf{e}_k \omega_k \times \mathbf{e}_j \otimes \mathbf{e}_j = \boldsymbol{\omega} \times \mathbf{e}_j \otimes \mathbf{e}_j \end{cases},$$

using equations (1.32) and (1.201) shown later.

It follows in view of equation $(1.48)_2$ that

$$(\mathbf{u} \otimes \mathbf{v} - \mathbf{v} \otimes \mathbf{u})\mathbf{a} = (\mathbf{v} \cdot \mathbf{a})\mathbf{u} - (\mathbf{u} \cdot \mathbf{a})\mathbf{v} = (\mathbf{v} \times \mathbf{u}) \times \mathbf{a}. \tag{1.145}$$

Thus, by virtue of equation (1.138), $\mathbf{v} \times \mathbf{u}$ is the axial vector of the anti-symmetric second-order tensor $\mathbf{u} \otimes \mathbf{v} - \mathbf{v} \otimes \mathbf{u}$.

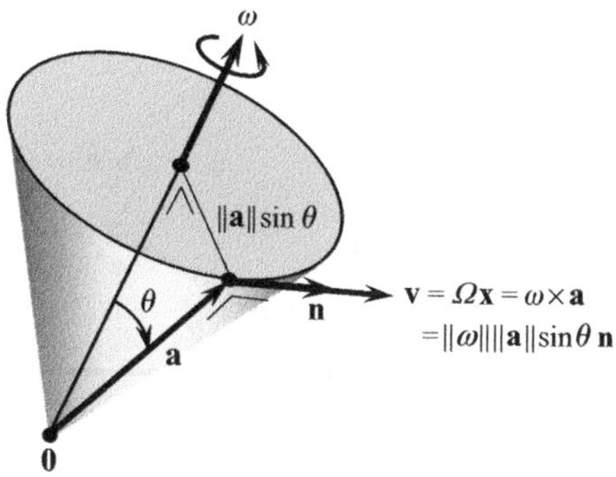

Figure 1.3 Axial vector in case of rotation

The relation of the skew-symmetric tensor and the axial vector is illustrated in Figure 1.3 for the rotation phenomenon, regarding $\boldsymbol{\omega}$ as the angular velocity and $\mathbf{v}$ as the peripheral velocity.

Now consider rotation in the Cartesian coordinate system. The time-differentiation of the base vector $\mathbf{e}_i^*$ in equation $(1.109)_2$ leads to

$$\overset{\bullet}{\mathbf{e}}{}_i^* = \overset{\bullet}{\mathbf{Q}}{}^T \mathbf{e}_i + \mathbf{Q}^T \overset{\bullet}{\mathbf{e}}_i = \overset{\bullet}{\mathbf{Q}}{}^T \mathbf{Q}\mathbf{e}_i^* + \mathbf{Q}^T \overset{\bullet}{\mathbf{e}}_i. \tag{1.146}$$

If we put $\overset{\bullet}{\mathbf{e}}_i = \mathbf{0}$ in equation (1.146), we have the variation of $\mathbf{e}_i^*$ observed from the coordinate system with the base $\{\mathbf{e}_i\}$, that is,

$$\overset{\bullet}{\mathbf{e}}{}_i^* = \boldsymbol{\Omega}\mathbf{e}_i^*, \tag{1.147}$$

where

$$\boldsymbol{\Omega} \equiv \overset{\bullet}{\mathbf{Q}}{}^T \mathbf{Q} = \overset{\bullet}{\mathbf{e}}{}_r^* \otimes \mathbf{e}_r^*, \quad \Omega_{ij} = (\overset{\bullet}{\mathbf{e}}{}_r^* \cdot \mathbf{e}_i)(\mathbf{e}_r^* \cdot \mathbf{e}_j). \tag{1.148}$$

$\boldsymbol{\Omega}$ designates the spin of the base vector $\mathbf{e}_i^*$ observed from the coordinate system with the base $\{\mathbf{e}_i\}$, bearing in mind that only the direction is changeable in the base vector because of the unit vector. $\overset{\bullet}{\mathbf{e}}{}_i^*$ can be rewritten as

$$\overset{\bullet}{\mathbf{e}}{}_r^* = \boldsymbol{\omega} \times \mathbf{e}_r^* \tag{1.149}$$

by virtue of the property in equation (1.138), where $\boldsymbol{\omega}$ is the angular velocity of $\mathbf{e}_i^*$, which is related to $\boldsymbol{\Omega}$ as

$$\boldsymbol{\omega} = -\frac{1}{2}\boldsymbol{\varepsilon}\boldsymbol{\Omega} = -\frac{1}{2}\Omega_{ij}\mathbf{e}_i \times \mathbf{e}_j = -\frac{1}{2}\varepsilon_{irs}\Omega_{rs}\mathbf{e}_i, \tag{1.150}$$

noting equation (1.143), where Ω_{rs} is specified by equation $(1.148)_2$ (see Figure 1.4).

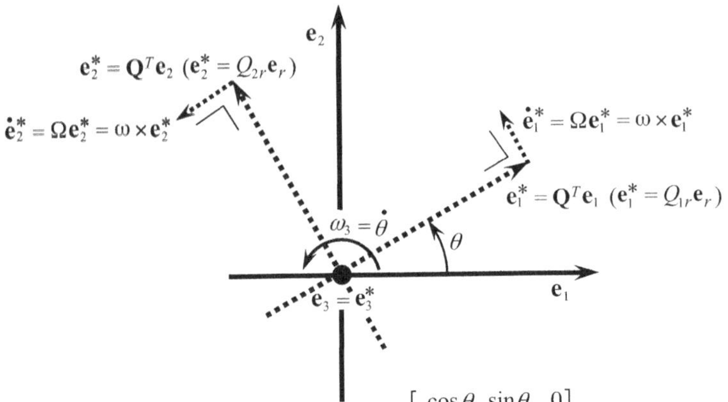

$$\mathbf{Q} = [\mathbf{e}_i^* \cdot \mathbf{e}_j] = \begin{bmatrix} \cos\theta & \sin\theta & 0 \\ -\sin\theta & \cos\theta & 0 \\ 0 & 0 & 1 \end{bmatrix}$$

$$\mathbf{\Omega} = \dot{\mathbf{Q}}^T\mathbf{Q} = \begin{bmatrix} 0 & -1 & 0 \\ 1 & 0 & 0 \\ 0 & 0 & 1 \end{bmatrix}\dot{\theta}, \quad \boldsymbol{\omega} = \left\{-\frac{1}{2}\varepsilon_{rsi}\Omega_{rs}\right\} = \begin{Bmatrix} 0 \\ 0 \\ \dot{\theta} \end{Bmatrix}$$

Figure 1.4 Rotation of coordinate system with base $\{\mathbf{e}_i^*\}$ (illustrated in two-dimensional state $\mathbf{e}_3 = \mathbf{e}_3^*$)

1.4.8 Determinant

The determinant is defined by

$$\boxed{\det\mathbf{T} = |T_{ij}| = \begin{vmatrix} T_{11} & T_{12} & T_{13} \\ T_{21} & T_{22} & T_{23} \\ T_{31} & T_{32} & T_{33} \end{vmatrix} = \begin{Bmatrix} \varepsilon_{pqr}T_{1p}T_{2q}T_{3r} \\ \varepsilon_{pqr}T_{p1}T_{q2}T_{r3} \end{Bmatrix} = \frac{1}{3!}\varepsilon_{abc}\varepsilon_{pqr}T_{ap}T_{bq}T_{cr}} \tag{1.151}$$

where the last term is divided by $3!(= 6)$ which is the total number of permutations of thee numbers. Equation (1.151) can be rewritten as

$$\det\mathbf{T} = \frac{1}{3}T_{ap}(\mathrm{cof}\,\mathbf{T})_{ap}, \quad \det\mathbf{T} = \frac{1}{3}\mathrm{tr}\{\mathbf{T}(\mathrm{cof}\,\mathbf{T})^T\}, \tag{1.152}$$

where $\mathrm{cof}\,\mathbf{T}$ is the *cofactor* defined by

$$\boxed{(\mathrm{cof}\,\mathbf{T})_{ap} = \frac{1}{2}\varepsilon_{abc}\varepsilon_{pqr}T_{bq}T_{cr}} \tag{1.153}$$

The cofactor is defined through multiplying the minor determinant lacking the ith row and jth colum components by the sign $(-1)^{i+j}$.

Now, note that the following equation satisfies equation (1.152).

$$\frac{1}{3}(\det\mathbf{T})\mathbf{I} = \frac{1}{3}\mathbf{T}(\mathrm{cof}\,\mathbf{T})^T, \quad \frac{1}{3}(\det\mathbf{T})\delta_{ij} = \frac{1}{3}T_{ip}(\mathrm{cof}\,\mathbf{T})_{jp}. \tag{1.154}$$

Equation (1.154) can be rewritten as

$$\mathbf{T}(\mathrm{cof}\,\mathbf{T})^T = (\det\mathbf{T})\mathbf{I}, \quad T_{ip}(\mathrm{cof}\,\mathbf{T})_{jp} = (\det\mathbf{T})\delta_{ij} \tag{1.155}$$

from which we have

$$\mathbf{T} = (\det\mathbf{T})(\mathrm{cof}\,\mathbf{T})^{-T}, \quad \mathbf{T}_{ij} = (\det\mathbf{T})(\mathrm{cof}\,\mathbf{T})^{-1}_{ji}, \tag{1.156}$$

$$\mathrm{cof}\,\mathbf{T} = (\det\mathbf{T})\mathbf{T}^{-T}, \quad (\mathrm{cof}\,\mathbf{T})_{ij} = (\det\mathbf{T})T^{-1}_{ji} \tag{1.157}$$

and

$$\boxed{\mathbf{T}^{-1} = \frac{(\mathrm{cof}\,\mathbf{T})^T}{\det\mathbf{T}}}, \quad T^{-1}_{ij} = \frac{(\mathrm{cof}\,\mathbf{T})_{ji}}{\det\mathbf{T}}. \tag{1.158}$$

Equation (1.158) is represented for the 2 × 2 and 3 × 3 matrices as follows:

$$[\mathbf{T}^{-1}] = \begin{bmatrix} T_{22} & -T_{21} \\ -T_{12} & T_{11} \end{bmatrix} \Big/ \begin{vmatrix} T_{11} & T_{12} \\ T_{21} & T_{22} \end{vmatrix},$$

$$[\mathbf{T}^{-1}] = \begin{bmatrix} T_{22}T_{33} - T_{23}T_{32} & T_{32}T_{13} - T_{33}T_{12} & T_{12}T_{23} - T_{13}T_{22} \\ T_{23}T_{31} - T_{21}T_{33} & T_{33}T_{11} - T_{31}T_{13} & T_{13}T_{21} - T_{11}T_{23} \\ T_{21}T_{32} - T_{22}T_{31} & T_{31}T_{12} - T_{32}T_{11} & T_{11}T_{22} - T_{12}T_{21} \end{bmatrix} \Big/ \begin{vmatrix} T_{11} & T_{12} & T_{12} \\ T_{21} & T_{22} & T_{23} \\ T_{31} & T_{32} & T_{33} \end{vmatrix}. \tag{1.158'}$$

We obtain the following equations:

$$\left.\begin{array}{l} \det[-\mathbf{T}] = -\det\mathbf{T}, \quad \det(s\mathbf{T}) = s^3\det\mathbf{T}, \quad \det\mathbf{T} = \det\mathbf{T}^T \\ \det[\mathbf{AB}] = \det\mathbf{A}\det\mathbf{B}, \quad \det[\mathbf{T}^n] = (\det\mathbf{T})^n, \quad \det(\exp\mathbf{T}) = \exp(\mathrm{tr}\mathbf{T}) \end{array}\right\}, \tag{1.159}$$

$$\det[\mathbf{a}\otimes\mathbf{b}] = 0, \tag{1.160}$$

$$\det[\mathbf{T}^{-1}] = (\det\mathbf{T})^{-1} \tag{1.161}$$

by virtue of the following:

$$\left.\begin{array}{l} \det[-\mathbf{T}] = \varepsilon_{pqr}(-1)T_{1p}(-1)T_{2q}(-1)T_{3r} = -\varepsilon_{pqr}T_{1p}T_{2q}T_{3r} \\ \det[s\mathbf{T}] = \varepsilon_{pqr}sT_{1p}sT_{2q}sT_{3r} = s^3\varepsilon_{pqr}T_{1p}T_{2q}T_{3r} \\ \det\mathbf{T} = \dfrac{1}{3!}\varepsilon_{abc}\varepsilon_{pqr}T_{ap}T_{bq}T_{cr} = \dfrac{1}{3!}\varepsilon_{pqr}\varepsilon_{abc}T_{pa}T_{qb}T_{rc} = \dfrac{1}{3!}\varepsilon_{abc}\varepsilon_{pqr}T_{pa}T_{qb}T_{rc} \\ \det[\mathbf{AB}] = \varepsilon_{pqr}(A_{1a}B_{ap})(A_{2b}B_{bq})(A_{3c}B_{cr}) = \varepsilon_{pqr}A_{1a}A_{2b}A_{3c}B_{ap}B_{bq}B_{cr} = A_{1a}A_{2b}A_{3c}\varepsilon_{abc}\det\mathbf{B} \end{array}\right\},$$

$$\det(\exp\mathbf{T}) = \begin{vmatrix} \exp(T_{11}) & 0 & 0 \\ 0 & \exp(T_{22}) & 0 \\ 0 & 0 & \exp(T_{33}) \end{vmatrix} = \exp(T_{11})\exp(T_{22})\exp(T_{33})$$
$$= \exp(T_{11} + T_{22} + T_{33})$$

(in spectral representation),

$$\det(\mathbf{a}\otimes\mathbf{b}) = \varepsilon_{ijk}(a_1b_i)(a_2b_j)(a_3b_k) = a_1a_2a_3\varepsilon_{ijk}b_ib_jb_k = (a_1a_2a_3)(\mathbf{b}\times\mathbf{b})\cdot\mathbf{b},$$

$$\det\mathbf{T}\,\det[\mathbf{T}^{-1}] = \det[\mathbf{TT}^{-1}] = 1.$$

An alternative proof of equation $(1.159)_6$ will be given later by applying equation (1.371) for the derivative of the determinant.

The following equations hold for the cofactor, noting equation (1.157) together with equation (1.161).

$$\left.\begin{aligned}&\mathrm{cof}(s\mathbf{T}) = s^2\mathrm{cof}\,\mathbf{T}\\&(\mathrm{cof}\,\mathbf{T})^T = \mathrm{cof}(\mathbf{T}^T),\quad (\mathrm{cof}\,\mathbf{T})^{-1} = \mathrm{cof}(\mathbf{T}^{-1}),\quad (\mathrm{cof}\,\mathbf{T})^{-T} = \mathrm{cof}(\mathbf{T}^{-T})\\&\mathrm{cof}(\mathbf{AB}) = \mathrm{cof}(\mathbf{A})\mathrm{cof}(\mathbf{B})\\&\mathrm{tr}(\mathrm{cof}\,\mathbf{T}) = \frac{1}{2}(\mathrm{tr}^2\mathbf{T} - \mathrm{tr}\mathbf{T}^2) = \mathrm{II}\\&\det(\mathrm{cof}\,\mathbf{T}) = 1\end{aligned}\right\}, \tag{1.162}$$

noting

$$\mathrm{tr}(\mathrm{cof}\,\mathbf{T}) = \frac{1}{2}\varepsilon_{abc}\varepsilon_{aqr}T_{bq}T_{cr} = \frac{1}{2}(\delta_{bq}\delta_{cr} - \delta_{br}\delta_{cq})T_{bq}T_{cr} = \frac{1}{2}(T_{qq}T_{rr} - T_{rq}T_{qp}).$$

II is called the second principal invariant as will be described in Section 1.6.1.

The vector product in equation (1.31) and the scalar triple product in equation (1.39) are described in terms of the determinant as follows:

$$\mathbf{a}\times\mathbf{b} = \begin{vmatrix} a_2 & a_3\\ b_2 & b_3\end{vmatrix}\mathbf{e}_1 + \begin{vmatrix} a_3 & a_1\\ b_3 & b_1\end{vmatrix}\mathbf{e}_2 + \begin{vmatrix} a_1 & a_2\\ b_1 & b_2\end{vmatrix}\mathbf{e}_3 = \begin{vmatrix}\mathbf{e}_1 & \mathbf{e}_2 & \mathbf{e}_3\\ a_1 & a_2 & a_3\\ b_1 & b_2 & b_3\end{vmatrix}, \tag{1.163}$$

$$\|\mathbf{a}\times\mathbf{b}\| = \sqrt{(a_2b_3 - a_3b_2)^2 + (a_3b_1 - a_1b_3)^2 + (a_1b_2 - a_2b_1)^2}, \tag{1.164}$$

$$[\mathbf{abc}] = \begin{vmatrix} a_1 & a_2 & a_3\\ b_1 & b_2 & b_3\\ c_1 & c_2 & c_3\end{vmatrix} = \begin{vmatrix}\mathbf{a}\cdot\mathbf{e}_1 & \mathbf{a}\cdot\mathbf{e}_2 & \mathbf{a}\cdot\mathbf{e}_3\\ \mathbf{b}\cdot\mathbf{e}_1 & \mathbf{b}\cdot\mathbf{e}_2 & \mathbf{b}\cdot\mathbf{e}_3\\ \mathbf{c}\cdot\mathbf{e}_1 & \mathbf{c}\cdot\mathbf{e}_2 & \mathbf{c}\cdot\mathbf{e}_3\end{vmatrix}, \tag{1.165}$$

$$(\mathbf{a}\times\mathbf{b})\cdot(\mathbf{c}\times\mathbf{d}) = \begin{vmatrix}\mathbf{a}\cdot\mathbf{c} & \mathbf{b}\cdot\mathbf{c}\\ \mathbf{a}\cdot\mathbf{d} & \mathbf{b}\cdot\mathbf{d}\end{vmatrix}, \tag{1.166}$$

$$[\mathbf{abc}][\mathbf{pqr}] = \begin{vmatrix} a_1 & a_2 & a_3\\ b_1 & b_2 & b_3\\ c_1 & c_2 & c_3\end{vmatrix}\begin{vmatrix} p_1 & q_2 & r_3\\ p_2 & q_2 & r_3\\ p_3 & q_2 & r_3\end{vmatrix} = \begin{vmatrix} a_ip_i & a_ip_i & a_iq_i\\ b_ip_i & b_ip_i & b_iq_i\\ c_ip_i & c_ip_i & c_iq_i\end{vmatrix} = \begin{vmatrix}\mathbf{a}\cdot\mathbf{p} & \mathbf{a}\cdot\mathbf{q} & \mathbf{a}\cdot\mathbf{r}\\ \mathbf{b}\cdot\mathbf{p} & \mathbf{b}\cdot\mathbf{q} & \mathbf{b}\cdot\mathbf{r}\\ \mathbf{c}\cdot\mathbf{p} & \mathbf{c}\cdot\mathbf{q} & \mathbf{c}\cdot\mathbf{r}\end{vmatrix}, \tag{1.167}$$

noting equation $(1.159)_4$. The following equation is derived as a special case of equation (1.167) for the three vectors $\mathbf{v}_1, \mathbf{v}_2, \mathbf{v}_3$:

$$v^2 = [\mathbf{v}_1\mathbf{v}_2\mathbf{v}_3]^2 = \begin{vmatrix}\mathbf{v}_1\cdot\mathbf{v}_1 & \mathbf{v}_1\cdot\mathbf{v}_2 & \mathbf{v}_1\cdot\mathbf{v}_3\\ & \mathbf{v}_2\cdot\mathbf{v}_2 & \mathbf{v}_2\cdot\mathbf{v}_3\\ \text{sym.} & & \mathbf{v}_3\cdot\mathbf{v}_3\end{vmatrix}, \tag{1.168}$$

where

$$v = [\mathbf{v}_1\mathbf{v}_2\mathbf{v}_3] = \varepsilon_{ijk}(\mathbf{v}_1)_i(\mathbf{v}_2)_j(\mathbf{v}_2)_k (= \varepsilon_{ijk}(\mathbf{v}_1\cdot\mathbf{e}_i)(\mathbf{v}_2\cdot\mathbf{e}_j)(\mathbf{v}_3\cdot\mathbf{e}_k)) \tag{1.169}$$

is the volume of the parallelepiped formed by the line-elements $\mathbf{v}_1, \mathbf{v}_2, \mathbf{v}_3$. Introducing the symbol

$$v_{ij} \equiv \mathbf{v}_i \cdot \mathbf{v}_j, \tag{1.170}$$

equation (1.168) can be written as

$$v^2 = \begin{vmatrix} v_{11} & v_{12} & v_{13} \\ & v_{22} & v_{23} \\ \text{sym.} & & v_{33} \end{vmatrix} = \det[v_{ij}]. \tag{1.171}$$

It follows from equation (1.167) that

$$\det\mathbf{T} = [\mathbf{Te}_1\ \mathbf{Te}_2\ \mathbf{Te}_3] = \mathbf{Te}_1 \cdot (\mathbf{Te}_2 \times \mathbf{Te}_2), \tag{1.172}$$

noting that

$$\det\mathbf{T} = \begin{vmatrix} T_{11} & T_{12} & T_{13} \\ T_{21} & T_{22} & T_{23} \\ T_{31} & T_{32} & T_{33} \end{vmatrix} = \begin{vmatrix} (\mathbf{e}_1 \cdot \mathbf{Te}_1) & (\mathbf{e}_1 \cdot \mathbf{Te}_2) & (\mathbf{e}_1 \cdot \mathbf{Te}_3) \\ (\mathbf{e}_2 \cdot \mathbf{Te}_1) & (\mathbf{e}_2 \cdot \mathbf{Te}_2) & (\mathbf{e}_2 \cdot \mathbf{Te}_3) \\ (\mathbf{e}_3 \cdot \mathbf{Te}_1) & (\mathbf{e}_3 \cdot \mathbf{Te}_2) & (\mathbf{e}_3 \cdot \mathbf{Te}_3) \end{vmatrix}.$$

$$= [\mathbf{e}_1\ \mathbf{e}_2\ \mathbf{e}_3][\mathbf{Te}_1\ \mathbf{Te}_2\ \mathbf{Te}_3] = [\mathbf{Te}_1\ \mathbf{Te}_2\ \mathbf{Te}_3]$$

Applying equations (1.159) to equations (1.102) and $(1.121)_2$, the determinants of the orthogonal tensor and the skew-symmetric tensor are given as

$$\boxed{\det\boldsymbol{Q} = \det\boldsymbol{Q}^T = \pm 1} \tag{1.173}$$

$$\boxed{\det\boldsymbol{\Omega} = 0} \tag{1.174}$$

1.4.9 On Solutions of Simultaneous Equation

The linear inhomogeneous simultaneous equation in the unknown variables x_r,

$$T_{ir}x_r = y_i \tag{1.175}$$

(where the y_i are not all zero), possesses a unique solution only if the determinant of the coefficient T_{ij} is not zero, that is,

$$\det[T_{ij}] \neq 0, \tag{1.176}$$

noting equation (1.158).

On the other hand, the necessary and sufficient condition for the linear homogeneous simultaneous equation

$$T_{ir}x_r = 0 \tag{1.177}$$

to possess a non-zero (non-trivial) solution is given by

$$\boxed{\det[T_{ij}] = 0} \tag{1.178}$$

as is known from $\mathbf{x} = \mathbf{T}^{-1}\mathbf{0} = (\text{cof}\,\mathbf{T})^T\mathbf{0}/\det\mathbf{T}$ obtained by virtue of equation (1.158); this solution is not unique because all scalar multiplications of any solution are also solutions.

1.4.10 Scalar Triple Products with Invariants

The following formulae of the scalar triple products related to the principal invariants hold:

$$\boxed{\begin{aligned}&[\mathbf{Ta\ b\ c}]+[\mathbf{a\ Tb\ c}]+[\mathbf{a\ b\ Tc}]=\mathrm{tr}\mathbf{T}[\mathbf{abc}]=\mathrm{I}[\mathbf{abc}]\\&[\mathbf{a\ Tb\ Tc}]+[\mathbf{Ta\ b\ Tc}]+[\mathbf{Ta\ Tb\ c}]=\mathrm{tr}(\mathrm{cof}\,\mathbf{T})[\mathbf{abc}]=\mathrm{II}[\mathbf{abc}]\\&[\mathbf{Ta\ Tb\ Tc}]=\det\mathbf{T}[\mathbf{abc}]=\mathrm{III}[\mathbf{abc}]\end{aligned}} \tag{1.179}$$

where I, II, III are *principal invariants* of $\mathbf{T}$, that is,

$$\mathrm{I}\equiv\mathrm{tr}\mathbf{T},\quad \mathrm{II}\equiv(\mathrm{tr}^2\mathbf{T}-\mathrm{tr}\mathbf{T}^2)/2,\quad \mathrm{III}\equiv\det\mathbf{T}$$

(see Section 1.6.1).

The proof for equation (1.179) is given below (cf. Chadwick, 1976; Kyoya, 2008).

Proof of equation (1.179)$_1$:
We have the relation

$$\begin{aligned}[\mathbf{Ta},\mathbf{b},\mathbf{c}]+[\mathbf{a},\mathbf{Tb},\mathbf{c}]+[\mathbf{a},\mathbf{b},\mathbf{Tc}]&=a_ib_jc_k([\mathbf{Te}_i,\mathbf{e}_j,\mathbf{e}_k]+[\mathbf{e}_i,\mathbf{Te}_j,\mathbf{e}_k]+[\mathbf{e}_i,\mathbf{e}_j,\mathbf{Te}_k])\\&=\varepsilon_{ijk}a_ib_jc_k([\mathbf{Te}_1,\mathbf{e}_2,\mathbf{e}_3]+[\mathbf{e}_1,\mathbf{Te}_2,\mathbf{e}_3]+[\mathbf{e}_1,\mathbf{e}_2,\mathbf{Te}_3])\\&=[\mathbf{abc}]([\mathbf{Te}_1,\mathbf{e}_2,\mathbf{e}_3]+[\mathbf{e}_1,\mathbf{Te}_2,\mathbf{e}_3]+[\mathbf{e}_1,\mathbf{e}_2,\mathbf{Te}_3]),\end{aligned} \tag{1.180}$$

making use of equation (1.43) and noting that

$$[\mathbf{Te}_i,\mathbf{e}_j,\mathbf{e}_k]+[\mathbf{e}_i,\mathbf{Te}_j,\mathbf{e}_k]+[\mathbf{e}_i,\mathbf{e}_j,\mathbf{Te}_k]=\varepsilon_{ijk}([\mathbf{Te}_1,\mathbf{e}_2,\mathbf{e}_3]+[\mathbf{e}_1,\mathbf{Te}_2,\mathbf{e}_3]+[\mathbf{e}_1,\mathbf{e}_2,\mathbf{Te}_3]). \tag{1.181}$$

Further, the terms in parentheses in the last equation in equation (1.180) are given by

$$\begin{aligned}[\mathbf{Te}_1,\mathbf{e}_2,\mathbf{e}_3]+[\mathbf{e}_1,\mathbf{Te}_2,\mathbf{e}_3]+[\mathbf{e}_1,\mathbf{e}_2,\mathbf{Te}_3]&=[T_{r1}\mathbf{e}_r,\mathbf{e}_2,\mathbf{e}_3]+[\mathbf{e}_1,T_{r2}\mathbf{e}_r,\mathbf{e}_3]+[\mathbf{e}_1,\mathbf{e}_2,T_{r3}\mathbf{e}_r]\\&=T_{r1}[\mathbf{e}_r\mathbf{e}_2\mathbf{e}_3]+T_{r2}[\mathbf{e}_1\mathbf{e}_r\mathbf{e}_3]+T_{r3}[\mathbf{e}_1\mathbf{e}_2\mathbf{e}_r]\\&=T_{11}[\mathbf{e}_1\mathbf{e}_2\mathbf{e}_3]+T_{22}[\mathbf{e}_1\mathbf{e}_2\mathbf{e}_3]+T_{33}[\mathbf{e}_1\mathbf{e}_2\mathbf{e}_3]\\&=T_{11}+T_{22}+T_{33}=\mathrm{tr}\mathbf{T}.\end{aligned} \tag{1.182}$$

Equation (1.179)$_1$ is obtained by substituting equation (1.182) into equation (1.180).

Proof of equation (1.179)$_2$:
Similarly to equation (1.180), we have first

$$\begin{aligned}&[\mathbf{a},\mathbf{Tb},\mathbf{Tc}]+[\mathbf{Ta},\mathbf{b},\mathbf{Tc}]+[\mathbf{Ta},\mathbf{Tb},\mathbf{c}]\\&\quad=\varepsilon_{ijk}a_ib_jc_k([\mathbf{e}_1,\mathbf{Te}_2,\mathbf{Te}_3]+[\mathbf{Te}_1,\mathbf{e}_2,\mathbf{Te}_3]+[\mathbf{Te}_1,\mathbf{Te}_2,\mathbf{e}_3])\\&\quad=[\mathbf{abc}]([\mathbf{e}_1,\mathbf{Te}_2,\mathbf{Te}_3]+[\mathbf{Te}_1,\mathbf{e}_2,\mathbf{Te}_3]+[\mathbf{Te}_1,\mathbf{Te}_2,\mathbf{e}_3]).\end{aligned} \tag{1.183}$$

Here, applying equations (1.43) and (1.181), the terms in parentheses are given by

$$
\begin{aligned}
&[\mathbf{e}_1, \mathbf{Te}_2, \mathbf{Te}_3] + [\mathbf{Te}_1, \mathbf{e}_2, \mathbf{Te}_3] + [\mathbf{Te}_1, \mathbf{Te}_2, \mathbf{e}_3] \\
&= T_{r2}T_{s3}[\mathbf{e}_1\mathbf{e}_r\mathbf{e}_s] + T_{r1}T_{s3}[\mathbf{e}_r\mathbf{e}_2\mathbf{e}_s] + T_{r1}T_{s2}[\mathbf{e}_r\mathbf{e}_s\mathbf{e}_3] \\
&= T_{r2}T_{s3}\varepsilon_{1rs} + T_{r1}T_{s3}\varepsilon_{r2s} + T_{r1}T_{s2}\varepsilon_{rs3} \\
&= (T_{22}T_{33} - T_{23}T_{32}) + (T_{11}T_{33} - T_{13}T_{31}) + (T_{11}T_{22} - T_{12}T_{12}) \\
&= T_{11}T_{22} + T_{22}T_{33} + T_{33}T_{11} - (T_{12}T_{12} + T_{23}T_{32} + T_{31}T_{13}) \\
&= \{T_{11}^2 + T_{22}^2 + T_{33}^2 + 2(T_{11}T_{22} + T_{22}T_{33} + T_{33}T_{11})\}/2 \\
&\quad - \{T_{11}^2 + T_{22}^2 + T_{33}^2 + 2(T_{12}T_{12} + T_{23}T_{32} + T_{31}T_{13})\}/2 \\
&= (T_{rr}T_{ss} - T_{rs}T_{sr})/2,
\end{aligned}
\qquad (1.184)
$$

leading to

$$
[\mathbf{e}_1, \mathbf{Te}_2, \mathbf{Te}_3] + [\mathbf{Te}_1, \mathbf{e}_2, \mathbf{Te}_3] + [\mathbf{Te}_1, \mathbf{Te}_2, \mathbf{e}_3] = \{(\mathrm{tr}\mathbf{T})^2 - \mathrm{tr}^2\mathbf{T}\}/2. \qquad (1.185)
$$

Equation $(1.179)_2$ is obtained by substituting equation (1.185) into equation (1.183).

Proof of equation $(1.179)_3$:
Changing to the component-based representations of the three vectors and then applying equations (1.43) and (1.181), we have

$$
[\mathbf{Ta}, \mathbf{Tb}, \mathbf{Tc}] = \varepsilon_{ijk}a_ib_jc_k[\mathbf{Te}_1, \mathbf{Te}_2, \mathbf{Te}_3] = [\mathbf{abc}][\mathbf{Te}_1, \mathbf{Te}_2, \mathbf{Te}_3]. \qquad (1.186)
$$

Equation $(1.179)_3$ is obtained by substituting equation (1.172) into equation (1.186).

1.4.11 Orthogonal Transformation of Scalar Triple Product

Consider the set of three vectors $\mathbf{a}, \mathbf{b}, \mathbf{c}$ which are arrayed as a right-handed set in this order fulfilling $[\mathbf{a}, \mathbf{b}, \mathbf{c}] > 0$ and the vectors $\mathbf{a}^*, \mathbf{b}^*, \mathbf{c}^*$ transformed from $\mathbf{a}, \mathbf{b}, \mathbf{c}$, respectively, by arbitrary orthogonal tensor $\boldsymbol{Q}$, fulfilling

$$
\left.
\begin{aligned}
\mathbf{a}^* = \boldsymbol{Q}\mathbf{a}, \quad \mathbf{b}^* = \boldsymbol{Q}\mathbf{b}, \quad \mathbf{c}^* = \boldsymbol{Q}\mathbf{c} \\
\mathbf{a} = \boldsymbol{Q}^T\mathbf{a}^*, \quad \mathbf{b} = \boldsymbol{Q}^T\mathbf{b}^*, \quad \mathbf{c} = \boldsymbol{Q}^T\mathbf{c}^*
\end{aligned}
\right\}. \qquad (1.187)
$$

By virtue of equation $(1.179)_3$, we find that

$$
[\mathbf{a}^*\mathbf{b}^*\mathbf{c}^*] = [\boldsymbol{Q}\mathbf{a}\boldsymbol{Q}\mathbf{b}\boldsymbol{Q}\mathbf{c}] = \det\boldsymbol{Q}[\mathbf{abc}] \qquad (1.188)
$$

which gives

$$
\left.
\begin{aligned}
&[\mathbf{abc}] = [\mathbf{a}^*\mathbf{b}^*\mathbf{c}^*] \quad \text{for } \det\boldsymbol{Q} = 1 \\
&[\mathbf{abc}] = -[\mathbf{a}^*\mathbf{b}^*\mathbf{c}^*] \quad \text{for } \det\boldsymbol{Q} = -1
\end{aligned}
\right\}, \qquad (1.189)
$$

noting equation (1.173). Therefore, the signs of $[\mathbf{abc}]$ and $[\mathbf{a}^*\mathbf{b}^*\mathbf{c}^*]$ are same and opposite for $\det\boldsymbol{Q} = 1$ and $\det\boldsymbol{Q} = -1$, respectively. In other words, the scalar triple product is transformed to same as and opposite of the right- and left-handed set for $\det\boldsymbol{Q} = 1$ and $\det\boldsymbol{Q} = -1$, respectively. The orthogonal tensors with the plus and the minus determinants are called the *proper orthogonal tensor* and the *improper orthogonal tensor*, respectively.

The transformation of the scalar triple product of the orthogonal base vectors is described from equation (1.188) as

$$\boxed{[\mathbf{e}_i^*\mathbf{e}_j^*\mathbf{e}_k^*] = \det\mathbf{Q}[\mathbf{e}_i\mathbf{e}_j\mathbf{e}_k]} \tag{1.190}$$

for the orthogonal tensor $\mathbf{Q}$ in equation (1.109) with equation $(1.159)_3$.

1.4.12 Pseudo Scalar, Vector and Tensor

Directions and senses of quantities such as force, displacement and force are determined naturally. They obey the transformation rule in equation (1.9) and possess a plane of symmetry perpendicular to their direction such that quantities with different signs possess opposite senses with respect to the plane. On the other hand, quantities such as a surface, a volume, a angular moment and a spin do not possess such a sense. These quantities do not obey the transformation rule in equation (1.9) and possess a plane of symmetry parallel to their direction. Their sense is chosen by the right-hand screw rule. These quantities are called pseudo scalars, vectors, or tensors. Several examples are shown below.

The transformation rule for the scalar triple product is given by

$$[\mathbf{a}^*\mathbf{b}^*\mathbf{c}^*] = [\mathbf{Qa}\ \mathbf{Qb}\ \mathbf{Qc}] = \det\mathbf{Q}[\mathbf{abc}], \tag{1.191}$$

noting equation (1.188). The scalar product changes its sign for $\det\mathbf{Q} = -1$ and thus it is called the *pseudo scalar*. Volume is regarded as a pseudo scalar.

The transformation of the vector product is given by

$$\mathbf{a}^* \times \mathbf{b}^* = (\det\mathbf{Q})\mathbf{Q}(\mathbf{a} \times \mathbf{b}), \tag{1.192}$$

noting that

$$\begin{aligned}(\mathbf{a}^*\times \mathbf{b}^*)_i &= \varepsilon_{ijk}a_j^*b_k^* = \varepsilon_{ijk}Q_{jp}Q_{kq}a_pb_q = Q_{is}Q_{us}\varepsilon_{ujk}Q_{jp}Q_{kq}a_pb_q = Q_{is}(\varepsilon_{ujk}Q_{us}Q_{jp}Q_{kq})a_pb_q\\ &= Q_{is}\varepsilon_{spq}(\varepsilon_{ujk}Q_{u1}Q_{j2}Q_{k3})a_pb_q = \det\mathbf{Q}Q_{is}\varepsilon_{spq}a_pb_q = \det\mathbf{Q}Q_{is}(\mathbf{a}\times\mathbf{b})_s.\end{aligned}$$

The vector obeying the transformation rule in equation (1.192) is called a *pseudo vector* or *axial vector*. A surface vector is regarded as an axial vector.

The transformation rule for the permutation symbol $\boldsymbol{\varepsilon}$ is given from equation (1.190) as follows:

$$\varepsilon_{ijk}^* = (\det\mathbf{Q})\varepsilon_{ijk}. \tag{1.193}$$

Thus, $\boldsymbol{\varepsilon}$ is a third-order pseudo tensor and plays the role of the third-order isotropic tensor as will be described later in equation (1.211).

1.5 Tensor Representations

Various notations are used to represent tensors in continuum mechanics. They are collectively shown in this section.

1.5.1 Tensor Notations

The tensor $\mathbf{T}$ can be expressed as

$$\mathbf{T} = T_{p_1 p_2 \cdots p_m} \mathbf{e}_{p_1} \otimes \mathbf{e}_{p_2} \cdots \otimes \mathbf{e}_{p_m} \tag{1.194}$$

which is called the *indicial* (or *component*) *notation with base*, where $\mathbf{e}_1, \mathbf{e}_2, \ldots, \mathbf{e}_m$ are the unit base vectors of the coordinate axes $x_1, x_2, \ldots, x_m$. The pertinence of the transformation of $\mathbf{T}$ to another coordinate system with the base vectors $\mathbf{e}_1^*, \mathbf{e}_2^*, \ldots, \mathbf{e}_m^*$ of the coordinate axes $x_1^*, x_2^*, \ldots, x_m^*$ is confirmed as follows:

$$\mathbf{T} = T_{p_1 p_2 \cdots p_m} \mathbf{e}_{p_1} \otimes \mathbf{e}_{p_2} \cdots \otimes \mathbf{e}_{p_m} = Q_{r_1 p_1} Q_{r_2 p_2} \cdots Q_{r_m p_m} T_{p_1 p_2 \cdots p_m} \mathbf{e}_{r_1}^* \otimes \mathbf{e}_{r_2}^* \otimes \cdots \otimes \mathbf{e}_{r_m}^*, \tag{1.195}$$

noting that

$$\begin{aligned} T_{p_1 p_2 \cdots p_m} \mathbf{e}_{p_1} \otimes \mathbf{e}_{p_2} \cdots \otimes \mathbf{e}_{p_m} &= T_{p_1 p_2 \cdots p_m} Q_{r_1 p_1} \mathbf{e}_{r_1}^* \otimes Q_{r_2 p_2} \mathbf{e}_{r_2}^* \otimes \cdots \otimes Q_{r_m p_m} \mathbf{e}_{r_m}^* \\ &= Q_{r_1 p_1} Q_{r_2 p_2} \cdots Q_{r_m p_m} T_{p_1 p_2 \cdots p_m} \mathbf{e}_{r_1}^* \otimes \mathbf{e}_{r_2}^* \otimes \cdots \otimes \mathbf{e}_{r_m}^* \end{aligned}$$

by substituting equation (1.17), where equation (1.195) designates equation (1.9).

In addition to the component notation with base (equation (1.194)), tensors can be written in *indicial notation* $T_{p_1 p_2 \cdots p_m}$; in *matrix notation* $[T]$, $[T_{ij}]$; or in *symbolic* (or *direct*) *notation* $\mathbf{T}$.

The matrix notation holds only for a vector or a second-order tensor or for a fourth-order tensor if it is formally expressed by two indices. For instance, the stress–strain relation can be represented in matrix notation by expressing the stress and the strain of second-order tensors as a form of vector and the stiffness coefficient in a fourth-order tensor as a form of second-order tensor, which is called the *Voigt notation*. Various contractions exist in the operation of higher-order tensors and thus the symbolic notation is not useful in general. In other words, the symbolic notation is limited to the representation of tensors of order lower than fourth. On the other hand, the component notation holds always without adopting special rules. However, it should be emphasized that equations described in symbolic notation hold in any coordinate system including not only Cartesian but also curvilinear coordinate systems, even if they are formulated or derived in the Cartesian coordinate system.

1.5.2 Tensor Components and Transformation Rule

The components of vector $\mathbf{v}$, second-order tensor $\mathbf{T}$ and fourth-order tensor $\mathfrak{T}$ can be written as follows:

$$v_i = \mathbf{v} \cdot \mathbf{e}_i, \quad T_{ij} = \mathbf{e}_i \cdot \mathbf{T} \mathbf{e}_j, \quad \mathfrak{T}_{ijkl} = (\mathbf{e}_i \otimes \mathbf{e}_j) : \mathfrak{T} : (\mathbf{e}_k \otimes \mathbf{e}_l), \tag{1.196}$$

noting that

$$v_i = v_r \mathbf{e}_r \cdot \mathbf{e}_i, \quad T_{ij} = \mathbf{e}_i \cdot T_{rs} \mathbf{e}_r \otimes \mathbf{e}_s \mathbf{e}_j,$$

$$\begin{aligned}(\mathbf{e}_i \otimes \mathbf{e}_j) &\mathfrak{T}_{pqrs} \mathbf{e}_p \otimes \mathbf{e}_q \otimes \mathbf{e}_r \otimes \mathbf{e}_s (\mathbf{e}_k \otimes \mathbf{e}_l) \\ &= (\mathbf{e}_i \cdot \mathbf{e}_p)(\mathbf{e}_j \cdot \mathbf{e}_q) \mathfrak{T}_{rspq} (\mathbf{e}_r \cdot \mathbf{e}_k)(\mathbf{e}_s \cdot \mathbf{e}_l) = \delta_{ip} \delta_{jq} \mathfrak{T}_{pqrs} \delta_{rk} \delta_{sl}.\end{aligned}$$

Noting equation (1.195) and then introducing the notation

$$\left.\begin{aligned}(\mathbf{Q}\,[\![\mathbf{T}]\!])_{p_1 p_2 \cdots p_m} &\equiv Q_{p_1 q_1} Q_{p_2 q_2} \cdots Q_{p_m q_m} T_{q_1 q_2 \cdots q_m} \\ (\mathbf{Q}^T\,[\![\mathbf{T}]\!])_{p_1 p_2 \cdots p_m} &\equiv Q_{p_1 q_1} Q_{p_2 q_2} \cdots Q_{p_m q_m} T_{q_1 q_2 \cdots q_m}\end{aligned}\right\} \tag{1.197}$$

for the component transformation, let equations (1.9) and (1.11) be expressed formally in symbolic notation for convenience as follows:

$$\boxed{\mathbf{T}^* = \mathbf{Q}\,[\![\mathbf{T}]\!], \quad \mathbf{T} = \mathbf{Q}^T\,[\![\mathbf{T}^*]\!]} \tag{1.198}$$

The objective transformation rules for vector and the second-order tensor are as follows:

$$\boxed{\mathbf{v}^* = \mathbf{Q}\mathbf{v}(\mathbf{v} = Q^T \mathbf{v}^*), \quad \mathbf{T}^* = \mathbf{Q}\mathbf{T}\mathbf{Q}^T\,(\mathbf{T} = \mathbf{Q}^T \mathbf{T}^* \mathbf{Q})} \tag{1.199}$$

Here, it should be emphasized that these equations do not express relations between different vectors or tensors but should be regarded as relations between components of vectors or tensors described by the two coordinate systems with bases $\{\mathbf{e}_i\}$ and $\{\mathbf{e}_i^*\}$.

1.5.3 Notations of Tensor Operations

The notation for first- to fourth-order tensors presented in this section will be used throughout this book. For vectors **a**, **b**, **c** and **v**, second-order tensors **A**, **B** and **T**, the third-order tensor Ξ and fourth-order tensor $\mathfrak{T}$, we have:

$$\left.\begin{aligned}&\mathbf{a} \cdot \mathbf{b} = \mathrm{tr}(\mathbf{a} \otimes \mathbf{b}) \text{ for } a_i b_i \\ &\mathbf{a} \times \mathbf{b} \text{ for } \varepsilon_{ijk} a_j b_k \\ &\mathbf{a} \otimes \mathbf{b} \text{ for } a_i b_j \text{ with } \mathbf{a} \otimes \mathbf{b} \cdot \mathbf{c} \text{ for } a_i b_j c_j\end{aligned}\right\}, \tag{1.200}$$

$$\left.\begin{aligned}&\mathbf{T}\mathbf{v} \text{ for } T_{ij} v_j, \quad \mathbf{v}\mathbf{T} \text{ for } v_i T_{ij} \\ &\mathbf{v} \times \mathbf{T} \text{ for } (\mathbf{v} \times \mathbf{T})_{ij} = \varepsilon_{ikl} v_k T_{lj}, \quad \mathbf{T} \times \mathbf{v} \text{ for } (\mathbf{T} \times \mathbf{v})_{ij} = \varepsilon_{jkl} T_{ik} v_l\end{aligned}\right\} \tag{1.201}$$

$$\left(\begin{aligned}\mathbf{v} \times \mathbf{T} = v_k \mathbf{e}_k \times T_{lj} \mathbf{e}_l \otimes \mathbf{e}_j = v_k T_{lj} \varepsilon_{kli} \mathbf{e}_i \otimes \mathbf{e}_j = \varepsilon_{kli} v_k T_{lj} \mathbf{e}_i \otimes \mathbf{e}_j = \varepsilon_{ikl} v_k T_{lj} \mathbf{e}_i \otimes \mathbf{e}_j \\ \mathbf{T} \times \mathbf{v} = T_{ik} \mathbf{e}_i \otimes \mathbf{e}_k \times v_l \mathbf{e}_l = T_{ik} v_l \mathbf{e}_i \otimes \varepsilon_{klj} \mathbf{e}_j = \varepsilon_{klj} T_{ik} v_l \mathbf{e}_i \otimes \mathbf{e}_j = \varepsilon_{jkl} T_{ik} v_l \mathbf{e}_i \otimes \mathbf{e}_j\end{aligned}\right),$$

$$\left.\begin{aligned}&\mathbf{A}\mathbf{B} \text{ for } A_{ir} B_{rj} \\ &\mathbf{A} : \mathbf{B} = \mathrm{tr}(\mathbf{A}\mathbf{B}^T) = \mathrm{tr}(\mathbf{A}^T \mathbf{B}) \text{ for } A_{ij} B_{ij} \\ &\mathbf{T} : \mathbf{I} = \mathbf{I} : \mathbf{T} = \mathrm{tr}\mathbf{T} \text{ for } T_{ij} \delta_{ij} = T_{ii} \\ &\mathbf{A} \otimes \mathbf{B} \text{ for } A_{ij} B_{kl} \\ &\mathbf{T} : (\mathbf{a} \otimes \mathbf{b}) = \mathbf{T}\mathbf{a} \cdot \mathbf{b} \text{ for } T_{ij} a_i b_j \\ &(\mathbf{a} \otimes \mathbf{b}) : (\mathbf{c} \otimes \mathbf{d}) = (\mathbf{a} \cdot \mathbf{c})(\mathbf{b} \cdot \mathbf{d}) \text{ for } a_i c_i b_j d_j\end{aligned}\right\}, \tag{1.202}$$

$$\left.\begin{array}{ll}\boldsymbol{\Xi}\mathbf{T} \text{ for } \Xi_{ijk}T_{kl}, & \mathbf{T}\boldsymbol{\Xi} \text{ for } T_{ij}\Xi_{jkl} \\ \boldsymbol{\Xi} : \mathbf{T} \text{ for } \Xi_{ijk}T_{jk}, & \mathbf{T} : \boldsymbol{\Xi} \text{ for } T_{ij}\Xi_{ijk}\end{array}\right\}, \tag{1.203}$$

$$\left.\begin{array}{ll}\boldsymbol{\mathfrak{T}}\mathbf{T} \text{ for } \mathfrak{T}_{ijkl}T_{ln}, & \mathbf{T}\boldsymbol{\mathfrak{T}} \text{ for } T_{ij}\mathfrak{T}_{jkln} \\ \boldsymbol{\mathfrak{T}} : \mathbf{T} \text{ for } \mathfrak{T}_{ijkl}T_{kl}, & \mathbf{T} : \boldsymbol{\mathfrak{T}} \text{ for } T_{ij}\mathfrak{T}_{ijkl}\end{array}\right\}, \tag{1.204}$$

$$\left.\begin{array}{l}(\boldsymbol{\mathfrak{T}} : \mathbf{A}) : \mathbf{B} = \mathbf{B} : \boldsymbol{\mathfrak{T}} : \mathbf{A} \text{ for } \mathfrak{T}_{ijkl}A_{kl}B_{ij} = B_{ij}\mathfrak{T}_{ijkl}A_{kl} \\ (\boldsymbol{\mathfrak{T}}^T : \mathbf{A}) : \mathbf{B} = \mathbf{A} : \boldsymbol{\mathfrak{T}} : \mathbf{B} \text{ for } \mathfrak{T}_{klij}A_{kl}B_{ij} = A_{kl}\mathfrak{T}_{klij}B_{ij}\end{array}\right\}, \tag{1.205}$$

$$\left.\begin{array}{l}||\mathbf{v}|| = \sqrt{\mathbf{v}\cdot\mathbf{v}} \text{ for } \sqrt{v_i v_i}, \\ ||\mathbf{T}|| = \sqrt{\mathbf{T} : \mathbf{T}} = \sqrt{\mathrm{tr}(\mathbf{T}\mathbf{T}^T)} \text{ for } \sqrt{T_{ij}T_{ij}}\end{array}\right\}, \tag{1.206}$$

$$\frac{\partial f^2(\mathbf{T})}{\partial \mathbf{T} \otimes \partial \mathbf{T}} \text{ for } \frac{\partial f^2(\mathbf{T})}{\partial T_{ij}\partial T_{kl}}. \tag{1.207}$$

The following symbols for the tensor products of second-order tensors **A**, **B**, **C** are defined (del Piero, 1979; Steinmann et al., 1997; Kintzel and Başar, 2006; Wang and Dui, 2008):

$$\left.\begin{array}{ll}\boxed{(\mathbf{A}\otimes\mathbf{B})_{ijkl} = A_{ij}B_{kl}} & \text{with } \mathbf{A}\otimes\mathbf{B} : \mathbf{C} = \mathbf{A}(\mathbf{B} : \mathbf{C}) \ ((\mathbf{A}\otimes\mathbf{B}\mathbf{C})_{ij} = A_{ij}(B_{kl}C_{kl})) \text{ and} \\ & \mathbf{C} : \mathbf{A}\otimes\mathbf{B} = (\mathbf{C} : \mathbf{A})\mathbf{B} \ ((\mathbf{C}\mathbf{A}\otimes\mathbf{B})_{kl} = C_{ij}A_{ij}B_{kl}) \\ \boxed{(\mathbf{A}\overline{\otimes}\mathbf{B})_{ijkl} = A_{ik}B_{jl}} & \text{with } \mathbf{A}\overline{\otimes}\mathbf{B} : \mathbf{C} = \mathbf{A}\mathbf{C}\mathbf{B}^T \ ((\mathbf{A}\overline{\otimes}\mathbf{B}\mathbf{C})_{ij} = A_{ik}B_{jl}C_{kl} = A_{ik}C_{kl}B_{jl}) \text{ and} \\ & \mathbf{C} : \mathbf{A}\overline{\otimes}\mathbf{B} = \mathbf{A}^T\mathbf{C}\mathbf{B} \ ((\mathbf{C}\mathbf{A}\overline{\otimes}\mathbf{B})_{kl} = C_{ij}A_{ik}B_{jl} = A_{ik}C_{ij}B_{jl}) \\ \boxed{(\mathbf{A}\underline{\otimes}\mathbf{B})_{ijkl} = A_{il}B_{jk}} & \text{with } \mathbf{A}\underline{\otimes}\mathbf{B} : \mathbf{C} = \mathbf{A}\mathbf{C}^T\mathbf{B}^T \ ((\mathbf{A}\underline{\otimes}\mathbf{B}\mathbf{C})_{ij} = A_{il}B_{jk}C_{kl} = A_{il}C_{kl}B_{jk}) \text{ and} \\ & \mathbf{C} : \mathbf{A}\underline{\otimes}\mathbf{B} = \mathbf{B}^T\mathbf{C}^T\mathbf{A} \ ((\mathbf{C}\mathbf{A}\underline{\otimes}\mathbf{B})_{kl} = C_{ij}A_{il}B_{jk} = B_{jk}C_{ij}A_{il}) \\ \boxed{(\mathbf{A}\tilde{\otimes}\mathbf{B})_{ijkl} = A_{ik}B_{lj}} & \text{with } \mathbf{A}\tilde{\otimes}\mathbf{B} : \mathbf{C} = \mathbf{A}\mathbf{C}\mathbf{B} \ ((\mathbf{A}\tilde{\otimes}\mathbf{B}\mathbf{C})_{ij} = A_{ik}B_{lj}C_{kl} = A_{ik}C_{kl}B_{lj}) \text{ and} \\ & \mathbf{C} : \mathbf{A}\tilde{\otimes} : \mathbf{B} = \mathbf{A}^T\mathbf{C}\mathbf{B}^T \ ((\mathbf{C}\mathbf{A}\tilde{\otimes}\mathbf{B})_{kl} = C_{ij}A_{ik}B_{lj} = A_{ik}C_{ij}B_{lj}) \\ \boxed{(\mathbf{A}\underset{\sim}{\otimes}\mathbf{B})_{ijkl} = A_{il}B_{kj}} & \text{with } \mathbf{A}\underset{\sim}{\otimes}\mathbf{B} : \mathbf{C} = \mathbf{A}\mathbf{C}^T\mathbf{B} \ ((\mathbf{A}\underset{\sim}{\otimes}\mathbf{B}\mathbf{C})_{ij} = A_{il}B_{kj}C_{kl} = A_{il}C_{kl}B_{kj}) \text{ and} \\ & \mathbf{C} : \mathbf{A}\underset{\sim}{\otimes}\mathbf{B} = \mathbf{B}\mathbf{C}^T\mathbf{A} \ ((\mathbf{C}\mathbf{A}\underset{\sim}{\otimes}\mathbf{B})_{kl} = C_{ij}A_{il}B_{kj} = B_{kj}C_{ij}A_{il})\end{array}\right\} \tag{1.208}$$

by which the following expressions hold.

$$\left.\begin{array}{l}\mathbf{A}\mathbf{B} = \mathbf{A}\mathbf{B}\mathbf{I} = \mathbf{A}\overline{\otimes}\mathbf{I} : \mathbf{B} \\ \mathbf{B}\mathbf{A}^T = \mathbf{I}\mathbf{B}\mathbf{A}^T = \mathbf{I}\overline{\otimes}\mathbf{A} : \mathbf{B} \\ \mathbf{A}\mathbf{B}^T = \mathbf{A}\mathbf{B}^T\mathbf{I} = \mathbf{A}\underline{\otimes}\mathbf{I} : \mathbf{B} \\ \mathbf{A}^T\mathbf{B}^T = \mathbf{I}\mathbf{A}^T\mathbf{B}^T = \mathbf{I}\underline{\otimes}\mathbf{B} : \mathbf{A}\end{array}\right\}. \tag{1.209}$$

1.5.4 Operational Tensors

The second-order *identity tensor* is defined by the following equation, which involves the Kronecker delta:

$$\boxed{\mathbf{I} = \delta_{ij}\mathbf{e}_i \otimes \mathbf{e}_j = \mathbf{e}_i \otimes \mathbf{e}_i}, \quad \delta_{ij} = \mathbf{e}_i \cdot \mathbf{e}_j. \tag{1.210}$$

The *alternating tensor* is defined by the following equation, which involves the alternating symbol:

$$\boxed{\boldsymbol{\varepsilon} = \varepsilon_{ijk}\mathbf{e}_i \otimes \mathbf{e}_j \otimes \mathbf{e}_k}, \quad \varepsilon_{ijk} = [\mathbf{e}_i\mathbf{e}_j\mathbf{e}_k] \tag{1.211}$$

The *fourth-order identity tensor* $\overline{\mathcal{I}}$ and the *transposing tensor* $\underline{\mathcal{I}}$ are defined by

$$\left.\begin{aligned} \overline{\mathcal{I}} &= \delta_{ik}\delta_{jl}\mathbf{e}_i \otimes \mathbf{e}_j \otimes \mathbf{e}_k \otimes \mathbf{e}_l = \mathbf{e}_i \otimes \mathbf{e}_j \otimes \mathbf{e}_i \otimes \mathbf{e}_j = \mathbf{I}\overline{\otimes}\mathbf{I} \\ \underline{\mathcal{I}} &= \delta_{il}\delta_{jk}\mathbf{e}_i \otimes \mathbf{e}_j \otimes \mathbf{e}_k \otimes \mathbf{e}_l = \mathbf{e}_i \otimes \mathbf{e}_j \otimes \mathbf{e}_j \otimes \mathbf{e}_i = \mathbf{I}\underline{\otimes}\mathbf{I} \end{aligned}\right\}. \tag{1.212}$$

Note that $\overline{\mathcal{I}} : \mathbf{T} = \mathbf{T} : \overline{\mathcal{I}} = \mathbf{T}, \underline{\mathcal{I}} : \mathbf{T} = \mathbf{T} : \underline{\mathcal{I}} = \mathbf{T}^T, \overline{\mathcal{I}} : \mathfrak{T} = \mathfrak{T} : \overline{\mathcal{I}} = \mathfrak{T}$ and $\partial\mathbf{T}/\partial\mathbf{T} = \overline{\mathcal{I}}$, $\partial\mathbf{T}^T/\partial\mathbf{T} = \underline{\mathcal{I}}$. Moreover, the fourth-order tracing identity tensor defined by

$$\boxed{\mathcal{I} \equiv \delta_{ij}\delta_{kl}\mathbf{e}_i \otimes \mathbf{e}_j \otimes \mathbf{e}_k \otimes \mathbf{e}_l = \mathbf{I} \otimes \mathbf{I}} \tag{1.213}$$

has the operation

$$\mathcal{I} : \mathbf{T} = (\mathrm{tr}\mathbf{T})\mathbf{I} \tag{1.214}$$

The *symmetrizing tensor* $\boldsymbol{\mathcal{S}}$ ant the *skew*-(or *anti-*) *symmetrizing tensor* $\boldsymbol{\mathcal{A}}$ are given by

$$\boxed{\begin{aligned} \boldsymbol{\mathcal{S}} &\equiv \frac{1}{2}(\delta_{ik}\delta_{jl} + \delta_{il}\delta_{jk})\mathbf{e}_i \otimes \mathbf{e}_j \otimes \mathbf{e}_k \otimes \mathbf{e}_l = \frac{1}{2}(\overline{\mathcal{I}} + \underline{\mathcal{I}}) \\ \boldsymbol{\mathcal{A}} &\equiv \frac{1}{2}(\delta_{ik}\delta_{jl} - \delta_{il}\delta_{jk})\mathbf{e}_i \otimes \mathbf{e}_j \otimes \mathbf{e}_k \otimes \mathbf{e}_l = \frac{1}{2}(\overline{\mathcal{I}} - \underline{\mathcal{I}}) \end{aligned}} \tag{1.215}$$

leading to $\boldsymbol{\mathcal{S}} : \mathbf{T} = (\mathbf{T} + \mathbf{T}^T)/2, \boldsymbol{\mathcal{A}} : \mathbf{T} = (\mathbf{T} - \mathbf{T}^T)/2$. The following notation is also used:

$$\boldsymbol{\mathcal{S}} : \mathbf{T} = \mathrm{sym}[\mathbf{T}] = (\mathbf{T})_s, \boldsymbol{\mathcal{A}} : \mathbf{T} = \mathrm{ant}[\mathbf{T}] = (\mathbf{T})_a \tag{1.216}$$

The *deviatoric projection tensor* is defined by

$$\boxed{\boldsymbol{\mathcal{D}} \equiv \left(\delta_{ik}\delta_{jl} - \frac{1}{3}\delta_{ij}\delta_{kl}\right)\mathbf{e}_i \otimes \mathbf{e}_j \otimes \mathbf{e}_k \otimes \mathbf{e}_l = \overline{\mathcal{I}} - \frac{1}{3}\mathcal{I}} \tag{1.217}$$

leading to $\boldsymbol{\mathcal{D}} : \mathbf{T} = \mathbf{T} - \{(\mathrm{tr}\mathbf{T})/3\}\mathbf{I}$.

The following relations hold in these tensors:

$$\left.\begin{aligned} &\overline{\mathcal{I}} : \overline{\mathcal{I}} = \overline{\mathcal{I}},\ \underline{\mathcal{I}} : \underline{\mathcal{I}} = \overline{\mathcal{I}},\ \overline{\mathcal{I}} : \underline{\mathcal{I}} = \underline{\mathcal{I}} : \overline{\mathcal{I}} = \underline{\mathcal{I}} \\ &\mathcal{I} : \mathcal{I} = 3\mathcal{I} \\ &\overline{\mathcal{I}} : \mathcal{I} = \mathcal{I} : \overline{\mathcal{I}} = \underline{\mathcal{I}} : \mathcal{I} = \mathcal{I} : \underline{\mathcal{I}} = \mathcal{I} \end{aligned}\right\}, \tag{1.218}$$

$$\boldsymbol{\mathcal{S}} : \boldsymbol{\mathcal{S}} = \boldsymbol{\mathcal{S}},\ \boldsymbol{\mathcal{D}} : \boldsymbol{\mathcal{D}} = \boldsymbol{\mathcal{D}},\ \boldsymbol{\mathcal{D}} : \boldsymbol{\mathcal{S}} = \boldsymbol{\mathcal{S}} : \boldsymbol{\mathcal{D}} = \mathbf{O}, \tag{1.219}$$

noting that

$$\begin{aligned} \overline{\mathcal{I}} : \overline{\mathcal{I}} &= \{(\mathbf{e}_i \otimes \mathbf{e}_j) \otimes (\mathbf{e}_i \otimes \mathbf{e}_j)\} : \{(\mathbf{e}_p \otimes \mathbf{e}_q) \otimes (\mathbf{e}_p \otimes \mathbf{e}_q)\} \\ &= \mathbf{e}_i \otimes \mathbf{e}_j\delta_{ip}\delta_{jq} \otimes \mathbf{e}_p \otimes \mathbf{e}_q = \mathbf{e}_i \otimes \mathbf{e}_j\mathbf{e}_i \otimes \mathbf{e}_j = \overline{\mathcal{I}}, \\ \underline{\mathcal{I}} : \underline{\mathcal{I}} &= \{(\mathbf{e}_i \otimes \mathbf{e}_j) \otimes (\mathbf{e}_j \otimes \mathbf{e}_i)\} : \{(\mathbf{e}_p \otimes \mathbf{e}_q) \otimes (\mathbf{e}_q \otimes \mathbf{e}_p)\} \\ &= \mathbf{e}_i \otimes \mathbf{e}_j\delta_{jp}\delta_{iq} \otimes \mathbf{e}_q \otimes \mathbf{e}_p = \mathbf{e}_i \otimes \mathbf{e}_j \otimes \mathbf{e}_i \otimes \mathbf{e}_j = \overline{\mathcal{I}}, \end{aligned}$$

$$
\begin{aligned}
\overline{\mathcal{I}} : \underline{\mathcal{I}} &= \{(\mathbf{e}_i \otimes \mathbf{e}_j) \otimes (\mathbf{e}_i \otimes \mathbf{e}_j)\} : \{(\mathbf{e}_r \otimes \mathbf{e}_s) \otimes (\mathbf{e}_s \otimes \mathbf{e}_r)\} \\
&= \mathbf{e}_i \otimes \mathbf{e}_j \delta_{ir} \delta_{js} \otimes \mathbf{e}_s \otimes \mathbf{e}_r = \mathbf{e}_i \otimes \mathbf{e}_j \mathbf{e}_j \otimes \mathbf{e}_i = \underline{\mathcal{I}}, \\
\mathcal{I} : \mathcal{I} &= (\mathbf{I} \otimes \mathbf{I}) : (\mathbf{I} \otimes \mathbf{I}) = (\mathbf{I} : \mathbf{I}) \mathbf{I} \otimes \mathbf{I} = 3 \mathbf{I} \otimes \mathbf{I} = 3 \mathcal{I}, \\
\overline{\mathcal{I}} : \mathcal{I} &= \{(\mathbf{e}_i \otimes \mathbf{e}_j) \otimes (\mathbf{e}_i \otimes \mathbf{e}_j)\} : \{(\mathbf{e}_r \otimes \mathbf{e}_r) \otimes (\mathbf{e}_s \otimes \mathbf{e}_s)\} \\
&= \mathbf{e}_i \otimes \mathbf{e}_j \delta_{ir} \delta_{jr} \otimes \mathbf{e}_s \otimes \mathbf{e}_s = \mathbf{e}_r \otimes \mathbf{e}_r \otimes \mathbf{e}_s \otimes \mathbf{e}_s = \mathcal{I}, \\
\underline{\mathcal{I}} : \mathcal{I} &= \{(\mathbf{e}_i \otimes \mathbf{e}_j) \otimes (\mathbf{e}_j \otimes \mathbf{e}_i)\} : \{(\mathbf{e}_r \otimes \mathbf{e}_r) \otimes (\mathbf{e}_s \otimes \mathbf{e}_s)\} \\
&= \mathbf{e}_i \otimes \mathbf{e}_j \delta_{jr} \delta_{ir} \mathbf{e}_s \otimes \mathbf{e}_s = \mathbf{e}_r \otimes \mathbf{e}_r \mathbf{e}_s \otimes \mathbf{e}_s = \mathcal{I}; \\
\mathcal{S} : \mathcal{S} &= \frac{1}{2}(\overline{\mathcal{I}} + \underline{\mathcal{I}}) : \frac{1}{2}(\overline{\mathcal{I}} + \underline{\mathcal{I}}) = \frac{1}{4}(\overline{\mathcal{I}} + 2\underline{\mathcal{I}} + \overline{\mathcal{I}}) = \mathcal{S}, \\
\mathcal{D} : \mathcal{D} &= \frac{1}{2}(\overline{\mathcal{I}} - \underline{\mathcal{I}}) : \frac{1}{2}(\overline{\mathcal{I}} - \underline{\mathcal{I}}) = \frac{1}{4}(\overline{\mathcal{I}} - 2\underline{\mathcal{I}} + \overline{\mathcal{I}}) = \mathcal{D}, \\
\mathcal{S} : \mathcal{D} &= \frac{1}{2}(\overline{\mathcal{I}} + \underline{\mathcal{I}}) : \frac{1}{2}(\overline{\mathcal{I}} - \underline{\mathcal{I}}) = \frac{1}{4}(\overline{\mathcal{I}} - \overline{\mathcal{I}}) = \mathbf{O}.
\end{aligned}
$$

Note that the symmetrizing tensor $\mathcal{S}$ can be used instead of the identity tensor $\overline{\mathcal{I}}$ for symmetric tensors.

1.5.5 Isotropic Tensors

An isotropic tensor is defined as a tensor possessing components which are unchanged by arbitrary rotation of the coordinate system and thus it must satisfy

$$
\mathbf{T} = \mathbf{Q} \, [\![\mathbf{T}]\!] \, , \tag{1.220}
$$

where use is made of the notation for objective transformation in equation (1.198) for general tensor. As a trivial case, all tensors possessing zero components are isotropic tensors. We consider non-trivial isotropic tensors for the first- to fourth-order tensors for which equation (1.220) is described as

$$
\left.
\begin{aligned}
&S = S \\
&v_i = Q_{ir} v_r \\
&T_{ij} = Q_{ir} Q_{js} T_{rs} \\
&T_{ijk} = Q_{ir} Q_{js} Q_{kt} T_{rst} \\
&T_{ijkl} = Q_{ir} Q_{js} Q_{kt} Q_{lu} T_{rstu}
\end{aligned}
\right\} . \tag{1.221}
$$

Here, note that the permutation of indices does not influence on the values of components by virtue of the isotropy.

Now consider a small rotation of coordinate system given by the following equation (Jeffreys, 1931):

$$
Q_{ij} = \delta_{ij} + \Omega_{ij} \tag{1.222}
$$

with an infinitesimal anti-symmetric tensor Ω_{ij}, for which one has

$$
Q_{ir} Q_{jr} = (\delta_{ir} + \Omega_{ir})(\delta_{jr} + \Omega_{jr}) = \delta_{ij} + \Omega_{ij} + \Omega_{ji} + \Omega_{ir}\Omega_{jr} = \delta_{ij} + \Omega_{ir}\Omega_{jr} \cong 0,
$$

exhibiting the property of the orthogonal tensor, while $\Omega_{ir}\Omega_{jr}$ is infinitesimal in the second order.

Substituting equation (1.222) into equation (1.221), we have

$$\begin{aligned}
v_i &= (\delta_{ir} + \Omega_{ir})v_r = v_i + \Omega_{ir}v_r, \\
T_{ij} &= (\delta_{ir} + \Omega_{ir})(\delta_{js} + \Omega_{js})T_{rs} \cong T_{ij} + \Omega_{ir}T_{rj} + \Omega_{js}T_{is}, \\
T_{ijk} &= (\delta_{ir} + \Omega_{ir})(\delta_{js} + \Omega_{js})(\delta_{kt} + \Omega_{kt})T_{rst} \\
&\cong (\delta_{ir}\delta_{js}\delta_{kt} + \delta_{js}\delta_{kt}\Omega_{ir} + \Omega_{ir}\delta_{kt}\Omega_{js} + \delta_{ir}\delta_{js}\Omega_{kt})T_{rst} \\
&= T_{ijk} + \Omega_{ir}T_{rjk} + \Omega_{js}T_{isk} + \Omega_{kt}T_{ijt}, \\
T_{ijkl} &= (\delta_{ir} + \Omega_{ir})(\delta_{js} + \Omega_{js})(\delta_{kt} + \Omega_{kt})(\delta_{lu} + \Omega_{lu})T_{rstu} \\
&\cong (\delta_{ir}\delta_{js}\delta_{kt}\delta_{lu} + \delta_{js}\delta_{kt}\delta_{lu}\Omega_{ir} + \delta_{ir}\delta_{kt}\delta_{lu}\Omega_{js} + \delta_{ir}\delta_{js}\delta_{lu}\Omega_{kt} + \delta_{ir}\delta_{js}\delta_{kt}\Omega_{lu})T_{rstu} \\
&= T_{ijkl} + \Omega_{ir}T_{rjkl} + \Omega_{js}T_{iskl} + \Omega_{kt}T_{ijtl} + \Omega_{lu}T_{ijku},
\end{aligned}$$

and thus

$$\left.\begin{aligned}
&\Omega_{ir}v_r = 0 \\
&\Omega_{ir}T_{rj} + \Omega_{js}T_{si} = 0 \\
&\Omega_{ir}T_{rjk} + \Omega_{jr}T_{irk} + \Omega_{kr}T_{ijr} = 0 \\
&\Omega_{ir}T_{rjkl} + \Omega_{jr}T_{irkl} + \Omega_{kr}T_{ijrl} + \Omega_{lr}T_{ijkr} = 0
\end{aligned}\right\} \tag{1.223}$$

for arbitrary Ω_{ij}, where $\Omega_{ij} = 0$ for $i = j$ and $\Omega_{ij} = -\Omega_{ji}$ for $i \neq j$.

Zero-order (scalar) isotropic tensor

Any scalar satisfies equation $(1.221)_1$ and thus is a zero-order isotropic tensor.

First-order (vector) isotropic tensor

The linear homogeneous simultaneous equation $(1.223)_1$ possesses a non-trivial (non-zero) solution of v_i for a given (fixed) Ω_{ir} because the necessary and sufficient condition, equation (1.178), for the existence of a non-trivial solution is satisfied by virtue of the property of the anti-symmetric tensor shown in equation (1.174), that is, $\det[\Omega_{ir}] = 0$. However, equation $(1.223)_1$ leads to

$$\left.\begin{aligned}
&\Omega_{12}v_2 - \Omega_{31}v_3 = 0 \\
&-\Omega_{12}v_1 + \Omega_{23}v_3 = 0 \\
&\Omega_{31}v_1 - \Omega_{23}v_2 = 0
\end{aligned}\right\}$$

which reduces to $\Omega_{12}(v_2 - v_1) + \Omega_{23}(v_3 - v_2) + \Omega_{31}(v_1 - v_3) = 0$. In order that this equation is satisfied for arbitrary Ω_{ij}, only the trivial solution $v_1 = v_2 = v_3 = 0$ can exist. Consequently, there is no first-order isotropic tensor (vector).

Second-order isotropic tensor

If $i = 1, j = 1$ in equation $(1.223)_2$, then

$$2(\Omega_{12}T_{21} + \Omega_{13}T_{31}) = 0, \tag{1}$$

and thus it follows that

$$T_{ij} = 0 \text{ for } i \neq j. \tag{2}$$

Further, if $i = 1, j = 2$ in equation $(1.223)_2$, then

$$\Omega_{12}(T_{22} - T_{11}) = 0, \tag{3}$$

noting equation (2), and thus it must follow that

$$T_{ii} = T_{jj} \text{ (no sum)}. \tag{4}$$

The second-order isotropic tensor satisfying equations (2) and (4) is generally given by

$$\boxed{s\delta_{ij}\mathbf{e}_i \otimes \mathbf{e}_j = s\mathbf{I}} \tag{1.224}$$

Third-order isotropic tensor

If $i = 1, j = 1$ in equation $(1.223)_3$, then

$$\Omega_{12}T_{21k} + \Omega_{13}T_{31k} + \Omega_{12}T_{12k} + \Omega_{13}T_{13k} + \Omega_{k1}T_{111} + \Omega_{k2}T_{112} + \Omega_{k3}T_{113} = 0. \tag{1}$$

Further, putting $k = 2$ and noting that $\Omega_{k2} = 0$ in this equation, one has

$$\Omega_{12}(T_{212} + T_{122} - T_{111}) + \Omega_{13}(T_{312} + T_{132}) + \Omega_{23}T_{113} = 0$$

and thus

$$\left.\begin{aligned} &T_{212} + T_{122} = T_{111} \\ &T_{312} + T_{132} = 0 \\ &T_{113} = 0 \end{aligned}\right\}. \tag{2}$$

Noting the symmetry, from the last equation, $T_{ijk} = 0$ if two of i, j, k are equal and the third unequal, that is,

$$T_{iij} = T_{iji} = T_{ijj} = 0 \text{ (no sum, } i \neq j). \tag{3}$$

Then, substituting this result into the first equation, it follows that

$$T_{iii} = 0 \text{ (no sum)}. \tag{4}$$

Further, noting that the same permutation must not influence the value of components by the isotropy of T_{ijk}, it follows from equation $(2)_2$ that

$$T_{ijk} = T_{jki} = T_{kij} = -T_{jik} = -T_{kji} = -T_{ikj} (i \neq j \neq k \neq i). \tag{5}$$

The third-order isotropic tensor satisfying equations (3), (4) and (5) is given generally is given generally by the permutation symbol multiplied by arbitrary scalar, that is,

$$\boxed{s\varepsilon_{ijk}\mathbf{e}_i \otimes \mathbf{e}_j \otimes \mathbf{e}_k = s\boldsymbol{\varepsilon}} \tag{1.225}$$

Fourth-order isotropic tensor

For a fourth-order isotropic tensor, with its four indices which can only take values $1, 2, 3$, at least two of the indices must be are equal. We can separate the components into four classes: (a) two are unequal and the other two equal; (b) three equal; (c) two equal and the other two equal; and (d) all four equal.

If $i = j = 1, k = 2, l = 3$ in class (a), equation $(1.223)_4$ leads to

$$\begin{aligned}&\Omega_{12}T_{2123} + \Omega_{13}T_{3123} + \Omega_{12}T_{1223} + \Omega_{13}T_{1323}\\&\quad + \Omega_{21}T_{1113} + \Omega_{23}T_{1133} + \Omega_{31}T_{1123} + \Omega_{32}T_{1122} = 0,\end{aligned}$$

that is,

$$\Omega_{12}(T_{2123} + T_{1223} - T_{1113}) + \Omega_{13}(T_{3123} + T_{1323} - T_{1123}) + \Omega_{23}(T_{1133} - T_{1122}) = 0,$$

and thus

$$\left.\begin{array}{l}T_{2123} + T_{1223} - T_{1113} = 0\\ T_{3123} + T_{1323} - T_{1123} = 0\\ T_{1133} - T_{1122} = 0\end{array}\right\}. \tag{1}$$

If $i = j = k = 1, l = 2$ in class (b), equation $(1.223)_4$ yields

$$\begin{aligned}&\Omega_{12}T_{2112} + \Omega_{13}T_{3112} + \Omega_{12}T_{1212} + \Omega_{13}T_{1312}\\&\quad + \Omega_{12}T_{1122} + \Omega_{13}T_{1132} + \Omega_{21}T_{1111} + \Omega_{23}T_{1113} = 0,\end{aligned}$$

that is,

$$\Omega_{12}(T_{2112} + T_{1212} + T_{1122} - T_{1111}) + \Omega_{13}(T_{3112} + T_{1312} + T_{1132}) + \Omega_{23}T_{1113} = 0,$$

and thus

$$\left.\begin{array}{l}T_{2112} + T_{1212} + T_{1122} - T_{1111} = 0\\ T_{3112} + T_{1312} + T_{1132} = 0\\ T_{1113} = 0\end{array}\right\}. \tag{2}$$

The last equation leads to the fact that all components in class (b) are zero, that is,

$$T_{iiij} = T_{iiji} = T_{iiij} = T_{ijii} = T_{jiii} = 0 \text{ (no sum, } i \neq j). \tag{3}$$

Then equation $(1)_1$ results in

$$T_{2123} + T_{1223} = 0$$

which, by changing $1 \to 3, 2 \to 1, 3 \to 2$, leads to

$$T_{1312} + T_{3112} = 0.$$

Substituting this into equation $(2)_2$, one has

$$T_{1132} = 0,$$

so that all components in class (a) are also zero, that is,

$$T_{iijk} = T_{ijik} = T_{ijki} = T_{jiki} = T_{jkii} = 0 \,(\text{no sum}, i \neq j \neq k \neq i). \tag{4}$$

Taking account of equation (4) in equation $(2)_1$, the components in class (d) can be expressed by the components in class (c).

Eventually, we have only to take account of the components in class (c), which are classified into the following three types:

$$\left.\begin{aligned} T_{1122} = T_{2211} = T_{2233} = T_{3322} = T_{3311} = T_{1133} = \lambda \\ T_{1212} = T_{2121} = T_{2323} = T_{3232} = T_{3131} = T_{1313} = \mu + \nu \\ T_{1221} = T_{2112} = T_{2332} = T_{3223} = T_{3113} = T_{1331} = \mu - \nu \end{aligned}\right\}. \tag{5}$$

Here, the permutation of indices does not influence the values of components by virtue of the isotropy and thus all components in class (d) are identical so that they can be expressed collectively as

$$T_{1111} = T_{2222} = T_{3333} = \lambda + 2\mu \tag{6}$$

by substituting equation (5) into equation $(2)_1$.

Let us now formulate the tensor satisfying equation (5). Note the following facts:

1. The tensor with components for $i = j$ and $k = l$ in $(5)_1$ given by λ and other components zero is given by multiplication of the tracing identity tensor $\mathcal{I}_{ijkl} (= \delta_{ij}\delta_{kl})$ in equation (1.213) by λ.
2. The tensor possessing components such that the half of the sum of the components for $i = k, j = l \neq i$ in equation $(5)_2$ and $i = l, j = k \neq i$ in equation $(5)_3$ $((T_{1212} + T_{1221})/2$ is given by multiplication of the symmetrizing tensor $\mathcal{S}_{ijkl} (= \delta_{ik}\delta_{jl} + \delta_{il}\delta_{jk})/2)$ in equation $(1.215)_1$ by μ.
3. On the other hand, its subtraction $((T_{1212} - T_{1221})/2$, etc.) is given by multiplication of the anti-symmetrizing tensor $\mathcal{A}_{ijkl} (= \delta_{ik}\delta_{jl} - \delta_{il}\delta_{jk})/2)$ in equation $(1.215)_2$ by ν.

Finally, the fourth-order isotropic tensor is given generally by the sum of these independent tensors as follows:

$$\boxed{\mathcal{J} \equiv (\lambda\mathcal{I}_{ijkl} + \mu\mathcal{S}_{ijkl} + \nu\mathcal{A}_{ijkl})\mathbf{e}_i \otimes \mathbf{e}_j \otimes \mathbf{e}_k \otimes \mathbf{e}_l = \lambda\mathcal{I} + \mu\mathcal{S} + \nu\mathcal{A}} \tag{1.226}$$

The basic fourth-order tensors described in Sections 1.5.4 and 1.5.5 are gathered together in Table 1.1.

Table 1.1 Various fourth-order basic tensors

Fourth-order tensors	Remarks
Identity tensor $\overline{\overline{\mathcal{I}}} = \delta_{ik}\delta_{jl}\mathbf{e}_i \otimes \mathbf{e}_j \otimes \mathbf{e}_k \otimes \mathbf{e}_l = \mathbf{e}_i \otimes \mathbf{e}_j \otimes \mathbf{e}_i \otimes \mathbf{e}_j = \mathbf{I}\overline{\otimes}\mathbf{I}$	$\overline{\overline{\mathcal{I}}} : \mathbf{T} = \mathbf{T}$ $\dfrac{\partial \mathbf{T}}{\partial \mathbf{T}} = \overline{\overline{\mathcal{I}}}$
Transposing tensor $\underline{\underline{\mathcal{I}}} = \delta_{il}\delta_{jk}\mathbf{e}_i \otimes \mathbf{e}_j \otimes \mathbf{e}_k \otimes \mathbf{e}_l = \mathbf{e}_i \otimes \mathbf{e}_j \otimes \mathbf{e}_j \otimes \mathbf{e}_i = \mathbf{I}\underline{\otimes}\mathbf{I}$	$\underline{\underline{\mathcal{I}}} : \mathbf{T} = \mathbf{T}^T$ $\dfrac{\partial \mathbf{T}^T}{\partial \mathbf{T}} = \underline{\underline{\mathcal{I}}}$
Tracing identity tensor $\mathcal{I} \equiv \delta_{ij}\delta_{kl}\mathbf{e}_i \otimes \mathbf{e}_j \otimes \mathbf{e}_j \otimes \mathbf{e}_i = \mathbf{I} \otimes \mathbf{I}$	$\mathcal{I} : \mathbf{T} = (\mathrm{tr}\mathbf{T})\mathbf{I}$
Symmetrizing tensor $\mathcal{S} \equiv \dfrac{1}{2}(\delta_{ik}\delta_{jl} + \delta_{il}\delta_{jk})\mathbf{e}_i \otimes \mathbf{e}_j \otimes \mathbf{e}_k \otimes \mathbf{e}_l = \dfrac{1}{2}(\overline{\overline{\mathcal{I}}} + \underline{\underline{\mathcal{I}}})$	$\mathcal{S} : \mathbf{T} = \mathrm{sym}[\mathbf{T}]$
Skew-symmetrizing tensor $\mathcal{A} \equiv \dfrac{1}{2}(\delta_{ik}\delta_{jl} - \delta_{il}\delta_{jk})\mathbf{e}_i \otimes \mathbf{e}_j \otimes \mathbf{e}_k \otimes \mathbf{e}_l = \dfrac{1}{2}(\overline{\overline{\mathcal{I}}} - \underline{\underline{\mathcal{I}}})$	$\mathcal{A} : \mathbf{T} = \mathrm{ant}[\mathbf{T}]$
Deviatoric projection tensor $\mathcal{D} \equiv \left(\delta_{ik}\delta_{jl} - \dfrac{1}{3}\delta_{ij}\delta_{kl}\right)\mathbf{e}_i \otimes \mathbf{e}_j \otimes \mathbf{e}_k \otimes \mathbf{e}_l = \overline{\overline{\mathcal{I}}} - \dfrac{1}{3}\mathcal{I}$	$\mathcal{D} : \mathbf{T} = \mathbf{T} - \{(\mathrm{tr}\mathbf{T})/3\}\mathbf{I} = \mathbf{T}'$
Isotropic tensor $\mathcal{J} \equiv (\lambda\mathcal{I}_{ijkl} + \mu\mathcal{S}_{ijkl} + \nu\mathcal{A}_{ijkl})\mathbf{e}_i \otimes \mathbf{e}_j \otimes \mathbf{e}_k \otimes \mathbf{e}_l$ $= \lambda\mathcal{I} + \mu\mathcal{S} + \nu\mathcal{A}$	
$\left.\begin{array}{l} \overline{\overline{\mathcal{I}}} : \overline{\overline{\mathcal{I}}} = \overline{\overline{\mathcal{I}}},\ \underline{\underline{\mathcal{I}}} : \underline{\underline{\mathcal{I}}} = \overline{\overline{\mathcal{I}}},\ \overline{\overline{\mathcal{I}}} : \underline{\underline{\mathcal{I}}} = \underline{\underline{\mathcal{I}}} : \overline{\overline{\mathcal{I}}} = \underline{\underline{\mathcal{I}}} \\ \mathcal{I} : \mathcal{I} = 3\mathcal{I} \\ \overline{\overline{\mathcal{I}}} : \mathcal{I} = \mathcal{I} : \overline{\overline{\mathcal{I}}} = \underline{\underline{\mathcal{I}}} : \mathcal{I} = \mathcal{I} : \underline{\underline{\mathcal{I}}} = \mathcal{I} \end{array}\right\}$ $\mathcal{S} : \mathcal{S} = \mathcal{S},\ \mathcal{D} : \mathcal{D} = \mathcal{D},\ \mathcal{D} : \mathcal{S} = \mathcal{S} : \mathcal{D} = \mathbf{0}$	

1.6 Eigenvalues and Eigenvectors

A tensor is expressed in the component notation having only normal components by choosing coordinate base into special directions. In what follows, we consider these special directions for the second-order tensor.

1.6.1 Eigenvalues and Eigenvectors of Second-Order Tensors

The unit vector **e** fulfilling

$$\boxed{\mathbf{T}\mathbf{e} = T\mathbf{e}}, \quad T_{ij}e_j = Te_i, \tag{1.227}$$

that is,

$$(\mathbf{T} - T\mathbf{I})\mathbf{e} = \mathbf{0}, \quad (\mathrm{T}_{ij} - T\delta_{ij})e_j = 0, \tag{1.228}$$

for the second-order tensor is called the *eigenvector* (or *principal* or *characteristic* or *proper vector*) and the scalar T is called the *eigenvalue* (or *principal* or *characteristic* or *proper value*). The prefix 'eigen-' comes from German and was first used in this context by David Hilbert.

In order for the linear simultaneous equation (1.228) to possess a non-zero solution, the determinant of the coefficient must be zero as described in equations (1.177) and (1.178), that is,

$$\boxed{\det[\mathbf{T} - T\mathbf{I}] = 0}, \quad \left|T_{ij} - T\delta_{ij}\right| = \begin{vmatrix} T_{11} - T & T_{12} & T_{13} \\ T_{21} & T_{22} - T & T_{23} \\ T_{31} & T_{32} & T_{33} - T \end{vmatrix} = 0. \tag{1.229}$$

Equation (1.229), a cubic equation in T, is called the *characteristic equation* of the tensor. The three eigenvectors $\mathbf{e}_1, \mathbf{e}_2, \mathbf{e}_3$ are derived by solving equation (1.228) for each of three solutions T_1, T_2, T_3 of T obtained from equation (1.229).

In what follows, it is verified that the eigenvectors are mutually orthogonal for eigenvalues that differ from each other in the second-order real symmetric tensor fulfilling $\mathbf{T} = \mathbf{T}^T$.

It follows from equation (1.227) that

$$\mathbf{T}\mathbf{e}_\alpha = T_\alpha \mathbf{e}_\alpha \text{ (no sum)} \tag{1.230}$$

for the eigenvectors $\mathbf{e}_\alpha (\alpha = 1, 2, 3)$ of $\mathbf{T}$. Multiplying equation (1.230) by the eigenvectors, we have

$$\left.\begin{aligned} \mathbf{e}_\beta \cdot \mathbf{T}\mathbf{e}_\alpha &= T_\alpha \mathbf{e}_\alpha \cdot \mathbf{e}_\beta \\ \mathbf{e}_\alpha \cdot \mathbf{T}\mathbf{e}_\beta &= T_\beta \mathbf{e}_\beta \cdot \mathbf{e}_\alpha \end{aligned}\right\} \quad \text{(no sum)}.$$

Subtracting the second equation from the first yields

$$\mathbf{e}_\beta \cdot \mathbf{T}\mathbf{e}_\alpha - \mathbf{e}_\alpha \cdot \mathbf{T}\mathbf{e}_\beta = (T_\alpha - T_\beta)\mathbf{e}_\alpha \cdot \mathbf{e}_\beta, \tag{1.231}$$

that is,

$$(T_\alpha - T_\beta)\mathbf{e}_\alpha \cdot \mathbf{e}_\beta = 0 \text{ (no sum)}, \tag{1.232}$$

noting that $\mathbf{e}_\beta \cdot \mathbf{T}\mathbf{e}_\alpha - \mathbf{e}_\alpha \cdot \mathbf{T}\mathbf{e}_\beta = \mathbf{e}_\beta \cdot \mathbf{T}\mathbf{e}_\alpha - \mathbf{T}^T\mathbf{e}_\alpha \cdot \mathbf{e}_\beta = \mathbf{e}_\beta \cdot (\mathbf{T} - \mathbf{T}^T)\mathbf{e}_\alpha = 0$ by virtue of the symmetry of the tensor $\mathbf{T}$ on the left-hand side of equation (1.231). The following facts can be concluded from equation (1.232).

1. If the three eigenvalues are all different from each other, there will exist three principal directions perpendicular to each other.
2. If two of three eigenvalues are equal, all directions in the plane perpendicular to the principal direction for the other eigenvalue will be principal directions for the two equal eigenvalues.
3. If all three eigenvalues are equal, all directions in the space will be principal directions.

Returning to the second-order tensor including a skew-symmetric tensor, the expansion of the characteristic equation (1.229) of $\mathbf{T}$ leads to

$$\begin{aligned} \begin{vmatrix} T_{11} - T & T_{12} & T_{13} \\ T_{21} & T_{22} - T & T_{23} \\ T_{31} & T_{32} & T_{33} - T \end{vmatrix} &= (T_{11} - T)(T_{22} - T)(T_{33} - T) + T_{12}T_{23}T_{31} + T_{21}T_{32}T_{13} \\ &\quad - (T_{11} - T)T_{23}T_{32} - (T_{22} - T)T_{31}T_{13} - (T_{33} - T)T_{12}T_{21} \end{aligned}$$

$$
\begin{aligned}
&= -T^3 + (T_{11} + T_{22} + T_{33})T^2 - (T_{11}T_{22} + T_{22}T_{33} + T_{33}T_{11} \\
&\quad - T_{12}T_{21} - T_{23}T_{32} - T_{31}T_{13})T + T_{11}T_{22}T_{33} + T_{12}T_{23}T_{31} \\
&\quad + T_{13}T_{32}T_{21} - T_{11}T_{23}T_{32} - T_{22}T_{31}T_{13} - T_{33}T_{12}T_{21} \\
&= -T^3 + (T_{11} + T_{22} + T_{33})T^2 - \frac{1}{2}[(T_{11} + T_{22} + T_{33})^2 \\
&\quad - \{T_{11}^2 + T_{22}^2 + T_{33}^2 + 2(T_{12}T_{21} + T_{23}T_{32} + T_{31}T_{13})\}]T \\
&\quad + T_{11}T_{22}T_{33} + T_{12}T_{23}T_{31} + T_{13}T_{32}T_{21} - T_{11}T_{23}T_{32} - T_{22}T_{31}T_{13} \\
&\quad - T_{33}T_{12}T_{21} = 0
\end{aligned}
$$

from which the characteristic equation is expressed as

$$
(-\det[\mathbf{T} - T\mathbf{I}] =) \boxed{T^3 - \mathrm{I}T^2 + \mathrm{II}T - \mathrm{III} = 0} \tag{1.233}
$$

where

$$
\boxed{\mathrm{I} \equiv T_{11} + T_{22} + T_{33} = T_{ii} = \mathrm{tr}\mathbf{T}} \tag{1.234}
$$

$$
\boxed{
\begin{aligned}
\mathrm{II} &\equiv \begin{vmatrix} T_{22} & T_{23} \\ T_{32} & T_{33} \end{vmatrix} + \begin{vmatrix} T_{11} & T_{13} \\ T_{31} & T_{33} \end{vmatrix} + \begin{vmatrix} T_{11} & T_{12} \\ T_{21} & T_{22} \end{vmatrix} = \mathrm{tr}(\mathrm{cof}\,\mathbf{T}) \\
&= \frac{1}{2}(T_{rr}T_{ss} - T_{rs}T_{sr}) = \frac{1}{2}(\mathrm{tr}^2\mathbf{T} - \mathrm{tr}\mathbf{T}^2)
\end{aligned}
} \tag{1.235}
$$

$$
\boxed{
\mathrm{III} \equiv \begin{vmatrix} T_{11} & T_{12} & T_{13} \\ T_{21} & T_{22} & T_{23} \\ T_{31} & T_{32} & T_{33} \end{vmatrix} = \det\mathbf{T} = \varepsilon_{rst}T_{r1}T_{s2}T_{t3} = \frac{1}{6}\mathrm{tr}^3\mathbf{T} - \frac{1}{2}\mathrm{tr}\mathbf{T}\,\mathrm{tr}\mathbf{T}^2 + \frac{1}{3}\mathrm{tr}\mathbf{T}^3
} \tag{1.236}
$$

The direct notation by the traces in equation (1.236) will be shown later in equation (1.297) based on the Cayley–Hamilton theorem in 1.6.6.

The characteristic equation (1.233) is derived above by expanding the determinant for components. On the other hand, it can be derived in the direct notation as follows (Chadwick, 1976; Kyoya, 2008). Multiply equation (1.229) by [**abc**],

$$
\det(\mathbf{T} - \lambda\mathbf{I})[\mathbf{abc}] = 0,
$$

and transform it using equation $(1.179)_3$ into

$$
[(\mathbf{T} - \lambda\mathbf{I})\mathbf{a}(\mathbf{T} - \lambda\mathbf{I})\mathbf{b}(\mathbf{T} - \lambda\mathbf{I})\mathbf{c}] = 0.
$$

This equation leads to

$$
\begin{aligned}
[(\mathbf{T} - \lambda\mathbf{I})\mathbf{a}(\mathbf{T} - \lambda\mathbf{I})\mathbf{b}(\mathbf{T} - \lambda\mathbf{I})\mathbf{c}] &= -[\mathbf{abc}]\lambda^3 + ([\mathbf{Ta\,b\,c}] + [\mathbf{a\,Tb\,c}] + [\mathbf{a\,b\,Tc}])\lambda^2 \\
&\quad - ([\mathbf{a\,Tb\,Tc}] + [\mathbf{Ta\,b\,Tc}] + [\mathbf{Ta\,Tb\,c}])\lambda + ([\mathbf{Ta\,Tb\,Tc}] \\
&= -[\mathbf{abc}][\lambda^3 - (\mathrm{tr}\mathbf{T})\lambda^2 + \frac{1}{2}\{(\mathrm{tr}\mathbf{T})^2 - \mathrm{tr}\mathbf{T}^2)\lambda - \det\mathbf{T}\} = 0
\end{aligned}
$$

by virtue of equation (1.179). Then we have

$$\lambda^3 - (\mathrm{tr}\mathbf{T})\lambda^2 + \frac{1}{2}\{(\mathrm{tr}\mathbf{T})^2 - \mathrm{tr}\mathbf{T}^2)\lambda - \mathrm{det}\mathbf{T} = 0$$

which is nothing more than the characteristic equation in equation (1.233).

On the other hand, the characteristic equation (1.233) is expressed by means of the eigenvalues as follows:

$$(T - T_1)(T - T_2)(T - T_3) = 0. \tag{1.237}$$

Comparing equations (1.233) and (1.237), we can express the coefficients I, II, and III as

$$\boxed{\begin{aligned} &\mathrm{I} = T_1 + T_2 + T_3 \\ &\mathrm{II} = T_1T_2 + T_1T_2 + T_3T_1 \\ &\mathrm{III} = T_1T_2T_3 \end{aligned}} \tag{1.238}$$

Equation (1.238) can also be derived by inserting $T_{11} = T_1, T_{22} = T_2, T_{33} = T_3, T_{12} = T_{23} = T_{31} = 0$ into equations (1.234)–(1.236). Since I, II, and III are symmetric functions of eigenvalues, they are invariant under coordinate system rotation and are called the *principal invariants*.

The following invariants are called *moments*.

$$\boxed{\bar{\mathrm{I}} \equiv \mathrm{tr}\mathbf{T}, \quad \bar{\mathrm{II}} \equiv \mathrm{tr}\mathbf{T}^2, \quad \bar{\mathrm{III}} \equiv \mathrm{tr}\mathbf{T}^3} \tag{1.239}$$

The principal invariants are described in terms of these moments from equations (1.234)–(1.236) as follows:

$$\left.\begin{aligned} &\mathrm{I} = \bar{\mathrm{I}} \\ &\mathrm{II} = \frac{1}{2}(\bar{\mathrm{I}}^2 - \bar{\mathrm{II}}) \\ &\mathrm{III} = \frac{1}{6}\bar{\mathrm{I}}^3 - \frac{1}{2}\bar{\mathrm{I}}\,\bar{\mathrm{II}} - + \frac{1}{3}\bar{\mathrm{III}} \end{aligned}\right\}. \tag{1.240}$$

The characteristic equation (1.233) can be obtained directly exploiting the formula for the scalar triple product in equation (1.179). Multiplying equation (1.229) by the triple scalar product of arbitrary three independent vectors **a**, **b**, **c** yields (Chadwick, 1976; Kyoya, 2008; Bertram, 2008)

$$\det(\mathbf{T} - T\mathbf{I})[\mathbf{abc}] = 0, \tag{1.241}$$

which is transformed using equation $(1.179)_3$ to

$$[(\mathbf{T} - T\mathbf{I})\mathbf{a}, (\mathbf{T} - T\mathbf{I})\mathbf{b}, (\mathbf{T} - T\mathbf{I})\mathbf{c}] = 0. \tag{1.242}$$

Here, expanding equation (1.242) and applying equation (1.179), one has

$$\begin{aligned} &[(\mathbf{T} - T\mathbf{I})\mathbf{a}, (\mathbf{T} - T\mathbf{I})\mathbf{b}, (\mathbf{T} - T\mathbf{I})\mathbf{c}] \\ &\quad = -T^3[\mathbf{abc}] + T^2([\mathbf{Ta}, \mathbf{b}, \mathbf{c}] + [\mathbf{a}, \mathbf{Tb}, \mathbf{c}] + [\mathbf{a}, \mathbf{b}, \mathbf{Tc}]) \\ &\qquad - T([\mathbf{a}, \mathbf{Tb}, \mathbf{Tc}] + [\mathbf{Ta}, \mathbf{b}, \mathbf{Tc}] + [\mathbf{Ta}, \mathbf{Tb}, \mathbf{c}]) + [\mathbf{Ta}, \mathbf{Tb}, \mathbf{Tc}] \\ &\quad = -[\mathbf{abc}][T^3 - (\mathrm{tr}\mathbf{T})T^2 + \frac{1}{2}\{(\mathrm{tr}\mathbf{T})^2 - \mathrm{tr}\mathbf{T}^2\}T - \mathrm{det}\mathbf{T}\} = 0. \end{aligned}$$

Because of the arbitrariness of **a**, **b**, **c** the following equality must hold:

$$T^3 - (\mathrm{tr}\mathbf{T})T^2 + \frac{1}{2}\mathrm{tr}(\mathrm{cof}\,\mathbf{T})T - \det\mathbf{T} = 0 \tag{1.243}$$

which is nothing more than the characteristic equation (1.233).

Next, consider the deviatoric tensor $\mathbf{T}'$. The characteristic equation of $\mathbf{T}'$ is given by replacing $\mathbf{T}$ by $\mathbf{T}'$ in equation (1.233) as follows:

$$\boxed{T'^3 - \mathrm{II}'\mathrm{T}' - \mathrm{III}' = 0} \tag{1.244}$$

noting that

$$\mathrm{I}' \equiv \mathrm{tr}\mathbf{T}' = 0, \tag{1.245}$$

where

$$\begin{aligned}
\boxed{\mathrm{II}' \equiv \frac{1}{2}\mathrm{tr}\mathbf{T}'^2} &= \frac{1}{2}T'_{rs}T'_{sr} \\
&= \frac{1}{2}(T'^2_{11} + T'^2_{22} + T'^2_{33}) + T'^2_{12} + T'^2_{23} + T'^2_{31} \\
&= \frac{1}{6}\{(T_{11} - T_{22})^2 + (T_{22} - T_{33})^2 + (T_{33} - T_{11})^2\} + T'^2_{12} + T'^2_{23} + T'^2_{31} \\
&= \frac{1}{2}(T'^2_1 + T'^2_2 + T'^2_3) = \frac{1}{6}\{(T_1 - T_2)^2 + (T_2 - T_3)^2 + (T_3 - T_1)^2\},
\end{aligned} \tag{1.246}$$

$$\begin{aligned}
\boxed{\mathrm{III}' \equiv \det\mathbf{T}' = \frac{1}{3}\mathrm{tr}\mathbf{T}'^3} &= \frac{1}{3}T'_{rs}T'_{sr}T'_{rt} \\
&= T'_{11}T'_{22}T'_{33} - T'_{11}T'^2_{23} - T'_{22}T'^2_{31} - T'_{33}T'^2_{12} + 2T'_{12}T'_{23}T'_{31} \\
&= T'_1T'_2T'_3 = \frac{1}{3}(T'^3_1 + T'^3_2 + T'^3_3).
\end{aligned} \tag{1.247}$$

1.6.2 Spectral Representation and Elementary Tensor Functions

By choosing as base vectors the principal directions $\mathbf{e}_P$ of the tensor $\mathbf{T}$, the identity tensor is represented by $\mathbf{I} = \mathbf{e}_P \otimes \mathbf{e}_P$. $\mathbf{T}$ can be represented by multiplying it by $\mathbf{T}$ itself, that is, $\mathbf{T} = \mathbf{TI} = \mathbf{T}\mathbf{e}_P \otimes \mathbf{e}_P$ as follows:

$$\boxed{\mathbf{T} = \sum_{P=1}^{3} T_P\mathbf{e}_P \otimes \mathbf{e}_P} \tag{1.248}$$

where T_P and $\mathbf{e}_P$ are the eigenvalues and eigenvectors, respectively, fulfilling

$$\mathbf{T}\mathbf{e}_P = T_P\mathbf{e}_P \ (P = 1, 2, 3;\ \text{no sum}). \tag{1.249}$$

Equation (1.248) is called the *spectral decomposition* (or *spectral representation*).

If the tensor $\tilde{\mathbf{T}}$ with eigenvectors $\tilde{\mathbf{e}}_P$ possesses the same eigenvalues as those of the tensor $\mathbf{T}$, we may write

$$\tilde{\mathbf{T}}\tilde{\mathbf{e}}_P = T_P\tilde{\mathbf{e}}_P \ (\text{no sum}), \tag{1.250}$$

where the orthogonal tensor $\tilde{\mathbf{Q}}$ between the eigenvectors of these tensors is given by

$$\tilde{Q}_{PQ} = \tilde{\mathbf{e}}_P \cdot \mathbf{e}_Q, \quad \tilde{\mathbf{Q}} = \mathbf{e}_P \otimes \tilde{\mathbf{e}}_P, \quad \mathbf{e}_P = \tilde{\mathbf{Q}}\tilde{\mathbf{e}}_P, \quad \tilde{\mathbf{e}}_P = \tilde{\mathbf{Q}}^T\mathbf{e}_P. \tag{1.251}$$

Then

$$\mathbf{T} = \tilde{\mathbf{Q}}\tilde{\mathbf{T}}\tilde{\mathbf{Q}}^T, \quad \tilde{\mathbf{T}} = \tilde{\mathbf{Q}}^T\mathbf{T}\tilde{\mathbf{Q}} \tag{1.252}$$

because of

$$\mathbf{T} = \sum_{P=1}^{3} T_P\mathbf{e}_p \otimes \mathbf{e}_P = \sum_{P=1}^{3} T_P\tilde{\mathbf{Q}}\tilde{\mathbf{e}}_P \otimes \tilde{\mathbf{Q}}\tilde{\mathbf{e}}_P = \tilde{\mathbf{Q}}\sum_{P=1}^{3} T_P\tilde{\mathbf{e}}_P \otimes \tilde{\mathbf{e}}_P\tilde{\mathbf{Q}}^T, \tag{1.253}$$

noting equation (1.94). Therefore, tensors having identical eigenvalues can be related by an orthogonal transformation: they are called *similar tensors*. The orthogonal transformation rule (1.199) and relation (1.252) for similar tensors are of mutually opposite forms. The magnitudes of similar tensors are identical to each other by equation (1.107).

Furthermore, it follows for the nth power of the tensor $\mathbf{T}$ that

$$\mathbf{T}^n\mathbf{e}_P = T_P^n\mathbf{e}_P, \tag{1.254}$$

$$\mathbf{T}^n = \mathbf{T}^n\sum_{P=1}^{3} \mathbf{e}_P \otimes \mathbf{e}_P = \sum_{P=1}^{3} T_P^n\mathbf{e}_P \otimes \mathbf{e}_P \tag{1.255}$$

by repeated application of $\mathbf{T}$ to equation (1.227), that is,

$$T_P^n\mathbf{e} = T_P^{n-1}T_P\mathbf{e} = T_P^{n-1}\mathbf{T}\mathbf{e} = \mathbf{T}T_P^{n-2}T_P\mathbf{e} = \mathbf{T}T_P^{n-2}\mathbf{T}\mathbf{e} = \mathbf{T}^2T_P^{n-3}T_P\mathbf{e} = \cdots = \mathbf{T}^n\mathbf{e}.$$

Equation (1.254) means that the eigenvalues of $\mathbf{T}^n$ are T_P^n, where T_P are the eigenvalues of $\mathbf{T}$ and the principal directions of the tensor $\mathbf{T}^n$ are identical to those of $\mathbf{T}$. Tensors that have an identical set of principal directions are called *coaxial* or said to fulfill *coaxiality*. The second-order tensor function $\mathbf{f}(\mathbf{T})$ of only $\mathbf{T}$ is coaxial with $\mathbf{T}$ and the eigenvalues are given by $f(T_P)$, and thus it follows that

$$f(\mathbf{T}) = \sum_{P=1}^{3} f(T_P)\mathbf{e}_P \otimes \mathbf{e}_P. \tag{1.256}$$

The general second-order tensor function can be defined by extending equation (1.256) as follows. Applying the *Maclaurin expansion*

$$f(T_P) = \sum_{n=0}^{\infty} \frac{f^{(n)}(0)}{n!} T_P^n = f(0) + f'(0)T_P + \frac{f''(0)}{2!} T_P^2 + \frac{f'''(0)}{3!} T_P^3 + \cdots \quad (1.257)$$

to the scalar function $f(T_P)$ in equation (1.256), various second-order tensor functions are described explicitly as follows:

$$\exp \mathbf{T} = \sum_{n=0}^{\infty} \frac{1}{n!} \mathbf{T}^n = \mathbf{I} + \mathbf{T} + \frac{1}{2!} \mathbf{T}^2 + \frac{1}{3!} \mathbf{T}^3 + \cdots, \quad (1.258)$$

$$\ln(\mathbf{I} + \mathbf{T}) = \sum_{n=1}^{\infty} \frac{(-1)^{n+1}}{n} \mathbf{T}^n = \mathbf{T} - \frac{\mathbf{T}^2}{2} + \frac{\mathbf{T}^3}{3} - \cdots, \quad (1.259)$$

$$(\mathbf{I} + \mathbf{T})^m = \sum_{n=0}^{\infty} \frac{m!}{(m-n)!n!} \mathbf{T}^n = \mathbf{I} + m\mathbf{T} + \frac{m(m-1)}{2!} \mathbf{T}^2 + \frac{m(m-1)(m-2)}{3!} \mathbf{T}^3 + \cdots, \quad (1.260)$$

$$\mathbf{T}^{-1} = \{\mathbf{I} + (\mathbf{T} - \mathbf{I})\}^{-1} = \mathbf{I} - (\mathbf{T} - \mathbf{I}) + (\mathbf{T} - \mathbf{I})^2 - \cdots, \quad (1.261)$$

$$a^{\mathbf{T}} = \sum_{n=0}^{\infty} \frac{(\ln a)^n}{n!} \mathbf{T}^n = \mathbf{I} + (\ln a)\mathbf{T} + \frac{(\ln a)^2}{2!} \mathbf{T}^2 + \frac{(\ln a)^3}{3!} \mathbf{T}^3 + \cdots, \quad (1.262)$$

$$\sin \mathbf{T} = \sum_{n=0}^{\infty} \frac{(-1)^n}{(2n+1)!} \mathbf{T}^{2n+1} = \mathbf{T} - \frac{1}{3!} \mathbf{T}^3 + \frac{1}{5!} \mathbf{T}^5 - \cdots, \quad (1.263)$$

$$\cos \mathbf{T} = \sum_{n=0}^{\infty} \frac{(-1)^n}{(2n)!} \mathbf{T}^{2n} = \mathbf{I} - \frac{1}{2!} \mathbf{T}^2 + \frac{1}{4!} \mathbf{T}^4 - \cdots. \quad (1.264)$$

1.6.3 Calculation of Eigenvalues and Eigenvectors

Eigenvalues and eigenvectors of tensor must be calculated in order to obtain the spectral representation. Solutions (cf. Carlson and Hoger, 1986; Bruhns, 2003) are as follows.

Eigenvalues

In order to obtain eigenvalues, one must solve the characteristic equation, which is the cubic equation with coefficients which are functions of the invariants. First, let the following form of solution be postulated.

$$T' = \sqrt{\frac{4\mathrm{II}'}{3}} \cos \psi \quad (1.265)$$

for the eigenvalues of the deviatoric part of tensor **T**. Substituting equation (1.265) into equation (1.244), we have

$$\left(\frac{4\mathrm{II}'}{3}\right)^{3/2}\cos^3\psi - \mathrm{II}'\left(\frac{4\mathrm{II}'}{3}\right)^{1/2}\cos\psi - \mathrm{III}' = 0 \tag{1.266}$$

which reduces to

$$\frac{2}{3\sqrt{3}}\mathrm{II}'^{3/2}\cos 3\psi - \mathrm{III}' = 0, \tag{1.267}$$

using the trigonometric formula

$$\cos^3\psi = \frac{1}{4}(\cos 3\psi + 3\cos\psi). \tag{1.268}$$

It follows from (1.267) that

$$\cos 3\psi = \frac{3\sqrt{3}\mathrm{III}'}{2\mathrm{II}'^{3/2}}. \tag{1.269}$$

Bearing in mind that the cosine is a periodic function with the period 2π, the angle ψ is expressed by the equation

$$\psi_K = \frac{1}{3}\left\{\cos^{-1}\left(\frac{3\sqrt{3}\mathrm{III}'}{2\mathrm{II}'^{3/2}}\right) - 2\pi K\right\}, \tag{1.270}$$

where K is an integer. Substituting equation (1.270) into equation (1.265) and adding the spherical component I/3, the eigenvalues of **T** are as follows:

$$T_K = \frac{1}{3}\mathrm{I} + T'_K = \frac{1}{3}\mathrm{I} + \sqrt{\frac{4\mathrm{II}'}{3}}\cos\psi_K = \frac{1}{3}\mathrm{I} + \sqrt{\frac{4\mathrm{II}'}{3}}\cos\left[\frac{1}{3}\left\{\cos^{-1}\left(\frac{3\sqrt{3}\mathrm{III}'}{2\mathrm{II}'^{3/2}}\right) - 2\pi K\right\}\right]. \tag{1.271}$$

Eigenvectors

Equation (1.248) can be expressed as

$$\mathbf{T} = \sum_{P=1}^{3} T_P \mathbf{E}_P \tag{1.272}$$

in which the tensor

$$\mathbf{E}_P \equiv \mathbf{e}_P \otimes \mathbf{e}_P \;\; \text{(no sum)} \tag{1.273}$$

is called the *eigenprojection* of **T** and satisfies

$$\sum_{P=1}^{3} \mathbf{E}_P (= \mathbf{e}_1 \otimes \mathbf{e}_1 + \mathbf{e}_2 \otimes \mathbf{e}_2 + \mathbf{e}_3 \otimes \mathbf{e}_3) = \mathbf{I}, \tag{1.274}$$

$$\left.\begin{aligned}&\mathbf{E}_J\mathbf{E}_K = \begin{cases}\mathbf{E}_J & \text{for} \quad J = K\\ \mathbf{0} & \text{for} \quad J \neq K\end{cases}\\ &\mathrm{tr}(\mathbf{E}_J\mathbf{E}_K) = \delta_{IJ}\end{aligned}\right\}. \tag{1.275}$$

Here, we have

$$\left.\begin{aligned}\mathbf{T}\mathbf{E}_J &= \left(\sum_{K=1}^{3} T_K\mathbf{E}_K\right)\mathbf{e}_J\otimes\mathbf{e}_J = \left(\sum_{K=1}^{3} T_K\mathbf{e}_K\otimes\mathbf{e}_K\right)\mathbf{e}_J\otimes\mathbf{e}_J = T_J\mathbf{e}_J\otimes\mathbf{e}_J\\ \mathbf{E}_J\mathbf{T} &= \mathbf{e}_J\otimes\mathbf{e}_J\left(\sum_{K=1}^{3} T_K\mathbf{E}_K\right) = \mathbf{e}_J\otimes\mathbf{e}_J\left(\sum_{K=1}^{3} T_K\mathbf{e}_K\otimes\mathbf{e}_K\right) = \mathbf{e}_J\otimes\mathbf{e}_J T_J\end{aligned}\right\} \text{(no sum for } J\text{)},$$

and thus we have also

$$\mathbf{T}\mathbf{E}_J = \mathbf{E}_J\mathbf{T} = T_J\mathbf{E}_J \text{ (no sum)}. \tag{1.276}$$

On the other hand, it follows from equation (1.274) that

$$\mathbf{T} - T_K\mathbf{I} = \sum_{J=1}^{3} T_J\mathbf{E}_J - T_K\sum_{J=1}^{3}\mathbf{E}_J$$

and thus

$$\mathbf{T} - T_K\mathbf{I} = \sum_{J=1}^{3}(T_J - T_K)\mathbf{E}_J \tag{1.277}$$

from which we also have

$$\prod_{\substack{K\neq\theta\\K=1}}^{3}(\mathbf{T} - T_K\mathbf{I}) = \prod_{\substack{K\neq\theta\\K=1}}^{3}\left(\sum_{J=1}^{3}(T_J - T_K)\mathbf{E}_J\right), \tag{1.278}$$

where θ is 1, 2, or 3. The right-hand side in equation (1.278) can be rewritten as

$$\prod_{\substack{K\neq\theta\\K=1}}^{3}\sum_{J=1}^{3}(T_J - T_K)\mathbf{E}_J = \left\{\prod_{\substack{K\neq\theta\\K=1}}^{3}(T_\theta - T_K)\right\}\mathbf{E}_\theta, \tag{1.279}$$

noting equation (1.275). Substituting equation (1.279) into equation (1.278), *Sylvester's formula* for the eigenprojections is obtained:

$$\mathbf{E}_\theta = \prod_{\substack{K\neq\theta\\K=1}}^{3}\frac{\mathbf{T} - T_K\mathbf{I}}{T_\theta - T_K}. \tag{1.280}$$

For instance, the eigenprojection of $\mathbf{E}_2(\theta = 2)$ is obtained by the above-mentioned method as follows:

$$\mathbf{E}_2 = \prod_{\substack{K\neq 2\\K=1}}^{3}\frac{\mathbf{T} - T_K\mathbf{I}}{T_2 - T_K} = \frac{(\mathbf{T} - T_1\mathbf{I})(\mathbf{T} - T_3\mathbf{I})}{(T_2 - \mathrm{T}_1)(T_2 - T_3)}. \tag{1.281}$$

1.6.4 Eigenvalues and Vectors of Orthogonal Tensor

It follows from equation (1.102) that

$$\det[\boldsymbol{Q}^T(\boldsymbol{Q}-\mathbf{I})] = \begin{cases} \det[\mathbf{I}-\boldsymbol{Q}^T] = \det[-(\boldsymbol{Q}-\mathbf{I})^T] = -\det[\boldsymbol{Q}-\mathbf{I}] \\ \det[\boldsymbol{Q}^T]\det[\boldsymbol{Q}-\mathbf{I}] = \det[\boldsymbol{Q}-\mathbf{I}] \end{cases}$$

by virtue of equation $(1.159)_{1,3}$ and (1.173). Then, we have

$$\det(\boldsymbol{Q}-\mathbf{I}) = 0. \tag{1.282}$$

By comparing equation (1.282) with equation (1.229), it is evident that one of the eigenvalues of $\boldsymbol{Q}$ is 1.

Here, choosing $\mathbf{e}_{\mathrm{III}}$ as the unit principal vector possessing eigenvalue 1 in the orthogonal tensor $\boldsymbol{Q}$, it follows that

$$\mathbf{e}_{\mathrm{III}} = \boldsymbol{Q}\mathbf{e}_{\mathrm{III}}. \tag{1.283}$$

Further, denoting the orthonormal set of base vectors by $\{\mathbf{e}_3, \mathbf{e}_2, \mathbf{e}_{\mathrm{III}}\}$, we have

$$\mathbf{e}_{\mathrm{III}} \cdot (\boldsymbol{Q}^T\mathbf{e}_1) = \boldsymbol{Q}\mathbf{e}_{\mathrm{III}} \cdot \mathbf{e}_1 = \mathbf{e}_{\mathrm{III}} \cdot \mathbf{e}_1 = 0, \quad \mathbf{e}_{\mathrm{III}} \cdot (\boldsymbol{Q}^T\mathbf{e}_2) = \boldsymbol{Q}\mathbf{e}_{\mathrm{III}} \cdot \mathbf{e}_2 = \mathbf{e}_{\mathrm{III}} \cdot \mathbf{e}_2 = 0$$

by virtue of equations $(1.93)_2$ and (1.283). In other words, $\boldsymbol{Q}^T\mathbf{e}_1$ and $\boldsymbol{Q}^T\mathbf{e}_2$ are rotatiosn around the principal base vector $\mathbf{e}_{\mathrm{III}}$ (see equation (1.109)).

The principal invariants are given by incorporating equations (1.113) into equations (1.234)–(1.236) as follows:

$$\mathrm{I}_Q = \mathrm{II}_Q = 1 + 2\cos\theta, \quad \mathrm{III}_Q = 1, \tag{1.284}$$

where

$$\left.\begin{aligned} \mathrm{I}_Q &= \mathrm{tr}\boldsymbol{Q} = 1 + 2\cos\theta \\ \mathrm{II}_Q &= \frac{1}{2}(\mathrm{tr}^2\boldsymbol{Q} - \mathrm{tr}\boldsymbol{Q}^2) = \frac{1}{2}(Q_{rr}Q_{ss} - Q_{rs}Q_{sr}) \\ &= \frac{1}{2}[(2\cos\theta + 1)^2 - \{\cos^2\theta + \cos^2\theta + 1 + 2\sin\theta(-\sin\theta)\}] = 1 + 2\cos\theta \\ \mathrm{III}_Q &= \det\boldsymbol{Q} = 1 \end{aligned}\right\}. \tag{1.285}$$

An arbitrary vector rotated from $\mathbf{v}$ on the $(\mathbf{e}_1, \mathbf{e}_2)$ plane, fulfilling $\mathbf{v} \cdot \mathbf{e}_{\mathrm{III}} = 0$, is expressed by multiplying $\mathbf{v}$ by $\boldsymbol{Q}^T$ based on equation (1.112) as follows:

$$\boldsymbol{Q}^T\mathbf{v} = \mathbf{v}\cos\theta + \mathbf{e}_{\mathrm{III}} \times \mathbf{v}\sin\theta, \tag{1.286}$$

noting that

$$\begin{aligned} \boldsymbol{Q}^T\mathbf{v} &= \{(\mathbf{e}_1 \otimes \mathbf{e}_1 + \mathbf{e}_2 \otimes \mathbf{e}_2)\cos\theta - (\mathbf{e}_1 \otimes \mathbf{e}_2 - \mathbf{e}_2 \otimes \mathbf{e}_1)\sin\theta + \mathbf{e}_{\mathrm{III}} \otimes \mathbf{e}_{\mathrm{III}}\}\mathbf{v} \\ &= \{(\mathbf{e}_1(\mathbf{e}_1 \cdot \mathbf{v}) + \mathbf{e}_2(\mathbf{e}_2 \cdot \mathbf{v})\}\cos\theta - \{(\mathbf{e}_1(\mathbf{e}_2 \cdot \mathbf{v}) - \mathbf{e}_2(\mathbf{e}_1 \cdot \mathbf{v})\}\sin\theta \\ &= \mathbf{v}\cos\theta - (v_2\mathbf{e}_1 - v_1\mathbf{e}_2)\sin\theta, \end{aligned}$$

$$-(v_2\mathbf{e}_1 - v_1\mathbf{e}_2) = v_1\mathbf{e}_2 - v_2\mathbf{e}_1 = \mathbf{e}_{\mathrm{III}} \times (v_1\mathbf{e}_1 + v_2\mathbf{e}_2) = \mathbf{e}_{\mathrm{III}} \times \mathbf{v}.$$

1.6.5 Eigenvalues and Vectors of Skew-Symmetric Tensor and Axial Vector

The characteristic equation of the skew-symmetric tensor $\boldsymbol{\Omega}$ is given by equation (1.233) as

$$\Omega^3 - \mathrm{I}_\Omega \Omega^2 + \mathrm{II}_\Omega \Omega - \mathrm{III}_\Omega = 0, \tag{1.287}$$

denoting the eigenvalue by Ω, where the principal invariants are given as

$$\left.\begin{array}{l} \mathrm{I}_\Omega = \mathrm{tr}\boldsymbol{\Omega} = 0 \\ \mathrm{II}_\Omega = -\dfrac{1}{2}\,\mathrm{tr}\boldsymbol{\Omega}^2 = -\dfrac{1}{2}\Omega_{rs}\Omega_{sr} = \omega_1^2 + \omega_2^2 + \omega_3^2 \geq 0 \\ \mathrm{III}_\Omega = \det\boldsymbol{\Omega} = 0 \end{array}\right\}, \tag{1.288}$$

noting equation (1.141). Equation (1.287) reduces to

$$\Omega(\Omega^2 + \omega_1^2 + \omega_2^2 + \omega_3^2) = 0. \tag{1.289}$$

Therefore, the skew-symmetric tensor possesses only one real eigenvalue 0. This is self-evident because the diagonal components of a skew-symmetric tensor are zero. There exists only one principal direction having zero eigenvalue.

Choosing $\mathbf{e}_1$ as the principal direction of $\boldsymbol{\Omega}$ possessing zero eigenvalue, $\boldsymbol{\Omega}$ is represented as

$$\boldsymbol{\Omega} = \Omega_{32}(\mathbf{e}_3 \otimes \mathbf{e}_2 - \mathbf{e}_2 \otimes \mathbf{e}_3). \tag{1.290}$$

noting $\Omega_{11} = \Omega_{22} = \Omega_{33} = 0$ and

$$\left.\begin{array}{l} \Omega_{21}(= \mathbf{e}_2 \cdot (\boldsymbol{\Omega}\mathbf{e}_1) = \mathbf{e}_2 \cdot (0\mathbf{e}_1)) = \Omega_{12} = 0 \\ \Omega_{31}(= \mathbf{e}_3 \cdot (\boldsymbol{\Omega}\mathbf{e}_1) = \mathbf{e}_3 \cdot (0\mathbf{e}_1)) = \Omega_{13} = 0 \end{array}\right\}.$$

It follows that

$$\begin{aligned} \boldsymbol{\Omega}\mathbf{a} - \Omega_{23}\mathbf{e}_1 \times \mathbf{a} &= \Omega_{32}(\mathbf{e}_3 \otimes \mathbf{e}_2 - \mathbf{e}_2 \otimes \mathbf{e}_3)\mathbf{a} + \Omega_{32}\mathbf{e}_1 \times \mathbf{a} \\ &= \Omega_{32}[(\mathbf{a}\cdot\mathbf{e}_2)\mathbf{e}_3 - (\mathbf{a}\cdot\mathbf{e}_3)\mathbf{e}_2 + \mathbf{e}_1 \times \{(\mathbf{a}\cdot\mathbf{e}_1)\mathbf{e}_1 + (\mathbf{a}\cdot\mathbf{e}_2)\mathbf{e}_2 + (\mathbf{a}\cdot\mathbf{e}_3)\mathbf{e}_3\}] \\ &= \Omega_{23}[(\mathbf{a}\cdot\mathbf{e}_3)\mathbf{e}_2 - (\mathbf{a}\cdot\mathbf{e}_2)\mathbf{e}_3 + (\mathbf{a}\cdot\mathbf{e}_2)\mathbf{e}_3 - (\mathbf{a}\cdot\mathbf{e}_3)\mathbf{e}_2\}] = \mathbf{0}. \end{aligned} \tag{1.291}$$

Then the axial vector $\boldsymbol{\omega}$ for the skew-symmetric tensor $\boldsymbol{\Omega}$ can be represented as

$$\boldsymbol{\omega} = \Omega_{23}\mathbf{e}_1, \tag{1.292}$$

exploiting equations (1.138) and (1.291). It follows from equations (1.290) and (1.292) that

$$\boldsymbol{\Omega}\boldsymbol{\omega} = \mathbf{0}. \tag{1.293}$$

1.6.6 Cayley–Hamilton Theorem

Multiplication to the characteristic equation of equation (1.233) by the principal vector **e** with the aid of equation (1.254) yields

$$\boxed{\mathbf{T}^3 - \mathrm{I}\mathbf{T}^2 + \mathrm{II}\mathbf{T} - \mathrm{III}\mathbf{I} = \mathbf{O}} \tag{1.294}$$

Equation (1.294) is referred to as the *Cayley–Hamilton theorem.*

By virtue of equation (1.294), any power function of **T** can be expressed as a polynomial in $\mathbf{T}^2, \mathbf{T}, \mathbf{I}$ with coefficients consisting of the invariants. For instance, the fourth-power of **T** reduces to

$$\begin{aligned}\mathbf{T}^4 &= (\mathrm{I}\mathbf{T}^2 - \mathrm{II}\mathbf{T} + \mathrm{III}\mathbf{I})\mathbf{T} = \mathrm{I}\mathbf{T}^3 - \mathrm{II}\mathbf{T}^2 + \mathrm{III}\mathbf{T} = \mathrm{I}(\mathrm{I}\mathbf{T}^2 - \mathrm{II}\mathbf{T} + \mathrm{III}\mathbf{I}) - \mathrm{II}\mathbf{T}^2 + \mathrm{III}\mathbf{T}\\ &= (\mathrm{I}^2 - \mathrm{II})\mathbf{T}^2 - (\mathrm{I}\,\mathrm{II} - \mathrm{III})\mathbf{T} + \mathrm{I}\,\mathrm{III}\mathbf{I},\end{aligned} \tag{1.295}$$

and the inverse of **T** is expressed from equation (1.294) by

$$\mathbf{T}^{-1} = (\mathbf{T}^2 - \mathrm{I}\mathbf{T} + \mathrm{II}\mathbf{I})/\mathrm{III}. \tag{1.296}$$

It is concluded that the power of **T** is expressed by the linear combination of $\mathbf{T}^2, \mathbf{T}, \mathbf{I}$ with coefficients consisting of the eigenvalues. Furthermore, any function other than power function can be described by these, exploiting the Maclaurin expansion described in Section 1.6.2. Besides, third principal invariant III is described from Equation (1.294) with Equations (1.234) and (1.235) as follows:

$$\mathrm{III} = \frac{1}{6}\mathrm{tr}^3\mathbf{T} - \frac{1}{2}\mathrm{tr}\,\mathbf{T}\mathrm{tr}\,\mathbf{T}^2 + \frac{1}{3}\mathrm{tr}\,\mathbf{T}^3. \tag{1.297}$$

1.7 Polar Decomposition

A second-order tensor **P** which is symmetric and fulfills

$$\boxed{\mathbf{P}\mathbf{v}\cdot\mathbf{v} > 0} \tag{1.298}$$

for an arbitrary vector $\mathbf{v}(\neq \mathbf{0})$, is called a *positive-definite tensor*. For an eigenvalue and direction of **P** given by P_J and $\mathbf{e}_J$, respectively, it follows that

$$\mathbf{P}\mathbf{e}_J\cdot\mathbf{e}_J = P_J\mathbf{e}_J\cdot\mathbf{e}_J = P_J||\mathbf{e}_J||^2 > 0 \text{ (no sum)}. \tag{1.299}$$

Thus the eigenvalues of a positive-definite tensor are positive. Taking account of this fact in equation $(1.238)_3$, we have

$$\det\mathbf{P} = \mathrm{III} > 0. \tag{1.300}$$

Now, consider the non-singular second-order tensor **T** which fulfills $\det\mathbf{T} \neq 0$, and thus $\mathbf{T}\mathbf{v} \neq \mathbf{0}$ holds for an arbitrary vector $\mathbf{v}(\neq \mathbf{0})$, noting equations (1.175) and (1.176).

The symmetric tensors $\mathbf{T}^T\mathbf{T}$ and $\mathbf{T}\mathbf{T}^T$ fulfill

$$\left.\begin{array}{l}\mathbf{T}^T\mathbf{T}\mathbf{v}\cdot\mathbf{v}\\ \mathbf{T}\mathbf{T}^T\mathbf{v}\cdot\mathbf{v} = \mathbf{v}\cdot\mathbf{T}^T\mathbf{T}\mathbf{v}\end{array}\right\} = \mathbf{T}\mathbf{v}\cdot\mathbf{T}\mathbf{v} = ||\mathbf{T}\mathbf{v}||^2 > 0, \tag{1.301}$$

using equation (1.93). Thus $\mathbf{T}^T\mathbf{T}$ and $\mathbf{T}\mathbf{T}^T$ are positive-definite tensors. Therefore, they possess the same positive eigenvalues, and thus they are mutually similar tensors described in equation

(1.252). Denoting their eigenvalues by λ_α^2, and the unit principal direction vectors of $\mathbf{T}^T\mathbf{T}$ and $\mathbf{T}\mathbf{T}^T$ by $\boldsymbol{N}^{(\alpha)}$ and $\boldsymbol{n}^{(\alpha)}(\alpha = 1, 2, 3)$, respectively, we can write the spectral representations

$$\left.\begin{aligned} \mathbf{T}^T\mathbf{T} &= \sum_{\alpha=1}^{3} \lambda_\alpha^2 \boldsymbol{N}^{(\alpha)} \otimes \boldsymbol{N}^{(\alpha)} \\ \mathbf{T}\mathbf{T}^T &= \sum_{\alpha=1}^{3} \lambda_\alpha^2 \boldsymbol{n}^{(\alpha)} \otimes \boldsymbol{n}^{(\alpha)} \end{aligned}\right\}. \tag{1.302}$$

Further, let the new positive-definite tensors be defined as follows:

$$\left.\begin{aligned} \boldsymbol{U} &= (\mathbf{T}^T\mathbf{T})^{1/2}(= \boldsymbol{U}^T)(\boldsymbol{U}^2 = \mathbf{T}^T\mathbf{T}) \\ \boldsymbol{V} &= (\mathbf{T}\mathbf{T}^T)^{1/2}(= \boldsymbol{V}^T)(\boldsymbol{V}^2 = \mathbf{T}\mathbf{T}^T) \end{aligned}\right\}. \tag{1.303}$$

It follows that

$$\boldsymbol{U}\boldsymbol{N}^{(\alpha)} = \lambda_\alpha \boldsymbol{N}^{(\alpha)}, \quad \boldsymbol{V}\boldsymbol{n}^{(\alpha)} = \lambda_\alpha \boldsymbol{n}^{(\alpha)}(\text{no sum}), \tag{1.304}$$

and then **U** and **V** can be written in the spectral representation as follows:

$$\left.\begin{aligned} \boldsymbol{U} &= \sum_{\alpha=1}^{3} \lambda_\alpha \boldsymbol{N}^{(\alpha)} \otimes \boldsymbol{N}^{(\alpha)} \\ \boldsymbol{V} &= \sum_{\alpha=1}^{3} \lambda_\alpha \boldsymbol{n}^{(\alpha)} \otimes \boldsymbol{n}^{(\alpha)} \end{aligned}\right\}. \tag{1.305}$$

Now let the following tensor $\boldsymbol{R}$ be introduced.

$$\boldsymbol{R} = \mathbf{T}\boldsymbol{U}^{-1}, \quad \boldsymbol{R} = \boldsymbol{V}^{-1}\mathbf{T}, \tag{1.306}$$

for which, noting equation (1.98), one has

$$\left.\begin{aligned} \boldsymbol{R}\boldsymbol{R}^T &= (\mathbf{T}\boldsymbol{U}^{-1})(\mathbf{T}\boldsymbol{U}^{-1})^T = \mathbf{T}\boldsymbol{U}^{-1}\boldsymbol{U}^{-1}\mathbf{T}^T = \mathbf{T}(\boldsymbol{U}^2)^{-1}\mathbf{T}^T = \mathbf{T}\mathbf{T}^{-1}\mathbf{T}^{-T}\mathbf{T}^T = \mathbf{I} \\ \boldsymbol{R}^T\boldsymbol{R} &= (\boldsymbol{V}^{-1}\mathbf{T})^T(\boldsymbol{V}^{-1}\mathbf{T}) = \mathbf{T}^T\boldsymbol{V}^{-1}\boldsymbol{V}^{-1}\mathbf{T} = \mathbf{T}(\boldsymbol{V}^2)^{-1}\mathbf{T}^T = \mathbf{T}^T\mathbf{T}^{-T}\mathbf{T}^{-1}\mathbf{T} = \mathbf{I} \end{aligned}\right\}. \tag{1.307}$$

Therefore, $\boldsymbol{R}$ is an orthogonal tensor. It follows from equation (1.306) that

$$\boxed{\mathbf{T} = \boldsymbol{R}\boldsymbol{U} = \boldsymbol{V}\boldsymbol{R}} \tag{1.308}$$

Now, let us prove that the decomposition in equation (1.308) is unique. Assume that there exists a different decomposition

$$\mathbf{T} = \bar{\boldsymbol{R}}\bar{\boldsymbol{U}} = \bar{\boldsymbol{V}}\bar{\boldsymbol{R}}. \tag{1.309}$$

It follows from equations (1.308) and (1.309) that

$$\left.\begin{aligned} \boldsymbol{U}^2 &= \mathbf{T}^T\mathbf{T} = (\bar{\boldsymbol{R}}\bar{\boldsymbol{U}})^T(\bar{\boldsymbol{R}}\bar{\boldsymbol{U}}) = \bar{\boldsymbol{U}}^2 \\ \boldsymbol{V}^2 &= \mathbf{T}\mathbf{T}^T = (\bar{\boldsymbol{V}}\bar{\boldsymbol{R}})(\bar{\boldsymbol{V}}\bar{\boldsymbol{R}})^T = \bar{\boldsymbol{V}}^2 \end{aligned}\right\} \tag{1.310}$$

and

$$\bar{\boldsymbol{R}} = \mathbf{T}\bar{\boldsymbol{U}}^{-1} = \mathbf{T}\boldsymbol{U}^{-1} = \boldsymbol{R}. \tag{1.311}$$

Thus, we have $\bar{U} = U, \bar{V} = V, \bar{R} = R$. It is concluded that the decomposition in equation (1.308) is unique. Equation (1.308) is called the *polar decomposition*, and RU and VR are called the *right* and *left polar decompositions*, respectively. From equation (1.308) one has

$$U = R^T V R, \quad V = R U R^T. \tag{1.312}$$

Substituting equation $(1.305)_1$ into equation $(1.312)_2$, it follows that

$$V = RUR^T = R \sum_{\alpha=1}^{3} \lambda_\alpha N^{(\alpha)} \otimes N^{(\alpha)} R^T = \sum_{\alpha=1}^{3} \lambda_\alpha RN^{(\alpha)} \otimes RN^{(\alpha)}.$$

The last quantity in this equation must coincide with equation $(1.305)_2$ and thus one obtains

$$n^{(\alpha)} = RN^{(\alpha)}, \quad N^{(\alpha)} = R^T n^{(\alpha)} \tag{1.313}$$

from which R can be represented as

$$R = \sum_{\alpha=1}^{3} n^{(\alpha)} \otimes N^{(\alpha)}. \tag{1.314}$$

The following expressions are obtained by substituting equations (1.305) and (1.314) into equation (1.308):

$$\mathbf{T} = \sum_{\alpha=1}^{3} \lambda_\alpha n^{(\alpha)} \otimes N^{(\alpha)}, \quad \mathbf{T}^{-1} = \sum_{\alpha=1}^{3} \frac{1}{\lambda_\alpha} N^{(\alpha)} \otimes n^{(\alpha)}. \tag{1.315}$$

1.8 Isotropy

Isotropic materials are widely used in engineering practice. The definition of isotropic material and the representation theorem for isotropic tensor-valued tensor functions of single variables are given below.

1.8.1 Isotropic Material

A function $\mathbf{f}$ of tensors $\mathbf{T}, \mathbf{S}, \cdots$ is called an *isotropic function* if it satisfies the equation

$$\mathbf{Q}\,[\![\mathbf{f}(\mathbf{T}, \mathbf{S}, \cdots)]\!] = \mathbf{f}(\mathbf{Q}\,[\![\mathbf{T}]\!]\,, \mathbf{Q}\,[\![\mathbf{S}]\!]\,, \cdots), \tag{1.316}$$

where the symbol in equation (1.198) is used. If $\mathbf{f}$ is a scalar, it is an invariant, and if it is a tensor, it is called an *isotropic tensor-valued tensor function*.

Isotropic material is defined to exhibit an identical mechanical response that is independent of the chosen direction of material element or of the coordinate system by which the response is described. Here, the input/output tensor-valued variables related to the mechanical response are the stress and the strain in the elastic material and the stress rate and the strain rate in inelastic (viscoelastic, plastic and viscoplastic) materials which are described in the rate type.

The elastic constitutive equation is generally given by

$$\mathbf{f}(\boldsymbol{\sigma}, \boldsymbol{\varepsilon}, H_i) = \mathbf{O}, \tag{1.317}$$

where the stress tensor and strain tensor are denoted by the symbols $\boldsymbol{\sigma}$ and $\boldsymbol{\varepsilon}$, respectively, and the scalar- and tensor-valued internal variables are collectively denoted by the symbol H_i. When the following equation holds by giving coordinate transformations only for stress and strain tensors in the function $\mathbf{f}$, it describes the constitutive equation of isotropic elastic material:

$$\boxed{\mathbf{f}(\mathbf{Q}\boldsymbol{\sigma}\mathbf{Q}^T, \mathbf{Q}\boldsymbol{\varepsilon}\mathbf{Q}^T, H_i) = \mathbf{Q}\mathbf{f}(\boldsymbol{\sigma}, \boldsymbol{\varepsilon}, H_i)\mathbf{Q}^T} \tag{1.318}$$

For inelastic materials, the stress rate tensor $\overset{\circ}{\boldsymbol{\sigma}}$ and the strain rate tensor $\mathbf{d}$ are related through the stress and the internal variables so that the constitutive equation is generally given by

$$\mathbf{f}(\overset{\circ}{\boldsymbol{\sigma}}, \boldsymbol{\sigma}, H_i, \mathbf{d}) = \mathbf{O}, \tag{1.319}$$

When the following equation holds by giving coordinate transformations only for stress (rate) and strain rate tensors in the function $\mathbf{f}$, it describes constitutive equations of isotropic materials:

$$\boxed{\mathbf{f}(\mathbf{Q}\overset{\circ}{\boldsymbol{\sigma}}\mathbf{Q}^T, \mathbf{Q}\boldsymbol{\sigma}\mathbf{Q}^T, H_i, \mathbf{Q}\mathbf{d}\mathbf{Q}^T) = \mathbf{Q}\mathbf{f}(\overset{\circ}{\boldsymbol{\sigma}}, \boldsymbol{\sigma}, H_i, \mathbf{d})\mathbf{Q}^T} \tag{1.320}$$

In the plastic constitutive equation formulated incorporating the yield and/or plastic potential function, the isotropy holds if the yield and/or plastic potential function is given by the scalar function of stress invariants and scalar internal variables. Then, denoting the yield and plastic potential functions by f, the following equation must hold in isotropic materials:

$$\boxed{f(\mathbf{Q}\boldsymbol{\sigma}\mathbf{Q}^T, H_i) = \mathbf{Q}f(\boldsymbol{\sigma}, H_i)\mathbf{Q}^T} \tag{1.321}$$

Therefore, f is limited to a function of stress invariants and scalar variables H_i.

1.8.2 Representation Theorem of Isotropic Tensor-Valued Tensor Function

The general representation of the isotropic tensor-valued tensor function of a single second-order tensor is given below, to which isotropic elastic constitute equations belong.

Letting $\mathbf{A}$ and $\mathbf{T}$ be symmetric second-order tensors, consider the following isotropic tensor function $\mathbf{A}$ of $\mathbf{T}$:

$$\mathbf{A} = \mathbf{f}(\mathbf{T}), \tag{1.322}$$

where $\mathbf{f}$ fulfills

$$\mathbf{f}(\mathbf{Q}\mathbf{T}\mathbf{Q}^T) = \mathbf{Q}\mathbf{f}(\mathbf{T})\mathbf{Q}^T. \tag{1.323}$$

Introduce the coordinate system with the bases $\mathbf{e}_1, \mathbf{e}_2, \mathbf{e}_3$ which are the eigenvectors of the tensor $\mathbf{T}$ so that equations (1.248) and (1.249) hold. Then adopt the following three particular types of orthogonal tensor possessing only diagonal components:

$$\left.\begin{aligned} \mathbf{Q}_1 &= \mathbf{e}_1 \otimes \mathbf{e}_1 - \mathbf{e}_2 \otimes \mathbf{e}_2 - \mathbf{e}_3 \otimes \mathbf{e}_3 \\ \mathbf{Q}_2 &= -\mathbf{e}_1 \otimes \mathbf{e}_1 + \mathbf{e}_2 \otimes \mathbf{e}_2 - \mathbf{e}_3 \otimes \mathbf{e}_3 \\ \mathbf{Q}_3 &= -\mathbf{e}_1 \otimes \mathbf{e}_1 - \mathbf{e}_2 \otimes \mathbf{e}_2 + \mathbf{e}_3 \otimes \mathbf{e}_3 \end{aligned}\right\}, \tag{1.324}$$

where $\mathbf{Q}_1, \mathbf{Q}_2, \mathbf{Q}_3$ give π-rotations in the anticlockwise direction around the axes $\mathbf{e}_1, \mathbf{e}_2, \mathbf{e}_3$, respectively. Denoting these principal base vectors by $\mathbf{e}_P$ and the orthogonal tensors in equation (1.324) collectively by $\mathbf{Q}_P$, it follows that

$$\mathbf{Q}_P = \mathbf{Q}_P^T, \tag{1.325}$$

$$\mathbf{Q}_P \mathbf{e}_P = \mathbf{Q}_P^T \mathbf{e}_P = \mathbf{e}_P \quad \text{(no sum)}. \tag{1.326}$$

Equation (1.326) means that the fixed base vectors $\mathbf{e}_P$ are also eigenvectors of the orthogonal tensor $\mathbf{Q}_P$, keeping in mind that $\mathbf{e}_P$ are the eigenvectors of the tensor $\mathbf{T}$.

Furthermore, from equations (1.248) and (1.324) we have that

$$\mathbf{Q}_P \mathbf{T} \mathbf{Q}_P^T = \mathbf{T} \quad \text{(no sum)} \tag{1.327}$$

or

$$\mathbf{Q}_P \mathbf{T} = \mathbf{T} \mathbf{Q}_P, \tag{1.328}$$

noting that

$$\mathbf{Q}_P \sum_{S=1}^{3} \lambda_S \mathbf{e}_S \otimes \mathbf{e}_S \mathbf{Q}_P^T = \mathbf{Q}_P \sum_{S=1}^{3} \lambda_S \mathbf{e}_S \otimes \mathbf{e}_S \mathbf{Q}_P$$
$$= \lambda_1 \mathbf{e}_1 \otimes \mathbf{e}_1 + \lambda_2 \mathbf{e}_2 \otimes \mathbf{e}_2 + \lambda_3 \mathbf{e}_3 \otimes \mathbf{e}_3 = \sum_{P=1}^{3} \lambda_P \mathbf{e}_P \otimes \mathbf{e}_P.$$

Substituting equation (1.327) into equation (1.322) leads to

$$\mathbf{f}(\mathbf{Q}_P \mathbf{T} \mathbf{Q}_P^T) = \mathbf{f}(\mathbf{T}) = \mathbf{A}. \tag{1.329}$$

From equation (1.323), on the other hand, one has

$$\mathbf{f}(\mathbf{Q}_P \mathbf{T} \mathbf{Q}_P^T) = \mathbf{Q}_P \mathbf{A} \mathbf{Q}_P^T. \tag{1.330}$$

From equations (1.329) and (1.330) the commutative law

$$\mathbf{A} \mathbf{Q}_P = \mathbf{Q}_P \mathbf{A} \tag{1.331}$$

is derived, and further, noting equation (1.326), the following relation is obtained:

$$\mathbf{Q}_P \mathbf{A} \mathbf{e}_P = \mathbf{A} \mathbf{Q}_P \mathbf{e}_P = \mathbf{A} \mathbf{e}_P \text{ (no sum)}, \tag{1.332}$$

which means that $\mathbf{A}\mathbf{e}_P$ is the eigenvector of $\mathbf{Q}_P$ and thus has the same direction as $\mathbf{e}_P$. Then, denoting the eigenvalue of $\mathbf{A}$ for the eigenvector $\mathbf{e}_P$ by α_P, one can write

$$\mathbf{A} \mathbf{e}_P = \alpha_P \mathbf{e}_P \text{ (no sum)}. \tag{1.333}$$

It can be concluded that the tensor $\mathbf{A}$ has the same eigenvectors as the tensor $\mathbf{T}$, leading to *coaxiality*, so that they are described in the spectral representation in a common eigenvector basis. Therefore, the eigenvalues $\alpha_1, \alpha_2, \alpha_3$ of the tensor $\mathbf{A}$ can be represented by the eigenvalues $\lambda_1, \lambda_2, \lambda_3$ of the tensor $\mathbf{T}$, that is,

$$\alpha_P = \alpha_P(\lambda_1, \lambda_2, \lambda_3). \tag{1.334}$$

Regarding the eigenvalues, we may consider three cases. First, suppose all three eigenvalues of $\mathbf{T}$ are different: $\lambda_1 \neq \lambda_2 \neq \lambda_3 \neq \lambda_1$. Assume the following linear simultaneous equation:

$$\alpha_i = \phi_0 + \phi_1 \lambda_i + \phi_2 \lambda_i^2, \tag{1.335}$$

that is,

$$\begin{Bmatrix} \alpha_1 \\ \alpha_2 \\ \alpha_3 \end{Bmatrix} = \begin{bmatrix} 1 & \lambda_1 & \lambda_1^2 \\ 1 & \lambda_2 & \lambda_2^2 \\ 1 & \lambda_3 & \lambda_3^2 \end{bmatrix} \begin{Bmatrix} \phi_0 \\ \phi_1 \\ \phi_2 \end{Bmatrix}, \tag{1.336}$$

where ϕ_0, ϕ_1, ϕ_2 are scalar functions of $\lambda_1, \lambda_2, \lambda_3$. If we regard equation (1.335) as a simultaneous equation for the unknown values ϕ_0, ϕ_1, ϕ_2, the *Vandermonde determinant* of the coefficients $\lambda_1, \lambda_2, \lambda_3$ is not zero, that is,

$$\begin{vmatrix} 1 & \lambda_1 & \lambda_1^2 \\ 1 & \lambda_2 & \lambda_2^2 \\ 1 & \lambda_3 & \lambda_3^2 \end{vmatrix} = (\lambda_1 - \lambda_2)(\lambda_2 - \lambda_3)(\lambda_3 - \lambda_1) \neq 0. \tag{1.337}$$

Therefore, ϕ_0, ϕ_1, ϕ_2 are uniquely determined by $\lambda_1, \lambda_2, \lambda_3$ and thus $\alpha_1, \alpha_2, \alpha_3$, as was described in equations (1.175) and (1.176) in Section 1.4.9. Equation (1.335) thus provides the exact relation of the eigenvalues $\alpha_1, \alpha_2, \alpha_3$ to $\lambda_1, \lambda_2, \lambda_3$.

Since $\alpha_1, \alpha_2, \alpha_3$ and $\lambda_1, \lambda_2, \lambda_3$ are the eigenvalues of the tensor $\mathbf{A} = \mathbf{f}(\mathbf{T})$ and $\mathbf{T}$, respectively, and these tensors possess the same principal directions, the following representation holds from equation (1.335):

$$\begin{bmatrix} \alpha_1 & 0 & 0 \\ 0 & \alpha_2 & 0 \\ 0 & 0 & \alpha_3 \end{bmatrix} = \phi_0 \begin{bmatrix} 1 & 0 & 0 \\ 0 & 1 & 0 \\ 0 & 0 & 1 \end{bmatrix} + \phi_1 \begin{bmatrix} \lambda_1 & 0 & 0 \\ 0 & \lambda_2 & 0 \\ 0 & 0 & \lambda_3 \end{bmatrix} + \phi_2 \begin{bmatrix} \lambda_1^2 & 0 & 0 \\ 0 & \lambda_2^2 & 0 \\ 0 & 0 & \lambda_3^2 \end{bmatrix} \tag{1.338}$$

or

$$\sum_{P=1}^{3} \alpha_P \mathbf{e}_P \otimes \mathbf{e}_P = \phi_0 \sum_{P=1}^{3} \mathbf{e}_P \otimes \mathbf{e}_P + \phi_1 \sum_{P=1}^{3} \lambda_P \mathbf{e}_P \otimes \mathbf{e}_P + \phi_2 \left(\sum_{P=1}^{3} \lambda_P \mathbf{e}_P \otimes \mathbf{e}_P \right)^2, \tag{1.339}$$

namely,

$$\boxed{\mathbf{f}(\mathbf{T}) = \phi_0 \mathbf{I} + \phi_1 \mathbf{T} + \phi_2 \mathbf{T}^2} \tag{1.340}$$

where ϕ_0, ϕ_1, ϕ_2 can be expressed by the invariants I, II, III of $\mathbf{T}$ in equation (1.238), because they are scalar functions.

Now suppose two of three eigenvalues are same: $\lambda_1 \neq \lambda_2 = \lambda_3$. Assume the following linear simultaneous equation:

$$\left. \begin{aligned} \alpha_1(\lambda_1, \lambda_2) &= \phi_0(\lambda_1, \lambda_2) + \phi_1(\lambda_1, \lambda_2)\lambda_1 \\ \alpha_2(\lambda_1, \lambda_2) &= \phi_0(\lambda_1, \lambda_2) + \phi_1(\lambda_1, \lambda_2)\lambda_2 \end{aligned} \right\}. \tag{1.341}$$

The Vandermonde determinant is not zero, that is,

$$\begin{vmatrix} 1 & \lambda_1 \\ 1 & \lambda_2 \end{vmatrix} = -\lambda_1 + \lambda_2 \neq 0. \tag{1.342}$$

Equation (1.341) for the spectral expression is represented in the direct tensor notation as follows:

$$\mathbf{f}(\mathbf{T}) = \phi_0\mathbf{I} + \phi_1\mathbf{T}, \tag{1.343}$$

where ϕ_0, ϕ_1 are functions of the first and second invariants, I and II of $\mathbf{T}$.

Finally, suppode all three eigenvalues are equal: $\lambda_1 = \lambda_2 = \lambda_3$. Then

$$\alpha_1(\lambda_1) = \phi_0(\lambda_1), \tag{1.344}$$

or in direct tensor notation,

$$\mathbf{f}(\mathbf{T}) = \phi_0\mathbf{I}, \tag{1.345}$$

where ϕ_0 is function of the first invariant of $\mathbf{T}$.

For the case where $\mathbf{T}$ is invertible, substituting equation (1.296) into $\mathbf{T}^2$ in equation (1.340), we have

$$\mathbf{f}(\mathbf{T}) = \psi_0\mathbf{I} + \psi_1\mathbf{T} + \psi_{-1}\mathbf{T}^{-1}, \tag{1.346}$$

where

$$\psi_0 = \phi_0 - \phi_2\mathrm{II}, \quad \psi_1 = \phi_1 + \phi_2\mathrm{I}, \quad \psi_{-1} = \phi_2\mathrm{III}, \tag{1.347}$$

$$\phi_0 = \psi_0 + \frac{\mathrm{II}}{\mathrm{III}}\psi_{-1}, \quad \phi_1 = \psi_1 - \frac{\mathrm{I}}{\mathrm{III}}\psi_{-1}, \quad \phi_2 = \frac{\mathrm{I}}{\mathrm{III}}\psi_{-1}. \tag{1.348}$$

In the particular case where $\mathbf{f}$ is a linear function of $\mathbf{T}$, equation (1.340) reduces to

$$\boxed{\mathbf{f}(\mathbf{T}) = \lambda(\mathrm{tr}\mathbf{T})\mathbf{I} + 2\mu\mathbf{T}} \tag{1.349}$$

where λ and μ are the material constants, called the *Lamé constants*, regarding $\mathbf{f}$ and $\mathbf{T}$ as the stress and strain, repectively, in the linear elastic constitutive equation. Equation (1.349) can be rewritten as

$$\mathbf{f}(\mathbf{T}) = \mathbf{C}\mathbf{T}, \tag{1.350}$$

where

$$\mathbf{C} \equiv \lambda\boldsymbol{\mathcal{I}} + \mu\boldsymbol{\mathcal{S}} \quad (C_{ijkl} \equiv \lambda\delta_{ij}\delta_{kl} + \mu(\delta_{ik}\delta_{il} + \delta_{il}\delta_{jk})). \tag{1.351}$$

Equation (1.350) is also obtained by multiplying the fourth-order isotropic tensor in equation (1.226) by $\mathbf{T}$, noting that the term with the anti-symmetrizing tensor $\boldsymbol{\mathcal{A}}$ vanishes because of the symmetry of the tensor $\mathbf{T}$.

Equation (1.340) is expressed in the eigenvalues $\lambda_1, \lambda_2, \lambda_3$ of the tensor $\mathbf{T}$ as follows:

$$\left.\begin{aligned} f(\lambda_1) &= \phi_0 + \phi_1\lambda_1 + \phi_2\lambda_1^2 \\ f(\lambda_2) &= \phi_0 + \phi_1\lambda_2 + \phi_2\lambda_2^2 \\ f(\lambda_3) &= \phi_0 + \phi_1\lambda_3 + \phi_2\lambda_3^2 \end{aligned}\right\}. \tag{1.352}$$

By solving equation (1.352) for ϕ_0, ϕ_1, ϕ_2 and substituting the solutions into equation (1.340), the following equation, called the *Lagrange–Sylvester equation*, is obtained

(Tanahashi, 2004):

$$\mathbf{f}(\mathbf{T}) = f(\lambda_1)\frac{(\mathbf{T}-\lambda_2\mathbf{I})(\mathbf{T}-\lambda_3\mathbf{I})}{(\lambda_1-\lambda_2)(\lambda_1-\lambda_3)} + f(\lambda_2)\frac{(\mathbf{T}-\lambda_2\mathbf{I})(\mathbf{T}-\lambda_1\mathbf{I})}{(\lambda_2-\lambda_3)(\lambda_2-\lambda_1)} + f(\lambda_3)\frac{(\mathbf{T}-\lambda_2\mathbf{I})(\mathbf{T}-\lambda_3\mathbf{I})}{(\lambda_3-\lambda_1)(\lambda_3-\lambda_2)}. \tag{1.353}$$

For the second-order isotropic tensor-valued tensor function of a single tensor considered above, the *representation theorem* of the symmetric isotropic second-order tensor-valued tensor function **f** of two tensors **A** and **B** is shown below (Spencer, 1971):

$$\mathbf{f}(\mathbf{A},\mathbf{B}) = \varphi_0\mathbf{I} + \varphi_1\mathbf{A} + \varphi_2\mathbf{B} + \varphi_3\mathbf{A}^2 + \varphi_4\mathbf{B}^2 + \varphi_5(\mathbf{AB}+\mathbf{BA}) + \varphi_6(\mathbf{A}^2\mathbf{B}+\mathbf{BA}^2) + \varphi_7(\mathbf{AB}^2+\mathbf{B}^2\mathbf{A}) + \varphi_8(\mathbf{A}^2\mathbf{B}^2+\mathbf{B}^2\mathbf{A}^2), \tag{1.354}$$

where $\varphi_0, \varphi_1, \ldots, \varphi_8$ are scalar-valued homogeneous functions of invariants:

$$\left.\begin{array}{l} \mathrm{tr}\mathbf{A},\ \mathrm{tr}\mathbf{A}^2,\ \mathrm{tr}\mathbf{A}A^3,\ \mathrm{tr}\mathbf{B},\ \mathrm{tr}\mathbf{B}^2,\ \mathrm{tr}\mathbf{B}^3 \\ \mathrm{tr}(\mathbf{AB}),\ \mathrm{tr}(\mathbf{AB}^2),\ \mathrm{tr}(\mathbf{A}^2\mathbf{B}),\ \mathrm{tr}(\mathbf{A}^2\mathbf{B}^2) \end{array}\right\}. \tag{1.355}$$

1.9 Differential Formulae

Thi section provides various differential formulae which are often used in continuum mechanics.

1.9.1 Partial Derivatives

Several partial derivatives which will appear in later chapters are listed as follows.

1.

$$\frac{\partial\mathbf{T}^n}{\partial\mathbf{T}} = \sum_{k=1}^{n}\mathbf{T}^{k-1}\,\tilde{\otimes}\,\mathbf{T}^{n-k} = \mathbf{I}\,\tilde{\otimes}\,\mathbf{T}^{n-1} + \mathbf{T}\,\tilde{\otimes}\,\mathbf{T}^{n-2} + \cdots + \mathbf{T}^{n-2}\,\tilde{\otimes}\,\mathbf{T} + \mathbf{T}^{n-1}\,\tilde{\otimes}\,\mathbf{I},$$

$$\left(\frac{\partial\mathbf{T}^n}{\partial\mathbf{T}}\right)_{ijkl} = \sum_{k=1}^{n}(\mathbf{T}^{k-1})_{ik}(\mathbf{T}^{n-k})_{lj} \tag{1.356}$$

$$\left(\begin{array}{l} \dfrac{\partial(T_{ir_1}T_{r_1r_2}T_{r_2r_3}\cdots T_{r_{n-2}r_{n-1}}T_{r_{n-1}j})}{\partial T_{kl}} \\ = \delta_{ik}\delta_{r_1l}T_{r_1r_2}T_{r_2r_3}\cdots T_{r_{n-2}r_{n-1}}T_{r_{n-1}j} + T_{ir_1}\delta_{r_1k}\delta_{r_2l}T_{r_2r_3}\cdots T_{r_{n-2}r_{n-1}}T_{r_{n-1}j} + \cdots \\ \quad + T_{ir_1}T_{r_1r_2}T_{r_2r_3}\cdots T_{r_{n-3}r_{n-2}}\delta_{r_{n-2}k}\delta_{r_{n-1}l}T_{r_{n-1}j} + T_{ir_1}T_{r_1r_2}T_{r_2r_3}\cdots T_{r_{n-2}r_{n-1}}\delta_{r_{n-1}k}\delta_{jl} \\ = \delta_{ik}(T_{lr_2}T_{r_2r_3}\cdots T_{r_{n-2}r_{n-1}}T_{r_{n-1}j}) + T_{ik}(T_{lr_3}\cdots T_{r_{n-2}r_{n-1}}T_{r_{n-1}j}) + \cdots \\ \quad + (T_{ir_1}T_{r_1r_2}T_{r_2r_3}\cdots T_{r_{n-3}k})T_{lj} + (T_{ir_1}T_{r_1r_2}T_{r_2r_3}\cdots T_{r_{n-3}r_{n-2}}T_{r_{n-2}k})\delta_{lj} \\ = \delta_{ik}(\mathbf{T}^{n-1})_{lj} + T_{ik}(\mathbf{T}^{n-2})_{lj} + \cdots + (\mathbf{T}^{n-2})_{ik}T_{lj} + (\mathbf{T}^{n-1})_{ik}\delta_{lj} \end{array}\right)$$

which will be derived later by the directional derivative in Section 1.9.2.

Examples:

$$\left.\begin{aligned}&\frac{\partial \mathbf{T}}{\partial \mathbf{T}}=\mathbf{I}\tilde{\otimes}\mathbf{I},\quad \frac{\partial T_{ij}}{\partial T_{kl}}=\delta_{ik}\delta_{jl}=\delta_{ik}\delta_{lj}\\ &\frac{\partial \mathbf{T}^2}{\partial \mathbf{T}}=\mathbf{I}\tilde{\otimes}\mathbf{T}+\mathbf{T}\tilde{\otimes}\mathbf{I},\quad \frac{\partial T_{ir}T_{rj}}{\partial T_{kl}}=\delta_{ik}T_{lj}+T_{ik}\delta_{lj}\\ &\frac{\partial \mathbf{T}^3}{\partial \mathbf{T}}=\mathbf{I}\tilde{\otimes}\mathbf{T}^2+\mathbf{T}\tilde{\otimes}\mathbf{T}+\mathbf{T}^2\tilde{\otimes}\mathbf{I},\quad \frac{\partial (T_{ir}T_{rs}T_{sj})}{\partial T_{kl}}=\delta_{ik}T_{ls}T_{sj}+T_{ik}T_{lj}+T_{ir}T_{rk}\delta_{lj}\end{aligned}\right\}. \tag{1.357}$$

For a symmetric tensor $\mathbf{T}=\mathbf{T}^T$, one can write

$$\frac{\partial \mathbf{T}}{\partial \mathbf{T}}=\boldsymbol{\mathcal{S}},\quad \frac{\partial T_{ij}}{\partial T_{kl}}=\frac{1}{2}(\delta_{ik}\delta_{jl}+\delta_{il}\delta_{jk}). \tag{1.357$'$}$$

2.

$$\frac{\partial \mathrm{tr}\mathbf{T}^n}{\partial \mathbf{T}}=n(\mathbf{T}^{n-1})^T,$$

$$\frac{\partial (T_{r_0r_1}T_{r_1r_2}T_{r_2r_3}\cdots T_{r_{n-2}r_{n-1}}T_{r_{n-1}r_0})}{\partial T_{ij}}=nT_{jr_1}T_{r_1r_2}T_{r_2r_3}\cdots T_{r_{n-3}r_{n-2}}T_{r_{n-2}i} \tag{1.358}$$

$$\left(\begin{aligned}&\frac{\partial (T_{r_0r_1}T_{r_1r_2}T_{r_2r_3}\cdots T_{r_{n-2}r_{n-1}}T_{r_{n-1}r_0})}{\partial T_{ij}}\\ &=\delta_{ir_0}\delta_{jr_1}T_{r_1r_2}T_{r_2r_3}\cdots T_{r_{n-2}r_{n-1}}T_{r_{n-1}r_0}+T_{r_0r_1}\delta_{ir_1}\delta_{jr_2}T_{r_2r_3}\cdots T_{r_{n-2}r_{n-1}}T_{r_{n-1}r_0}+\cdots\\ &\quad+T_{r_0r_1}T_{r_1r_2}T_{r_2r_3}\cdots T_{r_{n-3}r_{n-2}}\delta_{ir_{n-2}}\delta_{jr_{n-1}}T_{r_{n-1}r_0}+T_{r_0r_1}T_{r_1r_2}T_{r_2r_3}\cdots T_{r_{n-2}r_{n-1}}\delta_{ir_{n-1}}\delta_{jr_0}\\ &=T_{jr_2}T_{r_2r_3}\cdots T_{r_{n-2}r_{n-1}}T_{r_{n-1}i}+T_{r_0i}T_{jr_3}\cdots T_{r_{n-2}r_{n-1}}T_{r_{n-1}r_0}+\cdots\\ &\quad+T_{r_0r_1}T_{r_1r_2}T_{r_2r_3}\cdots T_{r_{n-3}i}T_{jr_0}+T_{jr_1}T_{r_1r_2}T_{r_2r_3}\cdots T_{r_{n-3}r_{n-2}}T_{r_{n-2}i}\\ &=n(\mathbf{T}^{n-1})_{ji}\end{aligned}\right)$$

which will be derived later by the directional derivative in Section 1.9.2.

Examples:

$$\frac{\partial \mathrm{tr}\mathbf{T}}{\partial \mathbf{T}}=\frac{\partial \mathrm{I}}{\partial \mathbf{T}}=\frac{\partial \bar{\mathrm{I}}}{\partial \mathbf{T}}=\mathbf{I},\quad \frac{\partial \mathrm{tr}\mathbf{T}^2}{\partial \mathbf{T}}=\frac{\partial \overline{\mathrm{II}}}{\partial \mathbf{T}}=2\mathbf{T}^T,\quad \frac{\partial \mathrm{tr}\mathbf{T}^3}{\partial \mathbf{T}}=\frac{\partial \overline{\mathrm{III}}}{\partial \mathbf{T}}=3(\mathbf{T}^2)^T=3(\mathbf{T}^T)^2. \tag{1.359}$$

3.

$$\frac{\partial \mathbf{T}'}{\partial \mathbf{T}}=\boldsymbol{\mathcal{D}},\quad \frac{\partial T'_{ij}}{\partial T_{kl}}=\mathcal{D}_{ijkl} \tag{1.360}$$

$$\left(\frac{\partial T'_{ij}}{\partial T_{kl}}=\frac{\partial (T_{ij}-T_m\delta_{ij})}{\partial T_{kl}}=\delta_{ik}\delta_{jl}-\frac{1}{3}\delta_{ij}\delta_{kl}\right).$$

4.

$$\frac{\partial ||\mathbf{T}'||}{\partial \mathbf{T}}=\frac{\mathbf{T}'}{||\mathbf{T}'||}\equiv \mathbf{t}',\quad \frac{\partial \sqrt{T'_{rs}T'_{rs}}}{\partial T_{ij}}=t'_{ij} \tag{1.361}$$

$$\left(\frac{\partial ||\mathbf{T}'||}{\partial \mathbf{T}}=\frac{\partial \sqrt{\mathrm{tr}\mathbf{T}'^2}}{\partial \mathbf{T}}=\frac{1}{2}\frac{1}{\sqrt{\mathrm{tr}\mathbf{T}'^2}}\frac{\partial \mathrm{tr}\mathbf{T}'^2}{\partial \mathbf{T}}=\frac{1}{2}\frac{2\mathbf{T}'}{\sqrt{\mathrm{tr}\mathbf{T}'^2}}\right).$$

5.

$$\frac{\partial \mathbf{t}'}{\partial \mathbf{T}'} = \frac{1}{||\mathbf{T}'||}(\bar{\mathcal{I}} - \mathbf{t}' \otimes \mathbf{t}'), \quad \frac{\partial t'_{ij}}{\partial T'_{kl}} = \frac{1}{\sqrt{T'_{pq}T'_{pq}}}(\delta_{ik}\delta_{jl} - t'_{ij}t'_{kl}) \tag{1.362}$$

$$\left(\frac{\partial \mathbf{t}'}{\partial \mathbf{T}'} = \frac{\partial}{\partial \mathbf{T}'}\left(\frac{\mathbf{T}'}{||\mathbf{T}'||}\right) = \frac{\bar{\mathcal{I}}}{||\mathbf{T}'||} - \mathbf{T}' \otimes \frac{1}{||\mathbf{T}'||^2}\mathbf{t}'\right).$$

6.

$$\frac{\partial \mathbf{t}'}{\partial \mathbf{T}} = \frac{1}{||\mathbf{T}'||}\left(\bar{\mathcal{I}} - \frac{1}{3}\mathcal{I} - \mathbf{t}' \otimes \mathbf{t}'\right), \quad \frac{\partial t'_{ij}}{\partial T_{kl}} = \frac{1}{||\mathbf{T}'||}\left(\bar{\mathcal{I}}_{ijkl} - \frac{1}{3}\mathcal{I}_{ijkl} - t'_{ij}t'_{kl}\right) \tag{1.363}$$

$$\left(\frac{\partial \mathbf{t}'}{\partial \mathbf{T}'} = \frac{\partial \mathbf{t}'}{\partial \mathbf{T}'} : \frac{\partial \mathbf{T}'}{\partial \mathbf{T}} = \frac{1}{||\mathbf{T}'||}(\bar{\mathcal{I}} - \mathbf{t}' \otimes \mathbf{t}') : \mathcal{D} = \left\{\frac{1}{||\mathbf{T}'||}(\bar{\mathcal{I}} - \mathbf{t}' \otimes \mathbf{t}')\right\} : \left(\bar{\mathcal{I}} - \frac{1}{3}\mathcal{I}\right)\right).$$

7.

$$\frac{\partial \cos 3\theta}{\partial \mathbf{t}'} = 3\sqrt{6}\mathbf{t}'^2, \quad \frac{\partial \cos 3\theta}{\partial t'_{ij}} = 3\sqrt{6}t'_{ir}t'_{rj} \tag{1.364}$$

$$\left(\frac{\partial \cos 3\theta}{\partial \mathbf{t}'} = \sqrt{6}\frac{\partial \mathrm{tr}\mathbf{t}'^3}{\partial \mathbf{t}'}\right),$$

where

$$\cos 3\theta \equiv \sqrt{6}\mathrm{tr}\mathbf{t}'^3, \tag{1.365}$$

which often appears in yield conditions depending on the third invariant of deviatoric stress, that is, the intermediate principal stress as observed in frictional materials (Hashiguchi, 2009).

8.

$$\frac{\partial \cos 3\theta}{\partial \mathbf{T}} = -\frac{3\sqrt{6}}{||\mathbf{T}'||}\left(\frac{1}{3}||\mathbf{t}'||^2\mathbf{I} + \frac{1}{\sqrt{6}}\cos 3\theta\mathbf{t}' - \mathbf{t}'^2\right),$$

$$\frac{\partial \cos 3\theta}{\partial T_{ij}} = -\frac{3\sqrt{6}}{\sqrt{T'_{pq}T'_{pq}}}\left(\frac{1}{3}t'_{rt}t'_{tr}\delta_{ij} + \frac{1}{\sqrt{6}}\cos 3\theta t'_{ij} - t'_{ir}t'_{rj}\right) \tag{1.366}$$

$$\left(\begin{aligned}\frac{\partial \cos 3\theta}{\partial \mathbf{T}} &= \frac{\partial \cos 3\theta}{\partial \mathbf{t}'} : \frac{\partial \mathbf{t}'}{\partial \mathbf{T}} = 3\sqrt{6}\mathbf{t}'^2 : \frac{1}{||\mathbf{T}'||}\left(\bar{\mathcal{I}} - \frac{1}{3}\mathbf{I} \otimes \mathbf{I} - \mathbf{t}' \otimes \mathbf{t}'\right) \\ &= \frac{3\sqrt{6}}{||\mathbf{T}'||}\left\{\mathbf{t}'^2 - \frac{1}{3}(\mathrm{tr}\mathbf{t}'^2)\mathbf{I} - (\mathrm{tr}\mathbf{t}'^3)\mathbf{t}'\right\}\end{aligned}\right).$$

9.

$$\frac{\partial \mathbf{T}^{-1}}{\partial \mathbf{T}} = -\mathbf{T}^{-1} \tilde{\otimes} \mathbf{T}^{-1} = -\mathbf{T}^{-1}\bar{\mathcal{I}}\mathbf{T}^{-1}, \quad \frac{\partial T^{-1}_{ij}}{\partial T_{kl}} = -T^{-1}_{ik}T^{-1}_{lj} = -T^{-1}_{ir}\bar{\mathcal{I}}_{rskl}T^{-1}_{sj} \tag{1.367}$$

$$\left(\begin{array}{l}\dfrac{\partial(T_{ir}^{-1}T_{rs})}{\partial T_{kl}}=\dfrac{\partial T_{ir}^{-1}}{\partial T_{kl}}T_{rs}+T_{ir}^{-1}\delta_{rk}\delta_{sl}=0\rightsquigarrow\dfrac{\partial T_{ir}^{-1}}{\partial T_{kl}}T_{rs}T_{sj}^{-1}+T_{ir}^{-1}\delta_{rk}\delta_{sl}T_{sj}^{-1}=0\\ \qquad\rightsquigarrow\dfrac{\partial T_{ij}^{-1}}{\partial T_{kl}}+T_{ik}^{-1}T_{lj}^{-1}=0\end{array}\right).$$

which will be derived later by the directional deriavative in Section 1.9.2.

10.

$$\dot{\mathbf{T}}^{-1}=-\mathbf{T}^{-1}\dot{\mathbf{T}}\mathbf{T}^{-1},\quad \dot{T}_{ij}^{-1}=-T_{ir}^{-1}\dot{T}_{rs}T_{sj}^{-1} \tag{1.368}$$

$$\left(\dot{\mathbf{I}}=(\mathbf{T}^{-1}\mathbf{T})^{\bullet}=\dot{\mathbf{T}}^{-1}\mathbf{T}+\mathbf{T}^{-1}\dot{\mathbf{T}}=\mathbf{O}\right)\left(\frac{\partial T_{ij}^{-1}}{\partial T_{kl}}\dot{T}_{kl}=-T_{ik}^{-1}\dot{T}_{kl}T_{lj}^{-1}\right)$$

which will be derived later by the directional derivative in Section 1.9.2.

11.

$$\frac{\partial\mathrm{tr}(\mathrm{cof}\mathbf{T})}{\partial\mathbf{T}}=\frac{\partial\mathrm{II}}{\partial\mathbf{T}}=(\mathrm{tr}\mathbf{T})\mathbf{I}-\mathbf{T}^{T},\quad \frac{\partial\mathrm{tr}(\mathrm{cof}\mathbf{T})}{\partial T_{ij}}=T_{rr}\delta_{ij}-T_{ji} \tag{1.369}$$

$$\left(\begin{array}{l}\dfrac{\partial\mathrm{tr}(\mathrm{cof}\mathbf{T})}{\partial T_{ij}}=\dfrac{\partial\frac{1}{2}\varepsilon_{abc}\varepsilon_{aqr}T_{bq}T_{cr}}{\partial T_{ij}}=2\dfrac{1}{2}\varepsilon_{abc}\varepsilon_{aqr}\delta_{ib}\delta_{jq}T_{cr}=\varepsilon_{aic}\varepsilon_{ajr}T_{cr}\\ \qquad=(\delta_{ij}\delta_{cr}-\delta_{ir}\delta_{cj})T_{cr}=T_{rr}\delta_{ij}-T_{ji}\end{array}\right),$$

noting equations (1.56), (1.153) and (1.235).

12.

$$\frac{\partial(\mathrm{cof}\mathbf{T})_{ij}}{\partial T_{kl}}=\varepsilon_{ikq}\varepsilon_{jls}T_{qs} \tag{1.370}$$

$$\left(\frac{\partial(\mathrm{cof}\mathbf{T})_{ij}}{\partial T_{kl}}=\frac{\partial\frac{1}{2}\varepsilon_{ipq}\varepsilon_{jrs}T_{pr}T_{qs}}{\partial T_{kl}}=\frac{1}{2}(\varepsilon_{ipq}\varepsilon_{jrs}\delta_{kp}\delta_{lr}T_{qs}+\varepsilon_{ipq}\varepsilon_{jrs}T_{pr}\delta_{kq}\delta_{ls})\right).$$

13.

$$\frac{\partial\det\mathbf{T}}{\partial\mathbf{T}}=\frac{\partial\mathrm{III}}{\partial\mathbf{T}}=\mathrm{cof}\,\mathbf{T}=(\det\mathbf{T})\mathbf{T}^{-T},\quad \frac{\partial\det\mathbf{T}}{\partial T_{ij}}=\frac{\partial\mathrm{III}}{\partial T_{ij}}=(\mathrm{cof}\,\mathbf{T})_{ij}=(\det\mathbf{T})T_{ji}^{-1} \tag{1.371}$$

$$\left(\frac{\partial\frac{1}{3!}\varepsilon_{abc}\varepsilon_{pqr}T_{ap}T_{bq}T_{cr}}{\partial T_{ij}}=\frac{3}{3!}\varepsilon_{abc}\varepsilon_{pqr}\delta_{ia}\delta_{jp}T_{bq}T_{cr}=\frac{1}{2!}\varepsilon_{ibc}\varepsilon_{jqr}T_{bq}T_{cr}=(\mathrm{cof}\,\mathbf{T})_{ij}\right)$$

$$\left(\begin{array}{l}\dfrac{\partial\mathrm{III}}{\partial\mathbf{T}}:\mathbf{T}=\left\{\dfrac{\partial}{\partial\mathbf{T}}\left(\dfrac{1}{6}\mathrm{tr}^3\mathbf{T}-\dfrac{1}{2}\mathrm{tr}\mathbf{T}\mathrm{tr}\mathbf{T}^2+\dfrac{1}{3}\mathrm{tr}\mathbf{T}^3\right)\right\}:\mathbf{T}\\ \qquad=\left\{\dfrac{3}{6}(\mathrm{tr}^2\mathbf{T})\mathbf{I}-\dfrac{1}{2}\mathbf{I}\mathrm{tr}\mathbf{T}^2-(\mathrm{tr}\mathbf{T})\mathbf{T}^T+(\mathbf{T}^2)^T\right\}:\mathbf{T}\\ \qquad=\dfrac{3}{6}\mathrm{tr}^2\mathbf{T}\mathrm{tr}\mathbf{T}^T-\dfrac{1}{2}(\mathrm{tr}\mathbf{T})\mathrm{tr}\mathbf{T}^{T^2}-(\mathrm{tr}\mathbf{T})\mathrm{tr}\mathbf{T}^{T^2}+\mathrm{tr}\mathbf{T}^{T^3}\\ \qquad=\dfrac{3}{6}\mathrm{tr}^3\mathbf{T}-\dfrac{3}{2}(\mathrm{tr}\mathbf{T})\mathrm{tr}\mathbf{T}^2+\mathrm{tr}\mathbf{T}^3=3\mathrm{III}=\mathrm{III}\mathbf{T}^{-T}:\mathbf{T}\end{array}\right),$$

noting equations (1.151), (1.153), (1.297) with equation (1.157), while equation (1.371) will be also derived later by the directional derivative in 1.9.2. It follows from equation (1.371) that

$$\frac{\partial\sqrt{\det\mathbf{T}}}{\partial\mathbf{T}}=\frac{1}{2}\sqrt{\det\mathbf{T}}\mathbf{T}^{-T}. \tag{1.372}$$

Besides, applying equation (1.371), equation $(1.159)_6$ can be proved by the following alternative method:

$$\begin{aligned}(\det\mathrm{e}^{\mathbf{A}t})^{\bullet} &= (\det\mathrm{e}^{\mathbf{A}t})(\mathrm{e}^{\mathbf{A}t})^{-T}:(\mathrm{e}^{\mathbf{A}t})^{\bullet}=(\det\mathrm{e}^{\mathbf{A}t})(\mathrm{e}^{\mathbf{A}t})^{-T}:(\mathrm{e}^{\mathbf{A}t}\mathbf{A})\\
&=(\det\mathrm{e}^{\mathbf{A}t})\mathrm{tr}\{(\mathrm{e}^{\mathbf{A}t})^{-1}(\mathrm{e}^{\mathbf{A}t})\mathbf{A}\}=(\det\mathrm{e}^{\mathbf{A}t})\mathrm{tr}\mathbf{A}\\
&\rightsquigarrow\frac{(\det\mathrm{e}^{\mathbf{A}t})^{\bullet}}{\det\mathrm{e}^{\mathbf{A}t}}=\mathrm{tr}\mathbf{A}\rightsquigarrow\ln(\det\mathrm{e}^{\mathbf{A}t})=\mathrm{tr}(\mathbf{A}t)\rightsquigarrow\det\{\exp(\mathbf{A}t)\}=\exp\{\mathrm{tr}(\mathbf{A}t)\}\\
&\rightsquigarrow\det(\exp\mathbf{T})=\exp(\mathrm{tr}\mathbf{T}),\end{aligned} \tag{1.159$'_6$}$$

noting $(\mathrm{e}^{\mathbf{A}t})^{\bullet}=\mathrm{e}^{\mathbf{A}t}\mathbf{A}$ by $d[(1/n!)(\mathbf{T}t)^n]/dt=[\{1/(n-1)!\}(\mathbf{T}t)^{n-1}]\mathbf{T}$ for $n=0,1,2,\ldots,\infty$ in equation (1.258). This is one of the formulae concerning the *exponential map* or *tensor exponential function* used widely in numerical analysis for finite strain elastoplasticity. Other useful formulae are:

$$\left.\begin{array}{l}\exp(\mathbf{A}+\mathbf{B})=(\exp\mathbf{A})(\exp\mathbf{B})\ \text{if}\ \mathbf{AB}=\mathbf{BA},\quad \exp(m\mathbf{A})=(\exp\mathbf{A})^m\\ \exp(\mathbf{BAB}^{-1})=\mathbf{B}(\exp\mathbf{A})\mathbf{B}^{-1}\end{array}\right\}, \tag{1.372$'$}$$

where m is a positive or negative integer. The former two equations can be proved readily by the spectral representation used for equation $(1.159)_6$ or by substituting $\mathbf{A}+\mathbf{B}$ and $m\mathbf{A}$ into $\mathbf{T}$ in the series expansion equation (1.258) with the *binomial theorem* (Hirsch and Small, 1974):

$$(\mathbf{A}+\mathbf{B})^n=n!\left(\frac{\mathbf{A}^i}{i!}\frac{\mathbf{B}^j}{j!}\right)\rightsquigarrow\exp(\mathbf{A}+\mathbf{B})=\sum_{r=0}^{n}\sum_{i+j=r}\left(\frac{\mathbf{A}^i}{i!}\frac{\mathbf{B}^j}{j!}\right)$$

under the infinite series for the first equation. The last equation can be proved by substituting $\mathbf{BAB}^{-1}$ into $\mathbf{T}$ in equation (1.258), taking into account that $(\mathbf{BAB}^{-1})^n=\mathbf{BA}^n\mathbf{B}^{-1}$.

14.

$$\frac{\partial\ln\sqrt{\det\mathbf{T}}}{\partial\mathbf{T}}=\frac{1}{2}\mathbf{T}^{-1},\quad\frac{\partial\ln\sqrt{\det\mathbf{T}}}{\partial T_{ij}}=\frac{1}{2}T_{ij}^{-1} \tag{1.373}$$

$$\left(\frac{\partial\ln\sqrt{\det\mathbf{T}}}{\partial\mathbf{T}}=\frac{1}{\sqrt{\det\mathbf{T}}}\frac{\partial\sqrt{\det\mathbf{T}}}{\partial\mathbf{T}}=\frac{1}{\sqrt{\det\mathbf{T}}}\frac{1}{2\sqrt{\det\mathbf{T}}}(\det\mathbf{T})\mathbf{T}^{-1}\right).$$

15.

$$\frac{\partial\exp\mathbf{T}}{\partial\mathbf{T}}=\sum_{n=1}^{\infty}\frac{1}{n!}\sum_{k=1}^{n}\mathbf{T}^{k-1}\tilde{\otimes}\mathbf{T}^{n-k},\quad\left(\frac{\partial\exp\mathbf{T}}{\partial\mathbf{T}}\right)_{ijkl}=\sum_{n=1}^{\infty}\frac{1}{n!}\sum_{k=1}^{n}(\mathbf{T}^{k-1})_{ik}(\mathbf{T}^{n-k})_{lj} \tag{1.374}$$

$$\left(\begin{aligned}\frac{\partial(\mathbf{I}+\mathbf{T}+\frac{1}{2!}\mathbf{T}^2+\frac{1}{3!}\mathbf{T}^3+\ldots)}{\partial\mathbf{T}} &= \mathbf{I}\tilde{\otimes}\mathbf{I}+\frac{1}{2}(\mathbf{T}\tilde{\otimes}\mathbf{I}+\mathbf{I}\tilde{\otimes}\mathbf{T})\\ &+\frac{1}{3!}(\mathbf{T}^2\tilde{\otimes}\mathbf{I}+\mathbf{T}\tilde{\otimes}\mathbf{T}+\mathbf{I}\tilde{\otimes}\mathbf{T}^2)+\ldots\end{aligned}\right),$$

noting equations (1.258) and (1.356) ((1.357)).

For later use, the partial derivatives of a scalar function of various second-order tensor invariants are given below. Noting equations (1.235), (1.359), (1.361), (1.366), (1.369), (1.371) and (1.372) shown in Section 1.9.1, one obtains:

$$\begin{aligned}\frac{\partial f(\mathrm{I},\mathrm{II},\mathrm{III})}{\partial\mathbf{T}} &= \frac{\partial f}{\partial\mathrm{I}}\frac{\partial\mathrm{I}}{\partial\mathbf{T}}+\frac{\partial f}{\partial\mathrm{II}}\frac{\partial\mathrm{II}}{\partial\mathbf{T}}+\frac{\partial f}{\partial\mathrm{III}}\frac{\partial\mathrm{III}}{\partial\mathbf{T}}\\ &= \frac{\partial f}{\partial\mathrm{I}}\mathbf{I}+\frac{\partial f}{\partial\mathrm{II}}\{(\mathrm{tr}\mathbf{T})\mathbf{I}-\mathbf{T}^T\}+\frac{\partial f}{\partial\mathrm{III}}(\det\mathbf{T})\mathbf{T}^{-T}\\ &= \left(\frac{\partial f}{\partial\mathrm{I}}+\frac{\partial f}{\partial\mathrm{II}}\mathrm{tr}\mathbf{T}\right)\mathbf{I}-\frac{\partial f}{\partial\mathrm{II}}\mathbf{T}^T+\frac{\partial f}{\partial\mathrm{III}}(\det\mathbf{T})\mathbf{T}^{-T}\end{aligned}\tag{1.375}$$

which is rewritten in vew of equation (1.296) as

$$\begin{aligned}\frac{\partial f(\mathrm{I},\mathrm{II},\mathrm{III})}{\partial\mathbf{T}} &= \left(\frac{\partial f}{\partial\mathrm{I}}+\frac{\partial f}{\partial\mathrm{II}}\mathrm{tr}\mathbf{T}\right)\mathbf{I}-\frac{\partial f}{\partial\mathrm{II}}\mathbf{T}^T+\frac{\partial f}{\partial\mathrm{III}}(\mathrm{III}-\mathrm{I}\mathbf{T}+\mathbf{T}^2)^T\\ &= \left(\frac{\partial f}{\partial\mathrm{I}}+\frac{\partial f}{\partial\mathrm{II}}\mathrm{tr}\mathbf{T}+\frac{\partial f}{\partial\mathrm{III}}\mathrm{II}\right)\mathbf{I}-\left(\frac{\partial f}{\partial\mathrm{II}}+\frac{\partial f}{\partial\mathrm{III}}\mathrm{II}\right)\mathbf{T}^T+\frac{\partial f}{\partial\mathrm{III}}\mathbf{T}^{2^T},\end{aligned}\tag{1.376}$$

$$\frac{\partial f(\bar{\mathrm{I}},\bar{\mathrm{II}},\bar{\mathrm{III}})}{\partial\mathbf{T}} = \frac{\partial f}{\partial\bar{\mathrm{I}}}\mathbf{I}+2\frac{\partial f}{\partial\bar{\mathrm{II}}}\mathbf{T}^T+3\frac{\partial f}{\partial\bar{\mathrm{III}}}\mathbf{T}^{2^T},\tag{1.377}$$

$$\begin{aligned}\frac{\partial f(\mathrm{I},||\mathbf{T}'||,\cos 3\theta)}{\partial\mathbf{T}} &= \frac{\partial f}{\partial\mathrm{I}}\frac{\partial\mathrm{I}}{\partial\mathbf{T}}+\frac{\partial f}{\partial||\mathbf{T}'||}\frac{\partial||\mathbf{T}'||}{\partial\mathbf{T}}+\frac{\partial f}{\partial\cos 3\theta}\frac{\partial\cos 3\theta}{\partial\mathbf{T}}\\ &= \frac{\partial f}{\partial\mathbf{T}}\mathbf{I}+\frac{\partial f}{\partial||\mathbf{T}'||}\mathbf{t}'-\frac{3\sqrt{6}}{||\mathbf{T}'||}\frac{\partial f}{\partial\cos 3\theta}\left(\frac{1}{3}||\mathbf{t}'||^2\mathbf{I}+\frac{1}{\sqrt{6}}\cos 3\theta\mathbf{t}'-\mathbf{t}'^2\right).\end{aligned}\tag{1.378}$$

1.9.2 Directional Derivatives

The derivative of a scalar-valued function $f(x)$ of a scalar x is defined by

$$f'(x)\equiv\frac{df(x)}{dx}=\lim_{h\to 0}\frac{f(x+h)-f(x)}{h}.\tag{1.379}$$

Then, we can write formally as

$$f'(x;s)\equiv\lim_{h\to 0}\frac{f(x+hs)-f(x)}{h}=\left.\frac{\partial f(x+hs)}{\partial h}\right|_{h=0}=\left.\frac{\partial f(x+hs)}{\partial(x+hs)}\frac{\partial(x+hs)}{\partial h}\right|_{h=0}=\frac{\partial f(x)}{\partial x}s\tag{1.380}$$

Equation (1.380) can be extended for the scalar-valued vector function as follows:

$$f'(\mathbf{v};\mathbf{a}) \equiv \lim_{h\to 0}\frac{f(\mathbf{v}+h\mathbf{a})-f(\mathbf{v})}{h} = \left.\frac{\partial f(\mathbf{v}+h\mathbf{a})}{\partial h}\right|_{h=0}$$
$$= \left.\frac{\partial f(\mathbf{v}+h\mathbf{a})}{\partial(\mathbf{v}+h\mathbf{a})}\frac{\partial(\mathbf{v}+h\mathbf{a})}{\partial h}\right|_{h=0} = \frac{\partial f(\mathbf{v})}{\partial \mathbf{v}}\bullet\mathbf{a} \qquad (1.381)$$

which is referred to as the *directional derivative*. Analogously, we have the following equations for the vector $\mathbf{f}$, $\mathbf{v}$, $\mathbf{a}$ and the second-order tensors $\mathbf{F}$, $\mathbf{T}$, $\mathbf{A}$.

$$\mathbf{f}'(\mathbf{v};\mathbf{a}) \equiv \lim_{h\to 0}\frac{\mathbf{f}(\mathbf{v}+h\mathbf{a})-f(\mathbf{v})}{h} = \left.\frac{\partial \mathbf{f}(\mathbf{v}+h\mathbf{a})}{\partial h}\right|_{h=0} = \frac{\partial \mathbf{f}(\mathbf{v})}{\partial \mathbf{v}}\mathbf{a}, \qquad (1.382)$$

$$\mathbf{F}'(\mathbf{v};\mathbf{a}) \equiv \lim_{h\to 0}\frac{\mathbf{F}(\mathbf{v}+h\mathbf{a})-\mathbf{F}(\mathbf{v})}{h} = \left.\frac{\partial \mathbf{F}(\mathbf{v}+h\mathbf{a})}{\partial h}\right|_{h=0} = \frac{\partial \mathbf{F}(\mathbf{v})}{\partial \mathbf{v}}\mathbf{a}, \qquad (1.383)$$

$$\mathbf{F}'(\mathbf{T};\mathbf{A}) \equiv \lim_{h\to 0}\frac{\mathbf{F}(\mathbf{T}+h\mathbf{A})-\mathbf{F}(\mathbf{A})}{h} = \left.\frac{\partial \mathbf{F}(\mathbf{T}+h\mathbf{A})}{\partial h}\right|_{h=0} = \frac{\partial \mathbf{F}(\mathbf{T})}{\partial \mathbf{T}}:\mathbf{A}. \qquad (1.384)$$

Here, the following relations are fulfilled.

$$\lim_{h\to 0}\frac{\mathbf{f}(\mathbf{v}+hs\mathbf{a})-\mathbf{f}(\mathbf{v})}{h} = s\lim_{h\to 0}\frac{\mathbf{f}(\mathbf{v}+h\mathbf{a})-\mathbf{f}(\mathbf{v})}{h}, \qquad (1.385)$$

$$\lim_{h\to 0}\frac{\mathbf{F}(\mathbf{v}+hs\mathbf{a})-\mathbf{F}(\mathbf{v})}{h} = s\lim_{h\to 0}\frac{\mathbf{F}(\mathbf{v}+h\mathbf{a})-\mathbf{F}(\mathbf{v})}{h}, \qquad (1.386)$$

$$\lim_{h\to 0}\frac{\mathbf{F}(\mathbf{T}+hs\mathbf{A})-\mathbf{F}(\mathbf{T})}{h} = s\lim_{h\to 0}\frac{\mathbf{F}(\mathbf{T}+h\mathbf{A})-\mathbf{F}(\mathbf{T})}{h}, \qquad (1.387)$$

and

$$\lim_{h\to 0}\frac{\mathbf{f}(\mathbf{v}+h(\mathbf{a}+\mathbf{b}))-\mathbf{f}(\mathbf{v})}{h} = \lim_{h\to 0}\frac{\mathbf{f}(\mathbf{v}+h(\mathbf{a}+\mathbf{b}))-\mathbf{f}(\mathbf{v}+h\mathbf{a})}{h}$$
$$+\lim_{h\to 0}\frac{\mathbf{f}(\mathbf{v}+h\mathbf{a})-\mathbf{f}(\mathbf{v})}{h} = \frac{\partial \mathbf{f}(\mathbf{v})}{\partial \mathbf{v}}\mathbf{b}+\frac{\partial \mathbf{f}(\mathbf{v})}{\partial \mathbf{v}}\mathbf{a}, \qquad (1.388)$$

$$\lim_{h\to 0}\frac{\mathbf{F}(\mathbf{v}+h(\mathbf{a}+\mathbf{b}))-\mathbf{F}(\mathbf{v})}{h} = \lim_{h\to 0}\frac{\mathbf{F}(\mathbf{v}+h(\mathbf{a}+\mathbf{b}))-\mathbf{F}(\mathbf{v}+h\mathbf{a})}{h}$$
$$+\lim_{h\to 0}\frac{\mathbf{F}(\mathbf{v}+h\mathbf{a})-\mathbf{F}(\mathbf{v})}{h} = \frac{\partial \mathbf{F}(\mathbf{v})}{\partial \mathbf{v}}\mathbf{b}+\frac{\partial \mathbf{F}(\mathbf{v})}{\partial \mathbf{v}}\mathbf{a},$$
$$(1.389)$$

$$\lim_{h\to 0}\frac{\mathbf{F}(\mathbf{T}+h(\mathbf{A}+\mathbf{B}))-\mathbf{F}(\mathbf{T})}{h} = \lim_{h\to 0}\frac{\mathbf{F}(\mathbf{T}+h(\mathbf{A}+\mathbf{B}))-\mathbf{F}(\mathbf{T}+h\mathbf{A})}{h}$$
$$+\lim_{h\to 0}\frac{\mathbf{F}(\mathbf{T}+h\mathbf{A})-\mathbf{F}(\mathbf{T})}{h} = \frac{\partial \mathbf{F}(\mathbf{T})}{\partial \mathbf{T}}:\mathbf{B}+\frac{\partial \mathbf{F}(\mathbf{T})}{\partial \mathbf{T}}:\mathbf{A}.$$
$$(1.390)$$

Therefore, we can be convinced that the directional derivatives $\mathbf{f}'(\mathbf{v};\mathbf{a})$, $\mathbf{F}(\mathbf{v};\mathbf{a})$ and $\mathbf{F}'(\mathbf{T};\mathbf{A})$ are the linear transformations obeying the scalar multiplication and distribution laws. The above-mentioned equations for the directional derivatives are often used to derive the partial derivatives of tensor functions.

In what follows, the partial derivatives for some second-order tensor functions will be derived by exploiting the directional derivative in equation (1.384), which have been derived already along the ordinary tensor derivatives in 1.9.1.

$$\frac{\partial \mathbf{T}^n}{\partial \mathbf{T}}:\mathbf{A}=\left.\frac{\partial(\mathbf{T}+h\mathbf{A})^n}{\partial h}\right|_{h=0}=\frac{\partial}{\partial h}\left.\left(\mathbf{T}^n+h\sum_{i=1}^{n}\mathbf{T}^{i-1}\mathbf{A}\mathbf{T}^{n-i}+h^2\sum_{i=1}^{n}\cdots+\cdots\right)\right|_{h=0}$$

$$=\sum_{i=1}^{n}\mathbf{T}^{i-1}\mathbf{A}\mathbf{T}^{n-i}=\sum_{i=1}^{n}\mathbf{T}^{k-1}\tilde{\otimes}\,\mathbf{T}^{n-k}:\mathbf{A},\tag{1.391}$$

$$\frac{\partial \mathrm{tr}\mathbf{T}^n}{\partial \mathbf{T}}:\mathbf{A}=\left.\frac{\partial \mathrm{tr}(\mathbf{T}+h\mathbf{A})^n}{\partial h}\right|_{h=0}=\frac{\partial}{\partial h}\mathrm{tr}\{\mathbf{T}^n+h\underbrace{(\mathbf{A}\mathbf{T}^{n-1}+\mathbf{T}\mathbf{A}\mathbf{T}^{n-2}+\cdots+\mathbf{T}^{n-1}\mathbf{A})}_{n}$$

$$\left.+h^2(\cdots)+\cdots\}\right|_{h=0}=\mathrm{tr}(n\mathbf{T}^{n-1}\mathbf{A})=n(\mathbf{T}^{n-1})^T:\mathbf{A},\tag{1.392}$$

$$\left.\frac{\partial\{(\mathbf{T}+h\mathbf{A})^{-1}(\mathbf{T}+h\mathbf{A})\}}{\partial h}\right|_{h=0}=\left.\frac{\partial(\mathbf{T}+h\mathbf{A})^{-1}}{\partial h}(\mathbf{T}+h\mathbf{A})\right|_{h=0}+\left.(\mathbf{T}+h\mathbf{A})^{-1}\frac{\partial(\mathbf{T}+h\mathbf{A})}{\partial h}\right|_{h=0}$$

$$=\left.\frac{\partial(\mathbf{T}+h\mathbf{A})^{-1}}{\partial h}\right|_{h=0}\mathbf{T}+\mathbf{T}^{-1}\mathbf{A}=\mathbf{O}\to\frac{\partial\mathbf{T}^{-1}}{\partial\mathbf{T}}:\mathbf{A}$$

$$=\left.\frac{\partial(\mathbf{T}+h\mathbf{A})^{-1}}{\partial h}\right|_{h=0}=-\mathbf{T}^{-1}\mathbf{A}\mathbf{T}^{-1}=-\mathbf{T}^{-1}\tilde{\otimes}\,\mathbf{T}^{-1}:\mathbf{A},\tag{1.393}$$

$$\frac{\partial \det\mathbf{T}}{\partial\mathbf{T}}:\mathbf{A}=\left.\frac{\partial\det(\mathbf{T}+h\mathbf{A})}{\partial h}\right|_{h=0}=\left.\frac{\partial(\det\mathbf{T}\det[\mathbf{I}+h\mathbf{T}^{-1}\mathbf{A}])}{\partial h}\right|_{h=0}$$

$$=\left.\frac{\partial(\det\mathbf{T}h^3\det[\mathbf{T}^{-1}\mathbf{A}-(-h^{-1})\mathbf{I}])}{\partial h}\right|_{h=0}$$

$$=\left.\frac{\partial[\det\mathbf{T}h^3\{-(-h^{-1})^3+(-h^{-1})^2I-(-h^{-1})II+III])}{\partial h}\right|_{h=0}$$

$$=\left.\frac{\partial[\det\mathbf{T}\{1+hI+h^2II+h^3III])}{\partial h}\right|_{h=0}=(\det\mathbf{T})I$$

$$=(\det\mathbf{T})\mathrm{tr}(\mathbf{T}^{-1}\mathbf{A})=(\det\mathbf{T})\mathbf{T}^{-T}:\mathbf{A},\tag{1.394}$$

designating principal invariants of the tensor $\mathbf{T}^{-1}\mathbf{A}$ as I, II, III and noting equation (1.233) and that the eigenvalue of $\mathbf{T}^{-1}\mathbf{A}$ is $-h^{-1}$. We obtain equations (1.356), (1.358), (1.367) and (1.371) from equations (1.391), (1.392), (1.393) and (1.394), respectively. The directional derivative is useful to derive the partial derivatives of tensor functions more concisely compared with the method in component forms for some cases."

1.9.3 Taylor Expansion

The *Taylor expansion* of a scalar-valued scalar function is given by

$$f(x+h) = \sum_{k=0}^{\infty} \frac{h^k}{k!}\frac{d^k f(x)}{dx^k} = f(x) + h\frac{df(x)}{dx} + \frac{h^2}{2!}\frac{d^2 f(x)}{dx^2} + \cdots, \tag{1.395}$$

defining $0! = 1$. Choosing $x = x_0$ and $h = \alpha u$ in equation (1.395) and noting the relation

$$\frac{d^n f(x_0)}{dx_0^n} = \left.\frac{d^n f(x)}{dx^n}\right|_{x=x_0} = \left.\frac{d^n f(x_0+\varepsilon u)}{d(\varepsilon u)^k}\right|_{\varepsilon=0} = \frac{1}{u^k}\left.\frac{d^n f(x_0+\varepsilon u)}{d\varepsilon^k}\right|_{\varepsilon=0}$$

$$\left(\frac{d^n f(x+y)}{dx^n} = \frac{d^n f(x+y)}{dy^n}, \quad \frac{d^n f}{d(ax)^n} = \frac{d^n f}{dx^n}\frac{d^n x}{d(ax)^n} = \frac{d^n f}{a^n dx^n} = \frac{1}{a^n}\frac{d^n f}{dx^n}\right),$$

it follows that

$$\begin{aligned} f(x_0+\alpha u) &= \sum_{k=0}^{\infty}\frac{(\alpha u)^k}{k!}\frac{d^n f(x_0)}{dx_0^n} = \sum_{k=0}^{\infty}\frac{(\alpha u)^k}{k!}\frac{1}{u^k}\left(\left.\frac{d^n f(x_0+\varepsilon u)}{d\varepsilon^k}\right|_{\varepsilon=0}\right) \\ &= \sum_{k=0}^{\infty}\frac{\alpha^k}{k!}\left(\left.\frac{d^n f(x_0+\varepsilon u)}{d\varepsilon^k}\right|_{\varepsilon=0}\right) \\ &= f(x_0) + \alpha\left(\left.\frac{df(x_0+\varepsilon u)}{d\varepsilon}\right|_{\varepsilon=0}\right) + \frac{\alpha^2}{2!}\left(\left.\frac{d^2 f(x_0+\varepsilon u)}{d\varepsilon^2}\right|_{\varepsilon=0}\right) + \cdots. \end{aligned} \tag{1.396}$$

Based on equation (1.396), the Taylor expansion of the tensor-valued tensor function $\mathbf{F}(\mathbf{x}_0 + \alpha\mathbf{u})$ is as follows:

$$\begin{aligned} \mathbf{F}(\mathbf{x}_0+\alpha\mathbf{u}) &= \sum_{k=0}^{\infty}\frac{\alpha^k}{k!}\left(\left.\frac{d^k\mathbf{F}(\mathbf{x}_0+\varepsilon\mathbf{u})}{d\varepsilon^k}\right|_{\varepsilon=0}\right) \\ &= \mathbf{F}(\mathbf{x}_0) + \alpha\left(\left.\frac{d\mathbf{F}(\mathbf{x}_0+\varepsilon\mathbf{u})}{d\varepsilon}\right|_{\varepsilon=0}\right) + \frac{\alpha^2}{2!}\left(\left.\frac{d^2\mathbf{F}(\mathbf{x}_0+\varepsilon\mathbf{u})}{d\varepsilon^2}\right|_{\varepsilon=0}\right) + \cdots. \end{aligned} \tag{1.397}$$

The truncation of the Taylor expansion in equation (1.397) reduces to

$$\mathbf{F}(\mathbf{x}_0+\alpha\mathbf{u}) - \mathbf{F}(\mathbf{x}_0) \cong \alpha\left(\left.\frac{d\mathbf{F}(\mathbf{x}_0+\varepsilon\mathbf{u})}{d\varepsilon}\right|_{\varepsilon=0}\right), \tag{1.398}$$

which reduces in turn to the following equation by choosing $\alpha = 1$:

$$\mathbf{F}(\mathbf{x}_0+\mathbf{u}) - \mathbf{F}(\mathbf{x}_0) \cong \left.\frac{d\mathbf{F}(\mathbf{x}_0+\varepsilon\mathbf{u})}{d\varepsilon}\right|_{\varepsilon=0}. \tag{1.399}$$

The right-hand side of equation (1.399) is the directional derivative shown in equation (1.383), and thus it follows that

$$\mathbf{F}(\mathbf{x}_0+\mathbf{u}) - \mathbf{F}(\mathbf{x}_0) \cong \mathbf{F}'(\mathbf{x}_0; \mathbf{u}). \tag{1.400}$$

The following equation obtained by setting $\mathbf{F}(\mathbf{x}_0 + \mathbf{u}) = \mathbf{O}$ in equation (1.400) is used for a convergence process in numerical calculation of the nonlinear algebraic equation $\mathbf{F}(\mathbf{x}) = \mathbf{O}$ (Bonet and Wood, 2008):

$$\mathbf{F}(\mathbf{x}_0) + \mathbf{F}'(\mathbf{x}_0; \mathbf{u}) = \mathbf{O} \tag{1.401}$$

which is rewritten by equation (1.383) as follows:

$$\mathbf{F}(\mathbf{x}_0) + \frac{\partial \mathbf{F}}{\partial \mathbf{x}_0}\mathbf{u} = \mathbf{O}. \tag{1.402}$$

Analogously, the following equation holds for the vector function $\mathbf{f}$ by virtue of equation (1.382).

$$\mathbf{f}(\mathbf{x}_0) + \frac{\partial \mathbf{f}}{\partial \mathbf{x}_0}\mathbf{u} = \mathbf{O}. \tag{1.403}$$

1.9.4 Time Derivatives in Lagrangian and Eulerian Descriptions

A map that completely describes the positions of all the *material particles* (or *material points* or *material elements*) in a body is called a *configuration*. For some scalar- or tensor-valued physical quantity ψ, denoting the time by t, the position vector of a material particle in the *reference configuration* by $\mathbf{X}$ and the position vector of a material particle in the *current configuration* by $\mathbf{x}$, the distribution of ψ in the space can be formally described by $\psi(\mathbf{X}, t)$ or $\psi(\mathbf{x}, t)$, where the *mapping* from $\mathbf{x}$ to $\mathbf{X}$ is generally described as follows:

$$\mathbf{x} = \chi(\mathbf{X}, t), \quad \mathbf{X} = \chi^{-1}(\mathbf{x}, t) \tag{1.404}$$

which is taken to describe the *motion* of the material particle. The representations $\psi(\mathbf{X}, t)$ and $\psi(\mathbf{x}, t)$ are called the *Lagrangian (material) description* (or *representation*) and the *Eulerian (spatial) description* (or *representation*), respectively.

The velocity of the material particle $\mathbf{X}$ is given by

$$\mathbf{v} \equiv \frac{\partial \mathbf{x}(\mathbf{X}, t)}{\partial t} \tag{1.405}$$

in the Lagrangian description and

$$\mathbf{v} = \frac{\partial \mathbf{x}(\chi^{-1}(\mathbf{x}, t), t)}{\partial t} \tag{1.406}$$

in the Eulerian description.

The rate of the physical quantity ψ for a particular material particle is gained by observation, moving with material particle, and is referred to as the *material-time derivative* which is given by the total differentiation with respect to time. The material-time derivative in the Lagrangian description is defined by

$$\dot{\psi} \equiv \frac{D\psi}{Dt} \equiv \frac{\partial \psi(\mathbf{X}, t)}{\partial t} + \frac{\partial \psi(\mathbf{X}, t)}{\partial \mathbf{X}} \cdot \underbrace{\frac{\partial \mathbf{X}}{\partial t}}_{0} = \frac{\partial \psi(\mathbf{X}, t)}{\partial t} = \lim_{t \to 0} \frac{\psi(\mathbf{X}, t + \Delta t) - \psi(\mathbf{X}, t)}{\Delta t}. \tag{1.407}$$

In the Eulerian description it is given by

$$\boxed{\begin{aligned}\overset{\bullet}{\psi} \equiv \frac{D\psi}{Dt} \equiv \frac{d\psi(\mathbf{x},t)}{dt} &= \frac{\partial\psi(\mathbf{x},t)}{\partial t} + \frac{\partial\psi(\mathbf{x},t)}{\partial\mathbf{x}} \cdot \frac{\partial\mathbf{x}}{\partial t} \\ &= \frac{\partial\psi(\mathbf{x},t)}{\partial t} + \mathbf{v}(\mathbf{x},t) \cdot \operatorname{grad}\psi(\mathbf{x},t)\end{aligned}} \tag{1.408}$$

The first term in equation (1.408) signifies the *non-steady* (or *local time derivative*) *term* describing the variation of the quantity at a spatially fixed point $\mathbf{x}$ with time – the *spatial-time derivative*. The second term signifies the *convective term* describing the variation attributable to the movement of the material particle under the existence of the gradient $\partial\psi(\mathbf{x},t)/\partial\mathbf{x}$ due to the heterogeneity in the spatial distribution of ψ. Both symbols $(^\bullet)$ and $D(\ )/Dt$ are used widely to specify the material-time derivative.

The material-time derivatives of vectors and second-order tensors in the Eulerian and the Lagrangian description are given from equation (1.408) as

$$\overset{\bullet}{\mathbf{a}} = \frac{\partial\mathbf{a}(\mathbf{X},t)}{\partial t} = \frac{d\mathbf{a}(\mathbf{x},t)}{dt} = \frac{\partial\mathbf{a}(\mathbf{x},t)}{\partial t} + \mathbf{v} \cdot \frac{\partial\mathbf{a}(\mathbf{x},t)}{\partial\mathbf{x}} = \frac{\partial\mathbf{a}(\mathbf{x},t)}{\partial t} + \mathbf{v} \cdot \operatorname{grad}\mathbf{a}, \tag{1.409}$$

$$\overset{\bullet}{\mathbf{T}} = \frac{\partial\mathbf{T}(\mathbf{X},t)}{\partial t} = \frac{d\mathbf{T}(\mathbf{x},t)}{dt} = \frac{\partial\mathbf{T}(\mathbf{x},t)}{\partial t} + \mathbf{v} \cdot \frac{\partial\mathbf{T}(\mathbf{x},t)}{\partial\mathbf{x}} = \frac{\partial\mathbf{T}(\mathbf{x},t)}{\partial t} + \mathbf{v} \cdot \operatorname{grad}\mathbf{T}. \tag{1.410}$$

The motion of a perfect fluid is independent of the deformation history from the initial configuration and thus only a current spatial flow (motion) of a physical quantity is meaningful. The Eulerian description based on the spatial-time derivative $\partial\psi(\mathbf{x},t)/\partial t$ is usually employed in fluid mechanics. On the other hand, the constitutive equation of a solid depends on the deformation history except for a perfectly plastic material and thus one has to analyze the deformation of each material particle moving in a space. Then, in solid mechanics the material-time derivative $\overset{\bullet}{\psi}$ must be used. Further, as will be explained in Section 5.2, the *corotational derivative* based on the rate of a physical quantity observed by the coordinate system rotating with a material is often used in rate-type or incremental-type constitutive equations of solids, for example, the viscoelastic, the elasto-plastic and the viscoplastic deformation. The material-time and spatial-time derivatives are illustrated in Figure 1.5

In what follows, we examine the Lagrangian and Eulerian descriptions, the spatial-time and material-time derivatives by means of some simple examples.

Example 1.1: Lagrangian and Eulerian descriptions of velocity (Belytschko et al., 2000) Consider the Lagrangian and Eulerian descriptions of the velocity of a martial particle prior to the derivations of the spatial- and the material-time derivatives. Here, assume a one-dimension motion in which the mapping $\mathbf{x} = \chi(\mathbf{X},t)$ and $\mathbf{X} = \chi^{-1}(\mathbf{x},t)$ is given by

$$x(X,t) = (1-X)t + \frac{1}{2}Xt^2 + X, \quad X(x,t) = \frac{x-t}{\frac{1}{2}t^2 - t + 1}. \tag{1.411}$$

The Lagrangian description of velocity is given by

$$v(X,t) = \frac{\partial x(X,t)}{\partial t} = 1 + X(t-1). \tag{1.412}$$

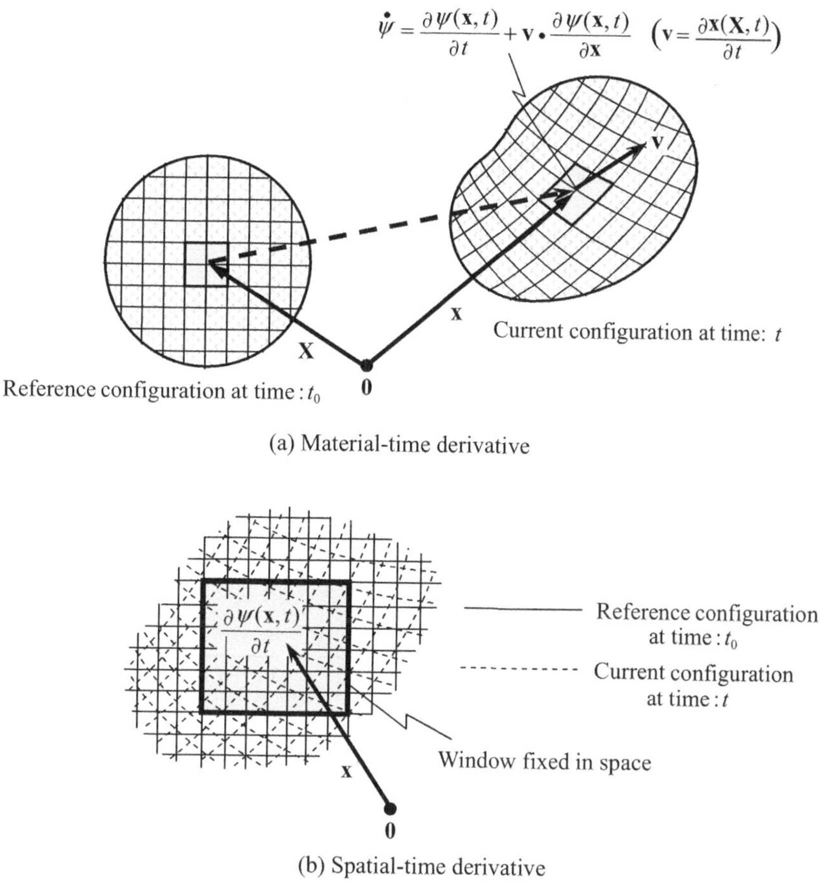

Figure 1.5 Material-time derivative and spatial-time derivative

On the other hand, the Eulrian description of the velocity of a material particle is given by substituting equation $(1.411)_2$ into equation (1.412) as follows:

$$v(x, t) = \frac{1 - x + xt - \frac{1}{2}t^2}{\frac{1}{2}t^2 - t + 1}. \tag{1.413}$$

The Eulrian description of velocity of material particle must be calculated though the Lagrangian description as shown above, since the velocity of the material particle has to be calculated for the condition $X = \text{const}$.

Example 1.2: Temperature distribution (Bonet and Wood, 2008) Consider the one-dimensional stretching of a bar in which the mappings $\mathbf{x} = \chi(\mathbf{X}, t)$ and $\mathbf{X} = \chi^{-1}(\mathbf{x}, t)$ are as follows:

$$x = (1 + t)X, \quad X = \frac{x}{1 + t}. \tag{1.414}$$

The bar experiences a temperature distribution given by

$$T = Xt^2, \quad T = xt^2/(1+t) \tag{1.415}$$

in the material and spatial descriptions, respectively. The velocity is

$$v = \frac{\partial x(X,t)}{\partial t} = X, \quad v = \frac{x}{1+t} \tag{1.416}$$

in the material and spatial descriptions, respectively. The material-time derivative of temperature is

$$\dot{T} = \frac{\partial T(X,t)}{\partial t} = 2Xt, \quad \dot{T} = \frac{2xt}{1+t} \tag{1.417}$$

in the material and spatial descriptions, respectively. On the other hand, the spatial-time derivative and the convective term are given by

$$\frac{\partial T(x,t)}{\partial t} = \frac{(2t+t^2)x}{(1+t)^2}, \quad \frac{\partial T(x,t)}{\partial x} = \frac{t^2}{1+t}, \quad \frac{\partial T(x,t)}{\partial x}v = \frac{t^2}{1+t}\frac{x}{1+t} = \frac{xt^2}{(1+t)^2}, \tag{1.418}$$

leading to

$$\dot{T} = \frac{\partial T(x,t)}{\partial t} + \frac{\partial T(x,t)}{\partial x}v = \frac{(2t+t^2)x}{(1+t)^2} + \frac{xt^2}{(1+t)^2} = \frac{2xt}{1+t} \tag{1.419}$$

which coincides with equation (1.417) so that the validity of equation (1.407) is confirmed.

Example 1.3: Infinitesimal shear strain field Consider the two-dimensional shear deformation in which the mapping $\mathbf{x} = \chi(\mathbf{X}, t)$ is given by

$$\left.\begin{aligned} x_1 &= X_1 + aX_2^2t^2 \\ x_2 &= X_2 + ct \end{aligned}\right\}, \tag{1.420}$$

where a and c are constants. The inverse mapping $\mathbf{X} = \chi^{-1}(\mathbf{x}, t)$ of equation (1.420) is given by

$$\left.\begin{aligned} X_1(= x_1 - aX_2^2t^2) &= x_1 + a(x_2 - ct)^2t^2 \\ X_2 &= x_2 - ct \end{aligned}\right\}. \tag{1.421}$$

The velocity of a material particle is given by

$$\left.\begin{aligned} v_1 &= \frac{\partial x_1(X_1, X_2, t)}{\partial t} = 2aX_2^2t \\ v_2 &= \frac{\partial x_2(X_1, X_2, t)}{\partial t} = c \end{aligned}\right\} \tag{1.422}$$

in the Lagrangian description, or by

$$\left.\begin{aligned} v_1 &= 2a(x_2 - ct)^2t \\ v_2 &= c \end{aligned}\right\} \tag{1.423}$$

in the Eulerian description.

Further, the infinitesimal shear strain γ_{12} is

$$\gamma_{12}(X_1, X_2, t) = \frac{x_1 - X_1}{X_2} = \frac{aX_2^2 t^2}{X_2} = aX_2 t^2 \tag{1.424}$$

in the Lagrangian description, or

$$\gamma_{12}(x_1, x_2, t) = a(x_2 - ct)t^2 \tag{1.425}$$

in the Eulerian description. Further, the material-time derivative of infinitesimal shear strain rate is given from equation (1.424) by

$$\dot{\gamma}_{12}(X_1, X_2, t) = \frac{\partial \gamma_{12}(X_1, X_2, t)}{\partial t} = 2aX_2 t \tag{1.426}$$

in the Lagrangian description, or

$$\dot{\gamma}_{12}(x_1, x_2, t) = 2a(x_2 - ct)t = 2ax_2 t - 2act^2 \tag{1.427}$$

in the Eulerian description, substituting equation (1.421) into equation (1.426).

On the other hand, the spatial-time derivative is given by

$$\frac{\partial \gamma_{12}(x_1, x_2, t)}{\partial t} = 2ax_2 t - 3act^2 \tag{1.428}$$

and the convective term is

$$\frac{\partial \gamma_{12}(x_1, x_2, t)}{\partial x_1} v_1 + \frac{\partial \gamma_{12}(x_1, x_2, t)}{\partial x_2} v_2 = 0 \cdot \{2a(x_2 - ct)^2 t\} + at^2 \cdot c = act^2. \tag{1.429}$$

Here, it can be seen that the material-time derivative in equation (1.427) is the sum of the spatial time-derivative in equation (1.428) and the convective term as described in equation (1.429).

Only Eulerian variables in the current configuration are usually used in fluid mechanics, but both Lagrangian variables based on the reference configuration and Eulerian variables based on the current configuration are used in solid mechanics. Some basic Lagrangian and the Eulerian variables used in solid mechanics are listed in Table 1.2. The exact definitions of the Lagrangian and Eulerian variables and their physical meanings will be explained in detail in later chapters. Here, note that there exist Eulerian–Lagrangian variables which are based on both the reference and current configurations, called two-point tensors, as represented by the deformation gradient.

The material-time derivative described in this section designates the rate of physical quantity observed by the observer translating in parallel with a certain material point. Further, a convected derivative will be introduced in Chapter 5, which designates the rate observed by the observer not only translating but also deforming and rotating with a certain infinitesimal region in material so that the adoption of the curvilinear coordinate system studied in Chapter 2 is required.

Table 1.2 Basic Lagrangian and Eulerian variables used in constitutive equations

	Lagrangian variables	Eulerian variables
Position vector	$\mathbf{X}$	$\mathbf{x}$
Strain tensor	Green strain $\mathbf{E} = \frac{1}{2}(\mathbf{F}^T\mathbf{F} - \mathbf{G})$	Almansi strain $\mathbf{e} = \frac{1}{2}(\mathbf{g} - \mathbf{F}^{-T}\mathbf{F}^{-1})$
Strain rate tensor	$\mathbf{D} = \mathrm{sym}[\mathbf{F}^T\dot{\mathbf{F}}] = \dot{\mathbf{E}}$	$\mathbf{d} = \mathrm{sym}[\dot{\mathbf{F}}\mathbf{F}^{-1}] \neq \dot{\mathbf{e}}$
Stress tensor	Second Piola–Kirchhoff stress $\mathbf{S}$ $\mathbf{F}^{-1}(\mathbf{t}da)/dA = \mathbf{SN}$	Cauchy stress $\boldsymbol{\sigma}$ $\mathbf{t} = \boldsymbol{\sigma}\mathbf{n}$

Note. $\mathbf{F} = \partial\mathbf{x}/\partial\mathbf{X}$: deformation gradient (Eulerian–Lagrangian two-point tensor). $\mathbf{G}$, $\mathbf{g}$: metric tensors based on reference and current configurations, which coincide with the identity tensor in the Cartesian coordinate system. $\mathbf{N}$, $\mathbf{n}$: unit outward-normal tensors of surface in reference and current configurations. A, a: surface areas in reference and current configurations. $\mathbf{t}$. force vector applying to current unit area, called the Cauchy stress vector.

1.9.5 Derivatives of Tensor Field

A field which is a function of a position vector $\mathbf{x}$ is called a tensor field. Various derivatives of the tensor field are shown in this section, where use is made of the operator

$$\nabla \equiv \frac{\partial}{\partial \mathbf{x}} = \frac{\partial}{\partial x_r}\mathbf{e}_r, \tag{1.430}$$

which is called the *nabla* or *del* or *Hamilton operator*.

1) Gradient

Scalar field:

$$\mathrm{grad}\, s = \nabla s = \frac{\partial s}{\partial x_i}\mathbf{e}_i. \tag{1.431}$$

Vector field:

$$\mathrm{grad}\,\mathbf{v} = \begin{cases} \mathbf{v} \otimes \nabla = v_i\mathbf{e}_i \otimes \dfrac{\partial}{\partial x_j}\mathbf{e}_j = \dfrac{\partial v_i}{\partial x_j}\mathbf{e}_i \otimes \mathbf{e}_j : \text{ right form} \\ \nabla \otimes \mathbf{v} = \dfrac{\partial}{\partial x_i}\mathbf{e}_i \otimes v_j\mathbf{e}_j = \dfrac{\partial v_j}{\partial x_i}\mathbf{e}_i \otimes \mathbf{e}_j : \text{ left form} \end{cases}. \tag{1.432}$$

Second-order tensor field:

$$\mathrm{grad}\,\mathbf{T} = \begin{cases} \mathbf{T} \otimes \nabla = T_{ij}\mathbf{e}_i \otimes \mathbf{e}_j \otimes \dfrac{\partial}{\partial x_k}\mathbf{e}_k = \dfrac{\partial T_{ij}}{\partial x_k}\mathbf{e}_i \otimes \mathbf{e}_j \otimes \mathbf{e}_k : \text{ right form} \\ \nabla \otimes \mathbf{T} = \dfrac{\partial}{\partial x_i}\mathbf{e}_i \otimes T_{jk}\mathbf{e}_j \otimes \mathbf{e}_k = \dfrac{\partial T_{jk}}{\partial x_i}\mathbf{e}_i \otimes \mathbf{e}_j \otimes \mathbf{e}_k : \text{ left form} \end{cases}. \tag{1.433}$$

2) Divergence
Vector field:

$$\operatorname{div}\mathbf{v} = \boldsymbol{\nabla}\cdot\mathbf{v}(=\mathbf{v}\cdot\boldsymbol{\nabla}) = v_i\mathbf{e}_i\cdot\frac{\partial}{\partial x_j}\mathbf{e}_j = \frac{\partial v_i}{\partial x_i}. \tag{1.434}$$

Second-order tensor field:

$$\operatorname{div}\mathbf{T} = \begin{cases} \mathbf{T}\boldsymbol{\nabla} = T_{ij}\mathbf{e}_i\otimes\mathbf{e}_j\dfrac{\partial}{\partial x_k}\mathbf{e}_k = \dfrac{\partial T_{ij}}{\partial x_j}\mathbf{e}_i : \text{ right form} \\ \boldsymbol{\nabla}\mathbf{T} = \dfrac{\partial}{\partial x_i}\mathbf{e}_i T_{jk}\mathbf{e}_j\otimes\mathbf{e}_k = \dfrac{\partial T_{ji}}{\partial x_j}\mathbf{e}_i : \text{ left form} \end{cases}. \tag{1.435}$$

3) Rotation (or **curl**)
Vector field:

$$\operatorname{rot}\mathbf{v} = \begin{cases} \mathbf{v}\times\boldsymbol{\nabla} = v_i\mathbf{e}_i\times\dfrac{\partial}{\partial x_j}\mathbf{e}_j = \varepsilon_{ijk}\dfrac{\partial v_i}{\partial x_j}\mathbf{e}_k : \text{ right form,} \\ \boldsymbol{\nabla}\times\mathbf{v} = \dfrac{\partial}{\partial x_i}\mathbf{e}_i\times v_j\mathbf{e}_j = \varepsilon_{jik}\dfrac{\partial v_i}{\partial x_j}\mathbf{e}_k = -\mathbf{v}\times\boldsymbol{\nabla} : \text{ left form.} \end{cases} \tag{1.436}$$

Second-order tensor field:

$$\operatorname{rot}\mathbf{T} = \begin{cases} \mathbf{T}\times\boldsymbol{\nabla} = T_{ij}\mathbf{e}_i\otimes\mathbf{e}_j\times\dfrac{\partial}{\partial x_k}\mathbf{e}_k \\ \qquad = \dfrac{\partial T_{ij}}{\partial x_k}\mathbf{e}_i\otimes(\mathbf{e}_j\times\mathbf{e}_k) = \varepsilon_{jkr}\dfrac{\partial T_{ij}}{\partial x_k}\mathbf{e}_i\otimes\mathbf{e}_r : \text{ right form} \\ \boldsymbol{\nabla}\times\mathbf{T} = \dfrac{\partial}{\partial x_i}\mathbf{e}_i\times T_{jk}\mathbf{e}_j\otimes\mathbf{e}_k \\ \qquad = \dfrac{\partial T_{jk}}{\partial x_i}(\mathbf{e}_i\times\mathbf{e}_j)\otimes\mathbf{e}_k = \varepsilon_{ijr}\dfrac{\partial T_{jk}}{\partial x_i}\mathbf{e}_r\otimes\mathbf{e}_k : \text{ left form} \end{cases}. \tag{1.437}$$

The symbol $\boldsymbol{\nabla}$ is regarded as a vector, and the scalar product of itself, that is,

$$\Delta \equiv \nabla^2 \equiv \boldsymbol{\nabla}\cdot\boldsymbol{\nabla} = \frac{\partial}{\partial x_r}\mathbf{e}_r\cdot\frac{\partial}{\partial x_s}\mathbf{e}_s = \frac{\partial^2}{\partial x_r\partial x_r} \tag{1.438}$$

has the meaning of $\nabla^2(\cdot) \equiv \operatorname{div}(\operatorname{grad}(\cdot))$. The symbol Δ is called the *Laplacian* or *Laplace operator*, which is often used for scalar or vector fields in the following sense:

$$\Delta s = \frac{\partial^2 s}{\partial x_r\partial x_r}, \qquad \Delta\mathbf{v} = \frac{\partial^2 v_s}{\partial x_r\partial x_r}\mathbf{e}_s. \tag{1.439}$$

Various formulae are derived from equations (1.431)–(1.437). Those often appearing in continuum mechanics are shown below. Note that equations (1.31), (1.32) and (1.56) are used for the derivations.

$$\left.\begin{array}{l}
\mathrm{grad}(s\mathbf{v}) = \begin{cases} \mathbf{v}\otimes \mathrm{grad}\, s + s\mathbf{v}\otimes\nabla : \text{ right form} \\ \mathrm{grad}\, s\otimes\mathbf{v} + s\nabla\otimes\mathbf{v} : \text{ left form} \end{cases} \\
\mathrm{grad}(\mathbf{u}\cdot\mathbf{v}) = (\mathrm{grad}\mathbf{u})\cdot\mathbf{v} + (\mathrm{grad}\mathbf{v})\cdot\mathbf{u} \\
\mathrm{div}(s\mathbf{v}) = s\mathrm{div}\,\mathbf{v} + \mathbf{v}\cdot\mathrm{grad}\, s \\
\mathrm{div}(\mathbf{u}\times\mathbf{v}) = \mathbf{v}\cdot(\nabla\times\mathbf{u}) + \mathbf{u}\cdot(\nabla\times\mathbf{v}) \\
\mathrm{div}(s\mathbf{T}) = \begin{cases} (s\mathbf{T})\nabla = \mathbf{T}\mathrm{grad}\, s + s(\mathbf{T}\nabla) : \text{ right form} \\ \nabla(s\mathbf{T}) = \mathbf{T}^T\mathrm{grad}\, s + s(\nabla\mathbf{T}) : \text{ left form} \end{cases} \\
\mathrm{div}(\mathbf{T}\mathbf{v}) = \mathbf{T}^T : \mathrm{grad}\,\mathbf{v} + (\nabla\mathbf{T})\cdot\mathbf{v} \\
\mathrm{rot}(\mathbf{u}\times\mathbf{v}) = \begin{cases} (\mathbf{u}\times\mathbf{v})\times\nabla \\ \quad = (\mathrm{div}\,\mathbf{u})\mathbf{v} - (\mathrm{div}\,\mathbf{v})\mathbf{u} - (\mathrm{grad}\,\mathbf{u})\mathbf{v} + (\mathrm{grad}\,\mathbf{v})\mathbf{u} : \text{ right form} \\ \nabla\times(\mathbf{u}\times\mathbf{v}) = -(\mathbf{u}\times\mathbf{v})\times\nabla \\ \quad = -(\mathrm{div}\,\mathbf{u})\mathbf{v} + (\mathrm{div}\,\mathbf{v})\mathbf{u} + (\mathrm{grad}\,\mathbf{u})\mathbf{v} - (\mathrm{grad}\,\mathbf{v})\mathbf{u} : \text{ left form} \end{cases}
\end{array}\right\} \quad (1.440)$$

because of

$$\mathrm{grad}(s\mathbf{v}) = \begin{cases} (s\mathbf{v})\otimes\nabla = \dfrac{\partial(sv_i)}{\partial x_j}\mathbf{e}_i\otimes\mathbf{e}_j = \dfrac{\partial s}{\partial x_j}v_i\mathbf{e}_i\otimes\mathbf{e}_j + s\dfrac{\partial v_i}{\partial x_j}\mathbf{e}_i\otimes\mathbf{e}_j \\ \quad = v_i\mathbf{e}_i\otimes\dfrac{\partial s}{\partial x_j}\mathbf{e}_j + s\dfrac{\partial v_i}{\partial x_j}\mathbf{e}_i\otimes\mathbf{e}_j = \mathbf{v}\otimes\mathrm{grad}\, s + s\mathbf{v}\otimes\nabla : \text{right form} \\ \nabla\otimes(s\mathbf{v}) = \dfrac{\partial(sv_j)}{\partial x_i}\mathbf{e}_i\otimes\mathbf{e}_j = \dfrac{\partial s}{\partial x_i}v_j\mathbf{e}_i\otimes\mathbf{e}_j + s\dfrac{\partial v_j}{\partial x_i}\mathbf{e}_i\otimes\mathbf{e}_j \\ \quad = \dfrac{\partial s}{\partial x_i}\mathbf{e}_i\otimes v_j\mathbf{e}_j + s\dfrac{\partial v_j}{\partial x_i}\mathbf{e}_i\otimes\mathbf{e}_j = \mathrm{grad}\, s\otimes\mathbf{v} + s\nabla\otimes\mathbf{v} : \text{left form} \end{cases}$$

$$\mathrm{grad}(\mathbf{u}\cdot\mathbf{v}) = \frac{\partial(\mathbf{u}\cdot\mathbf{v})}{\partial x_i}\mathbf{e}_i = \frac{\partial u_r v_r}{\partial x_i}\mathbf{e}_i = \frac{\partial u_r}{\partial x_i}v_r\mathbf{e}_i + \frac{\partial v_r}{\partial x_i}u_r\mathbf{e}_i$$

$$= \frac{\partial u_r}{\partial x_i}\mathbf{e}_i\otimes\mathbf{e}_r v_j\mathbf{e}_j + \frac{\partial v_r}{\partial x_i}\mathbf{e}_i\otimes\mathbf{e}_r u_j\mathbf{e}_j = (\mathrm{grad}\,\mathbf{u})\cdot\mathbf{v} + (\mathrm{grad}\,\mathbf{v})\cdot\mathbf{u},$$

$$\mathrm{div}(s\mathbf{v}) = \frac{\partial(sv_i)}{\partial x_i} = s\frac{\partial v_i}{\partial x_i} + \frac{\partial s}{\partial x_i}v_i = s\mathrm{div}\,\mathbf{v} + \mathrm{grad}\, s\cdot\mathbf{v},$$

$$\mathrm{div}(\mathbf{u}\times\mathbf{v}) = \frac{\partial(\mathbf{u}\times\mathbf{v})_i}{\partial x_i} = \frac{\partial\varepsilon_{ijk}u_jv_k}{\partial x_i} = \varepsilon_{ijk}\frac{\partial u_j}{\partial x_i}v_k + u_j\varepsilon_{ijk}\frac{\partial v_k}{\partial x_i} = \varepsilon_{ijk}\frac{\partial u_j}{\partial x_i}v_k + u_k\varepsilon_{jki}\frac{\partial v_i}{\partial x_j}$$

$$= \varepsilon_{ijk}\frac{\partial u_j}{\partial x_i}\mathbf{e}_k\cdot v_r\mathbf{e}_r + u_r\mathbf{e}_r\cdot\varepsilon_{jki}\frac{\partial v_i}{\partial x_j}\mathbf{e}_k = \mathbf{v}\cdot(\nabla\times\mathbf{u}) + \mathbf{u}\cdot(\nabla\times\mathbf{v}),$$

$$\operatorname{div}(s\mathbf{T}) = \begin{cases} (s\mathbf{T})\boldsymbol{\nabla} = \dfrac{\partial(sT_{ij})}{\partial x_j}\mathbf{e}_i = \dfrac{\partial s}{\partial x_j}T_{ij}\mathbf{e}_i + s\dfrac{\partial T_{ij}}{\partial x_j}\mathbf{e}_i = T_{ij}\mathbf{e}_i \otimes \mathbf{e}_j \dfrac{\partial s}{\partial x_r}\mathbf{e}_r + s\dfrac{\partial T_{ij}}{\partial x_j}\mathbf{e}_i \\ \quad = \mathbf{T}\operatorname{grad} s + s(\mathbf{T}\boldsymbol{\nabla}) : \text{ right form} \\ \boldsymbol{\nabla}(s\mathbf{T}) = \dfrac{\partial(sT_{ji})}{\partial x_j}\mathbf{e}_i = \dfrac{\partial s}{\partial x_j}T_{ji}\mathbf{e}_i + s\dfrac{\partial T_{ji}}{\partial x_j}\mathbf{e}_i = T_{ji}\mathbf{e}_i \otimes \mathbf{e}_j \dfrac{\partial s}{\partial x_r}\mathbf{e}_r + s\dfrac{\partial T_{ji}}{\partial x_j}\mathbf{e}_i \\ \quad = \mathbf{T}^T\operatorname{grad} s + s(\boldsymbol{\nabla}\mathbf{T}) : \text{ left form} \end{cases},$$

$$\begin{aligned} \operatorname{div}(\mathbf{T}\mathbf{v}) &= \frac{\partial(T_{ij}v_j)}{\partial x_i} = T_{ij}\frac{\partial v_j}{\partial x_i} + \frac{\partial T_{ij}}{\partial x_i}v_j = (T_{ij}\mathbf{e}_j \otimes \mathbf{e}_i) : \left(\frac{\partial v_s}{\partial x_r}\mathbf{e}_s \otimes \mathbf{e}_r\right) + \left(\frac{\partial T_{ir}}{\partial x_i}\mathbf{e}_r\right) \cdot (v_j\mathbf{e}_j) \\ &= \mathbf{T}^T : \operatorname{grad}\mathbf{v} + (\boldsymbol{\nabla}\mathbf{T}) \cdot \mathbf{v}, \end{aligned}$$

$$\operatorname{rot}(\mathbf{u} \times \mathbf{v}) = \begin{cases} (\mathbf{u} \times \mathbf{v}) \times \boldsymbol{\nabla} = \varepsilon_{ijk}\dfrac{\partial(\mathbf{u} \times \mathbf{v})_i}{\partial x_j}\mathbf{e}_k = \varepsilon_{ijk}\dfrac{\partial(\varepsilon_{irs}u_r v_s)}{\partial x_j}\mathbf{e}_k \\ \quad = (\delta_{jr}\delta_{ks} - \delta_{js}\delta_{kr})\left(\dfrac{\partial u_r}{\partial x_j}v_s\mathbf{e}_k + u_r\dfrac{\partial v_s}{\partial x_j}\mathbf{e}_k\right) \\ \quad = \dfrac{\partial u_j}{\partial x_j}v_k\mathbf{e}_k - \dfrac{\partial u_k}{\partial x_j}v_j\mathbf{e}_k + u_j\dfrac{\partial v_k}{\partial x_j}\mathbf{e}_k - u_k\dfrac{\partial v_j}{\partial x_j}\mathbf{e}_k \\ \quad = \dfrac{\partial u_j}{\partial x_j}v_k\mathbf{e}_k - \dfrac{\partial u_k}{\partial x_j}\mathbf{e}_k \otimes \mathbf{e}_j v_r\mathbf{e}_r + \dfrac{\partial v_k}{\partial x_j}\mathbf{e}_k \otimes \mathbf{e}_j u_r\mathbf{e}_r - \dfrac{\partial v_j}{\partial x_j}u_k\mathbf{e}_k \\ \quad = (\operatorname{div}\mathbf{u})\mathbf{v} - (\operatorname{div}\mathbf{v})\mathbf{u} - (\operatorname{grad}\mathbf{u})\mathbf{v} + (\operatorname{grad}\mathbf{v})\mathbf{u} : \text{ right form} \\ \boldsymbol{\nabla} \times (\mathbf{u} \times \mathbf{v}) = -(\mathbf{u} \times \mathbf{v}) \times \boldsymbol{\nabla} : \text{ left form} \end{cases}.$$

Hereinafter, only the symbols grad$(\cdot)$ and div$(\cdot)$ are used when there is no difference between the right and left forms.

1.9.6 Gauss's Divergence Theorem

Consider a scalar- or vector- or tensor-valued physical quantity $\psi(\mathbf{x})$ in the zone surrounded by a smooth surface inside a material. Imagine a thin prism cut by the four planes perpendicular to the x_2- and x_3-axes at infinitesimal intervals from a zone inside the material. The following equation holds for the prism:

$$\int_v \frac{\partial \psi}{\partial x_1}dv = \int_v \frac{\partial \psi}{\partial x_1}dx_1dx_2dx_3 = [\psi]_{x_1^-}^{x_1^+}dx_2dx_3 = (\psi^+ - \psi^-)dx_2dx_3, \tag{1.441}$$

where $(\cdot)^+$ and $(\cdot)^-$ denote the values of the physical quantity at the maximum and the minimum x_1-coordinates, respectively.

The neighborhood of the surface cut by the prism is magnified in Figure 1.6. Consider the infinitesimal rectangular surface ABCD of the prism exposed at the surface at the maximum x_1-coordinate and the infinitesimal rectangular section PB*C*D* cut by the plane passing through the point A and perpendicular to the x_1-axis. Then, denoting $\overline{\mathrm{BB}^*} = dx_Q$, $\overline{\mathrm{DD}^*} = dx_S$, the vectors $\overrightarrow{\mathrm{AB}}$, $\overrightarrow{\mathrm{AD}}$ are given by

$$\overrightarrow{\mathrm{AB}} = dx_2\mathbf{e}_2 + dx_Q\mathbf{e}_1, \quad \overrightarrow{\mathrm{AD}} = dx_3\mathbf{e}_3 + dx_S\mathbf{e}_1, \tag{1.442}$$

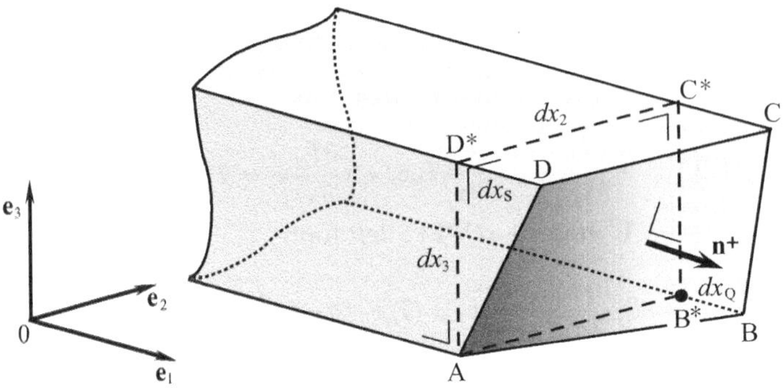

Figure 1.6 Material surface cut by prism with an infinitesimal square section

and thus it follows that

$$\begin{aligned}\mathbf{n}^+ da^+ &= \vec{\mathrm{AB}} \times \vec{\mathrm{AD}} = (dx_2\mathbf{e}_2 + dx_Q\mathbf{e}_1) \times (dx_3\mathbf{e}_3 + dx_S\mathbf{e}_1) \\ &= dx_2 dx_3 \mathbf{e}_2 \times \mathbf{e}_3 + dx_Q dx_3 \mathbf{e}_1 \times \mathbf{e}_3 + dx_S dx_2 \mathbf{e}_2 \times \mathbf{e}_1 \\ &= dx_2 dx_3 \mathbf{e}_1 - dx_Q dx_3 \mathbf{e}_2 - dx_S dx_2 \mathbf{e}_3. \end{aligned} \tag{1.443}$$

Comparing the components in the base $\mathbf{e}_1$ on the both sides in equation (1.443), one has

$$n_1^+ da^+ = dx_2 dx_3. \tag{1.444}$$

In a similar manner, for the surface of the prism exposed on the surface in the minimum x_1-coordinate, one has

$$n_1^- da^- = -dx_2 dx_3. \tag{1.445}$$

The general expression for projected area is given in Appendix A.

Adopting equations (1.444) and (1.445) in equation (1.441), the following expression holds for the prism:

$$\int_v \frac{\partial \psi}{\partial x_1} dv = \psi^+ n_1^+ da^+ - \psi^-(-n_1^- da^-) = \psi^+ n_1^+ da^+ + \psi^- n_1^- da^-. \tag{1.446}$$

Then the following equation holds for the whole zone:

$$\int_v \frac{\partial \psi}{\partial x_1} dv = \int_a \psi n_1 da. \tag{1.447}$$

Similar equations are obtained also for the x_2- and x_3-directions, and thus it follows that

$$\int_v \frac{\partial \psi}{\partial x_i} dv = \int_a \psi n_i da \tag{1.448}$$

which is called *Gauss's theorem* or *Gauss's divergence theorem*.

The following equations for the scalar s, the vector $\mathbf{v}$ and the tensor $\mathbf{T}$ hold from equation (1.448):

$$\boxed{\int_v \frac{\partial s}{\partial x_i} dv = \int_a s n_i da, \quad \int_v \operatorname{div} s dv = \int_a s\mathbf{n} da} \tag{1.449}$$

$$\boxed{\int_v \frac{\partial v_i}{\partial x_i} dv = \int_a v_i n_i da, \quad \int_v \operatorname{div} \mathbf{v} dv = \int_a \mathbf{v} \cdot \mathbf{n} da} \tag{1.450}$$

$$\boxed{\int_v \frac{\partial T_{ij}}{\partial x_i} dv = \int_a T_{ij} n_i da, \quad \int_v \nabla \mathbf{T} dv = \int_a \mathbf{T}^T \mathbf{n} da} \tag{1.451}$$

1.9.7 Material-Time Derivative of Volume Integration

Suppose that the zone of material occupying the volume v at the current moment ($t = t$) changes to occupy the volume $v + \delta v$ after an infinitesimal time ($t = t + \delta t$). The material-time derivative of the volume integration $\int_v \psi(\mathbf{x}, t) dv$ of the scalar- or tensor-valued physical quantity $\psi(\mathbf{x}, t)$ involved in the volume is given by

$$\begin{aligned}\left(\int_v \psi dv\right)^{\bullet} &= \lim_{\delta t \to 0} \frac{1}{\delta t} \left\{ \int_{v+\delta v} \psi(\mathbf{x}, t + \delta t) dv - \int_v \psi(\mathbf{x}, t) dv \right\} \\ &= \lim_{\delta t \to 0} \frac{1}{\delta t} \left[\int_v \{\psi(\mathbf{x}, t + \delta t) - \psi(\mathbf{x}, t)\} dv + \int_{\delta v} \psi(\mathbf{x}, t + \delta t) dv \right],\end{aligned} \tag{1.452}$$

The integration of the first term on the right-hand side in equation (1.452) is transformed as

$$\lim_{\delta t \to 0} \frac{1}{\delta t} \int_v \{\psi(\mathbf{x}, t + \delta t) - \psi(\mathbf{x}, t)\} dv = \int_v \frac{\partial \psi(\mathbf{x}, t)}{\partial t} dv. \tag{1.453}$$

The second term in equation (1.452) describes the effect of the change of volume during the infinite time. Here, the volume increment, δv, is given by subtracting the volume going out the boundary of the zone from the volume going into the boundary, which is the sum of $dv(= \mathbf{v} \cdot \mathbf{n} da \delta t)$ over the whole boundary surface (Figure 1.7). Therefore, using Gauss's divergence theorem (1.448) and ignoring the second-order infinitesimal quantity, the integration of the second term in the right-hand side of equation (1.452) is given by

$$\begin{aligned}&\lim_{\delta t \to 0} \frac{1}{\delta t} \int_{\delta v} \psi(\mathbf{x}, t + \delta t) dv \cong \lim_{\delta t \to 0} \frac{1}{\delta t} \int_{\delta v} \psi(\mathbf{x}, t) dv \\ &= \lim_{\delta t \to 0} \frac{1}{\delta t} \int_a \psi(\mathbf{x}, t) v_r n_r da \delta t = \int_a \psi(\mathbf{x}, t) v_r n_r da = \int_v \frac{\partial \psi(\mathbf{x}, t) v_r}{\partial x_r} dv.\end{aligned}$$

The sum of the first term on the right-hand side of this equation and equation (1.453) is equal to the material-time derivative of $\psi(\mathbf{x}, t)$, and thus equation (1.452) is given by

$$\boxed{\left(\int_v \psi(\mathbf{x}, t) dv\right)^{\bullet} = \int_v \{\dot{\psi}(\mathbf{x}, t) + \psi(\mathbf{x}, t) \operatorname{div} \mathbf{v}\} dv} \tag{1.454}$$

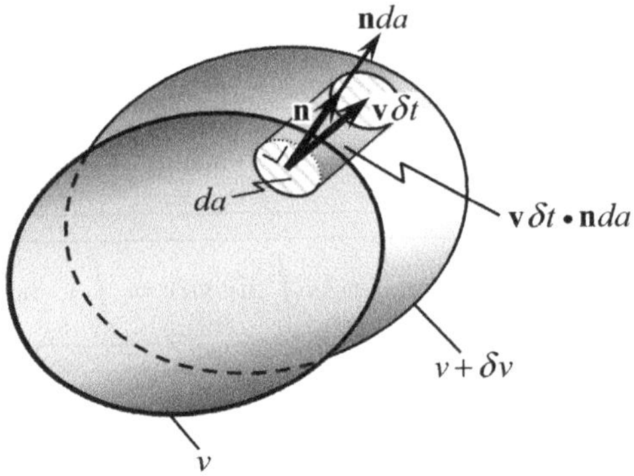

Figure 1.7 Transfer of volume element

which is called the *Reynolds transportation theorem*, noting

$$\left(\int_v \psi(\mathbf{x},t)dv\right)^{\bullet} = \int_v \left\{\frac{\partial \psi(\mathbf{x},t)}{\partial t} + \frac{\partial(\psi(\mathbf{x},t)v_r)}{\partial x_r}\right\}dv$$
$$= \int_v \left\{\frac{\partial \psi(\mathbf{x},t)}{\partial t} + \frac{\partial \psi(\mathbf{x},t)}{\partial x_r}v_r + \psi(\mathbf{x},t)\frac{\partial v_r}{\partial x_r}\right\}dv$$

and equation (1.408).

Equation (1.454) can be also obtained in the following manner:

$$\left(\int_v \psi(\mathbf{x},t)dv\right)^{\bullet} = \left(\int_V \psi(\mathbf{X},t)JdV\right)^{\bullet} = \int_V (\dot{\psi}(\mathbf{X},t)J + \psi(\mathbf{X},t)\dot{J})dV$$
$$= \int_v \left\{\dot{\psi}(\mathbf{x},t) + \psi(\mathbf{x},t)\frac{\partial v_r}{\partial x_r}\right\}dv,$$

where V is the initial volume of the zone and we set $J \equiv dv/dV$. Here, it follows that $\dot{J} = J(\partial v_r/\partial x_r)$ (see equation (1.475)).

For the physical quantity ψ kept constant in a volume element, equation (1.454) leads to

$$\int_v (\dot{\psi}(\mathbf{x},t) + \psi(\mathbf{x},t)\text{div}\,\mathbf{v})dv = 0. \tag{1.455}$$

The *local form* of equation (1.455) is given by

$$\dot{\psi}(\mathbf{x},t) + \psi(\mathbf{x},t)\text{div}\,\mathbf{v} = 0, \tag{1.456}$$

while equation (1.455) is called the *weak form*.

1.10 Variations and Rates of Geometrical Elements

Variations of line, surface and volume elements and their rates under the deformation are described in this section.

1.10.1 Variations of Line, Surface and Volume

The relation of the current infinitesimal line element $d\mathbf{x}$ to the initial infinitesimal line element $d\mathbf{X}$ is given by

$$d\mathbf{x} = \mathbf{F}d\mathbf{X}, \quad d\mathbf{X} = \mathbf{F}^{-1}d\mathbf{x}, \tag{1.457}$$

where

$$\boxed{\mathbf{F} \equiv \frac{\partial \mathbf{x}}{\partial \mathbf{X}}} \tag{1.458}$$

which is referred to as the *deformation gradient*. The determinant of the deformation gradient is called the *Jacobian* and is denoted by

$$J \equiv \det\mathbf{F} = \det\left[\frac{\partial \mathbf{x}}{\partial \mathbf{X}}\right]. \tag{1.459}$$

For the deformation gradient tensor $\mathbf{F}$, we comment briefly on the indices in its transpose, inverse and transposed inverse. The deformation gradient is expressed by $\mathbf{F} = \partial\mathbf{x}/\partial\mathbf{X} = \partial x_i/\partial X_J \mathbf{e}_i \otimes \mathbf{e}_J (F_{iJ} = \partial x_i/\partial X_J)$, where $\mathbf{X}(= X_I\mathbf{e}_I)$ and $\mathbf{x}(= x_i\mathbf{e}_i)$ are the position vectors of a material particle in the reference and current states, respectively. The transposed tensor is given by $\mathbf{F}^T = (\partial x_j/\partial X_I \mathbf{e}_j \otimes e_I)^T = \partial x_j/\partial X_I \mathbf{e}_I \otimes \mathbf{e}_j = (\mathbf{F}^T)_{Ij}\mathbf{e}_I \otimes \mathbf{e}_j$, leading to $F^T_{Ij} \equiv (\mathbf{F}^T)_{Ij} = \partial x_j/\partial X_I$. On the other hand, the inversed tensor is given by $\mathbf{F}^{-1} = \partial\mathbf{X}/\partial\mathbf{x} = (\partial X_I/\partial x_j)\mathbf{e}_I \otimes \mathbf{e}_j = (\mathbf{F}^{-1})_{Ij}\mathbf{e}_I \otimes \mathbf{e}_j$, leading to $F^{-1}_{Ij} \equiv (\mathbf{F}^{-1})_{Ij} = \partial X_I/\partial x_j$. Further, the transposed inverse tensor is given by $\mathbf{F}^{-T} = (\partial\mathbf{X}/\partial\mathbf{x})^T = ((\partial X_J/\partial x_i)\mathbf{e}_J \otimes \mathbf{e}_i)^T = (\partial X_J/\partial x_i)\mathbf{e}_i \otimes \mathbf{e}_J = (\mathbf{F}^{-T})_{iJ}\mathbf{e}_i \otimes \mathbf{e}_J$, leading to $F^{-T}_{iJ} \equiv (\mathbf{F}^{-T})_{iJ} = \partial X_J/\partial x_i$.

Then, it is summarized that

$$F_{iJ} = \partial x_i/\partial X_J,\ F^T_{Ij} \equiv (\mathbf{F}^T)_{Ij} = \partial x_j/\partial X_I,\ F^{-1}_{Ij} \equiv (\mathbf{F}^{-1})_{Ij} = \partial X_I/\partial x_j,\ F^{-T}_{iJ} \equiv (\mathbf{F}^{-T})_{iJ} = \partial X_J/\partial x_i$$

The current and the reference infinitesimal volume elements dv and dV formed by the infinitesimal line elements $d\mathbf{x}^a$, $d\mathbf{x}^b$, $d\mathbf{x}^c$ and $d\mathbf{X}^a$, $d\mathbf{X}^b$, $d\mathbf{X}^c$ are related by

$$dv = [d\mathbf{x}^a d\mathbf{x}^b d\mathbf{x}^c] = [\mathbf{F}d\mathbf{X}^a\, \mathbf{F}d\mathbf{X}^b \mathbf{F}d\mathbf{X}^c] = \det\mathbf{F}[d\mathbf{X}^a d\mathbf{X}^b d\mathbf{X}^c] = (\det\mathbf{F})dV \tag{1.460}$$

by virtue of equation $(1.179)_3$. Then, the relation of the current and the reference infinitesimal volume elements is given by

$$\boxed{J = \det\mathbf{F} = \det\mathbf{U} = \det\mathbf{V} = \sqrt{\det\mathbf{C}} = \sqrt{\det\mathbf{b}} = \frac{dv}{dV} = \frac{\rho_0}{\rho}} \tag{1.461}$$

noting equation (1.159), while $\mathbf{U} \equiv \sqrt{\mathbf{F}^T\mathbf{F}}$, $\mathbf{V} \equiv \sqrt{\mathbf{F}\mathbf{F}^T}$, $\mathbf{C} \equiv \mathbf{F}^T\mathbf{F}$ and $\mathbf{b} \equiv \mathbf{F}\mathbf{F}^T$ are described in Chapter 4. ρ_0 and ρ are the densities in the reference and current configurations, respectively.

The following equalities hold for the infinitesimal volume elements, denoting the infinitesimal reference and current surface element vectors as $d\mathbf{A} = d\mathbf{X}^a \times d\mathbf{X}^b$ and $d\mathbf{a} = d\mathbf{x}^a \times d\mathbf{x}^b$, respectively, and noting equation (1.93):

$$dv = \begin{cases} d\mathbf{a} \cdot d\mathbf{x}^c, \\ JdV = Jd\mathbf{A} \cdot d\mathbf{X}^c = Jd\mathbf{A} \cdot \mathbf{F}^{-1}d\mathbf{x}^c = J\mathbf{F}^{-T}d\mathbf{A} \cdot d\mathbf{x}^c. \end{cases}$$

Then the following relation holds:

$$\boxed{d\mathbf{a} = J\mathbf{F}^{-T}d\mathbf{A} = (\mathrm{cof}\,\mathbf{F})d\mathbf{A}, \quad d\mathbf{A} = \mathbf{F}^{T}d\mathbf{a}/J = (\mathrm{cof}\,\mathbf{F})^{-1}d\mathbf{a}} \tag{1.462}$$

noting that

$$\mathrm{cof}\,\mathbf{F} = J\mathbf{F}^{-T} \tag{1.463}$$

by virtue of equation (1.157), leading to

$$da = \mathbf{F}^{-T}\mathbf{N}\cdot\mathbf{n}J dA, \quad dA = \mathbf{F}^{T}\mathbf{n}\cdot\mathbf{N}da/J, \tag{1.464}$$

$$\mathbf{n} = \mathbf{F}^{-T}\mathbf{N}J dA/da, \quad \mathbf{N} = \mathbf{F}^{T}\mathbf{n}da/(dA/J). \tag{1.465}$$

Equation (1.464) is referred to as *Nanson's formula.*

Here, the following *Euler formula* holds for the cofactor

$$\nabla_X(\mathrm{cof}\,\mathbf{F}) = \frac{\partial\,\mathrm{cof}\,\mathbf{F}}{\partial\mathbf{X}} = \mathbf{0}, \qquad \frac{\partial(\mathrm{cof}\,\mathbf{F})_{ap}}{\partial X_p} = \frac{\partial\frac{1}{2}\varepsilon_{abc}\varepsilon_{pqr}F_{bq}F_{cr}}{\partial X_p} = 0, \tag{1.466}$$

noting equation (1.153) and that X_p is not contained in the deformation gradient.

Denote an arbitrary physical quantity in the reference and current states by T and t, respectively. Their relation can be written as

$$t(\mathbf{x},t)d\mathbf{a}(\mathbf{x},t) = T(\mathbf{X})d\mathbf{A}(\mathbf{X}). \tag{1.467}$$

The substitution of the Nanson's formula in equation (1.464) into equation (1.467) yields *Piola's transformation formula*

$$T = (\mathrm{cof}\,\mathbf{F})t. \tag{1.468}$$

Here, we have

$$\frac{\partial}{\partial\mathbf{x}} = \mathbf{F}^{-T}\frac{\partial}{\partial\mathbf{X}} \quad (\nabla_x = \mathbf{F}^{-T}\nabla_X), \tag{1.469}$$

noting that

$$\frac{\partial}{\partial x_i} = \frac{\partial}{\partial X_R}\frac{\partial X_R}{\partial x_i} = \frac{\partial X_R}{\partial x_i}\frac{\partial}{\partial X_R}.$$

It follows from equations (1.463), (1.466), (1.468) and (1.469) that

$$\frac{\partial T}{\partial\mathbf{X}} = J\frac{\partial t}{\partial\mathbf{x}}, \tag{1.470}$$

noting that

$$\frac{\partial T}{\partial\mathbf{X}} = \frac{\partial\{(\mathrm{cof}\,\mathbf{F})t\}}{\partial\mathbf{X}} = (\mathrm{cof}\,\mathbf{F})\frac{\partial t}{\partial\mathbf{X}} = J\mathbf{F}^{-T}\frac{\partial t}{\partial\mathbf{X}}.$$

Equation (1.470), which we will need in Chapter 6, is referred to as *Piola's formula.*

1.10.2 Rates of Changes of Surface and Volume

Differentiating equation (1.457), we have

$$(d\mathbf{x})^{\bullet} = \dot{\mathbf{F}}d\mathbf{X}, \quad (d\mathbf{x})^{\bullet} = \boldsymbol{l}d\mathbf{x}, \tag{1.471}$$

where

$$\boxed{\boldsymbol{l} = \dot{\mathbf{F}}\mathbf{F}^{-1} = \frac{\partial \mathbf{v}}{\partial \mathbf{x}}} \tag{1.472}$$

which is called the *velocity gradient*.

It follows from equations (1.434) and (1.461) that

$$(dv)^{\bullet} = (\mathrm{tr}\boldsymbol{l})dv = \dot{J}dV, \tag{1.473}$$

$$\mathrm{tr}\boldsymbol{l} = \mathrm{div}\,\mathbf{v}, \tag{1.474}$$

$$\boxed{\dot{J} = J\mathrm{tr}\boldsymbol{l}} \tag{1.475}$$

noting

$$\dot{J} = (\det\mathbf{F})^{\bullet} = \frac{\partial \det\mathbf{F}}{\partial \mathbf{F}} : \dot{\mathbf{F}} = (\det\mathbf{F})\mathbf{F}^{-T} : \dot{\mathbf{F}} = (\det\mathbf{F})\mathrm{tr}(\mathbf{F}^{-1}\dot{\mathbf{F}}) = (\det\mathbf{F})\mathrm{tr}(\dot{\mathbf{F}}\mathbf{F}^{-1})$$

with equation (1.371).

Further, it follows from equation (1.463) that

$$(\mathrm{cof}\,\mathbf{F})^{\bullet} = \tilde{\boldsymbol{l}}\mathrm{cof}\,\mathbf{F}, \tag{1.476}$$

where

$$\tilde{\boldsymbol{l}} \equiv (\mathrm{tr}\boldsymbol{l})\mathbf{I} - \boldsymbol{l}^{T}, \tag{1.477}$$

noting that

$$(\mathrm{cof}\,\mathbf{F})^{\bullet} = \dot{J}\mathbf{F}^{-T} + \dot{\mathbf{F}}^{-T}J = \frac{\dot{J}}{J}\mathrm{cof}\,\mathbf{F} + \dot{\mathbf{F}}^{-T}\mathbf{F}^{T}\mathrm{cof}\,\mathbf{F} = \frac{\dot{J}}{J}\mathrm{cof}\,\mathbf{F} - \mathbf{F}^{-T}\dot{\mathbf{F}}^{T}\mathrm{cof}\,\mathbf{F}.$$

$\tilde{\boldsymbol{l}}$ is referred to as the *surface strain rate tensor*. Further, differentiating equation (1.462) and noting equation (1.476), we have

$$(d\mathbf{a})^{\bullet} = (\mathrm{cof}\,\mathbf{F})^{\bullet}d\mathbf{A} = \tilde{\boldsymbol{l}}(\mathrm{cof}\,\mathbf{F})d\mathbf{A}$$

which is rewritten as

$$\boxed{(d\mathbf{a})^{\bullet} = \tilde{\boldsymbol{l}}d\mathbf{a}} \tag{1.478}$$

in the spatial description.

We now express the surface vectors as follows:

$$d\mathbf{a} \equiv \mathbf{n}da, \quad d\mathbf{A} \equiv \mathbf{N}dA, \tag{1.479}$$

where da and dA are infinitesimal areas and $\mathbf{n}$ and $\mathbf{N}$ are unit normal vectors of the current and the reference infinitesimal surface vectors, respectively.

Here, noting $\dot{\mathbf{n}} \cdot \mathbf{n} = 0$ from $\mathbf{n} \cdot \mathbf{n} = 1$ for the unit vector $\mathbf{n}$, it follows that

$$(da)^{\bullet} = \mathbf{n} \cdot \mathbf{n}(da)^{\bullet} = \mathbf{n} \cdot \{(\mathbf{n}da)^{\bullet} - \dot{\mathbf{n}}da\} = \mathbf{n} \cdot (d\mathbf{a})^{\bullet}. \tag{1.480}$$

Substituting equation (1.478) into equation (1.480), one obtains the rate of the current infinitesimal area as follows:

$$\boxed{(da)^{\bullet} = \mathbf{n} \cdot \tilde{\boldsymbol{l}}d\mathbf{a}} \tag{1.481}$$

or

$$(da)^{\bullet} = (\mathrm{tr}\boldsymbol{l} - \mathbf{n} \cdot \mathbf{dn})da, \tag{1.482}$$

where

$$\mathbf{d} \equiv \mathrm{sym}[\boldsymbol{l}], \tag{1.483}$$

which is called the *strain rate*, noting equation (1.93). Further, it follows from equations (1.477), (1.481) and (1.482) that

$$\begin{aligned}\dot{\mathbf{n}}da &= (\mathbf{n}da)^{\bullet} - \mathbf{n}(da)^{\bullet}\\ &= \{(\mathrm{tr}\boldsymbol{l})\mathbf{I} - \boldsymbol{l}^T\}\mathbf{n}da - \mathbf{n}\{(\mathrm{tr}\boldsymbol{l}) - \mathbf{n} \cdot \mathbf{dn}\}da.\end{aligned}$$

Then the rate of the unit normal of the current surface element is given by

$$\dot{\mathbf{n}} = \{(\mathbf{n} \cdot \mathbf{dn})\mathbf{I} - \boldsymbol{l}^T\}\mathbf{n} = \{(\mathbf{n} \cdot \boldsymbol{l}\mathbf{n})\mathbf{I} - \boldsymbol{l}^T\}\mathbf{n}. \tag{1.484}$$

The variations and rates of line, surface and volume elements are summarized in Table 1.3.

Table 1.3 Variations and rates of line, surface and volume elements

	Variations from reference to current states	**Rates**	
Elements		**Elements**	**Deformation gradient**
Line element	$d\mathbf{x} = \mathbf{F}d\mathbf{X}$	$(d\mathbf{x})^{\bullet} = \dot{\mathbf{F}}d\mathbf{X}$ $(d\mathbf{x})^{\bullet} = \boldsymbol{l}d\mathbf{x}$	$\dot{\mathbf{F}} = \boldsymbol{l}\mathbf{F}$
Surface element	$d\mathbf{a} = (\mathrm{cof}\,\mathbf{F})d\mathbf{A}$	$(d\mathbf{a})^{\bullet} = (\mathrm{cof}\,\mathbf{F})^{\bullet}d\mathbf{A}$ $(d\mathbf{a})^{\bullet} = \tilde{\boldsymbol{l}}d\mathbf{a}$ $(da)^{\bullet} = (\mathrm{tr}\boldsymbol{l} - \mathbf{n} \cdot \boldsymbol{l}\mathbf{n})da$ $\dot{\mathbf{n}} = \{(\mathbf{n} \cdot \boldsymbol{l}\mathbf{n})\mathbf{I} - \boldsymbol{l}^T\}\mathbf{n}$	$(\mathrm{cof}\,\mathbf{F})^{\bullet} = \tilde{\boldsymbol{l}}\mathrm{cof}\,\mathbf{F}$
Volume element	$dv = (\det\mathbf{F})dV$	$(dv)^{\bullet} = (\det\mathbf{F})^{\bullet}dV = (\mathrm{tr}\boldsymbol{l})dv$ $\dot{J} = J\mathrm{tr}\boldsymbol{l} = \dfrac{(dv)^{\bullet}}{dV}$	$(\det\mathbf{F})^{\bullet} = \mathrm{tr}\boldsymbol{l}\det\mathbf{F}$
$\mathbf{F} \equiv \dfrac{\partial \mathbf{x}}{\partial \mathbf{X}}$, $J \equiv \det\mathbf{F} = \dfrac{dv}{dV}$, $\mathrm{cof}\,\mathbf{F} = (\det\mathbf{F})\mathbf{F}^{-T} = J\mathbf{F}^{-T}$ $\boldsymbol{l} \equiv \dot{\mathbf{F}}\mathbf{F}^{-1}$, $\tilde{\boldsymbol{l}} \equiv (\mathrm{tr}\boldsymbol{l})\mathbf{I} - \boldsymbol{l}^T$			

1.11 Continuity and Smoothness Conditions

There exist various mechanical requirements, for example the thermodynamic restriction and the objectivity for constitutive equations. In addition, there exist the mechanical requirements observed in solid materials. Among them, the continuity and the smoothness conditions are violated in many elasto-plasticity models, while their importance for formulation of constitutive equations has not been sufficiently recognized to date. These requirements will be explained below (Hashiguchi, 1993a, 1993b, 1997, 2000).

1.11.1 Continuity Condition

It is observed in experiments that stress rate changes continuously for a continuous change of strain rate. This fact is called the *continuity condition* and is expressed mathematically as follows (Hashiguchi, 1993a, 1993b, 1997, 2000):

$$\boxed{\lim_{\delta \mathbf{d} \to \mathbf{0}} \overset{\circ}{\boldsymbol{\sigma}}(\boldsymbol{\sigma}, H_i; \mathbf{d} + \delta \mathbf{d}) = \overset{\circ}{\boldsymbol{\sigma}}(\boldsymbol{\sigma}, H_i; \mathbf{d})} \tag{1.485}$$

where H_i $(i = 1, 2, 3, \ldots)$ collectively denotes scalar-valued or tensor-valued internal state variables. In addition, $\delta(\cdot)$ stands for an infinitesimal variation. The response of the stress rate to the input of strain rate in the current state of stress and internal variables is denoted by $\overset{\circ}{\boldsymbol{\sigma}}(\boldsymbol{\sigma}, H_i; \mathbf{d})$. *Uniqueness* of the solution is not guaranteed in constitutive equations violating the continuity condition, predicting different stresses or strains. The violation of this condition is schematically shown in Figure 1.8. Ordinary elasto-plastic constitutive equations, in which the plastic strain rate is derived from the consistency condition, fulfill the continuity condition. As will be described later, however, no elasto-plastic constitutive equation satisfies it except for the subloading surface model when they are extended to describe the tangential inelastic strain rate (Hashiguchi, 2009).

The concept of the continuity condition was first advocated by Prager (1949). However, a mathematical expression for this condition was not given. The condition was defined as the continuity of strain rate with respect to the input of stress rate by Prager (1949) inversely to the definition given above. However, an identical stress rate directed into the yield surface can induce different strain rates in loading and unloading states in a softening material, as

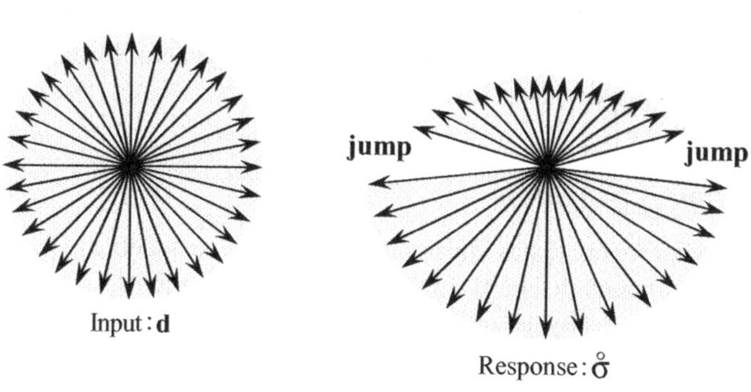

Figure 1.8 Violation of continuity condition

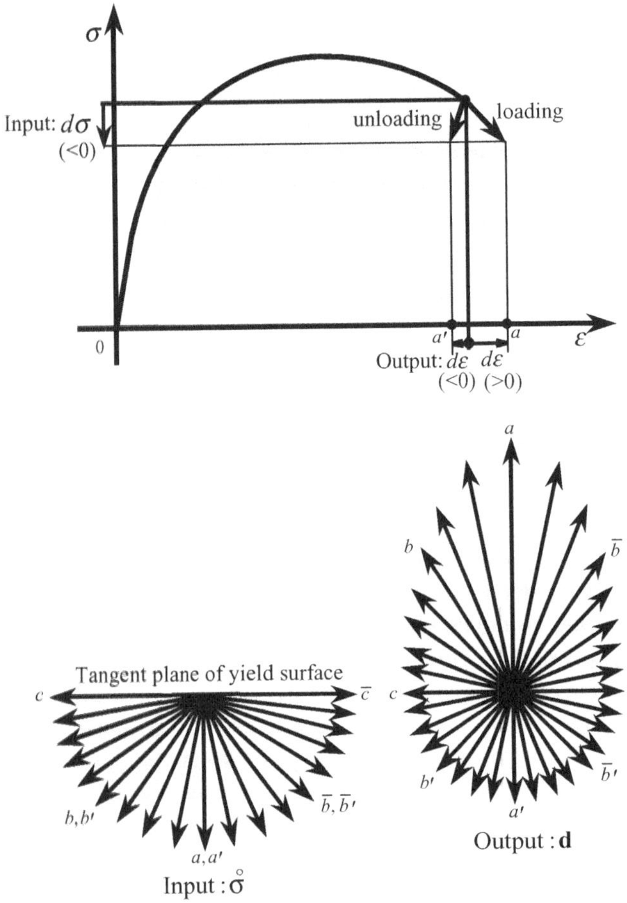

Figure 1.9 Impertinence of Prager 's(1949) continuity condition (illustrated for softening state)

shown in Figure 1.9. Here, it is noteworthy that a stress rate cannot be given arbitrarily since there exists a limitation in strength of materials, although a strain rate can be given arbitrarily. For that reason, Prager's (1949) notion does not hold in the general loading state, including softening and perfectly plastic states.

1.11.2 Smoothness Condition

It is observed in experiments that a stress rate induced by an identical strain rate changes continuously for a continuous change of stress state. This fact is called the *smoothness condition* and is expressed mathematically as follows (Hashiguchi, 1993a, 1993b, 1997, 2000):

$$\boxed{\lim_{\delta\boldsymbol{\sigma}\to\mathbf{0}} \overset{\circ}{\boldsymbol{\sigma}}(\boldsymbol{\sigma}+\delta\boldsymbol{\sigma}, H_i; \mathbf{d}) = \overset{\circ}{\boldsymbol{\sigma}}(\boldsymbol{\sigma}, H_i; \mathbf{d})} \tag{1.486}$$

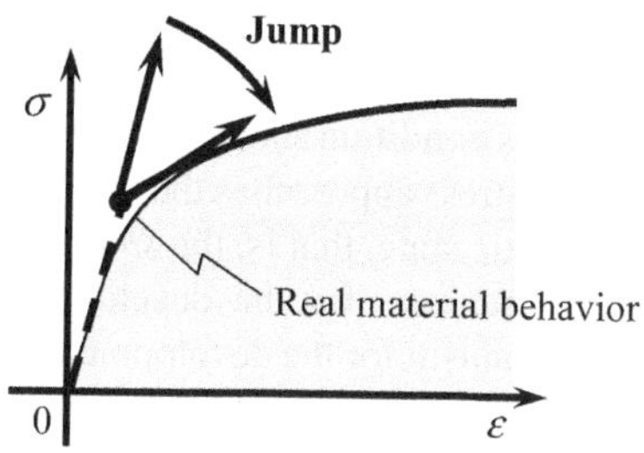

Figure 1.10 Violation of smoothness condition in conventional plasticity model and kinematic hardening cyclic plasticity models assuming a purely elastic domain

A smooth stress–strain response is not described in a constitutive equation violating the smoothness condition, causing a discontinuous change in the tangent modulus, as illustrated in Figure 1.10 for the elasto-plastic constitutive equations assuming the yield surface enclosing a purely elastic domain where the tangent modulus changes abruptly from the elastic to the elasto-plastic state.

The linear-rate constitutive equation is given by

$$\overset{\circ}{\boldsymbol{\sigma}} = \mathbf{M}^{ep}(\boldsymbol{\sigma}, H_i)\mathbf{d}, \tag{1.487}$$

where the fourth-order tensor $\mathbf{M}^{ep}$ is the elasto-plastic modulus, which is a function of the stress and internal variables and can be described generally as

$$\mathbf{M}^{ep} = \frac{\partial \overset{\circ}{\boldsymbol{\sigma}}}{\partial \mathbf{d}}. \tag{1.488}$$

Consequently, equation (1.486) can be rewritten as

$$\boxed{\lim_{\delta\boldsymbol{\sigma}\to\mathbf{0}} \mathbf{M}^{ep}(\boldsymbol{\sigma} + \delta\boldsymbol{\sigma}, H_i) = \mathbf{M}^{ep}(\boldsymbol{\sigma}, H_i)} \tag{1.489}$$

1.12 Unconventional Elasto-Plasticity Models

Ordinary elasto-plastic constitutive equations fulfill the continuity condition, in which plastic strain rate is derived from the consistency condition of yield condition. They are classified into conventional and the unconventional models by Drucker (1988). The interior of the yield surface is assumed to be a purely elastic domain in conventional models. In unconventional models, on the other hand, the interior of the yield surface is not assumed to be a purely elastic domain and thus the plastic strain rate induced by the rate of stress inside the yield surface is described. Unconventional models are classified further into models based on the notion of kinematic hardening, that is, the *multi-surface model* (Mroz, 1967; Iwan, 1967), the *two-surface model* (Dafalias and Popov, 1975; Krieg, 1975) and the *superposed kinematic hardening-single surface model* (Chaboche et al., 1979; Chaboche, 2008) based on the non-linear kinematic hardening rule (Armstrong and Frederick, 1966), and the model based on the notion that the plastic strain rate develops as a stress approaches the yield surface, that is, the *subloading surface model* (Hashiguchi, 1980, 1989). The former models based on the kinematic hardening models assume a small yield surface enclosing a purely lastic domain and thus they violate the smoothness condition. On the other hand, only the subloading surface model always satisfies the smoothness condition since it is based on quite natural postulate that the plastic strain rate develops as the stress approaches the yield surface, describing the smooth transition from the elastic to the plastic state, that is, the *smooth elastic–plastic transition*.

It should be noted that the mechanisms for the development of plastic deformation is substantially different from the mechanism for the development of kinematic hardening which is merely one of various mechanical behavior observed only in plastically-pressure independent metals. Here, remind that the kinematic hardening (Prager, 1956; Armstrong and Frederick, 1966) is merely the simple method proposed primarily to describe the induced anisotropy of plastically-pressure independent metals, while it is inapplicable to the description of the anisotropy of the plastically-pressure dependent, that is, frictional materials (for example, soils, rocks, concretes and friction behavior) to which the rotational hardening rule (Hashiguchi, 2001) has to be adopted. Furthermore, accumulation of plastic deformation cannot be predicted

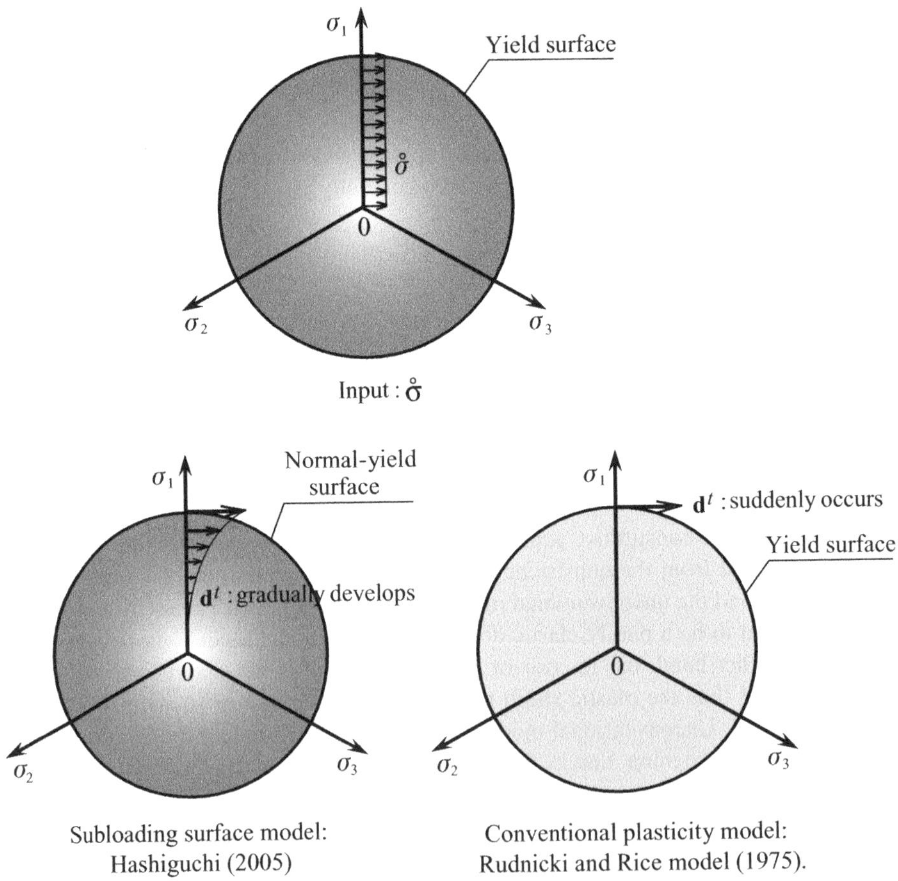

Figure 1.11 Incorporation of tangential inelastic strain rate

for cyclic loading of stress inside a small yield surface even at a high stress level near the overall yield state by the unconventional models based on the kinematic hardening, since a small yield surface enclosing a purely-elastic domain is assumed in them. Such cyclic loading is observed frequently in the practical engineering phenomena so that the risky mechanical prediction is given by the unconventional models based on the kinematic hardening rule.

Constitutive equations violating the smoothness condition cannot predict a smooth stress–strain curve. Therefore, they cannot pertinently describe softening behavior. In addition, the models violating the smoothness condition violate not only the smoothness but also the continuity conditions, predicting the sudden occurrence of tangential-inelastic strain rate, leading to the violation of the unique solution condition, when the tangential inelastic strain rate $\mathbf{d}^t$ (induced by the stress rate tangential to the yield surface) required for the analysis of plastic instability phenomena is incorporated (cf. Hashiguchi and Tsutsumi, 2003, 2007; Hashiguchi and Protasov, 2004; Hashiguchi, 2009) as shown in Figure 1.11. Among the existing constitutive models only the subloading surface model always fulfills the smoothness condition. The subloading surface model is capable of describing the wide classes of elasto-plastic deformation of metals (Hashiguchi et al., 2012), soils (Hashiguchi and Chen, 1998; Hashiguchi and Mase, 2007) and further sliding behavior between solids, that is, the friction phenomenon (Hashiguchi et al., 2005; Hashiguchi and Ozaki, 2008) with high accuracy. In addition, the subloading surface model does not require the judement of yielding, that is, whether or not the yield condition is satisfied and possesses the automatic controlling function to attract the stress to the yield surface in a plastic deformation process, so that numerical analyses would be enforced drastically as will be described in Chapter 8.

Finally, note that the superposed kinematic hardening-single surface model (Chaboche et al., 1979; Chaboche, 2008) nowadays is in fashion in the application to the deformation analysis of metals, and incorporated into commercial finite element method software, because it possesses quite primitive mathematical structures involving a lot of material parameters with ambiguous meanings. It can even be used by the beginners without expertise knowledge of elasto-plasticity. However, it lacks generality as it is applicable only to the limited deformation behavior of metals, possessing various insufficiencies in addition to the above-mentioned points. For instance, it abandoned the description of the stagnation of isotropic hardening during a strain cyclic loading observed widely in metals (Chaboche et al., 1979; Hashiguchi et al. 2012). Besides, the yield surface of geomaterials contains the origin of the stress space near its boundary and thus it goes away easily from the origin by the application of deviatoric stress or pressure if it translates, resulting in the invalidity. Needless to say, it cannot be applied to the description of the friction phenomenon. One should recognize these fundamental drawbacks for the steady development of elastoplasticity theory from the scientific aspect.

2

General (Curvilinear) Coordinate System

Knowledge of the curvilinear coordinate system is not required merely to apply finite elasto-plasticity theory since finite element analysis is usually performed in the rectangular coordinate system; therefore the curvilinear coordinate system is not considered in Chapters 9 to 10 on finite element analysis. However, the physical meanings of finite strain (rate) tensors, Lagrangian and Eulerian tensors, the pull-back and push-forward operations, the general objective derivatives, etc. appearing in the finite strain theory can be captured exactly as the geometrical variation of the coordinate system embedded in material itself. The embedded (convected) coordinate system changes to a curvilinear coordinate system even if the rectangular coordinate system is adopted at the initial state of the deformation process. Thus the mathematics for the general curvilinear coordinate system is addressed in this chapter before turning to the description of deformation in the convected coordinate system in later chapters.

2.1 Primary and Reciprocal Base Vectors

Let us consider the general curvilinear coordinate system $(\theta^1, \theta^2, \theta^3)$. The infinitesimal line element $d\mathbf{x}$ at the material point vector $\mathbf{x}$ is described as follows:

$$d\mathbf{x} = \frac{\partial \mathbf{x}}{\partial \theta^i} d\theta^i = \mathbf{a}_i d\theta^i, \tag{2.1}$$

where $\mathbf{a}_i$ is defined locally by the equation

$$\boxed{\mathbf{a}_i \equiv \frac{\partial \mathbf{x}}{\partial \theta^i}} \tag{2.2}$$

and is called the *natural base* or *primary base vector*. $\mathbf{a}_i$ describes the change in vector $\mathbf{x}$ per unit change along the coordinate axis θ^i and thus it is no more than the base vector of the coordinate axis which can be defined only locally (see Figure 2.1).

Introduction to Finite Strain Theory for Continuum Elasto-Plasticity, First Edition. Koichi Hashiguchi and Yuki Yamakawa.

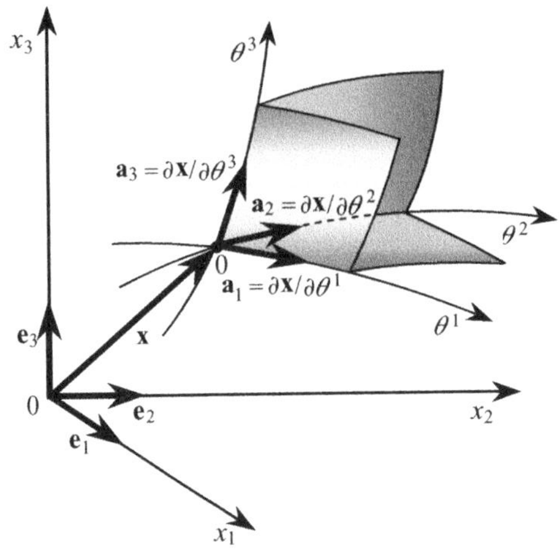

Figure 2.1 Curvilinear coordinate system $\{\theta^i\}$ with covariant base $\{\mathbf{a}_i\}$

Now, consider another curvilinear coordinate system $(\bar{\theta}^1, \bar{\theta}^2, \bar{\theta}^3)$ with the base $(\bar{\mathbf{a}}_1, \bar{\mathbf{a}}_2, \bar{\mathbf{a}}_3)$. The following coordinate transformation rule holds by the chain rule of differentiation:

$$\bar{\mathbf{a}}_i = \frac{\partial \mathbf{x}}{\partial \bar{\theta}^i} = \frac{\partial \mathbf{x}}{\partial \theta^j}\frac{\partial \theta^j}{\partial \bar{\theta}^i} = \frac{\partial \theta^j}{\partial \bar{\theta}^i}\mathbf{a}_j, \tag{2.3}$$

the inverse relation of which is given by

$$\mathbf{a}_i = \frac{\partial \bar{\theta}^j}{\partial \theta^i}\bar{\mathbf{a}}_j \left(= \frac{\partial \bar{\theta}^j}{\partial \theta^i}\frac{\partial \theta^r}{\partial \bar{\theta}^j}\mathbf{a}_r = \delta_i^r \mathbf{a}_r \right). \tag{2.4}$$

Vectors and tensors observed by the coordinate system $(\bar{\theta}^1, \bar{\theta}^2, \bar{\theta}^3)$ are denoted by adding an overbar (¯) for simplicity; elsewhere in the book they will be denoted by adding an asterisk.

Now, incorporate the reciprocal base vectors

$$\boxed{\mathbf{a}^i \equiv \frac{1}{2}\varepsilon_{ijk}\frac{\mathbf{a}_j \times \mathbf{a}_k}{V}}, \text{ that is, } \quad \mathbf{a}^1 \equiv \frac{\mathbf{a}_2 \times \mathbf{a}_3}{[\mathbf{a}_1\mathbf{a}_2\mathbf{a}_3]}, \quad \mathbf{a}^2 \equiv \frac{\mathbf{a}_3 \times \mathbf{a}_1}{[\mathbf{a}_1\mathbf{a}_2\mathbf{a}_3]}, \quad \mathbf{a}^3 \equiv \frac{\mathbf{a}_1 \times \mathbf{a}_2}{[\mathbf{a}_1\mathbf{a}_2\mathbf{a}_3]}, \tag{2.5}$$

noting equation (1.59), where V is the volume of parallelepiped made up of $\mathbf{a}_1, \mathbf{a}_2, \mathbf{a}_3$ in the right-hand coordinate system, that is,

$$V \equiv [\mathbf{a}_1\mathbf{a}_2\mathbf{a}_3] = \varepsilon_{pqr}(\mathbf{a}_1 \cdot \mathbf{e}_p)(\mathbf{a}_2 \cdot \mathbf{e}_q)(\mathbf{a}_3 \cdot \mathbf{e}_r). \tag{2.6}$$

The vectors $\mathbf{a}^1, \mathbf{a}^2, \mathbf{a}^3$ are nothing more than the reciprocal vectors of $\mathbf{a}_1, \mathbf{a}_2, \mathbf{a}_3$. Reciprocally, it follows that

$$\mathbf{a}_i \equiv \frac{1}{2}\varepsilon_{ijk} V \mathbf{a}^j \times \mathbf{a}^k, \text{ that is, } \mathbf{a}_1 \equiv \frac{\mathbf{a}^2 \times \mathbf{a}^3}{[\mathbf{a}^1\mathbf{a}^2\mathbf{a}^3]}, \quad \mathbf{a}_2 \equiv \frac{\mathbf{a}^3 \times \mathbf{a}^1}{[\mathbf{a}^1\mathbf{a}^2\mathbf{a}^3]}, \quad \mathbf{a}_3 \equiv \frac{\mathbf{a}^1 \times \mathbf{a}^2}{[\mathbf{a}^1\mathbf{a}^2\mathbf{a}^3]}, \tag{2.7}$$

noting equation (1.60), where

$$[\mathbf{a}_1\mathbf{a}_2\mathbf{a}_3][\mathbf{a}^1\mathbf{a}^2\mathbf{a}^3] = 1 \tag{2.8}$$

by equation (1.63), leading to

$$\boxed{[\mathbf{a}^1\mathbf{a}^2\mathbf{a}^3] = \frac{1}{V}} \tag{2.9}$$

It follows from equations (1.65) and (2.5) or (2.7) that

$$\mathbf{a}_i \cdot \mathbf{a}^j = \delta_i^j, \quad \mathbf{a}_i \cdot \mathbf{a}^i = 3, \tag{2.10}$$

$$\mathbf{a}_i \times \mathbf{a}^i = \mathbf{0}, \tag{2.11}$$

noting equations (1.61), (1.62) and (1.65).

The reciprocal base vector can be described as

$$\boxed{\mathbf{a}^i = \frac{\partial \theta^i}{\partial \mathbf{x}}} \tag{2.12}$$

by solving the equation

$$(\mathbf{a}_i \cdot \mathbf{a}^j =) \frac{\partial \mathbf{x}}{\partial \theta^i} \cdot \mathbf{a}^j = \delta_i^j$$

which is obtained by substituting equation (2.2) into equation (2.10).

Note that the coordinate system with base $\{\mathbf{a}^i\}$ formed by the reciprocal vectors does not exist, since it would not satisfy the relation (2.10) between the primary and its reciprocal base vectors. The reciprocal base vector $\mathbf{a}^k$ is always perpendicular to the coordinate plane formed by the primary base vectors $(\mathbf{a}_i, \mathbf{a}_j)(k \neq i, k \neq j)$. The reciprocal base vectors $\mathbf{a}^3$ is illustrated in Figure 2.2(a) and the relations of the base vectors are illustrated in Figure 2.2(b) in a two-dimensional representation where the base vectors $\mathbf{a}_3$ and $\mathbf{a}^3$ are taken to be perpendicular to the coordinate plane $(\mathbf{a}_1, \mathbf{a}_2)$ or $(\mathbf{a}^1, \mathbf{a}^2)$.

In the similar way to equations (2.3) and (2.4), the following transformation rule for the coordinate system is obtained:

$$\bar{\mathbf{a}}^i = \frac{\partial \bar{\theta}^i}{\partial \theta^j}\mathbf{a}^j, \quad \mathbf{a}^i = \frac{\partial \theta^i}{\partial \bar{\theta}^j}\bar{\mathbf{a}}^j \left(= \frac{\partial \theta^i}{\partial \bar{\theta}^j}\frac{\partial \bar{\theta}^j}{\partial \theta^r}\mathbf{a}^r = \delta_r^i \mathbf{a}^r \right). \tag{2.13}$$

In accordance with the transformation rules in equations (2.3), (2.4) and (2.13), $\mathbf{a}_i$ and $\mathbf{a}^i$ are called the *covariant base vector* and the *contravariant base vector*, respectively.

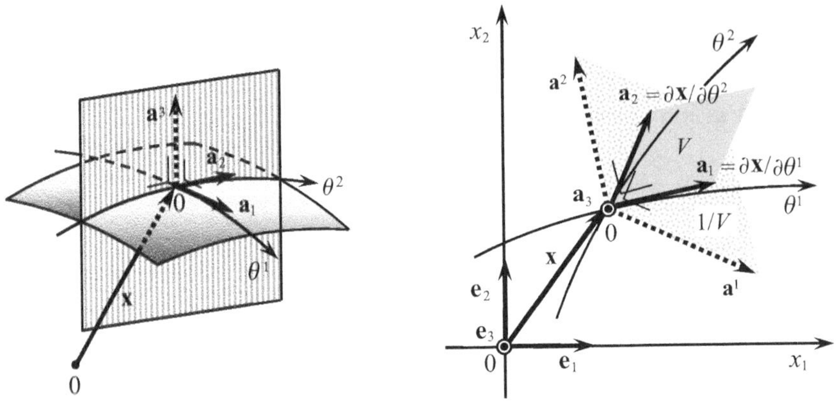

(a) Reciprocal base vector illustrated for $\mathbf{a}^3$.

(b) Two-dimensional representation. $(\mathbf{a}_3/\|\mathbf{a}_3\| = \mathbf{a}^3/\|\mathbf{a}^3\|)$

Figure 2.2 Primary and reciprocal base vectors and curvilinear coordinate system

Inserting the equation $\mathbf{x} = x_r\mathbf{e}_r$ in the rectangular coordinate system $\{\mathbf{e}_i\}$ into equations (2.2) and (2.12), the relation of $\mathbf{a}_i$ and $\mathbf{e}_r$ is given by

$$\left.\begin{aligned} \mathbf{a}_i &= \frac{\partial(x_r\mathbf{e}_r)}{\partial\theta^i} = \frac{\partial x_r}{\partial\theta^i}\mathbf{e}_r, \ \mathbf{e}_i = \frac{\partial\theta^r}{\partial x_i}\mathbf{a}_r \\ \mathbf{a}^i &= \frac{\partial\theta^i}{\partial(x_r\mathbf{e}_r)} = \frac{\partial\theta^i}{\partial x_r}\mathbf{e}_r, \ \mathbf{e}_i = \frac{\partial x_i}{\partial\theta^r}\mathbf{a}^r \end{aligned}\right\} \tag{2.14}$$

which is written as

$$\left.\begin{aligned} &\mathbf{a}_i = \underset{\sim}{a}_{ir}\mathbf{e}_r, \ \mathbf{a}^i = \tilde{\underset{\sim}{a}}^{ir}\mathbf{e}_r \\ &\mathbf{e}_i = \tilde{\underset{\sim}{a}}^{ri}\mathbf{a}_r = \underset{\sim}{a}_{ri}\mathbf{a}^r \end{aligned}\right\}, \tag{2.15}$$

adopting the notation

$$\left.\begin{aligned} &\tilde{\underset{\sim}{a}}^{ir} \equiv \mathbf{a}^i \cdot \mathbf{e}_r = \frac{\partial\theta^i}{\partial x_r}, \ \underset{\sim}{a}_{ir} \equiv \mathbf{a}_i \cdot \mathbf{e}_r = \frac{\partial x_r}{\partial\theta^i} \\ &\tilde{\underset{\sim}{a}}^{ir}\underset{\sim}{a}_{jr} = \delta^i_j, \ \underset{\sim}{a}_{ir}\tilde{\underset{\sim}{a}}^{jr} = \delta^j_i \end{aligned}\right\}. \tag{2.16}$$

Further, introducing the second-order tensor

$$\left.\begin{aligned} &\underset{\sim}{\mathbf{a}} \equiv \mathbf{a}_i \otimes \mathbf{e}_i = \underset{\sim}{a}_{ir}\mathbf{e}_r \otimes \mathbf{e}_i \\ &\underset{\sim}{\mathbf{a}}^{-1} = \mathbf{e}_i \otimes \mathbf{a}^i = \tilde{\underset{\sim}{a}}^{ir}\mathbf{e}_i \otimes \mathbf{e}_r \end{aligned}\right\}, \tag{2.17}$$

we may write

$$\left.\begin{aligned} &\mathbf{a}_i = \underset{\sim}{\mathbf{a}}\mathbf{e}_i, \quad \mathbf{e}_i = \underset{\sim}{\mathbf{a}}^{-1}\mathbf{a}_i \\ &\mathbf{a}^i = \mathbf{e}_i\underset{\sim}{\mathbf{a}}^{-1}, \quad \mathbf{e}_i = \mathbf{a}^i\underset{\sim}{\mathbf{a}} \end{aligned}\right\}. \tag{2.18}$$

2.2 Metric Tensors

We now introduce the scalar quantities

$$\left.\begin{aligned}
a^{ij} &\equiv \mathbf{a}^i \cdot \mathbf{a}^j = \frac{\partial\theta^i}{\partial x_r}\frac{\partial\theta^j}{\partial x_r} \\
a^i_{\bullet j} &\equiv \mathbf{a}^i \cdot \mathbf{a}_j = \frac{\partial\theta^i}{\partial x_r}\frac{\partial x_r}{\partial\theta^j} = \delta^i_j \\
a_i^{\bullet j} &\equiv \mathbf{a}_i \cdot \mathbf{a}^j = \frac{\partial x_r}{\partial\theta^i}\frac{\partial\theta^j}{\partial x_r} = \delta_i^j \\
a_{ij} &\equiv \mathbf{a}_i \cdot \mathbf{a}_j = \frac{\partial x_r}{\partial\theta^i}\frac{\partial x_r}{\partial\theta^j}
\end{aligned}\right\}, \tag{2.19}$$

noting equations (2.14) (also cf. equation $(2.10)_1$), where the symmetries $a_{ij} = a_{ji}$, $a^{ij} = a^{ji}$, $a^i_{\bullet j} = a^j_{\bullet i} = a_i^{\bullet j} = a_j^{\bullet i}$ hold. It follows from equation (2.19) that

$$a^{ir}a_{rj} = \delta^i_j,\ [a^{ir}][a_{rj}] = [a^{ir}a_{rj}] = [\delta^i_j] \tag{2.20}$$

$$\left(a^{ir}a_{rj} = \frac{\partial\theta^i}{\partial x_p}\frac{\partial\theta^r}{\partial x_p}\frac{\partial x_q}{\partial\theta^r}\frac{\partial x_q}{\partial\theta^j} = \frac{\partial\theta^i}{\partial\theta^j} = \delta^i_j\right).$$

Hence, we have

$$\boxed{a^{ij} = a_{ij}^{-1},\ [a^{ij}] = [a_{ij}]^{-1}} \tag{2.21}$$

from which it follows that

$$\det[a_{ij}]\det[a^{ij}] = 1\,(\text{no sum}),\ \det[a^{ij}] = 1/\det[a_{ij}], \tag{2.22}$$

and

$$a \equiv \det[a_{ij}] = [\mathbf{a}_1\ \mathbf{a}_2\ \mathbf{a}_3]^2,\ \det[a^{ij}] = 1/a, \tag{2.23}$$

noting equations $(1.159)_4$, (1.168) and (1.171). Equation (2.21) means that the matrices with the components a^{ij} and a_{ij} are mutually inverse, provided that $[a^{ij}] \neq 0$ and $[a_{ij}] \neq 0$. Furthermore, we have

$$\frac{\partial a}{\partial a_{ij}} = aa^{ij}, \tag{2.24}$$

noting that

$$\frac{\partial a}{\partial a_{ij}} = aa_{ij}^{-1}$$

by virtue of equations (1.371) with equation (2.21).

Using the notation in equation (2.19), the following transformation rules between the covariant and the contravariant base vectors hold:

$$\mathbf{a}^i = a^{ir}\mathbf{a}_r, \quad \mathbf{a}_i = a_{ir}\mathbf{a}^r, \tag{2.25}$$

noting that

$$\left.\begin{array}{l}\mathbf{a}^i=(\mathbf{a}^i\bullet\mathbf{e}_p)\mathbf{e}_p=\left(\mathbf{a}^i\bullet\dfrac{\partial\theta^s}{\partial x_p}\mathbf{a}_s\right)\dfrac{\partial\theta^r}{\partial x_p}\mathbf{a}_r=\dfrac{\partial\theta^i}{\partial x_p}\dfrac{\partial\theta^r}{\partial x_p}\mathbf{a}_r\\ \mathbf{a}_i=(\mathbf{a}_i\bullet\mathbf{e}_p)\mathbf{e}_p=\left(\mathbf{a}_i\bullet\dfrac{\partial x_p}{\partial\theta^s}a^s\right)\dfrac{\partial x_p}{\partial\theta^r}\mathbf{a}^r=\dfrac{\partial x_p}{\partial\theta^i}\dfrac{\partial x_p}{\partial\theta^r}\mathbf{a}^r\end{array}\right\}.$$

It is known from equation (2.25) that a^{ij} and a_{ij} play the role of the *shifter* which makes the index go up and down, respectively.

Substitution of equation (2.14) with equation (2.19) into the identity tensor **I** in equation (1.210) yields

$$\mathbf{e}_i\otimes\mathbf{e}_i=\left\{\begin{array}{l}\dfrac{\partial\theta^r}{\partial x_i}\mathbf{a}_r\otimes\dfrac{\partial\theta^s}{\partial x_i}\mathbf{a}_s=a^{rs}\mathbf{a}_r\otimes\mathbf{a}_s\\ \dfrac{\partial\theta^r}{\partial x_i}\mathbf{a}_r\otimes\dfrac{\partial x_i}{\partial\theta_s}\mathbf{a}^s=\delta^r_s\mathbf{a}_r\otimes\mathbf{a}^s=\mathbf{a}_r\otimes\mathbf{a}^r\\ \dfrac{\partial x_i}{\partial\theta_r}\mathbf{a}^r\otimes\dfrac{\partial\theta^s}{\partial x_i}\mathbf{a}_s=\delta^s_r\mathbf{a}^r\otimes\mathbf{a}_s=\mathbf{a}^r\otimes\mathbf{a}_r\\ \dfrac{\partial x_i}{\partial\theta_r}\mathbf{a}^r\otimes\dfrac{\partial x_i}{\partial\theta_s}\mathbf{a}^s=a_{rs}\mathbf{a}^r\otimes\mathbf{a}^s\end{array}\right..$$

Then, adopting the symbol **a** for the generalized identity tensor in the general coordinate system, we can write

$$\boxed{\mathbf{a}\equiv\mathbf{e}_i\otimes\mathbf{e}_i=\mathbf{a}_i\otimes\mathbf{a}^i=\mathbf{a}^i\otimes\mathbf{a}_i=a^{ij}\mathbf{a}_i\otimes\mathbf{a}_j=a_{ij}\mathbf{a}^i\otimes\mathbf{a}^j}\tag{2.26}$$

fulfilling

$$||\mathbf{a}||=\sqrt{3},\ \det\mathbf{a}=1,\ \mathbf{a}=\mathbf{a}^T=\mathbf{a}^{-1}.\tag{2.27}$$

The tensor **a** plays the role of identity tensor and its components shown in equation (2.26) are necessary for the description of the geometrical elements, for example, the length of a line element and the angle between line elements in the general space. It is therefore called a *metric tensor* as will be seen in Section 2.3.

The following equations hold:

$$\left.\begin{array}{l}\mathrm{sym}[\overset{\bullet}{\mathbf{a}}{}^i\otimes\mathbf{a}_i]=\mathrm{sym}[\mathbf{a}_i\otimes\overset{\bullet}{\mathbf{a}}{}^i]=\dfrac{1}{2}\overset{\bullet}{a}{}^{ij}\mathbf{a}_i\otimes\mathbf{a}_j\\ \mathrm{sym}[\overset{\bullet}{\mathbf{a}}_i\otimes\mathbf{a}^i]=\mathrm{sym}[\mathbf{a}^i\otimes\overset{\bullet}{\mathbf{a}}_i]=\dfrac{1}{2}\overset{\bullet}{a}_{ij}\mathbf{a}^i\otimes\mathbf{a}^j\end{array}\right\},\tag{2.28}$$

$$\left.\begin{array}{l}\mathrm{ant}[\overset{\bullet}{\mathbf{a}}{}^i\otimes\mathbf{a}_i]=-\mathrm{ant}[\mathbf{a}_i\otimes\overset{\bullet}{\mathbf{a}}{}^i]=\dfrac{1}{2}(\mathbf{a}^i\bullet\overset{\bullet}{\mathbf{a}}{}^j-\overset{\bullet}{\mathbf{a}}{}^i\bullet\mathbf{a}^j)\mathbf{a}_i\otimes\mathbf{a}_j\\ \mathrm{ant}[\overset{\bullet}{\mathbf{a}}_i\otimes\mathbf{a}^i]=-\mathrm{ant}[\mathbf{a}^i\otimes\overset{\bullet}{\mathbf{a}}_i]=\dfrac{1}{2}(\mathbf{a}_i\bullet\overset{\bullet}{\mathbf{a}}_j-\overset{\bullet}{\mathbf{a}}_i\bullet\mathbf{a}_j)\mathbf{a}^i\otimes\mathbf{a}^j\end{array}\right\},\tag{2.29}$$

noting that

$$2\text{sym}[\overset{\bullet}{\mathbf{a}}^i \otimes \mathbf{a}_i] = \overset{\bullet}{\mathbf{a}}^j \otimes \mathbf{a}_j + \mathbf{a}_i \otimes \overset{\bullet}{\mathbf{a}}^i = (\overset{\bullet}{\mathbf{a}}^j \cdot \mathbf{a}^i)\mathbf{a}_i \otimes \mathbf{a}_j + \mathbf{a}_i \otimes (\overset{\bullet}{\mathbf{a}}^i \cdot \mathbf{a}^j)\mathbf{a}_j = (\mathbf{a}^i \cdot \mathbf{a}^j)^\bullet \mathbf{a}_i \otimes \mathbf{a}_j,$$

$$2\text{sym}[\overset{\bullet}{\mathbf{a}}_i \otimes \mathbf{a}^i] = \overset{\bullet}{\mathbf{a}}_j \otimes \mathbf{a}^j + \mathbf{a}^i \otimes \overset{\bullet}{\mathbf{a}}_i = (\overset{\bullet}{\mathbf{a}}_j \cdot \mathbf{a}_i)\mathbf{a}^i \otimes \mathbf{a}^j + \mathbf{a}^i \otimes (\overset{\bullet}{\mathbf{a}}_i \cdot \mathbf{a}_j)\mathbf{a}^j = (\mathbf{a}^i \cdot \mathbf{a}_j)^\bullet \mathbf{a}^i \otimes \mathbf{a}^j,$$

$$2\text{ant}[\overset{\bullet}{\mathbf{a}}^i \otimes \mathbf{a}_i] = \overset{\bullet}{\mathbf{a}}^j \otimes \mathbf{a}_j - \mathbf{a}_i \otimes \overset{\bullet}{\mathbf{a}}^i = (\overset{\bullet}{\mathbf{a}}^j \cdot \mathbf{a}^i)\mathbf{a}_i \otimes \mathbf{a}_j - \mathbf{a}_i \otimes (\overset{\bullet}{\mathbf{a}}^i \cdot \mathbf{a}^j)\mathbf{a}_j,$$

$$2\text{ant}[\overset{\bullet}{\mathbf{a}}_i \otimes \mathbf{a}^i] = \overset{\bullet}{\mathbf{a}}_j \otimes \mathbf{a}^j - \mathbf{a}^i \otimes \overset{\bullet}{\mathbf{a}}_i = (\overset{\bullet}{\mathbf{a}}_j \cdot \mathbf{a}_i)\mathbf{a}^i \otimes \mathbf{a}^j - \mathbf{a}^i \otimes (\overset{\bullet}{\mathbf{a}}_i \cdot \mathbf{a}_j)\mathbf{a}^j.$$

By equations (2.6), (2.8) and (2.23), we may write

$$V = [\mathbf{a}_1\mathbf{a}_2\mathbf{a}_3] = \sqrt{a}, \quad [\mathbf{a}^1\mathbf{a}^2\mathbf{a}^3] = \frac{1}{V} = \frac{1}{\sqrt{a}}. \tag{2.30}$$

Then equations (2.5) and (2.7) can be written as

$$\mathbf{a}^i = \frac{1}{2}e^{ijk}\mathbf{a}_j \times \mathbf{a}_k, \quad \mathbf{a}_i = \frac{1}{2}e_{ijk}\mathbf{a}^j \times \mathbf{a}^k \tag{2.31}$$

and equation (2.30) leads to the general expression

$$[\mathbf{a}_i\ \mathbf{a}_j\mathbf{a}_k] = e_{ijk}, \ [\mathbf{a}^i\mathbf{a}^j\mathbf{a}^k] = e^{ijk}, \tag{2.32}$$

introducing the symbols

$$\boxed{e_{ijk} \equiv \varepsilon_{ijk}\sqrt{a}, \ e^{ijk} \equiv \frac{\varepsilon_{ijk}}{\sqrt{a}}} \tag{2.33}$$

which are related by the transformation rule

$$e^{ijk} = a^{ir}a^{js}a^{kt}e_{rst}, \ e_{ijk} = a_{ir}a_{js}a_{kt}e^{rst}, \tag{2.34}$$

noting that

$$\frac{\varepsilon_{ijk}}{\sqrt{a}} = \sqrt{a}\det(a^{pq})\varepsilon_{ijk} = \sqrt{a}\,a^{ir}a^{js}a^{kt}\varepsilon_{rst} = a^{ir}a^{js}a^{kt}\sqrt{a}\varepsilon_{rst},$$

$$\sqrt{a}\varepsilon_{ijk} = \frac{1}{\sqrt{a}}\det(a_{pq})\varepsilon_{ijk} = \frac{1}{\sqrt{a}}a_{ir}a_{js}a_{kt}\varepsilon_{rst} = a_{ir}a_{js}a_{kt}\frac{1}{\sqrt{a}}\varepsilon_{rst},$$

$$\det(a^{pq})\varepsilon_{ijk} = \varepsilon_{rst}a^{1r}a^{2s}a^{3t}\varepsilon_{ijk} = a^{ir}a^{js}a^{kt}\varepsilon_{rst}.$$

The following relations hold by virtue of equations (1.56)–(1.58):

$$\left.\begin{aligned} e^{ijk}e_{irs} &= e_{jki}e_{rsi} = \delta_{jr}\delta_{ks} - \delta_{js}\delta_{kr} \\ e^{ijk}e_{ijs} &= 2\delta_{ks} \\ e^{ijk}e_{ijk} &= 3! = 6 \end{aligned}\right\}. \tag{2.35}$$

Further, the vector products of base vectors are given by

$$\boxed{\mathbf{a}_i \times \mathbf{a}_j = e_{ijk}\mathbf{a}^k, \ \ \mathbf{a}^i \times \mathbf{a}^j = e^{ijk}\mathbf{a}_k} \tag{2.36}$$

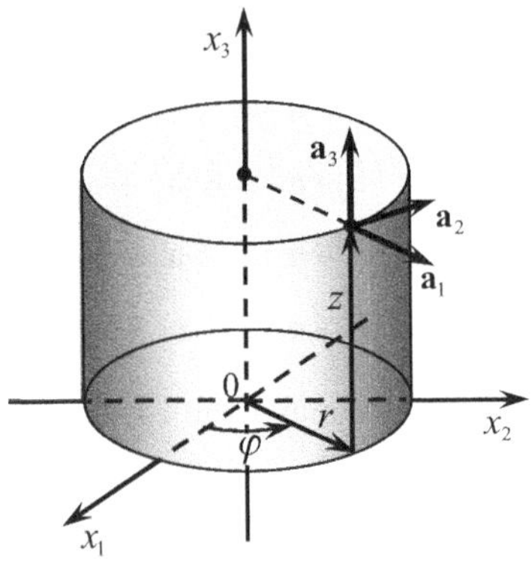

Figure 2.3 Oblique cylindrical polar coordinate system

noting that

$$\left.\begin{aligned}\mathbf{a}_r \times \mathbf{a}_s &= \frac{1}{2}(\mathbf{a}_r \times \mathbf{a}_s - \mathbf{a}_s \times \mathbf{a}_r) = \frac{1}{2}(\delta_{rj}\delta_{sk} - \delta_{rk}\delta_{sj})\mathbf{a}_j \times \mathbf{a}_k = \frac{1}{2}e_{irs}e^{ijk}\mathbf{a}_j \times \mathbf{a}_k = e_{irs}\mathbf{a}^i \\ \mathbf{a}^r \times \mathbf{a}^s &= \frac{1}{2}(\mathbf{a}^r \times \mathbf{a}^s - \mathbf{a}^s \times \mathbf{a}^r) = \frac{1}{2}(\delta_{rj}\delta_{sk} - \delta_{rk}\delta_{sj})\mathbf{a}^j \times \mathbf{a}^k = \frac{1}{2}e^{irs}e_{ijk}\mathbf{a}^j \times \mathbf{a}^k = e^{irs}\mathbf{a}_i\end{aligned}\right\}$$

with equations (1.34), (2.31) and (2.35).

In what follows, the base vectors in some typical curvilinear coordinate systems will be given.

Example 2.1: Cylindrical polar coordinate system (orthonormal coordinate system) In orthonormal coordinate systems we have $a_{ij} = 0$, $a^{ij} = 0$ for $i \neq j$ and thus $a_{ij} = 1/a^{ij}$ for $i = j$ due to equation (2.20).

It follows in the cylindrical polar coordinate system shown in Figure 2.3 that

$$x_1 = r\cos\varphi,\ x_2 = r\sin\varphi,\ x_3 = z, \tag{2.37}$$

leading to

$$\mathbf{x} = r\cos\varphi\mathbf{e}_1 + r\sin\varphi\mathbf{e}_2 + z\mathbf{e}_3. \tag{2.38}$$

Here, putting

$$\theta^1 = r,\ \theta^2 = \varphi,\ \theta^3 = z, \tag{2.39}$$

it follows from equation (2.14) for equation (2.38) that

$$\left.\begin{aligned}\mathbf{a}_1 &= \frac{\partial\mathbf{x}}{\partial r} = \cos\varphi\mathbf{e}_1 + \sin\varphi\mathbf{e}_2 \\ \mathbf{a}_2 &= \frac{\partial\mathbf{x}}{\partial\varphi} = -r\sin\varphi\mathbf{e}_1 + r\cos\varphi\mathbf{e}_2 \\ \mathbf{a}_3 &= \frac{\partial\mathbf{x}}{\partial z} = \mathbf{e}_3\end{aligned}\right\} \tag{2.40}$$

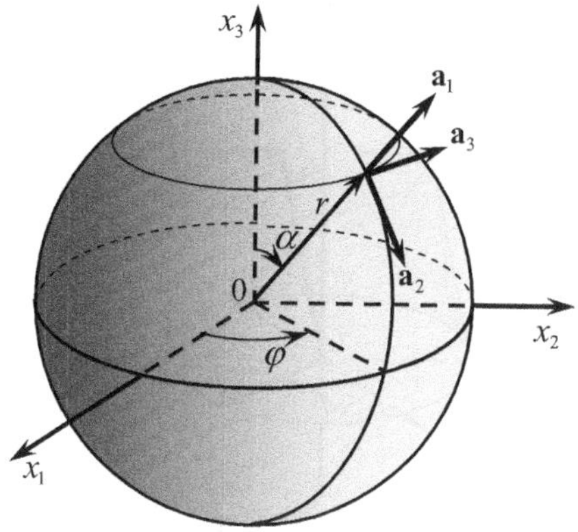

Figure 2.4 Spherical polar coordinate system

from which the metric tensor components are given by

$$[a_{ij}] = \begin{bmatrix} 1 & 0 & 0 \\ 0 & r^2 & 0 \\ 0 & 0 & 1 \end{bmatrix}, \quad [a^{ij}] = [a_{ij}]^{-1} = \begin{bmatrix} 1 & 0 & 0 \\ 0 & 1/r^2 & 0 \\ 0 & 0 & 1 \end{bmatrix}. \tag{2.41}$$

The reciprocal base vectors $\mathbf{a}^i$ are obtained by shifting the index of $\mathbf{a}_i$ from equation (2.25) with $a^{ij} = 0 (i \neq j)$ because of the orthogonal coordinate system as follows:

$$\left.\begin{aligned} \mathbf{a}^1 &= a^{11}\mathbf{a}_1 = \mathbf{a}_1 = \cos\varphi\mathbf{e}_1 + \sin\varphi\mathbf{e}_2 \\ \mathbf{a}^2 &= a^{22}\mathbf{a}_2 = (1/r^2)\mathbf{a}_2 = (-\sin\varphi\mathbf{e}_1 + \cos\varphi\mathbf{e}_2)/r \\ \mathbf{a}^3 &= a^{33}\mathbf{a}_3 = \mathbf{a}_3 = \mathbf{e}_3 \end{aligned}\right\}. \tag{2.42}$$

Example 2.2: Spherical polar coordinate system (orthonormal coordinate system) In the spherical polar coordinate system shown in Figure 2.4 we have

$$x_1 = r\sin\alpha\cos\varphi,\ x_2 = r\sin\alpha\sin\varphi,\ x_3 = r\cos\alpha \tag{2.43}$$

$$(r > 0, 0 < \alpha < \pi, 0 \leq \varphi < 2\pi),$$

leading to

$$\mathbf{x} = r\sin\alpha\cos\varphi\mathbf{e}_1 + r\sin\alpha\sin\varphi\mathbf{e}_2 + r\cos\alpha\mathbf{e}_3. \tag{2.44}$$

Here, putting

$$\theta^1 = r,\ \theta^2 = \alpha,\ \theta^3 = \varphi \tag{2.45}$$

we have from equation (2.14) for equation (2.44) that

$$\left.\begin{aligned} \mathbf{a}_1 &= \frac{\partial\mathbf{x}}{\partial r} = \sin\alpha\cos\varphi\mathbf{e}_1 + \sin\alpha\sin\varphi\mathbf{e}_2 + \cos\alpha\mathbf{e}_3 \\ \mathbf{a}_2 &= \frac{\partial\mathbf{x}}{\partial\alpha} = r\cos\alpha\cos\varphi\mathbf{e}_1 + r\cos\alpha\sin\varphi\mathbf{e}_2 - r\sin\alpha\mathbf{e}_3 \\ \mathbf{a}_3 &= \frac{\partial\mathbf{x}}{\partial\varphi} = -r\sin\alpha\sin\varphi\mathbf{e}_1 + r\sin\alpha\cos\varphi\mathbf{e}_2 \end{aligned}\right\} \tag{2.46}$$

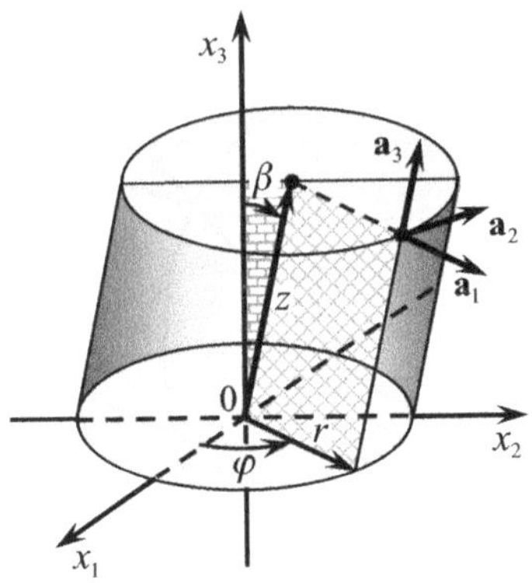

Figure 2.5 Oblique cylindrical polar coordinate system

from which the metric tensor components are given by

$$[a_{ij}] = \begin{bmatrix} 1 & 0 & 0 \\ 0 & r^2 & 0 \\ 0 & 0 & (r\sin\alpha)^2 \end{bmatrix}, \quad [a^{ij}] = [a_{ij}]^{-1} = \begin{bmatrix} 1 & 0 & 0 \\ 0 & 1/r^2 & 0 \\ 0 & 0 & 1/(r\sin\alpha)^2 \end{bmatrix}. \tag{2.47}$$

The reciprocal base vectors $\mathbf{a}^i$ are obtained by shifting the index of $\mathbf{a}_i$ from equation (2.25) with $a^{ij} = 0\ (i \neq j)$ because of the orthogonal coordinate system as follows:

$$\left.\begin{aligned} \mathbf{a}^1 &= \sin\alpha\cos\varphi\mathbf{e}_1 + \sin\alpha\sin\varphi\mathbf{e}_2 + \cos\alpha\mathbf{e}_3 \\ \mathbf{a}^2 &= \frac{1}{r}(\cos\alpha\cos\varphi\mathbf{e}_1 + \cos\alpha\sin\varphi\mathbf{e}_2 - \sin\alpha\mathbf{e}_3) \\ \mathbf{a}^3 &= \frac{1}{r\sin\alpha}(-\sin\varphi\mathbf{e}_1 + \cos\varphi\mathbf{e}_2) \end{aligned}\right\}. \tag{2.48}$$

Example 2.3: Skew cylindrical polar coordinate system (non-orthonormal coordinate system) The previous examples are concerned with orthogonal curvilinear coordinate systems. We now turn to the oblique cylindrical polar coordinate system ($a_{ij} \neq 0, a^{ij} \neq 0$ for $i \neq j$) shown in Figure 2.5 (Tanahashi, 1988) for which

$$x_1 = r\cos\varphi,\ x_2 = r\sin\varphi + z\sin\beta,\ x_3 = z\cos\beta, \tag{2.49}$$

leading to

$$\mathbf{x} = r\cos\varphi\mathbf{e}_1 + (r\sin\varphi + z\sin\beta)\mathbf{e}_2 + z\cos\beta\mathbf{e}_3. \tag{2.50}$$

Here, putting

$$\theta^1 = r,\ \theta^2 = \varphi,\ \theta^3 = z, \tag{2.51}$$

we have from equation (2.14) for equation (2.50) that

$$\left.\begin{aligned} \mathbf{a}_1 &= \frac{\partial \mathbf{x}}{\partial r} = \cos\varphi \mathbf{e}_1 + \sin\varphi \mathbf{e}_2 \\ \mathbf{a}_2 &= \frac{\partial \mathbf{x}}{\partial \varphi} = -r\sin\varphi \mathbf{e}_1 + r\cos\varphi \mathbf{e}_2 \\ \mathbf{a}_3 &= \frac{\partial \mathbf{x}}{\partial z} = \sin\beta \mathbf{e}_2 + \cos\beta \mathbf{e}_3 \end{aligned}\right\} \tag{2.52}$$

from which the metric tensor components are given by

$$\left.\begin{aligned} [a_{ij}] &= \begin{bmatrix} 1 & 0 & \sin\varphi\sin\beta \\ & r^2 & r\cos\varphi\sin\beta \\ \text{sym.} & & 1 \end{bmatrix} \\ [a^{ij}] = [a_{ij}]^{-1} &= \frac{1}{r^2\cos^2\beta}\begin{bmatrix} r^2(1-\cos^2\varphi\sin^2\beta) & r\sin\varphi\cos\varphi\sin^2\beta & -r^2\sin\varphi\sin\beta \\ r\sin\varphi\cos\varphi\sin^2\beta & 1-\sin^2\varphi\sin^2\beta & -r\cos\varphi\sin\beta \\ -r^2\sin\varphi\sin\beta & -r\cos\varphi\sin\beta & r^2 \end{bmatrix} \end{aligned}\right\}, \tag{2.53}$$

exploiting equation (1.158)′.

The reciprocal base vectors $\mathbf{a}^i$ are obtained by shifting the index of $\mathbf{a}_i$ from equation (2.25) with equation (2.53) as follows:

$$\left.\begin{aligned} \mathbf{a}^1 &= \frac{1}{\cos^2\beta}(\cos\varphi\cos^2\beta \mathbf{e}_1 + \sin\varphi\cos^2\beta \mathbf{e}_2 - \sin\varphi\sin\beta\cos\beta \mathbf{e}_3) \\ \mathbf{a}^2 &= \frac{1}{r}(-\sin\varphi \mathbf{e}_1 + \cos\varphi \mathbf{e}_2 - \cos\varphi\tan\beta \mathbf{e}_3) \\ \mathbf{a}^3 &= \frac{1}{\cos\beta}\mathbf{e}_3 \end{aligned}\right\}. \tag{2.54}$$

2.3 Representations of Vectors and Tensors

An arbitrary vector is described by equation (1.64) as

$$\mathbf{v} = (\mathbf{v}\cdot\mathbf{a}^i)\mathbf{a}_i = (\mathbf{v}\cdot\mathbf{a}_i)\mathbf{a}^i, \tag{2.55}$$

that is,

$$\boxed{\mathbf{v} = v_i\mathbf{e}_i = v^i\mathbf{a}_i = v_i\mathbf{a}^i} \tag{2.56}$$

where

$$v^i = \mathbf{v}\cdot\mathbf{a}^i, \quad v_i = \mathbf{v}\cdot\mathbf{a}_i \tag{2.57}$$

as shown in Figure 2.6, illustrated in the two-dimensional state. It is known from equation (2.56), based on equation (1.64), that a vector is represented by combinations of covariant base vectors with contravariant components or combinations of contravariant base vectors with covariant components.

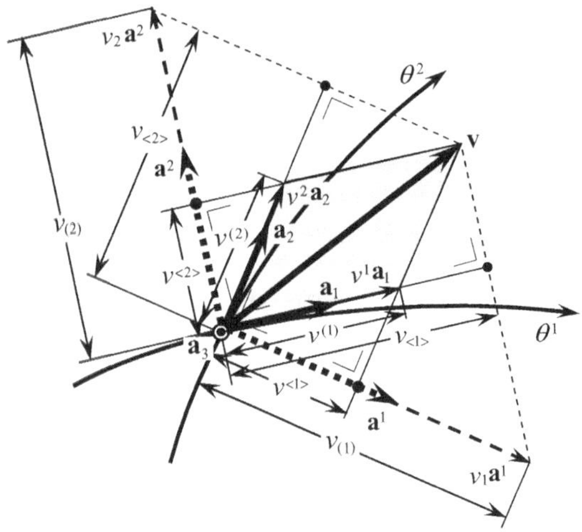

Figure 2.6 Vector representation in general coordinate system (two-dimensional observation for the state $\mathbf{a}_3/\|\mathbf{a}_3\| = \mathbf{a}^3/\|\mathbf{a}^3\|$)

By virtue of $\mathbf{v} = v^r\mathbf{a}_r = v_r\mathbf{a}^r$, we have

$$\left.\begin{aligned} v^i &= a^{ir}v_r \\ v_i &= a_{ir}v^r \end{aligned}\right\}, \tag{2.58}$$

noting that

$$\left.\begin{aligned} v^i &= \mathbf{v}\cdot\mathbf{a}^i = v_r\mathbf{a}^r\cdot\mathbf{a}^i \\ v_i &= \mathbf{v}\cdot\mathbf{a}_i = v^r\mathbf{a}_r\cdot\mathbf{a}_i \end{aligned}\right\},$$

where the metric tensor component plays the role of shifter.

The substitution of equation (2.15) into equation (2.56) leads to

$$\mathbf{v} = \mathrm{v}_i\mathbf{e}_i = v^i \underset{\sim}{g}_{ir}\mathbf{e}_r = v_i\tilde{a}^{ir}\mathbf{e}_r \tag{2.59}$$

from which it follows that

$$\mathrm{v}_i = \underset{\sim}{g}_{ri}v^r = \tilde{a}^{ri}v_r \tag{2.60}$$

On the other hand, substitution of equations (2.14) and (2.15) into equation (2.59) yields

$$\mathrm{v}_i\frac{\partial\theta^r}{\partial x_i}\mathbf{a}_r = v^i\mathbf{a}_i,\quad \mathrm{v}_i\frac{\partial x_i}{\partial\theta_r}\mathbf{a}^r = v_i\mathbf{a}^i$$

from which the inverse expression of equation (2.60) is given by

$$v^i = \tilde{a}^{ir}\mathrm{v}_r,\quad v_i = \underset{\sim}{g}_{ir}\mathrm{v}_r. \tag{2.61}$$

The vector $\mathbf{v}$ is described by the other covariant base $\{\bar{\mathbf{a}}_i\}(i = 1, 2, 3)$ as follows:

$$\mathbf{v} = \bar{v}^i\bar{\mathbf{a}}_i. \tag{2.62}$$

Substituting equation (2.3) into equation (2.62), one has

$$\mathbf{v} = \bar{v}^i \frac{\partial \theta^j}{\partial \bar{\theta}^i} \mathbf{a}_j \tag{2.63}$$

and, comparing equation (2.63) with equation (2.56), the following transformation rule is obtained:

$$v^i = \frac{\partial \theta^i}{\partial \bar{\theta}^j} \bar{v}^j \tag{2.64}$$

Further, inversely from equation (2.64), one obtains

$$\bar{v}^i = \frac{\partial \bar{\theta}^i}{\partial \theta^j} v^j \left(\bar{v}^i = \delta^{ir} \bar{v}^r = \frac{\partial \bar{\theta}^i}{\partial \theta^j} \frac{\partial \theta^j}{\partial \bar{\theta}^r} \bar{v}^r \right). \tag{2.65}$$

Similarly, we have

$$v_i = \frac{\partial \bar{\theta}^j}{\partial \theta^i} \bar{v}_j, \ \bar{v}_i = \frac{\partial \theta^j}{\partial \bar{\theta}^i} v_j. \tag{2.66}$$

$\mathbf{a}_i$ and v_i are transformed by the same rule as seen in equations (2.3) and (2.66), respectively, and $\mathbf{a}^i$ and v^i are also transformed by the same rule as seen in equations (2.13) and (2.65), respectively. v^i and v_i are called the *contravariant* and the *covariant components*, respectively.

The scalar, vector and the tensor products are given by equations (2.36) and (2.56) as

$$\mathbf{u} \cdot \mathbf{v} = a_{ij} u^i v^j = u^i v_i = u_i v^i = a^{ij} u_i v_j, \tag{2.67}$$

$$||\mathbf{v}|| = \sqrt{\mathbf{v} \cdot \mathbf{v}} = \sqrt{a_{ij} v^i v^j} = \sqrt{a^{ij} v_i v_j} = \sqrt{v^i v_i}, \tag{2.68}$$

$$\mathbf{u} \times \mathbf{v} = e_{ijk} u^i v^j \mathbf{a}^k = e^{ijk} u_i v_j \mathbf{a}_k, \tag{2.69}$$

$$(\mathbf{u} \times \mathbf{v}) \cdot \mathbf{w} = e_{ijk} u^i v^j w^k = e^{ijk} u_i v_j w_k, \tag{2.70}$$

$$\mathbf{u} \otimes \mathbf{v} = u^i v^j \mathbf{a}_i \otimes \mathbf{a}_j = u^i v_j \mathbf{a}_i \otimes \mathbf{a}^j = u_i v^j \mathbf{a}^i \otimes \mathbf{a}_j = u_i v_j \mathbf{a}^i \otimes \mathbf{a}^j, \tag{2.71}$$

noting that

$$\mathbf{u} \cdot \mathbf{v} = \begin{cases} u^i \mathbf{a}_i \cdot v^j \mathbf{a}_j = a_{ij} u^i v^j \\ u^i \mathbf{a}_i \cdot v_j \mathbf{a}^j = u^i v_i \\ u_i \mathbf{a}^i \cdot v^j \mathbf{a}_j = u_i v^i \\ u_i \mathbf{a}^i \cdot v_j \mathbf{a}^j = a^{ij} u_i v_j \end{cases},$$

$$\mathbf{u} \times \mathbf{v} = \begin{cases} u^i \mathbf{a}_i \times v^j \mathbf{a}_j = e_{ijk} u^i v^j \mathbf{a}^k \\ u_i \mathbf{a}^i \times v_j \mathbf{a}^j = e^{ijk} u_i v_j \mathbf{a}_k \end{cases},$$

$$(\mathbf{u} \times \mathbf{v}) \cdot \mathbf{w} = \begin{cases} \left(u^i \mathbf{a}_i \times v^j \mathbf{a}_j\right) \cdot w^k \mathbf{a}_k = \left(e_{ijr} u^i v^j \mathbf{a}^r\right) \cdot w^k \mathbf{a}_k \\ \left(u_i \mathbf{a}^i \times v_j \mathbf{a}^j\right) \cdot w_k \mathbf{a}^k = \left(e^{ijr} u_i v_j \mathbf{a}_r\right) \cdot w_k \mathbf{a}^k \end{cases},$$

$$\mathbf{u} \otimes \mathbf{v} = \begin{cases} u^i \mathbf{a}_i \otimes v^j \mathbf{a}_j = u^i v^j \mathbf{a}_i \otimes \mathbf{a}_j \\ u^i \mathbf{a}_i \otimes v_j \mathbf{a}^j = u^i v_j \mathbf{a}_i \otimes \mathbf{a}^j \\ u_i \mathbf{a}^i \otimes v^j \mathbf{a}_j = u_i v^j \mathbf{a}^i \otimes \mathbf{a}_j \\ u_i \mathbf{a}^i \otimes v_j \mathbf{a}^j = u_i v_j \mathbf{a}^i \otimes \mathbf{a}^j \end{cases}.$$

As seen in equation (2.67)–(2.71), the scalar product is described by a combination of contravariant and covariant components without the metric tensor components because the scalar product of base vectors of different variants reduces to the Kronecker delta.

The square of the length of an infinitesimal line-element vector $d\mathbf{x}$ is expressed using the metric tensor by equation (2.68) as follows:

$$||d\mathbf{x}||^2 = d\mathbf{x}\cdot d\mathbf{x} = a_{ij}dx^i dx^j = a^{ij}dx_i dx_j = dx^i dx_i, \tag{2.72}$$

noting that

$$d\mathbf{x}\cdot d\mathbf{x} = \begin{cases} dx^i\mathbf{a}_i \cdot dx^j\mathbf{a}_j = a_{ij}dx^i dx^j \\ dx_i\mathbf{a}^i \cdot dx_j\mathbf{a}^j = a^{ij}dx_i dx_j \\ dx^i\mathbf{a}_i \cdot dx_j\mathbf{a}^j = dx^i dx_i \end{cases}.$$

Further, the cosine of the angle between two infinitesimal line-element vectors $d\mathbf{x}$ and $d\mathbf{y}$ is expressed using the metric tensor by equations (2.67) and (2.72) as follows:

$$\cos(d\mathbf{x}\cdot d\mathbf{y}) = \frac{d\mathbf{x}\cdot d\mathbf{y}}{||d\mathbf{x}||\,||d\mathbf{y}||} = \begin{cases} \dfrac{a_{ij}dx^i dy^j}{\sqrt{a_{rs}dx^r dx^s}\sqrt{a_{pq}dy^p dy^q}} \\ \dfrac{a^{ij}dx_i dy_j}{\sqrt{a^{rs}dx_r dx_s}\sqrt{a^{pq}dy_p dy_q}} \\ \dfrac{dx^i v_i}{\sqrt{dx^r dx_r}\sqrt{dy^s dy_s}} \end{cases}, \tag{2.73}$$

It is seen that the tensor $\mathbf{a}$ transforms an arbitrary vector to itself, that is,

$$\mathbf{a}\mathbf{v} = \begin{Bmatrix} \mathbf{a}_i \otimes \mathbf{a}^i\mathbf{v} = (\mathbf{v}\cdot\mathbf{a}^i)\mathbf{a}_i \\ \mathbf{a}^i \otimes \mathbf{a}_i\mathbf{v} = (\mathbf{v}\cdot\mathbf{a}_i)\mathbf{a}^i \end{Bmatrix} = \mathbf{v}. \tag{2.74}$$

As will be evident from the above-mentioned facts, the tensor $\mathbf{a}$ plays a role in the representation of various geometrical elements in the Riemann space in addition to the role of identity tensor; it is thus called a *Riemann metric tensor* or simply *metric tensor*. The Riemann space is a more general space than the Eulerian space in which the parallel axiom holds as will be described in section 2.7.

The infinitesimal surface vector $d\mathbf{s}^k$ of the surface formed by the line elements $d\theta^i\mathbf{a}_i$ (no sum) and $d\theta^j\mathbf{a}_j$(no sum) along the primary base $\{\mathbf{a}_i\}$ is given from equation (2.36) by

$$\boxed{d\mathbf{s}^k = d\theta^i\mathbf{a}_i \times d\theta^j\mathbf{a}_j = d\theta^i d\theta^j\,\mathbf{a}^k \quad \text{(no sum)}} \tag{2.75}$$

where i, j, k are different from each other and are ordered in an even permutation, which is called the *surface element vector* (Figure 2.7).

Further, the volume dV of the infinitesimal parallelepiped (Figure 2.7) formed by the sides along the primary base $\{\mathbf{a}_i\}$ is given from equation (2.30) by

$$dV = [\mathbf{a}_1\ \mathbf{a}_2\ \mathbf{a}_3]d\theta^1 d\theta^2 d\theta^3 = \sqrt{a}d\theta^1 d\theta^2 d\theta^3 \tag{2.76}$$

which is called the *volume element*, noting that

$$[d\theta^1\mathbf{a}_1\ d\theta^2\mathbf{a}_2\ d\theta^3\mathbf{a}_3] = [\mathbf{a}_1\ \mathbf{a}_2\ \mathbf{a}_3]d\theta^1 d\theta^2 d\theta^3$$

with equation (1.43).

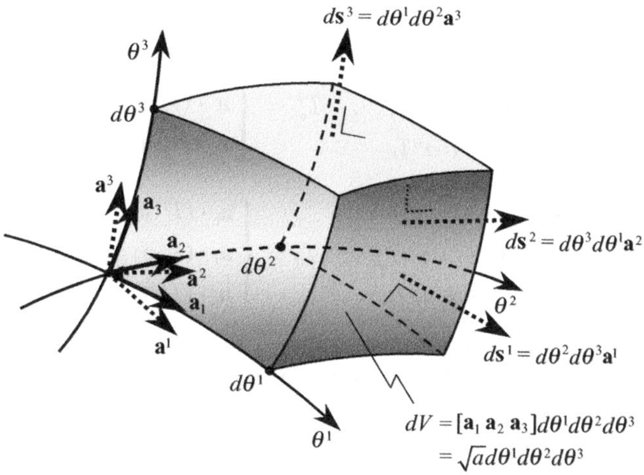

Figure 2.7 Surface and volume elements of parallelepiped formed by the side elements in the direction of primary base vectors

Following the expression in equation (2.71) for the tensor product, the second-order tensor **T** is expressed as follows:

$$\boxed{\mathbf{T} = \mathrm{T}_{ij}\mathbf{e}_i \otimes \mathbf{e}_j = T^{ij}\mathbf{a}_i \otimes \mathbf{a}_j = T^{i}_{\bullet j}\mathbf{a}_i \otimes \mathbf{a}^j = T_i^{\bullet j}\mathbf{a}^i \otimes \mathbf{a}_j = T_{ij}\mathbf{a}^i \otimes \mathbf{a}^j} \tag{2.77}$$

in the covariant, contravariant and mixed base vectors. Here, note that we cannot write T^i_j because of $T^i_{\bullet j} \neq T_i^{\bullet j}$ due to $\mathbf{a}_i \neq \mathbf{a}^j$ in general. The components of **T** are given as follows:

$$T^{ij} = \mathbf{a}^i \cdot \mathbf{T}\mathbf{a}^j, \quad T^i_{\bullet j} = \mathbf{a}^i \cdot \mathbf{T}\mathbf{a}_j, \quad T_i^{\bullet j} = \mathbf{a}_i \cdot \mathbf{T}\mathbf{a}^j, \quad T_{ij} = \mathbf{a}_i \cdot \mathbf{T}\mathbf{a}_j, \tag{2.78}$$

noting that

$$\left.\begin{aligned}
\mathbf{a}^i \cdot \mathbf{T}\mathbf{a}^j &= \mathbf{a}^i \cdot T^{rs}\mathbf{a}_r \otimes \mathbf{a}_s\mathbf{a}^j = \delta^i_r\delta^j_s T^{rs} \\
\mathbf{a}^i \cdot \mathbf{T}\mathbf{a}_j &= \mathbf{a}^i \cdot T^r_{\bullet s}\mathbf{a}_r \otimes \mathbf{a}^s\mathbf{a}_j = \delta^i_r\delta^s_j T^r_{\bullet s} \\
\mathbf{a}_i \cdot \mathbf{T}\mathbf{a}^j &= \mathbf{a}_i \cdot T_r^{\bullet s}\mathbf{a}^r \otimes \mathbf{a}_s\mathbf{a}^j = \delta^r_i\delta^j_s T_r^{\bullet s} \\
\mathbf{a}_i \cdot \mathbf{T}\mathbf{a}_j &= \mathbf{a}_i \cdot T_{rs}\mathbf{a}^r \otimes \mathbf{a}^s\mathbf{a}_j = \delta^r_i\delta^s_j T_{rs}
\end{aligned}\right\}$$

by virtue of equation (2.10). As is evident from equation (2.78), we obtain the components by applying the corresponding covariant and the contravariant base vectors. The relations between the components in the different variants are given by substituting equation (2.77) into equation (2.78) as follows:

$$\left.\begin{aligned}
T^{ij} &= T^i_{\bullet r}a^{rj} = a^{ir}T_r^{\bullet j} = a^{ir}T_{rs}a^{sj}, \quad T^i_{\bullet j} = T^{ir}a_{rj} = a^{ir}T_r^{\bullet s}a_{sj} = a^{ir}T_{rj} \\
T_i^{\bullet j} &= a_{ir}T^{rj} = a_{ir}T^r_{\bullet s}a^{sj} = T_{ir}a^{rj}, \quad T_{ij} = a_{ir}T^{rs}a_{sj} = a_{ir}T^r_{\bullet j} = T_i^{\bullet r}a_{rj}
\end{aligned}\right\}, \tag{2.79}$$

noting that

$$T^{ij} = \begin{cases} \mathbf{a}^i \cdot (T^r_{\bullet s}\mathbf{a}_r \otimes \mathbf{a}^s)\mathbf{a}^j = \delta^i_r a^{js} T^r_{\bullet s} \\ \mathbf{a}^i \cdot (T_r^{\bullet s}\mathbf{a}^r \otimes \mathbf{a}_s)\mathbf{a}^j = a^{ir}\delta^j_s T_r^{\bullet s} \\ \mathbf{a}^i \cdot (T_{rs}\mathbf{a}^r \otimes \mathbf{a}^s)\mathbf{a}^j = a^{ir}a^{sj}T_{rs} \end{cases}, \quad T^i_{\bullet j} = \begin{cases} \mathbf{a}^i \cdot (T^{rs}\mathbf{a}_r \otimes \mathbf{a}_s)\mathbf{a}_j = \delta^i_r a_{sj} T^{rs} \\ \mathbf{a}^i \cdot (T_r^{\bullet s}\mathbf{a}^r \otimes \mathbf{a}_s)\mathbf{a}_j = a^{ir}a_{sj}T_r^{\bullet s} \\ \mathbf{a}^i \cdot (T_{rs}\mathbf{a}^r \otimes \mathbf{a}^s)\mathbf{a}_j = a^{ir}\delta^s_j T_{rs} \end{cases},$$

$$T_i^{\bullet j} = \begin{cases} \mathbf{a}_i \cdot (T^{rs}\mathbf{a}_r \otimes \mathbf{a}_s)\mathbf{a}^j = a_{ir}\delta^j_s T^{rs} \\ \mathbf{a}_i \cdot (T^r_{\bullet s}\mathbf{a}_r \otimes \mathbf{a}^s)\mathbf{a}^j = a_{ir}a^{sj}T^r_{\bullet s} \\ \mathbf{a}_i \cdot (T_{rs}\mathbf{a}^r \otimes \mathbf{a}^s)\mathbf{a}^j = \delta^r_i a^{sj} T_{rs} \end{cases}, \quad T_{ij} = \begin{cases} \mathbf{a}_i \cdot (T^{rs}\mathbf{a}_r \otimes \mathbf{a}_s)\mathbf{a}_j = a_{ir}a^{sj}T^{rs} \\ \mathbf{a}_i \cdot (T^r_{\bullet s}\mathbf{a}_r \otimes \mathbf{a}^s)\mathbf{a}_j = a_{ir}\delta^s_j T^r_{\bullet s} \\ \mathbf{a}_i \cdot (T_r^{\bullet s}\mathbf{a}^r \otimes \mathbf{a}_s)\mathbf{a}_j = \delta^r_i a_{sj} T_r^{\bullet s} \end{cases}.$$

It is shown in equation (2.79) that the metric tensor plays the role of shifter of indices. One metric tensor component is required for each shifting operation, bearing in mind that the same index letters must be contained as the opposite combination of cotravariant and covariant. Any tensor can be expressed by the four types of variants in equation (2.77). Therefore, it would be inappropriate to call them contravariant or covariant or mixed variant tensors as some authors do (for example, Eringen, 1962; Bowen and Wang, 1976; Ogden, 1984; Kay, 1988; Belytschko et al., 2000). In other words, the terms ‘contravariant’, ‘covariant’ and ‘mixed’ should be taken to refer to tensor representation types rather than tensor classifications.

The scalar product of second-order tensors **A** and **B** is expressed in components as

$$\mathbf{A}:\mathbf{B} = A^{ij}B_{ij} = A^i_{\bullet j}B_i^{\bullet j} = A_i^{\bullet j}B^i_{\bullet j} = A_{ij}B^{ij}, \tag{2.80}$$

noting that

$$\mathrm{tr}(\mathbf{A}\mathbf{B}^T) = \mathbf{A}:\mathbf{B} = \begin{cases} (A^{ij}\mathbf{a}_i \otimes \mathbf{a}_j):(B_{rs}\mathbf{a}^r \otimes \mathbf{a}^s) = (\mathbf{a}_i \cdot \mathbf{a}^r)(\mathbf{a}_j \cdot \mathbf{a}^s)A^{ij}B_{rs} \\ (A^i_{\bullet j}\mathbf{a}_i \otimes \mathbf{a}^j):(B_r^{\bullet s}\mathbf{a}^r \otimes \mathbf{a}_s) = (\mathbf{a}_i \cdot \mathbf{a}^r)(\mathbf{a}^j \cdot \mathbf{a}_s)A^i_{\bullet j}B_r^{\bullet s} \\ (A_i^{\bullet j}\mathbf{a}^i \otimes \mathbf{a}_j)(B^r_{\bullet s}\mathbf{a}_r \otimes \mathbf{a}^s) = (\mathbf{a}^i \cdot \mathbf{a}_r)(\mathbf{a}_j \cdot \mathbf{a}^s)A_i^{\bullet j}B^r_{\bullet s} \\ (A_{ij}\mathbf{a}^i \otimes \mathbf{a}^j)(B^{rs}\mathbf{a}_r \otimes \mathbf{a}_s) = (\mathbf{a}^i \cdot \mathbf{a}_r)(\mathbf{a}_j \cdot \mathbf{a}^s)A_{ij}B^{rs} \end{cases}$$

The magnitude of the tensor **T** is given, by setting **A** = **B** = **T** in equation (2.80), as

$$||\mathbf{T}|| = \sqrt{\mathbf{T}:\mathbf{T}} = \sqrt{T^{ij}T_{ij}} = \sqrt{T^i_{\bullet j}T_i^{\bullet j}} = \sqrt{T_i^{\bullet j}T^i_{\bullet j}} = \sqrt{T_{ij}T^{ij}}. \tag{2.81}$$

Here, again we observe the contravariant and covariant combinations.

A metric tensor transforms an arbitrary second-order tensor to itself, that is,

$$\mathbf{Ta} = \left\{ \begin{array}{l} T^{ij}\mathbf{a}_i \otimes \mathbf{a}_j\mathbf{a}^r \otimes \mathbf{a}_r = T^{ij}\delta^r_j\mathbf{a}_i \otimes \mathbf{a}_r = T^{ij}\mathbf{a}_i \otimes \mathbf{a}_j \\ T^i_{\bullet j}\mathbf{a}_i \otimes \mathbf{a}^j\mathbf{a}_r \otimes \mathbf{a}^r = T^i_{\bullet j}\delta^j_r\mathbf{a}_i \otimes \mathbf{a}^r = T^i_{\bullet j}\mathbf{a}_i \otimes \mathbf{a}^j \\ T_i^{\bullet j}\mathbf{a}^i \otimes \mathbf{a}_j\mathbf{a}^r \otimes \mathbf{a}_r = T_i^{\bullet j}\delta^r_j\mathbf{a}^i \otimes \mathbf{a}_r = T_i^{\bullet j}\mathbf{a}^i \otimes \mathbf{a}_j \\ T_{ij}\mathbf{a}^i \otimes \mathbf{a}^j\mathbf{a}_r \otimes \mathbf{a}^r = T_{ij}\delta^j_r\mathbf{a}^i \otimes \mathbf{a}^r = T_{ij}\mathbf{a}^i \otimes \mathbf{a}^j \end{array} \right\} = \mathbf{T}. \tag{2.82}$$

The definition of the trace in equation (1.80) for the Cartesian space is extended as the scalar product with the metric tensor in the Riemann space, leading to the sum of the diagonal components of the mixed variant form, that is,

$$\mathrm{tr}\mathbf{T} \equiv \mathbf{T}:\mathbf{a} = \mathbf{a}:\mathbf{T} = \begin{cases} \mathbf{a}^i \cdot \mathbf{T}\mathbf{a}_i = T^i_{\bullet i} \\ \mathbf{a}_i \cdot \mathbf{T}\mathbf{a}^i = T_i^{\bullet i} \end{cases}, \tag{2.83}$$

where use is made of equation $(1.202)_{3,5}$.

The transpose of the tensor is given by

$$\mathbf{T}^T = T^{ij}\mathbf{a}_j \otimes \mathbf{a}_i = T^{i}_{\bullet j}\mathbf{a}^j \otimes \mathbf{a}_i = T_i^{\bullet j}\mathbf{a}_j \otimes \mathbf{a}^i = T_{ij}\mathbf{a}^j \otimes \mathbf{a}^i \tag{2.84}$$

by exchanging the base vectors of $\mathbf{T}$ in equation (2.77). Equation (2.84) can be rewritten by interchanging the indices i and j as

$$\mathbf{T}^T = T^{ji}\mathbf{a}_i \otimes \mathbf{a}_j = T_j^{\bullet i}\mathbf{a}_i \otimes \mathbf{a}^j = T^{j}_{\bullet i}\mathbf{a}^i \otimes \mathbf{a}_j = T_{ji}\mathbf{a}^i \otimes \mathbf{a}^j \tag{2.85}$$

in which the positions of front and rear indices in the component are interchanged in the expression of the original tensor. Here, note that the simple interchange of index letters in the mixed variant forms, that is, $T^{j}_{\bullet i}\mathbf{a}_i \otimes \mathbf{a}^j$ and $T_j^{\bullet i}\mathbf{a}^i \otimes \mathbf{a}_j$, is not permitted. Thus, the transposed tensors for the mixed variant forms are obtained by exchanging the base vectors, but cannot be obtained by exchanging the index letters in component.

If $\mathbf{T}$ is the symmetric tensor, it follows from equations (2.77) and (2.85) that

$$T^{ij} = T^{ji}, \quad T^{i}_{\bullet j} = T_j^{\bullet i}\,(T_i^{\bullet j} = T^{j}_{\bullet i}), \quad T_{ij} = T_{ji}. \tag{2.86}$$

However, the indices cannot be exchanged between the contravariant and the covariant ones, that is,

$$T^{i}_{\bullet j} \neq T^{j}_{\bullet i}, \quad T_i^{\bullet j} \neq T_j^{\bullet i}. \tag{2.87}$$

Therefore, the mixed components $T^{i}_{\bullet j}$ and $T_j^{\bullet i}$ possess nine independent values even in the symmetric tensor, which will often prove to be inconvenient.

The tensor representation in the Cartesian coordinate system is transformed to those in the curvilinear coordinate systems by equation (2.15) as follows:

$$\mathrm{T}_{ij}\mathbf{e}_i \otimes \mathbf{e}_j = T^{ij}\underset{\sim}{a}_{ir}\mathbf{e}_r \otimes \underset{\sim}{a}_{js}\mathbf{e}_s = T^{i}_{\bullet j}\underset{\sim}{a}_{ir}\mathbf{e}_r \otimes \tilde{a}^{js}\mathbf{e}_s = T_i^{\bullet j}\tilde{a}^{ir}\mathbf{e}_r \otimes \underset{\sim}{a}_{js}\mathbf{e}_s = T_{ij}\tilde{a}^{ir}\mathbf{e}_r \otimes \tilde{a}^{js}\mathbf{e}_s, \tag{2.88}$$

and thus the following relations hold between various components.

$$\mathrm{T}_{ij} = \underset{\sim}{a}_{ri}\underset{\sim}{a}_{sj}T^{rs} = \underset{\sim}{a}_{ri}\tilde{a}^{sj}T^{r}_{\bullet s} = \tilde{a}^{ri}\underset{\sim}{a}_{js}T_r^{\bullet s} = \tilde{a}^{ri}\tilde{a}^{sj}T_{rs}. \tag{2.89}$$

Inversely, from equation (2.89) with equation (2.16), one has

$$T^{ij} = \tilde{a}^{ir}\tilde{a}^{js}\mathrm{T}_{rs}, \quad T^{i}_{\bullet j} = \tilde{a}^{ir}\underset{\sim}{a}_{js}\mathrm{T}_{rs}, \quad T_i^{\bullet j} = \underset{\sim}{a}_{ir}\tilde{a}^{js}\mathrm{T}_{rs}, \quad T_{ij} = \underset{\sim}{a}_{ir}\underset{\sim}{a}_{js}\mathrm{T}_{rs}. \tag{2.90}$$

The following rule for transformation to the Cartesian coordinate system holds for the general mixed variant components:

$$T^{i_1\cdots i_m}_{j_1\cdots j_n} = \underset{\sim}{a}_{i_1 r_1}\cdots\underset{\sim}{a}_{i_m r_m}\tilde{a}^{j_1 s_1}\cdots\tilde{a}^{j_n s_n}\mathrm{T}_{r_1\cdots r_m s_1\cdots s_n}. \tag{2.91}$$

The representations of the tensor $\mathbf{T}$ in terms of the covariant base $\bar{\mathbf{a}}_i$ and the contravariant base $\bar{\mathbf{a}}^i$ are as follows:

$$\mathbf{T} = \bar{T}^{ij}\bar{\mathbf{a}}_i \otimes \bar{\mathbf{a}}_j = \bar{T}^{i}_{\bullet j}\bar{\mathbf{a}}_i \otimes \bar{\mathbf{a}}^j = \bar{T}_i^{\bullet j}\bar{\mathbf{a}}^i \otimes \bar{\mathbf{a}}_j = \bar{T}_{ij}\bar{\mathbf{a}}^i \otimes \bar{\mathbf{a}}^j. \tag{2.92}$$

Substitution of equation (2.4) into equation (2.77) leads to

$$\mathbf{T} = T^{ij}\frac{\partial\bar{\theta}^r}{\partial\theta^i}\bar{\mathbf{a}}_r \otimes \frac{\partial\bar{\theta}^s}{\partial\theta^j}\bar{\mathbf{a}}_s = T^i_{\bullet j}\frac{\partial\bar{\theta}^r}{\partial\theta^i}\bar{\mathbf{a}}_r \otimes \frac{\partial\theta^j}{\partial\bar{\theta}^s}\bar{\mathbf{a}}^s$$
$$= T_i^{\bullet j}\frac{\partial\theta^i}{\partial\bar{\theta}^r}\bar{\mathbf{a}}^r \otimes \frac{\partial\bar{\theta}^s}{\partial\theta^j}\bar{\mathbf{a}}_s = T_{ij}\frac{\partial\theta^i}{\partial\bar{\theta}^r}\bar{\mathbf{a}}^r \otimes \frac{\partial\theta^j}{\partial\bar{\theta}^s}\bar{\mathbf{a}}^s.$$

Comparing this with equation (2.92), the transformation rules of components are given by

$$\left.\begin{array}{ll}
\bar{T}^{ij} = \dfrac{\partial\bar{\theta}^i}{\partial\theta^r}\dfrac{\partial\bar{\theta}^j}{\partial\theta^s}T^{rs}, & T^{ij} = \dfrac{\partial\theta^i}{\partial\bar{\theta}^r}\dfrac{\partial\theta^j}{\partial\bar{\theta}^s}\bar{T}^{rs} \\
\bar{T}^i_{\bullet j} = \dfrac{\partial\bar{\theta}^i}{\partial\theta^r}\dfrac{\partial\theta^s}{\partial\bar{\theta}^j}T^r_{\bullet s}, & T^i_{\bullet j} = \dfrac{\partial\theta^i}{\partial\bar{\theta}^r}\dfrac{\partial\bar{\theta}^s}{\partial\theta^j}\bar{T}^r_{\bullet s} \\
\bar{T}_i^{\bullet j} = \dfrac{\partial\theta^r}{\partial\bar{\theta}^i}\dfrac{\partial\bar{\theta}^j}{\partial\theta^s}T_r^{\bullet s}, & T_i^{\bullet j} = \dfrac{\partial\bar{\theta}^r}{\partial\theta^i}\dfrac{\partial\theta^j}{\partial\bar{\theta}^s}\bar{T}_r^{\bullet s} \\
\bar{T}_{ij} = \dfrac{\partial\theta^r}{\partial\bar{\theta}^i}\dfrac{\partial\theta^s}{\partial\bar{\theta}^j}T_{rs}, & T_{ij} = \dfrac{\partial\bar{\theta}^r}{\partial\theta^i}\dfrac{\partial\bar{\theta}^s}{\partial\theta^j}\bar{T}_{rs}
\end{array}\right\}. \tag{2.93}$$

In general, the component $T^{q_1\cdots q_n}_{p_1\cdots p_m}$ fulfilling the transformation rule

$$\bar{T}^{i_1\cdots i_m}_{j_1\cdots j_n} = \frac{\partial\bar{\theta}^{i_1}}{\partial\theta^{r_1}}\cdots\frac{\partial\bar{\theta}^{i_m}}{\partial\theta^{r_m}}\frac{\partial\theta^{s_1}}{\partial\bar{\theta}^{j_1}}\cdots\frac{\partial\theta^{s_n}}{\partial\bar{\theta}^{j_n}}T^{r_1\cdots r_m}_{s_1\cdots s_n} \tag{2.94}$$

is called the $(m+n)$th-order *mixed variant* component, consisting of an mth-order *contravariant* and an nth-order *covariant*; note the superscript and subscript added to the contravariant and the covariant components, respectively.

2.4 Physical Components of Vectors and Tensors

The projections of the vector $\mathbf{v}$ onto the base vectors $\mathbf{a}_i$ and $\mathbf{a}^i$ are given by $v^i = \mathbf{v} \cdot \mathbf{a}^i$ and $v_i = \mathbf{v} \cdot \mathbf{a}_i$, respectively, and an arbitrary vector is described by equation (2.56). However, the base vectors $\mathbf{a}_i$ and $\mathbf{a}^i$ are not the unit vectors, possessing the lengths

$$||\mathbf{a}_i|| = \sqrt{a_{rr}}, \quad ||\mathbf{a}^i|| = \sqrt{a^{rr}}\ . \tag{2.95}$$

An arbitrary vector can be expressed in terms of the normalized base vectors as

$$\mathbf{v} = v^{(i)}\mathbf{a}_{(i)} = v_{(i)}\mathbf{a}^{(i)}, \tag{2.96}$$

where

$$v^{(i)} \equiv \sqrt{a_{rr}}v^i, \quad v_{(i)} \equiv \sqrt{a^{rr}}v_i\ , \tag{2.97}$$

$$\mathbf{a}_{(i)} \equiv \mathbf{a}_i/\sqrt{a_{rr}}, \quad \mathbf{a}^{(i)} \equiv \mathbf{a}^i/\sqrt{a^{rr}}\ , \tag{2.98}$$

noting that

$$\mathbf{v} = \left\{\begin{array}{l}
||\mathbf{a}_r||(\mathbf{v}\cdot\mathbf{a}^i)\dfrac{\mathbf{a}_i}{||\mathbf{a}_s||} = \sqrt{a_{rr}}v^i\dfrac{\mathbf{a}_i}{\sqrt{a_{ss}}} \\
||\mathbf{a}^r||(\mathbf{v}\cdot\mathbf{a}_i)\dfrac{\mathbf{a}^i}{||\mathbf{a}^s||} = \sqrt{a^{rr}}v_i\dfrac{\mathbf{a}^i}{\sqrt{a^{ss}}}
\end{array}\right. .$$

Here, note that $\mathbf{a}_{(i)}$ and $\mathbf{a}^{(i)}$ are the unit vectors and thus $v^{(i)}$ and $v_{(i)}$ are interpreted to possess the physical dimensions and the actual side lengths of the parallelogram which contains $\mathbf{v}$ and possesses the sides parallel to the coordinate axes (see Figure 2.6) as known from

$$||v^i\mathbf{a}_i|| = \sqrt{v^i\mathbf{a}_i \cdot v^i\mathbf{a}_i} = \sqrt{a_{ii}}v^i = v^{(i)},\ ||v_i\mathbf{a}^i|| = \sqrt{v_i\mathbf{a}^i \cdot v_i\mathbf{a}^i} = \sqrt{a^{ii}}v_i = v_{(i)}\ \text{(no sum)}. \tag{2.99}$$

$v^{(i)}$ and $v_{(i)}$ are referred to as the *physical components* of the vector $\mathbf{v}$. However, note that they do not obey the tensor transformation since they are not components of tensors.

On the other hand, the direct projections of $\mathbf{v}$ onto the unit vectors $\mathbf{a}^{(i)}$ and $\mathbf{a}_{(i)}$ are given by

$$v^{<i>} \equiv \mathbf{v} \cdot \mathbf{a}^{(i)} = \frac{v^i}{\sqrt{a^{rr}}},\ v_{<i>} \equiv \mathbf{v} \cdot \mathbf{a}_{(i)} = \frac{v_i}{\sqrt{a_{rr}}} \tag{2.100}$$

(see Figure 2.6), noting that

$$\mathbf{v} \cdot \mathbf{a}^{(i)} = v^j\mathbf{a}_j \cdot \frac{\mathbf{a}^i}{\sqrt{a^{rr}}},\ \mathbf{v} \cdot \mathbf{a}_{(i)} = v_j\mathbf{a}^j \cdot \frac{\mathbf{a}_i}{\sqrt{a_{rr}}}. \tag{2.101}$$

The expression

$$\mathbf{v} = v^{<i>}\mathbf{a}_{<i>} = v_{<i>}\mathbf{a}^{<i>} \tag{2.102}$$

holds, where

$$\mathbf{a}_{<i>} \equiv \sqrt{a^{rr}}\mathbf{a}_i,\ \mathbf{a}^{<i>} \equiv \sqrt{a_{rr}}\mathbf{a}^i, \tag{2.103}$$

noting that

$$\mathbf{v} = \begin{cases} \left(\mathbf{v} \cdot \dfrac{\mathbf{a}^i}{||\mathbf{a}^r||}\right) ||\mathbf{a}^s||\mathbf{a}_i = \dfrac{v^i}{\sqrt{a^{rr}}}\sqrt{a^{ss}}\mathbf{a}_i \\ \left(\mathbf{v} \cdot \dfrac{\mathbf{a}_i}{||\mathbf{a}_r||}\right) ||\mathbf{a}_s||\mathbf{a}^i = \dfrac{v_i}{\sqrt{a_{rr}}}\sqrt{a_{ss}}\mathbf{a}^i \end{cases}.$$

Similarly, the second-order tensor can be written from equation (2.88) as

$$\mathbf{T} = T^{(i)(j)}\mathbf{a}_{(i)} \otimes \mathbf{a}_{(j)} = T^{(i)}_{\bullet(j)}\mathbf{a}_{(i)} \otimes \mathbf{a}^{(j)} = T^{\bullet(j)}_{(i)}\mathbf{a}^{(i)} \otimes \mathbf{a}_{(j)} = T_{(i)(j)}\mathbf{a}^{(i)} \otimes \mathbf{a}^{(j)}, \tag{2.104}$$

where

$$\left.\begin{array}{l} T^{(i)(j)} \equiv a_{rr}T^{ij},\ T^{(i)}_{\bullet(j)} \equiv \sqrt{a_{rr}}\sqrt{a^{ss}}T^{i}_{\bullet j} \\ T^{\bullet(j)}_{(i)} \equiv \sqrt{a^{rr}}\sqrt{a_{ss}}T^{\bullet j}_{i},\ T_{(i)(j)} \equiv a^{rr}T_{ij} \end{array}\right\} \tag{2.105}$$

which are referred to as the *physical components* of the tensor $\mathbf{T}$.

2.5 Covariant Derivative of Base Vectors with Christoffel Symbol

The base vector in the rectangular Cartesian coordinate system with base $\{\mathbf{e}_i\}$ does not change but the base vectors $\mathbf{a}_i$ and $\mathbf{a}^i$ change with the coordinates and thus their derivatives with respect to the coordinate do not diminish, resulting in $\partial\mathbf{a}_i/\partial\theta^j \neq \mathbf{0}$, $\partial\mathbf{a}^i/\partial\theta^j \neq \mathbf{0}$.

From equation (2.2) we have

$$\frac{\partial\mathbf{a}_i}{\partial\theta^j} = \frac{\partial}{\partial\theta^j}\left\{\frac{\partial(x_r\mathbf{e}_r)}{\partial\theta^i}\right\} = \frac{\partial^2 x_r}{\partial\theta^i\partial\theta^j}\mathbf{e}_r. \tag{2.106}$$

Substituting equation (2.14) into equation (2.106) yields

$$\frac{\partial \mathbf{a}_i}{\partial \theta^j} = \frac{\partial^2 x_r}{\partial \theta^i \partial \theta^j} \frac{\partial \theta^s}{\partial x^r} \mathbf{a}_s = \frac{\partial \mathbf{a}_j}{\partial \theta^i}. \tag{2.107}$$

Let us rewrite equation (2.107) as

$$\frac{\partial \mathbf{a}_i}{\partial \theta^j} = \Gamma^s_{ij} \mathbf{a}_s, \ \Gamma^s_{ij} = \frac{\partial \mathbf{a}_i}{\partial \theta^j} \cdot \mathbf{a}^s, \tag{2.108}$$

where

$$\boxed{\Gamma^s_{ij} \equiv \left\{ \begin{matrix} s \\ ij \end{matrix} \right\} \equiv \frac{\partial^2 x_r}{\partial \theta^i \partial \theta^j} \frac{\partial \theta^s}{\partial x^r} \left(= \Gamma^s_{ji} \right)} \tag{2.109}$$

which is called the *Christoffel symbol of the second kind*. Further, setting

$$\boxed{\Gamma_{ijk} \equiv [ij, k] \equiv a_{ks} \Gamma^s_{ij} (= \Gamma_{jik})} \tag{2.110}$$

equation (2.108) is rewritten, by invoking equation (2.19), as

$$\frac{\partial \mathbf{a}_i}{\partial \theta^j} = \Gamma_{ijr} \mathbf{a}^r, \ \Gamma_{ijr} = \frac{\partial \mathbf{a}_i}{\partial \theta^j} \cdot \mathbf{a}_r. \tag{2.111}$$

Γ_{ijk} is called the *Christoffel symbol of the first kind*. Γ^s_{ij} and Γ_{ijk} do not obey the tensor transformation rule, as will be shown later, and thus they are not called 'tensors' but 'symbols'.

From the three types of derivatives

$$\left.\begin{aligned} \frac{\partial a_{ij}}{\partial \theta^k} &= \frac{\partial (\mathbf{a}_i \cdot \mathbf{a}_j)}{\partial \theta^k} = \frac{\partial \mathbf{a}_i}{\partial \theta^k} \cdot \mathbf{a}_j + \mathbf{a}_i \cdot \frac{\partial \mathbf{a}_j}{\partial \theta^k} \\ \frac{\partial a_{jk}}{\partial \theta^i} &= \frac{\partial (\mathbf{a}_j \cdot \mathbf{a}_k)}{\partial \theta^i} = \frac{\partial \mathbf{a}_j}{\partial \theta^i} \cdot \mathbf{a}_k + \mathbf{a}_j \cdot \frac{\partial \mathbf{a}_k}{\partial \theta^i} \\ \frac{\partial a_{ki}}{\partial \theta^j} &= \frac{\partial (\mathbf{a}_k \cdot \mathbf{a}_i)}{\partial \theta^j} = \frac{\partial \mathbf{a}_k}{\partial \theta^j} \cdot \mathbf{a}_i + \mathbf{a}_k \cdot \frac{\partial \mathbf{a}_i}{\partial \theta^j} \end{aligned}\right\}$$

we have

$$\boxed{\Gamma_{ijk} = \frac{1}{2} \left(\frac{\partial a_{jk}}{\partial \theta^i} + \frac{\partial a_{ik}}{\partial \theta^j} - \frac{\partial a_{ij}}{\partial \theta^k} \right)} \tag{2.112}$$

noting

$$\begin{aligned} &\frac{\partial a_{jk}}{\partial \theta^i} + \frac{\partial a_{ki}}{\partial \theta^j} - \frac{\partial a_{ij}}{\partial \theta^k} \\ &= \frac{\partial \mathbf{a}_j}{\partial \theta^i} \cdot \mathbf{a}_k + \mathbf{a}_j \cdot \frac{\partial \mathbf{a}_k}{\partial \theta^i} + \frac{\partial \mathbf{a}_k}{\partial \theta^j} \cdot \mathbf{a}_i + \mathbf{a}_k \cdot \frac{\partial \mathbf{a}_i}{\partial \theta^j} - \frac{\partial \mathbf{a}_i}{\partial \theta^k} \cdot \mathbf{a}_j - \mathbf{a}_i \cdot \frac{\partial \mathbf{a}_j}{\partial \theta^k} = 2 \frac{\partial \mathbf{a}_i}{\partial \theta^j} \cdot \mathbf{a}_k, \end{aligned}$$

where use is made of the symmetry shown in equation (2.107).

The gradient of the covariant component of a metric tensor is given from equation (2.112) as

$$\frac{\partial a_{ij}}{\partial \theta^k} = \Gamma_{ikj} + \Gamma_{kji} \tag{2.113}$$

by permuting the subscripts in equation (2.112) and summing them. By virtue of (2.110), equation (2.113) is rewritten as

$$\frac{\partial a_{ij}}{\partial \theta^k} = a_{jr}\Gamma^r_{ik} + a_{ir}\Gamma^r_{kj}. \tag{2.114}$$

Likewise, it follows that

$$\frac{\partial a^{ij}}{\partial \theta^k} = -a^{ir}\Gamma^j_{rk} - a^{jr}\Gamma^i_{rk}. \tag{2.115}$$

Substituting equation (2.112) into $\Gamma^k_{ij} = \delta^k_s\Gamma^s_{ij} = a^{kr}a_{rs}\Gamma^s_{ij} = a^{kr}\Gamma_{ijr}$ with $a^{kr}a_{rs} = \delta^k_s$ in equation (2.20), the inverse expression of equation (2.110) is obtained as follows:

$$\boxed{\left\{ {k \atop ij} \right\} \equiv \Gamma^k_{ij} = a^{kr}\Gamma_{ijr} = \frac{1}{2}a^{kr}\left(\frac{\partial a_{jr}}{\partial \theta^i} + \frac{\partial a_{ir}}{\partial \theta^j} - \frac{\partial a_{ij}}{\partial \theta^r}\right)} \tag{2.116}$$

The symbol on the left-hand side in equation (2.116) is simply called the *Christoffel symbol*; it is zero in the rectangular Cartesian coordinate system because of $a_{ij} = \delta_{ij} = \text{const}$.

It follows that

$$\frac{\partial a}{\partial \theta^i} = aa^{rs}\left(a_{st}\Gamma^t_{ri} + a_{rt}\Gamma^t_{is}\right), \tag{2.117}$$

noting

$$\frac{\partial a}{\partial \theta^i} = \frac{\partial \det[a_{pq}]}{\partial \theta^i} = \frac{\partial \det[a_{pq}]}{\partial a_{rs}}\frac{\partial a_{rs}}{\partial \theta^i} = \det[a_{pq}]a^{-1}_{rs}\frac{\partial a_{rs}}{\partial \theta^i} = aa^{rs}\frac{\partial a_{rs}}{\partial \theta^i} \tag{2.118}$$

with equations (1.371), (2.21) and (2.114).

The two kinds of Christoffel symbol Γ_{ijk} and Γ^k_{ij} mentioned above are symmetric with respect to the indices i and j. The upper and lower positions of the third index k can be exchanged by applying the metric tensor as the shifter, resulting in the exchange between the first and the second symbols as shown in equations (2.110) and (2.116).

Equation (2.108) is written in the coordinate system with the base $\{\bar{\mathbf{a}}_i\}$ as

$$\frac{\partial \bar{\mathbf{a}}_i}{\partial \bar{\theta}^j} = \bar{\Gamma}^k_{ij}\bar{\mathbf{a}}_k. \tag{2.119}$$

On the other hand, taking equations (2.3), (2.4), (2.108) into account, the following equation is obtained:

$$\begin{aligned}
\frac{\partial \bar{\mathbf{a}}_i}{\partial \bar{\theta}^j} = \frac{\partial}{\partial \bar{\theta}^j}\left(\frac{\partial \theta^r}{\partial \bar{\theta}^i}\mathbf{a}_r\right) &= \frac{\partial^2 \theta^r}{\partial \bar{\theta}^i \partial \bar{\theta}^j}\mathbf{a}_r + \frac{\partial \theta^r}{\partial \bar{\theta}^i}\frac{\partial \mathbf{a}_r}{\partial \bar{\theta}^j} \\
&= \frac{\partial^2 \theta^r}{\partial \bar{\theta}^i \partial \bar{\theta}^j}\mathbf{a}_r + \frac{\partial \theta^r}{\partial \bar{\theta}^i}\frac{\partial \theta^s}{\partial \bar{\theta}^j}\frac{\partial \mathbf{a}_r}{\partial \theta^s} \\
&= \frac{\partial^2 \theta^r}{\partial \bar{\theta}^i \partial \bar{\theta}^j}\mathbf{a}_r + \frac{\partial \theta^r}{\partial \bar{\theta}^i}\frac{\partial \theta^s}{\partial \bar{\theta}^j}\Gamma^t_{rs}\mathbf{a}_t \\
&= \frac{\partial^2 \theta^r}{\partial \bar{\theta}^i \partial \bar{\theta}^j}\frac{\partial \bar{\theta}^k}{\partial \theta^r}\bar{\mathbf{a}}_k + \frac{\partial \theta^r}{\partial \bar{\theta}^i}\frac{\partial \theta^s}{\partial \bar{\theta}^j}\Gamma^t_{rs}\frac{\partial \bar{\theta}^k}{\partial \theta^t}\bar{\mathbf{a}}_k.
\end{aligned}$$

Equating this equation to (2.119), one obtains the following coordinate transformation rule for Γ^k_{ij}:

$$\bar{\Gamma}^k_{ij} = \frac{\partial^2 \theta^r}{\partial \bar{\theta}^i \partial \bar{\theta}^j} \frac{\partial \bar{\theta}^k}{\partial \theta^r} + \frac{\partial \theta^r}{\partial \bar{\theta}^i} \frac{\partial \theta^s}{\partial \bar{\theta}^j} \frac{\partial \bar{\theta}^k}{\partial \theta^t} \Gamma^t_{rs}. \tag{2.120}$$

It is evident from equation (2.120) that the Christoffel symbol is not a tensor component due to the presence of the first term.

Substituting equation (2.108) into the relation

$$\frac{\partial \mathbf{a}^i}{\partial \theta^j} \cdot \mathbf{a}_s + \mathbf{a}^i \cdot \frac{\partial \mathbf{a}_s}{\partial \theta^j} = 0$$

obtained by differentiating $\mathbf{a}^i \cdot \mathbf{a}_s = \delta_{is}$ by θ^j, we have

$$\frac{\partial \mathbf{a}^i}{\partial \theta^j} \cdot \mathbf{a}_s = -\mathbf{a}^i \cdot \frac{\partial \mathbf{a}_s}{\partial \theta^j} = -\mathbf{a}^i \cdot \Gamma^r_{sj} \mathbf{a}_r = -\Gamma^i_{sj} = -\Gamma^i_{rj}(\mathbf{a}^r \cdot \mathbf{a}_s),$$

from which it follows that

$$\frac{\partial \mathbf{a}^i}{\partial \theta^j} = -\Gamma^i_{rj} \mathbf{a}^r = -\Gamma^i_{jr} \mathbf{a}^r. \tag{2.121}$$

We have

$$\begin{Bmatrix} i \\ ij \end{Bmatrix} = \frac{1}{2} a^{ir} \left(\frac{\partial a_{jr}}{\partial \theta^i} + \frac{\partial a_{ir}}{\partial \theta^j} - \frac{\partial a_{ij}}{\partial \theta^r} \right) = \frac{1}{2} a^{ir} \frac{\partial a_{ir}}{\partial \theta_j} \tag{2.122}$$

from equation (2.116), noting that

$$a^{ir} \frac{\partial a_{jr}}{\partial \theta^i} = a^{ri} \frac{\partial a_{ji}}{\partial \theta^r} = a^{ir} \frac{\partial a_{ij}}{\partial \theta^r}$$

($i \leftrightarrow r$ and $a_{ji} = a_{ij}$). Further, noting that

$$\frac{1}{2} a^{rs} \frac{\partial a_{rs}}{\partial \theta_i} = \frac{1}{2a} a a^{rs} \frac{\partial a_{rs}}{\partial \theta_i} = \frac{1}{2a} \frac{\partial a}{\partial \theta_i} = \frac{1}{\sqrt{a}} \frac{1}{2\sqrt{a}} \frac{\partial a}{\partial \theta_i} = \frac{\partial \log \sqrt{a}}{\partial \theta_i}$$

by virtue of (2.118), equation (2.122) can be rewritten as

$$\begin{Bmatrix} i \\ ij \end{Bmatrix} = \frac{1}{2a} \frac{\partial a}{\partial \theta_i} = \frac{1}{\sqrt{a}} \frac{\partial \sqrt{a}}{\partial \theta_i} = \frac{\partial \log \sqrt{a}}{\partial \theta^j}. \tag{2.123}$$

Further, it follows from equation (2.116) that

$$\begin{Bmatrix} k \\ ii \end{Bmatrix} = \frac{1}{2} a^{kr} \left(2 \frac{\partial a_{ir}}{\partial \theta^i} - \frac{\partial a_{ii}}{\partial \theta^r} \right) \quad \text{(no sum for } i\text{)}. \tag{2.124}$$

In the *orthogonal curvilinear coordinate system*, noting that $a_{ij} = a^{ij} = 0$ for $i \neq j$ and $a_{ii} = 1/a^{ii}$ (no sum), the components of the Christoffel symbol are given from equation (2.116) as follows (cf. Bowen and Wang, 1976):

$$\begin{Bmatrix} k \\ ij \end{Bmatrix} = \frac{1}{2a_{kk}}\left(\frac{\partial a_{jk}}{\partial \theta^i} + \frac{\partial a_{ik}}{\partial \theta^j} - \frac{\partial a_{ij}}{\partial \theta^r}\right) \text{ (no sum)}, \tag{2.125}$$

resulting in

$$\left.\begin{aligned} &\begin{Bmatrix} k \\ ij \end{Bmatrix} = 0 \quad \text{if} \quad i \neq j \neq k \neq i \\ &\begin{Bmatrix} k \\ ii \end{Bmatrix} = -\frac{1}{2a_{kk}}\frac{\partial a_{ii}}{\partial \theta^k} \quad \text{if} \quad i \neq k \\ &\begin{Bmatrix} i \\ ij \end{Bmatrix} = \begin{Bmatrix} i \\ ji \end{Bmatrix} = -\frac{1}{2a_{ii}}\frac{\partial a_{ii}}{\partial \theta^j} \quad \text{if} \quad i \neq j \\ &\begin{Bmatrix} i \\ ii \end{Bmatrix} = \frac{1}{2a_{ii}}\frac{\partial a_{ii}}{\partial \theta^i} \end{aligned}\right\} \text{(no sum)}. \tag{2.125)'$$

Example 2.3: Cylindrical polar coordinate system This is the orthogonal coordinate system considered in Example 2.1 and thus we can apply equation (2.125). The non-zero components of the Christoffel symbol are given by

$$\begin{Bmatrix} 1 \\ 2\,2 \end{Bmatrix} = -\frac{1}{2a_{11}}\frac{\partial a_{22}}{\partial \theta^1} = -\frac{1}{2}\frac{\partial r^2}{\partial r} = -r, \quad \begin{Bmatrix} 2 \\ 1\,2 \end{Bmatrix} = \frac{1}{2}\frac{1}{a_{22}}\frac{\partial a_{22}}{\partial \theta^1} = \frac{1}{2}\frac{1}{r^2}\frac{\partial r^2}{\partial r} = \frac{1}{r}. \tag{2.126}$$

Example 2.4: Spherical polar coordinate system This is also the orthogonal coordinate system considered in Example 2.2 and thus we can apply equation (2.125)′. The non-zero components of the Christoffel symbol are given by

$$\left.\begin{aligned} &\begin{Bmatrix} 1 \\ 2\,2 \end{Bmatrix} = -\frac{1}{2a_{11}}\frac{\partial a_{22}}{\partial \theta^1} = -\frac{1}{2}\frac{\partial r^2}{\partial r} = -r, \quad \begin{Bmatrix} 1 \\ 3\,3 \end{Bmatrix} = -\frac{1}{2a_{11}}\frac{\partial a_{33}}{\partial \theta^1} = -\frac{1}{2}\frac{\partial (r^2 \sin^2\alpha)}{\partial r} = -r\sin^2\alpha \\ &\begin{Bmatrix} 2 \\ 1\,2 \end{Bmatrix} = \frac{1}{2a_{22}}\frac{\partial a_{22}}{\partial \theta^1} = \frac{1}{2r^2}\frac{\partial r^2}{\partial r} = \frac{1}{r}, \quad \begin{Bmatrix} 2 \\ 3\,3 \end{Bmatrix} = -\frac{1}{2a_{22}}\frac{\partial a_{33}}{\partial \theta^2} = -\frac{1}{2r^2}\frac{\partial (r^2 \sin^2\alpha)}{\partial \alpha} = -\sin\alpha\cos\alpha \\ &\begin{Bmatrix} 3 \\ 1\,3 \end{Bmatrix} = \frac{1}{2a_{33}}\frac{\partial a_{33}}{\partial \theta^1} = \frac{1}{2r^2\sin^2\alpha}\frac{\partial (r^2 \sin^2\alpha)}{\partial r} = \frac{1}{2r^2}\frac{\partial r^2}{\partial r} = \frac{1}{r} \\ &\begin{Bmatrix} 3 \\ 2\,3 \end{Bmatrix} = \frac{1}{2a_{33}}\frac{\partial a_{33}}{\partial \theta^2} = \frac{1}{2r^2\sin^2\alpha}\frac{\partial (r^2 \sin^2\alpha)}{\partial \alpha} = \frac{\cos\alpha}{\sin\alpha} \end{aligned}\right\}. \tag{2.127}$$

2.6 Covariant Derivatives of Scalars, Vectors and Tensors

This section addresses the covariant derivative of scalar, vector and tensors, referring widely to Sokolnikoff (1951) and Hisada (1992).

The derivative of the scalar s by the contravariant coordinate, that is, $\partial s/\partial \theta^i$ obeys the covariant transformation

$$\frac{\partial s}{\partial \theta^i} = \frac{\partial \bar{\theta}^j}{\partial \theta^i}\frac{\partial s}{\partial \bar{\theta}^j}.$$

Then, we define it as the covariant derivative of scalar and denote it by the following symbol.

$$s|_i \equiv \frac{\partial s}{\partial \theta^i}. \tag{2.128}$$

Taking the partial derivative of the vector $\mathbf{v} = v^i \mathbf{a}_i = v_i \mathbf{a}^i$ and noting equations (2.108) and (2.121), one has

$$\frac{\partial \mathbf{v}}{\partial \theta^j} = \begin{cases} \dfrac{\partial (v^i \mathbf{a}_i)}{\partial \theta^j} = \dfrac{\partial v^i}{\partial \theta^j}\mathbf{a}_i + v^i \dfrac{\partial \mathbf{a}_i}{\partial \theta^j} = \dfrac{\partial v^i}{\partial \theta^j}\mathbf{a}_i + v^i \Gamma^r_{ij}\mathbf{a}_r = \left(\dfrac{\partial v^i}{\partial \theta^j} + v^r \Gamma^i_{rj}\right)\mathbf{a}_i \\ \dfrac{\partial (v_i \mathbf{a}^i)}{\partial \theta^j} = \dfrac{\partial v_i}{\partial \theta^j}\mathbf{a}^i + v_i \dfrac{\partial \mathbf{a}^i}{\partial \theta^j} = \dfrac{\partial v_i}{\partial \theta^j}\mathbf{a}^i - v_i \Gamma^i_{ir}\mathbf{a}^r = \left(\dfrac{\partial v_i}{\partial \theta^j} - v_r \Gamma^r_{ji}\right)\mathbf{a}^i \end{cases}.$$

The covariant derivative of vector is defined using the Christoffel symbol as

$$\boxed{\frac{\partial \mathbf{v}}{\partial \theta^j} = v^i|_j \mathbf{a}_i = v_i|_j \mathbf{a}^i} \tag{2.129}$$

where

$$v^i|_j \equiv \frac{\partial v^i}{\partial \theta^j} + v^r \Gamma^i_{rj}, \quad v_i|_j \equiv \frac{\partial v_i}{\partial \theta^j} - v_r \Gamma^r_{ji}. \tag{2.130}$$

The covariant derivative of vector requires a supplementary term in terms of the Christoffel symbol in addition to the usual derivative, while the scalar covariant derivative of scalar obeys the usual derivative. The supplementary term is interpreted to depend on the variation of the base vector along the coordinate axis as shown in equations (2.108) and (2.121).

Now, examine the coordinate transformation of $v^i|_j$. First, substituting equation (2.64) and the equation

$$\frac{\partial^2 \theta^t}{\partial \bar{\theta}^i \partial \bar{\theta}^j} = \bar{\Gamma}^k_{ij} \frac{\partial \theta^t}{\partial \bar{\theta}^k} - \frac{\partial \theta^p}{\partial \bar{\theta}^i} \frac{\partial \theta^q}{\partial \bar{\theta}^j} \Gamma^t_{pq} \tag{2.131}$$

based on equation (2.120) into

$$\frac{\partial v^i}{\partial \theta^j} = \frac{\partial}{\partial \theta^j}\left(\frac{\partial \theta^i}{\partial \bar{\theta}^r}\bar{v}^r\right) = \frac{\partial^2 \theta^i}{\partial \bar{\theta}^r \partial \theta^j}\bar{v}^r + \frac{\partial \theta^i}{\partial \bar{\theta}^r}\frac{\partial \bar{v}^r}{\partial \theta^j} = \frac{\partial^2 \theta^i}{\partial \bar{\theta}^r \partial \bar{\theta}^s}\frac{\partial \bar{\theta}^s}{\partial \theta^j}\bar{v}^r + \frac{\partial \theta^i}{\partial \bar{\theta}^r}\frac{\partial \bar{v}^r}{\partial \bar{\theta}^s}\frac{\partial \bar{\theta}^s}{\partial \theta^j},$$

we have

$$\frac{\partial v^i}{\partial \theta^j} = \left(\bar{\Gamma}^k_{rs}\frac{\partial \theta^i}{\partial \bar{\theta}^k} - \frac{\partial \theta^p}{\partial \bar{\theta}^r}\frac{\partial \theta^q}{\partial \bar{\theta}^s}\Gamma^i_{pq}\right)\frac{\partial \bar{\theta}^s}{\partial \theta^j}\bar{v}^r + \frac{\partial \theta^i}{\partial \bar{\theta}^r}\frac{\partial \bar{v}^r}{\partial \bar{\theta}^s}\frac{\partial \bar{\theta}^s}{\partial \theta^j}$$

which, considering equation (2.66), leads to

$$\frac{\partial v^i}{\partial \theta^j} + v^p \Gamma^t_{pj} = \left(\frac{\partial \bar{v}^r}{\partial \bar{\theta}^s} + \bar{v}^k \bar{\Gamma}^r_{ks}\right)\frac{\partial \theta^i}{\partial \bar{\theta}^r}\frac{\partial \bar{\theta}^s}{\partial \theta^j}.$$

It follows that

$$v^i|_j = \frac{\partial \theta^i}{\partial \bar{\theta}^r}\frac{\partial \bar{\theta}^s}{\partial \theta^j}\bar{v}^r|_s, \quad \bar{v}^i|_j = \frac{\partial \bar{\theta}^i}{\partial \theta^r}\frac{\partial \theta^s}{\partial \bar{\theta}^j}\bar{v}^r|_s. \tag{2.132}$$

$v^i|_j$ obeys the covariant transformation rule for the derivative coordinate θ_j, so that is called the *covariant derivative*.

Here, extending the Hamiltonian operator in equation (1.430), define the following generalized operator in the curvilinear coordinate system.

$$\nabla \equiv \frac{\partial}{\partial \theta^r}\mathbf{a}^r, \tag{2.133}$$

noting that the reciprocal base vector is used.

The divergence of a vector is given from equations (2.129) and (2.133) by

$$\mathrm{div}\mathbf{v} = \mathbf{v}\cdot\nabla = \mathbf{v}\cdot\frac{\partial}{\partial\theta^j}\mathbf{a}^j = \frac{\partial\mathbf{v}}{\partial\theta^j}\cdot\mathbf{a}^j = v^i|_j\mathbf{a}_i\cdot\mathbf{a}^j = v^j|_j. \tag{2.134}$$

The gradient of a vector is given by

$$\mathrm{grad}\mathbf{v} = \mathbf{v}\otimes\nabla = \mathbf{v}\otimes\frac{\partial}{\partial\theta^j}\mathbf{a}^j = \frac{\partial\mathbf{v}}{\partial\theta^j}\otimes\mathbf{a}^j = \begin{cases} v^i|_j\mathbf{a}_i\otimes\mathbf{a}^j \\ v_i|_j\mathbf{a}^i\otimes\mathbf{a}^j \end{cases} : \text{rear form}, \tag{2.135}$$

$$\mathrm{grad}\mathbf{v} = \nabla\otimes\mathbf{v} = \frac{\partial}{\partial\theta^i}\mathbf{a}^i\otimes\mathbf{v} = \mathbf{a}^i\otimes\frac{\partial\mathbf{v}}{\partial\theta^i} = \begin{cases} v^j|_i\mathbf{a}^i\otimes\mathbf{a}_j \\ v_j|_i\mathbf{a}^i\otimes\mathbf{a}^j \end{cases} : \text{front form}. \tag{2.136}$$

The divergence divv in equation (2.134) can be represented as

$$\mathrm{div}\mathbf{v} = \mathrm{grad}\mathbf{v} : \mathbf{a} \tag{2.137}$$

with the metric tensor $\mathbf{a}$, noting that

$$\mathrm{grad}\mathbf{v} : \mathbf{a} = (v^i|_j\mathbf{a}_i\otimes\mathbf{a}^j) : (\mathbf{a}^r\otimes\mathbf{a}_r) = \delta_i^r\delta_r^j v^i|_j = \delta_i^j v^i|_j = v^i|_i$$

with equation $(1.202)_6$.

Further, taking the partial derivative of the tensor $\mathbf{T} = T^{ij}\mathbf{a}_i\otimes\mathbf{a}_j$ with respect to θ^k, we have

$$\begin{aligned}
\frac{\partial\mathbf{T}}{\partial\theta^k} &= \frac{\partial}{\partial\theta^k}(T^{ij}\mathbf{a}_i\otimes\mathbf{a}_j) \\
&= \frac{\partial T^{ij}}{\partial\theta^k}\mathbf{a}_i\otimes\mathbf{a}_j + T^{ij}\frac{\partial\mathbf{a}_i}{\partial\theta^k}\otimes\mathbf{a}_j + T^{ij}\mathbf{a}_i\otimes\frac{\partial\mathbf{a}_j}{\partial\theta^k} \\
&= \frac{\partial T^{ij}}{\partial\theta^k}\mathbf{a}_i\otimes\mathbf{a}_j + T^{ij}\Gamma_{ik}^l\mathbf{a}_l\otimes\mathbf{a}_j + T^{ij}\mathbf{a}_i\otimes\Gamma_{jk}^l\mathbf{a}_l \\
&= \frac{\partial T^{ij}}{\partial\theta^k}\mathbf{a}_i\otimes\mathbf{a}_j + T^{rj}\Gamma_{rk}^i\mathbf{a}_i\otimes\mathbf{a}_j + T^{ir}\mathbf{a}_i\otimes\Gamma_{rk}^j\mathbf{a}_j \\
&= \left(\frac{\partial T^{ij}}{\partial\theta^k} + T^{rj}\Gamma_{rk}^i + T^{ir}\Gamma_{rk}^j\right)\mathbf{a}_i\otimes\mathbf{a}_j.
\end{aligned}$$

Performing similarly for the other variant forms, we have the following equation which is defined as the *covariant derivative* of the tensor.

$$\boxed{\frac{\partial \mathbf{T}}{\partial \theta^k} = T^{ij}|_k \mathbf{a}_i \otimes \mathbf{a}_j = T^i_{\bullet j}|_k \mathbf{a}_i \otimes \mathbf{a}^j = T_i^{\bullet j}|_k \mathbf{a}^i \otimes \mathbf{a}_j = T_{ij}|_k \mathbf{a}^i \otimes \mathbf{a}^j} \tag{2.138}$$

where

$$\left.\begin{aligned} T^{ij}|_k &\equiv \frac{\partial T^{ij}}{\partial \theta^k} + T^{rj}\Gamma^i_{kr} + T^{ir}\Gamma^j_{kr} \\ T^i_{\bullet j}|_k &\equiv \frac{\partial T^i_{\bullet j}}{\partial \theta^k} + T^r_{\bullet j}\Gamma^i_{kr} - T^i_{\bullet r}\Gamma^r_{jk} \\ T_i^{\bullet j}|_k &\equiv \frac{\partial T_i^{\bullet j}}{\partial \theta^k} - T_r^{\bullet j}\Gamma^r_{ik} + T_i^{\bullet r}\Gamma^j_{kr} \\ T_{ij}|_k &\equiv \frac{\partial T_{ij}}{\partial \theta^k} - T_{rj}\Gamma^r_{ik} - T_{ir}\Gamma^r_{jk} \end{aligned}\right\}. \tag{2.139}$$

Furthermore, the fact that the covariant derivative of the covariant component of a metric tensor is zero is derived from equation $(2.139)_4$ with equations (2.114) as

$$a_{ij}|_k = 0, \tag{2.140}$$

where

$$a_{ij}|_k = \frac{\partial a_{ij}}{\partial \theta^k} - a_{jr}\Gamma^r_{ik} - a_{ir}\Gamma^r_{kj}. \tag{2.141}$$

Equation (2.140) is referred to as *Ricci's lemma*.

Inner multiplication of equation $(2.93)_2$ by $\partial\theta^p/\partial\bar{\theta}^i$ yields

$$\bar{T}^i_{\bullet j}\frac{\partial \theta^p}{\partial \bar{\theta}^i}\left(= \frac{\partial \bar{\theta}^i}{\partial \theta^r}\frac{\partial \theta^s}{\partial \bar{\theta}^j}\frac{\partial \theta^p}{\partial \bar{\theta}^i}T^r_{\bullet s}\right) = \frac{\partial \theta^s}{\partial \bar{\theta}^j}T^p_{\bullet s}. \tag{2.142}$$

Differentiating equation (2.142) with respect to $\bar{\theta}^k$, that is,

$$\frac{\partial \bar{T}^i_{\bullet j}}{\partial \bar{\theta}^k}\frac{\partial \theta^p}{\partial \bar{\theta}^i} + \bar{T}^i_{\bullet j}\frac{\partial^2 \theta^p}{\partial \bar{\theta}^i \partial \bar{\theta}^k} = \frac{\partial^2 \theta^s}{\partial \bar{\theta}^j \partial \bar{\theta}^k}T^p_{\bullet s} + \frac{\partial \theta^s}{\partial \bar{\theta}^j}\frac{\partial T^p_{\bullet s}}{\partial \bar{\theta}^k},$$

and eliminating the second-order partial derivatives by means of equation (2.131) leads to

$$\begin{aligned} &\frac{\partial \bar{T}^i_{\bullet j}}{\partial \bar{\theta}^k}\frac{\partial \theta^p}{\partial \bar{\theta}^i} + \bar{T}^i_{\bullet j}\left(\bar{\Gamma}^l_{ik}\frac{\partial \theta^p}{\partial \bar{\theta}^l} - \frac{\partial \theta^a}{\partial \bar{\theta}^i}\frac{\partial \theta^b}{\partial \bar{\theta}^k}\Gamma^p_{ab}\right)\left(= \left(\bar{\Gamma}^l_{jk}\frac{\partial \theta^s}{\partial \bar{\theta}^l} - \frac{\partial \theta^a}{\partial \bar{\theta}^j}\frac{\partial \theta^b}{\partial \bar{\theta}^k}\Gamma^s_{ab}\right)T^p_{\bullet s} + \frac{\partial \theta^s}{\partial \bar{\theta}^j}\frac{\partial T^p_{\bullet s}}{\partial \bar{\theta}^k}\right) \\ &= \left(\bar{\Gamma}^l_{jk}\frac{\partial \theta^s}{\partial \bar{\theta}^l} - \frac{\partial \theta^a}{\partial \bar{\theta}^j}\frac{\partial \theta^b}{\partial \bar{\theta}^k}\Gamma^s_{ab}\right)T^p_{\bullet s} + \frac{\partial \theta^s}{\partial \bar{\theta}^j}\frac{\partial \theta^c}{\partial \bar{\theta}^k}\frac{\partial T^p_{\bullet s}}{\partial \theta^c}. \end{aligned}$$

Applying equation (2.142) to this equation, one has

$$\frac{\partial \bar{T}^i_{\bullet j}}{\partial \bar{\theta}^k}\frac{\partial \theta^p}{\partial \bar{\theta}^i} + \bar{T}^i_{\bullet j}\bar{\Gamma}^l_{ik}\frac{\partial \theta^p}{\partial \bar{\theta}^l} - \bar{T}^i_{\bullet j}\frac{\partial \theta^a}{\partial \bar{\theta}^i}\frac{\partial \theta^b}{\partial \bar{\theta}^k}\Gamma^p_{ab}$$

$$= T^p_{\bullet s}\bar{\Gamma}^l_{jk}\frac{\partial \theta^s}{\partial \bar{\theta}^l} - \frac{\partial \theta^a}{\partial \bar{\theta}^j}\frac{\partial \theta^b}{\partial \bar{\theta}^k}T^p_{\bullet s}\Gamma^s_{ab} + \frac{\partial \theta^s}{\partial \bar{\theta}^j}\frac{\partial \theta^c}{\partial \bar{\theta}^k}\frac{\partial T^p_{\bullet s}}{\partial \theta^c},$$

$$\frac{\partial \bar{T}^i_{\bullet j}}{\partial \bar{\theta}^k}\frac{\partial \theta^p}{\partial \bar{\theta}^i} + \bar{T}^i_{\bullet j}\bar{\Gamma}^l_{ik}\frac{\partial \theta^p}{\partial \bar{\theta}^l} - \frac{\partial \theta^s}{\partial \bar{\theta}^j}T^a_{\bullet s}\frac{\partial \theta^b}{\partial \bar{\theta}^k}\Gamma^p_{ab}$$

$$= \frac{\partial \theta^p}{\partial \bar{\theta}^i}\bar{T}^i_{\bullet l}\bar{\Gamma}^l_{jk} - \frac{\partial \theta^a}{\partial \bar{\theta}^j}\frac{\partial \theta^b}{\partial \bar{\theta}^k}T^p_{\bullet s}\Gamma^s_{ab} + \frac{\partial \theta^s}{\partial \bar{\theta}^j}\frac{\partial \theta^c}{\partial \bar{\theta}^k}\frac{\partial T^p_{\bullet s}}{\partial \theta^c},$$

$$\left(\frac{\partial \bar{T}^i_{\bullet j}}{\partial \bar{\theta}^k} + \bar{T}^m_{\bullet j}\bar{\Gamma}^i_{mk} - \bar{T}^i_{\bullet l}\bar{\Gamma}^l_{jk}\right)\frac{\partial \theta^p}{\partial \bar{\theta}^i} = \left(\frac{\partial T^p_{\bullet s}}{\partial \theta^c} + T^a_{\bullet s}\Gamma^p_{ac} - T^p_{\bullet t}\Gamma^t_{sc}\right)\frac{\partial \theta^s}{\partial \bar{\theta}^j}\frac{\partial \theta^c}{\partial \bar{\theta}^k}$$

which can be described by

$$\bar{T}^i_{\bullet j}|_k\frac{\partial \theta^p}{\partial \bar{\theta}^i} = \frac{\partial \theta^s}{\partial \bar{\theta}^j}\frac{\partial \theta^c}{\partial \bar{\theta}^k}T^p_{\bullet s}|_c.$$

Inner multiplication of this equation by $\partial\bar{\theta}^r/\partial\theta^p$ gives

$$\bar{T}^r_{\bullet j}|_k = \frac{\partial \theta^s}{\partial \bar{\theta}^j}\frac{\partial \bar{\theta}^r}{\partial \theta^p}\frac{\partial \theta^c}{\partial \bar{\theta}^k}T^p_{\bullet s}|_c \tag{2.143}$$

which obeys the transformation rule of the third-order tensor in equation (2.94), where the covariant transformation rule holds for the derivative coordinate θ_k.

Extending equations (2.130) for vector and equation (2.139) for second-order tensor, the covariant derivative of mixed variant component of tensor in general order is given by

$$T^{i_1\cdots i_m}_{j_1\cdots j_n}|_k = \frac{\partial}{\partial \theta^k}T^{i_1\cdots i_m}_{j_1\cdots j_n} + \sum_{r=1}^{m} T^{i_1\cdots i_{r-1}p_r i_{r+1}\cdots i_m}_{j_1\cdots j_n}\Gamma^{i_r}_{p_r k} - \sum_{s=1}^{n} T^{i_1\cdots i_m}_{j_1\cdots j_{s-1}q_s j_{s+1}\cdots j_n}\Gamma^{q_s}_{j_s k}. \tag{2.144}$$

The divergence of second-order tensor is given from equations (2.133) and (2.138) by

$$\mathrm{div}\mathbf{T} = \begin{cases} \mathbf{T}\nabla = T^{ij}|_j\mathbf{a}_i = T_i^{\bullet j}|_j\mathbf{a}^i\text{: right form} \\ \nabla\mathbf{T} = T^{ji}|_j\mathbf{a}_i = T^j_{\bullet i}|_j\mathbf{a}^i\text{: left form} \end{cases}, \tag{2.145}$$

noting that

$$\mathbf{T}\nabla = \frac{\partial \mathbf{T}}{\partial \theta^k}\mathbf{a}^k = \begin{cases} T^{ij}|_k\,\mathbf{a}_i\otimes\mathbf{a}_j\mathbf{a}^k \\ T_i^{\bullet j}|_k\,\mathbf{a}^i\otimes\mathbf{a}_j\mathbf{a}^k \end{cases}$$

$$\nabla\mathbf{T} = \mathbf{a}^k\frac{\partial \mathbf{T}}{\partial \theta^k} = \begin{cases} \mathbf{a}^kT^{ij}|_k\,\mathbf{a}_i\otimes\mathbf{a}_j = T^{ij}|_i\mathbf{a}_j \\ \mathbf{a}^kT^i_{\bullet j}|_k\,\mathbf{a}_i\otimes\mathbf{a}^j = T^i_{\bullet j}|_i\mathbf{a}^j \end{cases}.$$

The gradient of $\mathbf{T}$ is given by

$$\mathrm{grad}\mathbf{T} = \begin{cases} \mathbf{T}\otimes\nabla = \begin{cases} T^{ij}|_k \mathbf{a}_i \otimes \mathbf{a}_j \otimes \mathbf{a}^k \\ T^{i}_{\bullet j}|_k \mathbf{a}_i \otimes \mathbf{a}^j \otimes \mathbf{a}^k \\ T_i^{\bullet j}|_k \mathbf{a}^i \otimes \mathbf{a}_j \otimes \mathbf{a}^k \\ T_{ij}|_k \mathbf{a}^i \otimes \mathbf{a}^j \otimes \mathbf{a}^k \end{cases} & : \text{right form} \\ \nabla\otimes\mathbf{T} = \begin{cases} T^{ij}|_k \mathbf{a}^k \otimes \mathbf{a}_i \otimes \mathbf{a}_j \\ T^{i}_{\bullet j}|_k \mathbf{a}^k \otimes \mathbf{a}_i \otimes \mathbf{a}^j \\ T_i^{\bullet j}|_k \mathbf{a}^k \otimes \mathbf{a}^i \otimes \mathbf{a}_j \\ T_{ij}|_k \mathbf{a}^k \otimes \mathbf{a}^i \otimes \mathbf{a}^j \end{cases} & : \text{left form} \end{cases} \tag{2.146}$$

2.7 Riemann–Christoffel Curvature Tensor

A brief explanation of the curvature tensor in the general coordinate system is given in this section.

Regarding $v_i|_j$ in equation $(2.130)_2$ as the second-order tensor and then exploiting equation $(2.139)_4$, its covariant derivative with respect to θ^k is given by

$$\begin{aligned} v_i|_{jk} &= \frac{\partial v_i|_j}{\partial\theta^k} - v_r|_j\Gamma^r_{ik} - v_i|_r\Gamma^r_{jk} \\ &= \frac{\partial}{\partial\theta^k}\left(\frac{\partial v_i}{\partial\theta^j} - v_r\Gamma^r_{ij}\right) - \left(\frac{\partial v_r}{\partial\theta^j} - v_s\Gamma^s_{rj}\right)\Gamma^r_{ik} - \left(\frac{\partial v_i}{\partial\theta^r} - v_s\Gamma^s_{ir}\right)\Gamma^r_{jk}. \end{aligned} \tag{2.147}$$

Interchanging the order of differentiation, on the other hand, we have

$$v_i|_{kj} = \frac{\partial}{\partial\theta^j}\left(\frac{\partial v_i}{\partial\theta^k} - v_r\Gamma^r_{ik}\right) - \left(\frac{\partial v_r}{\partial\theta^k} - v_s\Gamma^s_{rk}\right)\Gamma^r_{ij} - \left(\frac{\partial v_i}{\partial\theta^r} - \Gamma^s_{ir}v_s\right)\Gamma^r_{kj}. \tag{2.148}$$

Carrying out the differentiations shown, equations (2.147) and (2.148) become

$$v_i|_{jk} = \frac{\partial^2 v_i}{\partial\theta^k\partial\theta^j} - v_r\frac{\partial\Gamma^r_{ij}}{\partial\theta^k} - \frac{\partial v_r}{\partial\theta^k}\Gamma^r_{ij} - \frac{\partial v_r}{\partial\theta^j}\Gamma^r_{ik} + v_s\Gamma^r_{ik}\Gamma^s_{rj} - \frac{\partial v_i}{\partial\theta^r}\Gamma^r_{jk} + v_s\Gamma^r_{jk}\Gamma^s_{ir}, \tag{2.149}$$

$$v_i|_{kj} = \frac{\partial^2 v_i}{\partial\theta^j\partial\theta^k} - v_r\frac{\partial\Gamma^r_{ik}}{\partial\theta^j} - \frac{\partial v_r}{\partial\theta^j}\Gamma^r_{ik} - \frac{\partial v_r}{\partial\theta^k}\Gamma^r_{ij} + v_s\Gamma^r_{ij}\Gamma^s_{rk} - \frac{\partial v_i}{\partial\theta^r}\Gamma^r_{kj} + v_s\Gamma^r_{kj}\Gamma^s_{ir}. \tag{2.150}$$

Subtracting equation (2.150) from equation (2.149) gives

$$v_i|_{jk} - v_i|_{kj} = v_s\Gamma^r_{ik}\Gamma^s_{rj} - v_r\frac{\partial\Gamma^r_{ij}}{\partial\theta^k} - v_s\Gamma^r_{ij}\Gamma^s_{rk} + v_r\frac{\partial\Gamma^r_{ik}}{\partial\theta^j}$$

which, by interchange of r and s, results in

$$v_i|_{jk} - v_i|_{kj} = \left(\frac{\partial\Gamma^r_{ik}}{\partial\theta^j} - \frac{\partial\Gamma^r_{ij}}{\partial\theta^k} + \Gamma^s_{ik}\Gamma^r_{sj} - \Gamma^s_{ij}\Gamma^r_{sk}\right)v_r,$$

that is,

$$v_i|_{jk} - v_i|_{kj} = R^r_{ijk}v_r, \tag{2.151}$$

where R^r_{ijk} is the mixed tensor of order four given by

$$\boxed{R^r_{ijk} \equiv \frac{\partial \Gamma^r_{ik}}{\partial \theta^j} - \frac{\partial \Gamma^r_{ij}}{\partial \theta^k} + \Gamma^s_{ik}\Gamma^r_{sj} - \Gamma^s_{ij}\Gamma^r_{sk}}$$

that is,

$$R^i_{jkl} \equiv \frac{\partial \Gamma^i_{jl}}{\partial \theta^k} - \frac{\partial \Gamma^i_{jk}}{\partial \theta^l} + \Gamma^r_{jl}\Gamma^i_{rk} - \Gamma^r_{jk}\Gamma^i_{rl}; \tag{2.152}$$

this can be rewritten as

$$R^i_{jkl} = \begin{vmatrix} \dfrac{\partial}{\partial \theta^k} & \dfrac{\partial}{\partial \theta^l} \\ \Gamma^i_{jk} & \Gamma^i_{jl} \end{vmatrix} + \begin{vmatrix} \Gamma^i_{rk} & \Gamma^i_{rl} \\ \Gamma^r_{jk} & \Gamma^r_{jl} \end{vmatrix}, \tag{2.153}$$

which is called the *mixed Riemann–Christoffel symbol of the second kind*. The associated tensor defined by

$$R_{ijkl} = a_{ir}R^r_{jkl} \tag{2.154}$$

is called the *covariant Riemann–Christoffel symbol of the first kind*.

Equation (2.154) can be transformed as follows:

$$\begin{aligned} R_{ijkl} = a_{is}R^s_{jkl} &= a_{is}\left(\frac{\partial \Gamma^s_{jl}}{\partial \theta^k} - \frac{\partial \Gamma^s_{jk}}{\partial \theta^l} + \Gamma^r_{jl}\Gamma^s_{rk} - \Gamma^r_{jk}\Gamma^s_{rl}\right) \\ &= a_{is}\frac{\partial \Gamma^s_{jl}}{\partial \theta^k} - a_{is}\frac{\partial \Gamma^s_{jk}}{\partial \theta^l} + a_{is}\Gamma^r_{jl}\Gamma^s_{rk} - a_{is}\Gamma^r_{jk}\Gamma^s_{rl} \\ &= \frac{\partial\left(a_{is}\Gamma^s_{jl}\right)}{\partial \theta^k} - \frac{\partial a_{is}}{\partial \theta^k}\Gamma^s_{jl} - \frac{\partial(a_{is}\Gamma^s_{jk})}{\partial \theta^l} + \frac{\partial a_{is}}{\partial \theta^l}\Gamma^s_{jk} + \Gamma^r_{jl}\Gamma_{rki} - \Gamma^r_{jk}\Gamma_{rli} \\ &= \frac{\partial \Gamma_{jli}}{\partial \theta^k} - \frac{\partial \Gamma_{jki}}{\partial \theta^l} + \Gamma^r_{jk}\left(\frac{\partial a_{ir}}{\partial \theta^l} - \Gamma_{rli}\right) - \Gamma^r_{jl}\left(\frac{\partial a_{ir}}{\partial \theta^k} - \Gamma_{rki}\right). \end{aligned}$$

Making use of equation (2.113), this reduces to

$$R_{ijkl} = \frac{\partial \Gamma_{jli}}{\partial \theta^k} - \frac{\partial \Gamma_{jki}}{\partial \theta^l} + \Gamma^r_{jk}\Gamma_{ilr} - \Gamma^r_{jl}\Gamma_{ikr} \tag{2.155}$$

or

$$R_{ijkl} = \begin{vmatrix} \dfrac{\partial}{\partial \theta^k} & \dfrac{\partial}{\partial \theta^l} \\ \Gamma_{jki} & \Gamma_{jli} \end{vmatrix} + \begin{vmatrix} \Gamma^r_{jk} & \Gamma^r_{jl} \\ \Gamma_{ikr} & \Gamma_{ilr} \end{vmatrix}. \tag{2.156}$$

Substitution of equations (2.110) into equation (2.155) yields

$$R_{ijkl} = \frac{\partial \Gamma_{jli}}{\partial \theta^k} - \frac{\partial \Gamma_{jki}}{\partial \theta^l} + a^{rs}(\Gamma_{jkr}\Gamma_{ils} - \Gamma_{jlr}\Gamma_{iks}), \tag{2.157}$$

noting that

$$\begin{aligned} \Gamma^r_{jk}\Gamma_{ilr} - \Gamma^r_{jl}\Gamma_{ikr} &= \delta^s_p\Gamma^p_{jk}\Gamma_{ils} - \delta^s_p\Gamma^p_{jl}\Gamma_{iks} = a^{rs}a_{rp}\Gamma^p_{jk}\Gamma_{ils} - a^{rs}a_{rp}\Gamma^p_{jl}\Gamma_{iks} \\ &= a^{rs}(\Gamma_{jkr}\Gamma_{ils} - \Gamma_{jlr}\Gamma_{iks}) \end{aligned}$$

Furthermore, equation (2.157) can be rewritten as

$$\boxed{R_{ijkl} = \frac{1}{2}\left(\frac{\partial^2 a_{il}}{\partial\theta^j\partial\theta^k} + \frac{\partial^2 a_{jk}}{\partial\theta^i\partial\theta^l} - \frac{\partial^2 a_{ik}}{\partial\theta^j\partial\theta^l} - \frac{\partial^2 a_{jl}}{\partial\theta^i\partial\theta^k} + \right) + a^{rs}(\Gamma_{jkr}\Gamma_{ils} - \Gamma_{jlr}\Gamma_{iks})} \quad (2.158)$$

noting that

$$\begin{aligned}\frac{\partial\Gamma_{jli}}{\partial\theta^k} - \frac{\partial\Gamma_{jki}}{\partial\theta^l} &= \frac{1}{2}\left(\frac{\partial^2 a_{li}}{\partial\theta^j\partial\theta^k} + \frac{\partial^2 a_{ji}}{\partial\theta^l\partial\theta^k} - \frac{\partial^2 a_{jl}}{\partial\theta^i\partial\theta^k}\right) - \frac{1}{2}\left(\frac{\partial^2 a_{ki}}{\partial\theta^j\partial\theta^l} + \frac{\partial^2 a_{ji}}{\partial\theta^k\partial\theta^l} - \frac{\partial^2 a_{jk}}{\partial\theta^i\partial\theta^l}\right)\\ &= \frac{1}{2}\left(\frac{\partial^2 a_{li}}{\partial\theta^j\partial\theta^k} + \frac{\partial^2 a_{jk}}{\partial\theta^i\partial\theta^l} - \frac{\partial^2 a_{jl}}{\partial\theta^i\partial\theta^k} - \frac{\partial^2 a_{ki}}{\partial\theta^j\partial\theta^l}\right)\end{aligned}$$

by equations (2.112).

R_{ijkl} fulfills the symmetry properties:

$$\left.\begin{array}{ll}\text{first skew symmetry,} & R_{ijkl} = -R_{jikl}\\ \text{second skew symmetry,} & R_{ijkl} = -R_{ijlk}\\ \text{block symmetry,} & R_{ijkl} = R_{klij}\\ \text{Bianchi's identity,} & R_{ijkl} + R_{iklj} + R_{iljk} = 0\end{array}\right\}. \quad (2.159)$$

In the Euclidian space in which the ordinary orthogonal coordinate system belongs, the order of partial derivatives in the left-hand side of equation (2.151) is immaterial so that the Riemann–Christoffel symbol is zero, that is,

$$R_{ijkl} = a_{ir}R^r_{jkl} = 0. \quad (2.160)$$

The spaces in which the Riemann–Christoffel symbol vanishes and does not vanish are defined as the Euclidian and the non-Euclidian space, respectively, both of which live in the Riemann space. Here, note that the Riemann–Christoffel symbol comprises the second-order partial derivatives of the metric tensor a_{ij} with respect to the coordinate θ_i as seen in equation (2.157) with equation (2.112). Therefore, it is related to the curvature of space, while the first derivative is related to the gradient. The Euclidian and non-Euclidian spaces can be also defined as the spaces in which the *parallel axiom* holds and does not hold, respectively. That is, the non-Euclidian space is curved such that the parallel axiom does not hold. Therefore, the length of a line element cannot be described by $||d\mathbf{x}|| = \sqrt{dx_i dx_i}$ but should be described by equation (2.72), that is, $||d\mathbf{x}|| = \sqrt{a_{ij}d\theta^i d\theta^j}$, etc. by use of the metric tensor a_{ij} so that the orthogonal coordinate system cannot be defined in the non-Euclidian space. This fact characterizes the Euclidian and non-Euclidian spaces such that the former space is flat but the latter space is not. Bearing in mind that a relation which holds in one coordinate system in a certain space always holds in that space regardless of the coordinate system, equation (2.160) holds in the Euclidian space in which the orthogonal coordinate system can be defined but does not hold in the non-Euclidian space in which the orthogonal coordinate system cannot be defined.

A surface in n-dimensional space is regarded as being $(n-1)$-dimensional. The simplest example is a surface in three-dimensional space, which is two-dimensional. If the surface can be smoothed out on a plane without stretching, one says that the surface is flat; otherwise the surface is curved. In this sense, for example, a cylindrical surface is flat but a spherical surface (for example, the Earth's surface) is curved. Mathematically, the parallel axiom holds so that the rectangular Cartesian coordinate system can be depicted on the cylindrical surface, but it cannot be copied onto the spherical surface. Therefore, the cylindrical surface belongs to the Euclidian space but the spherical surface belongs to the non-Euclidian space in two dimensions. Riemannian geometry has been developed to study the geometry of the non-Euclidian Riemannian spaces.

The equation $R_{ijkl} = 0$ gives rise to the compatibility condition in the Euclidian space. The compatibility condition of strain will be described in Section 4.3.

2.8 Relations of Convected and Cartesian Coordinate Descriptions

It should be emphasized that all tensors can be described by an arbitrary coordinate system and all tensor relations described in an arbitrary coordinate system hold in other coordinate systems in the Euclidian space, and thus such relations can be described in a unified form of symbolic notation. Tensor relations and calculations are often expressed more conveniently in the normalized rectangular (that is, Cartesian) coordinate system than in the curvilinear coordinate system, except for the deformation analysis of structures with a particular shape. Here, it should be noted that the symbolic notation is limited to the description of simple relations and we often encounter cases where we have to use a component description. The component description in the curvilinear coordinate system is obtained from the symbolic notation and the component description in the normalized-rectangular coordinate system by the following rules.

1. Indices are changed to pairs of super- and/or subscripts depending on the contravariant, covariant, and mixed variant forms.
2. A free index appears uniformly as either a super- or subscript in all terms.
3. The identity tensor **I** and the Kronecker delta δ_{ij} are replaced by the metric tensor **a** and its components a^{ij}, $a^{i}_{\bullet j}$, $a_{i}^{\bullet j}$, a_{ij}, and the permutation symbol $\boldsymbol{\varepsilon}$ and its components ε_{ijk} are replaced by $\underset{\sim}{\mathbf{e}} \equiv e_{ijk}\mathbf{a}^{i} \otimes \mathbf{a}^{j} \otimes \mathbf{a}^{k}$ or $\widehat{\mathbf{e}} \equiv e^{ijk}\mathbf{a}_{i} \otimes \mathbf{a}_{j} \otimes \mathbf{a}_{k}$ and e_{ijk} or e^{ijk}.
4. The partial derivative $\partial(\cdot)/\partial\theta^{i}$ is replaced by the covariant derivative $(\cdot)|_{i}$.

Several examples are given below:

$$\mathbf{u}\cdot\mathbf{v} = u^{i}v_{i} = u_{i}v^{i}, \tag{2.161}$$

$$\mathbf{u}\times\mathbf{v}, \quad \left.\begin{array}{l} e_{ijk}u^{i}v^{j} \\ e^{ijk}u_{i}v_{j} \end{array}\right\}, \tag{2.162}$$

$$\mathbf{T}:\mathbf{I} = T_{ij}\delta_{ij} \rightsquigarrow \mathbf{T}:\mathbf{a} = T^{ij}a_{ij} = T^{i}_{\bullet j}a_{i}^{\bullet j} = T_{i}^{\bullet j}a^{i}_{\bullet j} = T_{ij}a^{ij}, \tag{2.163}$$

$$\mathbf{A}:\mathbf{B} = A^{ij}B_{ij} = A^{i}_{\bullet j}B_{i}^{\bullet j} = A_{i}^{\bullet j}B^{i}_{\bullet j} = A_{ij}B^{ij}, \tag{2.164}$$

$$\nabla\mathbf{T} + \rho\mathbf{b} = \rho\dot{\mathbf{v}}, \quad \left.\begin{array}{ll} T^{ji}|_{j} + \rho b^{i} = \rho\dot{v}^{i}, & T^{j}_{\bullet i}|_{j} + \rho b_{i} = \rho\dot{v}_{i} \\ T_{j}^{\bullet i}|_{j} + \rho b^{i} = \rho\dot{v}^{i}, & T_{ji}|_{j} + \rho b_{i} = \rho\dot{v}_{i} \end{array}\right\}, \tag{2.165}$$

noting equation $(2.145)_2$, where only i in equation (2.165) is a free index. Equation (2.165) will appear as the equilibrium equation in Section 6.3. Equation (2.163) is derived from equation (2.80).

The Gauss divergence theorem in equations (1.449), (1.450) and (1.451) is described in the curvilinear coordinate system as follows:

$$\int_v s|_i dv = \int_a s n_i da, \tag{2.166}$$

$$\int_v v^i|_i dv = \int_a v^i n_i da, \tag{2.167}$$

$$\int_v T^{ij}|_i dv = \int_a T^{ij} n_i da. \tag{2.168}$$

Needless to say, the description of the tensor equation in the rectangular coordinate system can be obtained directly from the description in the curvilinear coordinate system by ignoring the distinction of variants in the components and the base vectors. The usual finite element analysis can be carried out in the rectangular coordinate system without the use of the curvilinear coordinate system, as will be seen in Chapters 9 and 10.

3

Description of Physical Quantities in Convected Coordinate System

The spatial-time derivative at a fixed point in space cannot be used, and the material-time derivative gained by the observer moving in parallel (translates) with a material is better than the spatial-time derivative in mechanics of solids the mechanical properties of which are influenced by the deformation history from the initial state as was explained in Section 1.9.4. However, the material-time derivative is still insufficient, and instead the convected (embedded) derivative gained by the observer not only translating but also deforming and rotating with the material has to be adopted in solid mechanics. The convected coordinate system has to be incorporated in order to describe exactly the variation of physical quantities under the deformation, in which the coordinate system is etched in the material and thus translates, deforms and rotates with it, falling within the framework of the curvilinear coordinate system outlined in Chapter 2. Deep insights into and clear physical interpretations of the deformation behavior of materials are gained by describing physical quantities such as stress (rate) and strain (rate) in the convected coordinate system. The pull-back and the push-forward operations leading to transformations between the Eulerian and the Lagrangian tensors can be interpreted systematically as exchanges of base vectors of the convected coordinate system in the reference and current configurations, while these operations can be substituted for multiplications by the deformation gradient tensor in the direct description independent of coordinate system.

It should be noted that the Cartesian coordinate system is usually convenient for describing the position of material particles as described in the beginning of the Chapter 2, but the convected coordinate system is more useful for the accurate description of the movement of material particles in the representative volume element.

3.1 Necessity for Description in Embedded Coordinate System

As described at the end of the previous chapter, a physical relation described in the normalized-rectangular (Cartesian) coordinate system holds in any coordinate system in the Euclidian space. Futhermore, deformations can be calculated using just the Cartesian coordinate system even if one does not possess any knowledge of the *convected* (*embedded* or *intrinsic*) coordinate

Introduction to Finite Strain Theory for Continuum Elasto-Plasticity, First Edition. Koichi Hashiguchi and Yuki Yamakawa.

system falling within the framework of the general curvilinear coordinate system. However, the representation of tensors in the convected coordinate system is indispensable in order to understand and describe exactly finite deformations by the following reasons.

1. All tensors describing deformation (rate) can be expressed exactly in terms of the variations of embedded coordinate base vectors. It is fundamentally noticeable that the deformation gradient tensor is shown in the identity tensor in the convected coordinate system. Then, the physical meaning of strain (rate) tensors can be captured clearly as a variation of the metric tensor in the embedded coordinate system. Thus, all strain and strain rate tensors will be shown in terms of the metric tensor in addition to the description in terms of the deformation or displacement gradient tensor in this book except for Chapters 9 and 10 for the application of the finite elasto-plastic constitutive equation.
2. Lagrangian and Eulerian tensors can be defined independently and their difference or mutual relation can be captured systematically and explicitly by representing them in terms of the reference base vectors and current base vectors, respectively, of the embedded coordinate system. In addition, the characteristics of the two-point tensor, extending over the reference and the current base vectors, appear explicitly in this representation. On the other hand, they cannot be captured by the representation in the fixed Cartesian coordinate system.
3. Transformations between Lagrangian and Eulerian tensors, that is, the pull-back and push-forward operations of tensors, can be performed by multiplying the tensor by the deformation gradient tensor and its inverse and transpose. Their mathematical and physical meanings can be captured clearly as the exchange of the reference and current embedded base vectors. There exist two and four types of transformation rules from the Eulerian vector and tensor to the Lagrangian vector and tensor (or their inverse transformation rules) for the vector and second-order tensor, respectively.
4. Relations between various stress tensors appear explicitly by the description in the convected coordinate system.
5. The variation of physical quantities can be described exactly in the convected coordinate system so that the rates of physical quantities have to be given by the convected time derivatives in general.
6. The general meaning of the Lie derivative of a tensor can be captured comprehensively as the convected derivative, that is, the rate of variation observed by the material element itself moving, deforming and rotating with material. The convected derivative can be attained by the push-forward operation of the material-time differentiation of the quantity pulled back from the current configuration.
7. The physical meaning of corotational derivatives can be captured clearly as the convected derivative taking account only of material element rotation.

It is noteworthy that the base vectors vary with a deformation but the components of material-line element vectors are kept constant in the embedded coordinate system, while the components change in the Cartesian coordinate system which possesses fixed base vectors.

3.2 Embedded Base Vectors

Consider the coordinate system $\{\Theta^I\}$ ($I = 1, 2, 3$) with the primary base $\{\mathbf{G}_I\}$, which is printed or etched in material in the reference state at time t_0, and which changes to the coordinate

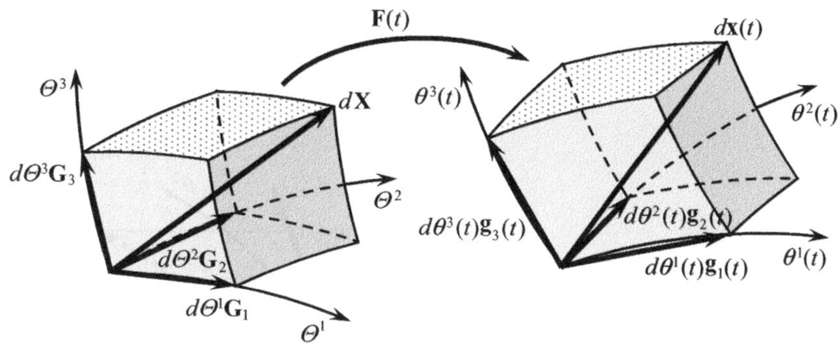

Figure 3.1 Deformation of parallelepiped element with sides parallel to the embedded coordinate

system $\{\theta^i(t)\}$ with the primary base $\{\mathbf{g}_i(t)\}$ at the current time t under the deformation of material as shown in Figure 3.1 for the volume element shown in equation (2.76). The primary base is termed the *convected* (or *embedded*) *base* and the coordinate system with the primary base is termed the *convected* (or *embedded*) *coordinate system*. On the other hand, the bases reciprocal to the primary bases $\{\mathbf{G}_I\}$ in the reference configuration and $\{\mathbf{g}_i(t)\}$ in the current configuration are denoted by $\{\mathbf{G}^I\}$ and $\{\mathbf{g}^i(t)\}$ respectively, the components of which are designated as $\{\Theta_I\}$ and $\{\theta_i(t)\}$ respectively, as shown by the dashed arrows in Figure 3.2. Here, it should be noted that the reciprocal base cannot be embedded and thus has to be reformed every moment with a deformation because the mathematical relation in equation (2.5) to the primary base cannot be kept if they are embedded. Needless to say, the contravariant components of infinitesimal material line-element vectors in the embedded coordinate system are kept constant, while the primary base $\{\mathbf{g}_i(t)\}$ changes. However, the covariant components change in general as known from $d\theta_i(t) = g_{ij}(t)d\theta^j(t) = \delta^j_J g_{ij}(t)d\Theta^J \neq G_{IJ}d\Theta^J = d\Theta_I$ because $g_{ij}(t) \neq G_{IJ}$, noting equation (2.58). Consequently, it follows that

$$d\theta^i(t) = \delta^i_I d\Theta^I \quad (d\theta_i(t) \neq \delta^I_i d\Theta_I). \tag{3.1}$$

The following relations hold for the primary and reciprocal base vectors by virtue of equation $(2.19)_{2,3}$:

$$\mathbf{G}_I \cdot \mathbf{G}^J = \delta^J_I, \quad \mathbf{g}_i(t) \cdot \mathbf{g}^j(t) = \delta^j_i, \tag{3.2}$$

$$\left.\begin{array}{ll} \mathbf{G}_I = G_{IR}\mathbf{G}^R, & \mathbf{G}^I = G^{IR}\mathbf{G}_R \\ \mathbf{g}_i(t) = g_{ir}(t)\mathbf{g}^r(t), & \mathbf{g}^i(t) = g^{ir}(t)\mathbf{g}_r(t) \end{array}\right\}, \tag{3.3}$$

where

$$\left.\begin{array}{ll} G^{IJ} = \mathbf{G}^I \cdot \mathbf{G}^J, & G_{IJ} = \mathbf{G}_I \cdot \mathbf{G}_J \\ g^{ij}(t) = \mathbf{g}^i(t) \cdot \mathbf{g}^j(t), & g_{ij}(t) = \mathbf{g}_i(t) \cdot \mathbf{g}_j(t) \end{array}\right\}. \tag{3.4}$$

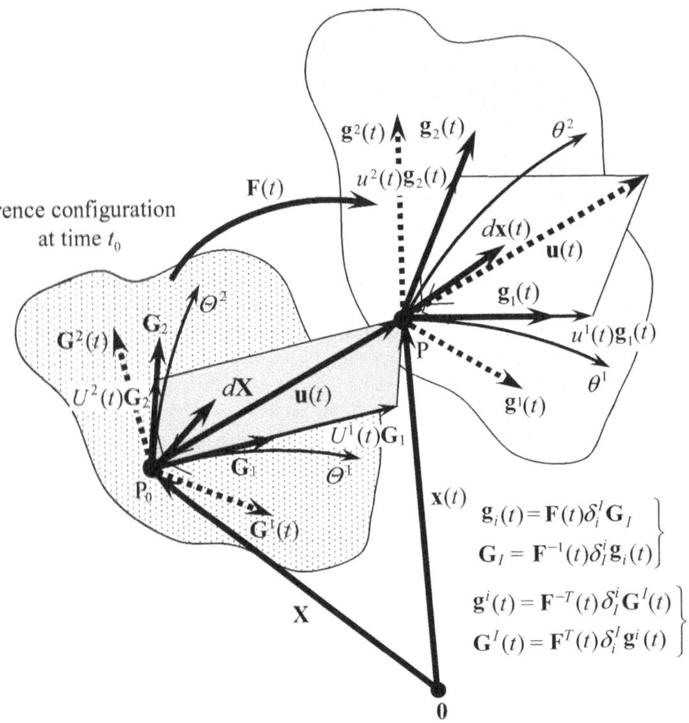

Figure 3.2 Embedded coordinate system with infinitesimal line-element and displacement vectors in the two-dimensional space

The following tensors are nothing more than the metric tensors defined by equation (2.26):

$$\left.\begin{aligned}\mathbf{G} &\equiv \mathbf{G}^I \otimes \mathbf{G}_I = G^{IJ}\mathbf{G}_I \otimes \mathbf{G}_J = G^I_{\bullet J}\mathbf{G}_I \otimes \mathbf{G}^J \\ &= \mathbf{G}_I^{\bullet J}\mathbf{G}^I \otimes \mathbf{G}_J = G_{IJ}\mathbf{G}^I \otimes \mathbf{G}^J \\ \mathbf{g}(t) &\equiv \mathbf{g}^i(t) \otimes \mathbf{g}_i(t) = g^{ij}(t)\mathbf{g}_i(t) \otimes \mathbf{g}_j(t) = g^i_{\bullet j}(t)\mathbf{g}_i(t) \otimes \mathbf{g}^j(t) \\ &= \mathbf{g}_i^{\bullet j}(t)\mathbf{g}^i(t) \otimes \mathbf{g}_j(t) = g_{ij}(t)\mathbf{g}^i(t) \otimes \mathbf{g}^j(t)\end{aligned}\right\}. \tag{3.5}$$

The vectors and the tensors based in the reference and the current configuration are called the *Lagrangian vectors* and *tensors* and the *Eulerian vectors* and *tensors*, respectively. In principle, the Lagrangian and the Eulrian vectors and tensors and their indices are denoted by uppercase and lowercase letters, respectively, throughout this book. Further, tensors based in both the reference and current configurations are called the *Lagrangian–Eulerian* or *Eulerian–Lagrangian two-point tensors* and are denoted by uppercase letters, and their indices are denoted by both uppercase and lowercase letters so as to specify the base vectors in which they are based.

3.3 Deformation Gradient Tensor

Suppose that the position vectors $\mathbf{X}$ and $\mathbf{X}+d\mathbf{X}$ of a material particle point and its neighborhood change to $\mathbf{x}(t)$ and $\mathbf{x}(t)+d\mathbf{x}(t)$, respectively, in the current configuration, where it follows from equation (2.2) that

$$d\mathbf{X}=\frac{\partial\mathbf{X}}{\partial\Theta^I}d\Theta^I=d\Theta^I\mathbf{G}_I,\quad d\mathbf{x}=\frac{\partial\mathbf{x}}{\partial\theta^i}d\theta^i=\delta^i_I d\Theta^I\mathbf{g}_i. \tag{3.6}$$

The deformation gradient tensor $\mathbf{F}(t)$ in equation (1.458) transforms the initial infinitesimal line-element vector $d\mathbf{X}$ to the current one $d\mathbf{x}(t)$ as described in equation $(1.432)_1$, that is,

$$\left.\begin{aligned}\mathbf{F}(t)&=\mathbf{x}(t)\otimes\nabla_X=\frac{\partial\mathbf{x}(t)}{\partial\mathbf{X}}=\frac{\partial(\mathbf{X}+\mathbf{u}(t))}{\partial\mathbf{X}}=\mathbf{G}+\frac{\partial\mathbf{u}(t)}{\partial\mathbf{X}}\\ \mathbf{F}(t)^{-1}&=\mathbf{X}\otimes\nabla_x=\frac{\partial\mathbf{X}}{\partial\mathbf{x}(t)}=\frac{\partial(\mathbf{x}(t)-\mathbf{u}(t))}{\partial\mathbf{x}(t)}=\mathbf{g}(t)-\frac{\partial\mathbf{u}(t)}{\partial\mathbf{x}(t)}\\ d\mathbf{x}(t)&=\mathbf{F}(t)d\mathbf{X},\ d\mathbf{X}=\mathbf{F}(t)^{-1}d\mathbf{x}(t)\end{aligned}\right\}, \tag{3.7}$$

where $\nabla_X=\partial(\cdot)/\partial\mathbf{X}$ and $\nabla_x=\partial(\cdot)/\partial\mathbf{x}(t)$, and $\mathbf{u}(t)$ is the displacement vector of the material point shown in Figure 3.2, that is,

$$\mathbf{x}(t)=\mathbf{X}+\mathbf{u}(t)=\mathbf{X}+\begin{cases}U^I(t)\mathbf{G}_I\\ u^i(t)\mathbf{g}_i(t)\end{cases}, \tag{3.8}$$

Hereinafter, the symbol (t) specifying quantities at time t will be omitted for the sake of simplicity.

Because $d\mathbf{X}$ and $d\mathbf{x}$ can be replaced by the embedded base vectors $\mathbf{G}_I$ and $\mathbf{g}_i$, respectively, in equation $(3.7)_3$, one has

$$\left.\begin{aligned}\mathbf{g}_i&=\frac{\partial\mathbf{x}}{\partial\theta^i}=\delta^I_i\frac{\partial\mathbf{x}}{\partial\Theta^I}=\partial^I_i\frac{\partial\mathbf{x}}{\partial\mathbf{X}}\frac{\partial\mathbf{X}}{\partial\Theta^I}=\mathbf{F}\delta^I_i\mathbf{G}_I\\ \mathbf{G}_I&=\frac{\partial\mathbf{X}}{\partial\Theta^I}=\partial^i_I\frac{\partial\mathbf{X}}{\partial\theta^i}=\partial^i_I\frac{\partial\mathbf{X}\partial\mathbf{x}}{\partial\mathbf{x}\partial\theta^i}=\mathbf{F}^{-1}\delta^i_I\mathbf{g}_i\end{aligned}\right\}, \tag{3.9}$$

noting equation (2.2) for the insides of the brackets, from which we have

$$\left.\begin{aligned}\mathbf{F}&=\delta^i_I\mathbf{g}_i\otimes\mathbf{G}^I,\quad F^i_{\bullet I}=\frac{\partial\theta^i}{\partial\Theta^I}=\delta^i_I\\ \mathbf{F}^{-1}&=\delta^I_i\mathbf{G}_I\otimes\mathbf{g}^i,\quad (\mathbf{F}^{-1})^I_{\bullet i}=\frac{\partial\Theta^I}{\partial\theta^i}=\delta^I_i\end{aligned}\right\} \tag{3.10}$$

because of

$$\left.\begin{aligned}\mathbf{F}\mathbf{G}&=\mathbf{F}\mathbf{G}_I\otimes\mathbf{G}^I=\delta^i_I\mathbf{F}\delta^J_i\mathbf{G}_J\otimes\mathbf{G}^I=\delta^i_I\mathbf{g}_i\otimes\mathbf{G}^I\\ \mathbf{F}^{-1}\mathbf{g}&=\mathbf{F}^{-1}\mathbf{g}_i\otimes\mathbf{g}^i=\delta^I_i\mathbf{F}^{-1}\delta^j_I\mathbf{g}_j\otimes\mathbf{g}^i=\delta^I_i\mathbf{G}_I\otimes\mathbf{g}^i\end{aligned}\right\}.$$

noting equation (3.1). It is noteworthy that the deformation gradient tensor is the idenity tensor in the convected coordinate system as known from equation (3.10). Here, it follows from equation (3.10) that

$$\left.\begin{aligned}\mathbf{F}^T&=\delta^i_I\mathbf{G}^I\otimes\mathbf{g}_i,\quad (\mathbf{F}^T)^{\bullet i}_I=\delta^i_I\\ \mathbf{F}^{-T}&=\delta^I_i\mathbf{g}^i\otimes\mathbf{G}_I,\quad (\mathbf{F}^{-T})^{\bullet I}_i=\delta^I_i\end{aligned}\right\} \tag{3.11}$$

and thus we have

$$\left.\begin{aligned} \mathbf{g}^i \left(= \delta^i_I \frac{\partial \Theta^I}{\partial \mathbf{x}}\right) &= \mathbf{F}^{-T} \delta^i_I \mathbf{G}^I \\ \mathbf{G}^I \left(= \frac{\partial \Theta^I}{\partial \mathbf{X}}\right) &= \mathbf{F}^T \delta^I_i \mathbf{g}^i \end{aligned}\right\} \tag{3.12}$$

for the reciprocal base vectors, noting that

$$\left.\begin{aligned} \mathbf{g}^i &= \delta^i_I \delta^I_j \mathbf{g}^j = \delta^i_I \mathbf{g}^j \otimes \delta^J_j \mathbf{G}_J \mathbf{G}^I = \delta^i_I \mathbf{F}^{-T} \mathbf{G}^I \\ \mathbf{G}^I &= \delta^I_i \delta^i_J \mathbf{G}^J = \delta^I_i \mathbf{G}^J \otimes \delta^j_J \mathbf{g}_j \mathbf{g}^i = \delta^I_i \mathbf{F}^T \mathbf{g}^i \end{aligned}\right\}$$

and equation (2.12) for the insides of the brackets.

The following expressions also hold:

$$d\mathbf{x} = d\mathbf{X}\mathbf{F}^T, \quad d\mathbf{X} = d\mathbf{x}\mathbf{F}^{-T}, \tag{3.13}$$

noting that

$$d\mathbf{x} = dx^i \mathbf{g}_i = \delta^i_I d\theta^I \mathbf{g}_i = d\theta^J \mathbf{G}_J \mathbf{G}^I \otimes \delta^i_I \mathbf{g}_i. \tag{3.14}$$

Substituting equations (3.9) and (3.12) into equation (3.5), one has

$$\left.\begin{aligned} \mathbf{g} &= \mathbf{F}^{-T} \mathbf{G} \mathbf{F}^T \\ \mathbf{G} &= \mathbf{F}^T \mathbf{g} \mathbf{F}^{-T} \end{aligned}\right\}, \tag{3.15}$$

noting that

$$\left.\begin{aligned} \mathbf{g} &= \mathbf{g}^i \otimes \mathbf{g}_i = \mathbf{F}^{-T} \delta^i_I \mathbf{G}^I \otimes \mathbf{F} \delta^J_i \mathbf{G}_J = \mathbf{F}^{-T} \mathbf{G}^I \otimes \mathbf{G}_I \mathbf{F}^T \\ \mathbf{G} &= \mathbf{G}^I \otimes \mathbf{G}_I = \mathbf{F}^T \delta^I_i \mathbf{g}^i \otimes \mathbf{F}^{-1} \delta^j_I \mathbf{g}_j = \mathbf{F}^T \mathbf{g}^i \otimes \mathbf{g}_i \mathbf{F}^{-T} \end{aligned}\right\}$$

by virtue of equation (1.94).

Further, using the orthogonal tensor $\mathbf{R}$ of material rotation instead of the deformation gradient tensor $\mathbf{F}$ in equations (3.9) and (3.12), adopt the base vectors $\widehat{\mathbf{G}}_{\overset{\frown}{I}}$ and $\widehat{\mathbf{G}}^{\overset{\frown}{I}}$, which are rotated inversely from $\mathbf{g}_i$ and $\mathbf{g}^i$, respectively, by $\mathbf{R}$ as

$$\left.\begin{aligned} \mathbf{g}_i &\equiv \mathbf{R} \delta_i^{\overset{\frown}{I}} \widehat{\mathbf{G}}_{\overset{\frown}{I}} \\ \widehat{\mathbf{G}}_{\overset{\frown}{I}} &= \mathbf{R}^T \delta^i_{\overset{\frown}{I}} \mathbf{g}_i \end{aligned}\right\}, \tag{3.16}$$

$$\left.\begin{aligned} \mathbf{g}^i &\equiv \mathbf{R} \delta^i_{\overset{\frown}{I}} \widehat{\mathbf{G}}^{\overset{\frown}{I}} \\ \widehat{\mathbf{G}}^{\overset{\frown}{I}} &= \mathbf{R}^T \delta_i^{\overset{\frown}{I}} \mathbf{g}^i \end{aligned}\right\} \tag{3.17}$$

and adopt the base vectors $\underset{\smile}{\mathbf{g}}_{\underset{\smile}{i}}$ and $\underset{\smile}{\mathbf{g}}^{\underset{\smile}{i}}$, which are rotated from $\mathbf{G}_I$ and $\mathbf{G}^I$, respectively, by $\mathbf{R}$ as follows:

$$\left.\begin{aligned} \underset{\smile}{\mathbf{g}}_{\underset{\smile}{i}} &\equiv \mathbf{R} \delta^I_{\underset{\smile}{i}} \mathbf{G}_I \\ \mathbf{G}_I &= \mathbf{R}^T \delta_I^{\overset{\frown}{i}} \underset{\smile}{\mathbf{g}}_{\underset{\smile}{i}} \end{aligned}\right\} \tag{3.18}$$

$$\left.\begin{aligned} \underset{\smile}{\mathbf{g}}^{\underset{\smile}{i}} &\equiv \mathbf{R}\delta_I^{\underset{\smile}{i}}\mathbf{G}^I \\ \mathbf{G}^I &= \mathbf{R}^T\delta^I_{\underset{\smile}{i}}\,\underset{\smile}{\mathbf{g}}^{\underset{\smile}{i}} \end{aligned}\right\}. \tag{3.19}$$

The convected coordinate systems which use the rectangular-normal coordinate system for the reference or current states, respectively, are called the *material* and *current coordinates*, respectively, in Başar and Weichert (2000). Needless to say, the relative position of other material points from the certain material point within a material is described by the convected coordinate system. On the other hand, the absolute position of a material point cannot be described by the convected coordinate system but has to be described by the coordinate system fixed in the space.

3.4 Pull-Back and Push-Forward Operations

The Eulerian tensor defined by the current base $\{\mathbf{g}_i\}$ or $\{\mathbf{g}^i\}$ is transformed to the Lagrangian tensor by replacing the current base by the reference base $\{\mathbf{G}_I\}$ or $\{\mathbf{G}^I\}$. This is called the *pull-back operation*. Its inverse is called the *push-forward operation*. These operations can also be performed by multiplication of the deformation gradient as will be explained below.

The Eulerian vector $\mathbf{v}$ is expressed by the following contravariant and covariant forms in the current configuration by equation (2.56):

$$\mathbf{v} = v^i\mathbf{g}_i = v_i\mathbf{g}^i. \tag{3.20}$$

The pull-back transformations from the Eulerian vector to the Lagrangian vectors are denoted by the symbols

$$\overset{\leftarrow}{\mathbf{v}}^G \equiv v^i\delta_i^I\mathbf{G}_I, \quad \overset{\leftarrow}{\mathbf{v}}_G \equiv v_i\delta_I^i\mathbf{G}^I, \tag{3.21}$$

where an over-arrow directed to the left ($^{\leftarrow}$) is added (the right arrow ($^{\rightarrow}$) for the push-forward operation is described later). Furthermore, the uppercase index G is added in order to stipulate the replacement of the current base by the reference base (the lowercase index g for the push-forward operation denoting replacement of the reference base by the current base is described later) and is put in the lower or upper position for the covariant or the contravariant component, respectively. These symbols were introduced by Hashiguchi (2011).

Equation (3.21) can be rewritten as

$$\overset{\leftarrow}{\mathbf{v}}^G = v^i\delta_i^I\mathbf{F}^{-1}\delta_I^j\mathbf{g}_j = \mathbf{F}^{-1}v^i\mathbf{g}_i, \quad \overset{\leftarrow}{\mathbf{v}}_G = v_i\delta_I^i\mathbf{F}^T\delta_j^I\mathbf{g}^j = \mathbf{F}^T v_i\mathbf{g}^i,$$

using equations (3.9) and (3.12), and thus the following direct notation is obtained:

$$\boxed{\overset{\leftarrow}{\mathbf{v}}^G = \delta_i^I v^i\mathbf{G}_I = \mathbf{F}^{-1}\mathbf{v}, \quad \overset{\leftarrow}{\mathbf{v}}_G = \delta_I^i v_i\mathbf{G}^I = \mathbf{F}^T\mathbf{v}} \tag{3.22}$$

Using the orthogonal tensor $\mathbf{R}$ instead of $\mathbf{F}$ in equation (3.22), the rotated pull-back of the vector $\mathbf{v}$ is defined by

$$\boxed{\widehat{\mathbf{V}} \equiv \overset{\leftarrow}{\mathbf{v}}^R = \mathbf{R}^T\mathbf{v} = \begin{cases} \delta_i^{\widehat{I}}\, v^i\widehat{\mathbf{G}}_{\widehat{I}} \\ \delta^i_{\widehat{I}}\, v_i\widehat{\mathbf{G}}^{\widehat{I}} \end{cases}} \tag{3.23}$$

noting equations (3.16) and (3.17), where the distinction of the contravariant and the covariant in the pull-back operations vanishes.

Inversely to the foregoing, the Lagrangian vector defined on the reference base $\{\mathbf{G}_I\}$ or $\{\mathbf{G}^I\}$, that is,

$$\mathbf{V} = V^I \mathbf{G}_I, \quad \mathbf{V} = V_I \mathbf{G}^I, \tag{3.24}$$

is transformed to the Eulerian vectors

$$\vec{\mathbf{V}}^g = V^I \delta_I^i \mathbf{g}_i, \quad \vec{\mathbf{V}}_g = V_I \delta_i^I \mathbf{g}^i \tag{3.25}$$

by the push-forward operations replacing the reference base $\{\mathbf{G}_I\}$ or $\{\mathbf{G}^I\}$ by the current base $\{\mathbf{g}_i\}$ or $\{\mathbf{g}^i\}$. Here, because of

$$\vec{\mathbf{V}}^g = V^I \mathbf{F} \mathbf{G}_I, \quad \vec{\mathbf{V}}_g = V_I \mathbf{F}^{-T} \mathbf{G}^I,$$

noting equations (3.9) and (3.12), the following direct notation holds:

$$\boxed{\vec{\mathbf{V}}^g = \delta_I^i V^I \mathbf{g}_i = \mathbf{F}\mathbf{V}, \quad \vec{\mathbf{V}}_g = \delta_i^I V_I \mathbf{g}^i = \mathbf{F}^{-T}\mathbf{V}} \tag{3.26}$$

Using the orthogonal tensor $\mathbf{R}$ instead of $\mathbf{F}$ in equation (3.22), the rotated pull-back of the vector $\mathbf{v}$ is defined by

$$\boxed{\underset{\smile}{\mathbf{v}} \equiv \vec{\mathbf{V}}^R = \mathbf{R}\mathbf{V} = \begin{cases} \delta_I^{\underset{\smile}{i}} V^I \underset{\smile}{\mathbf{g}}_{\underset{\smile}{i}} \\ \delta_{\underset{\smile}{i}}^I V_I \underset{\smile}{\mathbf{g}}^{\underset{\smile}{i}} \end{cases}} \tag{3.27}$$

noting equations (3.16) and (3.17), where the distinction of the contravariant and the covariant in the push-forward operations as well as the pull-back ones vanishes.

Next, the second-order tensor $\mathbf{t}$ in the current configuration is represented by the following four types from equation (2.77):

$$\mathbf{t} = t^{ij}\mathbf{g}_i \otimes \mathbf{g}_j = t^i_{\bullet j}\mathbf{g}_i \otimes \mathbf{g}^j = t_i^{\bullet j}\mathbf{g}^i \otimes \mathbf{g}_j = t_{ij}\mathbf{g}^i \otimes \mathbf{g}^j. \tag{3.28}$$

Here, when $t^{ij} = t^{ji}, t^i_{\bullet j} = t^j_{\bullet i}, t_i^{\bullet j} = t_j^{\bullet i}, t_{ij} = t_{ji}$, $\mathbf{t}$ is the symmetric tensor and thus $\mathbf{t} = \mathbf{t}^T$ holds by virtue of equation (2.86).

The following Lagrangian–Eulerian or Eulerian–Lagrangian two-point tensors are obtained by the pull-back operations replacing only one of the current base $\{\mathbf{g}_i\}$ or $\{\mathbf{g}^i\}$ by the reference base $\{\mathbf{G}_I\}$ or $\{\mathbf{G}^I\}$ in equation (3.28), noting equations (3.9) and (3.12):

$$\left.\begin{array}{l}
\overleftarrow{\mathbf{t}}^{G\hat{g}} \equiv t^{ij}\delta_i^I \mathbf{G}_I \otimes \mathbf{g}_j = t^{ij}\delta_i^I \mathbf{F}^{-1}\delta_I^r \mathbf{g}_r \otimes \mathbf{g}_j = \mathbf{F}^{-1} t^{ij}\mathbf{g}_i \otimes \mathbf{g}_j \\
\overleftarrow{\mathbf{t}}^{\hat{g}G} \equiv t^{ij}\mathbf{g}_i \otimes \delta_j^J \mathbf{G}_J = t^{ij}\mathbf{g}_i \otimes \delta_j^J \mathbf{F}^{-1}\delta_J^r \mathbf{g}_r = t^{ij}\mathbf{g}_i \otimes \mathbf{g}_j \mathbf{F}^{-T} \\
\overleftarrow{\mathbf{t}}^{G}_{\bullet\hat{g}} \equiv t^i_{\bullet j}\delta_i^I \mathbf{G}_I \otimes \mathbf{g}^j = t^i_{\bullet j}\delta_i^I \mathbf{F}^{-1}\delta_I^r \mathbf{g}_r \otimes \mathbf{g}^j = \mathbf{F}^{-1} t^i_{\bullet j}\mathbf{g}_i \otimes \mathbf{g}^j \\
\overleftarrow{\mathbf{t}}^{\hat{g}}_{\bullet G} \equiv t^i_{\bullet j}\mathbf{g}_i \otimes \delta_J^j \mathbf{G}^J = t^i_{\bullet j}\mathbf{g}_i \otimes \delta_J^j \mathbf{F}^T \delta_r^J \mathbf{g}^r = t^i_{\bullet j}\mathbf{g}_i \otimes \mathbf{g}^j \mathbf{F} \\
\overleftarrow{\mathbf{t}}_{G}^{\bullet\hat{g}} \equiv t_i^{\bullet j}\delta_I^i \mathbf{G}^I \otimes \mathbf{g}_j = t^i_{\bullet j}\delta_I^i \mathbf{F}^T \delta_r^I \mathbf{g}^r \otimes \mathbf{g}_j = \mathbf{F}^T t^i_{\bullet j}\mathbf{g}^i \otimes \mathbf{g}_j \\
\overleftarrow{\mathbf{t}}_{\hat{g}}^{\bullet G} \equiv t_i^{\bullet j}\mathbf{g}^i \otimes \delta_j^J \mathbf{G}_J = t_i^{\bullet j}\mathbf{g}^i \otimes \delta_j^J \mathbf{F}^{-1}\delta_J^r \mathbf{g}_r = t_i^{\bullet j}\mathbf{g}^i \otimes \mathbf{g}_j \mathbf{F}^{-T} \\
\overleftarrow{\mathbf{t}}_{G\hat{g}} \equiv t_{ij}\delta_I^i \mathbf{G}^I \otimes \mathbf{g}^j = t_{ij}\delta_I^i \mathbf{F}^T \delta_r^I \mathbf{g}^r \otimes \mathbf{g}^j = \mathbf{F}^T t_{ij}\mathbf{g}^i \otimes \mathbf{g}^j \\
\overleftarrow{\mathbf{t}}_{\hat{g}G} \equiv t_{ij}\mathbf{g}^i \otimes \delta_J^j \mathbf{G}^J = t_{ij}\mathbf{g}^i \otimes \delta_J^j \mathbf{F}^T \delta_r^J \mathbf{g}^r = t_{ij}\mathbf{g}^i \otimes \mathbf{g}^j \mathbf{F}
\end{array}\right\}. \tag{3.29}$$

This leads to

$$
\boxed{
\begin{array}{ll}
\text{Lagragian–Eulerian} & \text{Eulerian–Lagrangian} \\
\text{two-point tensors} & \text{two-point tensors} \\
\left.\begin{array}{l}
\overleftarrow{\mathbf{t}}^{G\hat{g}} = \delta_i^I t^{ij} \mathbf{G}_I \otimes \mathbf{g}_j \\
\overleftarrow{\mathbf{t}}^{G}_{\bullet\hat{g}} = t^i_{\bullet j} \delta_i^I \mathbf{G}_I \otimes \mathbf{g}^j
\end{array}\right\} = \mathbf{F}^{-1}\mathbf{t}, &
\left.\begin{array}{l}
\overleftarrow{\mathbf{t}}^{\hat{g}G} = \delta_j^J t^{ij} \mathbf{g}_i \otimes \mathbf{G}_J \\
\overleftarrow{\mathbf{t}}^{\bullet G}_{\hat{g}} = t_i^{\bullet j} \mathbf{g}^i \otimes \delta_j^J \mathbf{G}_J
\end{array}\right\} = \mathbf{t}\mathbf{F}^{-T} \\
\left.\begin{array}{l}
\overleftarrow{\mathbf{t}}_{G\hat{g}} = \delta_J^i t_{ij} \mathbf{G}^J \otimes \mathbf{g}^j \\
\overleftarrow{\mathbf{t}}^{\bullet\hat{g}}_{G} = t_i^{\bullet j} \delta_I^i \mathbf{G}^I \otimes \mathbf{g}_j
\end{array}\right\} = \mathbf{F}^{T}\mathbf{t}, &
\left.\begin{array}{l}
\overleftarrow{\mathbf{t}}_{\hat{g}G} = \delta_J^j t_{ij} \mathbf{g}^i \otimes \mathbf{G}^J \\
\overleftarrow{\mathbf{t}}^{\hat{g}}_{\bullet G} = t^i_{\bullet j} \mathbf{g}_i \otimes \delta_J^j \mathbf{G}^J
\end{array}\right\} = \mathbf{t}\mathbf{F}
\end{array}
}
\tag{3.30}
$$

where, in the same way as in the vectors, an over-arrow ($\overleftarrow{\ }$) and the index G are added. In addition, an over hat ($\hat{\ }$) is added in order to specify a base which is not replaced. A two-point tensor is not symmetric even if its original tensor $\mathbf{t}$ is symmetric.

The Eulerian tensor in equation (3.30) is replaced to the Lagrangian tensors by the pull-back operations replacing both of the current base vectors by reference base vectors as follows:

$$
\left.
\begin{array}{l}
\overleftarrow{\mathbf{t}}^{GG} \equiv t^{ij} \delta_i^I \mathbf{G}_I \otimes \delta_j^J \mathbf{G}_J = t^{ij} \delta_i^I \mathbf{F}^{-1} \delta_I^r \mathbf{g}_r \otimes \delta_j^J \mathbf{F}^{-1} \delta_J^s \mathbf{g}_s = \mathbf{F}^{-1} t^{ij} \mathbf{g}_i \otimes \mathbf{g}_j \mathbf{F}^{-T} \\
\overleftarrow{\mathbf{t}}^{G}_{\bullet G} \equiv t^i_{\bullet j} \delta_i^I \mathbf{G}_I \otimes \delta_J^j \mathbf{G}^J = t^i_{\bullet j} \delta_i^I \mathbf{F}^{-1} \delta_I^r \mathbf{g}_r \otimes \delta_J^j \mathbf{F}^{T} \delta_s^J \mathbf{g}^s = \mathbf{F}^{-1} t^i_{\bullet j} \mathbf{g}_i \otimes \mathbf{g}^j \mathbf{F} \\
\overleftarrow{\mathbf{t}}^{\bullet G}_{G} \equiv t_i^{\bullet j} \delta_I^i \mathbf{G}^I \otimes \delta_j^J \mathbf{G}_J = t_i^{\bullet j} \delta_I^i \mathbf{F}^{T} \delta_r^I \mathbf{g}^r \otimes \delta_j^J \mathbf{F}^{-1} \delta_J^s \mathbf{g}_s = \mathbf{F}^{T} t_i^{\bullet j} \mathbf{g}^i \otimes \mathbf{g}_j \mathbf{F}^{-T} \\
\overleftarrow{\mathbf{t}}_{GG} \equiv t_{ij} \delta_I^i \mathbf{G}^I \otimes \delta_J^j \mathbf{G}^J = t_{ij} \delta_I^i \mathbf{F}^{T} \delta_r^I \mathbf{g}^r \otimes \delta_J^j \mathbf{F}^{T} \delta_s^J \mathbf{g}^s = \mathbf{F}^{T} t_{ij} \mathbf{g}^i \otimes \mathbf{g}^j \mathbf{F}
\end{array}
\right\},
\tag{3.31}
$$

leading to

$$
\boxed{
\begin{array}{ll}
\overleftarrow{\mathbf{t}}^{GG} = \delta_i^I \delta_j^J t^{ij} \mathbf{G}_I \otimes \mathbf{G}_J = \mathbf{F}^{-1}\mathbf{t}\mathbf{F}^{-T}, & \overleftarrow{\mathbf{t}}^{G}_{\bullet G} = \delta_i^I \delta_J^j t^i_{\bullet j} \mathbf{G}_I \otimes \mathbf{G}^J = \mathbf{F}^{-1}\mathbf{t}\mathbf{F} \\
\overleftarrow{\mathbf{t}}^{\bullet G}_{G} = \delta_I^i \delta_j^J t_i^{\bullet j} \mathbf{G}^I \otimes \mathbf{G}_J = \mathbf{F}^{T}\mathbf{t}\mathbf{F}^{-T}, & \overleftarrow{\mathbf{t}}_{GG} = \delta_I^i \delta_J^j t_{ij} \mathbf{G}^I \otimes \mathbf{G}^J = \mathbf{F}^{T}\mathbf{t}\mathbf{F}
\end{array}
}
\tag{3.32}
$$

In the case where the original tensor $\mathbf{t}$ is symmetric, the pulled-back the tensors in the contravariant form and covariant form are symmetric, that is, $\overleftarrow{\mathbf{t}}^{GG} = \overleftarrow{\mathbf{t}}^{GG^T}$, $\overleftarrow{\mathbf{t}}_{GG} = \overleftarrow{\mathbf{t}}^{T}_{GG}$. On the other hand, the pull-back of the tensor in the mixed-variant form is not symmetric, that is, $\overleftarrow{\mathbf{t}}^{G}_{\bullet G} \neq \overleftarrow{\mathbf{t}}^{G}_{\bullet G}$ and $\overleftarrow{\mathbf{t}}^{\bullet G}_{G} \neq \overleftarrow{\mathbf{t}}^{\bullet G}_{G}$ but $\overleftarrow{\mathbf{t}}^{G}_{\bullet G} = \overleftarrow{\mathbf{t}}^{\bullet G^T}_{G}$ holds.

Further, replacing only one of the two base vectors by applying the rotation $\mathbf{F} = \mathbf{R}$, it follows from equation (3.30) that

$$
\boxed{
\begin{array}{ll}
\text{Rotated Lagrangian–Eulerian} & \text{Eulerian–rotated Lagrangian} \\
\text{two-point tensors} & \text{two-point tensors} \\
\left.\begin{array}{l}
\overleftarrow{\mathbf{t}}^{R\hat{g}} = \delta_i^{\widehat{I}} t^{ij} \widehat{\mathbf{G}}_{\widehat{I}} \otimes \mathbf{g}_j \\
\overleftarrow{\mathbf{t}}^{R}_{\bullet\hat{g}} = \delta_i^{\widehat{I}} t^i_{\bullet j} \widehat{\mathbf{G}}_{\widehat{I}} \otimes \mathbf{g}^j \\
\overleftarrow{\mathbf{t}}^{\bullet\hat{g}}_{R} = \delta^i_{\widehat{I}} t_i^{\bullet j} \widehat{\mathbf{G}}^{\widehat{I}} \otimes \mathbf{g}_j \\
\overleftarrow{\mathbf{t}}_{R\hat{g}} = \delta^i_{\widehat{I}} t_{ij} \widehat{\mathbf{G}}^{\widehat{I}} \otimes \mathbf{g}^j
\end{array}\right\} = \mathbf{R}^{T}\mathbf{t}, &
\left.\begin{array}{l}
\overleftarrow{\mathbf{t}}^{\hat{g}R} = \delta_j^{\widehat{J}} t^{ij} \mathbf{g}_i \otimes \widehat{\mathbf{G}}^{\widehat{J}} \\
\overleftarrow{\mathbf{t}}^{\hat{g}}_{\bullet R} = \delta^j_{\widehat{J}} t^i_{\bullet j} \mathbf{g}_i \otimes \widehat{\mathbf{G}}^{\widehat{J}} \\
\overleftarrow{\mathbf{t}}^{\bullet R}_{\hat{g}} = \delta_j^{\widehat{J}} t_i^{\bullet j} \mathbf{g}^i \otimes \widehat{\mathbf{G}}_{\widehat{J}} \\
\overleftarrow{\mathbf{t}}_{\hat{g}R} = \delta^j_{\widehat{J}} t_{ij} \mathbf{g}^i \otimes \widehat{\mathbf{G}}^{\widehat{J}}
\end{array}\right\} = \mathbf{t}\mathbf{R}
\end{array}
}
\tag{3.33}
$$

where the symbols for rotated base vectors in equations (3.16) and (3.17) are used. Here, even if the original tensor $\mathbf{t}$ is symmetric, the pulled-back tensors are not symmetric but $\overleftarrow{\mathbf{t}}^{R\hat{g}}(=\overleftarrow{\mathbf{t}}^{R}_{\bullet\hat{g}}=\overleftarrow{\mathbf{t}}^{\bullet\hat{g}}_{R}=\overleftarrow{\mathbf{t}}_{R\hat{g}})=\overleftarrow{\mathbf{t}}^{\hat{g}R^T}(=\overleftarrow{\mathbf{t}}^{\hat{g}T}_{\bullet R}=\overleftarrow{\mathbf{t}}^{\bullet RT}_{\hat{g}}=\overleftarrow{\mathbf{t}}^{T}_{\hat{g}R})$ holds.

Further, replacing one of the current base vectors by a reference base vector and rotating the other base with $\mathbf{F}=\mathbf{R}$, it follows from equation (3.32) that

$$\boxed{\begin{array}{ll}\overleftarrow{\mathbf{t}}^{GR}=\overleftarrow{\mathbf{t}}^{G}_{\bullet R}=\mathbf{F}^{-1}\mathbf{t}\mathbf{R}, & \overleftarrow{\mathbf{t}}^{\bullet R}_{G}=\overleftarrow{\mathbf{t}}_{GR}=\mathbf{F}^{T}\mathbf{t}\mathbf{R}\\ \overleftarrow{\mathbf{t}}^{RG}=\overleftarrow{\mathbf{t}}^{\bullet G}_{R}=\mathbf{R}^{T}\mathbf{t}\mathbf{F}^{-T}, & \overleftarrow{\mathbf{t}}^{R}_{\bullet G}=\overleftarrow{\mathbf{t}}_{RG}=\mathbf{R}^{T}\mathbf{t}\mathbf{F}\end{array}} \tag{3.34}$$

In this case also, even if the original tensor $\mathbf{t}$ is symmetric, the pulled-back tensors are not symmetric but $\overleftarrow{\mathbf{t}}^{GR}(=\overleftarrow{\mathbf{t}}^{G}_{\bullet R})=\overleftarrow{\mathbf{t}}^{RG^T}(=\overleftarrow{\mathbf{t}}^{\bullet G^T}_{R})$ and $\overleftarrow{\mathbf{t}}^{\bullet R}_{G}(=\overleftarrow{\mathbf{t}}_{GR})=\overleftarrow{\mathbf{t}}^{RT}_{\bullet G}(=\overleftarrow{\mathbf{t}}^{T}_{RG})$ hold. The descriptions in terms of the components with base vectors are omitted since they are self-evident but cumbersome.

Furthermore, when both base vectors are rotated, the pull-back tensors are described by the following unique equation in which the distinction between covariant, contravariant and mixed-variant operations vanishes:

$$\boxed{\overset{\frown}{\mathbf{T}}\equiv\overleftarrow{\mathbf{t}}^{R}=\mathbf{R}^{T}\mathbf{t}\mathbf{R}} \tag{3.35}$$

which can be described using only the rotated base vectors in equations (3.16) and (3.17). $\overset{\frown}{\mathbf{T}}$ is called the *corotational Lagrangian tensor*.

Inversely to the foregoing, the Lagrangian tensor defined by the reference bases $\{\mathbf{G}_I\}$, $\{\mathbf{G}^I\}$ is represented in the four types as

$$\mathbf{T}=T^{IJ}\mathbf{G}_I\otimes\mathbf{G}_J=T^{I}_{\bullet J}\mathbf{G}_I\otimes\mathbf{G}^J=T_I^{\bullet J}\mathbf{G}^I\otimes\mathbf{G}_J=T_{IJ}\mathbf{G}^I\otimes\mathbf{G}^J. \tag{3.36}$$

By the push-forward operations replacing only one of the two reference bases $\{\mathbf{G}_I\}$ or $\{\mathbf{G}^I\}$ by the current base $\{\mathbf{g}_i\}$ or $\{\mathbf{g}^i\}$, the Lagrangian tensor is transformed to the Eulerian–Lagrangian or the Lagrangian–Eulerian two-point tensors as follows:

$$\left.\begin{array}{l}
\vec{\mathbf{T}}^{g\hat{G}}\equiv T^{IJ}\delta^i_I\mathbf{g}_i\otimes\mathbf{G}_J=T^{IJ}\delta^i_I\mathbf{F}\delta^R_i\mathbf{G}_R\otimes\mathbf{G}_J=\mathbf{F}T^{IJ}\mathbf{G}_I\otimes\mathbf{G}_J\\
\vec{\mathbf{T}}^{g}_{\bullet\hat{G}}\equiv T^{I}_{\bullet J}\delta^i_I\mathbf{g}_i\otimes\mathbf{G}^J=T^{I}_{\bullet J}\delta^i_I\mathbf{F}\delta^R_i\mathbf{G}_R\otimes\mathbf{G}^J=\mathbf{F}T^{I}_{\bullet J}\mathbf{G}_I\otimes\mathbf{G}^J\\
\vec{\mathbf{T}}^{\hat{G}g}\equiv T^{IJ}\mathbf{G}_I\otimes\delta^j_J\mathbf{g}_j=T^{IJ}\mathbf{G}_J\otimes\delta^j_J\mathbf{F}\delta^R_j\mathbf{G}_R=T^{IJ}\mathbf{G}_I\otimes\mathbf{G}_J\mathbf{F}^T\\
\vec{\mathbf{T}}^{\bullet g}_{\hat{G}}\equiv T_I^{\bullet J}\mathbf{G}^I\otimes\delta^j_J\mathbf{g}_j=T_I^{\bullet J}\mathbf{G}^I\otimes\delta^j_J\mathbf{F}\delta^R_j\mathbf{G}_R=T_I^{\bullet J}\mathbf{G}^I\otimes\mathbf{G}_J\mathbf{F}^T\\
\vec{\mathbf{T}}_{g\hat{G}}\equiv T_{IJ}\delta^I_i\mathbf{g}^i\otimes\mathbf{G}^J=T_{IJ}\delta^I_i\mathbf{F}^{-T}\delta^i_R\mathbf{G}^R\otimes\mathbf{G}^J=\mathbf{F}^{-T}T_{IJ}\mathbf{G}^I\otimes\mathbf{G}^J\\
\vec{\mathbf{T}}^{\bullet\hat{G}}_{g}\equiv T_I^{\bullet J}\delta^I_i\mathbf{g}^i\otimes\mathbf{G}_J=T_I^{\bullet J}\delta^I_i\mathbf{F}^{-T}\delta^i_R\mathbf{G}^R\otimes\mathbf{G}_J=\mathbf{F}^{-T}T_I^{\bullet J}\mathbf{G}^I\otimes\mathbf{G}_J\\
\vec{\mathbf{T}}_{\hat{G}g}\equiv T_{IJ}\mathbf{G}_I\otimes\delta^J_j\mathbf{g}^j=T_{IJ}\mathbf{G}_I\otimes\delta^J_j\mathbf{F}^{-T}\delta^j_R\mathbf{G}^R=T^{IJ}\mathbf{G}^I\otimes\mathbf{G}^J\mathbf{F}^{-1}\\
\vec{\mathbf{T}}^{\hat{G}}_{\bullet g}\equiv T^{I}_{\bullet J}\mathbf{G}_I\otimes\delta^J_j\mathbf{g}^j=T^{I}_{\bullet J}\mathbf{G}_I\otimes\delta^J_j\mathbf{F}^{-T}\delta^j_R\mathbf{G}^R=T^{I}_{\bullet J}\mathbf{G}^I\otimes\mathbf{G}^J\mathbf{F}^{-1}
\end{array}\right\}, \tag{3.37}$$

leading to

$$
\begin{array}{ll}
\text{Eulerian–Lagrangian} & \text{Lagrangian–Eularian} \\
\text{two-point tensors} & \text{two-point tensors} \\
\left.\begin{array}{l}
\vec{\mathbf{T}}^{g\hat{G}} = \delta^i_I T^{IJ} \mathbf{g}_i \otimes \mathbf{G}_J \\
\vec{\mathbf{T}}^{g}_{\bullet\hat{G}} = \delta^i_I T^{I}_{\bullet J} \mathbf{g}_i \otimes \mathbf{G}^J
\end{array}\right\} = \mathbf{FT}, &
\left.\begin{array}{l}
\vec{\mathbf{T}}^{\hat{G}g} = \delta^i_J T^{IJ} \mathbf{G}_I \otimes \mathbf{g}_i \\
\vec{\mathbf{T}}^{\bullet g}_{\hat{G}} = \delta^j_J T_I^{\bullet J} \mathbf{G}^I \otimes \mathbf{g}_j
\end{array}\right\} = \mathbf{TF}^T \\
\left.\begin{array}{l}
\vec{\mathbf{T}}_{g\hat{G}} = \delta^I_i T_{IJ} \mathbf{g}^i \otimes \mathbf{G}^J \\
\vec{\mathbf{T}}^{\bullet\hat{G}}_{g} = \delta^I_i T_I^{\bullet J} \mathbf{g}^i \otimes \mathbf{G}_J
\end{array}\right\} = \mathbf{F}^{-T}\mathbf{T}, &
\left.\begin{array}{l}
\vec{\mathbf{T}}_{\hat{G}g} = \delta^J_i T_{IJ} \mathbf{G}_I \otimes \mathbf{g}^i \\
\vec{\mathbf{T}}^{\hat{G}}_{\bullet g} = \delta^J_j T^{I}_{\bullet J} \mathbf{G}_I \otimes \mathbf{g}^j
\end{array}\right\} = \mathbf{TF}^{-1}
\end{array}
\tag{3.38}
$$

by which the two-point tensors are induced from the Lagrangian tensors. Likewise, the push-forward of the two-point tensors becomes the Eulerian tensors as follows:

$$
\begin{array}{ll}
\text{From Lagrangian–Eulerian} & \text{From Eulerian–Lagrangian} \\
\text{two-point tensors} & \text{two-point tensors} \\
\left.\begin{array}{l}
\vec{\mathbb{T}}^{g\hat{g}} = \delta^i_I \mathbb{T}^{Ij} \mathbf{g}_i \otimes \mathbf{g}_j \\
\vec{\mathbb{T}}^{g}_{\bullet\hat{g}} = \delta^i_I \mathbb{T}^{I}_{\bullet j} \mathbf{g}_i \otimes \mathbf{g}^j
\end{array}\right\} = \mathbf{FT}, &
\left.\begin{array}{l}
\vec{\mathbb{T}}^{\hat{g}g} = \delta^j_J \mathbb{T}^{iJ} \mathbf{g}_i \otimes \mathbf{g}_j \\
\vec{\mathbb{T}}^{\bullet g}_{\hat{g}} = \delta^j_J \mathbb{T}_i^{\bullet J} \mathbf{g}^i \otimes \mathbf{g}_j
\end{array}\right\} = \mathbf{TF}^T \\
\left.\begin{array}{l}
\vec{\mathbb{T}}_{g\hat{g}} = \delta^I_i \mathbb{T}_{Ij} \mathbf{g}^i \otimes \mathbf{g}^j \\
\vec{\mathbb{T}}^{\bullet\hat{g}}_{g} = \delta^I_i \mathbb{T}_I^{\bullet j} \mathbf{g}^i \otimes \mathbf{g}_j
\end{array}\right\} = \mathbf{F}^{-T}\mathbf{T}, &
\left.\begin{array}{l}
\vec{\mathbb{T}}_{\hat{g}g} = \delta^J_j \mathbb{T}_{iJ} \mathbf{g}^i \otimes \mathbf{g}^j \\
\vec{\mathbb{T}}^{\hat{g}}_{\bullet g} = \delta^J_j \mathbb{T}^{i}_{\bullet J} \mathbf{g}_i \otimes \mathbf{g}^j
\end{array}\right\} = \mathbf{TF}^{-1}
\end{array}
\tag{3.39}
$$

where the hat is added in order to specify the base vector which is not replaced.

When replacing both of the reference base vectors by the current base vectors, the following relations are obtained by giving the combination of the transformations of equation (3.38) in front and rear:

$$
\begin{array}{ll}
\vec{\mathbf{T}}^{gg} = \delta^i_I \delta^j_J T^{IJ} \mathbf{g}_i \otimes \mathbf{g}_j = \mathbf{FTF}^T, &
\vec{\mathbf{T}}^{g}_{\bullet g} = \delta^i_I \delta^J_j T^{I}_{\bullet J} \mathbf{g}_i \otimes \mathbf{g}^j = \mathbf{FTF}^{-1} \\
\vec{\mathbf{T}}^{\bullet g}_{g} = \delta^I_i \delta^j_J T_I^{\bullet J} \mathbf{g}^i \otimes \mathbf{g}_j = \mathbf{F}^{-T}\mathbf{TF}^T, &
\vec{\mathbf{T}}_{gg} = \delta^I_i \delta^J_j T_{IJ} \mathbf{g}^i \otimes \mathbf{g}^j = \mathbf{F}^{-T}\mathbf{TF}^{-1}
\end{array}
\tag{3.40}
$$

In the case where the original tensor $\mathbf{T}$ is symmetric, the pushed-forward tensors in the contravariant and covariant forms are symmetric as $\vec{\mathbf{T}}^{gg} = \vec{\mathbf{T}}^{gg^T}$, $\vec{\mathbf{T}}_{gg} = \vec{\mathbf{T}}^T_{gg}$. On the other had, the pushed-forward mixed-variant tensors are not the symmetric but $\vec{\mathbf{T}}^{g}_{\bullet g} = \vec{\mathbf{T}}^{\bullet g^T}_{g}$ holds.

Further, replacing only one of the two base vectors and rotating the other base vector, that is, $\mathbf{F} = \mathbf{R}$, it is obtained from equation (3.38) that

$$
\begin{array}{l}
\vec{\mathbf{T}}^{R\hat{G}} = \vec{\mathbf{T}}^{R}_{\bullet\hat{G}} = \vec{\mathbf{T}}_{R\hat{G}} = \vec{\mathbf{T}}^{\bullet\hat{G}}_{R} = \mathbf{RT} \\
\vec{\mathbf{T}}^{\hat{G}R} = \vec{\mathbf{T}}^{\bullet R}_{\hat{G}} = \vec{\mathbf{T}}_{\hat{G}R} = \vec{\mathbf{T}}^{\hat{G}}_{\bullet R} = \mathbf{TR}^T
\end{array}
\tag{3.41}
$$

Here, even if the original tensor $\mathbf{T}$ is symmetric, the pushed-forward tensors are not symmetric but $\vec{\mathbf{T}}^{R\hat{G}} (= \vec{\mathbf{T}}^{R}_{\bullet\hat{G}} = \vec{\mathbf{T}}_{R\hat{G}} = \vec{\mathbf{T}}^{\bullet\hat{G}}_{R}) = \vec{\mathbf{T}}^{\hat{G}R^T} (= \vec{\mathbf{T}}^{\bullet RT}_{\hat{G}} = \vec{\mathbf{T}}^{T}_{\hat{G}R} = \vec{\mathbf{T}}^{\hat{G}T}_{\bullet R})$ holds.

Replacing one of the reference bases by the current base and rotating the other base, that is, $\mathbf{F} = \mathbf{R}$, it follows from equation (3.40) that

$$\boxed{\begin{aligned} \vec{\mathbf{T}}^{gR} &= \vec{\mathbf{T}}^{g}_{\bullet R} = \vec{\mathbf{T}}^{Rg^{T}} = \vec{\mathbf{T}}^{\bullet gT}_{R} = \mathbf{FTR}^{T} \\ \vec{\mathbf{T}}_{gR} &= \vec{\mathbf{T}}^{\bullet R}_{g} = \vec{\mathbf{T}}^{T}_{Rg} = \vec{\mathbf{T}}^{RT}_{\bullet g} = \mathbf{F}^{-T}\mathbf{TR}^{T} \end{aligned}} \tag{3.42}$$

In this case also, even if the original tensor $\mathbf{T}$ is symmetric, the pushed-forward tensors are not symmetric but $\vec{\mathbf{T}}^{gR}(=\vec{\mathbf{T}}^{g}_{\bullet R}) = \vec{\mathbf{T}}^{Rg^{T}}\ (=\vec{\mathbf{T}}^{\bullet gT}_{R})$ and $\vec{\mathbf{T}}_{gR}(=\vec{\mathbf{T}}^{\bullet R}_{g}) = \vec{\mathbf{T}}^{T}_{Rg}(=\vec{\mathbf{T}}^{RT}_{\bullet g})$ hold.

Furthermore, when both bases are rotated, the pushed-forward tensors are described by the following equation in which the distinction between contravariant, covariant and mixed-variant vanishes:

$$\boxed{\underset{\smile}{\mathbf{t}} \equiv \vec{\mathbf{T}}^{R} = \mathbf{RTR}^{T}} \tag{3.43}$$

which is called the *R-rotated Eulerian tensor*.

As described above, the pull-back and the push-forward operations can be realized by exchanging the base vectors between the reference and the current configurations but they are replaced by the multiplication of the deformation gradient $\mathbf{F}$, where there exist four types of operation depending on the covariant or contravariant or the mixed-variant forms. We can set $\mathbf{F} = \mathbf{R}$ when only rotation is induced or taken into account. The following rules exist for the multiplication of the deformation gradient by equations (3.9), (3.12), (3.22) and (3.40).

1. The deformation gradients multiplied from the left side and the right side of the tensor are related to each other by transposition, that is,

$$\boxed{(\,)\otimes\mathcal{F}\mathfrak{g} = (\,)\otimes\mathfrak{g}\mathcal{F}^{T}, \quad (\,)\otimes\mathcal{F}\mathfrak{G} = (\,)\otimes\mathfrak{G}\mathcal{F}^{T}} \tag{3.44}$$

   noting equation (1.94), where $\mathbf{F}$, $\mathbf{F}^{T}$, $\mathbf{F}^{-1}$, $\mathbf{F}^{-T}$ are denoted collectively by $\mathcal{F}$, $\mathbf{g}_i$, $\mathbf{g}^i$ by $\mathfrak{g}$ and $\mathbf{G}_I$, $\mathbf{G}^I$ by $\mathfrak{G}$.
2. The deformation gradients multiplied in the contravariant and the covariant operations are related to each other by transposition plus inversion, that is,

$$\boxed{\overleftarrow{\mathcal{F}}^{g}_{\bullet} = \overleftarrow{\mathcal{F}}^{\bullet}_{g}{}^{-T}, \quad \overrightarrow{\mathcal{F}}^{G}_{\bullet} = \overrightarrow{\mathcal{F}}^{\bullet}_{G}{}^{-T}} \tag{3.45}$$

   where the deformation gradient multiplied in the contravariant and the covariant operations are specified by the addtion of a superscript and a subscript, respectively. This property is captured by comparing the transformation rules of the covariant and the contravariant base vectors in equations (3.9) and (3.12).
3. The deformation gradients multiplied in the pull-back operation and the push-forward operation are inversely related to each other as shown in equations (3.9) and (3.12) so that

$$\boxed{\overleftarrow{\mathcal{F}} = \overrightarrow{\mathcal{F}}^{-1}} \tag{3.46}$$

In addition, the contravariant and the covariant pull-back or push-forward operations of a metric tensor yield the inverse relations to each other, namely

$$\begin{aligned} \overleftarrow{\mathbf{g}}^{GG} &= \mathbf{F}^{-1}\mathbf{g}\mathbf{F}^{-T} = (\mathbf{F}^{T}\mathbf{g}\mathbf{F})^{-1} = \overleftarrow{\mathbf{g}}_{GG}^{-1} \\ \overrightarrow{\mathbf{G}}^{gg} &= \mathbf{F}\mathbf{G}\mathbf{F}^{T} = (\mathbf{F}^{-T}\mathbf{G}\mathbf{F}^{-1})^{-1} = \overrightarrow{\mathbf{G}}_{gg}^{-1} \end{aligned} \tag{3.47}$$

On the other hand, of the cotravariant–covariant and covariant–cotravariant pull-back or push-forward operations of the metric tensor yield mutually transposed tensors, namely

$$\begin{aligned} \overleftarrow{\mathbf{g}}^{G}_{\bullet G} &= \mathbf{F}^{-1}\mathbf{g}\mathbf{F} = (\mathbf{F}^{T}\mathbf{g}\mathbf{F}^{-T})^{T} = \overleftarrow{\mathbf{g}}^{\bullet G^{T}}_{G} \\ \overrightarrow{\mathbf{G}}^{g}_{\bullet g} &= \mathbf{F}\mathbf{G}\mathbf{F}^{-1} = (\mathbf{F}^{-T}\mathbf{G}\mathbf{F}^{T})^{T} = \overrightarrow{\mathbf{G}}^{\bullet g^{T}}_{g} \end{aligned} \tag{3.48}$$

Further, the covariant pull-back and the contravariant push-forward of the metric tensor yield the transposed relation to each other, namely

$$\overleftarrow{\mathbf{g}}_{GG} = \mathbf{F}^{T}\mathbf{g}\mathbf{F} = (\mathbf{F}\mathbf{G}\mathbf{F}^{T})^{T} = \overrightarrow{\mathbf{G}}^{gg} \tag{3.49}$$

The pull-back and the push-forward operations are illustrated in Figure 3.3 and these operations for typical mechanical variables are listed in Table 3.1 which contains variables explained in the later chapters.

Variant forms	Pull-back	Push-forward
Contravariant	$\overleftarrow{\mathbf{t}}^{GG} = \mathbf{F}^{-1}\mathbf{t}\mathbf{F}^{-T}$	$\overrightarrow{\mathbf{T}}^{gg} = \mathbf{F}\mathbf{T}\mathbf{F}^{T}$
Covariant–contravariant	$\overleftarrow{\mathbf{t}}^{G}_{\bullet G} = \mathbf{F}^{-1}\mathbf{t}\mathbf{F}$	$\overrightarrow{\mathbf{T}}^{g}_{\bullet g} = \mathbf{F}\mathbf{T}\mathbf{F}^{-1}$
Contravariant–covariant	$\overleftarrow{\mathbf{t}}^{\bullet G}_{G} = \mathbf{F}^{T}\mathbf{t}\mathbf{F}^{-T}$	$\overrightarrow{\mathbf{T}}^{\bullet g}_{g} = \mathbf{F}^{-T}\mathbf{T}\mathbf{F}^{T}$
Covariant	$\overleftarrow{\mathbf{t}}_{GG} = \mathbf{F}^{T}\mathbf{t}\mathbf{F}$	$\overrightarrow{\mathbf{T}}_{gg} = \mathbf{F}^{-T}\mathbf{T}\mathbf{F}^{-1}$

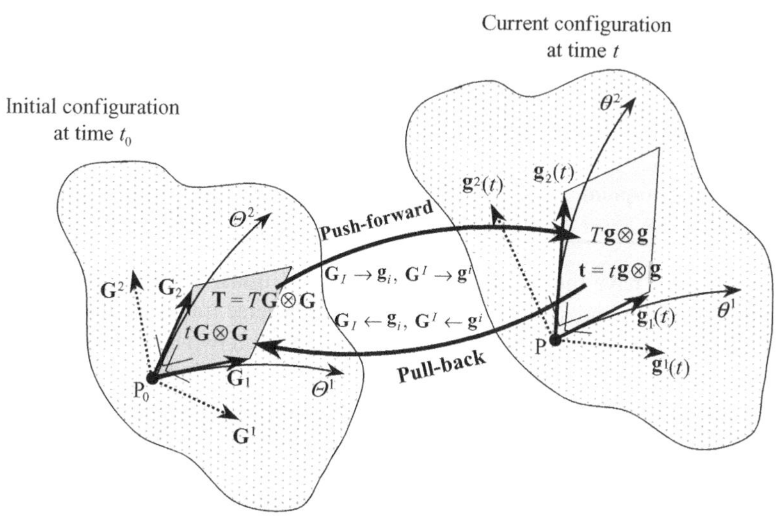

Figure 3.3 Pull-back and push-forward operations illustrated in the two-dimensional space

Table 3.1 Pull-back and push-forward operations

Variant types	Pull-back	Push-forward
Contravariant	$\overleftarrow{\mathbf{t}}^{GG} = \mathbf{F}^{-1}\mathbf{t}\mathbf{F}^{-T}$	$\overrightarrow{\mathbf{T}}^{gg} = \mathbf{F}\mathbf{T}\mathbf{F}^{T}$
	$\mathbf{G} = \mathbf{F}^{-1}\mathbf{b}\mathbf{F}^{-T}$ $\mathbf{S} = \mathbf{F}^{-1}\boldsymbol{\tau}\mathbf{F}^{-T}$ $\dot{\mathbf{S}} = \mathbf{F}^{-1}\overset{\circ}{\boldsymbol{\tau}}{}^{Ol}\mathbf{F}^{-T}$	$\mathbf{b} = \mathbf{F}\mathbf{G}\mathbf{F}^{T} = \mathbf{F}\mathbf{F}^{T}$ $\boldsymbol{\tau} = \mathbf{F}\mathbf{S}\mathbf{F}^{T}$ $\overset{\circ}{\boldsymbol{\tau}}{}^{Ol} = \mathbf{F}\dot{\mathbf{S}}\mathbf{F}^{T} = \dot{\boldsymbol{\tau}} - \boldsymbol{l}\boldsymbol{\tau} - \boldsymbol{\tau}\boldsymbol{l}^{T}$
Contravariant-covariant	$\overleftarrow{\mathbf{t}}^{G}_{\bullet G} = \mathbf{F}^{-1}\mathbf{t}\mathbf{F}$	$\overrightarrow{\mathbf{T}}^{g}_{\bullet g} = \mathbf{F}\mathbf{T}\mathbf{F}^{-1}$
	$\mathbf{L} = \mathbf{F}^{-1}\boldsymbol{l}\mathbf{F}$	$\boldsymbol{l} = \mathbf{F}\mathbf{L}\mathbf{F}^{-1}$
Covariant-contravariant	$\overleftarrow{\mathbf{t}}_{G}^{\bullet G} = \mathbf{F}^{T}\mathbf{t}\mathbf{F}^{-T}$ $\mathbf{M} = \mathbf{F}^{T}\boldsymbol{\tau}\mathbf{F}^{-T}\,(= \mathbf{C}\mathbf{S})$ $\dot{\mathbf{M}} = \mathbf{F}^{T}\overset{\circ}{\mathbf{t}}\mathbf{F}^{-T}$	$\overrightarrow{\mathbf{T}}_{g}^{\bullet g} = \mathbf{F}^{-T}\mathbf{T}\mathbf{F}^{T}$ $\boldsymbol{\tau} = \mathbf{F}^{-T}\mathbf{M}\mathbf{F}^{T}$ $\overset{\circ}{\mathbf{t}} = \mathbf{F}^{-T}\dot{\mathbf{M}}\mathbf{F}^{T} = \dot{\boldsymbol{\tau}} + \boldsymbol{l}^{T}\boldsymbol{\tau} - \boldsymbol{\tau}\boldsymbol{l}^{T}$
Covariant	$\overleftarrow{\mathbf{t}}_{GG} = \mathbf{F}^{T}\mathbf{t}\mathbf{F}$	$\overrightarrow{\mathbf{T}}_{gg} = \mathbf{F}^{-T}\mathbf{T}\mathbf{F}^{-1}$
	$\mathbf{C} = \mathbf{F}^{T}\mathbf{g}\mathbf{F}$ $\mathbf{E} = \mathbf{F}^{T}\mathbf{e}\mathbf{F}$ $\dot{\mathbf{E}} = \mathbf{F}^{T}\mathbf{d}\mathbf{F}$ $\mathsf{L} = \mathbf{F}^{T}\boldsymbol{l}\mathbf{F}$ $\mathfrak{S} = \mathbf{F}^{T}\boldsymbol{\sigma}\mathbf{F}$ $\dot{\mathfrak{S}} = \mathbf{F}^{T}\overset{\circ}{\boldsymbol{\sigma}}{}^{CR}\mathbf{F}$ $\mathfrak{T} = \mathbf{F}^{T}\boldsymbol{\tau}\mathbf{F}$ $\dot{\mathfrak{T}} = \mathbf{F}^{T}\overset{\circ}{\boldsymbol{\tau}}{}^{CR}\mathbf{F}$	$\mathbf{g} = \mathbf{F}^{-T}\mathbf{C}\mathbf{F}^{-1}$ $\mathbf{e} = \mathbf{F}^{-T}\mathbf{E}\mathbf{F}^{-1}$ $\mathbf{d} = \mathbf{F}^{-T}\dot{\mathbf{E}}\mathbf{F}^{-1}$ $\boldsymbol{l} = \mathbf{F}^{-T}\mathsf{L}\mathbf{F}^{-1}$ $\boldsymbol{\sigma} = \mathbf{F}^{-T}\mathfrak{S}\mathbf{F}^{-1}$ $\overset{\circ}{\boldsymbol{\sigma}}{}^{CR} = \mathbf{F}^{-T}\dot{\mathfrak{S}}\mathbf{F}^{-1} = \dot{\boldsymbol{\sigma}} + \boldsymbol{l}^{T}\boldsymbol{\sigma} + \boldsymbol{\sigma}\boldsymbol{l}$ $\boldsymbol{\tau} = \mathbf{F}^{-T}\mathfrak{T}\mathbf{F}^{-1}$ $\overset{\circ}{\boldsymbol{\tau}}{}^{CR} = \mathbf{F}^{-T}\dot{\mathfrak{T}}\mathbf{F}^{-1} = \dot{\boldsymbol{\tau}} + \boldsymbol{l}^{T}\boldsymbol{\tau} + \boldsymbol{\tau}\boldsymbol{l}$
Contravariant-current Two-point	$\overleftarrow{\mathbf{t}}^{G\hat{g}} = \mathbf{F}^{-1}\mathbf{t}$	$\overrightarrow{\mathbf{T}}^{g\hat{g}} = \mathbf{F}\mathbf{T}$
	$\mathbf{P} = \mathbf{F}^{-1}\boldsymbol{\tau}$	$\boldsymbol{\tau} = \mathbf{F}\mathbf{P}$
Current-contravariant two-point	$\overleftarrow{\mathbf{t}}^{\hat{g}G} = \mathbf{t}\mathbf{F}^{-T}$	$\overrightarrow{\mathbb{T}}^{\hat{g}g} = \mathbb{T}\mathbf{F}^{T}$
	$\boldsymbol{\Pi} = \boldsymbol{\tau}\mathbf{F}^{-T}$ $\dot{\boldsymbol{\Pi}} = {}_{\Pi}\overset{\circ}{\boldsymbol{\tau}}\mathbf{F}^{-T}$	$\boldsymbol{\tau} = \boldsymbol{\Pi}\mathbf{F}^{T}$ ${}_{\Pi}\overset{\circ}{\boldsymbol{\tau}} = \dot{\boldsymbol{\Pi}}\mathbf{F}^{T}$
Covariant-current Two-point	$\overleftarrow{\mathbf{t}}_{\hat{g}G} = \mathbf{t}\mathbf{F}$	$\overrightarrow{\mathbb{T}}_{\hat{g}g} = \mathbb{T}\mathbf{F}^{-1}$
	$\dot{\mathbf{F}} = \boldsymbol{l}\mathbf{F}$	$\boldsymbol{l} = \dot{\mathbf{F}}\mathbf{F}^{-1}$

It is shown that we have the two and the four types of transformations from the Eulerian to the Lagrangian vector and second-order tensor or its inverse transformations, respectively, since the vector and the second-order tensor possesses one and two base vectors, respectively. In general, there exist $n \times 2$ types of transformations for the n-th order tensor.

Finally, it should be noticed from the physical point of view that practically the Eulerian tensor is defined first and thereafter the Lagrangian tensors are produced from the Eulerian tensor by the pull-back operation in most cases. In addition, various Lagrangian tensors can be produced by the combination of pulled-back tensors operations shown in this chapter as will be seen in the Biot stress described in Section 6.2.3.

4

Strain and Strain Rate Tensors

In this chapter various deformation, strain, and rotation tensors and strain rate and spin tensors often used in continuum mechanics are described in detail. They are formulated in terms of the pull-back or push-forward operation of the metric tensors in the reference and current configurations. The relations of their Lagrangian and Eulerian measures are represented as pull-back and push-forward operations, that is, the exchange of reference and current base vectors in the convected coordinate system and multiplication of deformation gradient tensors. Thus, the geometrical meanings of deformation, rotation, strain rate and spin tensors and the mutual relations of Lagrangian and Eulerian deformation measures will be captured clearly. Furthermore, the compatibility condition is shown in the curvilinear coordinate system.

4.1 Deformation Tensors

Consider the covariant pull-back (equation $(3.32)_4$) of the metric tensor $\mathbf{g}(t)$ in the current configuration and the contravariant push-forward (equation $(3.40)_1$) of the metric tensor $\mathbf{G}$ in the reference configuration as follows:

$$\boxed{\begin{aligned}&\mathbf{C} \equiv \mathbf{U}^2 = \mathbf{F}^T\mathbf{F} = \mathbf{F}^T\mathbf{g}\mathbf{F} = \overleftarrow{\mathbf{g}}_{GG} = \delta^i_I\delta^j_J g_{ij}\mathbf{G}^I \otimes \mathbf{G}^J (= \mathbf{b}^T)\\ &\mathbf{b} \equiv \mathbf{V}^2 = \mathbf{F}\mathbf{F}^T = \mathbf{F}\mathbf{G}\mathbf{F}^T = \overrightarrow{\mathbf{G}}^{gg} = \delta^i_I\delta^j_J G^{IJ}\mathbf{g}_i \otimes \mathbf{g}_j (= \mathbf{C}^T)\end{aligned}} \tag{4.1}$$

as ascertained by

$$\left.\begin{aligned}\mathbf{F}^T\mathbf{F} &= \delta^i_I\mathbf{G}^I \otimes \mathbf{g}_i\delta^j_J\mathbf{g}_j \otimes \mathbf{G}^J = \delta^i_I\delta^j_J g_{ij}\mathbf{G}^I \otimes \mathbf{G}^J\\ \mathbf{F}\mathbf{F}^T &= \delta^i_I\mathbf{g}_i \otimes \mathbf{G}^I\delta^j_J\mathbf{G}^J \otimes \mathbf{g}_j = \delta^i_I\delta^j_J G^{IJ}\mathbf{g}_i \otimes \mathbf{g}_j\end{aligned}\right\}$$

so that the components of $\mathbf{C}$ and $\mathbf{b}$ are given by

$$C_{IJ} = \delta^i_I\delta^j_J g_{ij}, \quad b^{ij} = \delta^i_I\delta^j_J G^{IJ}. \tag{4.2}$$

Introduction to Finite Strain Theory for Continuum Elasto-Plasticity, First Edition. Koichi Hashiguchi and Yuki Yamakawa.

Table 4.1 Right and left Cauchy–Green deformation tensors

Operations	$\mathcal{K}_0$	Cauchy–Green deformation tensors	$\mathcal{K}$
Contravariant pull-back		$\mathbf{C}^{-1}$ $\leftarrow$	$\mathbf{g}$
Covariant pull-back		$\mathbf{C}$ $\leftarrow$	$\mathbf{g}$
Contravariant push-forward	$\mathbf{G}$	$\rightarrow$ $\mathbf{b}$	
Covariant push-forward	$\mathbf{G}$	$\rightarrow$ $\mathbf{b}^{-1}$	

$\mathbf{C}$ and $\mathbf{b}$ are referred to as the *right* and *left Cauchy–Green deformation tensors*, respectively. Their inverses are given by the following equation, noting equation (3.47).

$$\boxed{\begin{aligned}\mathbf{C}^{-1} &= \mathbf{F}^{-1}\mathbf{F}^{-T} = \mathbf{F}^{-1}\mathbf{g}\mathbf{F}^{-T} = \overleftarrow{\mathbf{g}}^{GG} = \delta^I_i\delta^J_j g^{ij}\mathbf{G}_I \otimes \mathbf{G}_J (\neq \mathbf{C}^{-T})\\ \mathbf{b}^{-1} &= \mathbf{F}^{-T}\mathbf{F}^{-1} = \mathbf{F}^{-T}\mathbf{G}\mathbf{F}^{-1} = \overrightarrow{\mathbf{G}}_{gg} = \delta^I_i\delta^J_j G_{IJ}\mathbf{g}^i \otimes \mathbf{g}^j (\neq \mathbf{b}^{-T})\end{aligned}} \tag{4.3}$$

where $\mathbf{C}^{-1}$ is sometimes called the *Piola deformation tensor* and $\mathbf{b}^{-1}$ the *Finger deformation tensor*.

Equations (4.1) and (4.3) show that the right and left Cauchy–Green deformation tensors and their inverses are regarded as transformations of the reference and the current metric tensors as follows:

1. The right Cauchy–Green deformation tensor $\mathbf{C}$ is defined by replacing the reference components G_{IJ} of the metric tensor $G_{IJ}\mathbf{G}^I \otimes \mathbf{G}^J$ of the covariant form in the reference configuration by the current components g_{ij}. Therefore, it is derived by the covariant pull-back of the current metric tensor $\mathbf{g}$.
2. Inversely, the left Cauchy–Green deformation tensor $\mathbf{b}$ is defined by replacing the current components g^{ij} of the metric tensor $g^{ij}\mathbf{g}_i \otimes \mathbf{g}_j$ of the contravariant form in the current configuration by the reference components G^{IJ}. Therefore, it is derived by the contravariant push-forward of the reference metric tensor $\mathbf{G}$.
3. $\mathbf{C}^{-1}$ is derived by replacing the reference components G^{IJ} of the metric tensor $G^{IJ}\mathbf{G}_I \otimes \mathbf{G}_J$ of the contravariant form in the reference configuration by the current components g^{ij}. Therefore, it is derived by the contravariant pull-back of the current metric tensor $\mathbf{g}$.
4. Inversely, $\mathbf{b}^{-1}$ is defined by replacing the current components g_{ij} of the metric tensor $g_{ij}\mathbf{g}^i \otimes \mathbf{g}^j$ of the covariant form in the current configuration by the reference components G_{IJ}. Therefore, it is derived by the covariant push-forward of the reference metric tensor $\mathbf{G}$.

These facts are shown in Table 4.1, where $\leftarrow$ and $\rightarrow$ designate the pull-back and the push-forward operations, respectively, and $\mathcal{K}_0$ and $\mathcal{K}$ imply the reference and the current configurations.

The application of equation (1.309) in the polar decomposition in Section 1.7 to the deformation gradient $\mathbf{F}$ reads

$$\boxed{\mathbf{F} = \mathbf{R}\mathbf{U} = \mathbf{V}\mathbf{R}} \tag{4.4}$$

where

$$\mathbf{U} = \mathbf{C}^{1/2} \ (\mathbf{U} = \mathbf{U}^T), \tag{4.5}$$

$$\mathbf{V} = \mathbf{b}^{1/2} \ (\mathbf{V} = \mathbf{V}^T), \tag{4.6}$$

$$\mathbf{R} = \mathbf{F}\mathbf{U}^{-1} = \mathbf{F}(\mathbf{F}^T\mathbf{F})^{-1/2}, \ \ \mathbf{R} = \mathbf{V}^{-1}\mathbf{F} = (\mathbf{F}\mathbf{F}^T)^{-1/2}\mathbf{F} \quad (\det\mathbf{R} = 1), \tag{4.7}$$

$$\mathbf{U} = \mathbf{R}^T\mathbf{V}\mathbf{R} = \mathbf{R}^T\mathbf{F} = \mathbf{F}^T\mathbf{R}, \ \ \mathbf{V} = \mathbf{R}\mathbf{U}\mathbf{R}^T = \mathbf{F}\mathbf{R}^T = \mathbf{R}\mathbf{F}^T, \tag{4.8}$$

$$\det\mathbf{F} = \det\mathbf{U} = \det\mathbf{V} = J, \det\mathbf{C} = \det\mathbf{b} = J^2. \tag{4.9}$$

U and **V** are called the right and the left stretch tensor, respectively.

Now, we designate the deformation gradient induced in the duration from the initial time t_0 to the current time t by $\mathbf{F}_0(t)$ and the one induced in the duration from t to an arbitrary time $\tau(> t)$ by $\mathbf{F}_t(\tau)$. Further, we designate the deformation gradient induced in the duration from t_0 to τ by $\mathbf{F}_0(\tau)$. Then, it follows that

$$\mathbf{F} = \mathbf{F}_0(t) = \frac{\partial\mathbf{x}(t)}{\partial\mathbf{X}}, \ \ \mathbf{F}_0(\tau) = \frac{\partial\mathbf{x}(\tau)}{\partial\mathbf{X}}, \ \ \mathbf{F}_t(\tau) = \frac{\partial\mathbf{x}(\tau)}{\partial\mathbf{x}(t)} \tag{4.10}$$

which fulfill

$$\mathbf{F}_0(\tau) = \mathbf{F}_t(\tau)\mathbf{F}_0(t) \tag{4.11}$$

because of

$$\frac{\partial\mathbf{x}(\tau)}{\partial\mathbf{X}} = \frac{\partial\mathbf{x}(t)}{\partial\mathbf{x}(\tau)}\frac{\partial\mathbf{x}(\tau)}{\partial\mathbf{X}}.$$

The generalized deformation gradient $\mathbf{F}_t(\tau)$ is referred to as the *relative deformation gradient* and its polar decomposition is given as follows:

$$\mathbf{F}_t(\tau) = \mathbf{R}_t(\tau)\mathbf{U}_t(\tau) = \mathbf{V}_t(\tau)\mathbf{R}_t(\tau). \tag{4.12}$$

The rate of deformation gradient $\mathbf{F}_t(\tau)$ in which τ is chosen as the current time t, called the *updated Lagrangian description*, is given by

$$\dot{\mathbf{F}}_t(t) = \left.\frac{\partial\mathbf{F}_t(\tau)}{\partial\tau}\right|_{\tau=t} = \left.\frac{\partial(\partial\mathbf{x}(\tau)/\partial\mathbf{x}(t))}{\partial\tau}\right|_{\tau=t} = \left.\frac{\partial\mathbf{v}(\tau)}{\partial\mathbf{x}(t)}\right|_{\tau=t} = \frac{\partial\mathbf{v}(t)}{\partial\mathbf{x}(t)}. \tag{4.13}$$

Since equation (4.8) holds for the orthogonal tensor **R**, the tensors **U** and **V** are similar tensors to each other as described for equation (1.252). Therefore, they possess the same eigenvalues. Denoting their eigenvalues and the eigenvectors of **U** and **V** by $\lambda_\alpha(\alpha = 1, 2, 3)$ and $\{\mathbf{N}^\alpha\}$ and $\{\mathbf{n}^{(\alpha)}\}$, respectively, the following relations hold from equations (1.304), (1.305), (1.313), and (1.314):

$$\mathbf{U}\mathbf{N}^{(\alpha)} = \lambda_\alpha\mathbf{N}^{(\alpha)}, \ \ \mathbf{V}\mathbf{n}^{(\alpha)} = \lambda_\alpha\mathbf{n}^{(\alpha)}, \tag{4.14}$$

$$\boxed{\mathbf{U} = \sum_{\alpha=1}^{3}\lambda_\alpha\mathbf{N}^{(\alpha)}\otimes\mathbf{N}^{(\alpha)}, \ \ \mathbf{V} = \sum_{\alpha=1}^{3}\lambda_\alpha\mathbf{n}^{(\alpha)}\otimes\mathbf{n}^{(\alpha)}} \tag{4.15}$$

$$\mathbf{n}^{(\alpha)} = \mathbf{R}\mathbf{N}^{(\alpha)}, \quad \mathbf{N}^{(\alpha)} = \mathbf{R}^T\mathbf{n}^{(\alpha)}, \tag{4.16}$$

$$\boxed{\mathbf{R} = \sum_{\alpha=1}^{3} \mathbf{n}^{(\alpha)} \otimes \mathbf{N}^{(\alpha)}} \tag{4.17}$$

$\mathbf{N}^{(\alpha)}$ and $\mathbf{n}^{(\alpha)}$ are called the *Lagrangian triad* and the *Eulerian triad*, respectively.

The deformation gradient $\mathbf{F}$ and its inverse are described by equation (1.315) as follows:

$$\mathbf{F} = \sum_{\alpha=1}^{3} \lambda_\alpha \mathbf{n}^{(\alpha)} \otimes \mathbf{N}^{(\alpha)}, \quad \mathbf{F}^{-1} = \sum_{\alpha=1}^{3} \frac{1}{\lambda_\alpha} \mathbf{N}^{(\alpha)} \otimes \mathbf{n}^{(\alpha)}. \tag{4.18}$$

We examine the mechanical meanings of $\mathbf{U}$, $\mathbf{V}$ and $\mathbf{R}$ below.

The change of an infinitesimal line element is given by the polar decomposition $\mathbf{F} = \mathbf{R}\mathbf{U}$, noting that $\mathbf{U}\mathbf{N}^{(\alpha)} = \lambda_\alpha \mathbf{N}^{(\alpha)}$(no sum), as follows:

$$d\mathbf{x} = \mathbf{F}d\mathbf{X} = \mathbf{R}\mathbf{U}d\mathbf{X} = \mathbf{R}\sum_{\alpha=1}^{3} \lambda_\alpha \mathbf{N}^{(\alpha)} \otimes \mathbf{N}^{(\alpha)} \sum_{\beta=1}^{3} dX_\beta \mathbf{N}^{(\beta)} = \mathbf{R}\sum_{\alpha=1}^{3} \lambda_\alpha dX_\alpha \mathbf{N}^{(\alpha)}. \tag{4.19}$$

Equation (4.19) means that the infinitesimal line elements $dX_\alpha \mathbf{N}^{(\alpha)}$ (no sum) in the principal directions $\mathbf{N}^{(\alpha)}$ are first stretched by a factor λ_α to $\lambda_\alpha dX_\alpha \mathbf{N}^{(\alpha)}$ (no sum) and then undergo the rotation $\mathbf{R}$ as shown in Figure 4.1.

On the other hand, the change of the infinitesimal line element is described by the polar decomposition $\mathbf{V}\mathbf{R}$ as

$$d\mathbf{x} = \mathbf{V}\mathbf{R}d\mathbf{X} = \sum_{\alpha=1}^{3} \lambda_\alpha \mathbf{n}^{(\alpha)} \otimes \mathbf{n}^{(\alpha)}\mathbf{R}\sum_{\beta=1}^{3} dX_\beta \mathbf{N}^{(\beta)} = \sum_{\alpha=1}^{3} \lambda_\alpha \mathbf{n}^{(\alpha)} \otimes \mathbf{n}^{(\alpha)} \sum_{\alpha=1}^{3} dX_\beta \mathbf{n}^{(\beta)}$$

$$= \sum_{\alpha=1}^{3} \lambda_\alpha dX_\alpha \mathbf{n}^{(\alpha)} = \sum_{\alpha=1}^{3} \lambda_\alpha \mathbf{R} dX_\alpha \mathbf{N}^{(\alpha)}. \tag{4.20}$$

Equation (4.20) means that the infinitesimal line elements $dX_\alpha \mathbf{N}^{(\alpha)}$ (no sum) in the principal directions $\mathbf{N}^{(\alpha)}$ are first rotated to $dX_\alpha \mathbf{n}^{(\alpha)}$ (no sum) by $\mathbf{R}$ and then are stretched by a factor λ_α to $\lambda_\alpha dX_\alpha \mathbf{n}^{(\alpha)}$ (no sum).

As described above, $\mathbf{U}$ and $\mathbf{V}$ designate deformations and $\mathbf{R}$ rotation. $\mathbf{U}$ and $\mathbf{V}$ are called *right* and *left stretch tensors*, respectively, and λ_α is called the *principal stretch*.

Denoting by $\mathbf{R}^L$ and $\mathbf{R}^E$ the rotations of the Lagrangian triad $\{\mathbf{N}^{(\alpha)}\}$ and the Eulerian triad $\{\mathbf{n}^{(\alpha)}\}$, respectively, from the fixed base $\{\mathbf{e}_\alpha\}$ $(\alpha = 1, 2, 3)$, we define

$$\boxed{\mathbf{R}^L \equiv \sum_{\alpha=1}^{3} \mathbf{N}^{(\alpha)} \otimes \mathbf{e}_\alpha, \quad \mathbf{R}^E \equiv \sum_{\alpha=1}^{3} \mathbf{n}^{(\alpha)} \otimes \mathbf{e}_\alpha} \tag{4.21}$$

where

$$\mathbf{N}^{(\alpha)} = \mathbf{R}^L \mathbf{e}_\alpha, \quad \mathbf{n}^{(\alpha)} = \mathbf{R}^E \mathbf{e}_\alpha, \tag{4.22}$$

$$\mathbf{R}^E = \mathbf{R}\mathbf{R}^L. \tag{4.23}$$

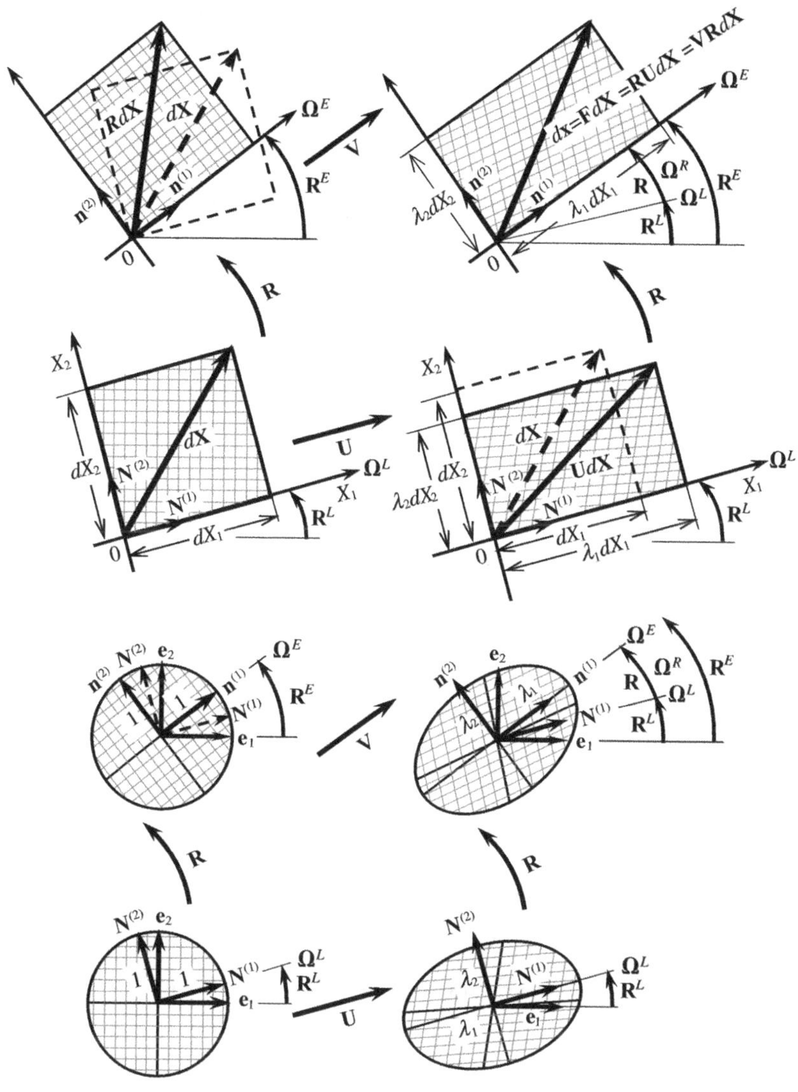

Figure 4.1 Polar decomposition of the deformation gradient

Noting equations (4.5), (4.6) and (4.15), the right and left Cauchy–Green deformation tensors are described as follows:

$$\begin{aligned} \mathbf{C} &= \mathbf{U}^2 = \mathbf{R}^T\mathbf{V}^2\mathbf{R} = \sum_{\alpha=1}^{3} \lambda_\alpha^2 \mathbf{N}^{(\alpha)} \otimes \mathbf{N}^{(\alpha)} \\ \mathbf{b} &= \mathbf{V}^2 = \mathbf{R}\mathbf{U}^2\mathbf{R}^T = \sum_{\alpha=1}^{3} \lambda_\alpha^2 \mathbf{n}^{(\alpha)} \otimes \mathbf{n}^{(\alpha)} \end{aligned} \tag{4.24}$$

The eigenvalues λ_a are calculated by solving the characteristic equation based on equation (1.233)

$$\lambda^3 - \mathrm{I_C}\lambda^2 - \mathrm{II_C}\lambda + \mathrm{III_C} = 0, \tag{4.25}$$

where

$$\mathrm{I_C} \equiv \mathrm{tr}\mathbf{C}, \mathrm{II_C} \equiv \frac{1}{2}(\mathrm{tr}^2\mathbf{C} - \mathrm{tr}\mathbf{C}^2), \mathrm{III_C} \equiv \frac{1}{6}\mathrm{tr}^3\mathbf{C} - \frac{1}{2}\mathrm{tr}\mathbf{C}\mathrm{tr}\mathbf{C}^2 + \frac{1}{3}\mathrm{tr}\mathbf{C}^3. \tag{4.26}$$

Eigenvectors and eigenvalues are calculated by the method described in Subsection 1.6.3.

In order to calculate $\mathbf{U}, \mathbf{V}$ and $\mathbf{R}$, it is first necessary to calculate the eigenvalues λ_α by equation (4.25) and eigenvectors $\mathbf{N}^{(\alpha)}$ and $\mathbf{n}^{(\alpha)} (\alpha = 1, 2, 3)$ by equation (1.227) for the known deformation gradient $\mathbf{F}$ and then $\mathbf{R}$ can be calculated from equation (4.17) as will be described for the simple shear deformation in Section 6.9.2.

4.2 Strain Tensors

The Lagrangian and Eulerian strain tensors described in this section are often used in continuum mechanics.

4.2.1 *Green and Almansi Strain Tensors*

The square of the length of a line element is described by the following equation, noting equations (2.19), (2.56), (4.1) and (4.3) with $d\theta^i = \delta^i_I d\Theta^I, d\Theta^I = \delta^I_i d\theta^i$:

$$\left.\begin{aligned} d\mathbf{x}\cdot d\mathbf{x} &= d\theta^i\mathbf{g}_i \cdot d\theta^j\mathbf{g}_j = g_{ij}d\theta^i d\theta^j = \delta^i_I\delta^j_J g_{ij}d\Theta^I d\Theta^J = C_{IJ}d\Theta^I d\Theta^J \\ d\mathbf{X}\cdot d\mathbf{X} &= d\Theta^I\mathbf{G}_I \cdot d\Theta^J\mathbf{G}_J = G_{IJ}d\Theta^I d\Theta^J = \delta^I_i\delta^J_j G_{IJ}d\theta^i d\theta^j = b^{-1}_{ij}d\theta^i d\theta^j \end{aligned}\right\} \tag{4.27}$$

or, further noting equation (1.93) and (3.7), as follows:

$$\left.\begin{aligned} d\mathbf{x}\cdot d\mathbf{x} &= \mathbf{F}d\mathbf{X}\cdot\mathbf{F}d\mathbf{X} = d\mathbf{X}\cdot\mathbf{F}^T\mathbf{F}d\mathbf{X} = d\mathbf{X}\cdot\mathbf{C}d\mathbf{X} \\ d\mathbf{X}\cdot d\mathbf{X} &= \mathbf{F}^{-1}d\mathbf{x}\cdot\mathbf{F}^{-1}d\mathbf{x} = d\mathbf{x}\cdot\mathbf{F}^{-T}\mathbf{F}^{-1}d\mathbf{x} = d\mathbf{x}\cdot\mathbf{b}^{-1}d\mathbf{x} \end{aligned}\right\}. \tag{4.28}$$

Then, from equations (4.27) or (4.28), the change in the square of the length of a line element is described as follows:

$$d\mathbf{x}\cdot d\mathbf{x} - d\mathbf{X}\cdot d\mathbf{X} = \begin{cases} (\delta^i_I\delta^j_J g_{ij} - G_{IJ})d\Theta^I d\Theta^J = (C_{IJ} - G_{IJ})d\Theta^I d\Theta^J = 2E_{IJ}d\Theta^I d\Theta^J \\ \quad = d\Theta^R\mathbf{G}_R\cdot\{(2E_{IJ}\mathbf{G}^I\otimes\mathbf{G}^J)(d\Theta^S\mathbf{G}_S)\} = d\mathbf{X}\cdot\{(2E_{IJ}\mathbf{G}^I\otimes\mathbf{G}^J)d\mathbf{X}\} \\ (g_{ij} - \delta^I_i\delta^J_j G_{IJ})d\theta^i d\theta^j = (g_{ij} - b^{-1}_{ij})d\theta^i d\theta^j = 2e_{ij}d\theta^i d\theta^j \\ \quad = d\theta^r\mathbf{g}_r\cdot\{(2e_{ij}\mathbf{g}^i\otimes\mathbf{g}^j)(d\theta^s\mathbf{g}_s)\} = d\mathbf{x}\cdot\{(2e_{ij}\mathbf{g}^i\otimes\mathbf{g}^j)d\mathbf{x}\} \end{cases} \tag{4.29}$$

or

$$d\mathbf{x}\cdot d\mathbf{x} - d\mathbf{X}\cdot d\mathbf{X} = \begin{cases} d\mathbf{X}\cdot(\mathbf{C} - \mathbf{G})d\mathbf{X} = d\mathbf{X}\cdot 2\mathbf{E}d\mathbf{X} \\ d\mathbf{x}\cdot(\mathbf{g} - \mathbf{b}^{-1})d\mathbf{x} = d\mathbf{x}\cdot 2\mathbf{e}d\mathbf{x} \end{cases} \tag{4.30}$$

where

$$\left.\begin{aligned}\mathbf{E} = E_{IJ}\mathbf{G}^I \otimes \mathbf{G}^J = \frac{1}{2}(\delta_I^i\delta_J^j g_{ij} - G_{IJ})\mathbf{G}^I \otimes \mathbf{G}^J = \frac{1}{2}(C_{IJ} - G_{IJ})\mathbf{G}^I \otimes \mathbf{G}^J \\ \mathbf{e} = e_{ij}\mathbf{g}^i \otimes \mathbf{g}^j = \frac{1}{2}(g_{ij} - \delta_i^I\delta_j^J G_{IJ})\mathbf{g}^i \otimes \mathbf{g}^j = \frac{1}{2}(g_{ij} - b_{ij}^{-1})\mathbf{g}^i \otimes \mathbf{g}^j\end{aligned}\right\} \tag{4.31}$$

that is,

$$\boxed{\begin{aligned}\mathbf{E} &= \frac{1}{2}(\mathbf{F}^T\mathbf{F} - \mathbf{G}) = \frac{1}{2}(\mathbf{C} - \mathbf{G}) = \frac{1}{2}(\overleftarrow{\mathbf{g}}_{GG} - \mathbf{G}) \\ \mathbf{e} &= \frac{1}{2}(\mathbf{g} - \mathbf{F}^{-T}\mathbf{F}^{-1}) = \frac{1}{2}(\mathbf{g} - \mathbf{b}^{-1}) = \frac{1}{2}(\mathbf{g} - \overrightarrow{\mathbf{G}}_{gg})\end{aligned}} \tag{4.32}$$

where $\mathbf{E}$ and $\mathbf{e}$ are called the *Green–Lagrangian strain tensor* and the *Almansi–Eulerian strain tensor*, respectively. These strain tensors are represented in terms of the position vectors $\mathbf{X}$ and $\mathbf{x}$ and the displacement vector $\mathbf{u}$ by equation (3.7) with equation (3.8) as follows:

$$\left.\begin{aligned}\mathbf{E} &= \frac{1}{2}\left\{\left(\frac{\partial \mathbf{x}}{\partial \mathbf{X}}\right)^T \frac{\partial \mathbf{x}}{\partial \mathbf{X}} - \mathbf{G}\right\} \\ &= \frac{1}{2}\left\{\left(\frac{\partial (\mathbf{X}+\mathbf{u})}{\partial \mathbf{X}}\right)^T \frac{\partial (\mathbf{X}+\mathbf{u})}{\partial \mathbf{X}} - \mathbf{G}\right\} = \frac{1}{2}\left\{\frac{\partial \mathbf{u}}{\partial \mathbf{X}} + \left(\frac{\partial \mathbf{u}}{\partial \mathbf{X}}\right)^T + \left(\frac{\partial \mathbf{u}}{\partial \mathbf{X}}\right)^T \frac{\partial \mathbf{u}}{\partial \mathbf{X}}\right\} \\ \mathbf{e} &= \frac{1}{2}\left\{\mathbf{g} - \left(\frac{\partial \mathbf{X}}{\partial \mathbf{x}}\right)^T \frac{\partial \mathbf{X}}{\partial \mathbf{x}}\right\} \\ &= \frac{1}{2}\left\{\mathbf{g} - \left(\frac{\partial (\mathbf{x}-\mathbf{u})}{\partial \mathbf{x}}\right)^T \frac{\partial (\mathbf{x}-\mathbf{u})}{\partial \mathbf{x}}\right\} = \frac{1}{2}\left\{\frac{\partial \mathbf{u}}{\partial \mathbf{x}} + \left(\frac{\partial \mathbf{u}}{\partial \mathbf{x}}\right)^T - \left(\frac{\partial \mathbf{u}}{\partial \mathbf{x}}\right)^T \frac{\partial \mathbf{u}}{\partial \mathbf{x}}\right\}\end{aligned}\right\}. \tag{4.33}$$

The geometrical interpretations of the strain tensors $\mathbf{E}$ and $\mathbf{e}$ are given in terms of the rectangular coordinate system, based on equation (4.33), in Appendix B. However, the physical meanings and the definite physical backgrounds and necessity of these strains are not clear in these interpretations. However, they can be captured clearly from the expressions in terms of the metric tensor of the embedded coordinate system as shown in equation (4.31) or (4.32). That is, the Green strain tensor $\mathbf{E}$ is regarded as half of the subtraction of the reference metric tensor $\mathbf{G}$ from the covariant pull-back of the current metric tensor $\mathbf{g}$, that is, $\mathbf{C}$. On the other hand, the Almansi strain tensor $\mathbf{e}$ is regarded as half of the subtraction of the covariant push-forward of the reference metric tensor $\mathbf{G}$, that is, $\mathbf{b}^{-1}$, from the current metric tensor $\mathbf{g}$, noting the relations described in Section 4.1.

It follows from equations (3.7) and (3.9) with equation (3.8) that

$$\left.\begin{aligned}\delta_I^i\mathbf{g}_i = \mathbf{F}\mathbf{G}_I = \frac{\partial (\mathbf{X}+\mathbf{u})}{\partial \mathbf{X}}\mathbf{G}_I = \mathbf{G}_I + \frac{\partial \mathbf{u}}{\partial \mathbf{X}}\frac{\partial \mathbf{X}}{\partial \Theta^I} = \mathbf{G}_I + \frac{\partial \mathbf{u}}{\partial \Theta^I} = \left(\delta_I^R + \frac{\partial U^R}{\partial \Theta^I}\right)\mathbf{G}_R \\ \delta_i^I\mathbf{G}_I = \mathbf{F}^{-1}\mathbf{g}_i = \frac{\partial (\mathbf{x}-\mathbf{u})}{\partial \mathbf{x}}\mathbf{g}_i = \mathbf{g}_i - \frac{\partial \mathbf{u}}{\partial \mathbf{x}}\frac{\partial \mathbf{x}}{\partial \theta^i} = \mathbf{g}_i - \frac{\partial \mathbf{u}}{\partial \theta^i} = \left(\delta_i^r + \frac{\partial u^r}{\partial \theta^i}\right)\mathbf{g}_r\end{aligned}\right\} \tag{4.34}$$

from which one has

$$\left.\begin{aligned}\delta_I^i\delta_J^j g_{ij} &= \delta_I^i\mathbf{g}_i\cdot\delta_J^j\mathbf{g}_j = \left(\mathbf{G}_I+\frac{\partial\mathbf{u}}{\partial\Theta^I}\right)\cdot\left(\mathbf{G}_J+\frac{\partial\mathbf{u}}{\partial\Theta^J}\right)\\ &= G_{IJ}+\mathbf{G}_I\cdot\frac{\partial\mathbf{u}}{\partial\Theta^J}+\mathbf{G}_J\cdot\frac{\partial\mathbf{u}}{\partial\Theta^I}+\frac{\partial\mathbf{u}}{\partial\Theta^I}\cdot\frac{\partial\mathbf{u}}{\partial\Theta^J}\\ \delta_i^I\delta_j^J G_{IJ} &= \delta_i^I\mathbf{G}_I\cdot\delta_j^J\mathbf{G}_J = \left(\mathbf{g}_i-\frac{\partial\mathbf{u}}{\partial\theta^i}\right)\cdot\left(\mathbf{g}_j-\frac{\partial\mathbf{u}}{\partial\theta^j}\right)\\ &= g_{ij}-\mathbf{g}_i\cdot\frac{\partial\mathbf{u}}{\partial\theta^j}-\mathbf{g}_j\cdot\frac{\partial\mathbf{u}}{\partial\theta^i}+\frac{\partial\mathbf{u}}{\partial\theta^i}\cdot\frac{\partial\mathbf{u}}{\partial\theta^j}\end{aligned}\right\}. \tag{4.35}$$

Substitution of equation (4.35) into equation (4.31) leads to

$$\left.\begin{aligned}\mathbf{E} &= \frac{1}{2}\left(\mathbf{G}_I\cdot\frac{\partial\mathbf{u}}{\partial\Theta^J}+\mathbf{G}_J\cdot\frac{\partial\mathbf{u}}{\partial\Theta^I}+\frac{\partial\mathbf{u}}{\partial\Theta^I}\cdot\frac{\partial\mathbf{u}}{\partial\Theta^J}\right)\mathbf{G}^I\otimes\mathbf{G}^J\\ \mathbf{e} &= \frac{1}{2}\left(\mathbf{g}_i\cdot\frac{\partial\mathbf{u}}{\partial\theta^j}+\mathbf{g}_j\cdot\frac{\partial\mathbf{u}}{\partial\theta^i}-\frac{\partial\mathbf{u}}{\partial\theta^i}\cdot\frac{\partial\mathbf{u}}{\partial\theta^j}\right)\mathbf{g}^i\otimes\mathbf{g}^j\end{aligned}\right\}$$

which, substituting the covariant derivative of vector in equation (2.129) with equation (3.8), become

$$\left.\begin{aligned}\mathbf{E} &= \frac{1}{2}(\mathbf{G}_I\cdot U_R|_J\mathbf{G}^R+\mathbf{G}_J\cdot U_R|_I\mathbf{G}^R+U^R|_I\mathbf{G}_R\cdot U_S|_J\mathbf{G}^S)\mathbf{G}^I\otimes\mathbf{G}^J\\ \mathbf{e} &= \frac{1}{2}(\mathbf{g}_i\cdot u_r|_j\mathbf{g}^r+\mathbf{g}_j\cdot u_s|_i\mathbf{g}^s-u^r|_i\mathbf{g}_r\cdot u_s|_j\mathbf{g}^s)\mathbf{g}^i\otimes\mathbf{g}^j\end{aligned}\right\}.$$

Then, component descriptions of **E** and **e** in terms of displacements with the embedded base vectors are given as follows:

$$\left.\begin{aligned}\mathbf{E} &= \frac{1}{2}(U_I|_J+U_J|_I+U^R|_IU_R|_J)\mathbf{G}^I\otimes\mathbf{G}^J\\ \mathbf{e} &= \frac{1}{2}(u_i|_j+\delta_i^I u_j|_i-u^r|_iu_r|_j)\mathbf{g}^i\otimes\mathbf{g}^j\end{aligned}\right\}. \tag{4.36}$$

In the fixed normalized-rectangular coordinate system we have

$$\left.\begin{aligned}E_{IJ} &= \frac{1}{2}(u_{I,J}+u_{J,I}+u_{R,I}\,u_{R,J})\\ \mathbf{e}_{ij} &= \frac{1}{2}(u_{i,j}+u_{j,i}-u_{r,i}\,u_{r,j})\end{aligned}\right\} \tag{4.37}$$

because the covariant derivatives $(\cdot)|_i$ and $(\cdot)|_I$ reduce to the ordinary partial derivatives $(\cdot)_I \equiv \partial(\cdot)/\partial X_I$ and $(\cdot)_{,i} \equiv \partial(\cdot)/\partial x_i$ respectively. The components of $\mathbf{e}$ are given by

$$\left.\begin{aligned}
e_{11} &= u_{1,1} - \frac{1}{2}\left\{(u_{1,1})^2 + (u_{2,1})^2 + (u_{3,1})^2\right\}\\
e_{22} &= u_{2,2} - \frac{1}{2}\left\{(u_{1,2})^2 + (u_{2,2})^2 + (u_{3,2})^2\right\}\\
e_{33} &= u_{3,3} - \frac{1}{2}\left\{(u_{1,3})^2 + (u_{2,3})^2 + (u_{3,3})^2\right\}\\
e_{12} &= e_{21} = \frac{1}{2}(u_{1,2} + u_{2,1} - u_{1,1}u_{1,2} - u_{2,1}u_{2,2} - u_{3,1}u_{3,2})\\
e_{23} &= e_{32} = \frac{1}{2}(u_{2,3} + u_{3,2} - u_{1,2}u_{1,3} - u_{2,2}u_{2,3} - u_{3,2}u_{3,3})\\
e_{31} &= e_{13} = \frac{1}{2}(u_{3,1} + u_{1,3} - u_{1,3}u_{1,1} - u_{2,3}u_{2,1} - u_{3,3}u_{3,1})
\end{aligned}\right\}. \tag{4.38}$$

$\mathbf{E}$ and $\mathbf{e}$ are derived by the covariant pull-back operation in equation $(3.32)_4$ and the push-forward operation in equation $(3.40)_4$ from each other as follows:

$$\mathbf{E} = \overleftarrow{\mathbf{e}}_{GG} = \mathbf{F}^T\mathbf{e}\mathbf{F},\ \ \mathbf{e} = \overrightarrow{\mathbf{E}}_{gg} = \mathbf{F}^{-T}\mathbf{E}\mathbf{F}^{-1} \tag{4.39}$$

The following relation holds for infinitesimal volumes dV and dv in the reference and the current configurations, respectively, by virtue of equation (2.76):

$$\left.\begin{aligned}
dV &\equiv [\mathbf{G}_1\ \mathbf{G}_2\ \mathbf{G}_3]d\Theta^1 d\Theta^2 d\Theta^3\\
dv &\equiv [\mathbf{g}_1\ \mathbf{g}_2\ \mathbf{g}_3]d\theta^1 d\theta^2 d\theta^3 = [\mathbf{g}_1\ \mathbf{g}_2\ \mathbf{g}_3]d\Theta^1 d\Theta^2 d\Theta^3
\end{aligned}\right\} \tag{4.40}$$

and thus

$$\boxed{\frac{dv}{dV} = \det\mathbf{F} = \sqrt{\frac{g}{G}}} \tag{4.41}$$

where

$$G \equiv \det[G_{IJ}] = [\mathbf{G}_1\ \mathbf{G}_2\ \mathbf{G}_3]^2,\ \ g \equiv \det[g_{ij}] = [\mathbf{g}_1\ \mathbf{g}_2\ \mathbf{g}_3]^2, \tag{4.42}$$

noting equation (2.23).

The relation between the initial area vector $\mathbf{A}$ formed by the initial base vectors $\mathbf{G}_1$, $\mathbf{G}_2$ and the initial volume V formed by the initial bases $\mathbf{G}_1$, $\mathbf{G}_2$, $\mathbf{G}_3$ is given by

$$\mathbf{A} = \mathbf{G}_1 \times \mathbf{G}_2 = V\mathbf{G}^3, \tag{4.43}$$

from equation (2.5). Similarly, the relation between the current area vector $\mathbf{a}$ formed by the current base vectors $\mathbf{g}_1$, $\mathbf{g}_2$ and the current volume v is given by

$$\mathbf{a} = v\mathbf{g}^3. \tag{4.44}$$

On the other hand, substituting $\mathbf{g}^3 = \mathbf{F}^{-T}\mathbf{G}^3$ due to equation $(3.12)_1$ into equations (4.43), we have

$$\mathbf{a} = \nu\mathbf{F}^{-T}\mathbf{G}^3 = J\mathbf{F}^{-T}\mathbf{A}, \tag{4.45}$$

while the third term is based on equation (1.462).

In what follows, geometrical interpretations of the Green and the Almansi strain tensors are given.

Let the angles formed by the reference primary base vectors $\mathbf{G}_I$ and $\mathbf{G}_J$ and by the current primary base vectors $\mathbf{g}_i$ and $\mathbf{g}_j$ be denoted by Ψ_{IJ} and ϕ_{ij}, respectively. Then it follows that

$$\left.\begin{aligned} G_{IJ} &= \mathbf{G}_I \cdot \mathbf{G}_J = ||\mathbf{G}_I||\,||\mathbf{G}_J||\cos\Psi_{IJ} \\ g_{ij} &= \mathbf{g}_i \cdot \mathbf{g}_j = ||\mathbf{g}_i||\,||\mathbf{g}_j||\cos\phi_{ij} \end{aligned}\right\}\text{(no sum)}. \tag{4.46}$$

Substituting equation (4.46) into equation (4.31), the components of the Green and the Almansi strains are described as

$$\left.\begin{aligned} E_{IJ} &= \frac{1}{2}\left\{\left(1+\frac{||\mathbf{g}_i||-||\mathbf{G}_I||}{||\mathbf{G}_I||}\right)\left(1+\frac{||\mathbf{g}_j||-||\mathbf{G}_J||}{||\mathbf{G}_J||}\right)\cos\phi_{ij} - \cos\Psi_{IJ}\right\}||\mathbf{G}_I||\,||\mathbf{G}_J|| \\ e_{ij} &= \frac{1}{2}\left\{\cos\phi_{ij} - \left(1-\frac{||\mathbf{g}_i||-||\mathbf{G}_I||}{||\mathbf{g}_i||}\right)\left(1-\frac{||\mathbf{g}_j||-||\mathbf{G}_J||}{||\mathbf{g}_j||}\right)\cos\Psi_{IJ}\right\}||\mathbf{g}_i||\,||\mathbf{g}_j|| \end{aligned}\right\}\text{(no sum)}. \tag{4.47}$$

For $\Psi_{IJ} = 0(\phi_{ij} = 0, \mathbf{G}_I = \mathbf{G}_J, \mathbf{g}_i = \mathbf{g}_j)$, equation (4.47) leads to

$$\left.\begin{aligned} E_{II} &= \frac{1}{2}\left\{\left(1+\frac{||\mathbf{g}_i||-||\mathbf{G}_I||}{||\mathbf{G}_I||}\right)^2 - 1\right\}||\mathbf{G}_I||^2 \\ e_{ii} &= \frac{1}{2}\left\{1-\left(1-\frac{||\mathbf{g}_i||-||\mathbf{G}_I||}{||\mathbf{g}_i||}\right)^2\right\}||\mathbf{g}_i||^2 \end{aligned}\right\}\text{(no sum)}, \tag{4.48}$$

which reduces to

$$\left.\begin{aligned} E_{II} &\cong \frac{||\mathbf{g}_i||-||\mathbf{G}_I||}{||\mathbf{G}_I||} \\ e_{ii} &\cong \frac{||\mathbf{g}_i||-||\mathbf{G}_I||}{||\mathbf{g}_i||} \end{aligned}\right\}\text{(no sum)} \tag{4.49}$$

for an infinitesimal deformation $||\mathbf{g}_i|| \cong ||\mathbf{G}_I||$ under the choice $||\mathbf{G}_I|| = 1$. E_{II} and e_{ii} in equation (4.49) denote the Lagrangian and Eulerian infinitesimal longitudinal strains, respectively.

For $\Psi_{IJ} = \pi/2$, equation (4.47) leads to

$$\left.\begin{aligned} E_{IJ} &= \frac{1}{2}\left(1+\frac{||\mathbf{g}_i||-||\mathbf{G}_I||}{||\mathbf{G}_I||}\right)\left(1+\frac{||\mathbf{g}_j||-||\mathbf{G}_J||}{||\mathbf{G}_J||}\right)\sin\gamma_{ij}||\mathbf{G}_I||\,||\mathbf{G}_J|| \\ e_{ij} &= \frac{1}{2}\sin\gamma_{ij}||\mathbf{g}_i||\,||\mathbf{g}_j|| \end{aligned}\right\}\text{(no sum)} \tag{4.50}$$

by introducing the angle γ_{ij} satisfying $\cos\phi_{ij} = \cos(\pi/2 - \gamma_{ij}) = \sin\gamma_{ij}$. Equation (4.50) reduces to

$$\left.\begin{array}{l} E_{IJ} \\ e_{ij} \end{array}\right\} \cong \frac{1}{2}\sin\gamma_{ij} \cong \frac{1}{2}\gamma_{ij} \tag{4.51}$$

for the infinitesimal deformation $||\mathbf{g}_i|| \cong ||\mathbf{G}_I||$ under the choice $||\mathbf{G}_I|| = 1$. Equation (4.51) denotes the deviatoric components of the infinitesimal strain.

4.2.2 General Strain Tensors

The second-order tensor function $\mathbf{f}(\mathbf{U})$ of $\mathbf{U}$ is co-axial to $\mathbf{U}$, and likewise the function $\mathbf{f}(\mathbf{V})$ of $\mathbf{V}$ is co-axial to $\mathbf{V}$ as known from the fact described in section 1.8. Noting equations (4.15), it therefore follows that

$$\left.\begin{array}{l} \mathsf{E} \equiv \mathbf{f}(\mathbf{U}) = \sum_{\alpha=1}^{3} f(\lambda_\alpha)\mathbf{N}^{(\alpha)} \otimes \mathbf{N}^{(\alpha)} \\ \mathsf{e} \equiv \mathbf{f}(\mathbf{V}) = \sum_{\alpha=1}^{3} f(\lambda_\alpha)\mathbf{n}^{(\alpha)} \otimes \mathbf{n}^{(\alpha)} \end{array}\right\}. \tag{4.52}$$

These functions are related by

$$\mathsf{E} = \mathbf{R}^T\mathbf{e}\mathbf{R}, \quad \mathsf{e} = \mathbf{R}\mathbf{E}\mathbf{R}^T, \tag{4.53}$$

noting that

$$\mathsf{E} = \sum_{\alpha=1}^{3} f(\lambda_\alpha)\mathbf{R}^T\mathbf{n}^{(\alpha)} \otimes \mathbf{R}^T\mathbf{n}^{(\alpha)} = \mathbf{R}^T \sum_{\alpha=1}^{3} f(\lambda_\alpha)\mathbf{n}^{(\alpha)} \otimes \mathbf{n}^{(\alpha)}\mathbf{R}.$$

E and e are respectively referred to as the *generalized strain measure* or *Hill's strain measure* (Hill, 1968, 1978).

Seth (1964) proposed the following particular forms of $\mathbf{f}(\mathbf{U})$ and $\mathbf{f}(\mathbf{v})$ in equation (4.52):

$$\boxed{\mathbf{E}^{(m)} = \mathbf{f}(\mathbf{U}) = \begin{cases} \dfrac{1}{2m}(\mathbf{U}^{2m} - \mathbf{G}) & \text{for } m \neq 0 \\ \ln\mathbf{U} & \text{for } m = 0 \end{cases}} \tag{4.54}$$

and

$$\boxed{\mathbf{e}^{(m)} = \mathbf{f}(\mathbf{V}) = \begin{cases} \dfrac{1}{2m}(\mathbf{V}^{2m} - \mathbf{g}) & \text{for } m \neq 0 \\ \ln\mathbf{V} & \text{for } m = 0 \end{cases}} \tag{4.55}$$

where $2m$ is a positive or negative integer. The strain tensors in equations (4.54) and (4.55) are related as follows:

$$\mathbf{E}^{(m)} = \mathbf{R}^T\mathbf{e}^{(m)}\mathbf{R} = \overleftarrow{\mathbf{e}}^{(m)R}, \quad \mathbf{e}^{(m)} = \mathbf{R}\mathbf{E}^{(m)}\mathbf{R}^T = \overleftarrow{\mathbf{E}}^{(m)R}, \tag{4.56}$$

noting equations (4.53) with equations (4.54) and (4.55).

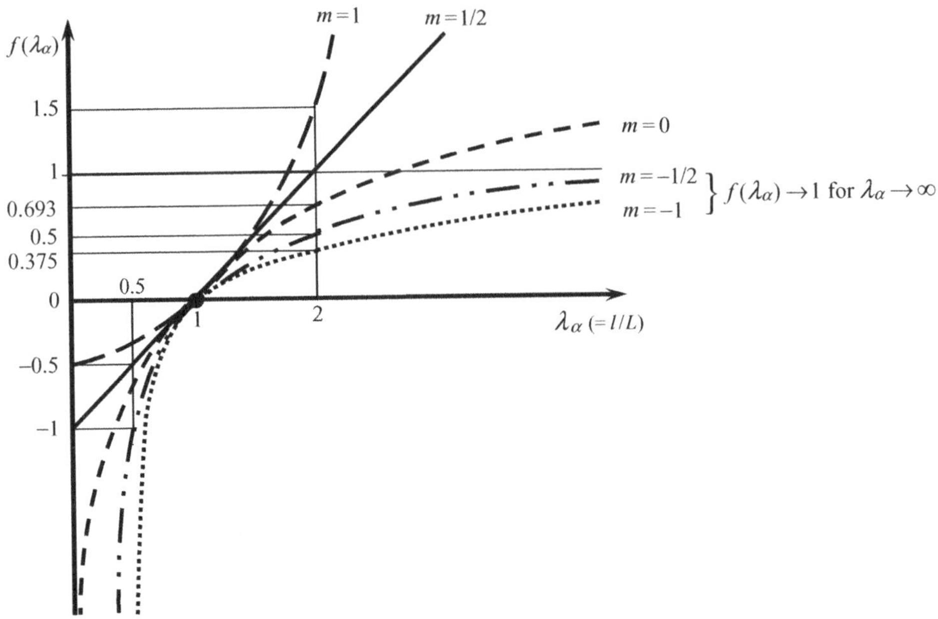

Figure 4.2 Properties of various strain measures

The eigenvalues of the general strain tensors $\mathbf{E}^{(m)}$ and $\mathbf{e}^{(m)}$ in equations (4.54) and (4.55) are identical to each other, and are given by

$$f(\lambda_\alpha) = \begin{cases} \dfrac{1}{2m}(\lambda_\alpha^{2m} - 1) & \text{for } m \neq 0, \\ \ln \lambda_\alpha & \text{for } m = 0. \end{cases} \tag{4.57}$$

The function $f(\lambda_\alpha)$ fulfills

$$f(1) = 0, f'(1) = 1, \tag{4.58}$$

and

$$f'(s) > 0 \tag{4.59}$$

for an arbitrary positive scalar s. The function $f(\lambda_\alpha)$ is shown in Figure 4.2, where L and l are the lengths of infinitesimal line elements in the reference and the current configurations. Note that equation $(4.57)_2$ for $m = 0$ is derived, with the aid of l'Hôpital's rule, as follows:

$$\lim_{m\to 0} \frac{1}{2m}(\lambda_\alpha^{2m} - 1) = \lim_{m\to 0} \frac{\exp(2m \ln \lambda_\alpha) - 1}{2m} = \lim_{m\to 0} \frac{\partial \{\exp(2m \ln \lambda_\alpha) - 1\} / \partial m}{\partial (2m)/\partial m}$$
$$= \lim_{m\to 0} \frac{\exp(2m \ln \lambda_\alpha)(2 \ln \lambda_\alpha)}{2} = \ln \lambda_\alpha.$$

Substituting equation (4.57) into equation (4.52), Seth's strains in equations (4.54) and (4.55) are described in the spectral decompositions as follows:

$$\mathbf{E}^{(m)} = \begin{cases} \sum_{\alpha=1}^{3} \frac{1}{2m}(\lambda_\alpha^{2m} - 1)\mathbf{N}^{(\alpha)} \otimes \mathbf{N}^{(\alpha)} & \text{for } m \neq 0 \\ \sum_{\alpha=1}^{3} \ln \lambda_\alpha \mathbf{N}^{(\alpha)} \otimes \mathbf{N}^{(\alpha)} & \text{for } m = 0 \end{cases}, \tag{4.60}$$

$$\mathbf{e}^{(m)} = \begin{cases} \sum_{\alpha=1}^{3} \frac{1}{2m}(\lambda_\alpha^{2m} - 1)\mathbf{n}^{(\alpha)} \otimes \mathbf{n}^{(\alpha)} & \text{for } m \neq 0 \\ \sum_{\alpha=1}^{3} \ln \lambda_\alpha \mathbf{n}^{(\alpha)} \otimes \mathbf{n}^{(\alpha)} & \text{for } m = 0 \end{cases}. \tag{4.61}$$

The following strain measures are used often in continuum mechanics.

$$\begin{array}{lll} \mathbf{E}^{(1)} & = \mathbf{E} = \frac{1}{2}(\mathbf{U}^2 - \mathbf{G}) = \frac{1}{2}(\mathbf{C} - \mathbf{G}) = \frac{1}{2}(\mathbf{F}^T\mathbf{F} - \mathbf{G}) & \text{Green strain} \\ \mathbf{E}^{(1/2)} & = \mathbf{B} = \mathbf{U} - \mathbf{G} = \mathbf{C}^{1/2} - \mathbf{G} = (\mathbf{F}^T\mathbf{F})^{1/2} - \mathbf{G} & \text{material Biot strain} \\ \mathbf{E}^{(0)} & = \ln \mathbf{U} = \frac{1}{2}\ln \mathbf{C} = \frac{1}{2}\ln(\mathbf{F}^T\mathbf{F}) & \text{material Hencky strain} \\ \mathbf{E}^{(-1)} & = \frac{1}{2}(\mathbf{G} - \mathbf{U}^{-2}) = \frac{1}{2}(\mathbf{G} - \mathbf{C}^{-1}) = \frac{1}{2}(\mathbf{G} - \mathbf{F}^{-1}\mathbf{F}^{-T}) & \text{material Piola strain} \end{array} \tag{4.62}$$

$$\begin{array}{lll} \mathbf{e}^{(1)} & = \frac{1}{2}(\mathbf{V}^2 - \mathbf{g}) = \frac{1}{2}(\mathbf{b} - \mathbf{g}) = \frac{1}{2}(\mathbf{F}\mathbf{F}^T - \mathbf{g}) & \text{Finger strain} \\ \mathbf{e}^{(0)} & = \ln \mathbf{V} = \frac{1}{2}\ln \mathbf{b} = \frac{1}{2}\ln(\mathbf{F}\mathbf{F}^T) & \text{spatial Hencky strain} \\ \mathbf{e}^{(-1/2)} & = \mathbf{g} - \mathbf{V}^{-1} = \mathbf{g} - \mathbf{b}^{-1/2} = \mathbf{g} - (\mathbf{F}\mathbf{F}^T)^{-1/2} & \text{spatial Biot strain} \\ \mathbf{e}^{(-1)} & = \mathbf{e} = \frac{1}{2}(\mathbf{g} - \mathbf{V}^{-2}) = \frac{1}{2}(\mathbf{g} - \mathbf{b}^{-1}) = \frac{1}{2}(\mathbf{g} - \mathbf{F}^{-T}\mathbf{F}^{-1}) & \text{Almansi strain} \end{array} \tag{4.63}$$

The generalized strain tensors in equations (4.54) and (4.55) can be represented in terms of the metric tensors, noting equations (4.1) and (4.3), as follows:

$$\mathbf{E}^{(m)} = \begin{cases} \frac{1}{2m}((\overleftarrow{\mathbf{g}}_{GG})^m - \mathbf{G}) & \text{for } m > 0 \\ \frac{1}{2}\ln \overleftarrow{\mathbf{g}}_{GG} & \text{for } m = 0 \\ \frac{1}{2m}((\overleftarrow{\mathbf{g}}^{GG})^{-m} - \mathbf{G}) & \text{for } m < 0 \end{cases} \tag{4.64}$$

and

$$\mathbf{e}^{(m)} = \begin{cases} \dfrac{1}{2m}((\vec{\mathbf{G}}^{gg})^m - \mathbf{g}) & \text{for } m > 0 \\ \dfrac{1}{2}\ln \vec{\mathbf{G}}^{gg} & \text{for } m = 0 \ , \\ \dfrac{1}{2m}((\vec{\mathbf{G}}_{gg})^{-m} - \mathbf{g}) & \text{for } m < 0 \end{cases} \tag{4.65}$$

noting equation (4.3). Then, the following expressions hold for the popular strain tensors:

$$\left.\begin{array}{ll} \mathbf{E}^{(1)} = \dfrac{1}{2}(\overleftarrow{\mathbf{g}}_{GG} - \mathbf{G}) = \dfrac{1}{2}(\delta^i_I\delta^j_J g_{ij} - G_{IJ})\mathbf{G}^I \otimes \mathbf{G}^J & \text{Green strain} \\ \mathbf{E}^{(1/2)} = (\overleftarrow{\mathbf{g}}_{GG})^{1/2} - \mathbf{G} & \text{material Biot strain} \\ \mathbf{E}^{(-1)} = \dfrac{1}{2}(\mathbf{G} - \overleftarrow{\mathbf{g}}^{GG}) = \dfrac{1}{2}(G^{IJ} - \delta^I_i\delta^J_j g^{ij})\mathbf{G}_I \otimes \mathbf{G}_J & \text{material Piola strain} \end{array}\right\} , \tag{4.66}$$

$$\left.\begin{array}{ll} \mathbf{e}^{(1)} = \dfrac{1}{2}(\vec{\mathbf{G}}^{gg} - \mathbf{g}) = \dfrac{1}{2}(\delta^i_I\delta^j_J G^{IJ} - g^{ij})\mathbf{g}_i \otimes \mathbf{g}_j & \text{Finger strain} \\ \mathbf{e}^{(-1/2)} = \mathbf{g} - (\vec{\mathbf{G}}_{gg})^{1/2} & \text{spatial Biot strain} \\ \mathbf{e}^{(-1)} = \dfrac{1}{2}(\mathbf{g} - \vec{\mathbf{G}}_{gg}) = \dfrac{1}{2}(g_{ij} - \delta^I_i\delta^J_j G_{IJ})\mathbf{g}^i \otimes \mathbf{g}^j & \text{Almansi strain} \end{array}\right\} , \tag{4.67}$$

while strains for $m = 0$ are shown in equations (4.64) and (4.65).

Any strain tensor can be described in terms of variations of the metric tensor in the convected coordinate system. This shows the importance of the incorporation of the convected coordinate system for the interpretation of physical meanings of strains.

4.2.3 *Hencky Strain Tensor*

In the particular case of $m = 0$ in the general strain tensor of equation (4.60) and (4.61), noting $\lambda_\alpha \geq 0$, the following strain measures are called *Hencky strain tensors*:

$$\begin{array}{l} \text{right Hencky strain tensor, } \mathbf{E}^{(0)} = \sum_{\alpha=1}^{3} \ln \lambda_\alpha \mathbf{N}^{(\alpha)} \otimes \mathbf{N}^{(\alpha)} \equiv \ln \mathbf{U} = \dfrac{1}{2}\ln \mathbf{C} \\ \text{left Hencky strain tensor, } \mathbf{e}^{(0)} = \sum_{\alpha=1}^{3} \ln \lambda_\alpha \mathbf{n}^{(\alpha)} \otimes \mathbf{n}^{(\alpha)} \equiv \ln \mathbf{V} = \dfrac{1}{2}\ln \mathbf{b} \end{array} \tag{4.68}$$

where $\lambda_\alpha\,(\alpha = 1, 2, 3)$ are the eigenvalues of $\mathbf{U}$ and $\mathbf{V}$, that is,

$$\lambda_\alpha \equiv U_\alpha = \sqrt{C_\alpha} = V_\alpha = \sqrt{b_\alpha}. \tag{4.69}$$

$\mathbf{E}^{(0)}$ and $\mathbf{e}^{(0)}$ are mutually related as follows.

$$\mathbf{E}^{(0)} = \mathbf{R}^T\mathbf{e}^{(0)}\mathbf{R},\ \mathbf{e}^{(0)} = \mathbf{R}\mathbf{E}^{(0)}\mathbf{R}^T. \tag{4.70}$$

When the principal directions of $\mathbf{C}$ and $\mathbf{b}$ are fixed, the following equations hold:

$$(\mathbf{E}^{(0)})_\alpha = (\mathbf{e}^{(0)})_\alpha = \ln \lambda_\alpha = \ln \frac{\partial x_\alpha}{\partial X_\alpha} \text{ (no sum)}, \tag{4.71}$$

$$\left\{(\mathbf{E}^{(0)})_\alpha\right\}^\bullet = \left\{(\mathbf{e}^{(0)})_\alpha\right\}^\bullet = (\ln \lambda_\alpha)^\bullet = \left(\frac{\partial x_\alpha}{\partial X_\alpha}\right)^\bullet \Big/ \left(\frac{\partial x_\alpha}{\partial X_\alpha}\right) = \frac{\partial \dot{x}_\alpha}{\partial x_\alpha} = d_{\alpha\alpha} \text{ (no sum)}, \tag{4.72}$$

where $\ln \lambda_\alpha$ in equation (4.71) is the *logarithmic strain* and $d_{\alpha\alpha}$(no sum) is the normal component in the strain rate tensor $\mathbf{d}$ defined in section 4.4. From equation (4.71) we obtain

$$\left.\begin{aligned} \mathrm{tr}\mathbf{E}^{(0)} &= \sum_{\alpha=1}^{3} \ln \lambda_\alpha = \frac{1}{2}\sum_{\alpha=1}^{3} \ln C_\alpha = \sum_{\alpha=1}^{3} \ln U_\alpha = \ln(U_1 U_2 U_3) \\ \mathrm{tr}\mathbf{e}^{(0)} &= \sum_{\alpha=1}^{3} \ln \lambda_\alpha = \frac{1}{2}\sum_{\alpha=1}^{3} \ln b_\alpha = \sum_{\alpha=1}^{3} \ln V_\alpha = \ln(V_1 V_2 V_3) \end{aligned}\right\} \tag{4.73}$$

which is identical to the *logarithmic volumetric strain*, that is,

$$\mathrm{tr}\mathbf{E}^{(0)} = \mathrm{tr}\mathbf{e}^{(0)} = \ln(\lambda_1\lambda_2\lambda_3) = \sum_{\alpha=1}^{3} \ln \frac{\partial x_\alpha}{\partial X_\alpha} = \ln J = \ln \frac{dv}{dV} \equiv \varepsilon_v. \tag{4.74}$$

4.3 Compatibility Condition

The Riemann–Christoffel symbol in equation (2.158) vanishes, that is, $R_{ijkl} = 0$ in the Euclidian space as shown in equation (2.160), which is described in terms of the metric tensors g_{ij} and G_{IJ} as follows (Dimitrienko, 2010):

$$\frac{1}{2}\left(\frac{\partial^2 g_{il}}{\partial\theta^j\partial\theta^k} + \frac{\partial^2 g_{jk}}{\partial\theta^i\partial\theta^l} - \frac{\partial^2 g_{ik}}{\partial\theta^j\partial\theta^l} - \frac{\partial^2 g_{jl}}{\partial\theta^i\partial\theta^k} +\right) + g^{rs}\left(\Gamma^g_{jkr}\Gamma^g_{ils} - g_{rp}\Gamma^g_{jlr}\Gamma^g_{iks}\right) = 0,$$

$$\frac{1}{2}\left(\frac{\partial^2 G_{il}}{\partial\theta^j\partial\theta^k} + \frac{\partial^2 G_{jk}}{\partial\theta^i\partial\theta^l} - \frac{\partial^2 G_{ik}}{\partial\theta^j\partial\theta^l} - \frac{\partial^2 G_{jl}}{\partial\theta^i\partial\theta^k} +\right) + G^{rs}\left(\Gamma^G_{jkr}\Gamma^G_{ils} - G_{rp}\Gamma^G_{jlr}\Gamma^G_{iks}\right) = 0,$$

where Γ^g_{ijk} and Γ^G_{ijk} denote the Christoffel symbols for g_{ij} or and G_{IJ}, respectively, and the indices in G_{IJ} are written in lowercase for the sake of simplicity. The subtraction of the lower equation from the upper equation yields the following *compatibility condition* in the general coordinate system, which must be satisfied in the strain components:

$$\begin{aligned} &\frac{1}{2}\left(\frac{\partial^2 e_{il}}{\partial\theta^j\partial\theta^k} + \frac{\partial^2 e_{jk}}{\partial\theta^i\partial\theta^l} - \frac{\partial^2 e_{ik}}{\partial\theta^j\partial\theta^l} - \frac{\partial^2 e_{jl}}{\partial\theta^i\partial\theta^k}\right) \\ &\qquad + g^{rs}(\Gamma^g_{jkr}\Gamma^g_{ils} - \Gamma^g_{jlr}\Gamma^g_{iks}) - G^{rs}(\Gamma^G_{jkr}\Gamma^G_{ils} - \Gamma^G_{jlr}\Gamma^G_{iks}) = 0 \end{aligned} \tag{4.75}$$

by virtue of equation $(4.31)_2$, where one has the relations

$$\Gamma^g_{ijk} = \Gamma^G_{ijk} + \frac{\partial e_{ik}}{\partial\theta^j} + \frac{\partial e_{jk}}{\partial\theta^i} - \frac{\partial e_{ij}}{\partial\theta^k} \tag{4.76}$$

and

$$g^{ir}(2e_{rj} + G_{rj}) = \delta^i_j, \tag{4.77}$$

noting that

$$\Gamma^g_{ijk} = \frac{1}{2}\left\{\frac{\partial(2e_{jk} + G_{jk})}{\partial\theta^i} + \frac{\partial(2e_{ik} + G_{ik})}{\partial\theta^j} - \frac{\partial(2e_{ij} + G_{ij})}{\partial\theta^k}\right\},$$

$$g^{ir}(2e_{rj} + G_{rj}) = g^{ir}\left\{2\frac{1}{2}(g_{ij} - G_{rj}) + G_{rj}\right\} = \delta^i_j$$

with $g_{ij} = 2e_{ij} + G_{ij}$.

The compatibility condition can be derived in a different way as follows (cf. Eringen, 1962). The Riemann–Christoffel symbol must hold for the metric tensors g_{ij} and G_{IJ} which are related to the Cauchy–Green deformation tensors C_{IJ} and b^{-1}_{ij} through the Green and Eulerian strain tensors E_{IJ} and e_{ij} as shown in equation (4.31). Substitution of equation (4.31) into equation (2.158) with equation (2.160) leads to the following compatibility conditions for e_{ij} and E_{IJ}. The compatibility condition for e_{ij} is given as follows:

$$\begin{aligned}
&\frac{\partial^2 e_{kn}}{\partial\theta^l\partial\theta^m} + \frac{\partial^2 e_{lm}}{\partial\theta^k\partial\theta^n} - \frac{\partial^2 e_{km}}{\partial\theta^l\partial\theta^n} - \frac{\partial^2 e_{ln}}{\partial\theta^k\partial\theta^m}\\
&\quad + 2g^{rs}\{(\Gamma_{knr})_e(\Gamma_{lms})_g + (\Gamma_{knr})_g(\Gamma_{lms})_e - 2(\Gamma_{knr})_e(\Gamma_{lms})_e\\
&\quad - (\Gamma_{kmr})_e(\Gamma_{lns})_g - (\Gamma_{kmr})_g(\Gamma_{lns})_e + 2(\Gamma_{kmr})_e(\Gamma_{lns})_e\}\\
&\quad + 2e^{rs}\{(\Gamma_{knr})_g(\Gamma_{lms})_g - (\Gamma_{kmr})_g(\Gamma_{lns})_g - 2(\Gamma_{knr})_e(\Gamma_{lms})_e\\
&\quad - 2(\Gamma_{knr})_g(\Gamma_{lms})_e + 4(\Gamma_{knr})_e(\Gamma_{lms})_e + 2(\Gamma_{kmr})_e(\Gamma_{lns})_g\\
&\quad + 2(\Gamma_{kmr})_g(\Gamma_{lns})_e - 4(\Gamma_{kmr})_e(\Gamma_{lns})_e\} = 0,
\end{aligned} \tag{4.78}$$

where $(\cdot)_g$ and $(\cdot)_e$ denote the Christoffel symbols of the first kind based on g_{ij} and e_{ij}, respectively.

If e_{ij} is replaced by the infinitesimal strain $\tilde{e}_{ij} \equiv (\partial u_i/\partial\theta_j + \partial u_j/\partial\theta_i)/2$ and then the terms containing the second powers of the Christoffel symbols of first kind are eliminated in equation (4.75) or (4.78), the following compatibility condition for the infinitesimal deformation is obtained:

$$\frac{\partial^2\tilde{e}_{il}}{\partial\theta^j\partial\theta^k} + \frac{\partial^2\tilde{e}_{jk}}{\partial\theta^i\partial\theta^l} - \frac{\partial^2\tilde{e}_{ik}}{\partial\theta^j\partial\theta^l} - \frac{\partial^2\tilde{e}_{jl}}{\partial\theta^i\partial\theta^k} = 0. \tag{4.79}$$

4.4 Strain Rate and Spin Tensors

Various strain rate and spin tensors used in rate (incremental) type constitutive equations are described in this section.

4.4.1 Strain Rate and Spin Tensors Based on Velocity Gradient Tensor

The velocity gradient tensor $\boldsymbol{l}$ is defined as follows:

$$\boldsymbol{l} = \text{grad } \mathbf{v} = \mathbf{v} \otimes \nabla_x = \frac{\partial \mathbf{v}}{\partial \mathbf{x}}, \quad d\mathbf{v} = d\overset{\bullet}{\mathbf{x}} = \frac{\partial \overset{\bullet}{\mathbf{x}}}{\partial \mathbf{x}} d\mathbf{x} = \frac{\partial \mathbf{v}}{\partial \mathbf{x}} d\mathbf{x} = \boldsymbol{l} d\mathbf{x} \tag{4.80}$$

which is described in terms of the deformation gradient as

$$\boxed{\boldsymbol{l} = \overset{\bullet}{\mathbf{F}}\mathbf{F}^{-1} = \overset{\rightarrow}{\overset{\bullet}{\mathbf{F}}}{}^{\hat{g}g}} \tag{4.81}$$

noting equation (3.7) and

$$\frac{\partial \mathbf{v}}{\partial \mathbf{x}} = \left(\frac{\partial \mathbf{x}}{\partial \mathbf{X}}\right)^{\bullet} \frac{\partial \mathbf{X}}{\partial \mathbf{x}},$$

where $\mathbf{v}$ is the velocity of material particle. Then $\overset{\bullet}{\mathbf{F}}$ is described as

$$\overset{\bullet}{\mathbf{F}} = \overset{\leftarrow}{\boldsymbol{l}}_{\hat{g}G} (= l_{ij}\mathbf{g}^i \otimes \delta^j_J \mathbf{G}^J = (l_{ij}\mathbf{g}^i \otimes \mathbf{g}^j)(\delta^r_R \mathbf{g}_r \otimes \mathbf{G}^R)) = \boldsymbol{l}\mathbf{F}, \ \overset{\bullet}{\mathbf{F}}{}^{-1} = -\mathbf{F}^{-1}\boldsymbol{l} \tag{4.82}$$

noting equation (3.10) and $\overset{\bullet}{\mathbf{F}}\mathbf{F}^{-1} + \mathbf{F}\overset{\bullet}{\mathbf{F}}{}^{-1} = \mathbf{O}$.

Further, noting that

$$\overset{\bullet}{\mathbf{F}}\mathbf{F}^{-1} = (\delta^i_I \mathbf{g}_i \otimes \mathbf{G}^I)^{\bullet} \delta^J_j \mathbf{G}_J \otimes \mathbf{g}^j = \delta^I_J \delta^i_I \overset{\bullet}{\mathbf{g}}_i \delta^J_j \otimes \mathbf{g}^j = \overset{\bullet}{\mathbf{g}}_i \otimes \mathbf{g}^j$$

based on equation (3.10), the velocity gradient can be rewritten as

$$\boldsymbol{l} = \overset{\bullet}{\mathbf{g}}_i \otimes \mathbf{g}^i = (\overset{\bullet}{\mathbf{g}}_i \cdot \mathbf{g}_r)\mathbf{g}^r \otimes \mathbf{g}^i, \tag{4.83}$$

from which we have

$$\overset{\bullet}{\mathbf{g}}_i = \boldsymbol{l}\mathbf{g}_i \tag{4.84}$$

and

$$\boldsymbol{l}^T = \mathbf{g}^i \otimes \overset{\bullet}{\mathbf{g}}_i = -\overset{\bullet}{\mathbf{g}}{}^i \otimes \mathbf{g}_i, \ \overset{\bullet}{\mathbf{g}}{}^i = -\boldsymbol{l}^T \boldsymbol{g}^i. \tag{4.85}$$

Then

$$\text{tr}\boldsymbol{l} = \frac{(\sqrt{g})^{\bullet}}{\sqrt{g}} = \frac{\overset{\bullet}{v}}{v} \tag{4.86}$$

noting that

$$\begin{aligned}(\sqrt{g})^{\bullet} = ([\mathbf{g}_1\mathbf{g}_2\mathbf{g}_3])^{\bullet} &= \overset{\bullet}{\mathbf{g}}_1 \cdot (\mathbf{g}_2 \times \mathbf{g}_3) + \mathbf{g}_1 \cdot (\overset{\bullet}{\mathbf{g}}_2 \times \mathbf{g}_3) + \mathbf{g}_1 \cdot (\mathbf{g}_2 \times \overset{\bullet}{\mathbf{g}}_3) \\ &= \sqrt{g}\,(\overset{\bullet}{\mathbf{g}}_1 \cdot \mathbf{g}^1 + \overset{\bullet}{\mathbf{g}}_2 \cdot \mathbf{g}^2 + \overset{\bullet}{\mathbf{g}}_3 \cdot \mathbf{g}^3) = \sqrt{g}\,\overset{\bullet}{\mathbf{g}}_i \cdot \mathbf{g}^i = \sqrt{g}\,\boldsymbol{l}\mathbf{g}_i \cdot \mathbf{g}^i = \sqrt{g}\,\text{tr}\boldsymbol{l} \\ &= \sqrt{g}\,\boldsymbol{l}{:}\mathbf{g}\end{aligned}$$

with the aid of equations (1.40), (2.36), (1.475), (2.83) and (4.84).

Comparison of equation (4.13) with equation (4.80) leads to the expression of the velocity gradient tensor by the relative deformation gradient tensor,

$$\boldsymbol{l} = \overset{\bullet}{\mathbf{F}}_t(t). \tag{4.87}$$

Exploiting equation (4.12), one can obtain the expression in the relative polar decomposition:

$$\boldsymbol{l} = \left.\frac{\partial(\mathbf{R}_t(\tau)\mathbf{U}_t(\tau))}{\partial\tau}\right|_{\tau=t} = (\mathbf{R}_t(\tau)\dot{\mathbf{U}}_t(\tau) + \dot{\mathbf{R}}_t(\tau)\mathbf{U}_t(\tau))|_{\tau=t} = \dot{\mathbf{U}}_t(t) + \dot{\mathbf{R}}_t(t) \tag{4.88}$$

noting $\mathbf{R}_t(t) = \mathbf{G}$, $\mathbf{U}_t(t) = \mathbf{G}$. Likewise, one has

$$\boldsymbol{l} = \dot{\mathbf{V}}_t(t) + \dot{\mathbf{R}}_t(t). \tag{4.89}$$

On the other hand, the following velocity gradient tensor can be defined:

$$\boldsymbol{L} \equiv \mathrm{Grad}\mathbf{v} = \mathbf{v} \otimes \nabla_X = \frac{\partial \mathbf{v}}{\partial \mathbf{X}} = \frac{\partial \mathbf{v}}{\partial \mathbf{x}}\frac{\partial \mathbf{x}}{\partial \mathbf{X}} = \boldsymbol{l}\mathbf{F} = \overleftarrow{\boldsymbol{l}}_{gG}, \tag{4.90}$$

which is an Eulerian–Lagrangian two-point tensor.

Further, we have the following contravariant-covariant pull-back of $\boldsymbol{l}$ by equation $(3.32)_2$:

$$\mathbf{L} \equiv \overleftarrow{\boldsymbol{l}}^{G}_{\bullet G} = \mathbf{F}^{-1}\boldsymbol{l}\mathbf{F} \tag{4.91}$$

which is a Lagrangian tensor.

Furthermore, the covariant pull-back of $\boldsymbol{l}$ by equation $(3.32)_4$ is given by

$$\mathsf{L} \equiv \overleftarrow{\boldsymbol{l}}_{GG} = \mathbf{F}^T\boldsymbol{l}\mathbf{F} = \mathbf{F}^T\dot{\mathbf{F}}\mathbf{F}^{-1}\mathbf{F} = \mathbf{F}^T\dot{\mathbf{F}} = \left(\frac{\partial \mathbf{x}}{\partial \mathbf{X}}\right)^T \frac{\partial \mathbf{v}}{\partial \mathbf{X}} \tag{4.92}$$

which is also a Lagrangian tensor.

We find from equations (4.31) and (4.32) that

$$\dot{\mathbf{E}} = \mathrm{sym}[\mathsf{L}] = \mathsf{D} = \frac{1}{2}\delta^i_I\delta^j_J\dot{g}_{ij}\mathbf{G}^I \otimes \mathbf{G}^J, \tag{4.93}$$

where we set

$$\mathsf{D} \equiv \overleftarrow{\mathbf{d}}_{GG} = \mathbf{F}^T\mathbf{d}\mathbf{F}, \tag{4.94}$$

noting

$$\dot{\mathbf{E}} = \frac{1}{2}(\dot{\mathbf{F}}^T\mathbf{F} + \mathbf{F}^T\dot{\mathbf{F}}) = \mathbf{F}^T\frac{1}{2}(\dot{\mathbf{F}}\mathbf{F}^{-1} + (\dot{\mathbf{F}}\mathbf{F}^{-1})^T)\mathbf{F}.$$

Here, the strain rate tensor $\mathbf{d}$ is the symmetric part of $\boldsymbol{l}$, while the *continuum spin* $\mathbf{w}$ is the anti-symmetric part of $\boldsymbol{l}$, that is,

$$\boxed{\boldsymbol{l} = \mathbf{d} + \mathbf{w}} \tag{4.95}$$

where

$$\left.\begin{aligned} \mathbf{d} &= \frac{1}{2}(\boldsymbol{l} + \boldsymbol{l}^T) = \frac{1}{2}(\mathbf{v} \otimes \nabla_x + \nabla_x \otimes \mathbf{v}) = \mathrm{sym}(\mathrm{grad}\mathbf{v}) \\ \mathbf{w} &= \frac{1}{2}(\boldsymbol{l} - \boldsymbol{l}^T) = \frac{1}{2}(\mathbf{v} \otimes \nabla_x - \nabla_x \otimes \mathbf{v}) = \mathrm{ant}(\mathrm{grad}\mathbf{v}) \end{aligned}\right\} \tag{4.96}$$

which can be rewritten by equation (4.83) as follows:

$$\left.\begin{aligned} \mathbf{d} &= \mathrm{sym}[\dot{\mathbf{g}}_i \otimes \mathbf{g}^i] = \frac{1}{2}(\mathbf{g}_i \cdot \dot{\mathbf{g}}_j + \dot{\mathbf{g}}_i \cdot \mathbf{g}_j)\mathbf{g}^i \otimes \mathbf{g}^j = \frac{1}{2}\dot{g}_{ij}\mathbf{g}^i \otimes \mathbf{g}^j \\ \mathbf{w} &= \mathrm{ant}[\dot{\mathbf{g}}_i \otimes \mathbf{g}^i] = \frac{1}{2}(\mathbf{g}_i \cdot \dot{\mathbf{g}}_j - \dot{\mathbf{g}}_i \cdot \mathbf{g}_j)\mathbf{g}^i \otimes \mathbf{g}^j \end{aligned}\right\}, \tag{4.97}$$

resulting in $d_{ij} = \dot{g}_{ij}/2$, noting equations $(2.28)_2$ and $(2.29)_2$.

Inserting equations (4.88) and (4.89) into equation (4.96), one has the expression

$$\mathbf{d} = \overset{\bullet}{\mathbf{U}}_t(t) = \overset{\bullet}{\mathbf{V}}_t(t), \quad \mathbf{w} = \overset{\bullet}{\mathbf{R}}_t(t). \tag{4.98}$$

The current embedded base vector is given from equation (2.2) by

$$\mathbf{g}_i = \frac{\partial \mathbf{x}}{\partial \theta^i}. \tag{4.99}$$

Noting that the components of a material particle position vector in the embedded coordinate system do not change, that is, $\overset{\bullet}{\theta}{}^i = 0$, the rate of the covariant base vector is given by

$$\overset{\bullet}{\mathbf{g}}_i = \frac{\partial \overset{\bullet}{\mathbf{x}}}{\partial \theta^i} = \frac{\partial \mathbf{v}}{\partial \theta^i} \tag{4.100}$$

which, using equation (2.129), can be rewritten as

$$\overset{\bullet}{\mathbf{g}}_j = v^i|_j \mathbf{g}_i = v_i|_j \mathbf{g}^i. \tag{4.101}$$

Substituting equation (4.101) into equations (4.83), that is, $\boldsymbol{l} = \overset{\bullet}{\mathbf{g}}_j \otimes \mathbf{g}^j$ and (4.97), one has

$$\boldsymbol{l} = v^i|_j \mathbf{g}_i \otimes \mathbf{g}^j = v_i|_j \mathbf{g}^i \otimes \mathbf{g}^j, \tag{4.102}$$

$$\mathbf{d} = \frac{1}{2}(v_i|_j \mathbf{g}^i \otimes \mathbf{g}^j + v_i|_j \mathbf{g}^j \otimes \mathbf{g}^i) = \frac{1}{2}(v_i|_j + v_j|_i)\mathbf{g}^i \otimes \mathbf{g}^j, \tag{4.103}$$

$$\mathbf{w} = \frac{1}{2}(v_i|_j - v_j|_i)\mathbf{g}^i \otimes \mathbf{g}^j. \tag{4.104}$$

A simple geometrical interpretation of the strain rate is now given. Substituting equation $(4.46)_2$ into equation $(4.97)_1$, it follows that

$$d_{ij} = \frac{1}{2}\left\{(||\mathbf{g}_i||)^{\bullet}||\mathbf{g}_j|| + ||\mathbf{g}_i||(||\mathbf{g}_j||)^{\bullet}\right\}\cos\phi_{ij} - \frac{1}{2}||\mathbf{g}_i||||\mathbf{g}_j||\overset{\bullet}{\phi}_{ij}\sin\phi_{ij}. \tag{4.105}$$

For $\mathbf{g}_i = \mathbf{g}_j (\phi_{ij} = 0)$, equation (4.105) leads to

$$d_{ii} = (||\mathbf{g}_i||)^{\bullet}||\mathbf{g}_i|| \text{ (no sum)} \tag{4.106}$$

which expresses the rate of elongation when we choose the magnitude of the base vectors to be unity, that is, $||\mathbf{g}_i|| = 1$. For $\phi_{ij} = \pi/2$,

$$d_{ij} = -\frac{1}{2}||\mathbf{g}_i||||\mathbf{g}_j||\overset{\bullet}{\phi}_{ij} \text{ (no sum)} \tag{4.107}$$

which is half of the reducing rate of the angle between the line elements perpendicular to each other when we choose the magnitude of the base vectors to be unity, that is, $||\mathbf{g}_i|| = 1$. More detailed geometrical interpretations of the strain rate tensor **d** and the continuum spin tensor **w** are given in Appendix B.

The relation between the components of the rate of the Green strain tensor $\overset{\bullet}{\mathbf{E}}$ and **d** is given by $\overset{\bullet}{E}_{IJ} = D_{IJ} = \delta^i_I \delta^j_J d_{ij}$ from equations (4.93) and (4.94) and thus it is known that their components have same values, while $\overset{\bullet}{\mathbf{E}}$ is the covariant pull-back of the strain rate tensor **d** in equation $(3.32)_4$.

Substituting equation (4.4) into equation (4.81), $\boldsymbol{l}$ is described by **U**, **V**, **R** as follows:

$$\boldsymbol{l} = \begin{cases} \mathbf{R}\,\overset{\bullet}{\tilde{\mathbf{U}}}\,\mathbf{R}^T + \boldsymbol{\Omega}^R \\ \overset{\bullet}{\tilde{\mathbf{V}}} + \mathbf{V}\boldsymbol{\Omega}^R\mathbf{V}^{-1} \end{cases} \tag{4.108}$$

where

$$\overset{\bullet}{\tilde{\mathbf{U}}} \equiv \dot{\mathbf{U}}\mathbf{U}^{-1},\ \overset{\bullet}{\tilde{\mathbf{V}}} \equiv \dot{\mathbf{V}}\mathbf{V}^{-1}, \tag{4.109}$$

$$\boxed{\boldsymbol{\Omega}^R \equiv \dot{\mathbf{R}}\mathbf{R}^T} \tag{4.110}$$

noting that

$$\boldsymbol{l} = \begin{cases} (\mathbf{RU})^{\bullet}(\mathbf{RU})^{-1} = (\dot{\mathbf{R}}\mathbf{U} + \mathbf{R}\dot{\mathbf{U}})\mathbf{U}^{-1}\mathbf{R}^T \\ (\mathbf{VR})^{\bullet}(\mathbf{VR})^{-1} = (\dot{\mathbf{V}}\mathbf{R} + \mathbf{V}\dot{\mathbf{R}})\mathbf{R}^T\mathbf{V}^{-1} \end{cases}. \tag{4.111}$$

$\boldsymbol{\Omega}^R$ is called the *relative spin* because it describes the difference of the spin of the Eulerian triad $\{\mathbf{n}^{(\alpha)}\}$ from the spin of the Lagrangian triad $\{\mathbf{N}^{(\alpha)}\}$, as will be shown in equation (4.119).

We can describe **d** and **w** in terms of **U**, **V**, **R** from equation (4.108) as follows:

$$\left.\begin{aligned} \mathbf{d} &= \mathbf{R}\mathrm{sym}[\overset{\bullet}{\tilde{\mathbf{U}}}]\mathbf{R}^T \\ \mathbf{w} &= \boldsymbol{\Omega}^R + \mathbf{R}\mathrm{ant}[\overset{\bullet}{\tilde{\mathbf{U}}}]\mathbf{R}^T \end{aligned}\right\}, \tag{4.112}$$

$$\left.\begin{aligned} \mathbf{d} &= \mathrm{sym}[\overset{\bullet}{\tilde{\mathbf{V}}}] + \mathrm{sym}[\mathbf{V}\boldsymbol{\Omega}^R\mathbf{V}^{-1}] \\ \mathbf{w} &= \mathrm{ant}[\mathbf{V}\boldsymbol{\Omega}^R\mathbf{V}^{-1}] + \mathrm{ant}[\overset{\bullet}{\tilde{\mathbf{V}}}] \end{aligned}\right\}. \tag{4.113}$$

The following strain rate derived from equation (4.112) with equation (4.109) is called the *rotationless* (or *corotational*) *strain rate* (Storen and Rice, 1975):

$$\widehat{\mathbf{D}} \equiv \overleftarrow{\mathbf{d}}^R = \mathbf{R}^T\mathbf{d}\mathbf{R} = \mathrm{sym}[\overset{\bullet}{\tilde{\mathbf{U}}}] = \frac{1}{2}(\dot{\mathbf{U}}\mathbf{U}^{-1} + \mathbf{U}^{-1}\dot{\mathbf{U}}) = \frac{1}{2}\mathbf{C}^{-1/2}\dot{\mathbf{C}}\mathbf{C}^{-1/2} \tag{4.114}$$

which depends only on the right stretch tensor **U**.

Noting that

$$\dot{\mathbf{R}}^L\mathbf{R}^{L^T} = \sum_{\alpha,\beta=1}^{3}(\mathbf{N}^{(\alpha)}\otimes\mathbf{e}_\alpha)^{\bullet}(\mathbf{N}^{(\beta)}\otimes\mathbf{e}_\beta)^T = \sum_{\alpha,\beta=1}^{3}\dot{\mathbf{N}}^{(\alpha)}\otimes\mathbf{e}^{(\alpha)}\mathbf{e}^{(\beta)}\otimes\mathbf{N}^{(\beta)},$$

it follows that

$$\boxed{\boldsymbol{\Omega}^L \equiv \dot{\mathbf{R}}^L\mathbf{R}^{L^T} = \sum_{\alpha=1}^{3}\dot{\mathbf{N}}^{(\alpha)}\otimes\mathbf{N}^{(\alpha)}} \tag{4.115}$$

and thus

$$\dot{\mathbf{N}}^{(\alpha)} = \boldsymbol{\Omega}^L\mathbf{N}^{(\alpha)} = \sum_{\beta=1}^{3}\Omega^L_{\beta\alpha}\mathbf{N}^{(\beta)} \tag{4.116}$$

noting that $\overset{\bullet}{\mathbf{e}}^{(\alpha)} = \mathbf{0}$ since $\{\mathbf{e}^{(\alpha)}\}$ is a fixed base. Therefore, $\mathbf{\Omega}^L$ describes the spin of the Lagrangian triad $\{\mathbf{N}^{(\alpha)}\}$ of the right stretch tensor $\mathbf{U}$ and is called the *Lagrangian spin tensor*. On the other hand, noting that

$$\overset{\bullet}{\mathbf{R}}^E\mathbf{R}^{E^T} = \sum_{\alpha,\beta=1}^{3} (\mathbf{n}^{(\alpha)} \otimes \mathbf{e}^{(\alpha)})^{\bullet}(\mathbf{n}^{(\beta)} \otimes \mathbf{e}^{(\beta)})^T = \sum_{\alpha,\beta=1}^{3} \overset{\bullet}{\mathbf{n}}^{(\alpha)} \otimes \mathbf{e}^{(\alpha)}\mathbf{e}^{(\beta)} \otimes \mathbf{n}^{(\beta)},$$

we have

$$\boxed{\mathbf{\Omega}^E \equiv \overset{\bullet}{\mathbf{R}}^E\mathbf{R}^{E^T} = \sum_{\alpha=1}^{3} \overset{\bullet}{\mathbf{n}}^{(\alpha)} \otimes \mathbf{n}^{(\alpha)}} = \sum_{\beta=1}^{3} \Omega^E_{\beta\alpha}\mathbf{N}^{(\beta)} \tag{4.117}$$

and thus

$$\overset{\bullet}{\mathbf{n}}^{(\alpha)} = \mathbf{\Omega}^E\mathbf{n}^{(\alpha)} = \sum_{\beta=1}^{3} \Omega^E_{\beta\alpha}\mathbf{n}^{(\beta)}. \tag{4.118}$$

Therefore, $\mathbf{\Omega}^E$ describes the spin of the Eulerian triad $\{\mathbf{n}^{(\alpha)}\}$ of the right stretch tensor $\mathbf{V}$ and is called the *Eulerian spin tensor*.

Furthermore, it follows that

$$\begin{aligned}
\overset{\bullet}{\mathbf{R}}\mathbf{R}^T &= \sum_{\alpha,\beta=1}^{3} (\mathbf{n}^{(\alpha)} \otimes \mathbf{N}^{(\alpha)})^{\bullet}(\mathbf{n}^{(\beta)} \otimes \mathbf{N}^{(\beta)})^T \\
&= \sum_{\alpha,\beta=1}^{3} (\overset{\bullet}{\mathbf{n}}^{(\alpha)} \otimes \mathbf{N}^{(\alpha)} + \mathbf{n}^{(\alpha)} \otimes \overset{\bullet}{\mathbf{N}}^{(\alpha)})\mathbf{N}^{(\beta)} \otimes \mathbf{n}^{(\beta)} \\
&= \sum_{\alpha,\beta=1}^{3} \left\{\overset{\bullet}{\mathbf{n}}^{(\alpha)} \otimes \mathbf{n}^{(\alpha)} + \mathbf{n}^{(\alpha)} \otimes \delta_{\alpha\gamma}\overset{\bullet}{\mathbf{N}}^{(\gamma)}\mathbf{N}^{(\beta)} \otimes \mathbf{n}^{(\beta)}\right\} \\
&= \sum_{\alpha,\beta=1}^{3} \left\{\overset{\bullet}{\mathbf{n}}^{(\alpha)} \otimes \mathbf{n}^{(\alpha)} + \mathbf{n}^{(\alpha)} \otimes \mathbf{N}^{(\alpha)}(\mathbf{N}^{(\gamma)} \otimes \overset{\bullet}{\mathbf{N}}^{(\gamma)})\mathbf{N}^{(\beta)} \otimes \mathbf{n}^{(\beta)}\right\} \\
&= \sum_{\alpha,\beta=1}^{3} \left\{\overset{\bullet}{\mathbf{n}}^{(\alpha)} \otimes \mathbf{n}^{(\alpha)} - \mathbf{n}^{(\alpha)} \otimes \mathbf{N}^{(\alpha)}(\overset{\bullet}{\mathbf{N}}^{(\gamma)} \otimes \mathbf{N}^{(\gamma)})\mathbf{N}^{(\beta)} \otimes \mathbf{n}^{(\beta)}\right\}
\end{aligned}$$

and thus the following relations hold:

$$\mathbf{\Omega}^R = \mathbf{\Omega}^E - \mathbf{R}\mathbf{\Omega}^L\mathbf{R}^T, \quad \mathbf{\Omega}^E = \mathbf{\Omega}^R + \mathbf{R}\mathbf{\Omega}^L\mathbf{R}^T, \quad \mathbf{\Omega}^L = \mathbf{R}^T(\mathbf{\Omega}^E - \mathbf{\Omega}^R)\mathbf{R}. \tag{4.119}$$

In the rigid-body rotation ($\overset{\bullet}{\mathbf{U}} = \overset{\bullet}{\tilde{\mathbf{U}}} = \overset{\bullet}{\mathbf{V}} = \overset{\bullet}{\tilde{\mathbf{V}}} = \mathbf{O}$, $\overset{\bullet}{\mathbf{N}}^{(\alpha)} = \mathbf{0}$), it follows from equations (4.98), (4.108), (4.109), (4.110), (4.112)$_2$, (4.115) and (4.119) that

$$\mathbf{d} = \mathbf{O}, \ \boldsymbol{l} = \mathbf{w} = \mathbf{\Omega}^R = \mathbf{\Omega}^E, \ \ \mathbf{\Omega}^L = \mathbf{O}, \tag{4.120}$$

noting that the Lagrangian triad $\{\mathbf{N}^{(\alpha)}\}$ is fixed during a rigid-body rotation. In the general deformation process, on the other hand, we can write

$$\mathbf{w} = \dot{\mathbf{R}}_t(t) = \boldsymbol{\Omega}_t^R(t) \tag{4.121}$$

from equations (4.98) and (4.110).

4.4.2 Strain Rate Tensor Based on General Strain Tensor

We now turn to the strain rate tensors based on the general strain described in Section 4.2.2. The rate of the general Lagrangian and Eulerian strain tensors are given by

$$\left.\begin{aligned}\dot{\mathbf{E}}^{(m)} &= \frac{1}{2m}(\mathbf{U}^{2m})^{\bullet} = \frac{1}{2m}(\dot{\mathbf{U}}\mathbf{U}^{2m-1} + \mathbf{U}\dot{\mathbf{U}}\mathbf{U}^{2m-2} + \cdots + \mathbf{U}^{2m-1}\dot{\mathbf{U}})\\ \dot{\mathbf{e}}^{(m)} &= \frac{1}{2m}(\mathbf{V}^{m})^{\bullet} = \frac{1}{2m}(\dot{\mathbf{V}}\mathbf{V}^{2m-1} + \mathbf{V}\dot{\mathbf{V}}\mathbf{V}^{2m-2} + \cdots + \mathbf{V}^{2m-1}\dot{\mathbf{V}})\end{aligned}\right\} \tag{4.122}$$

for which we have

$$\left.\begin{aligned}\dot{\mathbf{E}}^{(-m)} &= \mathbf{U}^{-2m}\dot{\mathbf{E}}^{(m)}\ \mathbf{U}^{-2m}\\ \dot{\mathbf{e}}^{(-m)} &= \mathbf{V}^{-2m}\dot{\mathbf{e}}^{(m)}\ \mathbf{V}^{-2m}\end{aligned}\right\}, \tag{4.123}$$

noting that

$$\begin{aligned}(\mathbf{U}^{-1})^{\bullet} &= -\mathbf{U}^{-1}\,\dot{\mathbf{U}}\,\mathbf{U}^{-1},\\ (\mathbf{U}^{-2m})^{\bullet} &= (\mathbf{U}^{-1})^{\bullet}\mathbf{U}^{-2m+1} + \mathbf{U}^{-1}(\mathbf{U}^{-1})^{\bullet}\mathbf{U}^{-2m+2} + \cdots + \mathbf{U}^{-2m+1}(\mathbf{U}^{-1})^{\bullet}\\ &= (-\mathbf{U}^{-1}\,\dot{\mathbf{U}}\,\mathbf{U}^{-1})\mathbf{U}^{-2m+1} + \mathbf{U}^{-1}(-\mathbf{U}^{-1}\,\dot{\mathbf{U}}\,\mathbf{U}^{-1})\mathbf{U}^{-2m+2} + \cdots + \mathbf{U}^{-2m+1}(-\mathbf{U}^{-1}\,\dot{\mathbf{U}}\,\mathbf{U}^{-1})\\ &= -\mathbf{U}^{-2m}(\dot{\mathbf{U}}\,\mathbf{U}^{2m-1} + \mathbf{U}\,\dot{\mathbf{U}}\,\mathbf{U}^{2m-2} + \cdots + \mathbf{U}^{2m-1}\,\dot{\mathbf{U}})\mathbf{U}^{-2m},\\ \dot{\mathbf{E}}^{(-m)} &= \frac{1}{-2m}(\mathbf{U}^{-2m})^{\bullet} = \frac{1}{-2m}(-\mathbf{U}^{-2m}2m\dot{\mathbf{E}}^{(m)}\mathbf{U}^{-2m}).\end{aligned}$$

Further, it follows that

$$\left.\begin{aligned}\mathbf{R}\dot{\mathbf{E}}^{(m)}\mathbf{R}^T &= \overset{\bullet}{\vec{\mathbf{E}}}{}^{(m)^R} = \overset{\bullet}{\overleftarrow{\mathbf{e}}}{}^{(m)^R} = \dot{\mathbf{e}}^{(m)} + \mathbf{e}^{(m)}\boldsymbol{\Omega}^R - \boldsymbol{\Omega}^R\mathbf{e}^{(m)}\\ \overset{\bullet}{\overleftarrow{\mathbf{e}}}{}^{(m)^w} &= \dot{\mathbf{e}}^{(m)} + \mathbf{e}^{(m)}\mathbf{w} - \mathbf{w}\mathbf{e}^{(m)}\,(\boldsymbol{\Omega}^R = \mathbf{w}\ \text{for}\ \mathbf{R} = \mathbf{R}_t(t))\end{aligned}\right\}, \tag{4.124}$$

noting the time-derivative of equation $(4.56)_1$:

$$\begin{aligned}\overset{\bullet}{\overleftarrow{\mathbf{e}}}{}^{(m)^R} &= \mathbf{R}(\mathbf{R}^T\mathbf{e}^{(m)}\mathbf{R})^{\bullet}\mathbf{R}^T = \mathbf{R}(\dot{\mathbf{R}}^T\mathbf{e}^{(m)}\mathbf{R} + \mathbf{R}^T\dot{\mathbf{e}}^{(m)}\mathbf{R} + \mathbf{R}^T\mathbf{e}^{(m)}\dot{\mathbf{R}})\mathbf{R}^T\\ &= \dot{\mathbf{e}}^{(m)} + \mathbf{e}^{(m)}\dot{\mathbf{R}}\mathbf{R}^T + \mathbf{R}\dot{\mathbf{R}}^T\mathbf{e}^{(m)} = \dot{\mathbf{e}}^{(m)} + \mathbf{e}^{(m)}\dot{\mathbf{R}}\mathbf{R}^T - \dot{\mathbf{R}}\mathbf{R}^T\mathbf{e}^{(m)},\end{aligned} \tag{4.125}$$

while the symbols with arrows in equation (4.124) denote the corotational rate tensors described in Section 5.2, which are based on the convected derivative as will be explained in Chapter 5.

For example, the rates of the general strain for $m = \pm 1$ are given by

$$\left.\begin{aligned} \overset{\bullet}{\mathbf{E}}^{(1)} &= \overset{\bullet}{\mathbf{E}} = \frac{1}{2}\overset{\bullet}{\mathbf{C}} = \mathbf{F}^T\mathbf{dF} = \mathbf{D} = \overset{\leftarrow}{\mathbf{d}}_{GG} = \overset{\bullet}{\overset{\leftarrow}{\mathbf{g}}}_{GG}/2 \\ \overset{\bullet}{\mathbf{E}}^{(-1)} &= \mathbf{F}^{-1}\mathbf{dF}^{-T} = \overset{\leftarrow}{\mathbf{d}}^{GG} \end{aligned}\right\}, \tag{4.126}$$

$$\left.\begin{aligned} \overset{\bullet}{\mathbf{e}}^{(1)} &= \frac{1}{2}\overset{\bullet}{\mathbf{b}} = \mathrm{sym}[\boldsymbol{l}\mathbf{b}] = \frac{1}{2}(\boldsymbol{l}\mathbf{b} + \mathbf{b}\boldsymbol{l}^T) \\ \overset{\bullet}{\mathbf{e}}^{(-1)} &= \overset{\bullet}{\mathbf{e}} = \mathbf{d} - \mathbf{e}\boldsymbol{l} - \boldsymbol{l}^T\mathbf{e} = \overset{\bullet}{\overset{\rightarrow}{\mathbf{G}}}_{gg}/2 \end{aligned}\right\}, \tag{4.127}$$

noting that

$$\left.\begin{aligned} \overset{\bullet}{\mathbf{E}}^{(1)} &= (\mathbf{U}^2)^\bullet/2 = (\mathbf{F}^T\mathbf{F})^\bullet/2 = (\mathbf{F}^T\overset{\bullet}{\mathbf{F}} + \overset{\bullet}{\mathbf{F}}^T F)/2 = \mathbf{F}^T(\overset{\bullet}{\mathbf{F}}\mathbf{F}^{-1} + \mathbf{F}^{-T}\overset{\bullet}{\mathbf{F}}^T)\mathbf{F}/2 \\ &= \mathbf{F}^T(\boldsymbol{l} + \boldsymbol{l}^T)\mathbf{F}/2 \\ \overset{\bullet}{\mathbf{E}}^{(-1)} &= -(\mathbf{U}^{-2})^\bullet/2 = -(\mathbf{F}^{-1}\mathbf{F}^{-T})^\bullet/2 = -(\overset{\bullet}{\mathbf{F}}^{-1}\mathbf{F}^{-T} + \mathbf{F}^{-1}\overset{\bullet}{\mathbf{F}}^{-T})/2 \\ &= -\{(-\mathbf{F}^{-1}\boldsymbol{l})\mathbf{F}^{-T} + \mathbf{F}^{-1}(-\mathbf{F}^{-1}\boldsymbol{l})^T)/2 \\ &= \mathbf{F}^{-1}(\boldsymbol{l} + \boldsymbol{l}^T)\mathbf{F}^{-T}/2 \end{aligned}\right\},$$

$$\left.\begin{aligned} \overset{\bullet}{\mathbf{e}}^{(1)} &= (\mathbf{V}^2)^\bullet/2 = (\mathbf{F}\mathbf{F}^T)^\bullet/2 = \frac{1}{2}(\overset{\bullet}{\mathbf{F}}\mathbf{F}^T + \mathbf{F}\overset{\bullet}{\mathbf{F}}^T) = \frac{1}{2}(\overset{\bullet}{\mathbf{F}}\mathbf{F}^{-1}\mathbf{F}\mathbf{F}^T + \mathbf{F}\mathbf{F}^T\mathbf{F}^{-T}\overset{\bullet}{\mathbf{F}}^T) \\ \overset{\bullet}{\mathbf{e}}^{(-1)} &= -(\mathbf{V}^{-2})^\bullet/2 = -(\mathbf{F}^{-T}\mathbf{F}^{-1})^\bullet/2 = -(\mathbf{F}^{-T}\overset{\bullet}{\mathbf{F}}^{-1} + \overset{\bullet}{\mathbf{F}}^{-T}\mathbf{F}^{-1})/2 \\ &= \left\{\mathbf{F}^{-T}\mathbf{F}^{-1}\boldsymbol{l} + (\mathbf{F}^{-1}\boldsymbol{l})^T\mathbf{F}^{-1}\right\}/2 \\ &= (\mathbf{b}^{-1}\boldsymbol{l} + \boldsymbol{l}^T\mathbf{b}^{-1})/2 = \{(\mathbf{g} - 2\mathbf{e})\boldsymbol{l} + \boldsymbol{l}^T(\mathbf{g} - 2\mathbf{e})\}/2 \end{aligned}\right\}$$

with equations $(4.63)_4$ and (4.82).

Basic tensors describing deformation shown in the foregoing are listed collectively in Table 4.2, in which double arrows over a tensor denote the convected derivative described in Chapter 5.

4.5 Representations of Strain Rate and Spin Tensors in Lagrangian and Eulerian Triads

Deformation analysis can often readily be carried out in the principal direction of the right and left stretch tensors **U** and **V**. Expressions for various strain rates and spins are shown in this section, following the explanation by Lubarda (2002) with the reference to Hill (1970, 1978).

The time-differentiation of **U** in equation (4.15) leads to

$$\overset{\bullet}{\mathbf{U}} = \sum_{\alpha=1}^{3}\left\{\overset{\bullet}{\lambda}_\alpha \mathbf{N}^{(\alpha)} \otimes \mathbf{N}^{(\alpha)} + \lambda_\alpha(\overset{\bullet}{\mathbf{N}}^{(\alpha)} \otimes \mathbf{N}^{(\alpha)} + \mathbf{N}^{(\alpha)} \otimes \overset{\bullet}{\mathbf{N}}^{(\alpha)})\right\}. \tag{4.128}$$

Table 4.2 Basic tensors describing deformation

Base vectors	Reference	$\mathbf{G}_I = \mathbf{F}^{-1}\delta_I^i\mathbf{g}_i$
		$\mathbf{G}^I = \mathbf{F}^T\delta_i^I\mathbf{g}^i$
	Current	$\mathbf{g}_i = \mathbf{F}\delta_i^I\mathbf{G}_I$
		$\mathbf{g}^i = \mathbf{F}^{-T}\delta_I^i\mathbf{G}^I$
Metric tensors	Reference	$\mathbf{G} \equiv \mathbf{G}^I \otimes \mathbf{G}_I = G^{IJ}\mathbf{G}_I \otimes \mathbf{G}_J = G_{\bullet J}^I\mathbf{G}_I \otimes \mathbf{G}^J$ $= G_I^{\bullet J}\mathbf{G}^I \otimes \mathbf{G}_J = G_{IJ}\mathbf{G}^I \otimes \mathbf{G}^J$
	Current	$\mathbf{g} \equiv \mathbf{g}^i \otimes \mathbf{g}_i = g^{ij}\mathbf{g}_i \otimes \mathbf{g}_j = g_{\bullet j}^i\mathbf{g}_i \otimes \mathbf{g}^j$ $= g_i^{\bullet j}\mathbf{g}^i \otimes \mathbf{g}_j = g_{ij}\mathbf{g}^i \otimes \mathbf{g}^j$
Deformation gradient		$\mathbf{F} = \delta_{\bullet I}^i\mathbf{g}_i \otimes \mathbf{G}^I,\ \mathbf{F}^T = \delta_I^{\bullet i}\mathbf{G}^I \otimes \mathbf{g}_i$
		$\mathbf{F}^{-1} = \delta_{\bullet i}^I\mathbf{G}_I \otimes \mathbf{g}^i,\ \mathbf{F}^{-T} = \delta_i^{\bullet I}\mathbf{g}^i \otimes \mathbf{G}_I$
Cauchy–Green deformation tensors	Right	$\mathbf{C} \equiv \overleftarrow{\mathbf{g}}_{GG} = \mathbf{F}^T\mathbf{F}$
	Left	$\mathbf{b} \equiv \overrightarrow{\mathbf{G}}^{gg} = \mathbf{F}\mathbf{F}^T$
Strains	Green	$\mathbf{E} = \frac{1}{2}(\overleftarrow{\mathbf{g}}_{GG} - \mathbf{G}) = \frac{1}{2}(\mathbf{C} - \mathbf{G})$
	Almansi	$\mathbf{e} = \frac{1}{2}(\mathbf{g} - \overrightarrow{\mathbf{G}}_{gg}) = \frac{1}{2}(\mathbf{g} - \mathbf{b}^{-1})$
Vecocity gradient		$\boldsymbol{l} = \dot{\mathbf{g}}_i \otimes \mathbf{g}^i = \dot{\mathbf{F}}\mathbf{F}^{-1}$
Strain rate		$\mathbf{d} = \mathrm{sym}[\boldsymbol{l}] = \mathrm{sym}[\dot{\mathbf{g}}_i \otimes \mathbf{g}^i] = \overset{\rightleftarrows}{\dot{\mathbf{g}}}_{gg} = \overset{\rightleftarrows}{\dot{\mathbf{e}}}_{gg}$
Continuum spin		$\mathbf{w} = \mathrm{ant}[\boldsymbol{l}] = \mathrm{ant}[\dot{\mathbf{g}}_i \otimes \mathbf{g}^i]$
Rate of Green strain tensor		$\dot{\mathbf{E}} = \overleftarrow{\mathbf{d}}_{GG} = \mathbf{F}^T\mathbf{d}\mathbf{F} = \overleftarrow{\dot{\mathbf{g}}}_{GG}/2$
Rate of Almansi tensor		$\dot{\mathbf{e}} = \mathbf{d} - \mathbf{e}\boldsymbol{l} - \boldsymbol{l}^T\mathbf{e} = \overrightarrow{\dot{\mathbf{G}}}_{gg}/2$

Substituting equation (4.116) into equation (4.128), one has

$$\dot{\mathbf{U}} = \sum_{\alpha=1}^{3} \dot{\lambda}_\alpha \mathbf{N}^{(\alpha)} \otimes \mathbf{N}^{(\alpha)} + \boldsymbol{\Omega}^L\mathbf{U} - \mathbf{U}\boldsymbol{\Omega}^L. \tag{4.129}$$

Here, the Lagrangian spin tensor can be described on the Lagrangian triad as follows:

$$\boldsymbol{\Omega}^L = \sum_{\alpha,\beta=1}^{3} \Omega_{\alpha\beta}^L \mathbf{N}^{(\alpha)} \otimes \mathbf{N}^{(\beta)} \tag{4.130}$$

and thus

$$\left.\begin{aligned}\boldsymbol{\Omega}^L\mathbf{U} &= \sum_{\alpha,\beta=1}^{3}\Omega^L_{\alpha\beta}\mathbf{N}^{(\alpha)}\otimes\mathbf{N}^{(\beta)}\sum_{\gamma=1}^{3}\lambda_\gamma\mathbf{N}^{(\gamma)}\otimes\mathbf{N}^{(\gamma)} = \sum_{\alpha,\beta=1}^{3}\Omega^L_{\alpha\beta}\lambda_\beta\mathbf{N}^{(\alpha)}\otimes\mathbf{N}^{(\beta)}\\ \mathbf{U}\boldsymbol{\Omega}^L &= \sum_{\gamma=1}^{3}\lambda_\gamma\mathbf{N}^{(\gamma)}\otimes\mathbf{N}^{(\gamma)}\sum_{\alpha,\beta=1}^{3}\Omega^L_{\alpha\beta}\mathbf{N}^{(\alpha)}\otimes\mathbf{N}^{(\beta)} = \sum_{\alpha,\beta=1}^{3}\Omega^L_{\alpha\beta}\lambda_\alpha\mathbf{N}^{(\alpha)}\otimes\mathbf{N}^{(\beta)}\end{aligned}\right\}. \tag{4.131}$$

By subtraction,

$$\boldsymbol{\Omega}^L\mathbf{U}-\mathbf{U}\boldsymbol{\Omega}^L = \boldsymbol{\Omega}^L\mathbf{U}+(\mathbf{U}\boldsymbol{\Omega}^L)^T = \sum_{\alpha,\beta=1}^{3}\Omega^L_{\alpha\beta}(\lambda_\beta-\lambda_\alpha)\mathbf{N}^{(\alpha)}\otimes\mathbf{N}^{(\beta)}. \tag{4.132}$$

Substituting equation (4.132) into equation (4.129), it follows that

$$\dot{\mathbf{U}} = \sum_{\alpha=1}^{3}\dot{\lambda}_\alpha\mathbf{N}^{(\alpha)}\otimes\mathbf{N}^{(\alpha)} + \sum_{\alpha,\beta=1}^{3}\Omega^L_{\alpha\beta}(\lambda_\beta-\lambda_\alpha)\mathbf{N}^{(\alpha)}\otimes\mathbf{N}^{(\beta)}. \tag{4.133}$$

Similarly, one gets

$$\dot{\mathbf{E}}^{(m)} = \begin{cases}\displaystyle\sum_{\alpha=1}^{3}\lambda_\alpha^{2m-1}\dot{\lambda}_\alpha\mathbf{N}^{(\alpha)}\otimes\mathbf{N}^{(\alpha)} - \sum_{\alpha,\beta=1}^{3}\Omega^L_{\alpha\beta}\frac{\lambda_\alpha^{2m}-\lambda_\beta^{2m}}{2m}\mathbf{N}^{(\alpha)}\otimes\mathbf{N}^{(\beta)}\ (m\neq 0)\\ \displaystyle\sum_{\alpha=1}^{3}\frac{1}{\lambda_\alpha}\dot{\lambda}_\alpha\mathbf{N}^{(\alpha)}\otimes\mathbf{N}^{(\alpha)} - \sum_{\alpha,\beta=1}^{3}\Omega^L_{\alpha\beta}\ln\frac{\lambda_\alpha}{\lambda_\beta}\mathbf{N}^{(\alpha)}\otimes\mathbf{N}^{(\beta)}\ (m=0)\end{cases}, \tag{4.134}$$

noting equation (4.116) and

$$\left.\begin{aligned}&\left(\sum_{\alpha=1}^{3}\frac{1}{2m}\lambda_\alpha^{2m}\mathbf{N}^{(\alpha)}\otimes\mathbf{N}^{(\alpha)}\right)^{\bullet} = \sum_{\alpha=1}^{3}\dot{\lambda}_\alpha^{2m-1}\mathbf{N}^{(\alpha)}\otimes\mathbf{N}^{(\alpha)} + \sum_{\alpha=1}^{3}\frac{1}{2m}\lambda_\alpha^{2m}(\dot{\mathbf{N}}^{(\alpha)}\otimes\mathbf{N}^{(\alpha)}+\mathbf{N}^{(\alpha)}\otimes\dot{\mathbf{N}}^{(\alpha)})\\ &= \sum_{\alpha=1}^{3}\dot{\lambda}_\alpha^{2m-1}\mathbf{N}^{(\alpha)}\otimes\mathbf{N}^{(\alpha)} + \sum_{\alpha=1}^{3}\frac{1}{2m}\lambda_\alpha^{2m}(\Omega^L_{\beta\alpha}\mathbf{N}^{(\beta)}\otimes\mathbf{N}^{(\alpha)}+\mathbf{N}^{(\alpha)}\otimes\Omega^L_{\beta\alpha}\mathbf{N}^{(\beta)})\ (m\neq 0)\\ &\left(\sum_{\alpha=1}^{3}\ln\lambda_\alpha\mathbf{N}^{(\alpha)}\otimes\mathbf{N}^{(\alpha)}\right)^{\bullet} = \sum_{\alpha=1}^{3}\frac{\dot{\lambda}_\alpha}{\lambda_\alpha}\mathbf{N}^{(\alpha)}\otimes\mathbf{N}^{(\alpha)} + \sum_{\alpha=1}^{3}\ln\lambda_\alpha(\dot{\mathbf{N}}^{(\alpha)}\otimes\mathbf{N}^{(\alpha)}+\mathbf{N}^{(\alpha)}\otimes\dot{\mathbf{N}}^{(\alpha)})\\ &= \sum_{\alpha=1}^{3}\frac{\dot{\lambda}_\alpha}{\lambda_\alpha}\mathbf{N}^{(\alpha)}\otimes\mathbf{N}^{(\alpha)} + \sum_{\alpha=1}^{3}\ln\lambda_\alpha(\Omega^L_{\beta\alpha}\mathbf{N}^{(\beta)}\otimes\mathbf{N}^{(\alpha)}+\mathbf{N}^{(\alpha)}\otimes\Omega^L_{\beta\alpha}\mathbf{N}^{(\beta)})\ (m=0)\end{aligned}\right\}. \tag{4.135}$$

On the other hand, the time-differentiation of $\mathbf{V}$ in equation (4.15) leads to

$$\dot{\mathbf{V}} = \sum_{\alpha=1}^{3} \left\{ \dot{\lambda}_\alpha \mathbf{n}^{(\alpha)} \otimes \mathbf{n}^{(\alpha)} + \lambda_\alpha (\dot{\mathbf{n}}^{(\alpha)} \otimes \mathbf{n}^{(\alpha)} + \mathbf{n}^{(\alpha)} \otimes \dot{\mathbf{n}}^{(\alpha)}) \right\}. \tag{4.136}$$

Substituting equation (4.118) into equation (4.136), one has

$$\dot{\mathbf{V}} = \sum_{\alpha=1}^{3} \dot{\lambda}_\alpha \mathbf{n}^{(\alpha)} \otimes \mathbf{n}^{(\alpha)} + \mathbf{\Omega}^E \mathbf{V} - \mathbf{V}\mathbf{\Omega}^E. \tag{4.137}$$

Here, the Lagrangian spin tensor can be described on the axes of Eulerian triad as follows:

$$\mathbf{\Omega}^E = \sum_{\alpha,\beta=1}^{3} \Omega^E_{\alpha\beta} \mathbf{n}^{(\alpha)} \otimes \mathbf{n}^{(\beta)} \tag{4.138}$$

and thus

$$\left.\begin{aligned} \mathbf{\Omega}^E \mathbf{V} &= \sum_{\alpha,\beta=1}^{3} \Omega^E_{\alpha\beta} \mathbf{n}^{(\alpha)} \otimes \mathbf{n}^{(\beta)} \sum_{\gamma=1}^{3} \lambda_\gamma \mathbf{n}^{(\gamma)} \otimes \mathbf{n}^{(\gamma)} = \sum_{\alpha,\beta=1}^{3} \Omega^E_{\alpha\beta} \lambda_\beta \mathbf{n}^{(\alpha)} \otimes \mathbf{n}^{(\beta)} \\ \mathbf{V}\mathbf{\Omega}^E &= \sum_{\gamma=1}^{3} \lambda_\gamma \mathbf{n}^{(\gamma)} \otimes \mathbf{n}^{(\gamma)} \sum_{\alpha,\beta=1}^{3} \Omega^L_{\alpha\beta} \mathbf{n}^{(\alpha)} \otimes \mathbf{n}^{(\beta)} = \sum_{\alpha,\beta=1}^{3} \Omega^L_{\alpha\beta} \lambda_\alpha \mathbf{n}^{(\alpha)} \otimes \mathbf{n}^{(\beta)} \end{aligned}\right\}. \tag{4.139}$$

By subtraction,

$$\mathbf{\Omega}^E \mathbf{V} - \mathbf{V}\mathbf{\Omega}^E = \mathbf{\Omega}^E \mathbf{V} + (\mathbf{V}\mathbf{\Omega}^E)^T = \sum_{\alpha,\beta=1}^{3} \Omega^E_{\alpha\beta} (\lambda_\beta - \lambda_\alpha) \mathbf{n}^{(\alpha)} \otimes \mathbf{n}^{(\beta)}. \tag{4.140}$$

Substituting equation (4.140) into equation (4.137), it follows that

$$\dot{\mathbf{V}} = \sum_{\alpha=1}^{3} \dot{\lambda}_\alpha \mathbf{n}^{(\alpha)} \otimes \mathbf{n}^{(\alpha)} + \sum_{\alpha,\beta=1}^{3} \Omega^E_{\alpha\beta} (\lambda_\beta - \lambda_\alpha) \mathbf{n}^{(\alpha)} \otimes \mathbf{n}^{(\beta)}. \tag{4.141}$$

Analogously to equation (4.134), one obtains

$$\dot{\mathbf{e}}^{(m)} = \begin{cases} \displaystyle\sum_{\alpha=1}^{3} \lambda_\alpha^{2m-1} \dot{\lambda}_\alpha \mathbf{n}^{(\alpha)} \otimes \mathbf{n}^{(\alpha)} - \sum_{\alpha,\beta=1}^{3} \Omega^E_{\alpha\beta} \frac{\lambda_\alpha^{2m} - \lambda_\beta^{2m}}{2m} \mathbf{n}^{(\alpha)} \otimes \mathbf{n}^{(\beta)} & (m \neq 0) \\ \displaystyle\sum_{\alpha=1}^{3} \frac{1}{\lambda_\alpha} \dot{\lambda}_\alpha \mathbf{n}^{(\alpha)} \otimes \mathbf{n}^{(\alpha)} - \sum_{\alpha,\beta=1}^{3} \Omega^E_{\alpha\beta} \ln \frac{\lambda_\alpha}{\lambda_\beta} \mathbf{n}^{(\alpha)} \otimes \mathbf{n}^{(\beta)} & (m = 0) \end{cases}. \tag{4.142}$$

The time-differentiation of equation (4.17) gives

$$\dot{\mathbf{R}} = \sum_{\alpha=1}^{3} (\dot{\mathbf{n}}^{(\alpha)} \otimes \mathbf{N}^{(\alpha)} + \mathbf{n}^{(\alpha)} \otimes \dot{\mathbf{N}}^{(\alpha)})$$

$$= \sum_{\alpha,\beta=1}^{3} (\dot{\mathbf{n}}^{(\alpha)} \otimes \mathbf{n}^{(\alpha)}\mathbf{n}^{(\beta)} \otimes \mathbf{N}^{(\beta)} + \mathbf{n}^{(\alpha)} \otimes \mathbf{N}^{(\alpha)}\mathbf{N}^{(\beta)} \otimes \dot{\mathbf{N}}^{(\beta)}) = \boldsymbol{\Omega}^E\mathbf{R} - \mathbf{R}\boldsymbol{\Omega}^L, \quad (4.143)$$

noting equations (4.115) and (4.117), and thus one can write

$$\dot{\mathbf{R}} = \sum_{\alpha,\beta=1}^{3} (\Omega^E_{\alpha\beta} - \Omega^L_{\alpha\beta})\mathbf{n}^{(\alpha)} \otimes \mathbf{N}^{(\beta)}. \quad (4.144)$$

Further, the time-differentiation of equation (4.18) leads to

$$\dot{\mathbf{F}} = \sum_{\alpha=1}^{3} \left\{ \dot{\lambda}_\alpha \mathbf{n}^{(\alpha)} \otimes \mathbf{N}^{(\alpha)} + \lambda_\alpha (\dot{\mathbf{n}}^{(\alpha)} \otimes \mathbf{N}^{(\alpha)} + \mathbf{n}^{(\alpha)} \otimes \dot{\mathbf{N}}^{(\alpha)}) \right\} \quad (4.145)$$

which is rewritten as

$$\dot{\mathbf{F}} = \sum_{\alpha=1}^{3} \dot{\lambda}_\alpha \mathbf{n}^{(\alpha)} \otimes \mathbf{N}^{(\alpha)} + \boldsymbol{\Omega}^E\mathbf{F} - \mathbf{F}\boldsymbol{\Omega}^L \quad (4.146)$$

and

$$\dot{\mathbf{F}} = \sum_{\alpha=1}^{3} \dot{\lambda}_\alpha \mathbf{n}^{(\alpha)} \otimes \mathbf{N}^{(\alpha)} + \sum_{\alpha,\beta=1}^{3} (\lambda_\beta \Omega^E_{\alpha\beta} - \lambda_\alpha \Omega^L_{\alpha\beta})\mathbf{n}^{(\alpha)} \otimes \mathbf{N}^{(\beta)}, \quad (4.147)$$

noting equations (4.115) and (4.117).

The velocity gradient tensor is described in terms of the Eulerian triad by substituting equations $(4.18)_2$ and (4.147) into equation (4.81) as

$$\boldsymbol{l} = \sum_{\alpha=1}^{3} \frac{\dot{\lambda}_\alpha}{\lambda_\alpha} \mathbf{n}^{(\alpha)} \otimes \mathbf{n}^{(\alpha)} + \sum_{\alpha,\beta=1}^{3} \left(\Omega^E_{\alpha\beta} - \frac{\lambda_\alpha}{\lambda_\beta} \Omega^L_{\alpha\beta} \right) \mathbf{n}^{(\alpha)} \otimes \mathbf{n}^{(\beta)}, \quad (4.148)$$

from which it follows that

$$\mathbf{d} = \sum_{\alpha=1}^{3} \frac{\dot{\lambda}_\alpha}{\lambda_\alpha} \mathbf{n}^{(\alpha)} \otimes \mathbf{n}^{(\alpha)} + \sum_{\alpha,\beta=1}^{3} \frac{\lambda_\alpha^2 - \lambda_\beta^2}{2\lambda_\alpha\lambda_\beta} \Omega^L_{\alpha\beta} \mathbf{n}^{(\alpha)} \otimes \mathbf{n}^{(\beta)}, \quad (4.149)$$

$$\mathbf{w} = \sum_{\alpha,\beta=1}^{3} \left(\Omega^E_{\alpha\beta} - \frac{\lambda_\alpha^2 + \lambda_\beta^2}{2\lambda_\alpha\lambda_\beta} \Omega^L_{\alpha\beta} \right) \mathbf{n}^{(\alpha)} \otimes \mathbf{n}^{(\beta)}. \quad (4.150)$$

The spin for the Lagrangian triad is described in terms of principal stretches and strain rate from equation (4.149) as follows:

$$\Omega^L_{\alpha\beta} = \frac{2\lambda_\alpha\lambda_\beta}{\lambda_\beta^2 - \lambda_\alpha^2} d_{\alpha\beta} \ (\alpha \neq \beta), \quad (4.151)$$

noting that

$$d_{\alpha\beta} = \frac{\lambda_\alpha^2 - \lambda_\beta^2}{2\lambda_\alpha \lambda_\beta} \Omega^L_{\alpha\beta}. \tag{4.152}$$

Substituting equation (4.151) into equation (4.150), the continuum spin is given by

$$w_{\alpha\beta} = \Omega^E_{\alpha\beta} + \frac{\lambda_\alpha^2 + \lambda_\beta^2}{\lambda_\alpha^2 - \lambda_\beta^2} d_{\alpha\beta} \ \ (\alpha \neq \beta). \tag{4.153}$$

Further, substitution of equation (4.144) and

$$\mathbf{R}^T = \sum_{\alpha=1}^{3} \mathbf{N}^{(\alpha)} \otimes \mathbf{n}^{(\alpha)} \tag{4.154}$$

into equation (4.110) yields

$$\boldsymbol{\Omega}^R = \sum_{\alpha \neq \beta} (\Omega^E_{\alpha\beta} - \Omega^L_{\alpha\beta}) \mathbf{n}^{(\alpha)} \otimes \mathbf{n}^{(\beta)}. \tag{4.155}$$

Thus, it follows that

$$\Omega^R_{\alpha\beta} = \Omega^E_{\alpha\beta} - \Omega^L_{\alpha\beta}. \tag{4.156}$$

Further, substituting equations (4.151) and (4.156) into equation (4.153), one obtains

$$w_{\alpha\beta} = \Omega^R_{\alpha\beta} + \frac{\lambda_\alpha + \lambda_\beta}{\lambda_\alpha - \lambda_\beta} d_{\alpha\beta} \ \ (\alpha \neq \beta). \tag{4.157}$$

4.6 Decomposition of Deformation Gradient Tensor into Isochoric and Volumetric Parts

The deformation gradient $\mathbf{F}$ is multiplicatively decomposed into a volume preserving, that is, *isochoric* (distortional) part $\underline{\mathbf{F}}$ and a volumetric part $\mathbf{F}_{vol}$ as follows:

$$\boxed{\mathbf{F} = \mathbf{F}_{vol}\underline{\mathbf{F}};\ \mathbf{F}_{vol} \equiv (\det\mathbf{F})^{1/3}\mathbf{G},\ \ \underline{\mathbf{F}} \equiv (\det\mathbf{F})^{-1/3}\mathbf{F}} \tag{4.158}$$

where

$$\det\mathbf{F}_{vol} = \det\mathbf{F} = J,\ \ \det\underline{\mathbf{F}} = 1, \tag{4.159}$$

exploiting the formula $\det[(\det\mathbf{T})^n\mathbf{G}] = (\det\mathbf{T})^{3n}$ and noting

$$\det\underline{\mathbf{F}} = \det[(\det\mathbf{F})^{-1/3}\mathbf{G}\mathbf{F}] = \det[(\det\mathbf{F})^{-1/3}\mathbf{G}]\det\mathbf{F} = (\det\mathbf{F})^{-1}\det\mathbf{F}.$$

Then, the tensor $\underline{\mathbf{F}}$ is a *unimodular tensor*.

The following partial derivatives hold for the unimodular tensor:

$$\frac{\partial \underline{\mathbf{F}}}{\partial \mathbf{F}} = (\det\mathbf{F})^{-1/3}\left(\mathcal{I} - \frac{1}{3}\mathbf{F}\otimes\mathbf{F}^{-T}\right) \tag{4.160}$$

$$\left(\begin{array}{l} \dfrac{\partial \underline{\mathbf{F}}}{\partial \mathbf{F}} = \dfrac{\partial\left\{\mathbf{F}(\det\mathbf{F})^{-1/3}\right\}}{\partial \mathbf{F}} = \dfrac{\partial \mathbf{F}}{\partial \mathbf{F}}(\det\mathbf{F})^{-1/3} + \mathbf{F}\otimes\left(-\dfrac{1}{3}(\det\mathbf{F})^{-4/3}\dfrac{\partial \det\mathbf{F}}{\partial \mathbf{F}}\right) \\ = \mathcal{I}(\det\mathbf{F})^{-1/3} - \mathbf{F}\otimes\dfrac{1}{3}(\det\mathbf{F})^{-4/3}(\det\mathbf{F})\mathbf{F}^{-T} \end{array}\right),$$

$$\frac{\partial \mathrm{tr}\underline{\mathbf{F}}}{\partial \mathbf{F}} = (\det\mathbf{F})^{-1/3}\left\{\mathbf{G} - \frac{1}{3}(\mathrm{tr}\mathbf{F})\mathbf{F}^{-T}\right\},\ \frac{\partial \mathrm{tr}\underline{\mathbf{F}}}{\partial F_{ij}} = (\det\mathbf{F})^{-1/3}\left\{\delta_{ij} - \frac{1}{3}(\mathrm{tr}\mathbf{F})F_{ji}^{-1}\right\} \tag{4.161}$$

$$\left(\frac{\partial \mathrm{tr}\underline{\mathbf{F}}}{\partial \mathbf{F}} = \frac{\partial \mathrm{tr}\underline{\mathbf{F}}}{\partial \underline{\mathbf{F}}} : \frac{\partial \underline{\mathbf{F}}}{\partial \mathbf{F}} = \mathbf{G} : \left\{(\det\mathbf{F})^{-1/3}\left(\mathcal{I} - \frac{1}{3}\mathbf{F}\otimes\mathbf{F}^{-T}\right)\right\}\right),$$

noting equation (1.371).

Further, define the decomposition

$$\boxed{\mathbf{C}_{vol} \equiv \mathbf{F}_{vol}^{T}\mathbf{F}_{vol} = (\det\mathbf{C})^{1/3}\mathbf{G},\ \ \underline{\mathbf{C}} \equiv \underline{\mathbf{F}}^{T}\underline{\mathbf{F}} = (\det\mathbf{C})^{-1/3}\mathbf{C}} \tag{4.162}$$

satisfying

$$\det\mathbf{C}_{vol} = \det\mathbf{C} = J^2,\ \ \det\underline{\mathbf{C}} = 1 \tag{4.163}$$

and the multiplicative decomposition

$$\mathbf{C}_{vol} = \mathbf{C}_{vol}\underline{\mathbf{C}} = \underline{\mathbf{C}}\mathbf{C}_{vol}. \tag{4.164}$$

Then the following partial derivatives hold for the unimodular tensor:

$$\frac{\partial \mathrm{tr}\underline{\mathbf{C}}}{\partial \mathbf{C}} = (\det\mathbf{C})^{-1/3}\left\{\mathbf{G} - \frac{1}{3}(\mathrm{tr}\mathbf{C})\mathbf{C}^{-1}\right\}, \tag{4.165}$$

noting that

$$\frac{\partial \mathrm{tr}\underline{\mathbf{C}}}{\partial \mathbf{C}} = \frac{\partial\left\{(\det\mathbf{C})^{-1/3}\mathrm{tr}\mathbf{C}\right\}}{\partial \mathbf{C}} = -\frac{1}{3}(\det\mathbf{C})^{-4/3}(\det\mathbf{C})\mathbf{C}^{-1}\mathrm{tr}\mathbf{C} + (\det\mathbf{C})^{-1/3}\mathbf{G}$$

by virtue of equations (1.359) and (1.371).

The decomposition of the deformation gradient described in this section is often used in formulations of the hyperelastic constitutive equation as will be addressed in Chapter 7.

5

Convected Derivative

Rates of tensors used in constitutive equations have to be independent of the rigid-body rotation, possessing an objectivity which is referred to as material-frame indifference. The convected tensor derivative, also called the Lie derivative, which is explained in this chapter, is the rate of tensor observed from the convected coordinate system moving, deforming, and rotating with a material. It is verified that corotational rate of a tensor can be regarded as the particular convected derivative observed from the coordinate system rotating with a material. In addition, the objectivities of tensors are examined, and the objectivity of the convected derivative is proved. The convected derivatives of stress tensors provide a systematic way to derive various stress rate tensors appearing in continuum mechanics.

5.1 Convected Derivative

The time-derivative of the vector $\mathbf{v}$ in equation (2.56), regarding the base vectors $\mathbf{a}_i$ and $\mathbf{a}^i$ as $\mathbf{g}_i$ and $\mathbf{g}^i$, is described as

$$\overset{\bullet}{\mathbf{v}} = \begin{cases} (v^r\mathbf{g}_r)^{\bullet} = \overset{\bullet}{v}{}^r\mathbf{g}_r + v^r\overset{\bullet}{\mathbf{g}}_r = \overset{\bullet}{v}{}^r\mathbf{g}_r + v^r\boldsymbol{l}\mathbf{g}_r, \\ (v_r\mathbf{g}^r)^{\bullet} = \overset{\bullet}{v}_r\mathbf{g}^r + v_r\overset{\bullet}{\mathbf{g}}{}^r = \overset{\bullet}{v}_r\mathbf{g}^r - v_r\boldsymbol{l}^T\mathbf{g}^r, \end{cases} \tag{5.1}$$

noting equations (4.84) and (4.85). The first terms of these equations represent the rates of the vector $\mathbf{v}$ observed from the embedded coordinate system, which are thus called *convected derivatives*. In other words, they represent the rate of a physical quantity observed from the embedded coordinate system having base vectors composed of line elements etched in a material (not material fiber). Also, they are interpreted as the rates observed from the material itself and thus they are independent of rigid-body rotation, possessing *objectivity*. They are expressed as

$$\boxed{\begin{aligned} \overset{\rightarrow}{\overset{\bullet}{\overleftarrow{\mathbf{v}}}}{}^{g} &\equiv \overset{\bullet}{v}{}^r\mathbf{g}_r = \overset{\bullet}{\mathbf{v}} - v^r\overset{\bullet}{\mathbf{g}}_r = \overset{\bullet}{\mathbf{v}} - \boldsymbol{l}\mathbf{v} = \mathbf{F}(\mathbf{F}^{-1}\mathbf{v})^{\bullet} = \mathbf{F}\overset{\bullet}{\overleftarrow{\mathbf{v}}}{}^{G} \equiv \overset{\rightarrow}{\overset{\bullet}{\mathbf{V}}}{}^{g} \\ \overset{\rightarrow}{\overset{\bullet}{\overleftarrow{\mathbf{v}}}}{}_{g} &\equiv \overset{\bullet}{v}_r\mathbf{g}^r = \overset{\bullet}{\mathbf{v}} - v_r\overset{\bullet}{\mathbf{g}}{}^r = \overset{\bullet}{\mathbf{v}} + \boldsymbol{l}^T\mathbf{v} = \mathbf{F}^{-T}(\mathbf{F}^T\mathbf{v})^{\bullet} = \mathbf{F}^{-T}\overset{\bullet}{\overleftarrow{\mathbf{v}}}{}_{G} \equiv \overset{\rightarrow}{\overset{\bullet}{\mathbf{V}}}{}_{g} \end{aligned}} \tag{5.2}$$

Introduction to Finite Strain Theory for Continuum Elasto-Plasticity, First Edition. Koichi Hashiguchi and Yuki Yamakawa.

It is known from equation (5.2) that the convected derivatives are obtained by time-differentiating the pulled-back quantity to the reference configuration and then pushing it forward to the current configuration. The symbols with double arrows in the first terms in equation (5.2) are introduced in order to stipulate this fact.

Similarly to the vector in equation (5.1), the material-time derivative of the second-order tensor $\mathbf{t}$ (equation (2.77)) in the current configuration is described by

$$
\dot{\mathbf{t}} = \begin{cases}
(t^{ij}\mathbf{g}_i \otimes \mathbf{g}_j)^{\bullet} = \dot{t}^{ij}\mathbf{g}_i \otimes \mathbf{g}_j + t^{ij}\dot{\mathbf{g}}_i \otimes \mathbf{g}_j + t^{ij}\mathbf{g}_i \otimes \dot{\mathbf{g}}_j, \\
(t^{i}_{\bullet j}\mathbf{g}_i \otimes \mathbf{g}^j)^{\bullet} = \dot{t}^{i}_{\bullet j}\mathbf{g}_i \otimes \mathbf{g}^j + t^{i}_{\bullet j}\dot{\mathbf{g}}_i \otimes \mathbf{g}^j + t^{i}_{\bullet j}\mathbf{g}_i \otimes \dot{\mathbf{g}}^j, \\
(t_{i}^{\bullet j}\mathbf{g}^i \otimes \mathbf{g}_j)^{\bullet} = \dot{t}_{i}^{\bullet j}\mathbf{g}^i \otimes \mathbf{g}_j + t_{i}^{\bullet j}\dot{\mathbf{g}}^i \otimes \mathbf{g}_j + t_{i}^{\bullet j}\mathbf{g}^i \otimes \dot{\mathbf{g}}_j, \\
(t_{ij}\mathbf{g}^i \otimes \mathbf{g}^j)^{\bullet} = \dot{t}_{ij}\mathbf{g}^i \otimes \mathbf{g}^j + t_{ij}\dot{\mathbf{g}}^i \otimes \mathbf{g}^j + t_{ij}\mathbf{g}^i \otimes \dot{\mathbf{g}}^j,
\end{cases} \tag{5.3}
$$

from which the convected derivatives, that is, the rates of the tensor $\mathbf{t}$ observed by the observer moving with the embedded coordinate system, are given by

$$
\left.\begin{array}{l}
\dot{t}^{ij}\mathbf{g}_i \otimes \mathbf{g}_j = (t^{ij}\mathbf{g}_i \otimes \mathbf{g}_j)^{\bullet} - t^{ij}\dot{\mathbf{g}}_i \otimes \mathbf{g}_j - t^{ij}\mathbf{g}_i \otimes \dot{\mathbf{g}}_j = \dot{\mathbf{t}} - t^{ij}\boldsymbol{l}\mathbf{g}_i \otimes \mathbf{g}_j - t^{ij}\mathbf{g}_i \otimes \mathbf{g}_j\boldsymbol{l}^T \\
\dot{t}^{i}_{\bullet j}\mathbf{g}_i \otimes \mathbf{g}^j = (t^{i}_{\bullet j}\mathbf{g}_i \otimes \mathbf{g}^j)^{\bullet} - t^{i}_{\bullet j}\dot{\mathbf{g}}_i \otimes \mathbf{g}^j - t^{i}_{\bullet j}\mathbf{g}_i \otimes \dot{\mathbf{g}}^j = \dot{\mathbf{t}} - t^{i}_{\bullet j}\boldsymbol{l}\mathbf{g}_i \otimes \mathbf{g}^j + t^{i}_{\bullet j}\mathbf{g}_i \otimes \mathbf{g}^j\boldsymbol{l} \\
\dot{t}_{i}^{\bullet j}\mathbf{g}^i \otimes \mathbf{g}_j = (t_{i}^{\bullet j}\mathbf{g}^i \otimes \mathbf{g}_j)^{\bullet} - t_{i}^{\bullet j}\dot{\mathbf{g}}^i \otimes \mathbf{g}_j - t_{i}^{\bullet j}\mathbf{g}^i \otimes \dot{\mathbf{g}}_j = \dot{\mathbf{t}} + t_{i}^{\bullet j}\boldsymbol{l}^T\mathbf{g}^i \otimes \mathbf{g}_j - t^{i}_{\bullet j}\mathbf{g}_i \otimes \mathbf{g}_j\boldsymbol{l}^T \\
\dot{t}_{ij}\mathbf{g}^i \otimes \mathbf{g}^j = (t_{ij}\mathbf{g}^i \otimes \mathbf{g}^j)^{\bullet} - t_{ij}\dot{\mathbf{g}}^i \otimes \mathbf{g}^j - t_{ij}\mathbf{g}^i \otimes \dot{\mathbf{g}}^j = \dot{\mathbf{t}} + t_{ij}\boldsymbol{l}^T\mathbf{g}^i \otimes \mathbf{g}^j + t_{ij}\mathbf{g}^i \otimes \mathbf{g}^j\boldsymbol{l}
\end{array}\right\}, \tag{5.4}
$$

noting equations (1.94), (4.84) and (4.85). The following four types of convected derivatives can be defined from equation (5.4):

$$
\begin{array}{|l|}
\hline
\overset{\rightarrow}{\overset{\bullet}{\overleftarrow{\mathbf{t}}}}{}^{gg} \equiv \dot{t}^{ij}\mathbf{g}_i \otimes \mathbf{g}_j = \dot{\mathbf{t}} - \boldsymbol{l}\mathbf{t} - \mathbf{t}\boldsymbol{l}^T = \mathbf{F}(\mathbf{F}^{-1}\mathbf{t}\mathbf{F}^{-T})^{\bullet}\mathbf{F}^T = \mathbf{F}\overset{\bullet}{\overleftarrow{\mathbf{t}}}{}^{GG}\mathbf{F}^T \equiv \overset{\rightarrow}{\dot{\mathbf{T}}}{}^{gg} \\
\overset{\rightarrow}{\overset{\bullet}{\overleftarrow{\mathbf{t}}}}{}^{g}_{\bullet g} \equiv \dot{t}^{i}_{\bullet j}\mathbf{g}_i \otimes \mathbf{g}^j = \dot{\mathbf{t}} - \boldsymbol{l}\mathbf{t} + \mathbf{t}\boldsymbol{l} = \mathbf{F}(\mathbf{F}^{-1}\mathbf{t}\mathbf{F})^{\bullet}\mathbf{F}^{-1} = \mathbf{F}\overset{\bullet}{\overleftarrow{\mathbf{t}}}{}^{G}_{\bullet G}\mathbf{F}^{-1} \equiv \overset{\rightarrow}{\dot{\mathbf{T}}}{}^{g}_{\bullet g} \\
\overset{\rightarrow}{\overset{\bullet}{\overleftarrow{\mathbf{t}}}}{}_{g}^{\bullet g} \equiv \dot{t}_{i}^{\bullet j}\mathbf{g}^i \otimes \mathbf{g}_j = \dot{\mathbf{t}} + \boldsymbol{l}^T\mathbf{t} - \mathbf{t}\boldsymbol{l}^T = \mathbf{F}^{-T}(\mathbf{F}^T\mathbf{t}\mathbf{F}^{-T})^{\bullet}\mathbf{F}^T = \mathbf{F}^{-T}\overset{\bullet}{\overleftarrow{\mathbf{t}}}{}_{G}^{\bullet G}\mathbf{F}^T \equiv \overset{\rightarrow}{\dot{\mathbf{T}}}{}_{g}^{\bullet g} \\
\overset{\rightarrow}{\overset{\bullet}{\overleftarrow{\mathbf{t}}}}{}_{gg} \equiv \dot{t}_{ij}\mathbf{g}^i \otimes \mathbf{g}^j = \dot{\mathbf{t}} + \boldsymbol{l}^T\mathbf{t} + \mathbf{t}\boldsymbol{l} = \mathbf{F}^{-T}(\mathbf{F}^T\mathbf{t}\mathbf{F})^{\bullet}\mathbf{F}^{-1} = \mathbf{F}^{-T}\overset{\bullet}{\overleftarrow{\mathbf{t}}}{}_{GG}\mathbf{F}^{-1} \equiv \overset{\rightarrow}{\dot{\mathbf{T}}}{}_{gg} \\
\hline
\end{array} \tag{5.5}
$$

$\overset{\rightarrow}{\overset{\bullet}{\overleftarrow{\mathbf{t}}}}{}^{gg}$ and $\overset{\rightarrow}{\overset{\bullet}{\overleftarrow{\mathbf{t}}}}{}_{gg}$ are the general forms of the *Oldroyd rate* and the *Cotter–Rivlin rate*, respectively.

For the case where one of the current base vectors is fixed so that the material-time derivation is performed through the two-point tensor, equation (5.5) reduces to

$$\boxed{\begin{array}{l} \left.\begin{array}{l} \overset{\rightleftarrows}{\dot{\mathbf{t}}}{}^{g\hat{g}} \equiv \dot{t}^{ij}\mathbf{g}_i \otimes \hat{\mathbf{g}}_j = \mathbf{F}\dot{\overleftarrow{\mathbf{t}}}{}^{G\hat{g}} \\ \overset{\rightleftarrows}{\dot{\mathbf{t}}}{}^{g}{}_{\bullet\hat{g}} \equiv \dot{t}^{i}{}_{\bullet j}\mathbf{g}_i \otimes \hat{\mathbf{g}}^j = \mathbf{F}\dot{\overleftarrow{\mathbf{t}}}{}^{G}{}_{\bullet\hat{g}} \end{array}\right\} = \mathbf{F}(\mathbf{F}^{-1}\mathbf{t})^{\bullet} = \dot{\mathbf{t}} - \mathbf{l}\mathbf{t} \equiv \left\{\begin{array}{l} \overset{\rightarrow}{\dot{\mathbb{T}}}{}^{g\hat{g}} \\ \overset{\rightarrow}{\dot{\mathbb{T}}}{}^{g}{}_{\bullet\hat{g}} \end{array}\right. \\ \left.\begin{array}{l} \overset{\rightleftarrows}{\dot{\mathbf{t}}}{}_{g}{}^{\bullet\hat{g}} \equiv \dot{t}_{i}{}^{\bullet j}\mathbf{g}^i \otimes \hat{\mathbf{g}}_j = \mathbf{F}^{-T}\dot{\overleftarrow{\mathbf{t}}}{}_{G}{}^{\bullet\hat{g}} \\ \overset{\rightleftarrows}{\dot{\mathbf{t}}}{}_{g\hat{g}} \equiv \dot{t}_{ij}\mathbf{g}^i \otimes \hat{\mathbf{g}}^j = \mathbf{F}^{-T}\dot{\overleftarrow{\mathbf{t}}}{}_{G\hat{g}} \end{array}\right\} = \mathbf{F}^{-T}(\mathbf{F}^{T}\mathbf{t})^{\bullet} = \dot{\mathbf{t}} + \mathbf{l}^{T}\mathbf{t} \equiv \left\{\begin{array}{l} \overset{\rightarrow}{\dot{\mathbb{T}}}{}_{g}{}^{\bullet\hat{g}} \\ \overset{\rightarrow}{\dot{\mathbb{T}}}{}_{g\hat{g}} \end{array}\right. \end{array}} \tag{5.6}$$

and

$$\boxed{\begin{array}{l} \left.\begin{array}{l} \overset{\rightleftarrows}{\dot{\mathbf{t}}}{}^{\hat{g}g} \equiv \dot{t}^{ij}\hat{\mathbf{g}}_i \otimes \mathbf{g}_j = \dot{\overleftarrow{\mathbf{t}}}{}^{\hat{g}G}\mathbf{F}^{T} \\ \overset{\rightleftarrows}{\dot{\mathbf{t}}}{}_{\hat{g}}{}^{\bullet g} \equiv \dot{t}_{i}{}^{\bullet j}\hat{\mathbf{g}}^i \otimes \mathbf{g}_j = \dot{\overleftarrow{\mathbf{t}}}{}_{\hat{g}}{}^{\bullet G}\mathbf{F}^{T} \end{array}\right\} = (\mathbf{t}\mathbf{F}^{-T})^{\bullet}\mathbf{F}^{T} = \dot{\mathbf{t}} - \mathbf{t}\mathbf{l}^{T} \equiv \left\{\begin{array}{l} \overset{\rightarrow}{\dot{\mathbb{T}}}{}^{\hat{g}g} \\ \overset{\rightarrow}{\dot{\mathbb{T}}}{}_{\hat{g}}{}^{\bullet g} \end{array}\right. \\ \left.\begin{array}{l} \overset{\rightleftarrows}{\dot{\mathbf{t}}}{}^{\hat{g}}{}_{\bullet g} \equiv \dot{t}^{i}{}_{\bullet j}\hat{\mathbf{g}}_i \otimes \mathbf{g}^j = \dot{\overleftarrow{\mathbf{t}}}{}^{\hat{g}}{}_{\bullet G}\mathbf{F}^{-1} \\ \overset{\rightleftarrows}{\dot{\mathbf{t}}}{}_{\hat{g}g} \equiv \dot{t}_{ij}\hat{\mathbf{g}}^i \otimes \mathbf{g}^j = \dot{\overleftarrow{\mathbf{t}}}{}_{\hat{g}G}\mathbf{F}^{-1} \end{array}\right\} = (\mathbf{t}\mathbf{F})^{\bullet}\mathbf{F}^{-1} = \dot{\mathbf{t}} + \mathbf{t}\mathbf{l} \equiv \left\{\begin{array}{l} \overset{\rightarrow}{\dot{\mathbb{T}}}{}^{\hat{g}}{}_{\bullet g} \\ \overset{\rightarrow}{\dot{\mathbb{T}}}{}_{\hat{g}g} \end{array}\right. \end{array}} \tag{5.7}$$

where a hat is added to specify the fixed base vector so that its material-time derivative is zero.

The convected derivatives of the two-point tensors are given by the equations similar to (5.6) and (5.7). The convected derivatives of the Eulerian–Lagrangian two-point tensor are given from equation (5.6) as follows:

$$\boxed{\begin{array}{l} \left.\begin{array}{l} \overset{\rightleftarrows}{\dot{\mathbb{T}}}{}^{g\hat{G}} \equiv \dot{\mathbb{T}}^{iJ}\mathbf{g}_i \otimes \mathbf{G}_J = \mathbf{F}\dot{\overleftarrow{\mathbb{T}}}{}^{G\hat{G}} \\ \overset{\rightleftarrows}{\dot{\mathbb{T}}}{}^{g}{}_{\bullet\hat{G}} \equiv \dot{\mathbb{T}}^{i}{}_{\bullet J}\mathbf{g}_i \otimes \mathbf{G}^J = \mathbf{F}\dot{\overleftarrow{\mathbb{T}}}{}^{G}{}_{\bullet\hat{G}} \end{array}\right\} = \mathbf{F}(\mathbf{F}^{-1}\mathbb{T})^{\bullet} = \dot{\mathbb{T}} - \mathbf{l}\mathbb{T} \equiv \left\{\begin{array}{l} \overset{\rightarrow}{\dot{\mathbf{T}}}{}^{g\hat{G}} \\ \overset{\rightarrow}{\dot{\mathbf{T}}}{}^{g}{}_{\bullet\hat{G}} \end{array}\right. \\ \left.\begin{array}{l} \overset{\rightleftarrows}{\dot{\mathbb{T}}}{}_{g}{}^{\bullet\hat{G}} \equiv \dot{\mathbb{T}}_{i}{}^{\bullet J}\mathbf{g}^i \otimes \mathbf{G}_J = \mathbf{F}^{-T}\dot{\overleftarrow{\mathbb{T}}}{}_{G}{}^{\bullet\hat{G}} \\ \overset{\rightleftarrows}{\dot{\mathbb{T}}}{}_{g\hat{G}} \equiv \dot{\mathbb{T}}_{iJ}\mathbf{g}^i \otimes \mathbf{G}^J = \mathbf{F}^{-T}\dot{\overleftarrow{\mathbb{T}}}{}_{G\hat{G}} \end{array}\right\} = \mathbf{F}^{-T}(\mathbf{F}^{T}\mathbb{T})^{\bullet} = \dot{\mathbb{T}} + \mathbf{l}^{T}\mathbb{T} \equiv \left\{\begin{array}{l} \overset{\rightarrow}{\dot{\mathbf{T}}}{}_{g}{}^{\bullet\hat{G}} \\ \overset{\rightarrow}{\dot{\mathbf{T}}}{}_{g\hat{G}} \end{array}\right. \end{array}} \tag{5.8}$$

and the convected derivatives of the Lagrangian–Eulerian two-point tensor are given from equation (5.7) as follows:

$$
\left.\begin{array}{l}
\overset{\rightarrow}{\overset{\bullet}{\overset{\leftarrow}{\mathbb{T}}}}{}^{\hat{G}g} \equiv \dot{\mathbb{T}}^{Ij}\mathbf{G}_I \otimes \mathbf{g}_j = \overset{\bullet}{\overset{\leftarrow}{\mathbb{T}}}{}^{\hat{G}G}\mathbf{F}^T \\
\overset{\rightarrow}{\overset{\bullet}{\overset{\leftarrow}{\mathbb{T}}}}{}_{\hat{G}}^{\;\bullet g} \equiv \dot{\mathbb{T}}_I^{\;\bullet j}\mathbf{G}^I \otimes \mathbf{g}_j = \overset{\bullet}{\overset{\leftarrow}{\mathbb{T}}}{}_{\hat{G}}^{\;\bullet G}\mathbf{F}^T
\end{array}\right\} = (\mathbb{T}\mathbf{F}^{-T})^{\bullet}\mathbf{F}^T = \dot{\mathbb{T}} - \mathbb{T}\boldsymbol{l}^T \equiv \left\{\begin{array}{l}
\overset{\rightarrow}{\dot{\mathbf{T}}}{}^{\hat{G}g} \\
\overset{\rightarrow}{\dot{\mathbf{T}}}{}_{\hat{G}}^{\;\bullet g}
\end{array}\right.
$$

$$
\left.\begin{array}{l}
\overset{\rightarrow}{\overset{\bullet}{\overset{\leftarrow}{\mathbb{T}}}}{}^{\hat{G}}_{\;\bullet g} \equiv \dot{\mathbb{T}}^I_{\;\bullet j}\mathbf{G}_I \otimes \mathbf{g}^j = \overset{\bullet}{\overset{\leftarrow}{\mathbb{T}}}{}^{\hat{G}}_{\;\bullet G}\mathbf{F}^{-1} \\
\overset{\rightarrow}{\overset{\bullet}{\overset{\leftarrow}{\mathbb{T}}}}{}_{\hat{G}g} \equiv \dot{\mathbb{T}}_{Ij}\mathbf{G}^I \otimes \mathbf{g}^j = \overset{\bullet}{\overset{\leftarrow}{\mathbb{T}}}{}_{\hat{G}G}\mathbf{F}^{-1}
\end{array}\right\} = (\mathbb{T}\mathbf{F})^{\bullet}\mathbf{F}^{-1} = \dot{\mathbb{T}} + \mathbb{T}\boldsymbol{l} \equiv \left\{\begin{array}{l}
\overset{\rightarrow}{\dot{\mathbf{T}}}{}^{\hat{G}}_{\;\bullet g} \\
\overset{\rightarrow}{\dot{\mathbf{T}}}{}_{\hat{G}g}
\end{array}\right.
\tag{5.9}
$$

As described above, the convected derivatives are the vector or tensor rates observed from the embedded coordinate system moving, deforming and rotating with material, which possess the objectivity as will be proved in Section 5.3. They are attained by first pulling back the tensor in the current configuration to the reference configuration, then taking its time-derivative, and finally returning it (pushing forward) to the current configuration as shown in Figure 5.1. The modes of the pull-back and push-forward operations depend on the variant of tensors. The convected derivative is often called the *Lie derivative* and denoted by the symbol $\mathcal{L}_v(\cdot)$ (for example, Truesdell and Toupin, 1960; Marsden and Hughes, 1983; Bonet and Wood, 2008; Simo, 1998; Belytschko et al., 2000; de Souza Neto et al., 2008). However, the symbol with the double arrows and the indices specifying the variants shown in this section express the operation and modes of the convected derivatives more clearly and intuitively. It was introduced by Hashiguchi (2011).

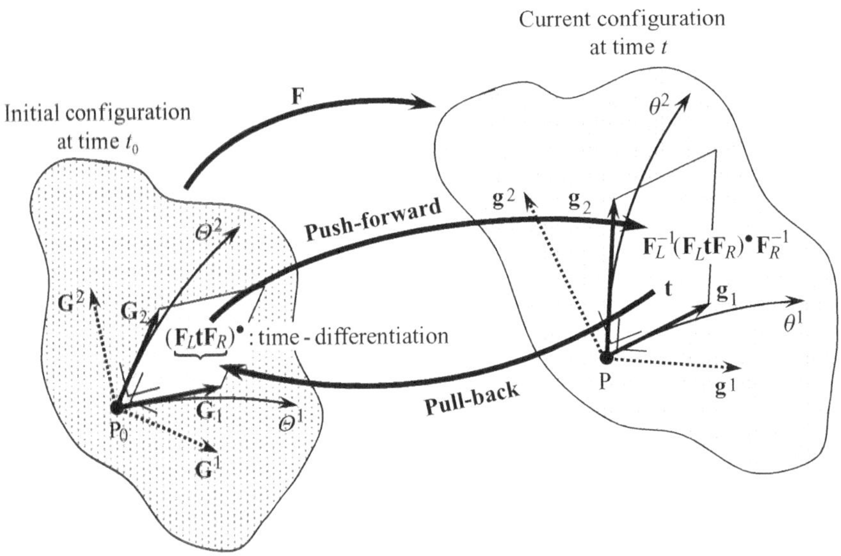

Figure 5.1 Convected derivative illustrated in the two-dimensional space

5.2 Corotational Rate

For the case of pure rotation $\mathbf{F} = \mathbf{R}$, all four kinds of derivatives in equation (5.5) reduce to the unique equation

$$\begin{aligned}\overset{\circ}{\mathbf{t}}^{R} &\equiv \overset{\rightarrow}{\overset{\bullet}{\overleftarrow{\mathbf{t}}}}{}^{gg} = \overset{\rightarrow}{\overset{\bullet}{\overleftarrow{\mathbf{t}}}}{}^{g}{}_{\bullet g} = \overset{\rightarrow}{\overset{\bullet}{\overleftarrow{\mathbf{t}}}}{}_{g}{}^{\bullet g} = \overset{\rightarrow}{\overset{\bullet}{\overleftarrow{\mathbf{t}}}}{}_{gg} \equiv \overset{\rightarrow}{\overset{\bullet}{\overleftarrow{\mathbf{t}}}}{}^{R} = \mathbf{R}(\mathbf{R}^{T}\mathbf{t}\mathbf{R})^{\bullet}\mathbf{R}^{T} = \mathbf{R}\overset{\bullet}{\hat{\mathbf{T}}}\mathbf{R}^{T} \\ &\equiv \overset{\rightarrow}{\overset{\bullet}{\hat{\mathbf{T}}}}{}^{R} = \dot{\mathbf{t}} - \boldsymbol{\Omega}^{R}\mathbf{t} + \mathbf{t}\boldsymbol{\Omega}^{R}\end{aligned} \tag{5.10}$$

The rate in equation (5.10) is called the *Green–Naghdi rate* (Green and Naghdi, 1965). It is zero when the quantity $(\mathbf{R}^{T}\mathbf{t}\mathbf{R})^{\bullet}$, which is observed by the coordinate system moving and rotating with a material line element, is kept constant.

Choosing the current state as the reference state leading to $\boldsymbol{\Omega}^{R} = \mathbf{w}$ as in equation (4.121), equation (5.10) results in

$$\overset{\circ}{\mathbf{t}}^{w} \equiv \overset{\rightarrow}{\overset{\bullet}{\overleftarrow{\mathbf{t}}}}{}^{w} \equiv \dot{\mathbf{t}} - \mathbf{w}\mathbf{t} + \mathbf{t}\mathbf{w} \tag{5.11}$$

which is called the *Zaremba–Jaumann rate* (Zaremba, 1903; Jaumann, 1911) and is related to the convected derivative in equation (5.5) as follows:

$$\overset{\rightarrow}{\overset{\bullet}{\overleftarrow{\mathbf{t}}}}{}^{w} = \overset{\rightarrow}{\overset{\bullet}{\overleftarrow{\mathbf{t}}}}{}^{gg} + \mathbf{t}\mathbf{d} + \mathbf{d}\mathbf{t} = \overset{\rightarrow}{\overset{\bullet}{\overleftarrow{\mathbf{t}}}}{}^{g}{}_{\bullet g} - \mathbf{t}\mathbf{d} + \mathbf{d}\mathbf{t} = \overset{\rightarrow}{\overset{\bullet}{\overleftarrow{\mathbf{t}}}}{}_{g}{}^{\bullet g} + \mathbf{t}\mathbf{d} - \mathbf{d}\mathbf{t} = \overset{\rightarrow}{\overset{\bullet}{\overleftarrow{\mathbf{t}}}}{}_{gg} - \mathbf{t}\mathbf{d} - \mathbf{d}\mathbf{t} \tag{5.12}$$

resulting in

$$\overset{\circ}{\mathbf{t}}^{w} = \overset{\rightarrow}{\overset{\bullet}{\overleftarrow{\mathbf{t}}}}{}^{w} = \frac{1}{2}(\overset{\rightarrow}{\overset{\bullet}{\overleftarrow{\mathbf{t}}}}{}^{gg} + \overset{\rightarrow}{\overset{\bullet}{\overleftarrow{\mathbf{t}}}}{}_{gg}) = \frac{1}{2}(\overset{\rightarrow}{\overset{\bullet}{\overleftarrow{\mathbf{t}}}}{}^{g}{}_{\bullet g} + \overset{\rightarrow}{\overset{\bullet}{\overleftarrow{\mathbf{t}}}}{}_{g}{}^{\bullet g}) = \frac{1}{2}\{\overset{\rightarrow}{\overset{\bullet}{\overleftarrow{\mathbf{t}}}}{}^{g}{}_{\bullet g} + (\overset{\rightarrow}{\overset{\bullet}{\overleftarrow{\mathbf{t}}}}{}^{g}{}_{\bullet g})^{T}\} = \frac{1}{2}\{\overset{\rightarrow}{\overset{\bullet}{\overleftarrow{\mathbf{t}}}}{}_{g}{}^{\bullet g} + (\overset{\rightarrow}{\overset{\bullet}{\overleftarrow{\mathbf{t}}}}{}_{g}{}^{\bullet g})^{T}\}. \tag{5.13}$$

The corotational rate has to be adopted for stress and stress-like internal variables in the framework of stress space formulation and for strain and strain-like internal variables in the framework of strain space formulation for the current configuration by the requirement of objectivity which will be explained in the next section. A brief comment on convected strain rates is given at the end of this section, whilst convected stress rates will be described in detail in the next chapter since the stress space formulation is much more popular than the strain space formulation.

The Green–Naghdi and Zaremba–Jaumann rates of the general Eulerian strain are given from equations (5.10) and (5.11) by

$$\left.\begin{aligned}\overset{\circ}{\mathbf{e}}^{(m)^{R}} &\equiv \mathbf{R}\dot{\mathbf{E}}^{(m)}\mathbf{R}^{T} = \mathbf{R}(\mathbf{R}^{T}\mathbf{e}^{(m)}\mathbf{R})^{\bullet}\mathbf{R}^{T} = \overset{\rightarrow}{\overset{\bullet}{\overleftarrow{\mathbf{e}}}}{}^{(m)R}_{gg} = \dot{\mathbf{e}}^{(m)} + \mathbf{e}^{(m)}\boldsymbol{\Omega}^{R} - \boldsymbol{\Omega}^{R}\mathbf{e}^{(m)} \\ \overset{\circ}{\mathbf{e}}^{(m)^{w}} &= \overset{\rightarrow}{\overset{\bullet}{\overleftarrow{\mathbf{e}}}}{}^{(m)^{w}}_{gg} = \dot{\mathbf{e}}^{(m)} + \mathbf{e}^{(m)}\mathbf{w} - \mathbf{w}\mathbf{e}^{(m)}\end{aligned}\right\} \tag{5.14}$$

which was already shown in equation (4.124). It follows for $m = -1$, that is, the Almansi strain $\mathbf{e}^{(-1)} = \mathbf{e}$, that

$$\left.\begin{aligned} \overset{\circ}{\mathbf{e}}{}^{R} &\equiv \mathbf{R}\dot{\mathbf{E}}^{(-1)}\mathbf{R}^{T} = \mathbf{R}(\mathbf{R}^{T}\mathbf{e}\mathbf{R})^{\bullet}\mathbf{R}^{T} = \overset{\leftrightarrows}{\dot{\mathbf{e}}}{}^{R} = \dot{\mathbf{e}} + \mathbf{e}\boldsymbol{\Omega}^{R} - \boldsymbol{\Omega}^{R}\mathbf{e} \\ \overset{\circ}{\mathbf{e}}{}^{W} &\equiv \overset{\leftrightarrows}{\dot{\mathbf{e}}}{}^{W} = \dot{\mathbf{e}} + \mathbf{e}\mathbf{w} - \mathbf{w}\mathbf{e} \end{aligned}\right\} \tag{5.15}$$

The covariant convected derivative of the current metric tensor $\mathbf{g}$ or the Almansi strain tensor $\mathbf{e}$ is identical to the strain rate as follows:

$$\left.\begin{aligned} \overset{\leftrightarrows}{\dot{\mathbf{g}}}{}_{gg} &= \mathbf{F}^{-T}\{\mathbf{F}^{T}(\mathbf{g}/2)\mathbf{F}\}^{\bullet}\mathbf{F}^{-1} = \frac{1}{2}(\dot{\mathbf{g}} + \boldsymbol{l}^{T}\mathbf{e} + \mathbf{e}\boldsymbol{l}) = \frac{1}{2}(\boldsymbol{l}^{T} + \boldsymbol{l}) \\ \overset{\leftrightarrows}{\dot{\mathbf{e}}}{}_{gg} &= \mathbf{F}^{-T}(\mathbf{F}^{T}\mathbf{e}\mathbf{F})^{\bullet}\mathbf{F}^{-1}(= \mathbf{F}^{-T}\dot{\mathbf{E}}\mathbf{F}^{-1}) = \dot{\mathbf{e}} + \boldsymbol{l}^{T}\mathbf{e} + \mathbf{e}\boldsymbol{l} \end{aligned}\right\} = \mathbf{d} \tag{5.16}$$

Equation $(5.16)_2$ was shown in equation $(4.127)_2$. Therefore, the strain rate $\mathbf{d}$ is half the covariant convected derivative, that is, the Cotter–Rivlin rate of the current metric tensor $\mathbf{g}$ itself. That is to say, $\mathbf{d}$ can be interpreted as half the rate of the current metric tensor $\mathbf{g}$ itself observed from the coordinate system with the current contravariant base $\mathbf{g}^{i} \otimes \mathbf{g}^{j}$. Needless to say, the strain rate $\mathbf{d}$ is a zero tensor in the case of pure rotation $\mathbf{F} = \mathbf{R}$ as can be confirmed by

$$\mathbf{d} = \overset{\leftrightarrows}{\dot{\mathbf{g}}}{}^{R} = \mathbf{R}\dot{\hat{\mathbf{G}}}\mathbf{R}^{T} = \mathbf{R}(\mathbf{R}^{T}\mathbf{g}\mathbf{R})^{\bullet}\mathbf{R}^{T}(= (\mathbf{R}\dot{\mathbf{R}}^{T} + \dot{\mathbf{R}}\mathbf{R}^{T})) = \mathbf{0}, \tag{5.17}$$

while $\dot{\mathbf{E}} = (\mathbf{R}^{T}\mathbf{R})^{\bullet}/2 = \mathbf{0}$ but $\dot{\mathbf{e}} = \boldsymbol{\Omega}^{R}\mathbf{e} - \mathbf{e}\boldsymbol{\Omega}^{R} \neq \mathbf{0}$. Note that $\dot{\mathbf{e}}$ is the material-time derivative of the Eulerian tensor based in the current configuration and thus is influenced by the rigid-body rotation. The general interpretation of this fact will be given in the next section.

5.3 Objectivity

Objectivity is the most basic concept in the description of deformation behavior of materials, and was the starting point of modern continuum mechanics. However, it is not a notion whose physical meaning can be easily captured, although it has been described in almost all books on continuum mechanics. An explanation is given in this section.

The constitutive properties of materials including deformation are independent of the observer. Therefore, any constitutive equation has to be formulated in a form which is not influenced by the superposition of rigid-body rotation. In other words, it has to be formulated in an identical form independent of coordinate systems such that included tensors obey a common coordinate transformation rule. This basic requirement for constitutive equations was advocated by Oldroyd (1950) as the *principle of material-frame indifference* and is usually referred to as the *objectivity principle*. A tensor that obeys the common transformation rule in equation (1.10) even if a mutual spin exists between the coordinate systems, that is, even if a rigid-body rotation is superposed on material, is defined as an *objective tensor*, and the common transformation rule in equation (1.10) is called the *objective transformation rule*. Here, an objective tensor has only to obey the objective transformation rule irrespective of the order of the tensor, that is, the number of the base vectors to which the tensor refers

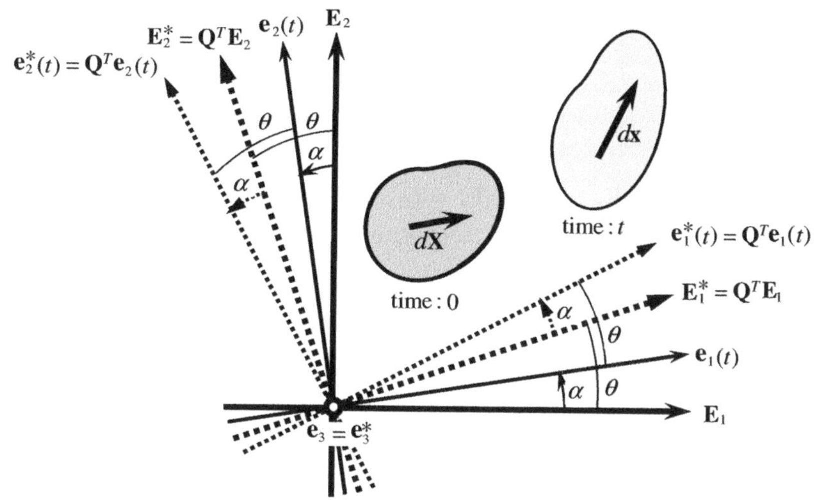

Figure 5.2 Coordinate systems possessing different directions without mutual rotation ($\dot{\mathbf{Q}} = \mathbf{O}$) (illustrated in two-dimensional state $\mathbf{e}_3 = \mathbf{e}_3^*$)

(is based on). A constitutive equation has to be formulated by objective tensors which obey an objective transformation rule in identical order (number of orthogonal tensors). In what follows, we examine the objectivity of various tensors.

Objectivity is concerned with how a tensor describing a physical quantity included in a material is observed from the outside of the material and thus we must use a coordinate system other than the convected coordinate system for the examination of objectivity. Therefore, we may adopt orthogonal coordinate systems as observers for examination of the objectivity of tensors.

First, consider the coordinate systems with the bases $\{\mathbf{e}_i(t)\}$ and $\{\mathbf{e}_i^*(t)\}$ rotating concurrently without a rate of mutual rotation ($\dot{\mathbf{Q}} = \mathbf{O}$) as shown in Figure 5.2, where α is the rotating angle of the coordinate systems from the reference coordinate systems with the bases $\{\mathbf{E}_I\}$ and $\{\mathbf{E}_I^*\}$. The following relations between these base vectors hold noting equations (1.17) and (1.109).

$$\left.\begin{array}{ll}\mathbf{E}_I^* = Q_{IR}\mathbf{E}_R, \quad \mathbf{E}_I = Q_{RI}\mathbf{E}_R^*; & \mathbf{E}_I^* = \mathbf{Q}^T\mathbf{E}_I, \quad \mathbf{E}_I = \mathbf{Q}\mathbf{E}_I^* \\ \mathbf{e}_i^* = Q_{ir}\mathbf{e}_r, \quad \mathbf{e}_i = Q_{ri}\mathbf{e}_r^*; & \mathbf{e}_i^* = \mathbf{Q}^T\mathbf{e}_i, \quad \mathbf{e}_i = \mathbf{Q}\mathbf{e}_i^*\end{array}\right\}, \tag{5.18}$$

where the base vectors in the reference states are specified by the uppercase letter **E**, while those in the current states are denoted by the lowercase letter **e** as usual.

Noting equation $(1.199)_1$, the following transformation rules between the components of the infinitesimal line-element vectors observed by these coordinate systems hold:.

$$\left.\begin{array}{l}d\mathbf{X}^* = \mathbf{Q}d\mathbf{X} \ \ (dX_I^* = Q_{IS}dX_S) \\ d\mathbf{x}^* = \mathbf{Q}d\mathbf{x} \ \ (dx_i^* = Q_{ir}dx_r)\end{array}\right\}, \tag{5.19}$$

and thus one has

$$\frac{\partial \mathbf{x}^*}{\partial \mathbf{X}^*} = \mathbf{Q}\frac{\partial \mathbf{x}}{\partial \mathbf{X}}\mathbf{Q}^T \left(\frac{\partial x_i^*}{\partial X_J^*} = Q_{ir}Q_{JS}\frac{\partial x_r}{\partial X_S}\right) \tag{5.20}$$

that is,

$$\mathbf{F}^* = \mathbf{Q}\mathbf{F}\mathbf{Q}^T \; (F^*_{iJ} = Q_{ir}Q_{JS}F_{rS}) \tag{5.21}$$

noting that

$$\left.\begin{aligned} dX_S\mathbf{E}_S &= dX_S Q_{IS}\mathbf{E}^*_I = Q_{IS}dX_S\mathbf{E}^*_I \equiv dX^*_I\mathbf{E}^*_I \\ dx_r\mathbf{e}_r &= dx_r Q_{ir}\mathbf{e}^*_i = Q_{ir}dx_r\mathbf{e}^*_i \equiv dx_i\mathbf{e}^*_i \end{aligned}\right\}, \tag{5.22}$$

$$\frac{\partial x_r}{\partial X_S}\mathbf{e}_r \otimes \mathbf{E}_S = \frac{\partial x_r}{\partial X_S}Q_{ir}\mathbf{e}^*_i \otimes Q_{JS}\mathbf{E}^*_J = Q_{ir}Q_{JS}\frac{\partial x_r}{\partial X_S}\mathbf{e}^*_i \otimes \mathbf{E}^*_J \equiv \frac{\partial x^*_i}{\partial X^*_J}\mathbf{e}^*_i \otimes \mathbf{E}^*_J, \tag{5.23}$$

$$\mathbf{F} = F_{rS}\mathbf{e}_r \otimes \mathbf{E}_S = F_{rS}Q_{ir}\mathbf{e}^*_i \otimes Q_{JS}\mathbf{E}^*_J = Q_{ir}Q_{JS}F_{rS}\mathbf{e}^*_i \otimes \mathbf{E}^*_J \equiv F^*_{iJ}\mathbf{e}^*_i \otimes \mathbf{E}^*_J. \tag{5.24}$$

By virtue of the transformation rules for the base vectors in equation (5.18), the vector obeys the objective transformation rule at first order, and all of the Eulerian, two-point and Lagrangian tensors obey the rule at second order as follows:

$$\mathbf{v}^* = \mathbf{Q}\mathbf{v}, \quad v^*_i = Q_{ir}v_r \tag{5.25}$$

$$\left.\begin{aligned} &\mathbf{t}^* = \mathbf{Q}\mathbf{t}\mathbf{Q}^T, \quad t^*_{ij} = Q_{ir}Q_{js}t_{rs} : \text{Eulerian tensor} \\ &\mathbb{T}^* = \begin{cases} \mathbf{Q}\mathbb{T}\mathbf{Q}^T, & \mathbb{T}^*_{iJ} = Q_{ir}Q_{JS}\mathbb{T}_{rS} : \text{Eulerian–Lagrangian two-point tensor} \\ \mathbf{Q}\mathbb{T}\mathbf{Q}^T, & \mathbb{T}^*_{Ij} = Q_{IR}Q_{js}\mathbb{T}_{Rs} : \text{Lagrangian–Eulerian two-point tensor} \end{cases} \\ &\mathbf{T}^* = \mathbf{Q}\mathbf{T}\mathbf{Q}^T, \quad T^*_{IJ} = Q_{IR}Q_{JS}T_{RS} : \text{Lagrangian tensor} \end{aligned}\right\}, \tag{5.26}$$

noting that

$$\mathbf{v} = v_i\mathbf{e}_i = Q_{ri}v_i\mathbf{e}^*_r = Q_{ir}v_r\mathbf{e}^*_i,$$

$$\left.\begin{aligned} &\mathbf{t} = t_{ij}\mathbf{e}_i \otimes \mathbf{e}_j = Q_{ri}Q_{sj}\tau_{ij}\mathbf{e}^*_r \otimes \mathbf{e}^*_s = Q_{ir}Q_{js}\tau_{rs}\mathbf{e}^*_i \otimes \mathbf{e}^*_j \\ &\mathbb{T} = \begin{cases} \mathbb{T}_{iJ}\mathbf{e}_i \otimes \mathbf{E}_J = Q_{ri}Q_{SJ}\mathbb{T}_{iJ}\mathbf{e}^*_r \otimes \mathbf{E}^*_S = Q_{ir}Q_{JS}\mathbb{T}_{rS}\mathbf{e}^*_i \otimes \mathbf{E}^*_J \\ \mathbb{T}_{Ij}\mathbf{E}_I \otimes \mathbf{e}_j = Q_{RI}Q_{sJ}\mathbb{T}_{Ij}\mathbf{E}^*_R \otimes \mathbf{e}^*_s = Q_{IR}Q_{js}\mathbb{T}_{Rs}\mathbf{E}^*_I \otimes \mathbf{e}^*_j \end{cases} \\ &\mathbf{T} = T_{IJ}\mathbf{E}_I \otimes \mathbf{E}_J = Q_{RI}Q_{SJ}T_{IJ}\mathbf{E}^*_R \otimes \mathbf{E}^*_S = Q_{RI}Q_{SJ}T_{RS}\mathbf{E}^*_I \otimes \mathbf{E}^*_J \end{aligned}\right\}.$$

Therefore, all vectors and tensors obey the objective transformation rule at an order identical to the order of the tensors themselves, that is, the number of base vectors. Several examples are shown below.

$$dv^* = dv,$$

$$d\mathbf{a}^* = \mathbf{Q}d\mathbf{a}, \quad da^*_i = Q_{ir}da_r,$$

$$\mathbf{C}^* = \mathbf{F}^{*T}\mathbf{F}^* = (\mathbf{Q}\mathbf{F}\mathbf{Q}^T)^T\mathbf{Q}\mathbf{F}\mathbf{Q}^T = \mathbf{Q}\mathbf{C}\mathbf{Q}^T,$$

$$\mathbf{b}^* = \mathbf{F}^*\mathbf{F}^{*T} = \mathbf{Q}\mathbf{F}\mathbf{Q}^T(\mathbf{Q}\mathbf{F}\mathbf{Q}^T)^T = \mathbf{Q}\mathbf{b}\mathbf{Q}^T,$$

$$\mathbf{S}^* = \mathbf{F}^{*-1}\boldsymbol{\tau}^*\mathbf{F}^{*-T} = (\mathbf{Q}\mathbf{F}\mathbf{Q}^T)^{-1}\mathbf{Q}\boldsymbol{\tau}\mathbf{Q}^T(\mathbf{Q}\mathbf{F}\mathbf{Q}^T)^{-T} = \mathbf{Q}\mathbf{F}^{-1}\boldsymbol{\tau}\mathbf{F}^{-T}\mathbf{Q}^T = \mathbf{Q}\mathbf{S}\mathbf{Q}^T,$$

$$\boldsymbol{\Pi}^* = \boldsymbol{\tau}^*\mathbf{F}^{*-T} = \mathbf{Q}\boldsymbol{\tau}\mathbf{Q}^T(\mathbf{Q}\mathbf{F}\mathbf{Q}^T)^{-T} = \mathbf{Q}\boldsymbol{\tau}\mathbf{F}^{-T}\mathbf{Q}^T = \mathbf{Q}\boldsymbol{\Pi}\mathbf{Q}^T.$$

Further, since there does not exist a rate of mutual rotation ($\dot{\mathbf{Q}} = \mathbf{O}$), the objective transformation rules hold also for rates of vectors and tensors as follows:

$$\dot{\mathbf{v}}^* = \mathbf{Q}\dot{\mathbf{v}}, \tag{5.27}$$

$$\dot{\mathbf{t}}^* = \mathbf{Q}\dot{\mathbf{t}}\mathbf{Q}^T,\ \dot{\mathbb{T}}^* = \mathbf{Q}\dot{\mathbb{T}}\mathbf{Q}^T,\ \dot{\mathbf{T}}^* = \mathbf{Q}\dot{\mathbf{T}}\mathbf{Q}^T. \tag{5.28}$$

For instance, the objective transformation rules at second order hold even for the velocity gradient tensor and the continuum spin tensor by virtue of $\dot{\mathbf{F}}^* = (\mathbf{QFQ}^T)^\bullet = \mathbf{Q}\dot{\mathbf{F}}\mathbf{Q}^T$ as follows:

$$\mathit{l}^* = \dot{\mathbf{F}}^*\mathbf{F}^{*-1} = \mathbf{Q}\dot{\mathbf{F}}\mathbf{Q}^T(\mathbf{QFQ}^T)^{-1} = \mathbf{Q}\dot{\mathbf{F}}\mathbf{F}^{-1}\mathbf{Q}^T = \mathbf{Q}\mathit{l}\mathbf{Q}^T, \tag{5.29}$$

$$\mathbf{w}^* = \mathrm{ant}[\mathbf{Q}\mathit{l}\mathbf{Q}^T] = \{\mathbf{Q}\mathit{l}\mathbf{Q}^T - (\mathbf{Q}\mathit{l}\mathbf{Q}^T)\}/2 = \mathbf{Q}\{(\mathit{l} - \mathit{l}^T)/2\}\mathbf{Q}^T = \mathbf{QwQ}^T. \tag{5.30}$$

Next, let us examine the observations in the situation where the two coordinate systems coincide with each other in the reference state and then rotate independently ($\dot{\mathbf{Q}} \neq \mathbf{O}$). Therefore, the components of the infinitesimal line-element vector observed by these coordinate systems in the initial state coincide with each other but the components of infinitesimal line-element vector observed by these systems are different at the current time as shown in Figure 5.3, where α and β are the rotating angles of these coordinate systems. This examination of the influence of the rotation of coordinate system on an observation is physically identical to the examination of the influence of rigid-body rotation on the observation. Consideration of this situation is indispensable in practical deformation analyses in which a rigid-body rotation of material takes place, while the fixed coordinate system is usually adopted for the description

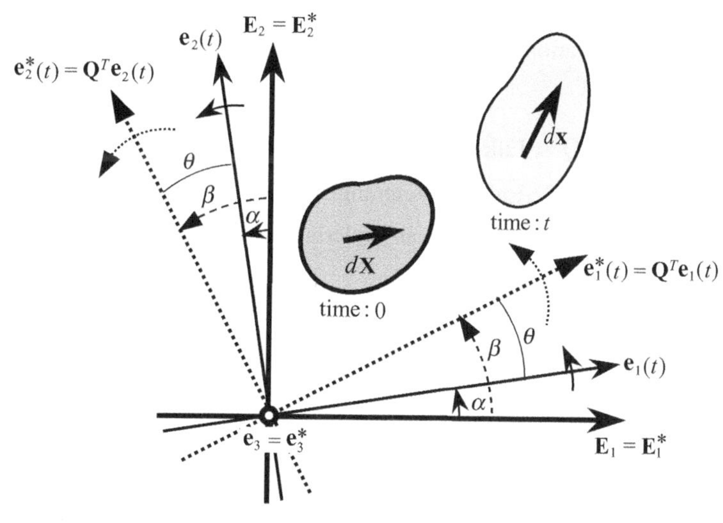

Figure 5.3 Coordinate systems coinciding at initial state and then rotating independently ($\dot{\mathbf{Q}} \neq \mathbf{O}$) (illustrated in two-dimensional state $\mathbf{e}_3 = \mathbf{e}_3^*$)

of the constitutive relation. The following relations of the base vectors hold:

$$\left.\begin{array}{l}\mathbf{E}_J^* = \mathbf{E}_J \\ \mathbf{e}_i^* = Q_{ir}\mathbf{e}_r, \quad \mathbf{e}_i = Q_{ri}\mathbf{e}_r^*; \quad \mathbf{e}_i^* = \mathbf{Q}^T\mathbf{e}_i, \quad \mathbf{e}_i = \mathbf{Q}\mathbf{e}_i^*\end{array}\right\}. \tag{5.31}$$

The following transformation rules between the components of the infinitesimal line-element vector observed by these coordinate systems hold:

$$\left.\begin{array}{l}d\mathbf{X}^* = d\mathbf{X}\ (dX_J^* = dX_J) \\ d\mathbf{x}^* = \mathbf{Q}d\mathbf{x}\ (dx_i^* = Q_{ir}dx_r)\end{array}\right\}, \tag{5.32}$$

and thus

$$\frac{\partial \mathbf{x}^*}{\partial \mathbf{X}} = \mathbf{Q}\frac{\partial \mathbf{x}}{\partial \mathbf{X}}\ \left(\frac{\partial x_i^*}{\partial X_J} = Q_{ir}\frac{\partial x_r}{\partial X_J}\right), \tag{5.33}$$

that is,

$$\mathbf{F}^* = \mathbf{Q}\mathbf{F}\ (F_{iJ}^* = Q_{ir}F_{rJ}), \tag{5.34}$$

noting that

$$\left.\begin{array}{l}dX_J\mathbf{E}_J \equiv dX_J^*\mathbf{E}_J^* \\ dx_r\mathbf{e}_r = dx_rQ_{ir}\mathbf{e}_i^* = Q_{ir}dx_r\mathbf{e}_i^* \equiv dx_i^*\mathbf{e}_i^*\end{array}\right\}, \tag{5.35}$$

$$\frac{\partial x_r}{\partial X_J}\mathbf{e}_r \otimes \mathbf{E}_J = \frac{\partial x_r}{\partial X_J}Q_{ir}\mathbf{e}_i^* \otimes \mathbf{E}_J^* = Q_{ir}\frac{\partial x_r}{\partial X_J}\mathbf{e}_i^* \otimes \mathbf{E}_J^* \equiv \frac{\partial x_i^*}{\partial X_J}\mathbf{e}_i^* \otimes \mathbf{E}_J^*, \tag{5.36}$$

and

$$\mathbf{F} = F_{rJ}\mathbf{e}_r \otimes \mathbf{E}_J = F_{rJ}Q_{ir}\mathbf{e}_i \otimes \mathbf{E}_J = Q_{ir}F_{rJ}\mathbf{e}_i \otimes \mathbf{E}_J^* \equiv F_{iJ}^*\mathbf{e}_i \otimes \mathbf{E}_J^*. \tag{5.37}$$

The following transformation rules hold in this situation:

$$\mathbf{v}^* = \mathbf{Q}\mathbf{v} \tag{5.38}$$

$$\left.\begin{array}{l}\mathbf{t}^* = \mathbf{Q}\mathbf{t}\mathbf{Q}^T\ : \text{Eulerian tensor} \\ \mathbb{T}^* = \begin{cases}\mathbf{Q}\mathbb{T} : \text{Eulerian–Lagrangian two-point tensor} \\ \mathbb{T}\mathbf{Q}^T\ : \text{Lagrangian–Eulerian two-point tensor}\end{cases} \\ \mathbf{T}^* = \mathbf{T} : \text{Lagrangian tensor}\end{array}\right\}, \tag{5.39}$$

noting that

$$\mathbf{v} = v_r\mathbf{e}_r = v_rQ_{ir}\mathbf{e}_i^* = Q_{ir}v_r\mathbf{e}_i^*,$$

$$\mathbf{t} = t_{rs}\mathbf{e}_r \otimes \mathbf{e}_s = t_{rs}Q_{ir}\mathbf{e}_i^* \otimes Q_{js}\mathbf{e}_j^* = Q_{ir}Q_{js}t_{rs}\mathbf{e}_i^* \otimes \mathbf{e}_j^* \equiv t_{ij}^*\mathbf{e}_i^* \otimes \mathbf{e}_j^*,$$

$$\left.\begin{array}{l}\mathbb{T} = \mathbb{T}_{rJ}\mathbf{e}_r \otimes \mathbf{E}_J = \mathbb{T}_{rJ}Q_{ir}\mathbf{e}_i^* \otimes \mathbf{E}_J^* = Q_{ir}\mathbb{T}_{rJ}\mathbf{e}_i^* \otimes \mathbf{E}_J^* \equiv \mathbb{T}_{iJ}^*\mathbf{e}_i^* \otimes \mathbf{E}_J^* \\ \mathbb{T} = \mathbb{T}_{Ir}\mathbf{E}_I \otimes \mathbf{e}_r = \mathbb{T}_{Ir}\mathbf{E}_I^* \otimes Q_{jr}\mathbf{e}_j^* = \mathbb{T}_{Ir}Q_{jr}\mathbf{E}_I^* \otimes \mathbf{e}_j^* \equiv \mathbb{T}_{Ij}^*\mathbf{E}_I^* \otimes \mathbf{e}_j^*\end{array}\right\},$$

$$\mathbf{T} = T_{IJ}\mathbf{E}_I \otimes \mathbf{E}_J = T_{IJ}\mathbf{E}_I^* \otimes \mathbf{E}_J^* \equiv T_{IJ}^*\mathbf{E}_I^* \otimes \mathbf{E}_J^*.$$

The transformations of basic tensors in continuum mechanics are listed below, as formulated directly by equation (5.39); the derivations based on the definitions of the Lagrangian and the two-point tensors by the pull-back operations from the Eulerian tensors are shown in brackets.

Infinitesimal current volume element (Eulerian) $dv^* = dv$

Infinitesimal reference volume element (Lagrangian) $dV^* = dV$

Infinitesimal current line element (Eulerian) $d\mathbf{x}^* = \mathbf{Q}d\mathbf{x}$

Infinitesimal reference line element (Lagrangian) $d\mathbf{X}^* = d\mathbf{X}$

Unit normal vector to current surface (Eulerian) $d\mathbf{n}^* = \mathbf{Q}d\mathbf{n}$

Unit normal vector to reference surface (Lagrangian) $d\mathbf{N}^* = d\mathbf{N}$

Infinitesimal current surface element (Eulerian) $d\mathbf{a}^* = \mathbf{Q}d\mathbf{a}$

Infinitesimal reference surface element (Lagrangian) $d\mathbf{A}^* = d\mathbf{A}$

Left stretch tensor (Eulerian) $\mathbf{V}^* = \mathbf{Q}\mathbf{V}\mathbf{Q}^T$

Right stretch tensor (Lagrangian) $\mathbf{U}^* = \mathbf{U}$

Rotation (Eulerian–Lagrangian two-point) $\mathbf{R}^* = \mathbf{Q}\mathbf{R}$

$$(\mathbf{R}^* = \mathbf{F}^*\mathbf{U}^{*-1} = (\mathbf{Q}\mathbf{F})\mathbf{U}^{*-1} = \mathbf{Q}\mathbf{F}\mathbf{U}^{-1} = \mathbf{Q}\mathbf{R})$$

Left Cauchy–Green deformation tensor (Eulerian) $\mathbf{b}^* = \mathbf{Q}\mathbf{b}\mathbf{Q}^T$

$$(\mathbf{b}^* = \mathbf{F}^*\mathbf{F}^{*T} = \mathbf{Q}\mathbf{F}(\mathbf{Q}\mathbf{F})^T = \mathbf{Q}(\mathbf{F}\mathbf{F}^T)\mathbf{Q}^T = \mathbf{Q}\mathbf{b}\mathbf{Q}^T)$$

Right Cauchy–Green deformation tensor (Lagrangian) $\mathbf{C}^* = \mathbf{C}$

$$(\mathbf{C}^* = \mathbf{F}^{*T}\mathbf{F}^* = (\mathbf{Q}\mathbf{F})^T\mathbf{Q}\mathbf{F} = \mathbf{F}^T\mathbf{F} = \mathbf{C})$$

General Eulerian strain (Eulerian) $\mathbf{e}^{(m)*} = \mathbf{Q}\mathbf{e}^{(m)}\mathbf{Q}^T$

General Lagrangian strain (Lagrangian) $\mathbf{E}^{(m)*} = \mathbf{E}^{(m)}\ (\dot{\mathbf{E}}^{(m)*} = \dot{\mathbf{E}}^{(m)})$

Cauchy stress vector (Eulerian) $\mathbf{t}^* = \mathbf{Q}\mathbf{t}$

Cauchy stress (Eulerian) $\boldsymbol{\sigma}^* = \mathbf{Q}\boldsymbol{\sigma}\mathbf{Q}^T$

Kirchhoff stress (Eulerian) $\boldsymbol{\tau}^* = \mathbf{Q}\boldsymbol{\tau}\mathbf{Q}^T$

First Piola–Kirchhoff stress (Eulerian–Lagrangian two point) $\boldsymbol{\Pi}^* = \mathbf{Q}\boldsymbol{\Pi}$

$$(\boldsymbol{\Pi}^* = \boldsymbol{\tau}^*\mathbf{F}^{*-T} = \mathbf{Q}\boldsymbol{\tau}\mathbf{Q}^T(\mathbf{Q}\mathbf{F})^{-T} = \mathbf{Q}\boldsymbol{\tau}\mathbf{Q}^T\mathbf{Q}\mathbf{F}^{-T} = \mathbf{Q}\boldsymbol{\tau}\mathbf{F}^{-T} = \mathbf{Q}\boldsymbol{\Pi})$$

Nominal stress (Lagrangian–Eulerian two-point) $\mathbf{P}^* = \mathbf{P}\mathbf{Q}^T$

$$(\mathbf{P}^* = \mathbf{F}^{*-1}\boldsymbol{\tau}^* = (\mathbf{Q}\mathbf{F})^{-1}\mathbf{Q}\boldsymbol{\tau}\mathbf{Q}^T = \mathbf{F}^{-1}\mathbf{Q}^T\mathbf{Q}\boldsymbol{\tau}\mathbf{Q}^T = \mathbf{F}^{-1}\boldsymbol{\tau}\mathbf{Q}^T = \mathbf{P}\mathbf{Q}^T)$$

Second Piola–Kirchhoff stress (Lagrangian) $\mathbf{S}^* = \mathbf{S}$

$$(\mathbf{S}^* = \mathbf{F}^{*-1}\boldsymbol{\tau}^*\mathbf{F}^{*-T} = (\mathbf{Q}\mathbf{F})^{-1}\mathbf{Q}\boldsymbol{\tau}\mathbf{Q}^T(\mathbf{Q}\mathbf{F})^{-T} = \mathbf{F}^{-1}\mathbf{Q}^T\mathbf{Q}\boldsymbol{\tau}\mathbf{Q}^T\mathbf{Q}\mathbf{F}^{-T} = \mathbf{F}^{-1}\boldsymbol{\tau}\mathbf{F}^{-T} = \mathbf{S})$$

Eshelby-like stress (Lagrangian) $\mathbf{M}^* = \mathbf{M}$

$$(\mathbf{M}^* = \mathbf{C}^*\mathbf{S}^* = \mathbf{CS} = \mathbf{M})$$

Convective stress (Lagrangian) $\mathfrak{T}^* = \mathfrak{T}$

$$(\mathbf{F}^{*T}\boldsymbol{\tau}^*\mathbf{F}^* = (\mathbf{QF})^T\mathbf{Q}\boldsymbol{\tau}\mathbf{Q}^T(\mathbf{QF}) = \mathbf{F}^T\mathbf{Q}^T\mathbf{Q}\boldsymbol{\tau}\mathbf{Q}^T\mathbf{QF} = \mathbf{F}^T\boldsymbol{\tau}\mathbf{F})$$

Corotational Kirchhoff stress (Lagrangian) $\widehat{\mathfrak{T}}^* = \widehat{\mathfrak{T}}$

$$(\widehat{\mathfrak{T}}^* = \mathbf{R}^{*T}\boldsymbol{\tau}^*\mathbf{R}^* = (\mathbf{QR})^T\mathbf{Q}\boldsymbol{\tau}\mathbf{Q}^T\mathbf{QR} = \mathbf{R}^T\boldsymbol{\tau}\mathbf{R} = \widehat{\mathfrak{T}})$$

Biot stress (Lagrangian) $\mathfrak{B}^* = \mathfrak{B}$

$$(\mathfrak{B}^* = \{\mathbf{S}^*\mathbf{U}^* + (\mathbf{S}^*\mathbf{U}^*)^T\}/2 = \{\mathbf{SU} + (\mathbf{SU})^T\}/2 = \mathfrak{B})$$

The Euler, two-point and Lagrangian tensors are all objective tensors obeying the objective transformation rules at second, first and zeroth orders, respectively. The stress tensors above will be described in Section 6.2.3. An objective tensor has only to obey the objective transformation rules irrespective of the order of the tensor itself. For instance, the second Piola–Kirchhoff stress $\mathbf{S}$ is regarded as an objective tensor, and it is a second-order tensor, but it obeys the objective transformation rule of a scalar under the superposition of rotation to the coordinate system in the deformation process, although it is written as a non-objective tensor in some literatures.

As described above, any tensor unrelated to the dimension of time can be regarded as an objective tensor. However, the objectivities of vectors and tensors describing the rate of physical quantities possess diversity as will be clarified in the following.

The time-differentiation of equation (1.199) leads to the fact that the material-time derivative of a vector does not obey the objective transformation rule but obeys

$$\dot{\mathbf{v}}^* = \mathbf{Q}\dot{\mathbf{v}} + \dot{\mathbf{Q}}\mathbf{v}, \tag{5.40}$$

and thus the material-time derivative of a vector is a non-objective tensor.

It follows from equation (5.39) that the material-time derivative of a Lagrangian tensor is the objective tensor obeying the objective transformation rule at zero order, that is,

$$\dot{\mathbf{T}}^* = \dot{\mathbf{T}}, \tag{5.41}$$

and thus the Lagrangian tensor is observed to be invariant. On the other hand, the material-time derivatives of Eulerian and two-point tensors are influenced by the rate of mutual rotation between the two coordinate systems so that they are non-objective tensors as follows:

$$\dot{\mathbf{t}}^* = \dot{\mathbf{Q}}\mathbf{t}\mathbf{Q}^T + \mathbf{Q}\dot{\mathbf{t}}\mathbf{Q}^T + \mathbf{Q}\mathbf{t}\dot{\mathbf{Q}}^T = \dot{\mathbf{Q}}\mathbf{t}\mathbf{Q}^T + \mathbf{Q}\dot{\mathbf{t}}\mathbf{Q}^T + \mathbf{Q}\mathbf{t}\dot{\mathbf{Q}}^T, \tag{5.42}$$

$$\left.\begin{array}{l} \dot{\mathbb{T}}^* = \mathbf{Q}\dot{\mathbb{T}} + \dot{\mathbf{Q}}\mathbb{T} \\ \dot{\mathbb{T}}^* = \dot{\mathbb{T}}\mathbf{Q}^T + \mathbb{T}\dot{\mathbf{Q}}^T \end{array}\right\}. \tag{5.43}$$

Having considered the objectivities of the material-time derivatives of vectors and tensors, we now examine the objectivities of original rate variables which are not derivatives of some tensors.

Because of

$$\overset{\bullet}{\mathbf{F}}^{*}\mathbf{F}^{*-1} = (\mathbf{QF})^{\bullet}(\mathbf{QF})^{-1} = (\overset{\bullet}{\mathbf{Q}}\mathbf{F} + \mathbf{Q}\overset{\bullet}{\mathbf{F}})\mathbf{F}^{-1}\mathbf{Q}^{-1} = \mathbf{Q}\overset{\bullet}{\mathbf{F}}\mathbf{F}^{-1}\mathbf{Q}^{T} + \overset{\bullet}{\mathbf{Q}}\mathbf{Q}^{T}, \tag{5.44}$$

the following transformation equations hold:

$$\boldsymbol{l}^{*} = \mathbf{Q}\boldsymbol{l}\mathbf{Q}^{T} + \overline{\boldsymbol{\Omega}}, \tag{5.45}$$

$$\mathbf{d}^{*} = \mathbf{QdQ}^{T}, \tag{5.46}$$

$$\mathbf{w}^{*} = \mathbf{QwQ}^{T} + \overline{\boldsymbol{\Omega}}, \tag{5.47}$$

where

$$\boxed{\overline{\boldsymbol{\Omega}} \equiv \overset{\bullet}{\mathbf{Q}}\mathbf{Q}^{T}} \tag{5.48}$$

Here, note that the material-time differentiation of equation (1.109) reads:

$$\overset{\bullet}{\mathbf{e}}_{i} = \overset{\bullet}{\mathbf{Q}}{}^{T}\mathbf{e}_{i}^{*} + \mathbf{Q}^{T}\overset{\bullet}{\mathbf{e}}{}_{i}^{*} = \overset{\bullet}{\mathbf{Q}}\mathbf{Q}^{T}\mathbf{e}_{i} + \mathbf{Q}^{T}\overset{\bullet}{\mathbf{e}}{}_{i}^{*}. \tag{5.49}$$

If we put $\overset{\bullet}{\mathbf{e}}{}_{i}^{*} = \mathbf{0}$ in equation (5.49), we have the variation of $\mathbf{e}_{i}$ observed from the coordinate system with the base $\{\mathbf{e}_{i}^{*}\}$, that is,

$$\overset{\bullet}{\mathbf{e}}_{i} = \overset{\bullet}{\mathbf{Q}}\mathbf{e}_{i}^{*} = \overline{\boldsymbol{\Omega}}\mathbf{e}_{i}. \tag{5.50}$$

$\overline{\boldsymbol{\Omega}}$ designates the spin of the base vector $\mathbf{e}_{i}$ observed from the coordinate system with the base $\{\mathbf{e}_{i}^{*}\}$, noting that only the direction is changeable in the base vector because of the unit vector. The tensor $\boldsymbol{\Omega}$ in equation (1.148) designating the spin of the base $\{\mathbf{e}_{i}^{*}\}$ observed from the coordinate system with the base $\{\mathbf{e}_{i}\}$ and the tensor $\overline{\boldsymbol{\Omega}}$ in equation (5.48) designating the spin of the base $\{\mathbf{e}_{i}\}$ observed from the coordinate system with the base $\{\mathbf{e}_{i}^{*}\}$ are obtained from each other by the replacement of $\mathbf{Q}$ by $\mathbf{Q}^{T}$ and vise versa. In the two-dimensional deformation perpendicular to the base vectors $\mathbf{e}_{3} = \mathbf{e}_{3}^{*}$ for instance, the relative angular velocities of the coordinate systems $\{\mathbf{e}_{i}\}$ and $\{\mathbf{e}_{i}^{*}\}$ around the base vector $\mathbf{e}_{3} = \mathbf{e}_{3}^{*}$ are given by $\overline{\omega}_{3} = -\overline{\Omega}_{12}$ and $\omega_{3} = -\Omega_{12}$, respectively, by virtue of equation (1.143). In addition, by setting $\mathbf{F} = \mathbf{R}$ in equation (5.44), the transformation equation of the relative spin $\boldsymbol{\Omega}^{R}$ in equation (4.110) is given as follows:

$$\boldsymbol{\Omega}^{R*} = \mathbf{Q}\boldsymbol{\Omega}^{R}\mathbf{Q}^{T} + \overline{\boldsymbol{\Omega}}. \tag{5.51}$$

As is evident from equations (5.45)–(5.47) and (5.51), the strain rate $\mathbf{d}$ is an objective tensor but the velocity gradient $\boldsymbol{l}$, the continuum spin $\mathbf{w}$, and the relative spin $\boldsymbol{\Omega}^{R}$ are non-objective tensors. Note that the rate tensors $\boldsymbol{l}$, $\mathbf{d}$, $\mathbf{w}$, $\boldsymbol{\Omega}^{P}$ are not derived by time-derivatives of state tensors but are defined independent of them. Among them, only $\mathbf{d}$ is an objective tensor, obtained by excluding the anti-symmetric part from the velocity gradient $\boldsymbol{l}$.

Let us now examine the objectivities of the convected derivatives of tensors. The material-time derivatives of tensors are non-objective tensosr as shown in equations (5.42) and (5.43), except for the Lagrangian tensors. Substitution of equation (5.45) with equation (5.48) into equations (5.2) and (5.5) leads to the objective transformation rules

$$\overset{\rightarrow}{\overset{\bullet}{\overset{\leftarrow}{\mathbf{v}}}}{}^{*} = \mathbf{Q}\overset{\rightarrow}{\overset{\bullet}{\overset{\leftarrow}{\mathbf{v}}}}, \quad \overset{\rightarrow}{\overset{\bullet}{\overset{\leftarrow}{\mathbf{t}}}}{}^{*} = \mathbf{Q}\overset{\rightarrow}{\overset{\bullet}{\overset{\leftarrow}{\mathbf{t}}}}\mathbf{Q}^{T} \tag{5.52}$$

for the same variant vectors and tensors, as proved for the contravariant vector and tensors as examples in the following:

$$\overset{\overrightarrow{\bullet}}{\underset{\leftarrow}{\mathbf{v}}}{}^{g^*} = \overset{\bullet}{\mathbf{v}}^* - l^*\mathbf{v}^* = (\mathbf{Q}\mathbf{v})^{\bullet} - (\mathbf{Q}l\mathbf{Q}^T + \dot{\mathbf{Q}}\mathbf{Q}^T)\mathbf{Q}\mathbf{v}$$
$$= \mathbf{Q}\dot{\mathbf{v}} + \dot{\mathbf{Q}}\mathbf{v} - \mathbf{Q}l\mathbf{v} + \mathbf{\Omega}\mathbf{Q}\mathbf{v} = \mathbf{Q}(\dot{\mathbf{v}} - l\mathbf{v}) + (\dot{\mathbf{Q}}\mathbf{Q}^T + \dot{\mathbf{Q}}\mathbf{Q}^T)\mathbf{Q}\mathbf{v} = \mathbf{Q}(\dot{\mathbf{v}} - l\mathbf{v}),$$

$$\overset{\overrightarrow{\bullet}}{\underset{\leftarrow}{\mathbf{t}}}{}^{gg^*} = \overset{\bullet}{\mathbf{t}}^* - l^*\mathbf{t}^* - \mathbf{t}^*l^{*T}$$
$$= (\mathbf{Q}\mathbf{t}\mathbf{Q}^T)^{\bullet} - (\mathbf{Q}l\mathbf{Q}^T + \dot{\mathbf{Q}}\mathbf{Q}^T)(\mathbf{Q}\mathbf{t}\mathbf{Q}^T) - (\mathbf{Q}\mathbf{t}\mathbf{Q}^T)(\mathbf{Q}l\mathbf{Q}^T + \dot{\mathbf{Q}}\mathbf{Q}^T)^T$$
$$= \mathbf{Q}\dot{\mathbf{t}}\mathbf{Q}^T + \dot{\mathbf{Q}}\mathbf{t}\mathbf{Q}^T + \mathbf{Q}\mathbf{t}\dot{\mathbf{Q}}^T - \mathbf{Q}l\mathbf{t}\mathbf{Q}^T - \dot{\mathbf{Q}}\mathbf{t}\mathbf{Q}^T - \mathbf{Q}\mathbf{t}l^T\mathbf{Q}^T - \mathbf{Q}\mathbf{t}\dot{\mathbf{Q}}^T$$
$$= \mathbf{Q}\dot{\mathbf{t}}\mathbf{Q}^T - \mathbf{Q}l\mathbf{t}\mathbf{Q}^T - \mathbf{Q}\mathbf{t}l^T\mathbf{Q}^T = \mathbf{Q}(\dot{\mathbf{t}} - l\mathbf{t} - \mathbf{t}l^T)\mathbf{Q}^T.$$

Consequently, it can be concluded that the convected derivatives of vectors and tensors possess objectivity. Therefore, they must be used instead of material-time derivatives in constitutive equations.

The objectivities of tensors were discussed above from the viewpoint of coordinate transformation. In what follows, let the objectivities be examined in the context of the influence of superposition of rigid-body rotation by use of the embedded coordinate system since the rigid-body rotation is added to material itself.

Superpose the pure rigid-body rotation as

$$\left.\begin{array}{l} \mathbf{F}_t(\tau) = \mathbf{R}_t(\tau) \\ \mathbf{g}_i(\tau) = \mathbf{R}_t(\tau)\mathbf{g}_i(t), \quad \mathbf{G}^I(\tau) = \mathbf{G}^I(t) = \mathbf{G}^I \\ \mathbf{F}_0(\tau) = \delta^i_I\mathbf{g}_i(\tau)\otimes\mathbf{G}^I = \mathbf{R}_t(\tau)\delta^i_I\mathbf{g}_i(t)\otimes\mathbf{G}^I = \mathbf{R}_t(\tau)\mathbf{F}_0(t) \end{array}\right\} \tag{5.53}$$

to the current state at the time t, where use is made of equations (3.9), (3.10) and (4.12), regarding $\mathbf{U}_t(\tau) = \mathbf{V}_t(\tau) = \mathbf{I}$. Substituting the relations $\mathbf{G}^I(\tau) = \mathbf{G}^I(t)$ and $\mathbf{g}_i(\tau) = \mathbf{R}_t(\tau)\mathbf{g}_i(t)$ shown in equation (5.53), the relations of the vector $\mathbf{v}$, the Eulerian tensor $\mathbf{t}$, the two-point tensor $\mathbb{T}$ and the Lagrangian tensor $\mathbf{T}$ at the current time t and the later time τ are given as

$$\left.\begin{array}{l} \mathbf{v}(\tau) = \mathbf{R}_t(\tau)\mathbf{v}(t): \text{ Eulerian vector} \\ \mathbf{V}(\tau) = \mathbf{V}(t): \text{ Lagrangian vector} \end{array}\right\}, \tag{5.54}$$

$$\left.\begin{array}{l} \mathbf{t}(\tau) = \mathbf{R}_t(\tau)\mathbf{t}(t)\mathbf{R}^T_t(\tau): \text{Eulerian tensor} \\ \mathbb{T}(\tau) = \left\{\begin{array}{l} \mathbf{R}_t(\tau)\mathbb{T}(t): \text{Eularian–Lagrangian two-point tensor} \\ \mathbb{T}(t)\mathbf{R}^T_t(\tau): \text{Lagrangian–Eularian two-point tensor} \end{array}\right. \\ \mathbf{T}(\tau) = \mathbf{T}(t): \text{Lagrangian tensor} \end{array}\right\}, \tag{5.55}$$

noting that

$$\mathbf{v}(\tau) = v^i(\tau)\mathbf{g}_i(\tau) = v^i(t)\mathbf{R}_t(\tau)\mathbf{g}_i(t) = \mathbf{R}_t(\tau)v^i(t)\mathbf{g}_i(t),$$
$$\mathbf{t}(\tau) = t^{ij}(\tau)\mathbf{g}_i(\tau)\otimes\mathbf{g}_j(\tau) = t^{ij}(t)\mathbf{R}_t(\tau)\mathbf{g}_i(t)\otimes\mathbf{R}_t(\tau)\mathbf{g}_i(t)$$
$$= \mathbf{R}_t(\tau)t^{ij}(t)\mathbf{g}_i(t)\otimes\mathbf{g}_i(t)\mathbf{R}^T_t(\tau),$$

$$\mathbb{T}(\tau) = \mathcal{T}^{iJ}(\tau)\mathbf{g}_i(\tau) \otimes \mathbf{G}_J(\tau) = \mathcal{T}^{iJ}(t)\mathbf{R}_t(\tau)\mathbf{g}_i(t) \otimes \mathbf{G}_J(t)$$
$$= \mathbf{R}_t(\tau)\mathcal{T}^{iJ}(t)\mathbf{g}_i(t) \otimes \mathbf{G}_J(t),$$
$$\mathbb{T}(\tau) = \mathcal{T}^{Ij}(\tau)\mathbf{G}_I(\tau) \otimes \mathbf{g}_j(\tau) = \mathcal{T}^{Ij}(t)\mathbf{G}_I(t) \otimes \mathbf{R}_t(\tau)\mathbf{g}_j(t)$$
$$= \mathcal{T}^{Ij}(t)\mathbf{G}_I(t) \otimes \mathbf{g}_j(t)\mathbf{R}_t^T(\tau),$$
$$\mathbf{T}(\tau) = T^{IJ}(\tau)\mathbf{G}_I(\tau) \otimes \mathbf{G}_J(\tau) = T^{IJ}(t)\mathbf{G}_I(t) \otimes \mathbf{G}_J(t),$$

in view of the fact that the components of tensors in the embedded coordinate system do not change during a pure rotation process. Then the Lagrangian tensor is independent of the superposition of rigid-body rotation, while the two-point and the Eulerian tensors are dependent.

The material-time derivatives of the tensors in equations (5.54) and (5.55) are given by

$$\left.\begin{array}{l} \dot{\mathbf{v}}(\tau) = \mathbf{R}_t(\tau)\dot{\mathbf{v}}(t) + \mathbf{\Omega}_t^R(\tau)\mathbf{v}(\tau): \text{ Eulerian vector} \\ \dot{\mathbf{V}}(\tau) = \dot{\mathbf{V}}(t) = \mathbf{0}: \text{ Lagrangian vector} \end{array}\right\}, \tag{5.56}$$

$$\left.\begin{array}{l} \dot{\mathbf{t}}(\tau) = \mathbf{R}_t(\tau)\dot{\mathbf{t}}(t)\mathbf{R}_t^T(\tau) + \mathbf{\Omega}_t^R(\tau)\mathbf{t}(\tau) - \mathbf{t}(\tau)\mathbf{\Omega}_t^R(\tau): \text{Eulerian tensor} \\ \dot{\mathbb{T}}(\tau) = \begin{cases} \mathbf{R}_t(\tau)\dot{\mathbb{T}}(t) + \mathbf{\Omega}_t^R(\tau)\mathbb{T}(\tau): \text{Eulerian–Lagrangian two-point tensor} \\ \dot{\mathbb{T}}(t)\mathbf{R}_t(\tau)^T - \mathbb{T}(t)\mathbf{\Omega}_t^R(\tau): \text{Lagrangian–Eulerian two-point tensor} \end{cases} \\ \dot{\mathbf{T}}(\tau) = \dot{\mathbf{T}}(t) = \mathbf{O}: \text{Lagrangian tensor} \end{array}\right\}, \tag{5.57}$$

where

$$\mathbf{\Omega}_t^R(\tau) \equiv \dot{\mathbf{R}}_t(\tau)\mathbf{R}_t^T(\tau), \tag{5.58}$$

noting that

$$\dot{\mathbf{v}}(\tau) = \{\mathbf{R}_t(\tau)\mathbf{v}(t)\}^\bullet = \dot{\mathbf{R}}_t(\tau)\mathbf{v}(t) + \mathbf{R}_t(\tau)\dot{\mathbf{v}}(t)$$
$$= \dot{\mathbf{R}}_t(\tau)\mathbf{R}_t^T(\tau)(\mathbf{R}_t(\tau)\mathbf{v}(t)) + \mathbf{R}_t(\tau)\dot{\mathbf{v}}(t),$$
$$\dot{\mathbf{t}}(\tau) = \{\mathbf{R}_t(\tau)\mathbf{t}(t)\mathbf{R}_t^T(\tau)\}^\bullet = \dot{\mathbf{R}}_t(\tau)\mathbf{t}(t)\mathbf{R}_t^T(\tau) + \mathbf{R}_t(\tau)\dot{\mathbf{t}}(t)\mathbf{R}_t^T(\tau) + \mathbf{R}_t(\tau)\mathbf{t}(t)\dot{\mathbf{R}}_t^T(\tau)$$
$$= \dot{\mathbf{R}}_t(\tau)\mathbf{R}_t^T(\tau)(\mathbf{R}_t(\tau)\mathbf{t}(t)\mathbf{R}_t^T(\tau)) + \mathbf{R}_t(\tau)\dot{\mathbf{t}}(t)\mathbf{R}_t^T(\tau) + (\mathbf{R}_t(\tau)\mathbf{t}(t)\mathbf{R}_t^T(\tau))\mathbf{R}_t(\tau)\dot{\mathbf{R}}_t^T(\tau),$$
$$\dot{\mathbb{T}}(\tau) = \{\mathbf{R}_t(\tau)\mathbb{T}(t)\}^\bullet = \dot{\mathbf{R}}_t(\tau)\mathbb{T}(t) + \mathbf{R}_t(\tau)\dot{\mathbb{T}}(t) = \dot{\mathbf{R}}_t(\tau)\mathbf{R}_t^T(\tau)(\mathbf{R}_t(\tau)\mathbb{T}(t)) + \mathbf{R}_t(\tau)\dot{\mathbb{T}}(t),$$
$$\dot{\mathbb{T}}(\tau) = \{\mathbb{T}(t)\mathbf{R}_t^T(\tau)\} = \dot{\mathbb{T}}(t)\mathbf{R}_t^T(\tau) + \mathbb{T}(t)\dot{\mathbf{R}}_t^T(\tau) = \dot{\mathbb{T}}(t)\mathbf{R}_t^T(\tau) + (\mathbb{T}(t)\mathbf{R}_t^T(\tau))\mathbf{R}_t(\tau)\dot{\mathbf{R}}_t^T(\tau).$$

The rates of the Eulerian and two-point tensors are influenced by the rigid-body rotation $\dot{\mathbf{R}}_t(\tau)$ but the Lagrangian tensor is independent of it.

Next, the rate tensors in equations (5.45)–(5.47) are influenced by the rigid-body rotation, regarding $\mathbf{Q}$ as $\mathbf{R}_t(\tau)$, as follows:

$$\boldsymbol{l}(\tau) = \mathbf{R}_t(\tau)\boldsymbol{l}(t)\mathbf{R}_t^T(\tau) + \mathbf{\Omega}_t^R(\tau), \tag{5.59}$$

$$\mathbf{w}(\tau) = \mathbf{R}_t(\tau)\mathbf{w}(t)\mathbf{R}_t^T(\tau) + \mathbf{\Omega}_t^R(\tau), \tag{5.60}$$

$$\mathbf{\Omega}_0^R(\tau) = \mathbf{R}_t(\tau)\mathbf{\Omega}_0^R(t)\mathbf{R}_t^T(\tau) + \mathbf{\Omega}_t^R(\tau), \tag{5.61}$$

$$(\mathbf{d}(\tau) = \mathbf{R}_t(\tau)\mathbf{d}(t)\mathbf{R}_t^T(\tau)),$$

noting that

$$\begin{aligned}
\dot{\mathbf{F}}_0(\tau)\mathbf{F}_0(\tau)^{-1} &= (\mathbf{R}_t(\tau)\mathbf{F}_0(t))^{\bullet}(\mathbf{R}_t(\tau)\mathbf{F}_0(t))^{-1} = (\dot{\mathbf{R}}_t(\tau)\mathbf{F}_0(t) + \mathbf{R}_t(\tau)\dot{\mathbf{F}}_0(t))\mathbf{F}_0(t)^{-1}\mathbf{R}_t^T(\tau)\\
&= \mathbf{R}_t(\tau)\dot{\mathbf{F}}_0(t)\mathbf{F}_0(t)^{-1}\mathbf{R}_t^T(\tau) + \dot{\mathbf{R}}_t(\tau)\mathbf{R}_t^T(\tau),\\
\dot{\mathbf{R}}_0(\tau)\mathbf{R}_0^T(\tau) &= (\mathbf{R}_t(\tau)\mathbf{R}_0(t))^{\bullet}(\mathbf{R}_t(\tau)\mathbf{R}_0(t))^T = (\dot{\mathbf{R}}_t(\tau)\mathbf{R}_0(t) + \mathbf{R}_t(\tau)\dot{\mathbf{R}}_0(t))\mathbf{R}_0^T(\tau)\mathbf{R}_t^T(\tau)\\
&= \mathbf{R}_t(\tau)\dot{\mathbf{R}}_0(t)\mathbf{R}_0^T(\tau)\mathbf{R}_t^T(\tau) + \dot{\mathbf{R}}_t(\tau)\mathbf{R}_t^T(\tau).
\end{aligned}$$

Further, the following equation holds for the convected derivatives:

$$\overset{\overrightarrow{\bullet}}{\overleftarrow{\mathbf{v}}}(\tau) = \mathbf{R}_t(\tau)\overset{\overrightarrow{\bullet}}{\overleftarrow{\mathbf{v}}}(t), \quad \overset{\overrightarrow{\bullet}}{\overleftarrow{\mathbf{t}}}(\tau) = \mathbf{R}_t(\tau)\overset{\overrightarrow{\bullet}}{\overleftarrow{\mathbf{t}}}(t)\mathbf{R}_t(\tau), \tag{5.62}$$

noting the following relations for the contravariant forms of vector and tensor as in the examples in equations (5.2) and (5.5):

$$\begin{aligned}
\dot{\mathbf{v}}(\tau) - \boldsymbol{l}(\tau)\mathbf{v}(\tau) &= \mathbf{R}_t(\tau)\dot{\mathbf{v}}(t) - \{\mathbf{R}_t(\tau)\boldsymbol{l}(t)\mathbf{R}_t^T(\tau) + \boldsymbol{\Omega}_t^R(\tau)\}\mathbf{R}_t(\tau)\mathbf{v}(t)\\
&= \mathbf{R}_t(\tau)(\dot{\mathbf{v}}(t) - \boldsymbol{l}(t)\mathbf{t}(t)),\\
\dot{\mathbf{t}}(\tau) - \boldsymbol{l}(\tau)\mathbf{t}(\tau) - \mathbf{t}(\tau)\boldsymbol{l}(\tau)^T &= \mathbf{R}_t(\tau)\dot{\mathbf{t}}(t)\mathbf{R}_t^T(\tau) - \{\mathbf{R}_t(\tau)\boldsymbol{l}(t)\mathbf{R}_t^T(\tau) + \boldsymbol{\Omega}_t^R(\tau)\}\mathbf{R}_t(\tau)\mathbf{t}(t)\\
&\quad \times \mathbf{R}_t^T(\tau) - \mathbf{R}_t(\tau)\mathbf{t}(t)\mathbf{R}_t^T(\tau)\{\mathbf{R}_t(\tau)\boldsymbol{l}(t)\mathbf{R}_t^T(\tau) + \boldsymbol{\Omega}_t^R(\tau)\}^T\\
&= \mathbf{R}_t(\tau)(\dot{\mathbf{t}}(t) - \boldsymbol{l}(t)\mathbf{t}(t) - \mathbf{t}(t)\boldsymbol{l}(t)^T)\mathbf{R}_t^T(\tau).
\end{aligned}$$

It is evident that the examination of the objectivity from the viewpoint of the transformation rule is substantially identical to that from the viewpoint of the influence of rigid-body rotation.

There remains the issue of the spin that correctly describes the rigid-body of material; various descriptions have been proposed to date. Here, it should be noticed that mechanically meaningful spin of material is not the rotation observed from the outer surface of the material but the rotation of the substructure (fiber) of the material, for example, crystals of metals and annual rings of trees. In other words, it is not given by the spin determined by the geometric variation of the outward appearance of the material but it depends on material properties reflecting substructures. For the hypoelastic-based formulation, the best-known idea from this physical background insists that the spin of substructure delays from the continuum spin $\mathbf{w}$ when accompanied with plastic deformation so that the Zaremba-Jaumann rate in equation (5.11) is modified to the rate $\overset{\circ}{\mathbf{t}}{}^{we}$ in the following.

$$\overset{\circ}{\mathbf{t}}{}^{we} \equiv \dot{\mathbf{t}} - \mathbf{w}^e\mathbf{t} + \mathbf{t}\mathbf{w}^e = \overset{\circ}{\mathbf{t}}{}^{w} + \mathbf{w}^p\mathbf{t} - \mathbf{t}\mathbf{w}^p, \tag{5.63}$$

$$\mathbf{w}^e \equiv \mathbf{w} - \mathbf{w}^p, \tag{5.64}$$

where $\mathbf{w}^e$ is the *substructure spin*, and $\mathbf{w}^p$ designates the spin restrained by plastic deformation and is referred to as *plastic spin* (Dafalias, 1985). This concept was suggested by Kroner (1959), Mandel (1972a, 1972b, 1973), and Kratochvil (1971), and the explicit formulation was studied first by Dafalias (1983) and Loret (1983). The simplest formulation of $\mathbf{w}^p$ was given by Zbib and Aifantis (1988) as follows:

$$\mathbf{w}^p = \mu(\boldsymbol{\sigma}\mathbf{d}^p - \mathbf{d}^p\boldsymbol{\sigma}), \tag{5.65}$$

where $\boldsymbol{\sigma}$ is the Cauchy stress, $\mathbf{d}^p$ is the plastic strain rate and μ is a material parameter. Because $\boldsymbol{\sigma}^* = \mathbf{Q}\boldsymbol{\sigma}\mathbf{Q}^T$ and $\mathbf{d}^{p*} = \mathbf{Q}\mathbf{d}^p\mathbf{Q}^T$, we have $\mathbf{w}^{p*} = \mathbf{Q}\mathbf{w}^p\mathbf{Q}^T$. Therefore, $\overset{\circ}{\mathbf{t}}{}^{we}$ fulfills the objective transformation, that is,

$$\overset{\circ}{\mathbf{t}}{}^{we*} = \mathbf{Q}\overset{\circ}{\mathbf{t}}{}^{we}\mathbf{Q}^T. \tag{5.66}$$

The deformation behavior with the corotational rate with the plastic spin is described in Dafalias (1985) and Zbib and Aifantis (1989). A more comment on the hypoelasticity and the plastic spin is given below.

It has been revealed by Dienes (1979) that a serious oscillatory stress–strain curve is predicted by the hypoelastic material based on the Zaremba–Jaumann rate since the continuum spin indicates the spin of the cross depicted momentarily in material. Further, it has been indicated by Nagtegaal (1981) that a serious oscillatory back stress–strain curve resulting in a serious oscillatory stress–strain curve is predicted by the hypoelastic-plastic material with the kinematic hardening based on the Zaremba–Jaumann rate. Then, it has been revealed by Dafalias (1985) that the serious oscillatory stress–strain stress curve generated in the hypoelastic-plastic material is remedied by incorporating the plastic spin. The issue on the plastic spin for the hyperelastic-based plasticity formulation will be discussed in Section 8.10.3.

6

Conservation Laws and Stress (Rate) Tensors

Conservation laws must be fulfilled for mass, momentum, angular momentum and energy during a deformation. These laws are outlined in this chapter. Concurrently, the Cauchy stress tensor is defined, and various Lagrangian and two-point stress tensors are derived from the Cauchy stress tensor in a systematic way by means of pull-back operations. Introducing these stress tensors, the equilibrium equations of force and angular moment are formulated based on the conservation laws. The virtual work principle required for the analyses of boundary-value problems is also described. Various objective stress rate tensors are derived systematically through the convected derivatives of the Eulerian stress tensors, that is, the Cauchy and Kirchhoff stresses. Further, the work-conjugacy concept of stress and strain rate tensors is described, and various conjugate pairs are derived systematically based on the pull-back operations.

6.1 Conservation Laws

Various conservation laws in the Eulerian and the Lagrangian descriptions are described in this section.

6.1.1 Basic Conservation Law

The conservation law for a scalar-, vector- or tensor-valued physical quantity ϕ is given in the Eulerian description as follows:

$$\left(\int_v \phi dv\right)^{\bullet} + \int_a \mathbf{v}_\phi \cdot \mathbf{n} da = \int_v s_\phi dv, \tag{6.1}$$

where the flow velocity of ϕ is denoted by $\mathbf{v}_\phi$ and the source strength of ϕ by s_ϕ. The first term in equation (6.1) is transformed by the Reynolds transportation theorem (equation (1.454)) as follows:

$$\left(\int_v \phi dv\right)^{\bullet} = \int_v \left(\overset{\bullet}{\phi} + \phi \mathrm{div}\mathbf{v}\right) dv, \tag{6.2}$$

where use is made of the symbol $\mathrm{div}(\cdot) \equiv (\partial/\partial \mathbf{x}) \cdot (\cdot) = (\cdot) \cdot (\partial/\partial \mathbf{x})$, where $(\cdot)$ is a Eulerian vector. Substituting equation (6.2) into equation (6.1) and applying Gauss's divergence theorem in equation (1.50) and $\overset{\bullet}{\phi} + \phi \mathrm{div}\mathbf{v} = \partial\phi/\partial t + (\partial\phi/\partial x_i)v_i + \phi \partial v_i/\partial x_i = \partial\phi/\partial t + \mathrm{div}(\phi \mathbf{v})$ due to equation (1.408), one has

$$\frac{\partial \phi}{\partial t} + \mathrm{div}(\phi \mathbf{v} + \mathbf{v}_{\phi}) = s_{\phi} \tag{6.3}$$

which is the *conservation law in the Eulerian description* in the local form.

Next, consider the conservation law of a scalar-, vector- or tensor-valued physical quantity Φ in the Lagrangian description. Denoting the flow velocity of Φ by $\mathbf{V}_{\Phi}$ and the source strength of Φ by S_{Φ}, the conservation law is given as

$$\left(\int_V \Phi dV\right)^{\bullet} + \int_A \mathbf{V}_{\Phi} \cdot \mathbf{N} dA = \int_V S_{\Phi} dV, \tag{6.4}$$

where V is the volume occupied in the reference configuration. Transforming equation (6.4) into a volume integration by use of Gauss's divergence theorem in equation (1.450), one has

$$\int_V \overset{\bullet}{\Phi} dV + \int_V \mathrm{Div}\mathbf{V}_{\Phi} dV = \int_V S_{\Phi} dV, \tag{6.5}$$

leading to

$$\overset{\bullet}{\Phi} + \mathrm{Div}\mathbf{V}_{\Phi} = S_{\Phi}, \tag{6.6}$$

which is the *conservation law in the Lagrangian description* in the local form, using the symbol $\mathrm{Div}(\cdot) \equiv (\partial/\partial \mathbf{X}) \cdot (\cdot) = (\cdot) \cdot (\partial/\partial \mathbf{X})$, where $(\cdot)$ is the Lagrangian vector.

6.1.2 *Conservation Law of Mass*

Denoting the field of mass density in a material as $\rho(\mathbf{x}, t)$, the mass in a region v is given as $m = \int_v \rho(\mathbf{x}, t) dv$. Therefore, the following *conservation law of mass* in the Eulerian description must hold:

$$(\overset{\bullet}{m} =) \left(\int_v \rho(\mathbf{x}, t) dv\right)^{\bullet} = 0. \tag{6.7}$$

From equation (6.7) the following relationship holds for an infinitesimal volume:

$$\rho(\mathbf{x}, t) dv = \rho_0(\mathbf{X}) dV, \tag{6.8}$$

where ρ_0 is the mass density in the reference state. Applying the Reynolds transportation theorem (equation (1.454)), one obtains the local form of equation (6.7):

$$\overset{\bullet}{\rho} + \rho \mathrm{div}\mathbf{v} = 0 \tag{6.9}$$

or

$$\frac{\partial \rho}{\partial t} + \text{div}(\rho \mathbf{v}) = 0, \quad \frac{\partial \rho}{\partial t} + (\rho v^i)|_i = 0 \tag{6.10}$$

which is referred to as the *continuity equation*. For the volume-preserving deformation process fulfilling $\rho = \text{const.}$, it follows from equation (6.9) or (6.10) that

$$\text{div}\mathbf{v} = \text{tr}\frac{\partial \mathbf{v}}{\partial \mathbf{x}} = \text{tr}\boldsymbol{l} = 0, \quad \frac{\partial v_r}{\partial x_r} = 0. \tag{6.11}$$

Further, setting $T(\mathbf{x}, t) \equiv \rho\phi$, where ϕ is a physical quantity per current unit mass, it follows from equation (6.2) that

$$\left(\int_v \rho\phi dv\right)^{\bullet} = \int_v \left\{(\rho\phi)^{\bullet} + \rho\phi\frac{\partial v_r}{\partial x_r}\right\}dv = \int_v \left(\rho\overset{\bullet}{\phi} + \overset{\bullet}{\rho}\phi + \rho\phi\frac{\partial v_r}{\partial x_r}\right)dv \tag{6.12}$$

and substituting equation (6.9) into equation (6.12), it follows that

$$\left(\int_v \rho\phi dv\right)^{\bullet} = \int_v \rho\overset{\bullet}{\phi} dv. \tag{6.13}$$

Here, choosing $\phi = 1/\rho$ and noting equation (6.9), equation (6.13) reduces to

$$\left(\int_v dv\right)^{\bullet} = \int_v \text{div}\mathbf{v} dv = \int_v \text{tr}\boldsymbol{l} dv, \tag{6.14}$$

noting that

$$\int_v \rho\overset{\bullet}{\phi} dv = \int_v \rho\frac{d(1/\rho)}{dt}dv = \int_v -\frac{1}{\rho}\overset{\bullet}{\rho} dv = \int_v -\frac{1}{\rho}(-\rho\text{div}\mathbf{v})dv.$$

The *conservation law of mass* is given in the Lagrangian description as follows:

$$(\overset{\bullet}{M} =)\left(\int_V \rho_0(\mathbf{X})dV\right)^{\bullet} = 0. \tag{6.15}$$

The local form of equation (6.15) is written as

$$\overset{\bullet}{\rho}_0(\mathbf{X}) = \frac{\partial \rho_0(\mathbf{X})}{\partial t} = 0 \tag{6.16}$$

descring that the reference density ρ_0 is independent of time.

6.1.3 Conservation Law of Linear Momentum

The *linear momentum* possessed by a region v in a current state is given by $\int_v \rho\mathbf{v}dv$. On the other hand, denoting the *traction* (or *stress*) *vector* applied to the current unit surface area of the region as $\mathbf{t}$, the total traction vector applied to the surface of the region is given as $\int_a \mathbf{t}da$ and, denoting the *body force* per current unit mass as $\mathbf{b}$, the body force applied to the region is given by $\int_v \rho\mathbf{b}dv$. The rate of momentum has to be equivalent to the sum of the traction and the body force applied to the region. Therefore, *Euler's first law of motion* (or *conservation law of linear momentum*) in the Eulerian description is given by

$$\left(\int_v \rho \mathbf{v} dv\right)^{\bullet} = \int_a \mathbf{t} da + \int_v \rho \mathbf{b} dv$$

or

$$\int_v \rho \overset{\bullet}{\mathbf{v}} dv = \int_a \mathbf{t} da + \int_v \rho \mathbf{b} dv \tag{6.17}$$

by virtue of equation (6.13).

The linear momentum possessed by a region V in the reference state is given by $\int_V \rho_0 \mathbf{v} dV$. On the other hand, denoting the traction vector applied to the reference unit surface area in the region as $\mathbf{T}$, the total stress vector applied to the surface of the region is given by $\int_A \mathbf{T} dA$, and the body force applied to the region is given by $\int_V \rho_0 \mathbf{b} dV$. The conservation law of linear momentum in the Lagrangian description is given by

$$\left(\int_V \rho_0 \mathbf{v} dV\right)^{\bullet} = \int_A \mathbf{T} dA + \int_V \rho_0 \mathbf{b} dV$$

or

$$\int_V \rho_0 \overset{\bullet}{\mathbf{v}} dV = \int_A \mathbf{T} dA + \int_V \rho_0 \mathbf{b} dV \tag{6.18}$$

by virtue of equation (6.13). The equivalence of equation (6.17) and (6.18) is ascertained by noting equation (6.8) and $\mathbf{T} dA \equiv \mathbf{t} da$.

6.1.4 Conservation Law of Angular Momentum

The *angular momentum* possessed by a region v in the current state is given as $\int_v \rho(\mathbf{x} \times \mathbf{v}) dv$. On the other hand, denoting the *angular moment of traction* and the *angular moment of body force* applied to the region as $\int_a (\mathbf{x} \times \mathbf{t}) da$ and $\int_v \rho(\mathbf{x} \times \mathbf{b}) dv$ respectively, *Euler's second law of motion* (or *conservation law of angular momentum*) in the Eulerian description is given by

$$\left(\int_v \rho \mathbf{x} \times \mathbf{v} dv\right)^{\bullet} = \int_a \mathbf{x} \times \mathbf{t} da + \int_v \rho \mathbf{x} \times \mathbf{b} dv$$

which reduces to

$$\int_v \rho \mathbf{x} \times \overset{\bullet}{\mathbf{v}} dv = \int_a \mathbf{x} \times \mathbf{t} da + \int_v \rho \mathbf{x} \times \mathbf{b} dv,$$

$$\int_v \rho \varepsilon_{ijk} x_j \overset{\bullet}{v}_k dv = \int_a \varepsilon_{ijk} x_j t_k da + \int_v \rho \varepsilon_{ijk} x_j b_k dv, \tag{6.19}$$

noting $(\mathbf{x} \times \mathbf{v})^{\bullet} = \mathbf{v} \times \mathbf{v} + \mathbf{x} \times \overset{\bullet}{\mathbf{v}} = \mathbf{x} \times \overset{\bullet}{\mathbf{v}}$ and equation (6.13).

The *angular momentum* possessed by the reference volume is given by $\int_V \rho_0(\mathbf{X} \times \mathbf{v}) dV$. On the other hand, the *angular moment of traction* and the *angular moment of body force* applied to that volume is given by $\int_A (\mathbf{X} \times \mathbf{T}) dA$ and $\int_V \rho_0(\mathbf{X} \times \mathbf{b}) dV$, respectively. Then the conservation law of angular momentum in the Lagrangian description is given by

$$\left(\int_V \rho_0 \mathbf{X} \times \mathbf{v} dV\right)^{\bullet} = \int_A \mathbf{X} \times \mathbf{T} dA + \int_V \rho_0 \mathbf{X} \times \mathbf{b} dV$$

which reduces to

$$\int_V \rho_0 \mathbf{X} \times \overset{\bullet}{\mathbf{v}} dV = \int_A \mathbf{X} \times \mathbf{T} dA + \int_V \rho_0 \mathbf{X} \times \mathbf{b} dV, \tag{6.20}$$

noting $(\mathbf{X} \times \mathbf{v})^{\bullet} = \mathbf{X} \times \overset{\bullet}{\mathbf{v}}$ and equation (6.13).

6.2 Stress Tensors

This section begins by defining the Cauchy stress tensor. Various stress tensors are defined from this tensor through various types of pull-back operation. The explanation in this section is the elaboration of the descriptions by Flügge (1972), Raniecki (2004), etc.

6.2.1 Cauchy Stress Tensor

Consider the equilibrium of force applied to the tetrahedron formed by the infinitesimal line vectors $d\theta^1\mathbf{g}_1$, $d\theta^2\mathbf{g}_2$ and $d\theta^3\mathbf{g}_3$ parallel to the current primary base vectors as shown in Figure 6.1. The surface vectors of the triangles ΔOAB, ΔOBCandΔOCA formed by these infinitesimal line vectors are given by

$$d\mathbf{a}^{(i)} = da_i \mathbf{g}^i \text{ (no sum)}, \tag{6.21}$$

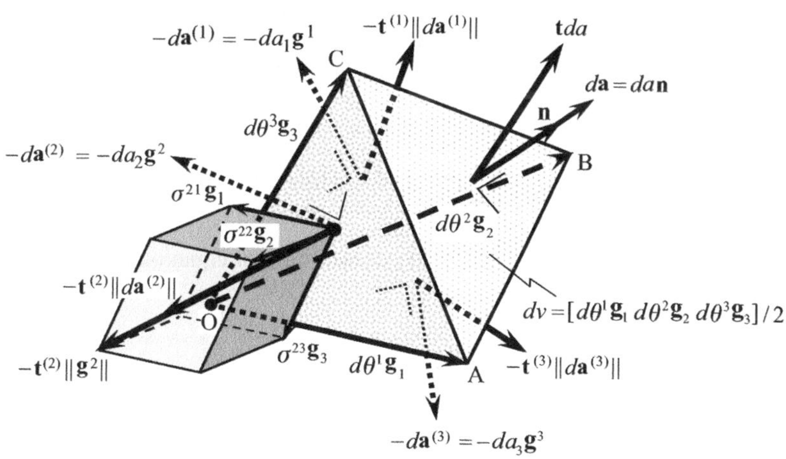

Figure 6.1 Equilibrium of force in tetrahedron formed by edges parallel to current primary base vectors

where

$$da_i \equiv \{([\mathbf{g}_1\,\mathbf{g}_2\,\mathbf{g}_3]d\theta^1 d\theta^2 d\theta^3)/d\theta^i\}/2 = (dv/d\theta^i)/2 = (\sqrt{g}d\theta^1 d\theta^2 d\theta^3/d\theta^i)/2, \quad (6.22)$$

$$dv = [d\theta^1\mathbf{g}_1\, d\theta^2\mathbf{g}_2\, d\theta^3\mathbf{g}_3] = \sqrt{g}d\theta^1 d\theta^2 d\theta^3, \quad (6.23)$$

noting that

$$d\mathbf{a}^{(1)} = d\theta^2\mathbf{g}_2 \times d\theta^3\mathbf{g}_3/2 = d\theta^2 d\theta^3[\mathbf{g}_1\,\mathbf{g}_2\,\mathbf{g}_3]\mathbf{g}^1/2 = [\{(d\theta^1 d\theta^2 d\theta^3[\mathbf{g}_1\,\mathbf{g}_2\,\mathbf{g}_3])/d\theta^1\}/2]\mathbf{g}^1$$

for instance, by virtue of equations (1.43), (2.5), (2.30) and (2.76). dv is twice the volume of the tetrahedron, that is, the volume of the parallelepiped formed by the infinitesimal line vectors $d\theta^1\mathbf{g}_1$, $d\theta^2\mathbf{g}_2$ and $d\theta^3\mathbf{g}_3$.

The surface vector of ΔABC is given by

$$\begin{aligned} d\mathbf{a} &= (d\theta^2\mathbf{g}_2 - d\theta^1\mathbf{g}_1) \times (d\theta^3\mathbf{g}_3 - d\theta^2\mathbf{g}_2)/2 \\ &= \sqrt{g}(d\theta^2 d\theta^3\mathbf{g}^1 + d\theta^3 d\theta^1\mathbf{g}^2 + d\theta^1 d\theta^2\mathbf{g}^3)/2 \\ &= da_1\mathbf{g}^1 + da_2\mathbf{g}^2 + da_3\mathbf{g}^3. \end{aligned} \quad (6.24)$$

It follows from equations (6.21) and (6.24) that

$$d\mathbf{a} = da\mathbf{n} = d\mathbf{a}^{(1)} + d\mathbf{a}^{(2)} + d\mathbf{a}^{(3)} = da_i\mathbf{g}^i, \quad (6.25)$$

where $da(= ||d\mathbf{a}||)$ and $\mathbf{n}(= d\mathbf{a}/||d\mathbf{a}||)$ are the area and the unit normal vector of the surface ΔABC. Further, it follows from equation (6.25) that

$$d\mathbf{a}\cdot\mathbf{g}_i = da\mathbf{n}\cdot\mathbf{g}_i = da_r\mathbf{g}^r\cdot\mathbf{g}_i = da_i. \quad (6.26)$$

Consequently, one obtains the following relation between the areas of the tetrahedron from equation (6.26):

$$da_i = da\mathbf{n}\cdot\mathbf{g}_i,\ \ ||d\mathbf{a}^{(i)}|| = da\mathbf{n}\cdot\mathbf{g}_i||\mathbf{g}^i||\ \text{(no sum)}. \quad (6.27)$$

Now, denoting the stress (unit force) vector applied to the surface ΔABC with the surface vector $d\mathbf{a}$ by $\mathbf{t}$ and the stress vectors applied to the surfaces ΔOAB, ΔOBC and ΔOCA with the surface vectors $d\mathbf{a}^{(i)} = da_i\mathbf{g}^i$ by $\mathbf{t}^{(i)}$ as shown in Figure 6.1, the following equilibrium equation for the forces in the tetrahedron must hold:

$$\mathbf{t}da - \sum_{i=1}^{3}\mathbf{t}^{(i)}||d\mathbf{a}^{(i)}|| + \mathbf{b}dv = \mathbf{0}. \quad (6.28)$$

For the infinitesimal tetrahedron ($dv/da \to 0$), substitution of equation (6.27) into equation (6.28) reduces to

$$\mathbf{t} = \sum_{i=1}^{3}\mathbf{t}^{(i)}||\mathbf{g}^i||(\mathbf{n}\cdot\mathbf{g}_i). \quad (6.29)$$

Here, we incorporate

$$\sigma^{ij} \equiv \mathbf{t}^{(i)}||\mathbf{g}^i||\cdot\mathbf{g}^j\ \text{(no sum)}, \quad (6.30)$$

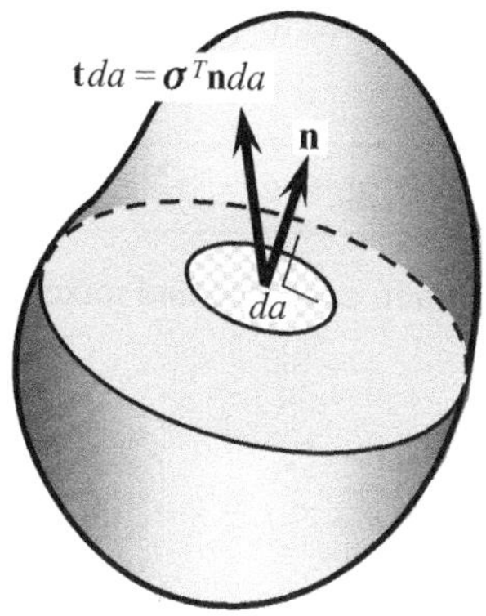

Figure 6.2 Stress vector acting to unit area of current surface: Cauchy's fundamental theorem

from which we have

$$\mathbf{t}^{(i)}||\mathbf{g}^i|| = (\mathbf{t}^{(i)}||\mathbf{g}^i|| \cdot \mathbf{g}^j)\mathbf{g}_j = \sigma^{ij}\mathbf{g}_j \text{ (no sum for } i). \tag{6.31}$$

Equation (6.29) is rewritten as

$$\boxed{\mathbf{t} = \boldsymbol{\sigma}^T\mathbf{n}} \tag{6.32}$$

noting that

$$\mathbf{t} = \sum_{i=1}^{3} \mathbf{t}^{(i)}(\mathbf{n} \cdot \mathbf{g}_i||\mathbf{g}^i||) = \sum_{i=1}^{3} (\mathbf{t}^{(i)} \cdot \mathbf{g}^j)\mathbf{g}_j(\mathbf{n} \cdot \mathbf{g}_i)||\mathbf{g}^i|| = \sigma^{ij}\mathbf{g}_j(\mathbf{n} \cdot \mathbf{g}_i) = \sigma^{ij}\mathbf{g}_j \otimes \mathbf{g}_i\mathbf{n}. \tag{6.33}$$

Equation (6.32) is referred to as *Cauchy's fundamental theorem* or *Cauchy's stress principle* and $\boldsymbol{\sigma}$ to as the *Cauchy stress tensor*. The Cauchy stress tensor plays the role of the linear mapping of the unit surface vector **n** to the *Cauchy stress vector* **t** (see Figure 6.2), while the fact that the Cauchy stress is a second-order tensor is confirmed by the quotient law described in Section 1.2.2.

The product of the Cauchy stress vector by the magnitude of the base vector, that is, $\mathbf{t}^{(i)}||\mathbf{g}^i||$(no sum), is decomposed into Cauchy stress tensor components in the direction of the base vectors, that is, $\sigma^{ij}\mathbf{g}_j$, as illustrated for the relation of $\mathbf{t}^{(2)}||\mathbf{g}^2||$ to $\sigma^{2j}\mathbf{g}_j$ in equation (6.30) (see Figure 6.1). The multiplication of the magnitude of base vector $||\mathbf{g}^i||$ is required in the curvilinear coordinate system with base vectors which are not unit vectors. Here, as is evident from equation (6.30), the first index in the components of the tensor $\boldsymbol{\sigma}$ denotes the direction of the surface to which the stress vector applies and the second index denotes the direction of action. The sign of stress component is defined to be positive when the stress component applies in the positive direction of the coordinate axis for the positive surface. Here, the sign of surface is defined to be positive when the outward-normal vector has the positive direction of the coordinate axis.

The Cauchy stress tensor $\boldsymbol{\sigma}$ was described for the contravariant form in the foregoing. It can be represented by equation (2.77) as

$$\boxed{\boldsymbol{\sigma} = \sigma^{ij}\mathbf{g}_i \otimes \mathbf{g}_j = \sigma^{i}_{\bullet j}\mathbf{g}_i \otimes \mathbf{g}^j = \sigma_i^{\bullet j}\mathbf{g}^i \otimes \mathbf{g}_j = \sigma_{ij}\mathbf{g}^i \otimes \mathbf{g}^j} \tag{6.34}$$

The relations between the contravariant, covariant and mixed variant components are given by equation (2.79) as follows:

$$\left.\begin{array}{l} \sigma^{ij} = \sigma^{i}_{\bullet r} g^{rj} = g^{ir}\sigma_r^{\bullet j} = g^{ir}\sigma_{rs}g^{sj}, \ \ \sigma^{i}_{\bullet j} = \sigma^{ir}g_{rj} = g^{ir}\sigma_r^{\bullet s}g_{sj} = g^{ir}\sigma_{rj} \\ \sigma_i^{\bullet j} = g_{ir}\sigma^{rj} = g_{ir}\sigma^{r}_{\bullet s}g^{sj} = \sigma_{ir}g^{rj}, \ \ \sigma_{ij} = g_{ir}\sigma^{rs}g_{sj} = g_{ir}\sigma^{r}_{\bullet j} = \sigma_i^{\bullet r}g_{rj} \end{array}\right\}. \tag{6.35}$$

All components are illustrated for the two-dimensional state in Figure 6.3, noting equation (6.31).

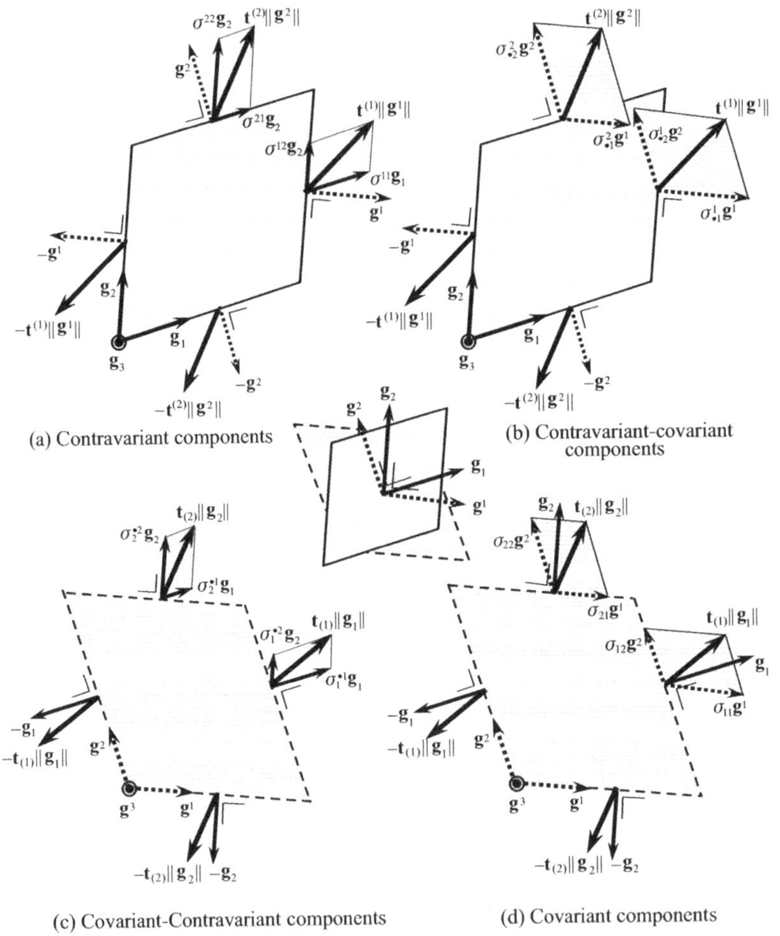

(a) Contravariant components

(b) Contravariant-covariant components

(c) Covariant-Contravariant components

(d) Covariant components

Figure 6.3 Load vectors and Cauchy stress components acting on parallelepiped formed by base vectors (illustrated in the two-dimensional state)

The equation $\mathbf{t}=\boldsymbol{\sigma}\mathbf{n}$ appears in many works (for example, Leigh, 1968; Truesdell, 1977; Bonet and Wood, 2008; Ogden, 1984; Haupt, 2000; Bertram, 2008; de Souza Neto et al., 2008) instead of equation (6.32). However, we have to accept only equation (6.32) before the verification of the symmetry $\boldsymbol{\sigma}=\boldsymbol{\sigma}^T$, as is confirmed by

$$\begin{aligned}\boldsymbol{\sigma}\mathbf{n} = \sigma^{ij}\mathbf{g}_i\otimes\mathbf{g}_j\mathbf{n} &= \sum_{i=1}^{3}(\mathbf{t}^{(i)}||\mathbf{g}^i||\cdot\mathbf{g}^j)\mathbf{g}_i\otimes\mathbf{g}_j\mathbf{n} \\ &\neq \sum_{i=1}^{3}(\mathbf{t}^{(i)}||\mathbf{g}^i||\cdot\mathbf{g}^j)\mathbf{g}_j\otimes\mathbf{g}_i\mathbf{n} = \sum_{i=1}^{3}\mathbf{t}^{(i)}||\mathbf{g}^i||(\mathbf{n}\cdot\mathbf{g}_i)=\mathbf{t}.\end{aligned}$$

6.2.2 Symmetry of Cauchy Stress Tensor

In order to verify the symmetry of the Cauchy stress tensor, consider the equilibrium of the angular moment applied to the infinitesimal parallelepiped element formed by the infinitesimal sides $d\theta^1\mathbf{g}_1$, $d\theta^2\mathbf{g}_2$, $d\theta^3\mathbf{g}_3$ parallel to the current primary base $\{\mathbf{g}_i\}$ (see Figure 6.4).

The load vector $d\mathbf{P}^{(i)}$ applied to the surface $2d\mathbf{a}^{(i)}$ of the infinitesimal parallelepiped element is given by

$$d\mathbf{P}^{(i)} \equiv \mathbf{t}^{(i)}||2d\mathbf{a}^{(i)}|| = 2da_i\sigma^{is}\mathbf{g}_s = (dv/d\theta^i)\sigma^{is}\mathbf{g}_s \text{ (no sum for } i), \tag{6.36}$$

noting that

$$\begin{aligned}\mathbf{t}^{(i)}||2d\mathbf{a}^{(i)}|| &= \boldsymbol{\sigma}^T(\mathbf{g}^i/||\mathbf{g}^i||)||2da_i\mathbf{g}^i||(=2\boldsymbol{\sigma}^T d\mathbf{a}^{(i)}) \\ &= \sigma^{rs}\mathbf{g}_s\otimes\mathbf{g}_r(2da_i)\mathbf{g}^i = \sigma^{rs}\mathbf{g}_s\otimes\mathbf{g}_r(dv/d\theta^i)\mathbf{g}^i \text{ (no sum for } i)\end{aligned}$$

with equations (6.21), (6.22), (6.23) and (6.32).

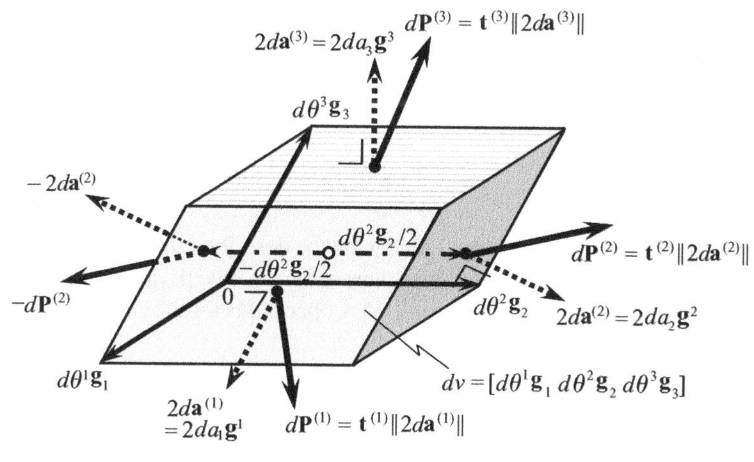

Figure 6.4 Load vectors acting on parallelepiped volume element formed by the infinitesimal sides parallel to current convected base vectors

Further, the angular moment caused by the force $d\mathbf{P}^{(i)}$ acting on the surface $d\mathbf{a}^{(i)}$ of the parallelepiped is given by

$$d\mathbf{P}^{(i)} \times d\theta^i \mathbf{g}_i = (dv/d\theta^i)\sigma^{is}\mathbf{g}_s \times d\theta^i \mathbf{g}_i = dv\sigma^{is}\mathbf{g}_s \times \mathbf{g}_i \text{ (no sum for } i), \tag{6.37}$$

noting equation (6.36) and that the load vectors form a pair of forces so that the angular moment is given by the vector product of the load vector and the distance vector between the load vectors as illustrated for $d\mathbf{P}^{(2)}$ in Figure 6.4.

The resultant of the angular moments in equation (6.37) has to be zero, that is,

$$\begin{aligned}\sum_{i=1}^{3} d\mathbf{P}^{(i)} \times d\theta^i \mathbf{g}_i &= dv\sigma^{is}\mathbf{g}_s \times \mathbf{g}_i \\ &= dv(-\sigma^{12}\mathbf{g}^3 + \sigma^{13}\mathbf{g}^2 - \sigma^{23}\mathbf{g}^1 + \sigma^{21}\mathbf{g}^3 - \sigma^{31}\mathbf{g}^2 + \sigma^{32}\mathbf{g}^1) \\ &= -dv\{(\sigma^{23} - \sigma^{32})\mathbf{g}^1 + (\sigma^{31} - \sigma^{13})\mathbf{g}^2 + (\sigma^{12} - \sigma^{21})\mathbf{g}^3\} = \mathbf{0}.\end{aligned} \tag{6.38}$$

For equation (6.38) to be satisfied it must follow that

$$\sigma^{12} = \sigma^{21}, \quad \sigma^{23} = \sigma^{32}, \quad \sigma^{31} = \sigma^{13}; \quad \sigma^{ij} = \sigma^{ji}, \tag{6.39}$$

leading to the symmetry of the Cauchy stress tensor, that is,

$$\boxed{\boldsymbol{\sigma} = \boldsymbol{\sigma}^T} \tag{6.40}$$

Therefore, from equations (2.86) and (2.87) one has

$$\sigma^{ij} = \sigma^{ji}, \quad \sigma^{i}_{\bullet j} = \sigma_j^{\bullet i}, \quad \sigma_{ij} = \sigma_{ji} \tag{6.41}$$

$$\sigma^{i}_{\bullet j} \neq \sigma^{j}_{\bullet i}, \quad \sigma_i^{\bullet j} \neq \sigma_j^{\bullet i}. \tag{6.42}$$

6.2.3 *Various Stress Tensors*

The above-mentioned Cauchy stress is concerned only with current state variables. The *Kirchhoff stress tensor* is defined via multiplying the Cauchy stress tensor by the ratio of the current volume to the reference volume as follows:

$$\boldsymbol{\tau} = \frac{dv}{dV}\boldsymbol{\sigma} = J\boldsymbol{\sigma}. \tag{6.43}$$

noting equation (1.461). Needless to say, a tensor can be described in any of the contravariant, covariant and mixed variant forms. Various Lagrangian or mixed variant stress tensors are derived from the Kirchhoff stress tensor by pull-back operations for the contravariant, covariant and mixed variants as described below.

Contravariant pull-back of Kirchhoff stress

The following tensor derived by the contravariant pull-back operation in equation $(3.32)_1$ from the Kirchhoff stress is referred to as the *second Piola–Kirchhoff stress* $\mathbf{S}$, which is the

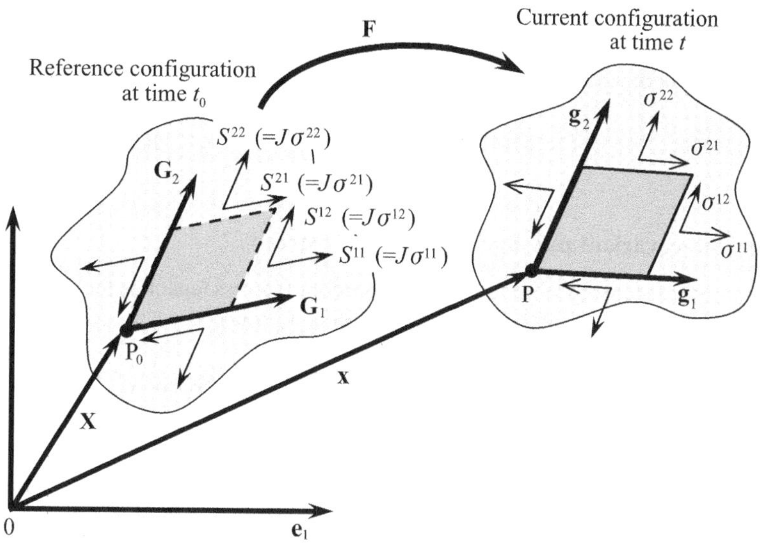

Figure 6.5 Relation of the contravariant Cauchy and the 2nd Piola-Kirchhoff stress components illustrated in the two-dimensional space

Lagrangian tensor

$$\mathbf{S} = S^{IJ}\mathbf{G}_I \otimes \mathbf{G}_J \equiv \delta^I_i \delta^J_j \tau^{ij} \mathbf{G}_I \otimes \mathbf{G}_J = \overset{\leftarrow}{\boldsymbol{\tau}}{}^{GG} = \mathbf{F}^{-1}\boldsymbol{\tau}\mathbf{F}^{-T} (= \mathbf{F}^{-1} J\boldsymbol{\sigma}\mathbf{F}^{-T}) \; (= \mathbf{S}^T). \tag{6.44}$$

The second Piola–Kirchhoff stress has components S^{IJ} which are obtained by multiplying the component σ^{ij} of the Cauchy stress in the current state by J, and the S^{IJ} act on the surface of the pallellepiped formed by the initial primary bases as shown in Figure 6.5. The relations of the components in different variants are given by equation (2.79), replacing $\mathbf{a}_i$ and $\mathbf{a}^i$ by the base vectors $\mathbf{G}_I$ and $\mathbf{G}^I$, for example, $S^I_{\bullet J} = G_{SJ} S^{IS}$. This relation of the second Piola–Kirchhoff stress tensor to the Cauchy or Kirchhoff stress tensor can be captured by describing them in the convected coordinate system.

Two-point pull-back of Kirchhoff stress

The first Piola–Kirchhoff stress $\boldsymbol{\Pi}$ as the Eulerian–Lagrangian two-point tensor and the nominal stress $\mathbf{P}$ as the Lagrangian–Eulerian two-point tensor are derived from the Kirchhoff stress tensor by pull-back operations as follows:

$$\boldsymbol{\Pi} = \left\{ \begin{array}{l} \Pi^{iJ}\mathbf{g}_i \otimes \mathbf{G}_J \equiv \delta^J_j \tau^{ij}\mathbf{g}_i \otimes \mathbf{G}_J = \overset{\leftarrow}{\boldsymbol{\tau}}{}^{\hat{g}G} \\ \Pi_i^{\bullet J}\mathbf{g}^i \otimes \mathbf{G}_J \equiv \delta^J_j \tau_i^{\bullet j}\mathbf{g}^i \otimes \mathbf{G}_J = \overset{\leftarrow}{\boldsymbol{\tau}}{}_{\hat{g}}^{\bullet G} \end{array} \right\} = \boldsymbol{\tau}\mathbf{F}^{-T} = \mathbf{P}^T \, (\neq \boldsymbol{\Pi}^T), \tag{6.45}$$

$$\mathbf{P} = \left\{ \begin{array}{l} P^{Ij}\mathbf{G}_I \otimes \mathbf{g}_j \equiv \delta^I_i \tau^{ij}\mathbf{G}_I \otimes \mathbf{g}_j = \overset{\leftarrow}{\boldsymbol{\tau}}{}^{G\hat{g}} \\ P^I_{\bullet j}\mathbf{G}_I \otimes \mathbf{g}^j \equiv \delta^I_i \tau^i_{\bullet j}\mathbf{G}_I \otimes \mathbf{g}^j = \overset{\leftarrow}{\boldsymbol{\tau}}{}^{G}_{\bullet\hat{g}} \end{array} \right\} = \mathbf{F}^{-1}\boldsymbol{\tau} = \boldsymbol{\Pi}^T \, (\neq \mathbf{P}^T), \tag{6.46}$$

noting equations (3.30).

Here, the relation $\mathbf{\Pi} = \boldsymbol{\tau}\mathbf{F}^{-T} = \boldsymbol{\tau}(1/J)\mathrm{cof}\mathbf{F} = \boldsymbol{\sigma}\mathrm{cof}\mathbf{F}$ in the identical form to equation (1.468) is derived by applying equation (1.463) into equation (6.45). Noting this fact, it follows from the Piola's formula in equation (1.470) that

$$\nabla_X \mathbf{\Pi} = J\nabla_x \boldsymbol{\sigma}. \tag{6.47}$$

Contravariant–covariant pull-back of Kirchhoff stress

The following stress tensor is derived from the Kirchhoff stress tensor as the Lagrangian tensor by the contravariant–covariant pull-back operation as follows:

$$\mathfrak{M} = \mathfrak{M}^{I}_{\bullet J}\mathbf{G}_I \otimes \mathbf{G}^J = \delta^I_i \delta^j_J \tau^i_{\bullet j}\mathbf{G}_I \otimes \mathbf{G}^J = \overleftarrow{\boldsymbol{\tau}}^{G}_{\bullet G} = \mathbf{F}^{-1}\boldsymbol{\tau}\mathbf{F} = \mathbf{SC} = \mathbf{M}^T\ (\neq \mathfrak{M}^T). \tag{6.48}$$

Covariant–contravariant pull-back of Kirchhoff stress

The *Eshelby-like stress* $\mathbf{M}$ (Maugin, 1994; Sansour et al., 2007) as the Lagrangian tensor is derived from the Kirchhoff stress tensor by the covariant–contravariant pull-back operation as follows:

$$\begin{aligned}\mathbf{M} &= M_I^{\bullet J}\mathbf{G}^I \otimes \mathbf{G}_J \equiv \delta^i_I \delta^J_j \tau_i^{\bullet j}\mathbf{G}^I \otimes \mathbf{G}_J = \overleftarrow{\boldsymbol{\tau}}^{\bullet G}_{G} = \mathbf{F}^T\boldsymbol{\tau}\mathbf{F}^{-T} = \mathfrak{M}^T \\ &= \mathbf{CS} = (\delta^i_I \delta^j_J g_{ij}\mathbf{G}^I \otimes \mathbf{G}^J)(S^{RS}\mathbf{G}_R \otimes \mathbf{G}_S) = \delta^i_I \delta^j_R g_{ij} S^{RJ}\mathbf{G}^I \otimes \mathbf{G}_J\ (\neq \mathbf{M}^T).\end{aligned} \tag{6.49}$$

Covariant pull-back of Kirchhoff stress

The *convected stress* $\mathfrak{T}$ is derived from the Kirchhoff stress tensor $\boldsymbol{\tau}$ as the Lagrangian tensor by the covariant pull-back operation as follows:

$$\mathfrak{T}(=J\mathfrak{S}) = \mathfrak{T}^{IJ}\mathbf{G}_I \otimes \mathbf{G}_J \equiv \delta^i_I \delta^j_J \tau_{ij}\mathbf{G}_I \otimes \mathbf{G}_J = \overleftarrow{\boldsymbol{\tau}}_{GG} = \mathbf{F}^T\boldsymbol{\tau}\mathbf{F} = \mathbf{CSC}\ (=\mathfrak{T}^T), \tag{6.50}$$

where $\mathfrak{S}(\equiv\overleftarrow{\boldsymbol{\sigma}}_{GG} = \mathbf{F}^T\boldsymbol{\sigma}\mathbf{F})$ is the covariant pull-back of the Cauchy stress $\boldsymbol{\sigma}$.

Rotational pull-back of Kirchhoff stress

The *corotational Cauchy stress* $\widehat{\mathbf{\Sigma}}$ and the *corotational* (or *rotationless*) *Kirchhoff stress* $\widehat{\mathfrak{T}}$ as the Lagrangian tensors are derived from the Cauchy and the Kirchhoff stress tensors by the rotational pull-back operation as follows:

$$\widehat{\mathbf{\Sigma}} \equiv \overleftarrow{\boldsymbol{\sigma}}^R = \mathbf{R}^T\boldsymbol{\sigma}\mathbf{R} = \mathbf{USU}/J\ (=\widehat{\mathbf{\Sigma}}^T), \tag{6.51}$$

$$\widehat{\mathfrak{T}} \equiv \overleftarrow{\boldsymbol{\tau}}^R = \mathbf{R}^T\boldsymbol{\tau}\mathbf{R} = \mathbf{USU}\ (=\widehat{\mathfrak{T}}^T) \tag{6.52}$$

which are independent of rigid-body rotation.

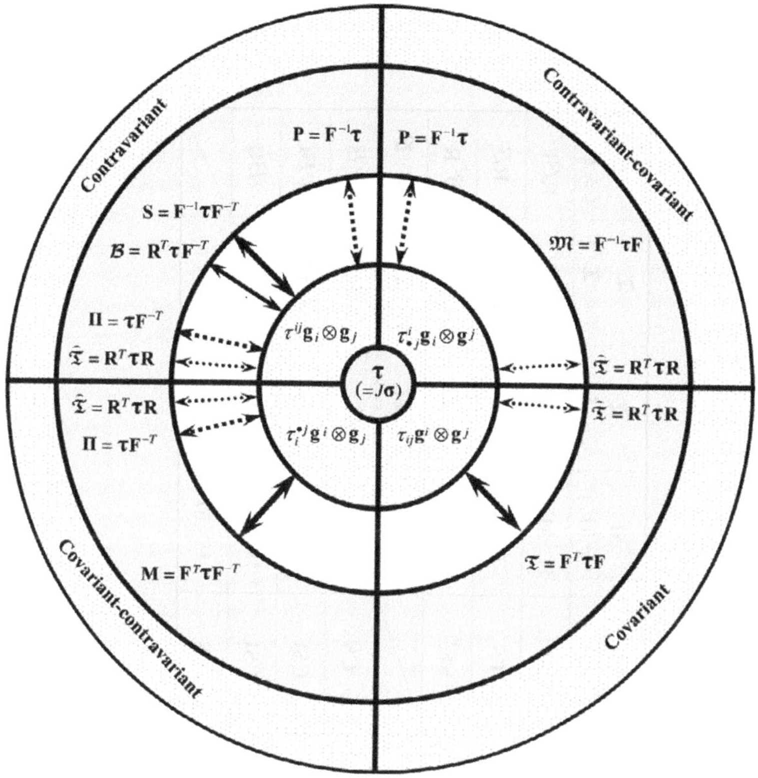

Figure 6.6 Classification of Lagrangian stress tensors obtained by the pull-back operations from the Kirchhoff stress tensor

Rotational–contravariant pull-back of Kirchhoff stress

The following stress tensor defined by the rotational–contravariant pull-back of Kirchhoff stress tensor as the Lagrangian tensor is due to Biot (1965):

$$\mathcal{B} \equiv \overleftarrow{\boldsymbol{\tau}}^{RG} = \mathbf{R}^T \boldsymbol{\tau} \mathbf{F}^{-T} = \mathbf{US}\ (\neq \mathcal{B}^T). \tag{6.53}$$

Further, the symmetric part of $\mathcal{B}$ is referred to as the *Biot* (or *Jaumann*) *stress tensor* $\mathfrak{B}$, that is,

$$\begin{aligned}\mathfrak{B} &\equiv (\overleftarrow{\boldsymbol{\tau}}^{RG} + \overleftarrow{\boldsymbol{\tau}}^{RGT})/2 = \{\mathbf{R}^T \boldsymbol{\tau} \mathbf{F}^{-T} + (\mathbf{R}^T \boldsymbol{\tau} \mathbf{F}^{-T})^T\}/2 \\ &= \{\mathbf{R}^T \boldsymbol{\Pi} + (\mathbf{R}^T \boldsymbol{\Pi})^T\}/2 = \{\mathbf{US} + (\mathbf{US})^T\}/2\ (= \mathfrak{B}^T).\end{aligned} \tag{6.54}$$

Obviously, the stress tensors $\mathcal{B}$ and $\mathfrak{B}$ are independent of the rigid-body rotation.

The relations of the above-mentioned Lagrangian stress tensors to the Eulerian tensors are shown in Figure 6.6, and various stress tensors appearing in continuum mechanics are collectively listed in Table 6.1.

Table 6.1 Relations of various stress tensors

	$\boldsymbol{\tau}$ $(=\boldsymbol{\tau}^T)$	$\boldsymbol{\Pi}$ $(\neq\boldsymbol{\Pi}^T)$	$\mathbf{P}$ $(\neq\mathbf{P}^T)$	$\mathbf{S}$ $(=\mathbf{S}^T)$	$\mathbf{M}$ $(\neq\mathbf{M}^T)$	$\mathfrak{T}$ $(=\mathfrak{T}^T)$	$\widehat{\mathfrak{T}}$ $(=\widehat{\mathfrak{T}}^T)$	$\mathcal{B}$ $(\neq\mathcal{B}^T)$
Kirchhoff $\boldsymbol{\tau}$ $(=J\sigma)$		$\mathbf{F}\boldsymbol{\Pi}^T$	$\mathbf{FP}$	$\mathbf{FSF}^T$	$\mathbf{F}^{-T}\mathbf{MF}^T$	$\mathbf{F}^{-T}\mathfrak{T}\mathbf{F}^{-1}$	$\mathbf{R}\widehat{\mathfrak{T}}\mathbf{R}^T$	$\mathbf{R}\mathcal{B}\mathbf{F}^T$
First Piola–Kirchhoff $\boldsymbol{\Pi}$	$\boldsymbol{\tau}\mathbf{F}^{-T}$		$\mathbf{P}^T$	$\mathbf{FS}$	$\mathbf{F}^{-T}\mathbf{M}$	$\mathbf{F}^{-T}\mathfrak{T}\mathbf{C}^{-1}$	$\mathbf{R}\widehat{\mathfrak{T}}\mathbf{U}^{-1}$	$\mathbf{R}\mathcal{B}$
Nominal $\mathbf{P}$	$\mathbf{F}^{-1}\boldsymbol{\tau}$	$\boldsymbol{\Pi}^T$		$\mathbf{SF}^T$	$\mathbf{M}^T\mathbf{F}^{-1}$	$\mathbf{C}^{-1}\mathfrak{T}\mathbf{F}^{-1}$	$\mathbf{U}^{-1}\widehat{\mathfrak{T}}\mathbf{R}^T$	$\mathcal{B}^T\mathbf{R}^T$
Second Piola–Kirchhoff $\mathbf{S}$	$\mathbf{F}^{-1}\boldsymbol{\tau}\mathbf{F}^{-T}$	$\mathbf{F}^{-1}\boldsymbol{\Pi}$	$\mathbf{PF}^{-T}$		$\mathbf{C}^{-1}\mathbf{M}$	$\mathbf{C}^{-1}\mathfrak{T}\mathbf{C}^{-1}$	$\mathbf{U}^{-1}\widehat{\mathfrak{T}}\mathbf{U}^{-1}$	$\mathbf{U}^{-1}\mathcal{B}$
Eshelby-like $\mathbf{M}$	$\mathbf{F}^T\boldsymbol{\tau}\mathbf{F}^{-T}$	$\mathbf{C}\boldsymbol{\Pi}^T$	$\mathbf{CPF}^{-T}$	$\mathbf{CS}$		$\mathfrak{T}\mathbf{C}^{-1}$	$\mathbf{U}\widehat{\mathfrak{T}}\mathbf{U}^{-1}$	$\mathbf{U}\mathcal{B}$
Convected $\mathfrak{T}$	$\mathbf{F}^T\boldsymbol{\tau}\mathbf{F}$	$\mathbf{C}\boldsymbol{\Pi}^T\mathbf{F}$	$\mathbf{CPF}$	$\mathbf{CSC}$	$\mathbf{MC}$		$\mathbf{U}\widehat{\mathfrak{T}}\mathbf{U}$	$\mathbf{U}\mathcal{B}\mathbf{C}$
Corotational Kirchhoff $\widehat{\mathfrak{T}}$	$\mathbf{R}^T\boldsymbol{\tau}\mathbf{R}$	$\mathbf{U}\boldsymbol{\Pi}^T\mathbf{R}$	$\mathbf{UPR}$	$\mathbf{USU}$	$\mathbf{U}^{-1}\mathbf{MU}$	$\mathbf{U}^{-1}\mathfrak{T}\mathbf{U}^{-1}$		$\mathcal{B}\mathbf{U}$
Original Biot $\mathcal{B}$	$\mathbf{R}^T\boldsymbol{\tau}\mathbf{F}^{-T}$	$\mathbf{U}\boldsymbol{\Pi}^T\mathbf{F}^{-T}$	$\mathbf{UPF}^{-T}$	$\mathbf{US}$	$\mathbf{U}^{-1}\mathbf{M}$	$\mathbf{U}^{-1}\mathfrak{T}\mathbf{C}^{-1}$	$\widehat{\mathfrak{T}}\mathbf{U}^{-1}$	

The relations of the stress tensors shown in Table 6.1 are derived as follows:

$$\boldsymbol{\Pi}=\boldsymbol{\tau}\mathbf{F}^{-T}=\begin{cases}(\mathbf{F}\mathbf{P})^T\mathbf{F}^{-T}=\mathbf{P}^T\mathbf{F}^T\mathbf{F}^{-T}=\mathbf{P}^T\\(\mathbf{F}\mathbf{S}\mathbf{F}^T)\mathbf{F}^{-T}=\mathbf{F}\mathbf{S}\\(\mathbf{F}^{-T}\mathbf{M}\mathbf{F}^T)\mathbf{F}^{-T}=\mathbf{F}^{-T}\mathbf{M}\\(\mathbf{F}^{-T}\mathfrak{T}\mathbf{F}^{-1})\mathbf{F}^{-T}=\mathbf{F}^{-T}\mathfrak{T}\mathbf{C}^{-1}\\(\mathbf{R}\hat{\mathfrak{T}}\mathbf{R}^T)\mathbf{F}^{-T}=\mathbf{R}\hat{\mathfrak{T}}\mathbf{U}^{-1}\\(\mathbf{R}\mathcal{B}\mathbf{F}^T)\mathbf{F}^{-T}=\mathbf{R}\mathcal{B}\end{cases}\quad \mathbf{P}=\mathbf{F}^{-1}\boldsymbol{\tau}=\begin{cases}\mathbf{F}^{-1}(\mathbf{F}\boldsymbol{\Pi}^T)=\boldsymbol{\Pi}^T\\\mathbf{F}^{-1}(\mathbf{F}\mathbf{S}\mathbf{F}^T)=\mathbf{S}\mathbf{F}^T\\\mathbf{F}^{-1}(\mathbf{F}^{-T}\mathbf{M}\mathbf{F}^T)^T=\mathbf{M}^T\mathbf{F}^{-1}\\\mathbf{F}^{-1}(\mathbf{F}^{-T}\mathfrak{T}\mathbf{F}^{-1})=\mathbf{C}^{-1}\mathfrak{T}\mathbf{F}^{-1}\\\mathbf{F}^{-1}(\mathbf{R}\hat{\mathfrak{T}}\mathbf{R}^T)=(\mathbf{R}\mathbf{U})^{-1}\mathbf{R}\hat{\mathfrak{T}}\mathbf{R}^T=\mathbf{U}^{-1}\hat{\mathfrak{T}}\mathbf{R}^T\\\mathbf{F}^{-1}(\mathbf{R}\mathcal{B}\mathbf{F}^T)^T=\mathcal{B}^T\mathbf{R}^T\end{cases}$$

$$\mathbf{S}=\mathbf{F}^{-1}\boldsymbol{\tau}\mathbf{F}^{-T}=\begin{cases}\mathbf{F}^{-1}(\mathbf{F}\boldsymbol{\Pi}^T)^T\mathbf{F}^{-T}=\mathbf{F}^{-1}\boldsymbol{\Pi}\\\mathbf{F}^{-1}(\mathbf{F}\mathbf{P})\mathbf{F}^{-T}=\mathbf{P}\mathbf{F}^{-T}\\\mathbf{F}^{-1}(\mathbf{F}^{-T}\mathbf{M}\mathbf{F}^T)\mathbf{F}^{-T}=\mathbf{C}^{-1}\mathbf{M}\\\mathbf{F}^{-1}(\mathbf{F}^{-T}\mathfrak{T}\mathbf{F}^{-1})\mathbf{F}^{-T}=\mathbf{C}^{-1}\mathfrak{T}\mathbf{C}^{-1}\\\mathbf{F}^{-1}(\mathbf{R}\hat{\mathfrak{T}}\mathbf{R}^T)\mathbf{F}^{-T}=\mathbf{U}^{-1}\hat{\mathfrak{T}}\mathbf{U}^{-1}\\\mathbf{F}^{-1}(\mathbf{R}\mathcal{B}\mathbf{F}^T)\mathbf{F}^{-T}=\mathbf{U}^{-1}\mathcal{B}\end{cases}\quad \mathbf{M}=\mathbf{F}^T\boldsymbol{\tau}\mathbf{F}^{-T}=\begin{cases}\mathbf{F}^T(\mathbf{F}\boldsymbol{\Pi}^T)\mathbf{F}^{-T}=\mathbf{C}\boldsymbol{\Pi}^T\\\mathbf{F}^T(\mathbf{F}\mathbf{P})\mathbf{F}^{-T}=\mathbf{C}\mathbf{P}\mathbf{F}^{-T}\\\mathbf{F}^T(\mathbf{F}\mathbf{S}\mathbf{F}^T)\mathbf{F}^{-T}=\mathbf{C}\mathbf{S}\\\mathbf{F}^T(\mathbf{F}^{-T}\mathfrak{T}\mathbf{F}^{-1})\mathbf{F}^{-T}=\mathfrak{T}\mathbf{C}^{-1}\\\mathbf{F}^T(\mathbf{R}\hat{\mathfrak{T}}\mathbf{R}^T)\mathbf{F}^{-T}=\mathbf{U}\hat{\mathfrak{T}}\mathbf{U}^{-1}\\\mathbf{F}^T(\mathbf{R}\mathcal{B}\mathbf{F}^T)\mathbf{F}^{-T}=\mathbf{U}\mathcal{B}\end{cases}$$

$$\mathfrak{T}=\mathbf{F}^T\boldsymbol{\tau}\mathbf{F}=\begin{cases}\mathbf{F}^T(\mathbf{F}\boldsymbol{\Pi}^T)\mathbf{F}=\mathbf{C}\boldsymbol{\Pi}^T\mathbf{F}\\\mathbf{F}^T(\mathbf{F}\mathbf{P})\mathbf{F}=\mathbf{C}\mathbf{P}\mathbf{F}\\\mathbf{F}^T(\mathbf{F}\mathbf{S}\mathbf{F}^T)\mathbf{F}=\mathbf{C}\mathbf{S}\mathbf{C}\\\mathbf{F}^T(\mathbf{F}^{-T}\mathbf{M}\mathbf{F}^T)\mathbf{F}=\mathbf{M}\mathbf{C}\\\mathbf{F}^T(\mathbf{R}\hat{\mathfrak{T}}\mathbf{R}^T)\mathbf{F}=\mathbf{U}\hat{\mathfrak{T}}\mathbf{U}\\\mathbf{F}^T(\mathbf{R}\mathcal{B}\mathbf{F}^T)\mathbf{F}=\mathbf{U}\mathcal{B}\mathbf{C}\end{cases}\quad \hat{\mathfrak{T}}=\mathbf{R}^T\boldsymbol{\tau}\mathbf{R}=\begin{cases}\mathbf{R}^T(\mathbf{F}\boldsymbol{\Pi}^T)\mathbf{R}=\mathbf{U}\boldsymbol{\Pi}^T\mathbf{R}\\\mathbf{R}^T(\mathbf{F}\mathbf{P})\mathbf{R}=\mathbf{U}\mathbf{P}\mathbf{R}\\\mathbf{R}^T(\mathbf{F}\mathbf{S}\mathbf{F}^T)\mathbf{R}=\mathbf{U}\mathbf{S}\mathbf{U}\\\mathbf{R}^T(\mathbf{F}^{-T}\mathbf{M}\mathbf{F}^T)\mathbf{R}=\mathbf{U}^{-1}\mathbf{M}\mathbf{U}\\\mathbf{R}^T(\mathbf{F}^{-T}\mathfrak{T}\mathbf{F}^{-1})\mathbf{R}=\mathbf{U}^{-1}\mathfrak{T}\mathbf{U}^{-1}\\\mathbf{R}^T(\mathbf{R}\mathcal{B}\mathbf{F}^T)\mathbf{R}=\mathcal{B}\mathbf{U}\end{cases}$$

$$\mathcal{B}=\mathbf{R}^T\boldsymbol{\tau}\mathbf{F}^{-T}=\begin{cases}\mathbf{R}^T(\mathbf{F}\boldsymbol{\Pi}^T)\mathbf{F}^{-T}=\mathbf{U}\boldsymbol{\Pi}^T\mathbf{F}^{-T}\\\mathbf{R}^T(\mathbf{F}\mathbf{P})\mathbf{F}^{-T}=\mathbf{U}\mathbf{P}\mathbf{F}^{-T}\\\mathbf{R}^T(\mathbf{F}\mathbf{S}\mathbf{F}^T)\mathbf{F}^{-T}=\mathbf{U}\mathbf{S}\\\mathbf{R}^T(\mathbf{F}^{-T}\mathbf{M}\mathbf{F}^T)\mathbf{F}^{-T}=\mathbf{U}^{-1}\mathbf{M}\\\mathbf{R}^T(\mathbf{F}^{-T}\mathfrak{T}\mathbf{F}^{-1})\mathbf{F}^{-T}=\mathbf{U}^{-1}\mathfrak{T}\mathbf{C}^{-1}\\\mathbf{R}^T(\mathbf{R}\hat{\mathfrak{T}}\mathbf{R}^T)\mathbf{F}^{-T}=\hat{\mathfrak{T}}\mathbf{U}^{-1}\end{cases}$$

where the symmetry of the Kirchhoff stress tensor is exploited.

Various Lagrangian and two-point stress tensors are derived from the pull-back operations from the Kirchhoff stress tensor. The relation of the Cauchy stress tensor to the Cauchy stress vector applied to the current surface vector is described by equation (6.32). In what follows, the relations of the above-mentioned stress tensors to the stress vectors applied to the reference surface vectors are described.

The stress vector $\mathbf{T}$ calculated by supposing that the Cauchy stress vector $\mathbf{t}$ applied to the unit reference area is given by

$$\mathbf{T}\equiv\mathbf{t}da/dA=\boldsymbol{\sigma}\mathbf{n}da/dA=\boldsymbol{\sigma}J\mathbf{F}^{-T}\mathbf{N}=(\boldsymbol{\tau}\mathbf{F}^{-T})\mathbf{N},\tag{6.55}$$

noting equations (1.464) and (6.32) with equation (6.40). Substituting equations (6.44), (6.45), (6.46), and (6.48)–(6.54) into equation (6.55), we have

$$\boxed{\begin{aligned}&\mathbf{T}=\boldsymbol{\tau}\mathbf{F}^{-T}\mathbf{N}=\boldsymbol{\Pi}\mathbf{N}=\mathbf{P}^T\mathbf{N}\\&\mathbf{F}^{-1}\mathbf{T}=\mathbf{S}\mathbf{N}\\&\mathbf{F}^T\mathbf{T}=\mathbf{M}\mathbf{N}=\mathfrak{T}\mathbf{C}^{-1}\mathbf{N}\\&\mathbf{R}^T\mathbf{T}=\mathcal{B}\mathbf{N}=\hat{\mathfrak{T}}\mathbf{U}^{-1}\mathbf{N}\end{aligned}}\tag{6.56}$$

noting that

$$\mathsf{T} = \begin{cases} (\mathbf{\Pi F}^T)\mathbf{F}^{-T}\mathbf{N} = \mathbf{\Pi N} \\ (\mathbf{FP})^T\mathbf{F}^{-T}\mathbf{N} = \mathbf{P}^T\mathbf{F}^T\mathbf{F}^{-T}\mathbf{N} \\ (\mathbf{FSF}^T)\mathbf{F}^{-T}\mathbf{N} = \mathbf{FSN} \\ \mathbf{F}^{-T}\mathbf{MF}^T\mathbf{F}^{-T}\mathbf{N} = \mathbf{F}^{-T}\mathbf{MN} \\ (\mathbf{F}^{-T}\mathfrak{T}\mathbf{F}^{-1})\mathbf{F}^{-T}\mathbf{N} = \mathbf{F}^{-T}\mathfrak{T}\mathbf{C}^{-1}\mathbf{N} \\ (\mathbf{R}\hat{\mathfrak{T}}\mathbf{R}^T)\mathbf{F}^{-T}\mathbf{N} = \mathbf{R}\hat{\mathfrak{T}}\mathbf{U}^{-1}\mathbf{N} \\ (\mathbf{R}\mathcal{B}\mathbf{F}^T)\mathbf{F}^{-T}\mathbf{N} = \mathbf{R}\mathcal{B}\mathbf{N} \end{cases} \tag{6.57}$$

It is shown in equation (6.56) that similar equations to Cauchy's fundamental law in equation (6.32) hold for these stress tensors, since stress tensors are second-order tensors transforming a vector linearly to another vector. Equation (6.56) is illustrated in Figure 6.7.

6.3 Equilibrium Equation

The equilibrium equation of force is obtained by substituting equation (6.32) into equation (6.17) as follows:

$$\int_v \rho\dot{\mathbf{v}}dv = \int_a \boldsymbol{\sigma}^T\mathbf{n}da + \int_v \rho\mathbf{b}dv. \tag{6.58}$$

This can be rewritten via Gauss's divergence theorem in equation (1.451) as follows:

$$\int_v \rho\dot{\mathbf{v}}dv = \int_v \nabla_x\boldsymbol{\sigma}dv + \int_v \rho\mathbf{b}dv, \tag{6.59}$$

where use is made of the symbol $\nabla_x(\cdot) \equiv \partial(\cdot)/\partial\mathbf{x}$, where $(\cdot)$ is the Eulerian second-order tensor. The local form of equation (6.59) is given as

$$\boxed{\nabla_x\boldsymbol{\sigma} + \rho\mathbf{b} = \rho\dot{\mathbf{v}}} \tag{6.60}$$

This equation is termed *Cauchy's first law of motion*, that is, the *equilibrium equation* in the Eulerian description. Equation (6.60) is represented in the general coordinate system as follows:

$$\left.\begin{array}{ll} \sigma^{ji}|_j + \rho b^i = \rho\dot{v}^i, & \sigma^{j}_{\bullet i}|_j + \rho b_i = \rho\dot{v}_i \\ \sigma_j^{\bullet i}|_j + \rho b^i = \rho\dot{v}^i, & \sigma_{ji}|_j + \rho b_i = \rho\dot{v}_i \end{array}\right\} \tag{6.61}$$

by virtue of equation $(2.145)_2$.

Further, substituting $\boldsymbol{\sigma}\mathbf{n}da = (\boldsymbol{\tau}/J)\mathbf{F}^{-T}\mathbf{N}JdA = \mathbf{\Pi N}dA$ obtained from equations (6.55) and $(6.56)_1$ into equation (6.58), one has

$$\int_V \rho_0\dot{\mathbf{v}}dV = \int_A \mathbf{\Pi N}dA + \int_V \rho_0\mathbf{b}dV, \tag{6.62}$$

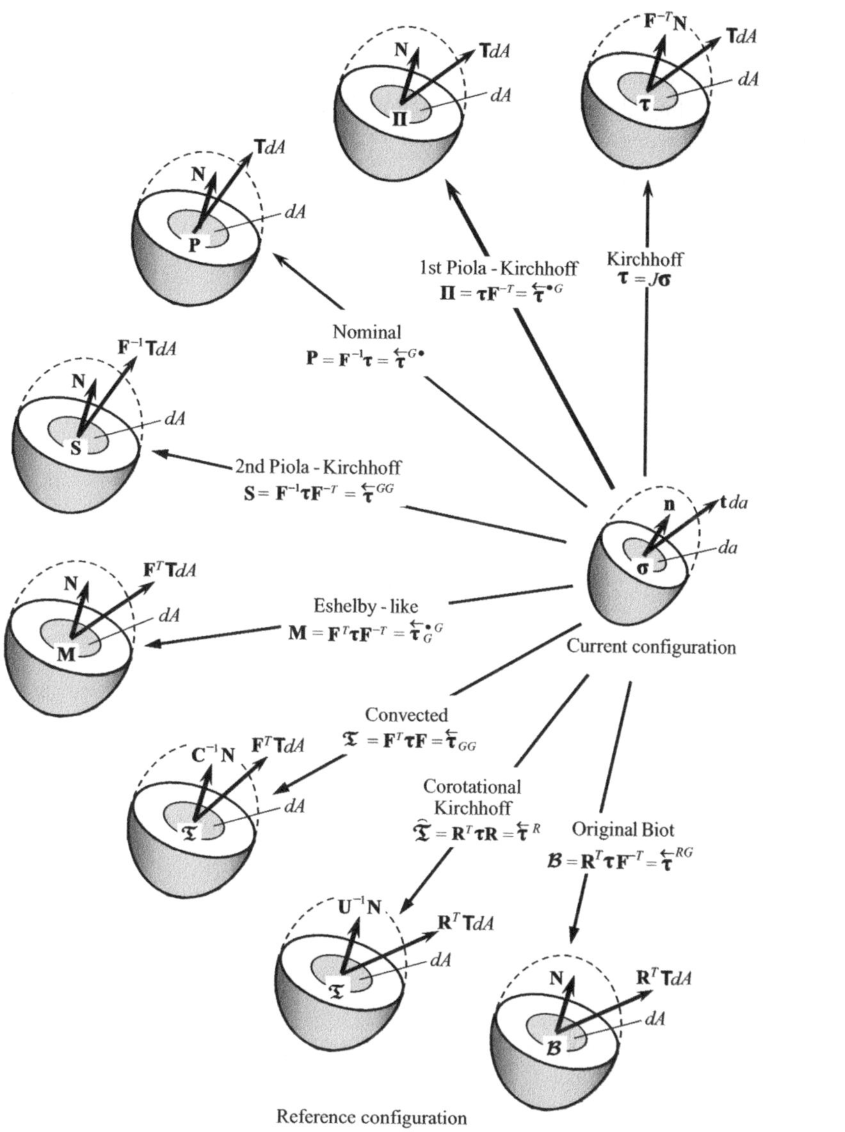

Figure 6.7 Relations of stress tensors and traction vectors

resulting in

$$\int_V \rho_0 \dot{\mathbf{v}} dV = \int_V \nabla_X \mathbf{\Pi} dV + \int_V \rho_0 \mathbf{b} dV, \tag{6.63}$$

where use is made of the symbol $\nabla_X(\cdot) \equiv \partial(\cdot)/\partial \mathbf{X}$, where $(\cdot)$ is the Lagrangian second-order tensor. The local form of equation (6.63) is given as

$$\boxed{\nabla_X \mathbf{\Pi} + \rho_0 \mathbf{b} = \rho_0 \dot{\mathbf{v}}} \tag{6.64}$$

Equation (6.64) can be obtained directly by substituting equation (6.47) into equation (6.60) with $J = \rho_0/\rho$.

The equilibrium equation in rate form is required in constitutive equations for irreversible deformation including elasto-plastic deformation. The time-differentiation of equation (6.64) engenders the following *rate-* (or *incremental-*) *type equilibrium equation* in the Lagrangian description, postulating that the acceleration does not change, that is, $\ddot{\mathbf{v}} = \mathbf{0}$:

$$\boxed{\nabla_X \overset{\bullet}{\boldsymbol{\Pi}} + \rho_0 \overset{\bullet}{\mathbf{b}} = \mathbf{0}} \tag{6.65}$$

On the other hand, the time-differentiation of equation (6.58) with the symmetry $\boldsymbol{\sigma} = \boldsymbol{\sigma}^T$ reads

$$\left(\int_v \rho\overset{\bullet}{\mathbf{v}}dv\right)^{\bullet} = \left(\int_a \boldsymbol{\sigma}\mathbf{n}da\right)^{\bullet} + \left(\int_v \rho\mathbf{b}dv\right)^{\bullet}$$

which can be rewritten via equation (6.13) as

$$\int_v \rho\,\ddot{\mathbf{v}}dv = \left(\int_a \boldsymbol{\sigma}\mathbf{n}da\right)^{\bullet} + \int_v \rho\overset{\bullet}{\mathbf{b}}dv.$$

This equation is further transformed into

$$\mathbf{0} = \int_a \overset{\bullet}{\boldsymbol{\sigma}}\,\mathbf{n}da + \int_a \boldsymbol{\sigma}(\mathbf{n}da)^{\bullet} + \int_v \rho\overset{\bullet}{\mathbf{b}}dv$$

by noting that $\ddot{\mathbf{v}} = \mathbf{0}$. Substituting equation (1.478) with equation (1.477) into this, one has

$$\int_a \overset{\bullet}{\boldsymbol{\sigma}}\,\mathbf{n}da + \int_a \boldsymbol{\sigma}\{(\mathrm{tr}\boldsymbol{l})\mathbf{I} - \boldsymbol{l}^T\}\mathbf{n}da + \int_v \rho\overset{\bullet}{\mathbf{b}}dv = \mathbf{0},$$

that is,

$$\int_v \nabla_x(\overset{\bullet}{\boldsymbol{\sigma}} + \boldsymbol{\sigma}\mathrm{tr}\boldsymbol{l} - \boldsymbol{\sigma}\boldsymbol{l}^T)^T dv + \int_v \rho\overset{\bullet}{\mathbf{b}}dv = \mathbf{0},$$

noting equation (1.451), from which one has

$$\boxed{\nabla_x {}_{\Pi}\overset{\circ}{\boldsymbol{\sigma}}{}^T + \overset{\bullet}{\mathbf{b}} = \mathbf{0}} \tag{6.66}$$

where

$${}_{\Pi}\overset{\circ}{\boldsymbol{\sigma}} \equiv \frac{1}{J}{}_{\Pi}\overset{\circ}{\boldsymbol{\tau}} = \frac{1}{J}\overset{\bullet}{\boldsymbol{\Pi}}\,\mathbf{F}^T = \overset{\bullet}{\boldsymbol{\sigma}} - \boldsymbol{\sigma}\boldsymbol{l}^T + \boldsymbol{\sigma}\,\mathrm{tr}\,\boldsymbol{l} \tag{6.67}$$

which is termed the *nominal stress rate*. Their definitions are given in equations (6.106) and (6.107).

6.4 Equilibrium Equation of Angular Moment

The symmetry property for the Cauchy stress tensor is proved in this section by considering the equilibrium of angular moment. Substituting equation (6.32) into equation (6.19) for the conservation law of angular momentum, one has

$$\begin{aligned}
&\int_v \rho \mathbf{x} \times \dot{\mathbf{v}} dv = \int_a \mathbf{x} \times \boldsymbol{\sigma}^T \mathbf{n} da + \int_v \rho \mathbf{x} \times \mathbf{b} dv, \\
&\int_v \rho \varepsilon_{ijk} x_j \dot{v}_k dv = \int_a \varepsilon_{ijk} x_j \sigma_{rk} n_r da + \int_v \rho \varepsilon_{ijk} x_j b_k dv.
\end{aligned} \tag{6.68}$$

Because the first term on the right-hand side of this equation is rewritten as

$$\begin{aligned}
&\int_a \mathbf{x} \times \boldsymbol{\sigma}^T \mathbf{n} da = \int_v \mathrm{div}(\mathbf{x} \times \boldsymbol{\nabla}_x \boldsymbol{\sigma}) dv = \int_v \boldsymbol{\varepsilon} : \boldsymbol{\sigma} dv + \int_v (\mathbf{x} \times \boldsymbol{\nabla}_x \boldsymbol{\sigma}) dv, \\
&\int_a \varepsilon_{ijk} x_j \sigma_{rk} n_r da = \int_a \frac{\partial(\varepsilon_{ijk} x_j \sigma_{rk})}{\partial x_r} dv = \int_a \varepsilon_{ijk} \frac{\partial x_j}{\partial x_r} \sigma_{rk} dv + \int_a \varepsilon_{ijk} x_j \frac{\partial \sigma_{rk}}{\partial x_r} dv \\
&\quad = \int_a \varepsilon_{ijk} \sigma_{jk} dv + \int_a \varepsilon_{ijk} x_j \frac{\partial \sigma_{rk}}{\partial x_r} dv,
\end{aligned}$$

equation (6.68) leads to

$$\begin{aligned}
&\int_v \boldsymbol{\varepsilon} : \boldsymbol{\sigma} dv + \int_v \mathbf{x} \times (\boldsymbol{\nabla}_x \boldsymbol{\sigma} + \rho \mathbf{b} - \rho \dot{\mathbf{v}}) dv = 0, \\
&\int_v \varepsilon_{ijk} \sigma_{jk} dv + \int_v \varepsilon_{ijk} x_j \left(\frac{\partial \sigma_{rk}}{\partial x_r} + \rho b_k - \rho \dot{v}_k \right) dv = 0.
\end{aligned} \tag{6.69}$$

Substituting the equilibrium equation (6.60) into equation (6.69) and noting equation (1.211), we obtain

$$\boldsymbol{\varepsilon} : \boldsymbol{\sigma} = \mathbf{0}, \quad \varepsilon_{ijk} \sigma_{jk} = 0, \tag{6.70}$$

leading to the symmetry of the Cauchy stress tensor described in equation (6.40) which was derived from the angular moment balance of the parallelepiped formed by the covariant base vectors in the curvilinear coordinate system.

6.5 Conservation Law of Energy

Forming the scalar product of the velocity $\mathbf{v}$ and the equilibrium equation (6.60), we obtain

$$\mathbf{v} \cdot \boldsymbol{\nabla}_x \boldsymbol{\sigma} + \rho \mathbf{v} \cdot \mathbf{b} = \rho \mathbf{v} \cdot \dot{\mathbf{v}}. \tag{6.71}$$

Noting that

$$\left. \begin{aligned}
&\mathrm{div}(\mathbf{v}\boldsymbol{\sigma}) = \boldsymbol{\sigma} : \mathrm{grad}\mathbf{v} + \mathbf{v} \cdot \boldsymbol{\nabla}_x \boldsymbol{\sigma} \\
&\boldsymbol{\sigma} : \mathrm{grad}\mathbf{v} = \boldsymbol{\sigma} : \boldsymbol{l} = \boldsymbol{\sigma} : \mathbf{d}
\end{aligned} \right\}, \tag{6.72}$$

equation (6.71) leads to

$$\boldsymbol{\sigma} : \mathbf{d} = \operatorname{div}(\mathbf{v}\boldsymbol{\sigma}) + \rho\mathbf{v} \cdot \mathbf{b} - \rho\left(\frac{1}{2}\mathbf{v} \cdot \mathbf{v}\right)^{\bullet} \tag{6.73}$$

which is the local form of the *equilibrium of kinetic energy*.

Now consider the integration of equation (6.73) over the current volume. The integration of the first term on the right-hand side in equation (6.73) yields

$$\int_v \operatorname{div}(\mathbf{v}\boldsymbol{\sigma})dv = \int_a (\mathbf{v}\boldsymbol{\sigma}) \cdot \mathbf{n}da = \int_a \mathbf{v} \cdot (\boldsymbol{\sigma}\mathbf{n})da = \int_a \mathbf{v} \cdot \mathbf{t}da \tag{6.74}$$

by virtue of equations (1.450), (6.32) and (6.40). Further noting the continuity equation (6.13), the integration of equation (6.73) is given by

$$\int_v \boldsymbol{\sigma} : \mathbf{d}dv = \int_a \mathbf{t} \cdot \mathbf{v}da + \int_v \rho\mathbf{b} \cdot \mathbf{v}dv - \left(\int_v \frac{1}{2}\rho\mathbf{v} \cdot \mathbf{v}dv\right)^{\bullet}. \tag{6.75}$$

By defining the stress power

$$P_\sigma \equiv \int_v \boldsymbol{\sigma} : \mathbf{d}dv, \tag{6.76}$$

the power of external forces

$$P_f \equiv \int_a \mathbf{t} \cdot \mathbf{v}da + \int_v \rho\mathbf{b} \cdot \mathbf{v}dv \tag{6.77}$$

and the kinetic energy

$$K \equiv \int_v \frac{1}{2}\rho\mathbf{v} \cdot \mathbf{v}dv, \tag{6.78}$$

equation (6.75) can be rewritten as

$$P_\sigma = P_f - \overset{\bullet}{K}. \tag{6.79}$$

Furthermore, equation (6.75) can be described in the reference configuration as follows:

$$\int_V \mathbf{S} : \overset{\bullet}{\mathbf{E}}dV = \int_A \mathbf{T} \cdot \mathbf{v}dA + \int_V \rho_0\mathbf{b} \cdot \mathbf{v}dV - \left(\int_V \frac{1}{2}\rho_0\mathbf{v} \cdot \mathbf{v}dV\right)^{\bullet}, \tag{6.80}$$

noting $\mathbf{t}da = \mathbf{T}dA$ in equation (6.55) and $\boldsymbol{\sigma} : \mathbf{d}dv = (\mathbf{S} : \overset{\bullet}{\mathbf{E}}/J)dv = \mathbf{S} : \overset{\bullet}{\mathbf{E}}dV$ in equation (6.100) which will be shown in Section 6.7 for the work conjugacy.

We now formulate the conservation law of energy. The *heat input* caused by the heat flux $\mathbf{q}$ per current unit area and the *internal heat source* r per unit mass is given by

$$Q \equiv -\int_a \mathbf{q} \cdot \mathbf{n}da + \int_v \rho r dv. \tag{6.81}$$

The *internal energy* E is defined by

$$E \equiv \int_v \rho\varepsilon dv, \tag{6.82}$$

where ε is the internal energy per unit mass. The energy equilibrium equation

$$\overset{\bullet}{K} + \overset{\bullet}{E} = P_f + Q \tag{6.83}$$

must hold, that is,

$$\left(\int_v \frac{1}{2}\rho\mathbf{v}\cdot\mathbf{v}dv\right)^{\bullet} + \left(\int_v \rho\varepsilon dv\right)^{\bullet} = \int_a \mathbf{t}\cdot\mathbf{v}da + \int_v \rho\mathbf{b}\cdot\mathbf{v}dv - \int_a \mathbf{q}\cdot\mathbf{n}da + \int_v \rho r dv. \tag{6.84}$$

Making use of equation (6.75) and noting the continuity equation (6.13), equation (6.84) leads to

$$\int_v (\mathrm{div}\mathbf{q} + \rho\overset{\bullet}{\varepsilon} - \boldsymbol{\sigma}:\mathbf{d} - \rho r)dv = 0, \tag{6.85}$$

from which the conservation law of energy in the local form is given as

$$\boxed{\rho\overset{\bullet}{\varepsilon} = \boldsymbol{\sigma}:\mathbf{d} + \rho r - \mathrm{div}\mathbf{q}} \tag{6.86}$$

Equation (6.86) is written in the reference configuration as

$$\boxed{\rho_0\overset{\bullet}{\varepsilon} = \mathbf{S}:\overset{\bullet}{\mathbf{E}} + \rho_0 r - \mathrm{Div}\mathbf{Q}} \tag{6.87}$$

where

$$\mathbf{Q} = J\mathbf{q}\mathbf{F}^{-T}, \tag{6.88}$$

noting

$$\int_v \mathrm{div}\mathbf{q}dv = \int_a \mathbf{q}\cdot\mathbf{n}da = \int_A \mathbf{q}\cdot J\mathbf{F}^{-T}\mathbf{N}dA = \int_V \mathrm{Div}(\mathbf{q}\cdot J\mathbf{F}^{-T})dV.$$

The explanation in this section follows Başar and Weichert (2000).

6.6 Virtual Work Principle

The stress (rate) field fulfilling the equilibrium equation and the boundary condition of stress is called a *statically admissible field*. On the other hand, the displacement (velocity) field fulfilling the geometrical requirement $F_{iA}\ (=\partial x_i/\partial X_A) = \delta_{iA} + \partial u_i/\partial X_A$ or $d_{ij} = (\partial v_i/\partial x_j + \partial v_j/\partial x_i)/2$ and the boundary condition of displacement (velocity) is called a *kinematically admissible field*. Denoting arbitrary statically admissible stress (rate) field and kinematically admissible displacement (velocity) field by $(\cdot)^{\Delta}$ and $(\cdot)^{\nabla}$, respectively, we have the following equation from equation (6.60):

$$\int_v (\boldsymbol{\nabla}_x\boldsymbol{\sigma}^{\Delta} + \rho\mathbf{b} - \rho\overset{\bullet}{\mathbf{v}})\cdot\mathbf{u}^{\nabla}dv = 0,\quad \int_v \left(\frac{\partial\sigma_{ij}^{\Delta}}{\partial x_j} + \rho b_i - \rho\overset{\bullet}{v}_i\right)u_i^{\nabla}dv = 0. \tag{6.89}$$

Making use of Gauss' divergence theorem (equation (1.450)), we have

$$\begin{aligned}\int_v \sigma_{ij}^{\Delta} \frac{\partial u_i^{\nabla}}{\partial x_j} dv &= \int_v \frac{\partial(\sigma_{ij}^{\Delta} u_i^{\nabla})}{\partial x_j} dv - \int_v \frac{\partial \sigma_{ij}^{\Delta}}{\partial x_j} u_i^{\nabla} dv \\ &= \int_{a_t} \bar{\sigma}_{ij} u_i^{\nabla} n_j da + \int_{a_v} \sigma_{ij}^{\Delta} \bar{u}_i n_j da - \int_v \frac{\partial \sigma_{ij}^{\Delta}}{\partial x_j} u_i^{\nabla} dv \\ &= \int_{a_t} \bar{t}_i u_i^{\nabla} da + \int_{a_v} \sigma_{ij}^{\Delta} n_j \bar{u}_i da - \int_v \frac{\partial \sigma_{ij}^{\Delta}}{\partial x_j} u_i^{\nabla} dv,\end{aligned}$$

where the overbar ($\bar{\ }$) denotes the given boundary condition, and a_t and a_v specify the surfaces on which the traction (rate) and the displacement (velocity) are given, respectively. Substituting equation (6.89) into this equation, the following *virtual work principle* described by the quantities in the current configuration is obtained:

$$\int_v \sigma_{ij}^{\Delta} \frac{\partial u_i^{\nabla}}{\partial x_j} dv = \int_{a_t} \bar{t}_i u_i^{\nabla} da + \int_{a_v} \sigma_{ij}^{\Delta} n_j \bar{u}_i da + \int_v \rho b_i u_i^{\nabla} dv - \int_v \rho \dot{v}_i u_i^{\nabla} dv. \tag{6.90}$$

Similarly, the following virtual work principle can be described by the quantities in the reference configuration from equation (6.64):

$$\int_V \Pi_{iJ}^{\Delta} \frac{\partial u_i^{\nabla}}{\partial X_J} dV = \int_{A_t} \bar{T}_i u_i^{\nabla} dA + \int_{A_v} \Pi_{iJ}^{\Delta} N_J \bar{u}_i dA + \int_V \rho_0 b_i u_i^{\nabla} dV - \int_V \rho_0 \dot{v}_i u_i^{\nabla} dV. \tag{6.91}$$

Furthermore, one has the following rate-type virtual work principle from equations (6.65) and (6.66):

$$\int_v \text{п}\overset{\circ}{\sigma}{}_{ij}^{\Delta} d_{ij}^{\nabla} dv = \int_{a_t} \dot{t}_i v_i^{\nabla} da + \int_{a_v} \text{п}\overset{\circ}{\sigma}{}_{ij}^{\Delta} n_j \bar{v}_i da + \int_v \rho \dot{b}_i v_i^{\nabla} dv, \tag{6.92}$$

$$\int_V \dot{\Pi}_{iJ}^{\Delta} \dot{F}_{iJ}^{\nabla} dV = \int_{A_t} \dot{\bar{T}}_i v_i^{\nabla} dA + \int_{A_v} \dot{\Pi}_{iJ}^{\Delta} N_J \bar{v}_i dA + \int_V \rho_0 \dot{b}_i v_i^{\nabla} dV. \tag{6.93}$$

6.7 Work Conjugacy

Constitutive relations must be considered for a particular volume element of material, as the material-time derivative for a particular volume element is adopted for the time-derivative of physical quantities in the formulation of constitutive relation as described in Section 1.9.4. Likewise, the rate of work (power) in a particular volume element which can be prescribed as a reference volume element is meaningful in constitutive relations. A pair of stress and strain rates which give rise to the rate of work is referred to as *work-conjugate pair* and this concept is referred to as *work conjugacy* (Hill, 1968).

Note that the rate of work done in a bar which has current length $d\theta^1\mathbf{g}_1$ and is subjected to a force $d\mathbf{P}^{(1)}$ (not perpendicular to the longitudinal direction in general) at the section surface $2d\mathbf{a}^1$ is given by $d\mathbf{P}^{(1)} \cdot (d\theta^1\mathbf{g}_1)^{\bullet}$. The total *rate of work* (or *stress power*) in the current

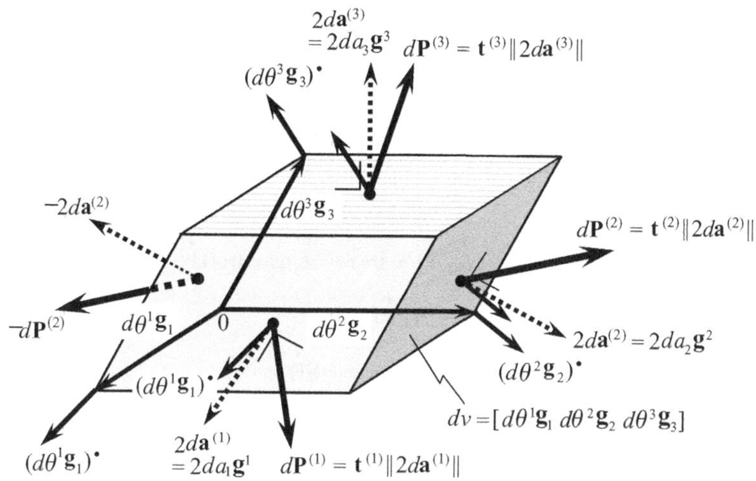

Figure 6.8 Load vectors and velocity vectors of surface movements in the parallelepiped formed by edges parallel to covariant base vectors in the current configuration

pallellepiped volume element $[d\theta^1\mathbf{g}_1\ d\theta^2\mathbf{g}_2\ d\theta^3\mathbf{g}_3]$ is given by $\sum_{i=1}^{3} d\mathbf{P}^{(i)}\cdot(d\theta^i\mathbf{g}_i)^\bullet$ (see Figure 6.8) and thus the work rate in the unit reference volume dV is given, noting equations (4.83) and (6.36), by

$$\begin{aligned}
\overset{\bullet}{w} &= \{d\mathbf{P}^{(1)}\cdot(d\theta^1\mathbf{g}_1)^\bullet + d\mathbf{P}^{(2)}\cdot(d\theta^2\mathbf{g}_2)^\bullet + d\mathbf{P}^{(3)}\cdot(d\theta^3\mathbf{g}_3)^\bullet\}/dV \\
&= \left\{(dv/d\theta^1)\sigma^{1j}\mathbf{g}_j\cdot d\theta^1\,\overset{\bullet}{\mathbf{g}}_1 + (dv/d\theta^2)\sigma^{2j}\mathbf{g}_j\cdot d\theta^2\,\overset{\bullet}{\mathbf{g}}_2 + (dv/d\theta^3)\sigma^{3j}\mathbf{g}_j\cdot d\theta^3\,\overset{\bullet}{\mathbf{g}}_3\right\}/dV \\
&= \frac{dv}{dV}\sigma^{rj}\mathbf{g}_j\cdot\overset{\bullet}{\mathbf{g}}_r = \frac{dv}{dV}(\sigma^{rj}\mathbf{g}_r\otimes\mathbf{g}_j):(\overset{\bullet}{\mathbf{g}}_p\cdot\mathbf{g}_q)\mathbf{g}^p\otimes\mathbf{g}^q = \frac{dv}{dV}\boldsymbol{\sigma}:\boldsymbol{l} \\
&\qquad = J\mathrm{tr}(\boldsymbol{\sigma}\boldsymbol{l}) = J\mathrm{tr}(\boldsymbol{\sigma}\mathbf{d}) = \mathrm{tr}(\boldsymbol{\tau}\boldsymbol{l}) = \mathrm{tr}(\boldsymbol{\tau}\mathbf{d}) \\
&\left(\text{or } \frac{dv}{dV}\sigma^{rj}\mathbf{g}_j\cdot\overset{\bullet}{\mathbf{g}}_r = \frac{dv}{dV}\sigma^{rj}\mathbf{g}_j\cdot(\overset{\bullet}{\mathbf{g}}_i\otimes\mathbf{g}^i)\mathbf{g}_r = \frac{dv}{dV}\sigma^{rj}\mathbf{g}_j\cdot\boldsymbol{l}\mathbf{g}_r = \frac{dv}{dV}\sigma^{jr}l_{jr}\right),
\end{aligned} \tag{6.94}$$

noting that the coordinate components of a material line element vector are kept constant in the convected coordinate system, leading to $(d\theta^i)^\bullet = 0$. It is verified directly in equation (6.94) that $\boldsymbol{\tau}:\mathbf{d}$ designates the work rate done by the stress in the unit reference volume. On the other hand, this fact has been explained by the indirect way through the energy balance requirement that the work rate done by the stress is equal to the sum of the work rates done by the traction and the body force minus the kinetic energy in the current unit mass as shown in equation (6.75) in various literatures (Ogden, 1984; Haupt, 2000; Lubarda, 2002; Bonet and Wood, 1997, etc.). The work rate done by the stress per reference unit mass, that is, per reference mass density is given from equation (6.94) as

$$\frac{1}{\rho_0}\overset{\bullet}{w} = \frac{1}{\rho_0}\boldsymbol{\tau}:\mathbf{d}, \tag{6.95}$$

which designates the stress power per unit mass density and thus is termed the *specific stress power*.

Various work-conjugate pairs are obtained from the Kirchhoff stress $\boldsymbol{\tau}$ and the strain rate $\mathbf{d}$ by applying the pull-back operations in equations (3.30), (3.32), (3.34) and (3.35) such that the base vectors disappear, resulting in scalar quantities, by inverse combinations of contravariant and covariant bases. Some typical examples are shown below.

$$\mathrm{tr}(\boldsymbol{\tau}\boldsymbol{l}) = \mathrm{tr}(\boldsymbol{\tau}\mathbf{d}) = \begin{cases} \mathrm{tr}\{(\boldsymbol{\tau}\mathbf{F}^{-T})(\boldsymbol{l}\mathbf{F})^T\} = \mathrm{tr}\{\overleftarrow{\boldsymbol{\tau}}^{\hat{g}G}(\overleftarrow{\boldsymbol{l}}_{\hat{g}G})^T\} = \mathrm{tr}(\overleftarrow{\boldsymbol{\tau}}^{\hat{g}G}\,\overleftarrow{\boldsymbol{l}}_{G\hat{g}}) = \mathrm{tr}(\boldsymbol{\Pi}\dot{\mathbf{F}}^T) \\ \mathrm{tr}\{(\mathbf{F}^{-1}\boldsymbol{\tau})(\boldsymbol{l}\mathbf{F})\} = \mathrm{tr}\{\overleftarrow{\boldsymbol{\tau}}^{G\hat{g}}\overleftarrow{\boldsymbol{l}}_{\hat{g}G}\} = \mathrm{tr}(\mathbf{P}\dot{\mathbf{F}}) \\ \mathrm{tr}\{(\mathbf{F}^{-1}\boldsymbol{\tau}\mathbf{F}^{-T})(\mathbf{F}^T\mathbf{d}\mathbf{F})\} = \mathrm{tr}(\overleftarrow{\boldsymbol{\tau}}^{GG}\,\overleftarrow{\mathbf{d}}_{GG}) = \mathrm{tr}(\mathbf{S}\dot{\mathbf{E}}) \\ \mathrm{tr}\{(\mathbf{F}^{-1}\boldsymbol{\tau}\mathbf{F})(\mathbf{F}^{-1}\boldsymbol{l}\mathbf{F})\} = \mathrm{tr}(\overleftarrow{\boldsymbol{\tau}}^{G}_{\bullet G}\,\overleftarrow{\mathbf{d}}^{G}_{\bullet G}) = \mathrm{tr}(\mathfrak{M}\mathbf{L}) \\ \mathrm{tr}\{(\mathbf{F}^T\boldsymbol{\tau}\mathbf{F}^{-T})(\mathbf{F}^{-1}\boldsymbol{l}\mathbf{F})^T\} = \mathrm{tr}\{\overleftarrow{\boldsymbol{\tau}}_G^{\bullet G}(\overleftarrow{\boldsymbol{l}}^{G}_{\bullet G})^T\} = \mathrm{tr}(\overleftarrow{\boldsymbol{\tau}}_G^{\bullet G}\,\overleftarrow{\boldsymbol{l}}_G^{\bullet G}) = \mathrm{tr}(\mathbf{M}\mathbf{L}^T) \\ \mathrm{tr}\{(\mathbf{F}^T\boldsymbol{\tau}\mathbf{F})(\mathbf{F}^{-1}\mathbf{d}\mathbf{F}^{-T})\} = \mathrm{tr}(\overleftarrow{\boldsymbol{\tau}}_{GG}\overleftarrow{\mathbf{d}}^{GG}) = \mathrm{tr}(\mathfrak{T}\dot{\mathbf{E}}^{(-1)}) \\ \mathrm{tr}\{(\mathbf{R}^T\boldsymbol{\tau}\mathbf{R})(\mathbf{R}^T\mathbf{d}\mathbf{R})\} = \mathrm{tr}(\overleftarrow{\boldsymbol{\tau}}^R\,\overleftarrow{\mathbf{d}}^R) = \mathrm{tr}(\widehat{\mathfrak{T}}\widehat{\mathbf{D}}) \\ \mathrm{tr}\{(\mathbf{R}^T\boldsymbol{\tau}\mathbf{F}^{-T})(\mathbf{R}^T\boldsymbol{l}\mathbf{F} + \mathbf{R}^T\boldsymbol{\Omega}^{RT}\mathbf{F})^T\} = \mathrm{tr}\{\overleftarrow{\boldsymbol{\tau}}^{RG}(\overleftarrow{\boldsymbol{l}}_{RG} + \mathbf{R}^T\boldsymbol{\Omega}^{RT}\mathbf{F})^T\} = \mathrm{tr}(\mathcal{B}\dot{\mathbf{B}}) = \mathrm{tr}(\mathfrak{B}\dot{\mathbf{B}}) \end{cases} \tag{6.96}$$

where $\mathbf{L}$ is the contravariant-covariant pull-back of $\boldsymbol{l}$ as shown in equation (4.91) and $\widehat{\mathbf{D}}$ is the rotationless strain rate described in equation (4.114). The relations $(\overleftarrow{\boldsymbol{l}}_{\hat{g}G})^T = \overleftarrow{\boldsymbol{l}}_{G\hat{g}}$ and $(\overleftarrow{\mathbf{t}}^{G}_{\bullet G})^T = \overleftarrow{\mathbf{t}}_G^{\bullet G}$ are used in equation (6.96), which are based on equations (3.30) and (3.32). The rate of the Biot strain tensor is given by

$$\begin{aligned} \dot{\mathbf{B}} = \dot{\mathbf{U}} &= (\mathbf{R}^T\mathbf{F})^{\bullet} = \mathbf{R}^T\dot{\mathbf{F}} + \dot{\mathbf{R}}^T\mathbf{F} = \mathbf{R}^T\dot{\mathbf{F}}\mathbf{F}^{-1}\mathbf{F} + \mathbf{R}^T\mathbf{R}\dot{\mathbf{R}}^T\mathbf{F} \\ &= \mathbf{R}^T\boldsymbol{l}\mathbf{F} + \mathbf{R}^T\boldsymbol{\Omega}^{RT}\mathbf{F} = \overleftarrow{\boldsymbol{l}}_{RG} + \mathbf{R}^T\boldsymbol{\Omega}^{RT}\mathbf{F}(=\dot{\mathbf{B}}^T). \end{aligned} \tag{6.97}$$

In addition, note that

$$\begin{aligned} 2\mathrm{tr}(\boldsymbol{\tau}\boldsymbol{l}) &= \mathrm{tr}(\boldsymbol{\tau}\boldsymbol{l}) + \mathrm{tr}(\boldsymbol{\tau}\boldsymbol{l}^T) + \mathrm{tr}(\boldsymbol{\tau}\boldsymbol{\Omega}^{RT}) + \mathrm{tr}(\boldsymbol{\tau}\boldsymbol{\Omega}^R) \\ &= \mathrm{tr}\{(\mathbf{F}^{-1}\boldsymbol{\tau}\mathbf{R})(\mathbf{R}^T\boldsymbol{l}\mathbf{F})\} + \mathrm{tr}\{(\mathbf{R}^T\boldsymbol{\tau}\mathbf{F}^{-T})(\mathbf{F}^T\boldsymbol{l}^T\mathbf{R})\} \\ &\quad + \mathrm{tr}\{(\mathbf{F}^{-1}\boldsymbol{\tau}\mathbf{R})(\mathbf{R}^T\boldsymbol{\Omega}^{RT}\mathbf{F})\} + \mathrm{tr}\{(\mathbf{R}^T\boldsymbol{\tau}\mathbf{F}^{-T})(\mathbf{F}^T\boldsymbol{\Omega}^R\mathbf{R})\} \\ &= \mathrm{tr}\{(\mathbf{F}^{-1}\boldsymbol{\tau}\mathbf{R})(\mathbf{R}^T\boldsymbol{l}\mathbf{F} + \mathbf{R}^T\boldsymbol{\Omega}^{RT}\mathbf{F})\} + \mathrm{tr}\{(\mathbf{R}^T\boldsymbol{\tau}\mathbf{F}^{-T})(\mathbf{R}^T\boldsymbol{l}\mathbf{F} + \mathbf{R}^T\boldsymbol{\Omega}^{RT}\mathbf{F})^T\} \\ &= \mathrm{tr}\{(\mathbf{F}^{-1}\boldsymbol{\tau}\mathbf{R})\dot{\mathbf{B}}\} + \mathrm{tr}\{(\mathbf{F}^{-1}\boldsymbol{\tau}\mathbf{R})^T\dot{\mathbf{B}}^T\} \\ &= 2\mathrm{tr}(\mathcal{B}\dot{\mathbf{B}}) = 2\mathrm{tr}(\mathfrak{B}\dot{\mathbf{B}}), \end{aligned}$$

while the work conjugacy of $\mathfrak{B}$ and $\dot{\mathbf{B}}$ can also be proven by

$$\mathbf{S}:\dot{\mathbf{E}} = \mathrm{tr}\left[\mathbf{S}\left\{\frac{1}{2}(\mathbf{U}\dot{\mathbf{U}} + \dot{\mathbf{U}}\mathbf{U})\right\}\right] = \frac{1}{2}\mathrm{tr}(\mathbf{S}\mathbf{U}\dot{\mathbf{U}} + \mathbf{U}\mathbf{S}\dot{\mathbf{U}}) = \frac{1}{2}(\mathbf{S}\mathbf{U} + \mathbf{U}\mathbf{S}):\dot{\mathbf{U}} = \mathfrak{B}:\dot{\mathbf{B}}, \tag{6.98}$$

noting equations $(4.62)_1$ and (6.54).

The work-conjugate pairs obtained above are shown collectively below:

$$\dot{w} = \begin{cases} \mathrm{tr}(\boldsymbol{\tau}\mathbf{d}) = \mathrm{tr}(\boldsymbol{\tau}\overset{\rightarrow}{\underset{\leftarrow}{\dot{\mathbf{e}}}}) = J\mathrm{tr}(\boldsymbol{\sigma}\mathbf{d}) \ \text{(Eulerian)}, \\ \mathrm{tr}(\boldsymbol{\Pi}\dot{\mathbf{F}}^T) = \mathrm{tr}(\mathbf{P}\dot{\mathbf{F}}) = \mathrm{tr}(\mathbf{S}\dot{\mathbf{E}}) = \mathrm{tr}(\mathfrak{M}\mathbf{L}) = \mathrm{tr}(\mathbf{M}\mathbf{L}^T) = \mathrm{tr}(\mathfrak{T}\dot{\mathbf{E}}^{(-1)}) \\ \quad = \mathrm{tr}(\widehat{\mathfrak{T}}\widehat{\mathbf{D}}) = \mathrm{tr}(\mathcal{B}\dot{\mathbf{B}}) = \mathrm{tr}(\mathfrak{B}\dot{\mathbf{B}}) \ \text{(Lagrangian)}, \end{cases} \tag{6.99}$$

that is,

$$\boxed{\dot{w} = \begin{cases} \boldsymbol{\tau}:\mathbf{d} = \boldsymbol{\tau}:\overset{\rightarrow}{\underset{\leftarrow}{\dot{\mathbf{e}}}} = J\boldsymbol{\sigma}:\mathbf{d} \quad \text{(Eulerian)} \\ \boldsymbol{\Pi}:\dot{\mathbf{F}} = \mathbf{P}:\dot{\mathbf{F}}^T = \mathbf{S}:\dot{\mathbf{E}} = \mathfrak{M}:\mathbf{L}^T = \mathbf{M}:\mathbf{L} = \mathfrak{T}:\dot{\mathbf{E}}^{(-1)} \\ \quad = \widehat{\mathfrak{T}}:\widehat{\mathbf{D}} = \mathcal{B}:\dot{\mathbf{B}} = \mathfrak{B}:\dot{\mathbf{B}} \ \text{(Lagrangian)} \end{cases}} \tag{6.100}$$

noting equation $(5.16)_2$.

6.8 Stress Rate Tensors

Various stress rates can be obtained from the Cauchy stress $\boldsymbol{\sigma}$ and the Kirchhoff stress $\boldsymbol{\tau}(=J\boldsymbol{\sigma})$ by the convected derivative described in Chapter 5. Here, by denoting the four types of convected derivatives collectively by $\overset{\rightarrow}{\underset{\leftarrow}{\dot{\boldsymbol{\sigma}}}}$ and $\overset{\rightarrow}{\underset{\leftarrow}{\dot{\boldsymbol{\tau}}}}$, their relations are described as

$$\overset{\rightarrow}{\underset{\leftarrow}{\dot{\boldsymbol{\tau}}}} = J(\overset{\rightarrow}{\underset{\leftarrow}{\dot{\boldsymbol{\sigma}}}} + \boldsymbol{\sigma}\mathrm{tr}\mathbf{d}), \quad \overset{\rightarrow}{\underset{\leftarrow}{\dot{\boldsymbol{\sigma}}}} = \frac{1}{J}\overset{\rightarrow}{\underset{\leftarrow}{\dot{\boldsymbol{\tau}}}} - \boldsymbol{\sigma}\mathrm{tr}\mathbf{d} \tag{6.101}$$

by virtue of $\dot{\boldsymbol{\tau}} = \dot{\boldsymbol{\sigma}}J + \boldsymbol{\sigma}\dot{J} = J(\dot{\boldsymbol{\sigma}} + \boldsymbol{\sigma}\mathrm{tr}\mathbf{d})$.

6.8.1 *Contravariant Convected Derivatives*

Based on equation $(5.5)_1$, the contravariant convected rate of the Cauchy stress $\boldsymbol{\sigma}$ is given by

$$\overset{\circ}{\boldsymbol{\sigma}}{}^{Ol} \equiv \overset{\rightarrow}{\underset{\leftarrow}{\dot{\boldsymbol{\sigma}}}}{}^{gg} = \mathbf{F}(\mathbf{F}^{-1}\boldsymbol{\sigma}\mathbf{F}^{-T})^{\bullet}\mathbf{F}^T\,(=\mathbf{F}(\dot{\overline{\mathbf{S}/J}})\mathbf{F}^T = (\overset{\rightarrow}{\dot{\overline{\mathbf{S}/J}}})^{gg}) = \dot{\boldsymbol{\sigma}} - \boldsymbol{l}\boldsymbol{\sigma} - \boldsymbol{\sigma}\boldsymbol{l}^T\,(=\overset{\circ}{\boldsymbol{\sigma}}{}^{OlT}) \tag{6.102}$$

which is termed the *Oldroyd rate of Cauchy stress*. Likewise, for the Kirchhoff stress we have

$$\overset{\circ}{\boldsymbol{\tau}}{}^{Ol} \equiv \overset{\rightarrow}{\underset{\leftarrow}{\dot{\boldsymbol{\tau}}}}{}^{gg} = \mathbf{F}(\mathbf{F}^{-1}\boldsymbol{\tau}\mathbf{F}^{-T})^{\bullet}\mathbf{F}^T\,(=\mathbf{F}\dot{\mathbf{S}}\mathbf{F}^T = \overset{\rightarrow}{\dot{\mathbf{S}}}{}^{gg}) = \dot{\boldsymbol{\tau}} - \boldsymbol{l}\boldsymbol{\tau} - \boldsymbol{\tau}\boldsymbol{l}^T\,(=\overset{\circ}{\boldsymbol{\tau}}{}^{OlT}) \tag{6.103}$$

which is termed the *Oldroyd rate of Kirchhoff stress*.

Further,

$$\begin{aligned} \overset{\circ}{\boldsymbol{\sigma}}{}^{Tr} &\equiv J^{-1}\,(\overset{\rightarrow}{\dot{\overline{J\boldsymbol{\sigma}}}})^{gg} = J^{-1}\mathbf{F}(\mathbf{F}^{-1}(J\boldsymbol{\sigma})\mathbf{F}^{-T})^{\bullet}\mathbf{F}^T\,(=J^{-1}\mathbf{F}\dot{\mathbf{S}}\mathbf{F}^T = J^{-1}\,\overset{\rightarrow}{\dot{\mathbf{S}}}{}^{gg}) \\ &= \overset{\circ}{\boldsymbol{\sigma}}{}^{Ol} + \boldsymbol{\sigma}\mathrm{tr}\mathbf{d} = \dot{\boldsymbol{\sigma}} - \boldsymbol{l}\boldsymbol{\sigma} - \boldsymbol{\sigma}\boldsymbol{l}^T + \boldsymbol{\sigma}\mathrm{tr}\mathbf{d}\ (=\overset{\circ}{\boldsymbol{\sigma}}{}^{TrT}) \end{aligned} \tag{6.104}$$

is termed the *Truesdell rate of Cauchy stress*.

6.8.2 Covariant–Contravariant Convected Derivatives

The covariant–contravariant convected rate of the Kirchhoff stress $\boldsymbol{\tau}$ is given from equation (5.5)$_3$ as

$$\overset{\circ}{\mathbf{t}} \equiv \overset{\rightarrow}{\overset{\bullet}{\overleftarrow{\boldsymbol{\tau}}}}{}_{g}^{\bullet g} = \mathbf{F}^{-T}(\mathbf{F}^{T}\boldsymbol{\tau}\mathbf{F}^{-T})^{\bullet}\mathbf{F}^{T}\,(= \mathbf{F}^{-T}\overset{\bullet}{\mathbf{M}}\mathbf{F}^{T} = \overset{\rightarrow}{\overset{\bullet}{\mathbf{M}}}{}_{g}^{\bullet g}) = \overset{\bullet}{\boldsymbol{\tau}} + \boldsymbol{l}^{T}\boldsymbol{\tau} - \boldsymbol{\tau}\boldsymbol{l}^{T}\,(\neq \overset{\circ}{\mathbf{t}}{}^{T}). \tag{6.105}$$

Based on equation (5.7), one has

$$_{\Pi}\overset{\circ}{\boldsymbol{\tau}} \equiv \overset{\rightarrow}{\overset{\bullet}{\overleftarrow{\boldsymbol{\tau}}}}{}_{\hat{g}}^{\bullet g} = (\boldsymbol{\tau}\mathbf{F}^{-T})^{\bullet}\mathbf{F}^{T}\,(= \overset{\bullet}{\boldsymbol{\Pi}}\mathbf{F}^{T} = \overset{\rightarrow}{\overset{\bullet}{\boldsymbol{\Pi}}}{}_{\hat{g}}^{\bullet g}) = \overset{\bullet}{\boldsymbol{\tau}} - \boldsymbol{\tau}\boldsymbol{l}^{T}\ \ (\neq {}_{\Pi}\overset{\circ}{\boldsymbol{\tau}}{}^{T}) \tag{6.106}$$

which is termed the *relative first Piola–Kirchhoff stress rate*. The *nominal stress rate*, given by

$$_{\Pi}\overset{\circ}{\boldsymbol{\sigma}} \equiv \frac{1}{J}\,{}_{\Pi}\overset{\circ}{\boldsymbol{\tau}} = \overset{\bullet}{\boldsymbol{\sigma}} - \boldsymbol{\sigma}\boldsymbol{l}^{T} + \boldsymbol{\sigma}\mathrm{tr}\boldsymbol{l}\ (\neq {}_{\Pi}\overset{\circ}{\boldsymbol{\sigma}}{}^{T}) \tag{6.107}$$

is used in equation (6.66) for the equilibrium equation of rate form in the current configuration.

6.8.3 Covariant Convected Derivatives

The *covariant convected rate* of Cauchy stress is given from equation (5.5)$_4$ as

$$\overset{\circ}{\boldsymbol{\sigma}}{}^{CR} \equiv \overset{\rightarrow}{\overset{\bullet}{\overleftarrow{\boldsymbol{\sigma}}}}{}_{gg} = \mathbf{F}^{-T}(\mathbf{F}^{T}\boldsymbol{\sigma}\mathbf{F})^{\bullet}\mathbf{F}^{-1}\,(= \mathbf{F}^{-T}\overset{\bullet}{\mathfrak{S}}\mathbf{F}^{-1} = \overset{\rightarrow}{\overset{\bullet}{\mathfrak{S}}}{}_{gg}) = \overset{\bullet}{\boldsymbol{\sigma}} + \boldsymbol{l}^{T}\boldsymbol{\sigma} + \boldsymbol{\sigma}\boldsymbol{l}\ (= \overset{\circ}{\boldsymbol{\sigma}}{}^{CRT}) \tag{6.108}$$

which is termed the *Cotter–Rivlin rate* of Cauchy stress. Likewise, the covariant convected rate of Kirchhoff stress is given by

$$\overset{\circ}{\boldsymbol{\tau}}{}^{CR} \equiv \overset{\rightarrow}{\overset{\bullet}{\overleftarrow{\boldsymbol{\tau}}}}{}_{gg} = \mathbf{F}^{-T}(\mathbf{F}^{T}\boldsymbol{\tau}\mathbf{F})^{\bullet}\mathbf{F}^{-1}\,(= \mathbf{F}^{-T}\overset{\bullet}{\mathfrak{T}}\mathbf{F}^{-1} = \overset{\rightarrow}{\overset{\bullet}{\mathfrak{T}}}{}_{gg}) = \overset{\bullet}{\boldsymbol{\tau}} + \boldsymbol{l}^{T}\boldsymbol{\tau} + \boldsymbol{\tau}\boldsymbol{l}\ (= \overset{\circ}{\boldsymbol{\tau}}{}^{CRT}). \tag{6.109}$$

6.8.4 Corotational Convected Derivatives

The following covariant convected derivative based on equation (5.10) is termed the *Green–Naghdi rate of Cauchy stress*:

$$\overset{\circ}{\boldsymbol{\sigma}}{}^{R} \equiv \overset{\rightarrow}{\overset{\bullet}{\overleftarrow{\boldsymbol{\sigma}}}}{}^{R} = \mathbf{R}(\mathbf{R}^{T}\boldsymbol{\sigma}\mathbf{R})^{\bullet}\mathbf{R}^{T}\ (= \mathbf{R}\overset{\bullet}{\hat{\boldsymbol{\Sigma}}}\mathbf{R}^{T} = \overset{\rightarrow}{\overset{\bullet}{\hat{\boldsymbol{\Sigma}}}}{}^{R}) = \overset{\bullet}{\boldsymbol{\sigma}} - \boldsymbol{\Omega}^{R}\boldsymbol{\sigma} + \boldsymbol{\sigma}\boldsymbol{\Omega}^{R}\ (= \overset{\circ}{\boldsymbol{\sigma}}{}^{GNT}). \tag{6.110}$$

Similarly, the *Green-Naghdi rate of Kirchhoff stress* is given by

$$\overset{\circ}{\boldsymbol{\tau}}{}^{R} \equiv \overset{\rightarrow}{\overset{\bullet}{\overleftarrow{\boldsymbol{\tau}}}}{}^{R} = \mathbf{R}(\mathbf{R}^{T}\boldsymbol{\tau}\mathbf{R})^{\bullet}\mathbf{R}^{T}\ (= \mathbf{R}\overset{\bullet}{\hat{\mathfrak{T}}}\mathbf{R}^{T} = \overset{\rightarrow}{\overset{\bullet}{\hat{\mathfrak{T}}}}{}^{R}) = \overset{\bullet}{\boldsymbol{\tau}} - \boldsymbol{\Omega}^{R}\boldsymbol{\tau} + \boldsymbol{\tau}\boldsymbol{\Omega}^{R}\ (= \overset{\circ}{\boldsymbol{\tau}}{}^{GNT}). \tag{6.111}$$

The contravariant convected rate based on equation (5.11) is given by

$$\overset{\circ}{\boldsymbol{\sigma}}{}^{W} \equiv \overset{\rightarrow}{\overset{\bullet}{\overleftarrow{\boldsymbol{\sigma}}}}{}^{W} = \dot{\boldsymbol{\sigma}} - \mathbf{w}\boldsymbol{\sigma} + \boldsymbol{\sigma}\mathbf{w} \,(= \overset{\circ}{\boldsymbol{\sigma}}{}^{JT}) \tag{6.112}$$

which is termed the *Zaremba–Jaumann rate of Cauchy stress*. Likewise, it follows for the Kirchhoff stress that

$$\overset{\circ}{\boldsymbol{\tau}}{}^{W} \equiv \overset{\rightarrow}{\overset{\bullet}{\overleftarrow{\boldsymbol{\tau}}}}{}^{W} = \dot{\boldsymbol{\tau}} - \mathbf{w}\boldsymbol{\tau} + \boldsymbol{\tau}\mathbf{w} \,(= \overset{\circ}{\boldsymbol{\tau}}{}^{JT}). \tag{6.113}$$

All the stress rate tensors described in this section are objective tensors obeying the objective transformation rule for second-order tensors in equation $(1.199)_2$, since they are derived from the convected derivatives of second-order tensors as was proved in equation (5.52).

The convected derivatives of the second-order Eulerian tensors in which both of the base vectors are based in the current configuration are explained above; they often appear in standard constitutive equations. However, we can consider the convected derivative of a two-point tensor. For instance, the convected derivative of the deformation gradient tensor $\mathbf{F}$ as an Eulerian–Lagrangian two-point tensor is given by equation (5.8) as follows:

$$\overset{\rightarrow}{\overset{\bullet}{\overleftarrow{\mathbf{F}}}}{}^{g}_{\bullet\hat{G}} = \mathbf{F}(\mathbf{F}^{-1}\mathbf{F})^{\bullet} = \dot{\mathbf{F}} - \boldsymbol{l}\mathbf{F} = \mathbf{O} \tag{6.114}$$

by virtue of equation $(4.82)_1$. On the other hand, the Zaremba–Jaumann rate of $\mathbf{F}$ is given by

$$\overset{\circ}{\mathbf{F}}{}^{W} = \overset{\rightarrow}{\overset{\bullet}{\overleftarrow{\mathbf{F}}}}{}^{W} = \dot{\mathbf{F}} - \mathbf{w}\mathbf{F} = \mathbf{d}\mathbf{F} \,(\neq \overset{\circ}{\mathbf{F}}{}^{JT}). \tag{6.115}$$

The convected derivatives of stress in this section and of strain described in Section 5.2 are shown in Table 6.2 and the stress rates are also shown in Figure 6.9.

Table 6.2 Convected derivatives of stresses (and strain)

Types of convected derivatives		Rates of tensors
Contravariant	$\overset{\rightarrow}{\overset{\bullet}{\overleftarrow{\mathbf{t}}}}{}^{gg} = \mathbf{F}(\mathbf{F}^{-1}\mathbf{t}\mathbf{F}^{-T})^{\bullet}\mathbf{F}^{T}$ $= \mathbf{F}\,\overset{\bullet}{\overleftarrow{\mathbf{t}}}{}^{GG}\mathbf{F}^{T}$	$\overset{\circ}{\boldsymbol{\sigma}}{}^{Ol}(= \overset{\rightarrow}{\overset{\bullet}{\overleftarrow{\boldsymbol{\sigma}}}}{}^{gg} = \overset{\rightarrow}{\overset{\bullet}{\overline{\mathbf{S}/J}}}{}^{gg}) = \dot{\boldsymbol{\sigma}} - \boldsymbol{l}\boldsymbol{\sigma} - \boldsymbol{\sigma}\boldsymbol{l}^{T}$ $\overset{\circ}{\boldsymbol{\tau}}{}^{Ol}(= \overset{\rightarrow}{\overset{\bullet}{\overleftarrow{\boldsymbol{\tau}}}}{}^{gg} = \overset{\rightarrow}{\dot{\mathbf{S}}}{}^{gg}) = \dot{\boldsymbol{\tau}} - \boldsymbol{l}\boldsymbol{\tau} - \boldsymbol{\tau}\boldsymbol{l}^{T}$ $\overset{\circ}{\boldsymbol{\sigma}}{}^{Tr}(= J^{-1}\,(\overset{\rightarrow}{\overset{\bullet}{\overleftarrow{J\boldsymbol{\sigma}}}})^{gg} = J^{-1}\,\overset{\rightarrow}{\dot{\mathbf{S}}}{}^{gg})$ $= \dot{\boldsymbol{\sigma}} - \boldsymbol{l}\boldsymbol{\sigma} - \boldsymbol{\sigma}\boldsymbol{l}^{T} + \boldsymbol{\sigma}\mathrm{tr}\mathbf{d}$
Covariant–contravariant	$\overset{\rightarrow}{\overset{\bullet}{\overleftarrow{\mathbf{t}}}}{}_{g}^{\bullet g} = \mathbf{F}^{-T}(\mathbf{F}^{T}\mathbf{t}\mathbf{F}^{-T})^{\bullet}\mathbf{F}^{T}$ $= \mathbf{F}^{-T}\,\overset{\bullet}{\overleftarrow{\mathbf{t}}}{}_{G}^{\bullet G}\mathbf{F}^{T}$	$\overset{\circ}{\mathfrak{T}}\,(= \overset{\rightarrow}{\overset{\bullet}{\overleftarrow{\boldsymbol{\tau}}}}{}_{g}^{\bullet g} = \overset{\rightarrow}{\dot{\mathbf{M}}}{}_{g}^{\bullet g}) = \dot{\boldsymbol{\tau}} + \boldsymbol{l}^{T}\boldsymbol{\tau} - \boldsymbol{\tau}\boldsymbol{l}^{T}$

(continued)

Table 6.2 *(Continued)*

Types of convected derivatives		Rates of tensors
(Contravariant/covariant)–contravariant	$\overset{\rightleftarrows}{\mathbf{t}}{}^{\bullet g}_{\hat{g}} = (\mathbf{t}\mathbf{F}^{-T})^{\bullet}\mathbf{F}^{T}$ $= \overset{\leftarrow}{\dot{\mathbf{t}}}{}^{\bullet G}_{\hat{g}}\mathbf{F}^{T} = \overrightarrow{\{(\mathbf{t}\mathbf{F}^{-T})^{\bullet}\}}{}^{\bullet g}_{\hat{g}}$	${}_{\Pi}\overset{\circ}{\boldsymbol{\tau}}\ (= \overset{\rightleftarrows}{\boldsymbol{\tau}}{}^{\bullet g}_{\hat{g}} = \overset{\rightleftarrows}{\boldsymbol{\Pi}}{}^{\bullet g}_{\hat{g}})$ $= \dot{\boldsymbol{\tau}} - \boldsymbol{\tau}\boldsymbol{l}^{T}\ (\neq {}_{\Pi}\overset{\circ}{\boldsymbol{\tau}}{}^{T})$ ${}_{\Pi}\overset{\circ}{\boldsymbol{\sigma}} \equiv \frac{1}{J}\,{}_{\Pi}\overset{\circ}{\boldsymbol{\tau}}$ $= \dot{\boldsymbol{\sigma}} - \boldsymbol{\sigma}\boldsymbol{l}^{T} + \boldsymbol{\sigma}\,\mathrm{tr}\boldsymbol{l}\,(\neq {}_{\Pi}\overset{\circ}{\boldsymbol{\sigma}}{}^{T})$
Covariant	$\overset{\rightleftarrows}{\mathbf{t}}_{gg} = \mathbf{F}^{-T}(\mathbf{F}^{T}\mathbf{t}\mathbf{F})^{\bullet}\mathbf{F}^{-1}$ $= \mathbf{F}^{-T}\,\overset{\leftarrow}{\dot{\mathbf{t}}}_{GG}\,\mathbf{F}^{-1}$	$\overset{\circ}{\boldsymbol{\sigma}}{}^{CR}(= \overset{\rightleftarrows}{\boldsymbol{\sigma}}_{gg} = \overset{\rightleftarrows}{\mathfrak{S}}_{gg})$ $= \dot{\boldsymbol{\sigma}} + \boldsymbol{l}^{T}\boldsymbol{\sigma} + \boldsymbol{\sigma}\boldsymbol{l}$ $\overset{\circ}{\boldsymbol{\tau}}{}^{CR}(= \overset{\rightleftarrows}{\boldsymbol{\tau}}_{gg} = \overset{\rightleftarrows}{\mathfrak{T}}_{gg}) = \dot{\boldsymbol{\tau}} + \boldsymbol{l}^{T}\boldsymbol{\tau} + \boldsymbol{\tau}\boldsymbol{l}$ $\mathbf{d} = \begin{cases} \overset{\rightleftarrows}{\mathbf{g}}_{gg} = \frac{1}{2}(\dot{\mathbf{g}} + \boldsymbol{l}^{T}\mathbf{g} + \mathbf{g}\boldsymbol{l}) \\ \overset{\rightleftarrows}{\mathbf{e}}_{gg} = \dot{\mathbf{e}} + \boldsymbol{l}^{T}\boldsymbol{e} + \boldsymbol{e}\boldsymbol{l} \end{cases}$
Green–Naghdi rate	$\overset{\rightleftarrows}{\mathbf{t}}{}^{R} = \mathbf{R}(\mathbf{R}^{T}\mathbf{t}\mathbf{R})^{\bullet}\mathbf{R}^{T}$ $= \mathbf{R}\,\overset{\leftarrow}{\dot{\mathbf{t}}}{}^{R}\mathbf{R}^{T}$	$\overset{\circ}{\boldsymbol{\sigma}}{}^{R}(= \overset{\rightleftarrows}{\boldsymbol{\sigma}}{}^{R} = \overset{\rightleftarrows}{\boldsymbol{\Sigma}}{}^{R}) = \dot{\boldsymbol{\sigma}} - \boldsymbol{\Omega}^{R}\boldsymbol{\sigma} + \boldsymbol{\sigma}\boldsymbol{\Omega}^{R}$ $\overset{\circ}{\boldsymbol{\tau}}{}^{R}(= \overset{\rightleftarrows}{\boldsymbol{\tau}}{}^{R} = \overset{\rightleftarrows}{\mathfrak{T}}{}^{R}) = \dot{\boldsymbol{\tau}} - \boldsymbol{\Omega}^{R}\boldsymbol{\tau} + \boldsymbol{\tau}\boldsymbol{\Omega}^{R}$ $\mathbf{R}\dot{\mathbf{E}}^{(m)}\mathbf{R}^{T} = \overset{\rightleftarrows}{\mathbf{e}}{}^{(m)}_{gg}$ $= \dot{\mathbf{e}}^{(m)} - \boldsymbol{\Omega}^{R}\mathbf{e}^{(m)} + \mathbf{e}^{(m)}\boldsymbol{\Omega}^{R}$
Zaremba–Jaumann rate	$\overset{\rightleftarrows}{\mathbf{t}}{}^{w}$	$\overset{\circ}{\boldsymbol{\sigma}}{}^{w}\ (= \overset{\rightleftarrows}{\boldsymbol{\sigma}}{}^{w}) = \dot{\boldsymbol{\sigma}} - \mathbf{w}\boldsymbol{\sigma} + \boldsymbol{\sigma}\mathbf{w}$ $\overset{\circ}{\boldsymbol{\tau}}{}^{w}\ (= \overset{\rightleftarrows}{\boldsymbol{\tau}}{}^{w}) = \dot{\boldsymbol{\tau}} - \mathbf{w}\boldsymbol{\tau} + \boldsymbol{\tau}\mathbf{w}$ $\dot{\mathbf{E}}^{(m)} = \overset{\rightleftarrows}{\mathbf{e}}{}^{(m)w}_{gg} = \dot{\mathbf{e}}^{(m)} - \mathbf{w}\mathbf{e}^{(m)} + \mathbf{e}^{(m)}\mathbf{w}$
Contravariant–(contravariant/covariant) of two-point tensor	$\overset{\rightleftarrows}{\mathbb{T}}{}^{g}_{\bullet\hat{G}} = \mathbf{F}(\mathbf{F}^{-1}\mathbb{T})^{\bullet} = \mathbf{F}(\overset{\leftarrow}{\mathbb{T}}{}^{G}_{\bullet\hat{G}})^{\bullet}$ $= \overrightarrow{\{(\mathbf{F}^{-1}\mathbb{T})^{\bullet}\}}{}^{g}_{\bullet\hat{G}}$	$\overset{\rightleftarrows}{\mathbf{F}}{}^{g}_{\bullet\hat{G}} = \dot{\mathbf{F}} - \boldsymbol{l}\mathbf{F} = \mathbf{O}$
Zaremba-Jaumann rate of two-point tensor	$\overset{\rightleftarrows}{\mathbb{T}}{}^{w}$	$\overset{\circ}{\mathbf{F}}{}^{w}(= \overset{\rightleftarrows}{\mathbf{F}}{}^{w}) = \dot{\mathbf{F}} - \mathbf{w}\mathbf{F} = \mathbf{d}\mathbf{F}$

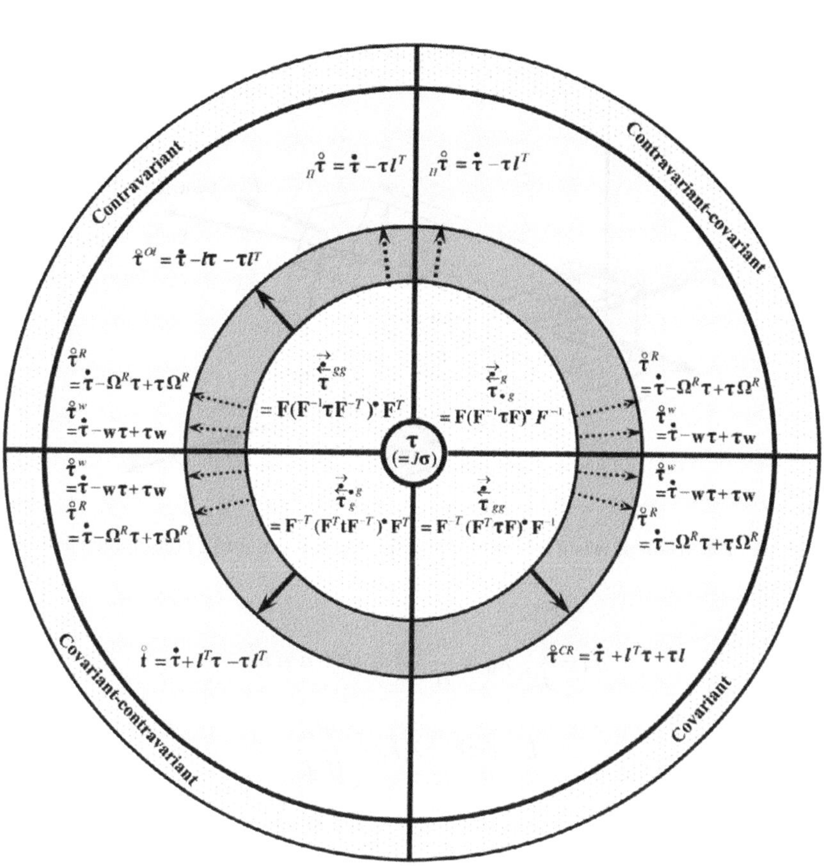

Figure 6.9 Various convective stress rates

6.9 Some Basic Loading Behavior

The various types of strain (rate) and stress (rate) described in the foregoing are shown explicitly in this section and their relations are described for various simple deformations in order to capture their physical meanings. These deformations are often observed in experiments for measurement of material properties. Homogeneous and isotropic deformation is assumed and the normalized-rectangular coordinate system is adopted for the representation of components.

6.9.1 Uniaxial Loading Followed by Rotation

Let a prism with initial length L and initial width H be first subjected to an extension, changing the length to l and the width to h, and then a rigid-body rotation under constant tensional force P as shown in Figure 6.10. This problem was taken up and analyzed by Hisada (1992).

At the current time t under extension, choosing the X-axis to the axial direction of cylinder, it follows that

$$x_{\mathrm{I}} = \lambda_{\mathrm{I}} X_{\mathrm{I}}, \ x_{\mathrm{II}} = \lambda_{\mathrm{II}} X_{\mathrm{II}}, \ x_{\mathrm{III}} = \lambda_{\mathrm{II}} X_{\mathrm{III}}, \tag{6.116}$$

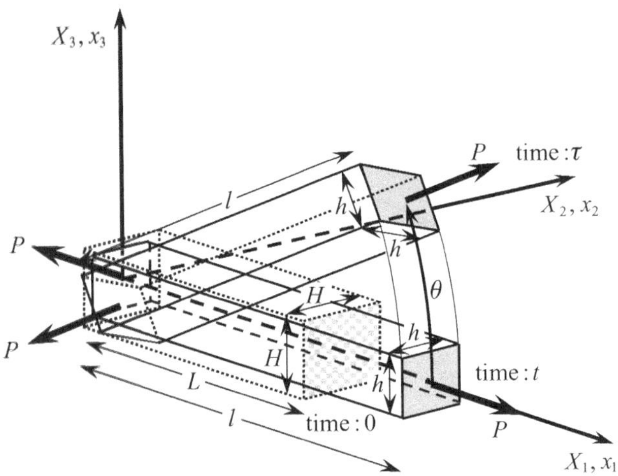

Figure 6.10 Extension of prism followed by rigid-body rotation under constant tensional force

where

$$\lambda_{\rm I} = \frac{l}{L},\ \lambda_{\rm II} = \frac{h}{H},\ A \equiv H^2,\ a \equiv h^2, \tag{6.117}$$

$$J = \det\mathbf{F} = \lambda_{\rm I}\lambda_{\rm II}^2 = \frac{h^2 l}{H^2 L} \tag{6.118}$$

for which we have

$$\mathbf{F}(t) = \mathbf{F}^T(t) = \mathbf{U}(t) = \mathbf{V}(t) = \begin{bmatrix} \lambda_{\rm I} & 0 & 0 \\ 0 & \lambda_{\rm II} & 0 \\ 0 & 0 & \lambda_{\rm II} \end{bmatrix} = \begin{bmatrix} l/L & 0 & 0 \\ 0 & h/H & 0 \\ 0 & 0 & h/H \end{bmatrix}, \tag{6.119}$$

$$\mathbf{F}^{-1}(t) = \mathbf{F}^{-T}(t) = \begin{bmatrix} \lambda_{\rm I}^{-1} & 0 & 0 \\ 0 & \lambda_{\rm II}^{-1} & 0 \\ 0 & 0 & \lambda_{\rm II}^{-1} \end{bmatrix} = \begin{bmatrix} L/l & 0 & 0 \\ 0 & H/h & 0 \\ 0 & 0 & H/h \end{bmatrix}, \tag{6.120}$$

$$\mathbf{R}(t) = \begin{bmatrix} 1 & 0 & 0 \\ 0 & 1 & 0 \\ 0 & 0 & 1 \end{bmatrix}, \tag{6.121}$$

$$\begin{aligned} \mathbf{C}(t) = \mathbf{F}^T(t)\mathbf{F}(t) &= \begin{bmatrix} l/L & 0 & 0 \\ 0 & h/H & 0 \\ 0 & 0 & h/H \end{bmatrix} \begin{bmatrix} l/L & 0 & 0 \\ 0 & h/H & 0 \\ 0 & 0 & h/H \end{bmatrix} \\ &= \begin{bmatrix} l^2/L^2 & 0 & 0 \\ 0 & h^2/H^2 & 0 \\ 0 & 0 & h^2/H^2 \end{bmatrix} = \mathbf{b}(t), \end{aligned} \tag{6.122}$$

$$\mathbf{U}(t) = \mathbf{C}^{1/2}(t) = \begin{bmatrix} l/L & 0 & 0 \\ 0 & h/H & 0 \\ 0 & 0 & h/H \end{bmatrix} = \mathbf{V}(t) = \mathbf{b}^{1/2}(t), \tag{6.123}$$

$$\mathbf{C}^{-1}(t) = \begin{bmatrix} L^2/l^2 & 0 & 0 \\ 0 & H^2/h^2 & 0 \\ 0 & 0 & H^2/h^2 \end{bmatrix} = \mathbf{b}^{-1}(t), \tag{6.124}$$

$$\mathbf{E}(t) = \frac{1}{2}(\mathbf{C}(t) - \mathbf{I}) = \frac{1}{2}\begin{bmatrix} l^2/L^2 - 1 & 0 & 0 \\ 0 & h^2/H^2 - 1 & 0 \\ 0 & 0 & h^2/H^2 - 1 \end{bmatrix}, \tag{6.125}$$

$$\mathbf{e}(t) = \frac{1}{2}(\mathbf{I} - \mathbf{b}^{-1}(t)) = \frac{1}{2}\begin{bmatrix} 1 - L^2/l^2 & 0 & 0 \\ 0 & 1 - H^2/h^2 & 0 \\ 0 & 0 & 1 - H^2/h^2 \end{bmatrix}, \tag{6.126}$$

$$\mathbf{E}^{(0)}(t) = \mathbf{e}^{(0)}(t) = \begin{bmatrix} \ln\lambda_{\mathrm{I}} & 0 & 0 \\ 0 & \ln\lambda_{\mathrm{II}} & 0 \\ 0 & 0 & \ln\lambda_{\mathrm{II}} \end{bmatrix} = \begin{bmatrix} \ln(l/L) & 0 & 0 \\ 0 & \ln(h/H) & 0 \\ 0 & 0 & \ln(h/H) \end{bmatrix}. \tag{6.127}$$

For the infinitesimal nominal strain $\hat{\mathbf{E}} = \{\partial\mathbf{u}/\partial\mathbf{X} + (\partial\mathbf{u}/\partial\mathbf{X})^T\}/2 = \{(\mathbf{F} + \mathbf{F}^T) - 2\mathbf{I}\}/2$, we have

$$\hat{\mathbf{E}}(t) = \begin{bmatrix} l/L - 1 & 0 & 0 \\ 0 & h/H - 1 & 0 \\ 0 & 0 & h/H - 1 \end{bmatrix}, \tag{6.128}$$

$$\begin{aligned} \boldsymbol{l}(t) = \dot{\mathbf{F}}(t)\mathbf{F}^{-1}(t) &= \begin{bmatrix} \dot{l}/L & 0 & 0 \\ 0 & \dot{h}/H & 0 \\ 0 & 0 & \dot{h}/H \end{bmatrix} \begin{bmatrix} L/l & 0 & 0 \\ 0 & H/h & 0 \\ 0 & 0 & H/h \end{bmatrix} = \begin{bmatrix} \dot{l}/l & 0 & 0 \\ 0 & \dot{h}/h & 0 \\ 0 & 0 & \dot{h}/h \end{bmatrix} \\ &= \mathbf{d}(t) = \dot{\mathbf{E}}^{(0)}(t) = \dot{\mathbf{e}}^{(0)}(t), \end{aligned} \tag{6.129}$$

$$\mathbf{w}(t) = \mathbf{0}, \tag{6.130}$$

$$\boldsymbol{\sigma}(t) = \frac{P}{h^2}\begin{bmatrix} 1 & 0 & 0 \\ 0 & 0 & 0 \\ 0 & 0 & 0 \end{bmatrix}, \quad \boldsymbol{\tau}(t) = J\boldsymbol{\sigma}(t) = \frac{h^2 l}{H^2 L}\frac{P}{h^2}\begin{bmatrix} 1 & 0 & 0 \\ 0 & 0 & 0 \\ 0 & 0 & 0 \end{bmatrix} = \frac{Pl}{H^2 L}\begin{bmatrix} 1 & 0 & 0 \\ 0 & 0 & 0 \\ 0 & 0 & 0 \end{bmatrix}, \tag{6.131}$$

$$\mathbf{\Pi}(t) = \mathbf{P}(t)^T = \boldsymbol{\tau}(t)\mathbf{F}^{-T}(t) = \frac{Pl}{H^2L}\begin{bmatrix} 1 & 0 & 0 \\ 0 & 1 & 0 \\ 0 & 0 & 1 \end{bmatrix}\begin{bmatrix} L/l & 0 & 0 \\ & H/h & 0 \\ \text{sym.} & & H/h \end{bmatrix} = \frac{P}{H^2}\begin{bmatrix} 1 & 0 & 0 \\ 0 & 1 & 0 \\ 0 & 0 & 1 \end{bmatrix}, \tag{6.132}$$

$$\mathbf{S}(t) = \mathbf{P}(t)\mathbf{F}(t)^{-T} = \frac{P}{H^2}\begin{bmatrix} 1 & 0 & 0 \\ 0 & 0 & 0 \\ 0 & 0 & 0 \end{bmatrix}\begin{bmatrix} L/l & 0 & 0 \\ & H/h & 0 \\ \text{sym.} & & H/h \end{bmatrix} = \frac{PL}{H^2 l}\begin{bmatrix} 1 & 0 & 0 \\ 0 & 0 & 0 \\ 0 & 0 & 0 \end{bmatrix}, \tag{6.133}$$

$$\mathfrak{M}(t) = \mathbf{S}(t)\mathbf{C}(t) = \frac{P}{H^2}\begin{bmatrix} L/l & 0 & 0 \\ 0 & 0 & 0 \\ 0 & 0 & 0 \end{bmatrix}\begin{bmatrix} l^2/L^2 & 0 & 0 \\ & h^2/H^2 & 0 \\ \text{sym.} & & h^2/H^2 \end{bmatrix} = \frac{Pl}{H^2L}\begin{bmatrix} 1 & 0 & 0 \\ 0 & 0 & 0 \\ 0 & 0 & 0 \end{bmatrix}, \tag{6.134}$$

$$\mathbf{M}(t) = \mathbf{C}(t)\mathbf{S}(t) = \mathfrak{M}(t), \tag{6.135}$$

$$\mathfrak{T}(t) = \mathbf{M}(t)\mathbf{C}(t) = \frac{P}{H^2}\begin{bmatrix} l/L & 0 & 0 \\ 0 & 0 & 0 \\ 0 & 0 & 0 \end{bmatrix}\begin{bmatrix} l^2/L^2 & 0 & 0 \\ 0 & h^2/H^2 & 0 \\ 0 & 0 & h^2/H^2 \end{bmatrix} = \frac{Pl^3}{H^2L^3}\begin{bmatrix} 1 & 0 & 0 \\ 0 & 0 & 0 \\ 0 & 0 & 0 \end{bmatrix}, \tag{6.136}$$

$$\begin{aligned} \hat{\mathfrak{T}}(t) = \mathbf{R}^T(t)\boldsymbol{\tau}(t)\mathbf{R}(t) &= \begin{bmatrix} 1 & 0 & 0 \\ 0 & 1 & 0 \\ 0 & 0 & 1 \end{bmatrix}\frac{Pl}{H^2L}\begin{bmatrix} 1 & 0 & 0 \\ 0 & 0 & 0 \\ 0 & 0 & 0 \end{bmatrix}\begin{bmatrix} 1 & 0 & 0 \\ 0 & 1 & 0 \\ 0 & 0 & 1 \end{bmatrix} \\ &= \frac{Pl}{H^2L}\begin{bmatrix} 1 & 0 & 0 \\ 0 & 0 & 0 \\ 0 & 0 & 0 \end{bmatrix} = \boldsymbol{\tau}(t), \end{aligned} \tag{6.137}$$

$$\begin{aligned} \mathfrak{B}(t) = \frac{1}{2}(\mathbf{S}(t)\mathbf{U}(t) + \mathbf{U}(t)\mathbf{S}(t)) = \frac{1}{2}\Bigg\{ &\frac{P}{H^2}\begin{bmatrix} L/l & 0 & 0 \\ 0 & 0 & 0 \\ 0 & 0 & 0 \end{bmatrix}\begin{bmatrix} l/L & 0 & 0 \\ 0 & h/H & 0 \\ 0 & 0 & h/H \end{bmatrix} \\ + \begin{bmatrix} l/L & 0 & 0 \\ 0 & h/H & 0 \\ 0 & 0 & h/H \end{bmatrix}\frac{P}{H^2}\begin{bmatrix} L/l & 0 & 0 \\ 0 & 0 & 0 \\ 0 & 0 & 0 \end{bmatrix}\Bigg\} &= \frac{P}{H^2}\begin{bmatrix} 1 & 0 & 0 \\ 0 & 1 & 0 \\ 0 & 0 & 1 \end{bmatrix} = \mathcal{B}(t). \end{aligned} \tag{6.138}$$

At time $\tau(> t)$ under a rigid-body rotation, we have

$$\mathbf{R}(\tau)=\begin{bmatrix}\cos\theta & -\sin\theta & 0\\ \sin\theta & \cos\theta & 0\\ 0 & 0 & 1\end{bmatrix},\qquad \mathbf{R}(\tau)^T=\begin{bmatrix}\cos\theta & \sin\theta & 0\\ -\sin\theta & \cos\theta & 0\\ 0 & 0 & 1\end{bmatrix},\tag{6.139}$$

$$\mathbf{F}(\tau)=\mathbf{R}(\tau)\mathbf{F}(t)=\begin{bmatrix}\cos\theta & -\sin\theta & 0\\ \sin\theta & \cos\theta & 0\\ 0 & 0 & 1\end{bmatrix}\begin{bmatrix}l/L & 0 & 0\\ & h/H & 0\\ \text{sym.} & & h/H\end{bmatrix}$$

$$=\begin{bmatrix}(l/L)\cos\theta & -(h/H)\sin\theta & 0\\ (l/L)\sin\theta & (h/H)\cos\theta & 0\\ 0 & 0 & h/H\end{bmatrix},\tag{6.140}$$

$$\mathbf{F}^{-1}(\tau)=\mathbf{F}^{-1}(t)\mathbf{R}^T(\tau)=\begin{bmatrix}L/l & 0 & 0\\ 0 & H/h & 0\\ 0 & 0 & H/h\end{bmatrix}\begin{bmatrix}\cos\theta & \sin\theta & 0\\ -\sin\theta & \cos\theta & 0\\ 0 & 0 & 1\end{bmatrix}$$

$$=\begin{bmatrix}(L/l)\cos\theta & (L/l)\sin\theta & 0\\ -(H/h)\sin\theta & (H/h)\cos\theta & 0\\ 0 & 0 & H/h\end{bmatrix},\tag{6.141}$$

$$\mathbf{C}(\tau)=(\mathbf{R}(\tau)\mathbf{F}(t))^T\mathbf{R}(\tau)\mathbf{F}(t)=\mathbf{F}^T(t)\mathbf{F}(t)=\mathbf{C}(t),\tag{6.142}$$

$$\mathbf{U}(\tau)=\mathbf{U}(t),\quad \mathbf{E}(\tau)=\mathbf{E}(t),\tag{6.143}$$

$$\begin{aligned}\mathbf{b}(\tau)&=\mathbf{F}(\tau)\mathbf{F}^T(\tau)\\ &=\begin{bmatrix}(l/L)\cos\theta & -(h/H)\sin\theta & 0\\ (l/L)\sin\theta & (h/H)\cos\theta & 0\\ 0 & 0 & h/H\end{bmatrix}\begin{bmatrix}(l/L\cos\theta) & (l/L)\sin\theta & 0\\ -(h/H\sin\theta) & (h/H)\cos\theta & 0\\ 0 & 0 & h/H\end{bmatrix}\\ &=\begin{bmatrix}(l/L)^2\cos^2\theta+(h/H)^2\sin^2\theta & \{(l/L)^2-(h/H)^2\}\sin\theta\cos\theta & 0\\ & (l/L)^2\sin^2\theta+(h/H)^2\cos^2\theta & 0\\ \text{sym.} & & (h/H)^2\end{bmatrix},\end{aligned}\tag{6.144}$$

$$\begin{aligned}\mathbf{b}^{-1}(\tau)&=\mathbf{F}^{-T}(\tau)\mathbf{F}^{-1}(\tau)\\ &=\begin{bmatrix}(l/L)\cos\theta & -(h/H)\sin\theta & 0\\ (l/L)\sin\theta & (h/H)\cos\theta & 0\\ 0 & 0 & h/H\end{bmatrix}\begin{bmatrix}(L/l)\cos\theta & (L/l)\sin\theta & 0\\ -(H/h)\sin\theta & (H/h)\cos\theta & 0\\ 0 & 0 & H/h\end{bmatrix}\\ &=\begin{bmatrix}(L/l)^2\cos^2\theta+(H/h)^2\sin^2\theta & \{(L/l)^2-(H/h)^2\}\sin\theta\cos\theta & 0\\ & (L/l)^2\sin^2\theta+(H/h)^2\cos^2\theta & 0\\ \text{sym.} & & (H/h)^2\end{bmatrix},\end{aligned}\tag{6.145}$$

$$\mathbf{e}(\tau)=\frac{1}{2}\begin{bmatrix} 1-(L/l)^2\cos^2\theta-(H/h)^2\sin^2\theta & \{(L/l)^2-(H/h)^2\}\sin\theta\cos\theta & 0 \\ & 1-(L/l)^2\sin^2\theta-(H/h)^2\cos^2\theta & 0 \\ \text{sym.} & & 1-(H/h)^2 \end{bmatrix}, \tag{6.146}$$

$$\mathbf{E}^{(0)}(t)=\mathbf{e}^{(0)}(t)=\mathbf{E}^{(0)}(\tau)=\mathbf{e}^{(0)}(\tau), \tag{6.147}$$

$$\begin{aligned}\hat{\mathbf{E}}(\tau) &= \frac{1}{2}\left\{\begin{bmatrix} (l/L)\cos\theta & -(h/H)\sin\theta & 0 \\ (l/L)\sin\theta & (h/H)\cos\theta & 0 \\ 0 & 0 & h/H \end{bmatrix}\right. \\ &\quad \left. + \begin{bmatrix} (l/L)\cos\theta & (l/L)\sin\theta & 0 \\ -(h/H)\sin\theta & (h/H)\cos\theta & 0 \\ 0 & 0 & h/H \end{bmatrix} - 2\begin{bmatrix} 1 & 0 & 0 \\ 0 & 1 & 0 \\ 1 & 0 & 1 \end{bmatrix}\right\} \\ &= \begin{bmatrix} (l/L)\cos\theta-1 & \{(l/L)-(h/H)\}\sin\theta/2 & 0 \\ & (h/H)\cos\theta-1 & 0 \\ \text{sym.} & & h/H-1 \end{bmatrix},\end{aligned} \tag{6.148}$$

$$\begin{aligned} \boldsymbol{l}(\tau)=\dot{\mathbf{F}}(\tau)\mathbf{F}^{-1}(\tau) &= \begin{bmatrix} (\dot{l}/L)\cos\theta-(l/L)\sin\theta\dot{\theta} & -(\dot{h}/H)\sin\theta-(l/L)\cos\theta\dot{\theta} & 0 \\ (\dot{l}/L)\sin\theta+(l/L)\cos\theta\dot{\theta} & (\dot{h}/H)\cos\theta-(h/H)\sin\theta\dot{\theta} & 0 \\ 0 & 0 & \dot{h}/H \end{bmatrix} \\ &\quad \begin{bmatrix} (L/l)\cos\theta & (L/l)\sin\theta & 0 \\ -(H/h)\sin\theta & (H/h)\cos\theta & 0 \\ 0 & 0 & H/h \end{bmatrix} \\ &= \begin{bmatrix} (\dot{l}/l)\cos^2\theta+(\dot{h}/h)\sin^2\theta & \{(\dot{l}/l)-(\dot{h}/h)\}\sin\theta\cos\theta & 0 \\ -\{1-(l/L)(H/h)\}\sin\theta\cos\theta\dot{\theta} & -\{\sin^2\theta+(l/L)(H/h)\cos^2\theta\}\dot{\theta} & \\ \{(\dot{l}/l)-(\dot{h}/h)\}\sin\theta\cos\theta+\dot{\theta} & (\dot{l}/l)\sin^2\theta+(\dot{h}/h)\cos^2\theta & 0 \\ 0 & 0 & \dot{h}/h \end{bmatrix}, \end{aligned} \tag{6.149}$$

$$\mathbf{d}(\tau)=\begin{bmatrix} (\dot{l}/l)\cos^2\theta+(\dot{h}/h)\sin^2\theta & \{(\dot{l}/l)-(\dot{h}/h)\}\sin\theta\cos\theta & 0 \\ -\{1-(l/L)(H/h)\}\sin\theta\cos\theta\dot{\theta} & +\{1-\sin^2\theta-(l/L)(H/h)\}\cos^2\theta\dot{\theta}/2 & \\ & (\dot{l}/l)\sin^2\theta+(\dot{h}/h)\cos^2\theta & 0 \\ \text{sym.} & 0 & \dot{h}/h \end{bmatrix}, \tag{6.150}$$

$$\mathbf{w}(\tau)=\begin{bmatrix} 0 & -\{1+\sin^2\theta+(l/L)(H/h)\cos^2\theta\}\dot{\theta}/2 & 0 \\ & 0 & 0 \\ \text{ant.} & & 0 \end{bmatrix}, \tag{6.151}$$

$$\begin{aligned}\boldsymbol{\sigma}(\tau)=\mathbf{R}(\tau)\boldsymbol{\sigma}(t)\mathbf{R}^T(\tau)&=\begin{bmatrix} \cos\theta & -\sin\theta & 0 \\ \sin\theta & \cos\theta & 0 \\ 0 & 0 & 1 \end{bmatrix}\frac{P}{h^2}\begin{bmatrix} 1 & 0 & 0 \\ & 0 & 0 \\ \text{sym.} & & 0 \end{bmatrix}\begin{bmatrix} \cos\theta & \sin\theta & 0 \\ -\sin\theta & \cos\theta & 0 \\ 0 & 0 & 1 \end{bmatrix} \\ &=\frac{P}{h^2}\begin{bmatrix} \cos^2\theta & \sin\theta\cos\theta & 0 \\ & \sin^2\theta & 0 \\ \text{sym.} & 0 & 0 \end{bmatrix}=\boldsymbol{\sigma}(t)\begin{bmatrix} \cos^2\theta & \sin\theta\cos\theta & 0 \\ & \sin^2\theta & 0 \\ \text{sym.} & 0 & 0 \end{bmatrix},\end{aligned} \tag{6.152}$$

$$\boldsymbol{\tau}(\tau)=J\boldsymbol{\sigma}(\tau)=\frac{h^2l}{H^2L}\frac{P}{h^2}\begin{bmatrix} \cos^2\theta & \sin\theta\cos\theta & 0 \\ & \sin^2\theta & 0 \\ \text{sym.} & 0 & 0 \end{bmatrix}=\frac{Pl}{H^2L}\begin{bmatrix} \cos^2\theta & \sin\theta\cos\theta & 0 \\ & \sin^2\theta & 0 \\ \text{sym.} & 0 & 0 \end{bmatrix}, \tag{6.153}$$

$$\begin{aligned}\boldsymbol{\Pi}(\tau)&=\mathbf{P}^T(\tau)=\boldsymbol{\tau}(\tau)\mathbf{F}^{-T}(\tau) \\ &=\frac{Pl}{H^2L}\begin{bmatrix} \cos^2\theta & \sin\theta\cos\theta & 0 \\ & \sin^2\theta & 0 \\ \text{sym.} & 0 & 0 \end{bmatrix}\begin{bmatrix} (L/l)\cos\theta & -(H/h)\sin\theta & 0 \\ (L/l)\sin\theta & (H/h)\cos\theta & 0 \\ 0 & 0 & H/h \end{bmatrix} \\ &=\frac{P}{H^2}\begin{bmatrix} \cos\theta & 0 & 0 \\ \sin\theta & 0 & 0 \\ 0 & 0 & 0 \end{bmatrix},\end{aligned} \tag{6.154}$$

$$\begin{aligned}\mathbf{S}(\tau)=\mathbf{F}^{-1}(\tau)\boldsymbol{\Pi}(\tau)&=\begin{bmatrix} (L/l)\cos\theta & (L/l)\sin\theta & 0 \\ -(H/h)\sin\theta & (H/h)\cos\theta & 0 \\ 0 & 0 & H/h \end{bmatrix}\frac{P}{H^2}\begin{bmatrix} \cos\theta & 0 & 0 \\ \sin\theta & 0 & 0 \\ 0 & 0 & 0 \end{bmatrix} \\ &=\frac{PL}{H^2l}\begin{bmatrix} 1 & 0 & 0 \\ 0 & 0 & 0 \\ 0 & 0 & 0 \end{bmatrix}=\mathbf{S}(t),\end{aligned} \tag{6.155}$$

$$\mathfrak{M}(\tau) = \mathbf{S}(\tau)\mathbf{C}(\tau) = \begin{bmatrix} PL/(H^2 l) & 0 & 0 \\ 0 & 0 & 0 \\ 0 & 0 & 0 \end{bmatrix} \begin{bmatrix} l^2/L^2 & 0 & 0 \\ 0 & h^2/H^2 & 0 \\ 0 & 0 & h^2/H^2 \end{bmatrix}$$

$$= \frac{Pl}{H^2 L} \begin{bmatrix} 1 & 0 & 0 \\ 0 & 0 & 0 \\ 0 & 0 & 0 \end{bmatrix} = \mathfrak{M}(t), \qquad (6.156)$$

$$\mathbf{M}(\tau) = \mathbf{C}(\tau)\mathbf{S}(\tau) = \mathfrak{M}(\tau) = \mathbf{M}(t), \qquad (6.157)$$

$$\mathfrak{T}(\tau) = \mathbf{M}(\tau)\mathbf{C}(\tau) = \frac{Pl}{H^2 L} \begin{bmatrix} 1 & 0 & 0 \\ 0 & 1 & 0 \\ 0 & 0 & 1 \end{bmatrix} \begin{bmatrix} l^2/L^2 & 0 & 0 \\ 0 & h^2/H^2 & 0 \\ 0 & 0 & h^2/H^2 \end{bmatrix}$$

$$= \frac{Pl^3}{H^2 L^3} \begin{bmatrix} 1 & 0 & 0 \\ 0 & 0 & 0 \\ 0 & 0 & 0 \end{bmatrix} = \mathfrak{T}(t), \qquad (6.158)$$

$$\widehat{\mathfrak{T}}(\tau) = \mathbf{U}(\tau)\mathbf{S}(\tau)\mathbf{U}(\tau) = \begin{bmatrix} l/L & 0 & 0 \\ 0 & h/H & 0 \\ 0 & 0 & h/H \end{bmatrix} \frac{PL}{H^2 l} \begin{bmatrix} 1 & 0 & 0 \\ 0 & 0 & 0 \\ 0 & 0 & 0 \end{bmatrix} \begin{bmatrix} l/L & 0 & 0 \\ 0 & h/H & 0 \\ 0 & 0 & h/H \end{bmatrix}$$

$$= \frac{Pl}{H^2 L} \begin{bmatrix} 1 & 0 & 0 \\ 0 & 0 & 0 \\ 0 & 0 & 0 \end{bmatrix} = \widehat{\mathfrak{T}}(t) = \boldsymbol{\tau}(t), \qquad (6.159)$$

where the corotational (rotationless) stress tensor $\widehat{\mathfrak{T}}$ is independent of the rigid-body rotation, and

$$\mathfrak{B}(\tau) = (\mathbf{S}(\tau)\mathbf{U}(\tau) + \mathbf{U}(\tau)\mathbf{S}(\tau))/2 = (\mathbf{S}(t)\mathbf{U}(t) + \mathbf{U}(t)\mathbf{S}(t))/2$$

$$= \frac{P}{H^2} \begin{bmatrix} 1 & 0 & 0 \\ 0 & 0 & 0 \\ 0 & 0 & 0 \end{bmatrix} = \mathfrak{B}(t) = \mathcal{B}(\tau) = \mathcal{B}(t). \qquad (6.160)$$

These results confirm that the Lagrangian tensors $\mathbf{S}$, $\mathfrak{M}$, $\mathbf{M}$, $\mathfrak{T}$, $\widehat{\mathfrak{T}}$, $\mathfrak{B}$ and $\mathcal{B}$ are independent of rigid-body rotation but the Eulerian tensors $\boldsymbol{\sigma}$, $\boldsymbol{\tau}$ and the two-point tensors $\mathbf{P}$, $\boldsymbol{\Pi}$ are influenced by it.

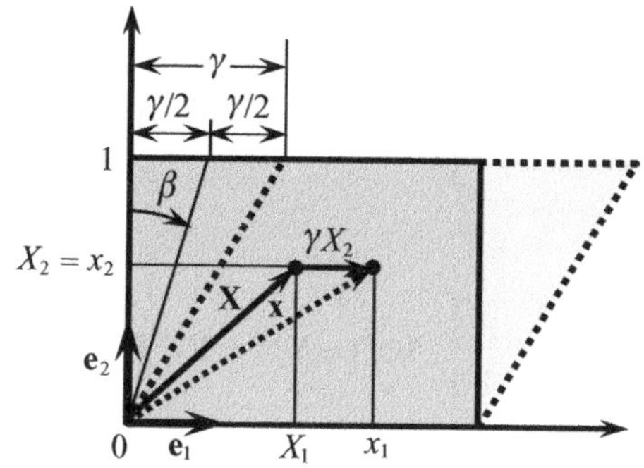

Figure 6.11 Simple shear

6.9.2 Simple Shear

As shown in Figure 6.11, consider the simple shear in which the material particle moves along the x_1-axis as follows:

$$\mathbf{x} = (X_1 + \gamma X_2)\mathbf{e}_1 + X_2\mathbf{e}_2 + X_3\mathbf{e}_3, \tag{6.161}$$

where γ is the engineering shear strain. We introduce the angle β defined by

$$\gamma = 2\tan\beta, \ \dot{\gamma} = 2\dot{\beta}\sec^2\beta. \tag{6.162}$$

It follows in this situation that

$$\mathbf{F} = \begin{bmatrix} 1 & \gamma & 0 \\ 0 & 1 & 0 \\ 0 & 0 & 1 \end{bmatrix}, \ \mathbf{F}^{-1} = \begin{bmatrix} 1 & -\gamma & 0 \\ 0 & 1 & 0 \\ 0 & 0 & 1 \end{bmatrix}, \ J = \det\mathbf{F} = 1, \tag{6.163}$$

where the inverse tensor $\mathbf{F}^{-1}$ is derived using equation $(1.158)'_1$. The components in the third column and those in the third row are zero except for unity in the third column and third row in all tensors appearing hereinafter for the simple shear deformation. For simplicity, let them be expressed by as a 2 × 2 matrix

$$\mathbf{F} = \begin{bmatrix} 1 & \gamma \\ 0 & 1 \end{bmatrix}, \ \mathbf{F}^{-1} = \begin{bmatrix} 1 & -\gamma \\ 0 & 1 \end{bmatrix}, \tag{6.164}$$

from which we obtain

$$\boldsymbol{l} = \dot{\mathbf{F}}\mathbf{F}^{-1} = \begin{bmatrix} 0 & \dot{\gamma} \\ 0 & 0 \end{bmatrix}\begin{bmatrix} 0 & -\gamma \\ 0 & 1 \end{bmatrix} = \begin{bmatrix} 0 & \dot{\gamma} \\ 0 & 0 \end{bmatrix}, \tag{6.165}$$

$$\mathbf{d} = \frac{\dot{\gamma}}{2}\begin{bmatrix} 0 & 1 \\ 1 & 0 \end{bmatrix}, \ \mathbf{w} = \frac{\dot{\gamma}}{2}\begin{bmatrix} 0 & 1 \\ -1 & 0 \end{bmatrix}. \tag{6.166}$$

Further, it follows that

$$\mathbf{C}(=\mathbf{U}^2=\mathbf{F}^T\mathbf{F})=\begin{bmatrix}1 & \gamma\\ \gamma & 1+\gamma^2\end{bmatrix},\ \mathbf{C}^{-1}(=\mathbf{U}^{-2}=\mathbf{F}^{-1}\mathbf{F}^{-T})=\begin{bmatrix}1+\gamma^2 & -\gamma\\ -\gamma & 1\end{bmatrix},\tag{6.167}$$

$$\mathbf{b}(=\mathbf{V}^2=\mathbf{F}\mathbf{F}^T)=\begin{bmatrix}1+\gamma^2 & \gamma\\ \gamma & 1\end{bmatrix},\ \mathbf{b}^{-1}(=\mathbf{V}^{-2}=\mathbf{F}^{-T}\mathbf{F}^{-1})=\begin{bmatrix}1 & -\gamma\\ -\gamma & 1+\gamma^2\end{bmatrix},\tag{6.168}$$

$$\mathbf{E}=\frac{1}{2}(\mathbf{C}-\mathbf{I})=\frac{1}{2}\begin{bmatrix}0 & \gamma\\ \gamma & \gamma^2\end{bmatrix},\ \mathbf{e}=\frac{1}{2}(\mathbf{I}-\mathbf{b}^{-1})=\frac{1}{2}\begin{bmatrix}0 & \gamma\\ \gamma & -\gamma^2\end{bmatrix},\tag{6.169}$$

where γ denotes twice the deviatoric component of the Green or Almansi strain tensor. The following more detailed analysis is based on Watanabe (1990, 1991).

The eigenvalues λ_α of $\mathbf{C}$ must fulfill the characteristic equation

$$\det[\mathbf{C}-\lambda_\alpha\mathbf{I}]=\det\begin{bmatrix}1-\lambda_\alpha & \gamma\\ \gamma & 1+\gamma^2-\lambda_\alpha\end{bmatrix}=\lambda_\alpha^2-(2+\gamma^2)\lambda_\alpha+1=0,$$

from which we obtain

$$\left.\begin{aligned}\lambda_{\mathrm{I}}&=\frac{2+\gamma^2+\gamma\sqrt{4+\gamma^2}}{2}=\left(\frac{1+\sin\beta}{\cos\beta}\right)^2\\ \lambda_{\mathrm{II}}&=\frac{2+\gamma^2-\gamma\sqrt{4+\gamma^2}}{2}=\left(\frac{1-\sin\beta}{\cos\beta}\right)^2\end{aligned}\right\}.\tag{6.170}$$

Equation (6.170) is illustrated in Figure 6.12.

We have $\mathbf{C}\mathbf{N}^{(\alpha)}=\lambda_\alpha\mathbf{N}^{(\alpha)}$, leading to

$$\left.\begin{aligned}\begin{bmatrix}1 & \gamma\\ \gamma & 1+\gamma^2\end{bmatrix}\begin{Bmatrix}N_1^{(\mathrm{I})}\\ N_2^{(\mathrm{I})}\end{Bmatrix}&=\left(\frac{\sin\beta+1}{\cos\beta}\right)^2\begin{Bmatrix}N_1^{(\mathrm{I})}\\ N_2^{(\mathrm{I})}\end{Bmatrix}\\ \begin{bmatrix}1 & \gamma\\ \gamma & 1+\gamma^2\end{bmatrix}\begin{Bmatrix}N_1^{(\mathrm{II})}\\ N_2^{(\mathrm{II})}\end{Bmatrix}&=\left(\frac{\sin\beta-1}{\cos\beta}\right)^2\begin{Bmatrix}N_1^{(\mathrm{II})}\\ N_2^{(\mathrm{II})}\end{Bmatrix}\end{aligned}\right\}\tag{6.171}$$

from which the following single independent equation holds for each of $\mathbf{N}^{(\mathrm{I})}$ and $\mathbf{N}^{(\mathrm{II})}$:

$$N_1^{(\mathrm{I})}+\gamma N_2^{(\mathrm{I})}=\left(\frac{\sin\beta+1}{\cos\beta}\right)^2N_1^{(\mathrm{I})},\ N_1^{(\mathrm{II})}+\gamma N_2^{(\mathrm{II})}=\left(\frac{\sin\beta-1}{\cos\beta}\right)^2N_1^{(\mathrm{II})},$$

that is,

$$\left.\begin{aligned}(\cos^2\beta-1-2\sin\beta-\sin^2\beta)N_1^{(\mathrm{I})}+2\sin\beta\cos\beta N_2^{(\mathrm{I})}=0\\ (\cos^2\beta-1+2\sin\beta-\sin^2\beta)N_1^{(\mathrm{II})}+2\sin\beta\cos\beta N_2^{(\mathrm{II})}=0\end{aligned}\right\}\tag{6.172}$$

Then, we have

$$\frac{N_1^{(\mathrm{I})}}{N_2^{(\mathrm{I})}} = \frac{\cos\beta}{\sin\beta + 1}, \quad \frac{N_1^{(\mathrm{II})}}{N_2^{(\mathrm{II})}} = \frac{\cos\beta}{\sin\beta - 1}, \tag{6.173}$$

Hence, the unit eigenvectors $\mathbf{N}^{(\mathrm{I})}$ and $\mathbf{N}^{(\mathrm{II})}$ of $\mathbf{C}$ are given by

$$\left.\begin{aligned} \mathbf{N}^{(\mathrm{I})} &= \frac{1}{\sqrt{2+2\sin\beta}}\{\cos\beta\mathbf{e}_1 + (\sin\beta + 1)\mathbf{e}_2\} \\ \mathbf{N}^{(\mathrm{II})} &= \frac{1}{\sqrt{2-2\sin\beta}}\{\cos\beta\mathbf{e}_1 + (\sin\beta - 1)\mathbf{e}_2\} \end{aligned}\right\} \tag{6.174}$$

from which the inclination angles of the principal directions in the plane ($\mathbf{e}_1$, $\mathbf{e}_2$) are given by

$$\theta_{\mathrm{I}} = \tan^{-1}\frac{\sin\beta + 1}{\cos\beta}, \quad \theta_{\mathrm{II}} = \tan^{-1}\frac{\sin\beta - 1}{\cos\beta}. \tag{6.175}$$

Equation (6.175) is depicted in Figure 6.12.

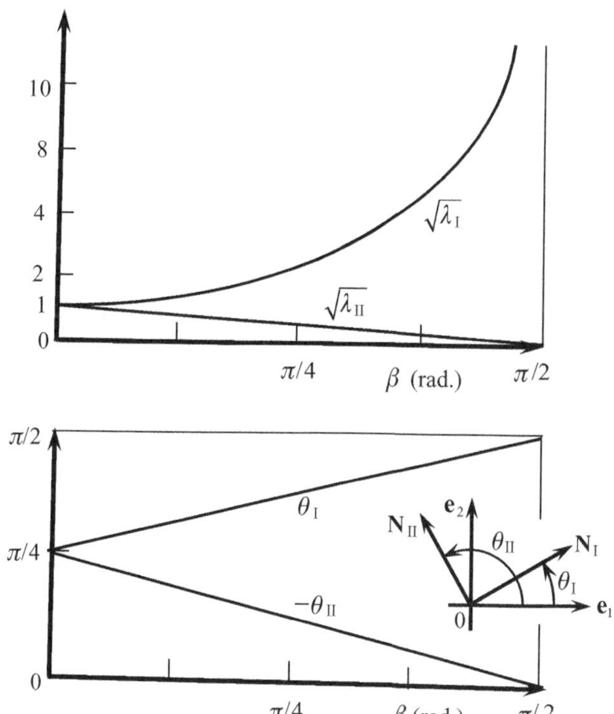

Figure 6.12 Variations of principal values and directions of right stretch tensor in simple shear (Watanabe, 1990)

$\mathbf{R}^L$ is given by equation (6.174) as follows:

$$\mathbf{R}^L = \frac{\cos\beta}{\sqrt{2+2\sin\beta}}\mathbf{e}_1\otimes\mathbf{e}_1 + \frac{\cos\beta}{\sqrt{2-2\sin\beta}}\mathbf{e}_1\otimes\mathbf{e}_2 + \frac{\sin\beta+1}{\sqrt{2+2\sin\beta}}\mathbf{e}_2\otimes\mathbf{e}_1 + \frac{\sin\beta-1}{\sqrt{2-2\sin\beta}}\mathbf{e}_2\otimes\mathbf{e}_2, \tag{6.176}$$

that is,

$$\mathbf{R}^L = \begin{bmatrix} \dfrac{\cos\beta}{\sqrt{2+2\sin\beta}} & \dfrac{\cos\beta}{\sqrt{2-2\sin\beta}} \\ \dfrac{\sin\beta+1}{\sqrt{2+2\sin\beta}} & \dfrac{\sin\beta-1}{\sqrt{2-2\sin\beta}} \end{bmatrix}, \tag{6.177}$$

noting that

$$\begin{aligned}\mathbf{R}^L &= \mathbf{N}^{(\mathrm{I})}\otimes\mathbf{e}_1 + \mathbf{N}^{(\mathrm{II})}\otimes\mathbf{e}_2 \\ &= \frac{1}{\sqrt{2+2\sin\beta}}[\{\cos\beta\mathbf{e}_1 + (\sin\beta+1)\mathbf{e}_2\}\otimes\mathbf{e}_1 + \{\cos\beta\mathbf{e}_1 + (\sin\beta-1)\mathbf{e}_2\}\otimes\mathbf{e}_2]\end{aligned} \tag{6.178}$$

from equation (4.21). Then $\mathbf{U}$ is expressed in the coordinate system $\{\mathbf{e}_i\}$ as follows:

$$\mathbf{U} = \begin{bmatrix} \cos\beta & \sin\beta \\ \sin\beta & \dfrac{1+\sin^2\beta}{\cos\beta} \end{bmatrix}, \tag{6.179}$$

noting that the inverse transformation of $\mathbf{U}$ from the coordinate system $\{\mathbf{N}^{(\alpha)}\}$ to $\{\mathbf{e}_i\}$ is given by

$$\begin{aligned}\mathbf{U} &= \mathbf{R}^{LT}(\sqrt{\lambda_{\mathrm{I}}}\mathbf{N}^{(\mathrm{I})}\otimes\mathbf{N}^{(\mathrm{I})} + \sqrt{\lambda_{\mathrm{II}}}\mathbf{N}^{(\mathrm{II})}\otimes\mathbf{N}^{(\mathrm{II})})\mathbf{R}^L = \mathbf{R}^{LT}\mathrm{diag}[\sqrt{\lambda_i}]\mathbf{R}^L \\ &= \begin{bmatrix} \dfrac{\cos\beta}{\sqrt{2+2\sin\beta}} & \dfrac{\sin\beta+1}{\sqrt{2+2\sin\beta}} \\ \dfrac{\cos\beta}{\sqrt{2-2\sin\beta}} & \dfrac{\sin\beta-1}{\sqrt{2-2\sin\beta}} \end{bmatrix} \begin{bmatrix} \dfrac{1+\sin\beta}{\cos\beta} & 0 \\ 0 & \dfrac{1-\sin\beta}{\cos\beta} \end{bmatrix} \begin{bmatrix} \dfrac{\cos\beta}{\sqrt{2+2\sin\beta}} & \dfrac{\cos\beta}{\sqrt{2-2\sin\beta}} \\ \dfrac{\sin\beta+1}{\sqrt{2+2\sin\beta}} & \dfrac{\sin\beta-1}{\sqrt{2-2\sin\beta}} \end{bmatrix} \\ &= \begin{bmatrix} \cos\beta & \sin\beta \\ \sin\beta & \dfrac{(1+\sin\beta)(1-\sin^2\beta)+(1-\sin\beta)^3}{2\cos\beta(1-\sin\beta)} \end{bmatrix} = \begin{bmatrix} \cos\beta & \sin\beta \\ \sin\beta & \dfrac{1+\sin^2\beta}{\cos\beta} \end{bmatrix},\end{aligned} \tag{6.180}$$

$$\mathbf{U}^{-1} = \begin{bmatrix} \dfrac{1+\sin^2\beta}{\cos\beta} & -\sin\beta \\ -\sin\beta & \cos\beta \end{bmatrix}. \tag{6.181}$$

The tensor $\mathbf{R}$ is given by

$$\mathbf{R}=\mathbf{F}\mathbf{U}^{-1}=\begin{bmatrix}1 & \gamma\\ 0 & 1\end{bmatrix}\begin{bmatrix}\dfrac{1+\sin^2\beta}{\cos\beta} & -\sin\beta\\ -\sin\beta & \cos\beta\end{bmatrix}$$

$$=\begin{bmatrix}\dfrac{1+\sin^2\beta}{\cos\beta}-\gamma\sin\beta & -\sin\beta+\gamma\cos\beta\\ -\sin\beta & \cos\beta\end{bmatrix}=\begin{bmatrix}\cos\beta & \sin\beta\\ -\sin\beta & \cos\beta\end{bmatrix}. \tag{6.182}$$

The relative spin is given by

$$\boldsymbol{\Omega}^R=\left(\begin{bmatrix}\cos\beta & \sin\beta\\ -\sin\beta & \cos\beta\end{bmatrix}\right)^{\bullet}\begin{bmatrix}\cos\beta & -\sin\beta\\ \sin\beta & \cos\beta\end{bmatrix}$$

$$=\dot{\beta}\begin{bmatrix}-\sin\beta & \cos\beta\\ -\cos\beta & -\sin\beta\end{bmatrix}\begin{bmatrix}\cos\beta & -\sin\beta\\ \sin\beta & \cos\beta\end{bmatrix}=\dot{\beta}\begin{bmatrix}0 & 1\\ -1 & 0\end{bmatrix}. \tag{6.183}$$

Then, we have

$$\mathbf{w}-\boldsymbol{\Omega}^R=\frac{\dot{\gamma}}{2}\begin{bmatrix}0 & 1\\ -1 & 0\end{bmatrix}-\dot{\beta}\begin{bmatrix}0 & 1\\ -1 & 0\end{bmatrix}$$

$$=\frac{\dot{\gamma}}{2}\begin{bmatrix}0 & 1\\ -1 & 0\end{bmatrix}-\dot{\beta}\begin{bmatrix}0 & 1\\ -1 & 0\end{bmatrix}=\dot{\beta}\tan^2\beta\begin{bmatrix}0 & 1\\ -1 & 0\end{bmatrix}. \tag{6.184}$$

Equations (6.183) and (6.184) are depicted in Figure 6.13.

Denoting $\tau=\sigma_{12}$, various stress tensors are described as follows:

$$\boldsymbol{\sigma}=\begin{bmatrix}\sigma_{11} & \tau\\ \tau & \sigma_{22}\end{bmatrix}, \tag{6.185}$$

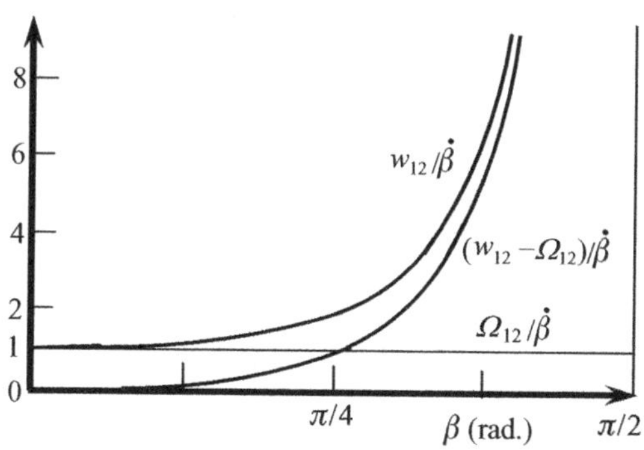

Figure 6.13 Relative and continuum spins in simple shear (Watanabe, 1990)

$$\boldsymbol{\Pi} = J\boldsymbol{\sigma}\mathbf{F}^{-T} = \begin{bmatrix} \sigma_{11} & \tau \\ \tau & \sigma_{22} \end{bmatrix} \begin{bmatrix} 1 & 0 \\ -\gamma & 1 \end{bmatrix} = \begin{bmatrix} \sigma_{11} - \gamma\tau & \tau \\ \tau - \gamma\sigma_{22} & \sigma_{22} \end{bmatrix}, \tag{6.186}$$

$$\begin{aligned} \mathbf{S} = J\mathbf{F}^{-1}\boldsymbol{\sigma}\mathbf{F}^{-T} = \mathbf{F}^{-1}\boldsymbol{\Pi} &= \begin{bmatrix} 1 & -\gamma \\ 0 & 1 \end{bmatrix} \begin{bmatrix} \sigma_{11} - \gamma\tau & \tau \\ \tau - \gamma\sigma_{22} & \sigma_{22} \end{bmatrix} \\ &= \begin{bmatrix} \sigma_{11} - \gamma^2\sigma_{22} - 2\gamma\tau & \tau - \gamma\sigma_{22} \\ \tau - \gamma\sigma_{22} & \sigma_{22} \end{bmatrix}. \end{aligned} \tag{6.187}$$

6.9.3 Combined Loading of Tension and Distortion

Consider a thin cylindrical specimen subjected to a combination of tension and distortion described by the following equation in the polar coordinate system (cf. Khan and Huang, 1995):

$$r = \alpha R, \quad \theta = \Theta + \omega Z, \quad z = \lambda Z, \tag{6.188}$$

where (R, Θ, Z) signifies the initial configuration, and α, ω and λ denote the proportionality factors depending on the deformation, while ω is described by the relative distortion angle ϕ between both ends as follows:

$$\omega \equiv \phi / L, \tag{6.189}$$

L being the length of the specimen.

Variables depending on the deformation are given as follows:

$$\begin{aligned} \mathbf{F} = \begin{bmatrix} F_{rR} & F_{r\Theta} & F_{rZ} \\ F_{\theta R} & F_{\theta\Theta} & F_{\theta Z} \\ F_{zR} & F_{z\Theta} & F_{zZ} \end{bmatrix} &= \begin{bmatrix} \dfrac{\partial r}{\partial R} & \dfrac{\partial r}{R\partial\Theta} & \dfrac{\partial r}{\partial Z} \\ \dfrac{r\partial\theta}{\partial R} & \dfrac{r\partial\theta}{R\partial\Theta} & \dfrac{r\partial\theta}{\partial Z} \\ \dfrac{\partial z}{\partial R} & \dfrac{\partial z}{R\partial\Theta} & \dfrac{\partial z}{\partial Z} \end{bmatrix} \\ &= \begin{bmatrix} \dfrac{\partial\alpha R}{\partial R} & \dfrac{\partial\alpha R}{R\partial\Theta} & \dfrac{\partial\alpha R}{\partial Z} \\ \dfrac{r\partial(\Theta + \omega Z)}{\partial R} & \dfrac{r\partial(\Theta + \omega Z)}{R\partial\Theta} & \dfrac{r\partial(\Theta + \omega Z)}{\partial Z} \\ \dfrac{\partial\lambda Z}{\partial R} & \dfrac{\partial\lambda Z}{R\partial\Theta} & \dfrac{\partial\lambda Z}{\partial Z} \end{bmatrix} = \begin{bmatrix} \alpha & 0 & 0 \\ 0 & \alpha & \alpha\omega\bar{R} \\ 0 & 0 & \lambda \end{bmatrix}, \end{aligned} \tag{6.190}$$

from which we have

$$\mathbf{F}^{-1}=\begin{bmatrix} 1/\alpha & 0 & 0 \\ 0 & 1/\alpha & -\omega\bar{R}/\lambda \\ 0 & 0 & 1/\lambda \end{bmatrix}, \tag{6.191}$$

$$\mathbf{C}=\mathbf{F}^T\mathbf{F}=\begin{bmatrix} \alpha^2 & 0 & 0 \\ 0 & \alpha^2 & \omega\alpha^2\bar{R} \\ 0 & \omega\alpha^2\bar{R} & \lambda^2+\omega^2\alpha^2\bar{R}^2 \end{bmatrix}, \tag{6.192}$$

$$\mathbf{b}=\mathbf{F}\mathbf{F}^T=\begin{bmatrix} \alpha & 0 & 0 \\ 0 & \alpha & \alpha\omega\bar{R} \\ 0 & 0 & \lambda \end{bmatrix}\begin{bmatrix} \alpha & 0 & 0 \\ 0 & \alpha & 0 \\ 0 & \alpha\omega\bar{R} & \lambda \end{bmatrix}=\begin{bmatrix} \alpha^2 & 0 & 0 \\ & \alpha^2+(\alpha\,\omega\bar{R})^2 & \lambda\alpha\omega\bar{R} \\ & \text{sym.} & \lambda^2 \end{bmatrix}, \tag{6.193}$$

$$\mathbf{b}^{-1}=\mathbf{F}^{-T}\mathbf{F}^{-1}=\begin{bmatrix} \frac{1}{\alpha^2} & 0 & 0 \\ 0 & \frac{1}{\alpha^2} & -\frac{\omega\bar{R}}{\alpha\lambda} \\ \text{sym.} & & \frac{1+\omega^2\bar{R}^2}{\lambda^2} \end{bmatrix}, \tag{6.194}$$

$$\mathbf{V}=\begin{bmatrix} \alpha & 0 & 0 \\ 0 & \alpha\cos\phi+\alpha\omega\bar{R}\sin\phi & \lambda\text{sins}\phi \\ \text{sym.} & & \lambda\cos\phi \end{bmatrix}, \tag{6.195}$$

$$\mathbf{U}=\begin{bmatrix} \alpha & 0 & 0 \\ 0 & \alpha\cos\phi & \alpha\sin\phi \\ \text{sym.} & & \alpha\omega\bar{R}\sin\phi+\lambda\cos\phi \end{bmatrix}, \tag{6.196}$$

$$\mathbf{E}=\frac{1}{2}\begin{bmatrix} \alpha-1 & 0 & 0 \\ 0 & \alpha\cos\phi-1 & \alpha\sin\phi \\ \text{sym.} & & \alpha\omega\bar{R}\sin\phi+\lambda\cos\phi-1 \end{bmatrix}, \tag{6.197}$$

$$\mathbf{e}=\frac{1}{2}\begin{bmatrix} 1-\frac{1}{\alpha^2} & 0 & 0 \\ & 1-\frac{1}{\alpha^2} & -\frac{\omega\bar{R}}{\alpha\lambda} \\ \text{sym.} & & 1-\frac{1+\omega^2\bar{R}^2}{\lambda^2} \end{bmatrix}, \tag{6.198}$$

where

$$\cos\phi = \frac{\lambda + \alpha}{\sqrt{(\alpha+\lambda)^2 + (\alpha\omega\bar{R})^2}}, \quad \sin\phi = \frac{\alpha\omega\bar{R}}{\sqrt{(\alpha+\lambda)^2 + (\alpha\omega\bar{R})^2}}, \tag{6.199}$$

$\bar{R}$ being the mean radius ($R \cong \bar{R}$) of the thin cylindrical specimen in the initial state, and

$$\boldsymbol{l} = \dot{\mathbf{F}}\mathbf{F}^{-1} = \begin{bmatrix} \dot{\alpha} & 0 & 0 \\ 0 & \dot{\alpha} & (\dot{\alpha}\omega + \dot{\omega}\alpha)\bar{R} \\ 0 & 0 & \dot{\lambda} \end{bmatrix} \begin{bmatrix} 1/\alpha & 0 & 0 \\ 0 & 1/\alpha & -\omega\bar{R}/\lambda \\ 0 & 0 & 1/\lambda \end{bmatrix} = \begin{bmatrix} \dfrac{\dot{\alpha}}{\alpha} & 0 & 0 \\ 0 & \dfrac{\dot{\alpha}}{\alpha} & \dfrac{\alpha\dot{\omega}\bar{R}}{\lambda} \\ 0 & 0 & \dfrac{\dot{\lambda}}{\lambda} \end{bmatrix}, \tag{6.200}$$

from which we have

$$\mathbf{d} = \begin{bmatrix} \dfrac{\dot{\alpha}}{\alpha} & 0 & 0 \\ & \dfrac{\dot{\alpha}}{\alpha} & \dfrac{\alpha\dot{\omega}\bar{R}}{2\lambda} \\ \text{sym.} & & \dfrac{\dot{\lambda}}{\lambda} \end{bmatrix}, \quad \mathbf{w} = \begin{bmatrix} 0 & 0 & 0 \\ & 0 & \dfrac{\alpha\dot{\omega}\bar{R}}{2\lambda} \\ \text{ant.} & & 0 \end{bmatrix}. \tag{6.201}$$

Stresses in various definitions are described by the following equations, designating the normal stress σ_{zz} and the shear stress $\sigma_{r\theta}$ applied to the traverse section of the cylinder by σ and τ, respectively:

$$\boldsymbol{\sigma} = \begin{bmatrix} 0 & 0 & 0 \\ 0 & 0 & \tau \\ 0 & \tau & \sigma \end{bmatrix}, \tag{6.202}$$

assuming compressibility. It follows from equations (6.191), (6.202) and $J = \det\mathbf{F} = \alpha^2\lambda$ that

$$\boldsymbol{\Pi} = J\boldsymbol{\sigma}\mathbf{F}^{-T} = \alpha^2\lambda \begin{bmatrix} 0 & 0 & 0 \\ 0 & 0 & \tau \\ 0 & \tau & \sigma \end{bmatrix} \begin{bmatrix} \dfrac{1}{\alpha} & 0 & 0 \\ 0 & \dfrac{1}{\alpha} & 0 \\ 0 & -\dfrac{\omega\bar{R}}{\lambda} & \dfrac{1}{\lambda} \end{bmatrix}$$

$$= \begin{bmatrix} 0 & 0 & 0 \\ 0 & -\alpha^2\omega\tau\bar{R} & \alpha^2\tau \\ 0 & \alpha\lambda\tau - \alpha^2\omega\sigma\bar{R} & \alpha^2\sigma \end{bmatrix}, \tag{6.203}$$

$$\mathbf{S} = \mathbf{F}^{-1}\mathbf{\Pi} = \begin{bmatrix} \dfrac{1}{\alpha} & 0 & 0 \\ 0 & \dfrac{1}{\alpha} & -\dfrac{\omega\bar{R}}{\lambda} \\ 0 & 0 & \dfrac{1}{\lambda} \end{bmatrix} \begin{bmatrix} 0 & 0 & 0 \\ 0 & -\alpha^2\omega\tau\bar{R} & \alpha^2\tau \\ 0 & \alpha\lambda\tau - \alpha^2\omega\sigma\bar{R} & \alpha^2\sigma \end{bmatrix}$$

$$= \begin{bmatrix} 0 & 0 & 0 \\ 0 & -2\alpha\omega\tau\bar{R} + \dfrac{\alpha^2\omega^2\sigma\bar{R}^2}{\lambda} & \alpha\tau - \dfrac{\alpha^2\omega\sigma\bar{R}}{\lambda} \\ 0 & \alpha\tau - \dfrac{\alpha^2\omega\sigma\bar{R}}{\lambda} & \dfrac{\alpha^2\sigma}{\lambda} \end{bmatrix}. \tag{6.204}$$

7

Hyperelasticity

Elastic constitutive equations can be classified into *hypoelastic*, *Cauchy elastic* and *hyperelastic* (or *Green elastic*) equations. Work is produced or dissipated during a closed loading (stress and/or strain) cycle in both the hypoelastic and Cauchy elastic equations; in the latter there is a one-to-one correspondence between stress and strain, while in the former there is not. The hyperelastic constitutive equation has to be used in the finite elasto-plastic constitutive equation, which exhibits a purely elastic deformation without the inexactness contained in the hypoelastic and the Cauchy elastic equations. In this chapter various types of hyperelastic constitutive equations based on St. Venant–Kirchhoff and neo-Hookean elastic materials are described.

7.1 Hyperelastic Constitutive Equation and Its Rate Form

The description of tensor quantities and equations in the convected coordinate system is useful to capture physical meanings of strain (rate) tensors, stress (rate) tensors and their relations as shown in the preceding chapters. Here, note that constitutive equation has to be formulated by stress (rate) and strain (rate) tensors in a work-conjugate pair. However, component descriptions of tensors in a work-conjugate pair possess different variant forms in general, as shown in Section 6.7. Therefore, component descriptions of constitutive equations in the convected coordinate system are not simple, usually containing a metric tensor component as a shifter since all tensors must be expressed in a common variant. Therefore, in this chapter constitutive equations are shown in the direct notation and the component descriptions in the rectangular coordinate system, and this should be sufficient for use in finite strain theory including the calculations in finite element analysis.

In a hyperelastic material, not only does a one-to-one correspondence between the stress and the strain exists but also the work done during the loading process from one strain to another strain is determined uniquely independent of the loading path in that process. The hyperelastic material must possess a *strain energy function* determined uniquely by a tensor describing the deformation of the material. For instance, let us adopt the deformation gradient tensor $\mathbf{F}$ as the tensor describing the deformation in the strain energy function; this is the most basic tensor describing the deformation of material. Denoting the deformation energy function per unit volume in the reference configuration by ψ, the work done in the deformation process

Introduction to Finite Strain Theory for Continuum Elasto-Plasticity, First Edition. Koichi Hashiguchi and Yuki Yamakawa.

from the reference time t_0 to the current time t must be described by the following equation, determined uniquely by the variation of the deformation gradient:

$$\int_{t_0}^{t} \overset{\bullet}{w} dt = \psi(\mathbf{F}) - \psi(\mathbf{F}_0) = \int_{t_0}^{t} \overset{\bullet}{\psi}(\mathbf{F}) dt = \int_{t_0}^{t} \left(\frac{\partial \psi(\mathbf{F})}{\partial \mathbf{F}} : \overset{\bullet}{\mathbf{F}} \right) dt. \tag{7.1}$$

Here $\mathbf{F}_0$ is the deformation gradient tensor in the reference configuration. Here, the first Piola–Kirchhoff stress tensor $\mathbf{\Pi}$ must be given by

$$\mathbf{\Pi} = \frac{\partial \psi(\mathbf{F})}{\partial \mathbf{F}} \tag{7.2}$$

by virtue of the work conjugacy in equation (6.100).

Substituting equation (7.2) into the stress tensors listed in Table 6.1, we obtain various expressions for hyperelasticity in terms of the deformation gradient as follows:

$$\left.\begin{aligned} \boldsymbol{\sigma} &= \frac{1}{J} \frac{\partial \psi(\mathbf{F})}{\partial \mathbf{F}} \mathbf{F}^T \\ \boldsymbol{\tau} &= \frac{\partial \psi(\mathbf{F})}{\partial \mathbf{F}} \mathbf{F}^T \\ \mathbf{\Pi} &= \frac{\partial \psi(\mathbf{F})}{\partial \mathbf{F}} \\ \mathbf{S} &= \mathbf{F}^{-1} \frac{\partial \psi(\mathbf{F})}{\partial \mathbf{F}} \end{aligned}\right\}. \tag{7.3}$$

Furthermore, noting that

$$\begin{aligned} \frac{\partial}{\partial F_{iA}} &= \frac{\partial}{\partial C_{PQ}} \frac{\partial C_{PQ}}{\partial F_{iA}} = \frac{\partial}{\partial C_{PQ}} \frac{\partial F_{rP} F_{rQ}}{\partial F_{iA}} = \frac{\partial}{\partial C_{PQ}} (\delta_{ri} \delta_{PA} F_{rQ} + F_{rP} \delta_{ri} \delta_{QA}) \\ &= \frac{\partial}{\partial C_{AQ}} F_{iQ} + \frac{\partial}{\partial C_{PA}} F_{iP} = 2 F_{iP} \frac{\partial}{\partial C_{PA}}, \end{aligned} \tag{7.4}$$

we have

$$\left.\begin{aligned} \frac{\partial \psi}{\partial \mathbf{F}} &= 2\mathbf{F} \frac{\partial \psi}{\partial \mathbf{C}} = \mathbf{F} \frac{\partial \psi}{\partial \mathbf{E}} \\ \frac{\partial \psi}{\partial \mathbf{C}} &= \frac{1}{2} \mathbf{F}^{-1} \frac{\partial \psi}{\partial \mathbf{F}} = \frac{1}{2} \frac{\partial \psi}{\partial \mathbf{E}} \\ \frac{\partial \psi}{\partial \mathbf{E}} &= \mathbf{F}^{-1} \frac{\partial \psi}{\partial \mathbf{F}} = 2 \frac{\partial \psi}{\partial \mathbf{C}} \end{aligned}\right\}. \tag{7.5}$$

Substituting equation (7.5) into equation (7.3), hyperelasticity is expressed in terms of the right Cauchy–Green deformation tensor **C** and the Green strain tensor **E** as follows:

$$\boxed{\begin{aligned}\boldsymbol{\sigma} &= 2\frac{1}{J}\mathbf{F}\frac{\partial\psi(\mathbf{C})}{\partial\mathbf{C}}\mathbf{F}^T = \frac{1}{J}\mathbf{F}\frac{\partial\psi(\mathbf{E})}{\partial\mathbf{E}}\mathbf{F}^T\\ \boldsymbol{\tau} &= 2\mathbf{F}\frac{\partial\psi(\mathbf{C})}{\partial\mathbf{C}}\mathbf{F}^T = \mathbf{F}\frac{\partial\psi(\mathbf{E})}{\partial\mathbf{E}}\mathbf{F}^T\\ \boldsymbol{\Pi} &= 2\mathbf{F}\frac{\partial\psi(\mathbf{C})}{\partial\mathbf{C}} = \mathbf{F}\frac{\partial\psi(\mathbf{E})}{\partial\mathbf{E}}\\ \mathbf{S} &= 2\frac{\partial\psi(\mathbf{C})}{\partial\mathbf{C}} = \frac{\partial\psi(\mathbf{E})}{\partial\mathbf{E}}\end{aligned}} \tag{7.6}$$

The general isotopic elastic constitutive equation is given, based on equation (1.340) with equation $(7.6)_4$, by

$$\mathbf{S} = \phi_0^C\mathbf{I} + \phi_1^C\mathbf{C} + \phi_2^C\mathbf{C}^2 \tag{7.7}$$

or

$$\mathbf{S} = \phi_0^E\mathbf{I} + \phi_1^E\mathbf{E} + \phi_2^E\mathbf{E}^2, \tag{7.8}$$

noting the symmetry properties of **C** and **E**, where $\phi_0^C, \phi_1^C, \phi_2^C$ and $\phi_0^E, \phi_1^E, \phi_2^E$ are the functions of invariants of **C** and **E**, respectively. Equations (7.7) and (7.8) reduce to the following equations for isotropic linear elastic material:

$$\mathbf{S} = \lambda^C(\mathrm{tr}\mathbf{C})\mathbf{I} + 2\mu^C\mathbf{C}, \tag{7.9}$$

$$\mathbf{S} = \lambda^E(\mathrm{tr}\mathbf{E})\mathbf{I} + 2\mu^E\mathbf{E}, \tag{7.10}$$

noting equation (1.349), where λ^C, μ^C and λ^E, μ^E are the material constants.

The strain energy function $\psi(\mathbf{C})$ for isotropic elastic materials has to be a function of the invariants of **C** and thus we obtain

$$\mathbf{S} = 2\frac{\partial\psi(\mathbf{C})}{\partial\mathbf{C}} = 2\left(\frac{\partial\psi(\mathbf{C})}{\partial\bar{\mathrm{I}}_C}\frac{\partial\bar{\mathrm{I}}_C}{\partial\mathbf{C}} + \frac{\partial\psi(\mathbf{C})}{\partial\overline{\mathrm{II}}_C}\frac{\partial\overline{\mathrm{II}}_C}{\partial\mathbf{C}} + \frac{\partial\psi(\mathbf{C})}{\partial\overline{\mathrm{III}}_C}\frac{\partial\overline{\mathrm{III}}_C}{\partial\mathbf{C}}\right), \tag{7.11}$$

where

$$\bar{\mathrm{I}}_C \equiv \mathrm{tr}\mathbf{C},\ \ \overline{\mathrm{II}}_C \equiv \mathrm{tr}\mathbf{C}^2,\ \ \overline{\mathrm{III}}_C \equiv \det\mathbf{C} = J^2. \tag{7.12}$$

Here, it follows that

$$\frac{\partial\bar{\mathrm{I}}_C}{\partial\mathbf{C}} = \mathbf{I},\ \ \frac{\partial\overline{\mathrm{II}}_C}{\partial\mathbf{C}} = 2\mathbf{C},\ \ \frac{\partial\overline{\mathrm{III}}_C}{\partial\mathbf{C}} = (\det\mathbf{C})\mathbf{C}^{-1} = \overline{\mathrm{III}}_C\mathbf{C}^{-1} = J^2\mathbf{C}^{-1}, \tag{7.13}$$

noting equations (1.359) and (1.371). Substitution of equation (7.13) into equation (7.11) yields

$$\mathbf{S} = 2\left(\frac{\partial\psi(\mathbf{C})}{\partial\bar{\mathrm{I}}_C}\mathbf{I} + 2\frac{\partial\psi(\mathbf{C})}{\partial\overline{\mathrm{II}}_C}\mathbf{C} + \overline{\mathrm{III}}_C\frac{\partial\psi(\mathbf{C})}{\partial\overline{\mathrm{III}}_C}\mathbf{C}^{-1}\right). \tag{7.14}$$

Taking account of

$$\mathbf{F}\mathbf{C}^n\mathbf{F}^T = \mathbf{F}\underbrace{(\mathbf{F}^T\mathbf{F})(\mathbf{F}^T\mathbf{F})\cdots(\mathbf{F}^T\mathbf{F})}_{n}\mathbf{F}^T = \underbrace{(\mathbf{F}\mathbf{F}^T)(\mathbf{F}\mathbf{F}^T)\cdots(\mathbf{F}\mathbf{F}^T)}_{n+1} = \mathbf{b}^{n+1} \tag{7.15}$$

in equation (7.14), we have the following spatial description of the hyperelastic constitutive equation:

$$\boldsymbol{\tau}(=\mathbf{F}\mathbf{S}\mathbf{F}^T) = 2\left(\frac{\partial\psi(\mathbf{b})}{\partial\overline{\mathrm{II}}_b}\mathbf{b} + 2\frac{\partial\psi(\mathbf{b})}{\partial\overline{\mathrm{II}}_b}\mathbf{b}^2 + \overline{\mathrm{III}}_b\frac{\partial\psi(\mathbf{b})}{\partial\overline{\mathrm{III}}_b}\mathbf{I}\right) \tag{7.16}$$

or

$$\boldsymbol{\sigma} = 2\left(J^{-1}\frac{\partial\psi(\mathbf{b})}{\partial\overline{\mathrm{II}}_b}\mathbf{b} + 2J^{-1}\frac{\partial\psi(\mathbf{b})}{\partial\overline{\mathrm{II}}_b}\mathbf{b}^2 + J\frac{\partial\psi(\mathbf{b})}{\partial\overline{\mathrm{III}}_b}\mathbf{I}\right), \tag{7.17}$$

where

$$\overline{\mathrm{I}}_b \equiv \mathrm{tr}\mathbf{b}(=\overline{\mathrm{I}}_C),\quad \overline{\mathrm{II}}_b \equiv \mathrm{tr}\mathbf{b}^2(=\overline{\mathrm{II}}_C),\quad \overline{\mathrm{III}}_b \equiv \det\mathbf{b} = J^2(=\overline{\mathrm{III}}_C). \tag{7.18}$$

Now the material-time derivative of equation $(7.6)_4$ is given by

$$\boxed{\overset{\bullet}{\mathbf{S}} = \frac{1}{2}\mathbb{C}^E : \overset{\bullet}{\mathbf{C}}} = \mathbb{C}^E : \overset{\bullet}{\mathbf{E}}, \tag{7.19}$$

where

$$\boxed{\mathbb{C}^E \equiv 2\frac{\partial\mathbf{S}}{\partial\mathbf{C}} = 4\frac{\partial^2\psi(\mathbf{C})}{\partial\mathbf{C}\otimes\partial\mathbf{C}}} = \frac{\partial^2\psi(\mathbf{E})}{\partial\mathbf{E}\otimes\partial\mathbf{E}}. \tag{7.20}$$

Substituting equations $(4.126)_1$ and (6.103) into equation (7.19) yields

$$\overset{\circ}{\tau}{}^{Ol} = \mathbf{F}\mathbb{C}^E : \mathbf{F}^T\mathbf{d}\mathbf{F}\mathbf{F}^T = \mathbb{c}^e : \mathbf{d} = \mathbb{c}^e : \boldsymbol{l} \tag{7.21}$$

$$\left(\begin{aligned}\overset{\circ}{\tau}{}^{Ol}_{ij} &= F_{iI}(\mathbb{C}^E_{IJKL}\,\overset{\bullet}{E}_{KL})F_{jJ}\\ &= F_{iI}(\mathbb{C}^E_{IJKL}F_{kK}d_{kl}F_{lL})F_{jJ} = F_{iI}F_{jJ}F_{kK}F_{lL}\mathbb{C}^E_{IJKL}d_{kl}\end{aligned}\right),$$

where

$$\mathbb{c}^e_{ijkl} = F_{iI}F_{jJ}F_{kK}F_{lL}\mathbb{C}^E_{IJKL}. \tag{7.22}$$

Here, the relation between $\overset{\bullet}{\mathbf{C}}$ and $\overset{\bullet}{J}$ is given from equation (7.2) by

$$\overset{\bullet}{J} = \frac{1}{2}J\mathbf{C}^{-1} : \overset{\bullet}{\mathbf{C}} \tag{7.23}$$

because of

$$\overset{\bullet}{J} = J\mathrm{tr}\mathbf{d} = J\mathrm{tr}(\mathbf{F}^{-T}\overset{\bullet}{\mathbf{E}}\mathbf{F}^{-1}) = J\mathrm{tr}(\mathbf{C}^{-1}\overset{\bullet}{\mathbf{E}}) = J\mathbf{C}^{-1} : \overset{\bullet}{\mathbf{E}}, \tag{7.24}$$

noting equations (1.475), (4.3) and $(4.126)_1$.

Further, the rate of the first Piola–Kirchhoff stress tensor $\mathbf{\Pi}$ is given from equation (7.2) by

$$\overset{\bullet}{\mathbf{\Pi}} = \mathbb{C}^F : \overset{\bullet}{\mathbf{F}}, \tag{7.25}$$

where

$$\mathbb{C}^F \equiv \frac{\partial \mathbf{\Pi}}{\partial \mathbf{F}} = \frac{\partial^2 \psi(\mathbf{F})}{\partial \mathbf{F} \otimes \partial \mathbf{F}}. \tag{7.26}$$

Substituting equation (7.25) into equation (6.106), we have

$$\begin{aligned}{}_{\Pi}\overset{\circ}{\boldsymbol{\tau}} &\equiv \overset{\bullet}{\mathbf{\Pi}}\mathbf{F}^T = \mathbb{C}^F : \overset{\bullet}{\mathbf{F}}\mathbf{F}^T = \mathbb{C}^F : \overset{\bullet}{\mathbf{F}}\mathbf{F}^{-1}\mathbf{F}\mathbf{F}^T = \mathbb{C}^F : \boldsymbol{l}\mathbf{F}\mathbf{F}^T = \mathbb{C}^F : (\mathbf{I}\boldsymbol{l}\mathbf{F})\mathbf{F}^T \\ &= \mathbf{I}\{\mathbb{C}^F : (\mathbf{I} \overline{\otimes} \mathbf{F}^T\boldsymbol{l})\}\mathbf{F}^T = \mathbf{I} \overline{\otimes} \mathbf{F} : \{\mathbb{C}^F : (\mathbf{I} \overline{\otimes} \mathbf{F}^T\boldsymbol{l})\} = (\mathbf{I} \overline{\otimes} \mathbf{F}) : \mathbb{C}^F : (\mathbf{I} \overline{\otimes} \mathbf{F}^T) : \boldsymbol{l},\end{aligned}$$

leading to

$$_{\Pi}\overset{\circ}{\boldsymbol{\tau}} = {}_{\Pi}\mathbb{c} : \boldsymbol{l}, \tag{7.27}$$

where

$$\begin{aligned}{}_{\Pi}\mathbb{c} &\equiv (\mathbf{I} \overline{\otimes} \mathbf{F}) : \mathbb{C}^F : (\mathbf{I} \overline{\otimes} \mathbf{F}^T) \\ (\{(\mathbf{I} \overline{\otimes} \mathbf{F}) &: \mathbb{C}^F : (\mathbf{I} \overline{\otimes} \mathbf{F}^T)\}_{ijkl} = (\mathbf{I} \overline{\otimes} \mathbf{F})_{ijpQ}\mathbb{C}^F_{pQrS}(\mathbf{I} \overline{\otimes} \mathbf{F}^T)_{rSkl}).\end{aligned} \tag{7.28}$$

In the above, the notation $\overline{\otimes}$ in equation $(1.208)_2$ is exploited.

Noting equation (7.21), the material-time derivative of the Kirchhoff stress is given from equation (6.103) as

$$\begin{aligned}\overset{\bullet}{\boldsymbol{\tau}} &= \overset{\circ}{\boldsymbol{\tau}}^{Ol} + \boldsymbol{l}\boldsymbol{\tau} + \boldsymbol{\tau}\boldsymbol{l}^T = \mathbb{c}^e : \boldsymbol{l} + \boldsymbol{l}\boldsymbol{\tau} + \boldsymbol{\tau}\boldsymbol{l}^T = \mathbb{c}^e : \boldsymbol{l} + \mathbf{I} \overline{\otimes} \boldsymbol{\tau} : \boldsymbol{l} + \boldsymbol{\tau} \underline{\otimes} \mathbf{I} : \boldsymbol{l}, \\ &= (\mathbb{c}^e + \mathbf{I} \overline{\otimes} \boldsymbol{\tau} + \boldsymbol{\tau} \underline{\otimes} \mathbf{I}) : \boldsymbol{l},\end{aligned}$$

leading to

$$\overset{\bullet}{\boldsymbol{\tau}} = {}_{\bullet}\mathbb{c} : \boldsymbol{l}, \tag{7.29}$$

where

$$_{\bullet}\mathbb{c} \equiv \mathbb{c}^e + \mathbf{I} \overline{\otimes} \boldsymbol{\tau} + \boldsymbol{\tau} \underline{\otimes} \mathbf{I}. \tag{7.30}$$

making use of the notations in equation $(1.208)_{2,3}$.

It follows from equations (6.106) and (7.21) that

$$_{\Pi}\overset{\circ}{\boldsymbol{\tau}} = \overset{\bullet}{\boldsymbol{\tau}} - \boldsymbol{\tau}\boldsymbol{l}^T = \overset{\circ}{\boldsymbol{\tau}}^{Ol} + \boldsymbol{l}\boldsymbol{\tau} = \mathbb{c}^e : \boldsymbol{l} + \mathbf{I} \overline{\otimes} \boldsymbol{\tau} : \boldsymbol{l} = (\mathbb{c}^e + \mathbf{I} \overline{\otimes} \boldsymbol{\tau}) : \boldsymbol{l}, \tag{7.31}$$

and thus $_{\Pi}\mathbb{c}$ is represented also by

$$_{\Pi}\mathbb{c} \equiv \mathbb{c}^e + \mathbf{I} \overline{\otimes} \boldsymbol{\tau}. \tag{7.32}$$

The convected derivative of the Kirchhoff stress in equation (6.105) is given as

$$\begin{aligned}\overset{\circ}{\mathbf{t}} &= \overset{\bullet}{\boldsymbol{\tau}} + \boldsymbol{l}^T\boldsymbol{\tau} - \boldsymbol{\tau}\boldsymbol{l}^T = \overset{\circ}{\boldsymbol{\tau}}^{Ol} + \boldsymbol{l}\boldsymbol{\tau} + \boldsymbol{\tau}\boldsymbol{l}^T + \boldsymbol{l}^T\boldsymbol{\tau} - \boldsymbol{\tau}\boldsymbol{l}^T = \overset{\circ}{\boldsymbol{\tau}}^{Ol} + \boldsymbol{l}\boldsymbol{\tau} + \boldsymbol{l}^T\boldsymbol{\tau} \\ &= \mathbb{c}^e : \boldsymbol{l} + \boldsymbol{l}\boldsymbol{\tau} + \boldsymbol{l}^T\boldsymbol{\tau} = \mathbb{c}^e : \boldsymbol{l} + \mathbf{I} \overline{\otimes} \boldsymbol{\tau} : \boldsymbol{l} + \mathbf{I} \underline{\otimes} \boldsymbol{\tau} : \boldsymbol{l} = (\mathbb{c}^e + \mathbf{I} \overline{\otimes} \boldsymbol{\tau} + \mathbf{I} \underline{\otimes} \boldsymbol{\tau}) : \boldsymbol{l},\end{aligned} \tag{7.33}$$

Table 7.1 Relations between various Eulerian elastic tangential modulus tensors

Elastic tangential modulus tensors	$\mathbb{c}^e$	$\cdot\mathbb{c}$	$_\Pi\mathbb{c}$	$_c\mathbb{c}$
$\mathbb{c}^e\ (\overset{\circ}{\boldsymbol{\tau}}{}^{Ol} = \mathbb{c}^e : \boldsymbol{l})$ $(\overset{\circ}{\boldsymbol{\tau}}{}^{Ol} = \dot{\boldsymbol{\tau}} - \boldsymbol{l}\boldsymbol{\tau} - \boldsymbol{\tau}\boldsymbol{l}^T)$		$\cdot\mathbb{c} - \mathbf{I}\,\overline{\otimes}\,\boldsymbol{\tau} - \boldsymbol{\tau}\,\underline{\otimes}\,\mathbf{I}$	$_\Pi\mathbb{c} - \mathbf{I}\,\overline{\otimes}\,\boldsymbol{\tau}$	$_c\mathbb{c} - \mathbf{I}\,\overline{\otimes}\,\boldsymbol{\tau} - \mathbf{I}\,\underline{\otimes}\,\boldsymbol{\tau}$
$\cdot\mathbb{c}\ (\dot{\boldsymbol{\tau}} = \cdot\mathbb{c} : \boldsymbol{l})$	$\mathbb{c}^e + \mathbf{I}\,\overline{\otimes}\,\boldsymbol{\tau} + \boldsymbol{\tau}\,\underline{\otimes}\,\mathbf{I}$		$_\Pi\mathbb{c} + \boldsymbol{\tau}\,\underline{\otimes}\,\mathbf{I}$	$_c\mathbb{c} - \mathbf{I}\,\underline{\otimes}\,\boldsymbol{\tau} + \boldsymbol{\tau}\,\underline{\otimes}\,\mathbf{I}$
$_\Pi\mathbb{c}\ ({}_\Pi\overset{\circ}{\boldsymbol{\tau}} = {}_\Pi\mathbb{c} : \boldsymbol{l})$ $({}_\Pi\overset{\circ}{\boldsymbol{\tau}} = \dot{\boldsymbol{\tau}} - \boldsymbol{\tau}\boldsymbol{l}^T)$	$\mathbb{c}^e + \mathbf{I}\,\overline{\otimes}\,\boldsymbol{\tau}$	$\cdot\mathbb{c} - \boldsymbol{\tau}\,\underline{\otimes}\,\mathbf{I}$		$_c\mathbb{c} - \mathbf{I}\,\underline{\otimes}\,\boldsymbol{\tau}$
$_c\mathbb{c}\ (\overset{\circ}{\mathbf{t}} = {}_c\mathbb{c} : \boldsymbol{l})$ $(\overset{\circ}{\mathbf{t}} = \dot{\boldsymbol{\tau}} + \boldsymbol{l}^T\boldsymbol{\tau} - \boldsymbol{\tau}\boldsymbol{l}^T)$	$\mathbb{c}^e + \mathbf{I}\,\overline{\otimes}\,\boldsymbol{\tau} + \mathbf{I}\,\underline{\otimes}\,\boldsymbol{\tau}$	$\cdot\mathbb{c} - \boldsymbol{\tau}\,\underline{\otimes}\,\mathbf{I} + \mathbf{I}\,\underline{\otimes}\,\boldsymbol{\tau}$	$_\Pi\mathbb{c} + \mathbf{I}\,\underline{\otimes}\,\boldsymbol{\tau}$	

Note: $\mathbb{c}^e_{ijkl} = F_{iI}F_{jJ}F_{kK}F_{lL}\mathbb{C}^E_{IJKL},\ \mathbb{C}^E_{IJKL} = 4\dfrac{\partial^2\psi}{\partial C_{IJ}\partial C_{KL}}.$

leading to

$$\overset{\circ}{\mathbf{t}} = {}_c\mathbb{c} : \boldsymbol{l}, \tag{7.34}$$

where

$$_c\mathbb{c} \equiv \mathbb{c}^e + \mathbf{I}\,\overline{\otimes}\,\boldsymbol{\tau} + \mathbf{I}\,\underline{\otimes}\,\boldsymbol{\tau} \tag{7.35}$$

in the notation of equation $(1.208)_{2,3}$.

The relations between various Eulerian elastic tangential modulus tensors are listed in Table 7.1.

7.2 Examples of Hyperelastic Constitutive Equations

Six types of hyperelastic constitutive equations are shown in this section, which are used for metals. The partial derivatives described in Section 1.9.1 are exploited, and the equation $J = \sqrt{\det\mathbf{C}} = \det\mathbf{U} = \det\mathbf{F}$ is often used.

7.2.1 St. Venant–Kirchhoff Elasticity

The strain energy of St. Venant–Kirchhoff elastic material is given by

$$\psi(\mathbf{E}) = \frac{1}{2}\lambda(\mathrm{tr}\mathbf{E})^2 + \mu\mathrm{tr}\mathbf{E}^2,\quad \psi(\mathbf{C}) = \frac{1}{2}\lambda[\mathrm{tr}\{(\mathbf{C}-\mathbf{I})/2\}]^2 + \mu\mathrm{tr}\{(\mathbf{C}-\mathbf{I})/2\}^2, \tag{7.36}$$

where the material constants λ and μ correspond to the Lamé constants, respectively, in the infinitesimal linear elasticity theory.

Substitution of equation (7.36) into equation (7.6) yields

$$\mathbf{S} = \lambda(\mathrm{tr}\mathbf{E})\mathbf{I} + 2\mu\mathbf{E} = \lambda[\mathrm{tr}\{(\mathbf{C}-\mathbf{I})/2\}]\mathbf{I} + \mu(\mathbf{C}-\mathbf{I}), \tag{7.37}$$

which can also be expressed, substituting equations (4.31) and (6.44), as

$$S^{IJ}\mathbf{G}_I \otimes \mathbf{G}_J = \{\lambda\delta^{IJ}\delta^{PQ} + \mu(\delta^{IP}\delta^{JQ} + \delta^{IQ}\delta^{JP})\mathbf{G}_I \otimes \mathbf{G}_J \otimes \mathbf{G}_P \otimes \mathbf{G}_Q)\} : E_{KL}\mathbf{G}^K \otimes \mathbf{G}^L$$
$$= \{\lambda\delta^{IJ}\delta^{KL} + \mu(\delta^{IK}\delta^{JL} + \delta^{IL}\delta^{JK})\}E_{KL}\mathbf{G}_I \otimes \mathbf{G}_J.$$

Thus we can write

$$\mathbf{S} = \mathbb{C}_{SV} : \mathbf{E}, \;\; S^{IJ} = \mathbb{C}_{SV}^{IJKL} E_{KL}, \tag{7.38}$$

where

$$\left.\begin{aligned} \mathbb{C}_{SV} &\equiv \lambda\mathcal{I} + 2\mu\mathcal{S}, \\ \mathbb{C}_{SV}^{IJKL} &\equiv \lambda\delta^{IJ}\delta^{KL} + \mu(\delta^{IK}\delta^{JL} + \delta^{IL}\delta^{JK}) \end{aligned}\right\}. \tag{7.39}$$

Equation (7.37) or (7.38) is nothing more than an extention of the isotropic linear elasticity in equations (1.349)–(1.351) in the infinitesimal theory to the finite strain, possessing a similar form to *Hooke's law*.

From equation (7.37) we have the *Eshelby-like stress*

$$\mathbf{M} \equiv \mathbf{CS} = \lambda[\mathrm{tr}\{(\mathbf{C} - \mathbf{I})/2\}]\mathbf{C} + \mu(\mathbf{C}^2 - \mathbf{C}). \tag{7.40}$$

and the Kirchhoff stress

$$\boldsymbol{\tau} = \frac{1}{2}\lambda(\mathrm{tr}\mathbf{b} - 3)\mathbf{b} + \mu\mathbf{b}(\mathbf{b} - \mathbf{I}), \tag{7.41}$$

noting that

$$\boldsymbol{\tau} = \mathbf{FSF}^T = \mathbf{F}\{\lambda(\mathrm{tr}\mathbf{E})\mathbf{I} + 2\mu\mathbf{E}\}\mathbf{F}^T$$
$$= \lambda(\mathrm{tr}\mathbf{E})\mathbf{FF}^T + 2\mu\mathbf{FEF}^T = \lambda\left\{\mathrm{tr}\frac{1}{2}(\mathbf{F}^T\mathbf{F} - \mathbf{I})\right\}\mathbf{FF}^T + 2\mu\mathbf{F}\left\{\frac{1}{2}(\mathbf{F}^T\mathbf{F} - \mathbf{I})\right\}\mathbf{F}^T.$$

The Lagrangian tangent modulus tensor in equation (7.19) is given from equation (7.38) by

$$\mathbb{C}^E = \mathbb{C}_{SV}, \tag{7.42}$$

since $\mathbb{C}_{SV}$ is constant in this model.

7.2.2 Modified St. Venant–Kirchhoff Elasticity

The following strain energy function is used by Wallin and Ristinmaa (2005), modifying equation (7.36) in the St. Venant–Kirchhoff elastic material:

$$\psi(\mathbf{C}) = \frac{1}{2}K(\ln J)^2 + 2GJ_2 \tag{7.43}$$

with

$$\ln J = \ln\sqrt{\det\mathbf{C}}), \;\; J_2 \equiv \frac{1}{2}\mathrm{tr}(\ln\mathbf{C}^{1/2})'^2, \tag{7.44}$$

where the material constants K and G correspond to the bulk and shear moduli, respectively, in the infinitesimal elasticity theory.

Differentiation of equation (7.43) with respect to $\mathbf{C}$ leads to

$$\mathbf{S} = K(\ln J)\mathbf{C}^{-1} + 2G(\ln\mathbf{C}^{1/2})'\mathbf{C}^{-1} \tag{7.45}$$

from which the Eshelby-like stress $\mathbf{M}$ is given as

$$\mathbf{M} = K(\ln J)\mathbf{I} + 2G(\ln \mathbf{C}^{1/2})', \tag{7.46}$$

making use of the following differential equations.

$$\left.\begin{aligned} \frac{\partial J}{\partial \mathbf{C}} &= \frac{\partial \sqrt{\det\mathbf{C}}}{\partial \mathbf{C}} = \frac{1}{2} J\mathbf{C}^{-1} \\ \frac{\partial \ln J}{\partial \mathbf{C}} &= \frac{\partial \ln \sqrt{\det\mathbf{C}}}{\partial \mathbf{C}} = \frac{1}{2}\mathbf{C}^{-1} \\ \frac{\partial J_2}{\partial \mathbf{C}} &= \frac{1}{2}(\ln \mathbf{C}^{1/2})'\mathbf{C}^{-1} \end{aligned}\right\}, \tag{7.47}$$

where equation $(7.47)_{1,2}$ are based on equations (1.372) and (1.373). The derivation of equation $(7.47)_3$ is given in Appendix C.

7.2.3 *Neo-Hookean Elasticity*

Elastic material possessing the following strain energy is called *neo-Hookean material*:

$$\psi(\mathbf{C}) = \frac{1}{2}\lambda(\ln J)^2 - \mu \ln J + \frac{1}{2}\mu(\mathrm{tr}\mathbf{C} - 3). \tag{7.48}$$

Here λ and μ are material constants, from which we have

$$\mathbf{S} = 2\left\{\frac{1}{2}\lambda 2(\ln J)\frac{1}{2}\mathbf{C}^{-1} - \mu\frac{1}{2}\mathbf{C}^{-1} + \frac{1}{2}\mu\mathbf{I}\right\},$$

leading to

$$\mathbf{S} = \lambda(\ln J)\mathbf{C}^{-1} + \mu(\mathbf{I} - \mathbf{C}^{-1}), \tag{7.49}$$

noting equation $(7.47)_2$. Then the Eshelby-like stress $\mathbf{M}$ is given by

$$\mathbf{M} = \lambda(\ln J)\mathbf{I} + \mu(\mathbf{C} - \mathbf{I}). \tag{7.50}$$

Equation (7.49) is described by the Kirchhoff stress as

$$\begin{aligned} \boldsymbol{\tau} = \mathbf{F}\mathbf{S}\mathbf{F}^T &= \mathbf{F}\{\lambda(\ln J)\mathbf{C}^{-1} + \mu(\mathbf{I} - \mathbf{C}^{-1})\}\mathbf{F}^T \\ &= \mathbf{F}\{\lambda(\ln J)\mathbf{F}^{-1}\mathbf{F}^{-T} + \mu(\mathbf{I} - \mathbf{F}^{-1}\mathbf{F}^{-T})\}\mathbf{F}^T, \end{aligned}$$

leading to

$$\boldsymbol{\tau} = \lambda(\ln J)\mathbf{I} + \mu(\mathbf{b} - \mathbf{I}). \tag{7.51}$$

The Lagrangian tangent modulus tensor in equation (7.19) is given for equation (7.49), noting equations (1.208), $(1.367)'$ and $(7.47)_{1,2}$, as follows:

$$\left.\begin{aligned} \mathbb{C}^E &= \lambda\mathbf{C}^{-1} \otimes \mathbf{C}^{-1} + \{\mu - \lambda(\ln J)\}(\mathbf{C}^{-1} \overline{\otimes} \mathbf{C}^{-1} + \mathbf{C}^{-1} \underline{\otimes} \mathbf{C}^{-1}) \\ \mathbb{C}^E_{IJKL} &= \lambda C^{-1}_{IJ} C^{-1}_{KL} + \{\mu - \lambda(\ln J)\}(C^{-1}_{IK} C^{-1}_{JL} + C^{-1}_{IL} C^{-1}_{JK}) \end{aligned}\right\} \tag{7.52}$$

because of

$$
\begin{aligned}
\mathbb{C}^E_{IJKL} &\equiv 4\frac{\partial^2\psi(\mathbf{C})}{\partial C_{IJ}\partial C_{KL}} = 2\frac{\partial S_{IJ}}{\partial C_{KL}} = 2\frac{\partial(\lambda(\ln J)C^{-1}_{IJ} + \mu(\delta_{IJ} - C^{-1}_{IJ}))}{\partial C_{KL}} \\
&= 2\left[\lambda\frac{1}{2}C^{-1}_{IJ}C^{-1}_{KL} + \lambda(\ln J)\left\{-\frac{1}{2}(C^{-1}_{IK}C^{-1}_{JL} + C^{-1}_{IL}C^{-1}_{JK})\right\}\right. \\
&\left.-\mu\left\{-\frac{1}{2}(C^{-1}_{IK}C^{-1}_{JL} + C^{-1}_{IL}C^{-1}_{JK})\right\}\right] \\
&= \lambda C^{-1}_{IJ}C^{-1}_{KL} + \{\mu - \lambda(\ln J)\}(C^{-1}_{IK}C^{-1}_{JL} + C^{-1}_{IL}C^{-1}_{JK}).
\end{aligned}
$$

Further, the Eulerian elastic tangent modulus is given by substituting equation (7.52) into equation (7.22) as follows:

$$
\mathbb{c}^e = \lambda\mathcal{I} + \{\mu - \lambda(\ln J)\}\mathcal{S}, \quad \mathbb{c}^e_{ijkl} = \lambda\delta_{ij}\delta_{kl} + \{\mu - \lambda(\ln J)\}(\delta_{ik}\delta_{jl} + \delta_{il}\delta_{jk}) \tag{7.53}
$$

because of

$$
\begin{aligned}
\mathbb{c}^e_{ijkl} &= F_{iI}F_{jJ}F_{kK}F_{lL}[\lambda C^{-1}_{IJ}C^{-1}_{KL} + \{\mu - \lambda(\ln J)\}(C^{-1}_{IK}C^{-1}_{JL} + C^{-1}_{IL}C^{-1}_{JK})] \\
&= F_{iI}F_{jJ}F_{kK}F_{lL}[\lambda F^{-1}_{Ir}F^{-1}_{Jr}F^{-1}_{Ks}F^{-1}_{Ls} + \{\mu - \lambda(\ln J)\}(F^{-1}_{Ir}F^{-1}_{Kr}F^{-1}_{Js}F^{-1}_{Ls} \\
&\quad + F^{-1}_{Ir}F^{-1}_{Lr}F^{-1}_{Js}F^{-1}_{Ks})] \\
&= \lambda\delta_{ir}\delta_{jr}\delta_{ks}\delta_{ls} + \{\mu - \lambda(\ln J)\}(\delta_{ir}\delta_{js}\delta_{kr}\delta_{ls} + \delta_{ir}\delta_{js}\delta_{ks}\delta_{lr}) \\
&= \lambda\delta_{ij}\delta_{kl} + \{\mu - \lambda(\ln J)\}(\delta_{ik}\delta_{jl} + \delta_{il}\delta_{jk}).
\end{aligned}
$$

7.2.4 *Modified Neo-Hookean Elasticity (1)*

By modifying equation (7.48) in the Neo-Hookean elasticity, let the following the strain energy function be adopted (Simo and Pister, 1984; Ciarlet, 1988):

$$
\psi(\mathbf{C}) = \frac{1}{4}\lambda(J^2 - 1) - \left(\frac{1}{2}\lambda + \mu\right)\ln J + \frac{1}{2}\mu(\mathrm{tr}\mathbf{C} - 3) \tag{7.54}
$$

from which, noting equations (6.44) and (6.49) with equation $(7.47)_{1,2}$, we have

$$
\mathbf{S} = \frac{1}{2}\lambda(J^2 - 1)\mathbf{C}^{-1} + \mu(\mathbf{I} - \mathbf{C}^{-1}), \tag{7.55}
$$

$$
\mathbf{M} = \frac{1}{2}\lambda(J^2 - 1)\mathbf{I} + \mu(\mathbf{C} - \mathbf{I}), \tag{7.56}
$$

$$
\boldsymbol{\tau} = \frac{1}{2}\lambda(J^2 - 1)\mathbf{I} + \mu(\mathbf{b} - \mathbf{I}). \tag{7.57}
$$

7.2.5 *Modified Neo-Hookean Elasticity (2)*

Vladimirov et al. (2008, 2010) used the following strain energy function for their finite elatoplastic constitutive relation:

$$\psi(\mathbf{C}) = \frac{\Lambda}{4}(J^2 - 1 - 2\ln J) + \frac{\mu}{2}(\mathrm{tr}\mathbf{C} - 3 - 2\ln J), \tag{7.58}$$

where μ and Λ are material constants, from which we have

$$\mathbf{S} = \frac{\Lambda}{2}(J^2 - 1)\mathbf{C}^{-1} + \mu(\mathbf{I} - \mathbf{C}^{-1}), \tag{7.59}$$

$$\mathbf{M} = \frac{\Lambda}{2}(J^2 - 1)\mathbf{I} + \mu(\mathbf{C} - \mathbf{I}), \tag{7.60}$$

$$\boldsymbol{\tau} = \frac{\Lambda}{2}(J^2 - 1)\mathbf{I} + \mu(\mathbf{b} - \mathbf{I}), \tag{7.61}$$

noting equation $(7.47)_{1,2}$. Equation (7.58) will be adopted for the formulation of the finite elasto-plastic constitutive equation in the next chapter.

7.2.6 *Modified Neo-Hookean Elasticity (3)*

The following strain energy function has been applied to the finite elasto-plasticity theory by Brunig (1998) and Helm (2001) and to the finite viscoplasticity theory by Shutov and Kreissing (2008a, b, 2010) and Shutov et al. (2011):

$$\psi(\mathbf{C}) = \frac{K}{2}(\ln J)^2 + \frac{\mu}{2}(\mathrm{tr}\underline{\mathbf{C}} - 3), \tag{7.62}$$

where K and μ are material constants. Here, $\underline{\mathbf{C}}$ is the unimodular tensor defined in equation (4.162).

It follows from equation (7.62), noting equations (4.165) and $(7.47)_{1,2}$, that

$$\mathbf{S} = k(\ln J)\mathbf{C}^{-1} + \mu J^{-2/3}\left\{\mathbf{I} - \frac{1}{3}(\mathrm{tr}\mathbf{C})\mathbf{C}^{-1}\right\}, \tag{7.63}$$

$$\mathbf{M} = k(\ln J)\mathbf{I} + \mu J^{-2/3}\mathbf{C}', \tag{7.64}$$

$$\boldsymbol{\tau} = k(\ln J)\mathbf{I} + \mu J^{-2/3}\mathbf{b}'. \tag{7.65}$$

7.2.7 *Modified Neo-Hookean Elasticity (4)*

By modifying equation (7.48) in the neo-Hookean elasticity, the following the strain energy function is assumed by Simo and Pister (1984), Ciarlet (1988), and others, separated into the two terms related to volume change J and the isochoric deformation $\underline{\mathbf{C}}$:

$$\psi(\mathbf{C}) = \frac{K}{2}\left(\frac{J^2 - 1}{2} - \ln J\right) + \frac{\mu}{2}(\mathrm{tr}\underline{\mathbf{C}} - 3), \tag{7.66}$$

where K and μ are the material constants, from which we have

$$\mathbf{S} = \frac{1}{2}K(J^2 - 1)\mathbf{C}^{-1} + \mu J^{-2/3}\left\{\mathbf{I} - \frac{1}{3}(\mathrm{tr}\mathbf{C})\mathbf{C}^{-1}\right\}, \tag{7.67}$$

$$\mathbf{M} = \frac{1}{2}K(J^2 - 1)\mathbf{I} + \mu J^{-2/3}\mathbf{C}' = \frac{1}{2}K(J^2 - 1)\mathbf{I} + \mu\underline{\mathbf{C}}', \tag{7.68}$$

$$\boldsymbol{\tau} = \frac{1}{2}K(J^2 - 1)\mathbf{I} + \mu\underline{\mathbf{b}}'. \tag{7.69}$$

8

Finite Elasto-Plastic Constitutive Equation

The *hypoelastic-based plastic constitutive equation*, used widely in elasto-plastic deformation analyses, is concerned with finite deformation and rotation under the limitation of infinitesimal elastic deformation by introducing appropriate corotational rates for stress and internal variables describing anisotropy. This chapter addresses the finite elasto-plastic constitutive equation concerned with finite deformation and finite rotation in both elastic and plastic parts. It is formulated as a *hyperelastic-based plastic constitutive equation*. The basic structure of the hyperelastic-based plastic constitutive equation was constructed after 2000, building on over a half century of pioneering studies, such as Kroner (1959), Green and Naghdi (1965), Lee and Liu (1967), Lee (1969), Kratochvil (1971), Mandel (1972a, b), and Hill (1978); other valuable work includes Dafalias (1985), Lubliner (1986), Simo (1992, 1998), Simo and Ortiz (1995), Miehe (1994), Steinmann et al. (1997), Lion (2000), Wallin et al. (2003), Wallin and Ristinmaa (2005), Dettmer and Reese (2004), Tsakmakis (2004), Menzel et al. (2005), Gurtin and Anand (2005), and Vladimirov et al. (2008, 2010). Historically, the finite elasto-plastic model in the current configuration described by the spectral representation has been in fashion and applied widely to the finite elasto-plastic deformation analysis (see Simo, 1992, 1998; Bonet and Wood, 2008; de Souza Neto et al., 2008). However, it requires the spectral decomposition of the stress and deformation tensors, postulating the elastic and plastic isotropy. It has recently been replaced by the finite elasto-plastic constitutive equation formulated in the intermediate configuration, which is free from the spectral decomposition. Besides, it should be noted that almost all finite elasto-plastic constitutive equations have been formulated based on the maximum dissipation principle and/or the Clausius–Duhem inequality. It would be impertinent at present to formulate plastic constitutive equation under the restriction by them as will be described in 8.14.

In this chapter, the finite elasto-plastic constitutive equation based in the intermediate configuration will be formulated based on past equations (Dettmer and Reese, 2004; Vladimirov et al., 2008, 2010) but in the general form which is not bound by the maximum dissipation theorem and the Clausius–Duhem inequality and, moreover, in the extended form by the unconventional elasto-plasticity based on the subloading surface model (Hashiguchi, 1980, 1989, 2009; Hashiguchi et al., 2012) capable of describing the plastic strain rate caused by the

Introduction to Finite Strain Theory for Continuum Elasto-Plasticity, First Edition. Koichi Hashiguchi and Yuki Yamakawa.

rate of stress inside the yield surface (Hashiguchi and Yamakawa, 2011, 2012), fulfilling the smoothness condition described in Section 1.11.2.

8.1 Basic Structures of Finite Elasto-Plasticity

The defects of infinitesimal elasto-plasticity and the hypoelastic-based plasticity are described in Introduction. There the basic structures of the hyperelastic-based plasticity without their defects are also described, which are outlined as follows:

1. The *multiplicative decomposition* as the decomposition of the deformation gradient tensor into elastic and plastic parts is adopted. This incorporates the *intermediate configuration* which is attained by unloading the material from the current stress to the stress-free state along the hyperelastic deformation.
2. The constitutive relation is formulated by tensor variables in the intermediate configuration.
3. The elastic deformation is described by the elastic part of deformation gradient tensor and is related to the stress by the hyperelastic constitutive equation.
4. The plastic part of deformation gradient tensor is further decomposed into an energy-storage part and an energy-dissipative part. Then, the kinematic hardening variable is also formulated in relation to the elastic strain-like variable induced by the energy-storage part as a hyperelastic-like equation possessing a potential energy function of the variable.
5. Stress and back stress tensors are calculated directly from the elastic deformation and the elastic strain-like deformation variable of the kinematic hardening.
6. The plastic flow rule is given by the relationship between the plastic velocity gradient tensor and the *Mandel stress tensor* (Mandel, 1972b) in the intermediate configuration, where these tensors satisfy the *plastic work conjugacy*.

An explicit formulation founded on these basic structures will be described in this chapter.

8.2 Multiplicative Decomposition

Elastic deformation is induced by the elastic deformation of material particles themselves and the plastic deformation by the mutual slips between them. Elasto-plasticity is premised on the decomposition of deformation into elastic and plastic parts. In particular, the decomposition of deformation itself is required to provide the hyperelasticity in an exact formulation of elasto-plasticity, although the strain rate is decomposed in the hypoelastic-based plasticity. The decomposition of deformation itself can be materialized by the *multiplicative decomposition* of the deformation gradient tensor into the elastic and the plastic deformation gradient tensors as will be described in the following.

The elastic deformation can be captured as the deformation which is removed in the unloading process from the current to the stress-free state, obeying the hyperelastic constitutive equation, which possesses the strain energy function so that a one-to-one correspondence between the stress and the elastic strain exists and energy is not produced or dissipated in the loading cycle as described in Chapter 7 for the hyperelastic constitutive equation. Thus the plastic deformation can be found by removing the elastic deformation from the total deformation. The configuration attained by removing the hyperelastic deformation from the current configuration is referred to as the *intermediate configuration*. Based on this notion, the following *multiplicative decomposition* of the deformation gradient tensor $\mathbf{F}$ into an elastic

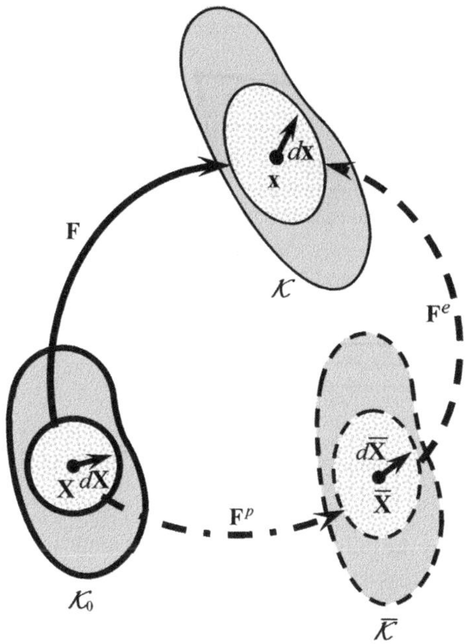

Figure 8.1 Multiplicative decomposition of deformation gradient

part $\mathbf{F}^e$ and a plastic part $\mathbf{F}^p$ was developed by Kroner (1959), Fox (1968), Lee and Liu (1967), Lee (1969), Mandel (1972a), and others as shown in Figure 8.1, where the initial, intermediate and current configurations are denoted by $\mathcal{K}_0$, $\bar{\mathcal{K}}$ and $\mathcal{K}$, respectively:

$$\boxed{\mathbf{F} = \mathbf{F}^e\mathbf{F}^p} \tag{8.1}$$

assuming

$$J = \det\mathbf{F} > 0,\ J^e = \det\mathbf{F}^e > 0,\ J^p = \det\mathbf{F}^p > 0, \tag{8.2}$$

so that not only $\mathbf{F}$ but also $\mathbf{F}^e$ and $\mathbf{F}^p$ are invertible. Here, denoting by $\bar{\mathbf{X}}$ the position vector of a material particle in the intermediate configuration, the mapping of infinitesimal material-line elements $d\mathbf{X}$ in the reference configuration to $d\bar{\mathbf{X}}$ in the intermediate configuration is defined as the plastic deformation gradient $\mathbf{F}^p$, and the mapping of $d\bar{\mathbf{X}}$ to $d\mathbf{x}$ in the current configuration is defined as the elastic deformation gradient $\mathbf{F}^e$, that is

$$d\mathbf{x} = \mathbf{F}^e d\bar{\mathbf{X}},\ d\bar{\mathbf{X}} = \mathbf{F}^p d\mathbf{X}. \tag{8.3}$$

Here, the following observations should be made. First, mutual slips between material particles are generated and thus plastic deformation is induced even in the unloading process because a purely elastic domain does not exist in real materials, as illustrated for the one-dimensional deformation in Figure 8.2. In general, plastic deformation develops gradually as the stress approaches the yield surface, which is described pertinently by the subloading surface model (Hashiguchi, 1980, 1989) in which the yield surface interior is not assumed to be a purely-elastic domain. Second, materials possess *statically indeterminate substructure* at a microscopic level so that they undergo heterogeneous plastic deformation. Most heterogeneous

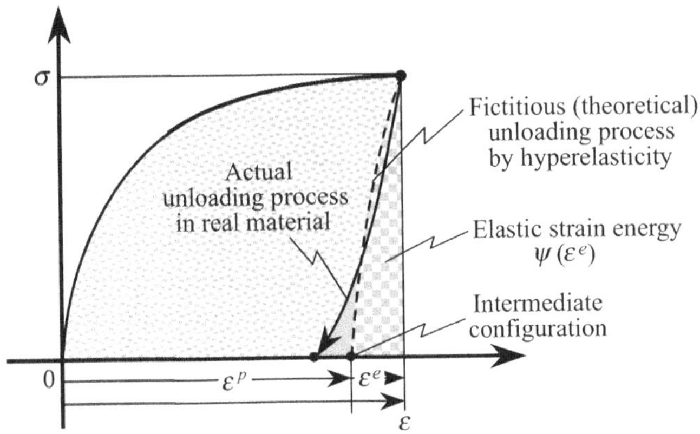

Figure 8.2 Intermediate configuration in one-dimensional deformation

deformation is not removed by the unloading. One must cut material into pieces in order to bring each material point to a real stress-free state. Therefore, there does not exist a continuous field of the position vector $\bar{\mathbf{X}}$ in the real stress-free state. Recall that the intermediate configuration is not real but merely fictitious. Besides, the tensor fields $\mathbf{F}^e$ and $\mathbf{F}^p$ do not satisfy the compatibility condition. Thus, we cannot write $\mathbf{F}^e = \partial\mathbf{x}/\partial\bar{\mathbf{X}}$ and $\mathbf{F}^p = \partial\bar{\mathbf{X}}/\partial\mathbf{X}$, although $\mathbf{F}^e$ and $\mathbf{F}^p$ are called deformation gradients. Here, on the other hand, the initial and the current position vectors of material points, that is $\mathbf{X}$ and $\mathbf{x}$, are actual material particle position vectors.

Chapter 3 discussed the Lagrangian tensor defined on the reference base $\{\mathbf{G}_I\}$ and the Eulerian tensor defined on the current base $\{\mathbf{g}_i\}$. Further, an intermediate tensor is defined on the intermediate base $\{\bar{\mathbf{G}}_{\bar{I}}\}$, while the reference base $\{\mathbf{G}_I\}$ changes to $\{\bar{\mathbf{G}}_{\bar{I}}\}$. In principle, the Lagrangian and the intermediate tensors and their indices are denoted by uppercase letters, and an overbar is added to the intermediate tensors and their indices. On the other hand, the Eulerian tensors and their indices are denoted by lowercase letters. Exploiting these base vectors, the deformation gradients in the multiplicative decomposition are described as

$$\left.\begin{array}{l}
\mathbf{F} = F^{i}_{\bullet I}\mathbf{g}_i \otimes \mathbf{G}^I = \delta^i_I\mathbf{g}_i \otimes \mathbf{G}^I = F^{ei}_{\bullet\bar{R}}F^{p\bar{R}}_{\bullet I}\mathbf{g}_i \otimes \mathbf{G}^I, \quad \mathbf{F}^{-1} = \delta^I_i\mathbf{G}_I \otimes \mathbf{g}^i \\
\mathbf{F}^e = F^{ei}_{\bullet\bar{I}}\mathbf{g}_i \otimes \bar{\mathbf{G}}^{\bar{I}} = \delta^i_{\bar{I}}\mathbf{g}_i \otimes \bar{\mathbf{G}}^{\bar{I}}, \quad \mathbf{F}^{e-1} = \delta^{\bar{I}}_i\bar{\mathbf{G}}_{\bar{I}} \otimes \mathbf{g}^i \\
\mathbf{F}^p = F^{p\bar{I}}_{\bullet I}\bar{\mathbf{G}}_{\bar{I}} \otimes \mathbf{G}^I = \delta^{\bar{I}}_I\bar{\mathbf{G}}_{\bar{I}} \otimes \mathbf{G}^I, \quad \mathbf{F}^{p-1} = \delta^I_{\bar{I}}\mathbf{G}_I \otimes \bar{\mathbf{G}}^{\bar{I}}
\end{array}\right\}, \tag{8.4}$$

noting equation (3.10). The bases $\{\bar{\mathbf{G}}_{\bar{I}}\}$ and $\{\bar{\mathbf{G}}^{\bar{I}}\}$ are obtained by the push-forward operation of the reference bases $\{\mathbf{G}_I\}$ and $\{\mathbf{G}^I\}$ by $\mathbf{F}^p$ or the pull-back operation of the current bases $\{\mathbf{g}_i\}$ and $\{\mathbf{g}^i\}$ by $\mathbf{F}^e$ to the intermediate configuration as follows:

$$\bar{\mathbf{G}}_{\bar{I}} = \left\{\begin{array}{l}
\mathbf{F}^p\delta^I_{\bar{I}}\mathbf{G}_I \\
\mathbf{F}^{e-1}\delta^i_{\bar{I}}\mathbf{g}_i
\end{array}\right., \tag{8.5}$$

$$\bar{\mathbf{G}}^{\bar{I}} = \left\{\begin{array}{l}
\mathbf{F}^{p-T}\delta^{\bar{I}}_I\mathbf{G}^I \\
\mathbf{F}^{eT}\delta^{\bar{I}}_i\mathbf{g}^i
\end{array}\right., \tag{8.6}$$

noting equation (3.9), (3.12) and regarding $\mathbf{G}_I$ and $\mathbf{G}^I$ or $\mathbf{g}_i$ and $\mathbf{g}^i$ as $\bar{\mathbf{G}}_{\bar{I}}$ and $\bar{\mathbf{G}}^{\bar{I}}$, respectively. Obviously, the two-point tensors $\mathbf{F}$, $\mathbf{F}^p$ and $\mathbf{F}^e$ live in different pairs of configurations from each other. Here, note that the embedded bases $\{\mathbf{G}_I\}$, $\{\bar{\mathbf{G}}_{\bar{I}}\}$, $\{\mathbf{g}_i\}$ and their reciprocal bases are different from the *director triad* (Mandel, 1972b) proposed for the crystal plasticity, which is composed of three unit orthogonal vectors, while the orthogonal line elements etched in the material are not kept orthogonal to each other under deformaton.

Other than the multiplicative decomposition in equation (8.1), the multiplicative decomposition in the reverse order, that is $\mathbf{F} = \mathbf{F}^p\mathbf{F}^e$ (Clifton, 1972; Nemat-Nasser, 1979; Zbib, 1993; Davidson, 1995, Shen, 1998; Lubarda, 1999, 2002; Sansour et al., 2007), and the additive decomposition of rate, that is $\dot{\mathbf{F}} = \dot{\mathbf{F}}^e + \dot{\mathbf{F}}^p$ (Clifton, 1972; Lubarda, 1999), have been proposed.

We now examine the superposition of rigid-body rotation described by the orthogonal tensor $\widehat{\mathbf{Q}}$ in the intermediate configuration. It follows from equation (8.1) that

$$\mathbf{F} = \mathbf{F}^e\mathbf{F}^p = \widehat{\mathbf{F}}^e\widehat{\mathbf{F}}^p, \tag{8.7}$$

where

$$\widehat{\mathbf{F}}^e \equiv \mathbf{F}^e\widehat{\mathbf{Q}}^T, \quad \widehat{\mathbf{F}}^p \equiv \widehat{\mathbf{Q}}\mathbf{F}^p. \tag{8.8}$$

Here, note that the intermediate configuration is not uniquely determined depending on the rigid-body rotation $\widehat{\mathbf{Q}}$ applied to the intermediate configuration. It would be natural since the intermediate configuration is actually fictitious. Therefore, a constitutive equation should be formulated such that the objectivity is not lost by the non-uniqueness. This problem will be discussed in Sections 8.8 and 8.10.3 after the introduction of various physical quantities.

The elastic and plastic deformation gradient tensors $\mathbf{F}^e$ and $\mathbf{F}^p$ are represented in the polar decompositions in terms of the positive-definite tensors $\mathbf{U}^e$, $\mathbf{V}^e$ and $\mathbf{U}^p$, $\mathbf{V}^p$ and the orthogonal tensors $\mathbf{R}^e$ and $\mathbf{R}^p$ by equations (1.308) as follows:

$$\left.\begin{aligned} \mathbf{F}^e &= \mathbf{R}^e\mathbf{U}^e = \mathbf{V}^e\mathbf{R}^e \\ \mathbf{F}^p &= \mathbf{R}^p\mathbf{U}^p = \mathbf{V}^p\mathbf{R}^p \end{aligned}\right\} \tag{8.9}$$

from which one has the following expressions.

$$\left.\begin{aligned} \mathbf{U}^e &= \mathbf{R}^{eT}\mathbf{V}^e\mathbf{R}^e, \quad \mathbf{V}^e = \mathbf{R}^e\mathbf{U}^e\mathbf{R}^{eT} \\ \mathbf{U}^p &= \mathbf{R}^{pT}\mathbf{V}^p\mathbf{R}^p, \quad \mathbf{V}^p = \mathbf{R}^p\mathbf{U}^p\mathbf{R}^{pT} \end{aligned}\right\}. \tag{8.10}$$

Further, the deformation gradient tensor $\mathbf{F}$ is represented by the spectral decomposition in equation (4.18) as follows:

$$\mathbf{F} = \mathbf{R}\mathbf{U} = \mathbf{V}\mathbf{R} = \sum_{\alpha=1}^{3} \lambda_\alpha \mathbf{n}^{(\alpha)} \otimes \mathbf{N}^{(\alpha)}. \tag{8.11}$$

Here, denoting the eigenvectors of $\mathbf{V}^p$ by $\bar{\mathbf{N}}^{(\alpha)}$, and the eigenvalues of $\mathbf{U}^e$, $\mathbf{V}^e$ and $\mathbf{U}^p$, $\mathbf{V}^p$ by λ^e_α and λ^p_α, respectively, the following relations hold:

$$\mathbf{F} = \sum_{\alpha=1}^{3} \lambda^e_\alpha\lambda^p_\alpha \mathbf{n}^{(\alpha)} \otimes \mathbf{N}^{(\alpha)} = \mathbf{F}^e\mathbf{F}^p = (\mathbf{V}^e\mathbf{R}^e)(\mathbf{R}^p\mathbf{U}^p) = \mathbf{V}^e\mathbf{R}\mathbf{U}^p, \tag{8.12}$$

$$\lambda_\alpha = \lambda^e_\alpha \lambda^p_\alpha, \tag{8.13}$$

$$\left.\begin{aligned} \mathbf{F}^e &= \sum_{\alpha=1}^{3} \lambda^e_\alpha \mathbf{n}^{(\alpha)} \otimes \bar{\mathbf{N}}^{(\alpha)} = \mathbf{R}^e\mathbf{U}^e = \mathbf{V}^e\mathbf{R}^e \\ \mathbf{F}^p &= \sum_{\alpha=1}^{3} \lambda^p_\alpha \bar{\mathbf{N}}^{(\alpha)} \otimes \mathbf{N}^{(\alpha)} = \mathbf{R}^p\mathbf{U}^p = \mathbf{V}^p\mathbf{R}^p \end{aligned}\right\}, \tag{8.14}$$

$$\left.\begin{aligned} \mathbf{U}^e &= \sum_{\alpha=1}^{3} \lambda^e_\alpha \bar{\mathbf{N}}^{(\alpha)} \otimes \bar{\mathbf{N}}^{(\alpha)}, \quad \mathbf{V}^e = \sum_{\alpha=1}^{3} \lambda^e_\alpha \mathbf{n}^{(\alpha)} \otimes \mathbf{n}^{(\alpha)} \\ \mathbf{U}^p &= \sum_{\alpha=1}^{3} \lambda^p_\alpha \mathbf{N}^{(\alpha)} \otimes \mathbf{N}^{(\alpha)}, \quad \mathbf{V}^p = \sum_{\alpha=1}^{3} \lambda^p_\alpha \bar{\mathbf{N}}^{(\alpha)} \otimes \bar{\mathbf{N}}^{(\alpha)} \end{aligned}\right\}, \tag{8.15}$$

$$\mathbf{R} = \sum_{\alpha=1}^{3} \mathbf{n}^{(\alpha)} \otimes \mathbf{N}^{(\alpha)} = \mathbf{R}^e\mathbf{R}^p, \tag{8.16}$$

$$\mathbf{R}^e = \sum_{\alpha=1}^{3} \mathbf{n}^{(\alpha)} \otimes \bar{\mathbf{N}}^{(\alpha)}, \quad \mathbf{R}^p = \sum_{\alpha=1}^{3} \bar{\mathbf{N}}^{(\alpha)} \otimes \mathbf{N}^{(\alpha)}. \tag{8.17}$$

Let the normalized-rectangular triads $\{\mathbf{N}^{(\alpha)}\}$, $\{\bar{\mathbf{N}}^{(\alpha)}\}$ and $\{\mathbf{n}^{(\alpha)}\}$ be called here the *reference*, *intermediate* and *current triad*, respectively (needless to say, they are irrelevant to the convected bases). $\{\mathbf{N}^{(\alpha)}\}$ and $\{\mathbf{n}^{(\alpha)}\}$ are nothing but the Lagrangian and the Eulerian triads defined in relation to equation (4.16). The effects of the superposition of rigid-body rotation on the intermediate configuration are given by

$$\left.\begin{aligned} \widehat{\mathbf{F}}^e &= \sum_{\alpha=1}^{3} \lambda^e_\alpha \mathbf{n}^{(\alpha)} \otimes \widehat{\bar{\mathbf{N}}}^{(\alpha)} = \widehat{\mathbf{R}}^e\widehat{\mathbf{U}}^e = \widehat{\mathbf{V}}^e\widehat{\mathbf{R}}^e \\ \widehat{\mathbf{F}}^p &= \sum_{\alpha=1}^{3} \lambda^p_\alpha \widehat{\bar{\mathbf{N}}}^{(\alpha)} \otimes \mathbf{N}^{(\alpha)} = \widehat{\mathbf{R}}^p\widehat{\mathbf{U}}^p = \widehat{\mathbf{V}}^p\widehat{\mathbf{R}}^p \end{aligned}\right\}, \tag{8.18}$$

$$\left.\begin{aligned} \widehat{\mathbf{U}}^e &= \sum_{\alpha=1}^{3} \lambda^e_\alpha \widehat{\bar{\mathbf{N}}}^{(\alpha)} \otimes \widehat{\bar{\mathbf{N}}}^{(\alpha)}, \quad \mathbf{V}^e = \sum_{\alpha=1}^{3} \lambda^e_\alpha \mathbf{n}^{(\alpha)} \otimes \mathbf{n}^{(\alpha)} \\ \mathbf{U}^p &= \sum_{\alpha=1}^{3} \lambda^p_\alpha \mathbf{N}^{(\alpha)} \otimes \mathbf{N}^{(\alpha)}, \quad \widehat{\mathbf{V}}^p = \sum_{\alpha=1}^{3} \lambda^p_\alpha \widehat{\bar{\mathbf{N}}}^{(\alpha)} \otimes \widehat{\bar{\mathbf{N}}}^{(\alpha)} \end{aligned}\right\}, \tag{8.19}$$

$$\widehat{\mathbf{R}}^e = \sum_{\alpha=1}^{3} \mathbf{n}^{(\alpha)} \otimes \widehat{\bar{\mathbf{N}}}^{(\alpha)}, \quad \widehat{\mathbf{R}}^p = \sum_{\alpha=1}^{3} \widehat{\bar{\mathbf{N}}}^{(\alpha)} \otimes \mathbf{N}^{(\alpha)}, \tag{8.20}$$

$$\left.\begin{aligned} &\widehat{\bar{\mathbf{N}}}^{(\alpha)} = \widehat{\mathbf{Q}}\bar{\mathbf{N}}^{(\alpha)} \\ &\widehat{\mathbf{U}}^e = \widehat{\mathbf{Q}}\mathbf{U}^e\widehat{\mathbf{Q}}^T, \quad \widehat{\mathbf{V}}^e = \mathbf{V}^e, \quad \widehat{\mathbf{R}}^e = \mathbf{R}^e\widehat{\mathbf{Q}}^T \\ &\widehat{\mathbf{U}}^p = \mathbf{U}^p, \quad \widehat{\mathbf{V}}^p = \widehat{\mathbf{Q}}\mathbf{V}^p\widehat{\mathbf{Q}}^T, \quad \widehat{\mathbf{R}}^p = \widehat{\mathbf{Q}}\mathbf{R}^p \end{aligned}\right\}, \tag{8.21}$$

noting that

$$\begin{aligned}\mathbf{F} = \widehat{\mathbf{F}}^e\widehat{\mathbf{F}}^p &= \left\{\left(\sum_{\alpha=1}^{3}\lambda_\alpha^e\mathbf{n}^{(\alpha)}\otimes\bar{\mathbf{N}}^{(\alpha)}\right)\widehat{\mathbf{Q}}^T\right\}\left\{\widehat{\mathbf{Q}}\left(\sum_{\beta=1}^{3}\lambda_\beta^p\bar{\mathbf{N}}^{(\beta)}\otimes\mathbf{N}^{(\beta)}\right)\right\}\\ &= \left\{\sum_{\alpha=1}^{3}\lambda_\alpha^e\mathbf{n}^{(\alpha)}\otimes\left(\widehat{\mathbf{Q}}\bar{\mathbf{N}}^{(\alpha)}\right)\right\}\left\{\sum_{\beta=1}^{3}\lambda_\beta^p\left(\widehat{\mathbf{Q}}\bar{\mathbf{N}}^{(\beta)}\right)\otimes\mathbf{N}^{(\beta)}\right\}\\ &= \left(\sum_{\alpha=1}^{3}\lambda_\alpha^e\mathbf{n}^{(\alpha)}\otimes\widehat{\bar{\mathbf{N}}}^{(\alpha)}\right)\left(\sum_{\beta=1}^{3}\lambda_\beta^p\widehat{\bar{\mathbf{N}}}^{(\beta)}\otimes\mathbf{N}^{(\beta)}\right).\end{aligned}$$

8.3 Stress and Deformation Tensors for Multiplicative Decomposition

Consider the contravariant pull-back of the Kirchhoff stress $\boldsymbol{\tau}$ or the contravariant push-forward of the second Piola–Kirchhoff stress $\mathbf{S}$ to the intermediate configuration as follows:

$$\bar{\mathbf{S}}(=\bar{\mathbf{S}}^T) = \bar{S}^{\bar{I}\bar{J}}\bar{\mathbf{G}}_{\bar{I}}\otimes\bar{\mathbf{G}}_{\bar{J}} \equiv \begin{cases}{}^e\overleftarrow{\boldsymbol{\tau}}^{\bar{G}\bar{G}} = \mathbf{F}^{e-1}\boldsymbol{\tau}\mathbf{F}^{e-T} = \delta_i^{\bar{I}}\delta_j^{\bar{J}}\tau^{ij}\bar{\mathbf{G}}_{\bar{I}}\otimes\bar{\mathbf{G}}_{\bar{J}}\\ {}^p\overrightarrow{\mathbf{S}}^{\bar{G}\bar{G}} = \mathbf{F}^p\mathbf{S}\mathbf{F}^{pT} = \delta_I^{\bar{I}}\delta_J^{\bar{J}}S^{IJ}\bar{\mathbf{G}}_{\bar{I}}\otimes\bar{\mathbf{G}}_{\bar{J}}\end{cases}, \tag{8.22}$$

by virtue of equations (3.32) and (3.40). Here, note $S^{IJ} = \delta_i^I\delta_j^J\tau^{ij}$ from equation (6.44). An index added to the left of symbols for the pull-back or push-forward operations specifies the type of the deformation gradient used in the operation. For instance, ${}^e\overleftarrow{\mathbf{t}}^{\bar{G}\bar{G}}$ and ${}^p\overrightarrow{\mathbf{T}}^{\bar{G}\bar{G}}$ designate the pull-back operation by $\mathbf{F}^e$ and the push-forward operation by $\mathbf{F}^p$, respectively.

Let the following elastic and plastic Cauchy–Green deformation tensors be defined noting equations (3.32), (3.40), (4.1) and (4.3):

$$\mathbf{C}^p = C_{IJ}^p\mathbf{G}^I\otimes\mathbf{G}^J \equiv \mathbf{F}^{pT}\mathbf{F}^p = \mathbf{F}^{pT}\bar{\mathbf{G}}\mathbf{F}^p = {}^p\overleftarrow{\bar{\mathbf{G}}}_{GG} = \delta_I^{\bar{I}}\delta_J^{\bar{J}}\bar{G}_{\bar{I}\bar{J}}\mathbf{G}^I\otimes\mathbf{G}^J, \tag{8.23}$$

$$\mathbf{C}^{p-1} = C^{p-1IJ}\mathbf{G}_I\otimes\mathbf{G}_J \equiv (\mathbf{F}^{pT}\mathbf{F}^p)^{-1} = \mathbf{F}^{p-1}\bar{\mathbf{G}}\mathbf{F}^{p-T} = {}^p\overleftarrow{\bar{\mathbf{G}}}^{GG} = \delta_{\bar{I}}^I\delta_{\bar{J}}^J\bar{G}^{\bar{I}\bar{J}}\mathbf{G}_I\otimes\mathbf{G}_J, \tag{8.24}$$

$$\bar{\mathbf{C}}^e = \bar{C}_{\bar{I}\bar{J}}^e\bar{\mathbf{G}}^{\bar{I}}\otimes\bar{\mathbf{G}}^{\bar{J}} \equiv \mathbf{F}^{eT}\mathbf{F}^e = \begin{cases}\mathbf{F}^{eT}\mathbf{g}\mathbf{F}^e = {}^e\overleftarrow{\mathbf{g}}_{\bar{G}\bar{G}} = \delta_{\bar{I}}^i\delta_{\bar{J}}^jg_{ij}\bar{\mathbf{G}}^{\bar{I}}\otimes\bar{\mathbf{G}}^{\bar{J}}\\ \mathbf{F}^{p-T}\mathbf{C}\mathbf{F}^{p-1} = {}^p\overrightarrow{\mathbf{C}}_{\bar{G}\bar{G}} = \delta_{\bar{I}}^I\delta_{\bar{J}}^JC_{IJ}\bar{\mathbf{G}}^{\bar{I}}\otimes\bar{\mathbf{G}}^{\bar{J}}\end{cases}, \tag{8.25}$$

$$\bar{\mathbf{C}}^{e-1} = \bar{C}^{e-1\bar{I}\bar{J}}\bar{\mathbf{G}}_{\bar{I}}\otimes\bar{\mathbf{G}}_{\bar{J}} \equiv (\mathbf{F}^{eT}\mathbf{F}^e)^{-1} = \begin{cases}\mathbf{F}^{e-1}\mathbf{g}\mathbf{F}^{e-T} = {}^e\overleftarrow{\mathbf{g}}^{\bar{G}\bar{G}} = \delta_i^{\bar{I}}\delta_j^{\bar{J}}g^{ij}\bar{\mathbf{G}}_{\bar{I}}\otimes\bar{\mathbf{G}}_{\bar{J}}\\ \mathbf{F}^p\mathbf{C}^{-1}\mathbf{F}^{pT} = {}^p\overrightarrow{\mathbf{C}}^{-1\bar{G}\bar{G}} = \delta_I^{\bar{I}}\delta_J^{\bar{J}}C^{-1IJ}\bar{\mathbf{G}}_{\bar{I}}\otimes\bar{\mathbf{G}}_{\bar{J}}\end{cases}, \tag{8.26}$$

$$\mathbf{b}^e = b_{ij}^e\mathbf{g}_i\otimes\mathbf{g}_j \equiv \mathbf{F}^e\mathbf{F}^{eT} = \mathbf{F}^e\bar{\mathbf{G}}\mathbf{F}^{eT} = {}^e\overrightarrow{\bar{\mathbf{G}}}^{gg} = \delta_{\bar{I}}^i\delta_{\bar{J}}^j\bar{G}^{\bar{I}\bar{J}}\mathbf{g}_i\otimes\mathbf{g}_j, \tag{8.27}$$

$$\mathbf{b}^{e-1} = b_{ij}^{e-1}\mathbf{g}^i\otimes\mathbf{g}^j \equiv (\mathbf{F}^e\mathbf{F}^{eT})^{-1} = \mathbf{F}^{e-T}\bar{\mathbf{G}}\mathbf{F}^{e-1} = {}^e\overrightarrow{\bar{\mathbf{G}}}_{gg} = \delta_i^{\bar{I}}\delta_j^{\bar{J}}\bar{G}_{\bar{I}\bar{J}}\mathbf{g}^i\otimes\mathbf{g}^j, \tag{8.28}$$

$$\bar{\mathbf{B}}^p = \bar{B}^{p\bar{I}\bar{J}}\bar{\mathbf{G}}_{\bar{I}} \otimes \bar{\mathbf{G}}_{\bar{J}} \equiv \mathbf{F}^p\mathbf{F}^{pT} = \begin{cases} \mathbf{F}^p\mathbf{G}\mathbf{F}^{pT} = {}^p\vec{\mathbf{G}}^{\bar{G}\bar{G}} = \delta_I^{\bar{I}}\delta_J^{\bar{J}}G^{IJ}\bar{\mathbf{G}}_{\bar{I}} \otimes \bar{\mathbf{G}}_{\bar{J}} \\ \mathbf{F}^{e-1}\mathbf{b}\mathbf{F}^{e-T} = {}^e\overleftarrow{\mathbf{b}}^{\bar{G}\bar{G}} = \delta_i^{\bar{I}}\delta_j^{\bar{J}}b^{ij}\bar{\mathbf{G}}_{\bar{I}} \otimes \bar{\mathbf{G}}_{\bar{J}} \end{cases}, \tag{8.29}$$

$$\bar{\mathbf{B}}^{p-1} = \bar{B}^{p-1}_{\bar{I}\bar{J}}\bar{\mathbf{G}}^{\bar{I}} \otimes \bar{\mathbf{G}}^{\bar{J}} \equiv (\mathbf{F}^p\mathbf{F}^{pT})^{-1} = \begin{cases} \mathbf{F}^{p-T}\mathbf{G}\mathbf{F}^{p-1} = {}^p\vec{\mathbf{G}}_{\bar{G}\bar{G}} = \delta_{\bar{I}}^{I}\delta_{\bar{J}}^{J}G_{IJ}\bar{\mathbf{G}}^{\bar{I}} \otimes \bar{\mathbf{G}}^{\bar{J}} \\ \mathbf{F}^{eT}\mathbf{b}^{-1}\mathbf{F}^e = {}^e\overleftarrow{\mathbf{b}}^{-1}_{\bar{G}\bar{G}} = \delta_{\bar{I}}^{i}\delta_{\bar{J}}^{j}b^{-1}_{ij}\bar{\mathbf{G}}^{\bar{I}} \otimes \bar{\mathbf{G}}^{\bar{J}} \end{cases} \tag{8.30}$$

because of

$$\left.\begin{array}{l} \bar{\mathbf{C}}^e = \mathbf{F}^{eT}\mathbf{F}^e = (\mathbf{F}\mathbf{F}^{p-1})^T\mathbf{F}\mathbf{F}^{p-1} = \mathbf{F}^{p-T}\mathbf{F}^T\mathbf{F}\mathbf{F}^{p-1} = \mathbf{F}^{p-T}\mathbf{C}\mathbf{F}^{p-1} = {}^p\vec{\mathbf{C}}_{\bar{G}\bar{G}} \\ \bar{\mathbf{C}}^{e-1} = (\mathbf{F}^{eT}\mathbf{F}^e)^{-1} = (\mathbf{F}^{p-T}\mathbf{C}\mathbf{F}^{p-1})^{-1} = \mathbf{F}^p\mathbf{C}^{-1}\mathbf{F}^{pT} = {}^p\vec{\mathbf{C}}^{-1\bar{G}\bar{G}} \\ \bar{\mathbf{B}}^p = \mathbf{F}^p\mathbf{F}^{pT} = \mathbf{F}^{e-1}\mathbf{F}(\mathbf{F}^{e-1}\mathbf{F})^T = \mathbf{F}^{e-1}\mathbf{F}\mathbf{F}^T\mathbf{F}^{e-T} = \mathbf{F}^{e-1}\mathbf{b}\mathbf{F}^{e-T} = {}^e\overleftarrow{\mathbf{b}}^{\bar{G}\bar{G}} \\ \bar{\mathbf{B}}^{p-1} = (\mathbf{F}^p\mathbf{F}^{pT})^{-1} = (\mathbf{F}^{e-1}\mathbf{b}\mathbf{F}^{e-T})^{-1} = \mathbf{F}^{eT}\mathbf{b}^{-1}\mathbf{F}^e = {}^e\overleftarrow{\mathbf{b}}^{-1}_{\bar{G}\bar{G}} \end{array}\right\}. \tag{8.31}$$

The following relations of the components hold from equations (8.23)–(8.30):

$$\left.\begin{array}{l} C^p_{IJ} = \delta_I^{\bar{I}}\delta_J^{\bar{J}}\bar{G}_{\bar{I}\bar{J}}, \;\; C^{p-1IJ} = \delta_{\bar{I}}^{I}\delta_{\bar{J}}^{J}\bar{G}^{\bar{I}\bar{J}} \\ \bar{C}^e_{\bar{I}\bar{J}} = \delta_{\bar{I}}^{i}\delta_{\bar{J}}^{j}g_{ij} = \delta_{\bar{I}}^{I}\delta_{\bar{J}}^{J}C_{IJ}, \;\; \bar{C}^{e-1\bar{I}\bar{J}} = \delta_i^{\bar{I}}\delta_j^{\bar{J}}g^{ij} = \delta_I^{\bar{I}}\delta_J^{\bar{J}}C^{-1IJ} \\ b^e_{ij} = \delta_{\bar{I}}^{i}\delta_{\bar{J}}^{j}\bar{G}^{\bar{I}\bar{J}}, \;\; b^{e-1}_{ij} = \delta_i^{\bar{I}}\delta_j^{\bar{J}}\bar{G}_{\bar{I}\bar{J}} \\ \bar{B}^{p\bar{I}\bar{J}} = \delta_I^{\bar{I}}\delta_J^{\bar{J}}\bar{G}^{IJ} = \delta_i^{\bar{I}}\delta_j^{\bar{J}}b^{ij}, \;\; \bar{B}^{p-1}_{\bar{I}\bar{J}} = \delta_{\bar{I}}^{I}\delta_{\bar{J}}^{J}G_{IJ} = \delta_{\bar{I}}^{i}\delta_{\bar{J}}^{j}b^{-1}_{ij} \end{array}\right\}. \tag{8.32}$$

8.4 Incorporation of Nonlinear Kinematic Hardening

Materials possess statically-indeterminate substructure, and thus heterogeneous elastic and plastic deformations are induced microscopically in general. Here, most of the plastic deformation is not removed even in the unloaded state of the external load. Thus, the heterogeneous elastic deformation remains, inducing the residual stress such that the yield stress region moves in the loading direction, resulting in the Bauschinger effect. This effect can be described by the movement of the yield surface, that is *kinematic hardening* for metals (Prager, 1956), and by the *rotational hardening* for frictional materials such as geomaterials (Hashiguchi and Chen, 1998; Hashiguchi, 2001a). Kinematic hardening can be formulated appropriately by the *nonlinear kinematic hardening rule* proposed by Armstrong and Frederick (1966), while a peculiar deformation behavior in the inverse loading process is predicted by the linear kinematic hardening rule proposed by Prager (1956).

Now, in order to obtain the exact finite elasto-plastic constitutive equation, we will formulate the nonlinear kinematic hardening rule such that it is derived based on a potential function of a pertinent strain variable similarly to the hyperelastic constitutive equation. To this end, a further multiplicative decomposition of the plastic deformation gradient into energy dissipative and storage parts will be introduced. This notion was advocated by Lion (2000) and incorporated into the finite elasto-plastic constitutive equation by Dettmer and Reese

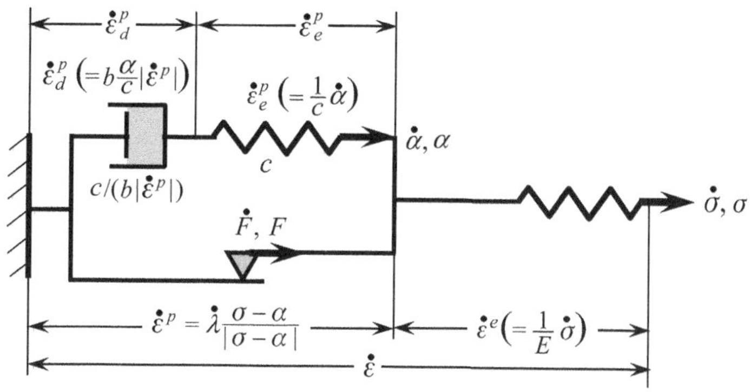

Figure 8.3 Rheological model for nonlinear kinematic hardening

(2004) and Vladimirov et al. (2008, 2010). Plastic deformation brings about the homogeneous plastic deformation induced by the slips between material particles and the heterogeneous plastic deformation induced by the development of obstacles of slips in the loading direction causing kinematic hardening. The former dissipates energy and the latter stores energy. Thus, the plastic deformation gradient tensor is further decomposed into energy dissipative and storage parts.

8.4.1 Rheological Model for Nonlinear Kinematic Hardening

The strain attained by the fictitious unloading operation obeying the hyperelastic equation is defined as the plastic strain and denoted by ε^p in the one-dimensional deformation. In accordance with the foregoing ideas, the plastic strain ε^p is decomposed into an energy-dissipating part ε_d^p caused by the slips between material particles and the energy-storage (free-energy) part ε_e^p caused by the development of obstacles of slips in the loading direction inducing the anisotropic hardening, as shown in Figure 8.3. Assume the following relations:

$$\dot{\varepsilon} = \dot{\varepsilon}^e + \dot{\varepsilon}^p, \tag{8.33}$$

$$\dot{\varepsilon}^p = \dot{\varepsilon}_e^p + \dot{\varepsilon}_d^p, \tag{8.34}$$

with

$$\dot{\varepsilon}^e = \frac{1}{E}\dot{\sigma}, \tag{8.35}$$

$$\dot{\varepsilon}^p = \dot{\lambda}\hat{N}, \hat{N} \equiv \frac{\sigma - \alpha}{|\sigma - \alpha|} = \pm 1, \tag{8.36}$$

$$\dot{\varepsilon}_e^p = \frac{1}{c}\dot{\alpha}, \quad \dot{\varepsilon}_d^p = b\frac{\alpha}{c}|\dot{\varepsilon}^p|, \tag{8.37}$$

where E is the elastic modulus. α is the kinematic hardening variable, that is back stress, which is derived from the strain energy function ψ_e^p of elastic-like strain variable ε_e^p for the kinematic hardening as follows:

$$\psi_e^p(\varepsilon_e^p) = c(\varepsilon_e^p)^2/2, \tag{8.38}$$

$$\alpha = \partial\psi_e^p/\partial\varepsilon_e^p = c\varepsilon_e^p. \tag{8.39}$$

$\overset{\bullet}{\lambda}$ is the positive plastic multiplier denoting the magnitude of the plastic strain rate $\overset{\bullet}{\varepsilon}^p$. $c\,(=\overset{\bullet}{\alpha}/\overset{\bullet}{\varepsilon}_e^p)$ is an elastic tangent modulus-like material constant for kinematic hardening and b is the dimensionless material constant, while $c/(b|\overset{\bullet}{\varepsilon}^p|)(=\alpha/\overset{\bullet}{\varepsilon}_d^p)$ is regarded as the pseudo-viscosity. The upper and lower signs in equations in this section correspond to the extension ($\overset{\bullet}{\varepsilon}^p > 0$) and the compression ($\overset{\bullet}{\varepsilon}^p < 0$), respectively.

It follows from equations (8.34), (8.36) and (8.37) that

$$\overset{\bullet}{\alpha} = (c\hat{N} - b\alpha)|\overset{\bullet}{\varepsilon}^p| = (\pm c - b\alpha)(\pm\overset{\bullet}{\varepsilon}^p), \tag{8.40}$$

leading to

$$\frac{\overset{\bullet}{\alpha}}{\overset{\bullet}{\varepsilon}^p} = c \mp b\alpha. \tag{8.41}$$

Equation (8.40) is the one-dimensional equation describing the nonlinear kinematic hardening rule of Armstrong and Frederick (1966). The variation of the kinematic hardening variable α in the uniaxial loading and reverse loading processes is shown in Figure 8.4.

A mechanism similar to that described above is induced also in the *rotational hardening* (Hashiguchi, 2001a, 2009), that is the rotation of the yield surface for frictional materials, while only the kinematic hardening for pressure-independent metals is discussed in this section. The deformation tensor for the kinematic hardening will be described in the next section.

8.4.2 *Multiplicative Decomposition of Plastic Deformation Gradient Tensor*

Based on the rheological model described in Section 8.4.1, the plastic deformation gradient $\mathbf{F}^p$ is further decomposed multiplicatively into the *kinematic hardening-elastic deformation gradient* $\mathbf{F}_e^p$ (energy storage part) and the *kinematic hardening-dissipative deformation gradient* $\mathbf{F}_d^p$ (energy dissipative part) as shown in Figure 8.5. The configuration attained by extracting $\mathbf{F}_e^p$ from the intermediate configuration $\bar{\mathcal{K}}$ is called the *kinematic hardening-intermediate configuration* and denoted by $\tilde{\mathcal{K}}$ (Lion, 2000). Thus, denoting the position vectors of material particles in the kinematic hardening-intermediate configuration $\tilde{\mathcal{K}}$ as $\tilde{\mathbf{X}}$, it follows that

$$\mathbf{F}^p = \mathbf{F}_e^p\mathbf{F}_d^p, \tag{8.42}$$

where

$$d\tilde{\mathbf{X}} = \mathbf{F}_d^p\,d\mathbf{X}, \quad d\bar{\mathbf{X}} = \mathbf{F}_e^p d\tilde{\mathbf{X}}. \tag{8.43}$$

Here, note that $\mathbf{F}^p$ is independent of the superposition of rigid-body rotation because it is irrelevant to the current configuration, so that $\mathbf{F}_e^p$ and $\mathbf{F}_d^p$ are also independent of that

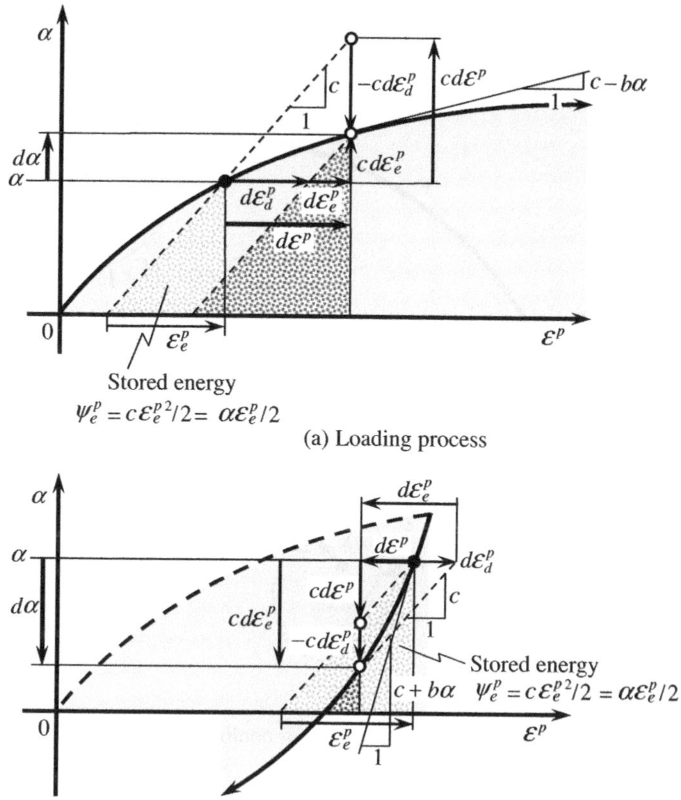

(b) Inverse loading process

Figure 8.4 Nonlinear kinematic hardening in uniaxial loading described by the rheological model

superposition. Their component descriptions in terms of the base vectors are as follows:

$$\left.\begin{array}{l}
\mathbf{F}^p = F^{p\bar{I}}_{\bullet J}\bar{\mathbf{G}}_{\bar{I}}\otimes\mathbf{G}^J = \delta^{\bar{I}}_I\bar{\mathbf{G}}_{\bar{I}}\otimes\mathbf{G}^I,\ \ \mathbf{F}^{p-1} = \delta^I_{\bar{I}}\mathbf{G}_I\otimes\bar{\mathbf{G}}^{\bar{I}} \\
\mathbf{F}^p_e = F^{p\,\bar{I}}_{e\,\bullet\tilde{I}}\bar{\mathbf{G}}_{\bar{I}}\otimes\tilde{\mathbf{G}}^{\tilde{I}} = \delta^{\bar{I}}_{\tilde{I}}\bar{\mathbf{G}}_{\bar{I}}\otimes\tilde{\mathbf{G}}^{\tilde{I}},\ \ \mathbf{F}^{p-1}_e = \delta^{\tilde{I}}_{\bar{I}}\tilde{\mathbf{G}}_{\tilde{I}}\otimes\bar{\mathbf{G}}^{\bar{I}} \\
\mathbf{F}^p_d = F^{p\,\tilde{I}}_{e\,\bullet I}\tilde{\mathbf{G}}_{\tilde{I}}\otimes\mathbf{G}^I = \delta^{\tilde{I}}_I\tilde{\mathbf{G}}_{\tilde{I}}\otimes\mathbf{G}^I,\ \ \mathbf{F}^{p-1}_d = \delta^I_{\tilde{I}}\mathbf{G}_I\otimes\tilde{\mathbf{G}}^{\tilde{I}}
\end{array}\right\}, \tag{8.44}$$

noting equation (3.10).

The base vectors $\tilde{\mathbf{G}}_{\tilde{I}}$ and $\tilde{\mathbf{G}}^{\tilde{I}}$ in the kinematic hardening-intermediate configuration are given by

$$\tilde{\mathbf{G}}_{\tilde{I}} = \left\{\begin{array}{l} \delta^I_{\tilde{I}}\mathbf{F}^p_d\mathbf{G}_I \\ \delta^{\bar{I}}_{\tilde{I}}\mathbf{F}^{p-1}_e\bar{\mathbf{G}}_{\bar{I}} \end{array}\right.,\quad \tilde{\mathbf{G}}^{\tilde{I}} = \left\{\begin{array}{l} \delta^{\tilde{I}}_I\mathbf{F}^{p-T}_d\mathbf{G}^I \\ \delta^{\tilde{I}}_{\bar{I}}\mathbf{F}^{pT}_e\bar{\mathbf{G}}^{\bar{I}} \end{array}\right., \tag{8.45}$$

noting equations (3.9), (3.12) and (8.44).

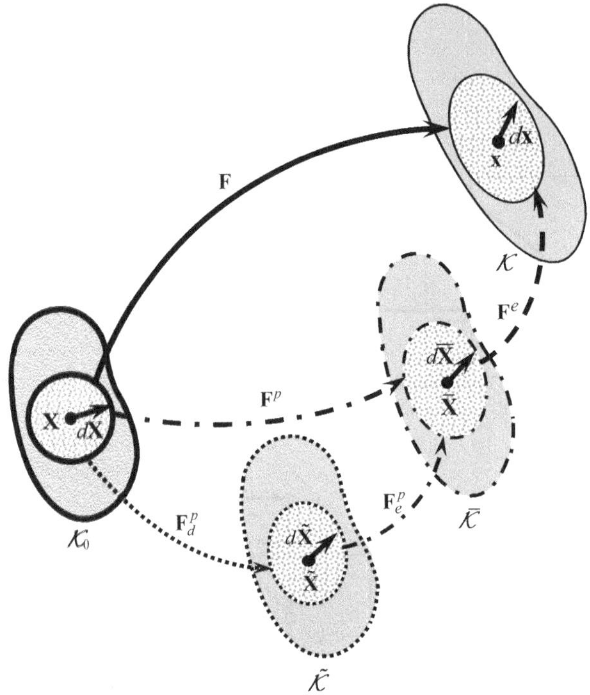

Figure 8.5 Multiplicative decomposition for nonlinear kinematic hardening

Let the following elastic and plastic Cauchy–Green deformation tensors for kinematic hardening be defined, noting equations (3.32), (3.40), (4.1) and (4.3):

$$\mathbf{C}_d^p = C^p_{dIJ}\mathbf{G}^I \otimes \mathbf{G}^J \equiv \mathbf{F}_d^{pT}\mathbf{F}_d^p = \mathbf{F}_d^{pT}\tilde{\mathbf{G}}\mathbf{F}_d^p = {}^p_d\overleftarrow{\tilde{\mathbf{G}}}_{\tilde{G}\tilde{G}} = \delta_I^{\tilde{I}}\delta_J^{\tilde{J}}\tilde{G}_{\tilde{I}\tilde{J}}\mathbf{G}^I \otimes \mathbf{G}^J, \tag{8.46}$$

$$\mathbf{C}_d^{p-1} = C_d^{p-1IJ}\mathbf{G}_I \otimes \mathbf{G}_J \equiv (\mathbf{F}_d^{pT}\mathbf{F}_d^p)^{-1} = \mathbf{F}_d^{p-1}\tilde{\mathbf{G}}\mathbf{F}_d^{p-T} = {}^p_d\overleftarrow{\tilde{\mathbf{G}}}^{\tilde{G}\tilde{G}} = \delta_{\tilde{I}}^{I}\delta_{\tilde{J}}^{J}\tilde{G}^{\tilde{I}\tilde{J}}\mathbf{G}_I \otimes \mathbf{G}_J, \tag{8.47}$$

$$\tilde{\mathbf{C}}_e^p = \tilde{C}^p_{e\tilde{I}\tilde{J}}\tilde{\mathbf{G}}^{\tilde{I}} \otimes \tilde{\mathbf{G}}^{\tilde{J}} \equiv \mathbf{F}_e^{pT}\mathbf{F}_e^p = \begin{cases} \mathbf{F}_e^{pT}\bar{\mathbf{G}}\mathbf{F}_e^p = {}^p_e\overleftarrow{\bar{\mathbf{G}}}_{\tilde{G}\tilde{G}} = \delta_{\tilde{I}}^{\bar{I}}\delta_{\tilde{J}}^{\bar{J}}\bar{G}_{\bar{I}\bar{J}}\tilde{\mathbf{G}}^{\tilde{I}} \otimes \tilde{\mathbf{G}}^{\tilde{J}} \\ \mathbf{F}_d^{p-T}\mathbf{C}^p\mathbf{F}_d^{p-1} = {}^p_d\overrightarrow{\mathbf{C}}_{\tilde{G}\tilde{G}} = \delta_{\tilde{I}}^{I}\delta_{\tilde{J}}^{J}C^p_{IJ}\tilde{\mathbf{G}}^{\tilde{I}} \otimes \tilde{\mathbf{G}}^{\tilde{J}} \end{cases}, \tag{8.48}$$

$$\tilde{\mathbf{C}}_e^{p-1} = \tilde{C}_e^{p-1\tilde{I}\tilde{J}}\tilde{\mathbf{G}}_{\tilde{I}} \otimes \tilde{\mathbf{G}}_{\tilde{J}} \equiv (\mathbf{F}_e^{pT}\mathbf{F}_e^p)^{-1} = \begin{cases} \mathbf{F}_e^{p-1}\bar{\mathbf{G}}\mathbf{F}_e^{p-T} = {}^p_e\overleftarrow{\bar{G}}^{\tilde{G}\tilde{G}} = \delta_{\bar{I}}^{\tilde{I}}\delta_{\bar{J}}^{\tilde{J}}\bar{G}^{\bar{I}\bar{J}}\tilde{\mathbf{G}}_{\tilde{I}} \otimes \tilde{\mathbf{G}}_{\tilde{J}} \\ \mathbf{F}_d^p\mathbf{C}^{p-1}\mathbf{F}_d^{pT} = {}^p_d\overrightarrow{\mathbf{C}}^{p-1\tilde{G}\tilde{G}} = \delta_I^{\tilde{I}}\delta_J^{\tilde{J}}C^{p-1IJ}\tilde{\mathbf{G}}_{\tilde{I}} \otimes \tilde{\mathbf{G}}_{\tilde{J}} \end{cases}, \tag{8.49}$$

$$\tilde{\mathbf{B}}_d^p = \tilde{B}^p_{d\tilde{I}\tilde{J}}\tilde{\mathbf{G}}^{\tilde{I}} \otimes \tilde{\mathbf{G}}^{\tilde{J}} = \mathbf{F}_d^p\mathbf{F}_d^{pT} = \begin{cases} \mathbf{F}_d^p\mathbf{G}\mathbf{F}_d^{pT} = {}^p_d\overrightarrow{\mathbf{G}}^{\tilde{G}\tilde{G}} = \delta_I^{\tilde{I}}\delta_J^{\tilde{J}}G^{IJ}\tilde{\mathbf{G}}_{\tilde{I}} \otimes \tilde{\mathbf{G}}_{\tilde{J}} \\ \mathbf{F}_e^{pT}\bar{\mathbf{B}}^p\mathbf{F}_e^p = {}^p_d\overleftarrow{\bar{\mathbf{B}}}^{p\tilde{G}\tilde{G}} = \delta_{\bar{I}}^{\tilde{I}}\delta_{\bar{J}}^{\tilde{J}}\bar{B}^{p\bar{I}\bar{J}}\tilde{\mathbf{G}}_{\tilde{I}} \otimes \tilde{\mathbf{G}}_{\tilde{J}} \end{cases}, \tag{8.50}$$

$$\tilde{\mathbf{B}}_d^{p-1} = \tilde{B}_{d\tilde{I}\tilde{J}}^{p-1}\tilde{\mathbf{G}}^{\tilde{I}} \otimes \tilde{\mathbf{G}}^{\tilde{J}} = (\mathbf{F}_d^p \mathbf{F}_d^{pT})^{-1} = \begin{cases} \mathbf{F}_d^{p-T}\mathbf{G}\mathbf{F}_d^{p-1} = {}_d^p\overrightarrow{\mathbf{G}}_{\tilde{G}\tilde{G}} = \delta_{\tilde{I}}^I\delta_{\tilde{J}}^J G_{IJ}\tilde{\mathbf{G}}^{\tilde{I}} \otimes \tilde{\mathbf{G}}^{\tilde{J}} \\ \mathbf{F}_e^{pT}\bar{\mathbf{B}}^{p-1}\mathbf{F}_e^p = {}_d^p\overleftarrow{\bar{\mathbf{B}}^{p-1}}_{\tilde{G}\tilde{G}} = \delta_{\tilde{I}}^{\bar{I}}\delta_{\tilde{J}}^{\bar{J}}\bar{B}_{\bar{I}\bar{J}}^{p-1}\tilde{\mathbf{G}}^{\tilde{I}} \otimes \tilde{\mathbf{G}}^{\tilde{J}} \end{cases}, \tag{8.51}$$

where we have

$$\left.\begin{aligned} \tilde{\mathbf{C}}_e^p &= \mathbf{F}_e^{pT}\mathbf{F}_e^p = (\mathbf{F}^p\mathbf{F}_d^{p-1})^T\mathbf{F}^p\mathbf{F}_d^{p-1} = \mathbf{F}_d^{p-T}\mathbf{F}^{pT}\mathbf{F}^p\mathbf{F}_d^{p-1} = \mathbf{F}_d^{p-T}\mathbf{C}^p\mathbf{F}_d^{p-1} = {}_d^p\overrightarrow{\mathbf{C}}_{\tilde{G}\tilde{G}} \\ \tilde{\mathbf{C}}_e^{p-1} &= (\mathbf{F}_e^{pT}\mathbf{F}_e^p)^{-1} = (\mathbf{F}_d^{p-T}\mathbf{C}^p\mathbf{F}_d^{p-1})^{-1} = \mathbf{F}_d^p\mathbf{C}^{p-1}\mathbf{F}_d^{pT} = {}_d^p\overrightarrow{\mathbf{C}}^{p-1\tilde{G}\tilde{G}} \\ \tilde{\mathbf{B}}_d^p &= \mathbf{F}_d^p\mathbf{F}_d^{pT} = \mathbf{F}_e^{p-1}\mathbf{F}^p(\mathbf{F}_e^{p-1}\mathbf{F}^p)^T = \mathbf{F}_e^{p-1}\mathbf{F}^p\mathbf{F}^{pT}\mathbf{F}_e^{p-T} = \mathbf{F}_e^{p-1}\bar{\mathbf{B}}^pF_e^{p-T} = {}_d^p\overleftarrow{\bar{\mathbf{B}}}^{\tilde{G}\tilde{G}} \\ \tilde{\mathbf{B}}_d^{p-1} &= (\mathbf{F}_d^p\mathbf{F}_d^{pT})^{-1} = (\mathbf{F}_e^{p-1}\bar{\mathbf{B}}^p\mathbf{F}_e^{p-T})^{-1} = \mathbf{F}_e^{pT}\bar{\mathbf{B}}^{p-1}\mathbf{F}_e^p = {}_d^p\overleftarrow{\bar{\mathbf{B}}^{p-1}}_{\tilde{G}\tilde{G}} \end{aligned}\right\}, \tag{8.52}$$

The following relations of the components hold from equations (8.46)–(8.51):

$$\left.\begin{aligned} &C_{dIJ}^p = \delta_I^{\tilde{I}}\delta_J^{\tilde{J}}\tilde{G}_{\tilde{I}\tilde{J}},\ C_d^{p-1IJ} = \delta_{\tilde{I}}^I\delta_{\tilde{J}}^J\tilde{G}^{\tilde{I}\tilde{J}} \\ &\tilde{C}_{e\tilde{I}\tilde{J}}^p = \delta_{\tilde{I}}^{\tilde{I}}\delta_{\tilde{J}}^{\tilde{J}}\tilde{G}_{\tilde{I}\tilde{J}} = \delta_{\tilde{I}}^I\delta_{\tilde{J}}^J C_{IJ}^p,\ \tilde{C}_e^{p-1\tilde{I}\tilde{J}} = \delta_I^{\tilde{I}}\delta_J^{\tilde{J}}\tilde{G}^{IJ} = \delta_{\tilde{I}}^{\bar{I}}\delta_{\tilde{J}}^{\bar{J}}C^{p-1IJ} \\ &\tilde{B}_{d\bar{I}\bar{J}}^p = \delta_I^{\tilde{I}}\delta_J^{\tilde{J}}G^{IJ} = \delta_{\bar{I}}^{\tilde{I}}\delta_{\bar{J}}^{\tilde{J}}\bar{B}^{p\bar{I}\bar{J}},\ \tilde{B}_{d\tilde{I}\tilde{J}}^{p-1} = \delta_{\tilde{I}}^I\delta_{\tilde{J}}^J G_{IJ} = \delta_{\tilde{I}}^{\bar{I}}\delta_{\tilde{J}}^{\bar{J}}\bar{B}_{\bar{I}\bar{J}}^{p-1} \end{aligned}\right\}. \tag{8.53}$$

Furthermore, the push-forward of $\tilde{\mathbf{C}}_e^p$ to the intermediate configuration is given by

$$\bar{\mathbf{C}}_e^p = \bar{C}_{e\bar{I}\bar{J}}^p\bar{\mathbf{G}}^{\bar{I}} \otimes \bar{\mathbf{G}}^{\bar{J}} = \mathbf{F}_e^{p-T}\tilde{\mathbf{C}}_e^p\mathbf{F}_e^{p-1} = {}_e^p\overrightarrow{\tilde{\mathbf{C}}_e^p}_{\bar{G}\bar{G}} = \bar{\mathbf{G}} = \bar{G}_{\bar{I}\bar{J}}\bar{\mathbf{G}}^{\bar{I}} \otimes \bar{\mathbf{G}}^{\bar{J}}, \tag{8.54}$$

noting that

$$\mathbf{F}_e^{p-T}\tilde{\mathbf{C}}_e^p\mathbf{F}_e^{p-1} = \mathbf{F}_e^{p-T}(\mathbf{F}_e^{pT}\mathbf{F}_e^p)\mathbf{F}_e^{p-1} = \bar{\mathbf{G}}.$$

$\tilde{\mathbf{C}}_e^p$ is the right Cauchy–Green deformation tensor induced in the process from the kinematic hardening-intermediate configuration $\tilde{\mathcal{K}}$ to the intermediate configuration $\bar{\mathcal{K}}$. On the other hand, $\bar{\mathbf{C}}_e^p$ is the right Cauchy–Green deformation tensor in the process from and to the intermediate configuration $\bar{\mathcal{K}}$ and thus coincides with the metric tensor $\bar{\mathbf{G}}$ in the intermediate configuration $\bar{\mathcal{K}}$.

As shown above, various elastic and plastic Cauchy–Green deformation tensors with the multiplicative decomposition accompanying the kinematic hardening decomposition are defined as the pull-back and the push-forward of the metric tensors in the current and the reference configurations. They are shown together in Table 8.1.

8.5 Strain Tensors

Various deformation (rate) tensors for describing the finite elasto-plastic deformation have been proposed, as will be described below.

The Green Lagrangian strain $\mathbf{E}$ based in the reference configuration, shown in equation $(4.31)_1$ and $(4.32)_1$, additively splits into elastic and plastic parts as follows:

$$\boxed{\mathbf{E} = \mathbf{E}^e + \mathbf{E}^p} \tag{8.55}$$

Table 8.1 Elastic and plastic right and left Cauchy–Green deformation tensors and their kinematic hardening parts

Deformation gradient	$\mathbf{F}$						
Multiplicative decomposition	$\mathbf{F}^p$					$\mathbf{F}^e$	
Multiplicative decomposition of kinematic hardening	$\mathbf{F}^p_d$		$\mathbf{F}^p_e$			$\mathbf{F}^e$	
Configurations	$\mathcal{K}_0$		$\tilde{\mathcal{K}}$		$\bar{\mathcal{K}}$		$\mathcal{K}$
Contravariant	$\mathbf{G}$	$\mathbf{C}^{-1}$ $\leftarrow$					$\mathbf{g}$
		$\mathbf{C}^{p-1}$ $\leftarrow$			$\bar{\mathbf{G}}$	$\bar{\mathbf{C}}^{e-1}$ $\leftarrow$	
		$\mathbf{C}^{p-1}_d$ $\leftarrow$	$\tilde{\mathbf{G}}$	$\tilde{\mathbf{C}}^{p-1}_e$ $\leftarrow$			
Covariant		$\mathbf{C}$ $\leftarrow$					
		$\mathbf{C}^p$ $\leftarrow$			$\bar{\mathbf{G}}$	$\bar{\mathbf{C}}^e$ $\leftarrow$	
		$\mathbf{C}^p_d$ $\leftarrow$	$\tilde{\mathbf{G}}$	$\tilde{\mathbf{C}}^p_e$ $\leftarrow$			
Cotravariant		$\rightarrow$ $\mathbf{b}$					
		$\rightarrow$ $\bar{\mathbf{B}}^p$			$\bar{\mathbf{G}}$	$\rightarrow$ $\mathbf{b}^e$	
		$\rightarrow$ $\tilde{\mathbf{B}}^p_d$	$\tilde{\mathbf{G}}$	$\rightarrow$ $\bar{\mathbf{B}}^p_e$			
Covariant		$\rightarrow$ $\mathbf{b}^{-1}$					
		$\rightarrow$ $\bar{\mathbf{b}}^{p-1}$			$\bar{\mathbf{G}}$	$\rightarrow$ $\mathbf{b}^{e-1}$	
		$\rightarrow$ $\tilde{\mathbf{B}}^{p-1}_d$	$\tilde{\mathbf{G}}$	$\rightarrow$ $\bar{\mathbf{B}}^{p-1}_e$			

Note: $\leftarrow$ pull-back, $\rightarrow$ push-forward.

where

$$\left.\begin{aligned} \mathbf{E}^e &\equiv (\mathbf{C}-\mathbf{C}^p)/2 \\ &= \{(C_{IJ}-C^p_{IJ})/2\}\mathbf{G}^I\otimes\mathbf{G}^J = \{(\delta^i_I\delta^j_J g_{ij}-\delta^{\bar{I}}_I\delta^{\bar{J}}_J\bar{G}_{\bar{I}\bar{J}})/2\}\mathbf{G}^I\otimes\mathbf{G}^J \\ \mathbf{E}^p &\equiv (\mathbf{C}^p-\mathbf{G})/2 \\ &= \{(C^p_{IJ}-G_{IJ})/2\}\mathbf{G}^I\otimes\mathbf{G}^J = \{(\delta^{\bar{I}}_I\delta^{\bar{J}}_J\bar{G}_{\bar{I}\bar{J}}-G_{IJ})/2\}\mathbf{G}^I\otimes\mathbf{G}^J \end{aligned}\right\}, \tag{8.56}$$

noting equations (4.1) and (8.23).

The Almansi Eulerian strain **e** based in the current configuration, shown in equation $(4.31)_2$ and $(4.32)_2$, additively splits into elastic and plastic parts as follows:

$$\boxed{\mathbf{e} = \mathbf{e}^e + \mathbf{e}^p} \tag{8.57}$$

where

$$\left.\begin{aligned} \mathbf{e}^e &\equiv (\mathbf{g}-\mathbf{b}^{e-1})/2 \\ &= \{(g_{ij}-b^{e-1}_{ij})/2\}\mathbf{g}^i\otimes\mathbf{g}^j = \{(g_{ij}-\delta^{\bar{I}}_i\delta^{\bar{J}}_j\bar{G}_{\bar{I}\bar{J}})/2\}\mathbf{g}^i\otimes\mathbf{g}^j \\ \mathbf{e}^p &\equiv (\mathbf{b}^{e-1}-\mathbf{b}^{-1})/2 \\ &= \{(b^{e-1}_{ij}-b^{-1}_{ij})/2\}\mathbf{g}^i\otimes\mathbf{g}^j = \{(\delta^{\bar{I}}_i\delta^{\bar{J}}_j\bar{G}_{\bar{I}\bar{J}}-\delta^I_i\delta^J_j G_{IJ})/2\}\mathbf{g}^i\otimes\mathbf{g}^j \end{aligned}\right\}, \tag{8.58}$$

noting equations (4.3) and (8.28). From equations (8.56) and (8.58), the following relations hold:

$$\left.\begin{aligned} \mathbf{E}^e &= \overleftarrow{\mathbf{e}}^e_{GG} = \mathbf{F}^T\mathbf{e}^e\mathbf{F} \\ \mathbf{e}^p &= \overrightarrow{\mathbf{E}}^p_{gg} = \mathbf{F}^{-T}\mathbf{E}^p\mathbf{F}^{-1} \end{aligned}\right\}. \tag{8.59}$$

The following strain tensor and its elastic and plastic parts are defined by the push-forward operations of $\mathbf{E}$, $\mathbf{E}^e$, $\mathbf{E}^p$ to the intermediate configuration:

$$\boxed{\bar{\mathbf{E}} = \bar{\mathbf{E}}^e + \bar{\mathbf{E}}^p} \tag{8.60}$$

where

$$\left.\begin{aligned} \bar{\mathbf{E}} &\equiv {}^p\overrightarrow{\mathbf{E}}_{\bar{G}\bar{G}} = \mathbf{F}^{p-T}\mathbf{E}\mathbf{F}^{p-1} = (\bar{\mathbf{C}}^e - \bar{\mathbf{B}}^{p-1})/2 \\ &= \{(\bar{C}^e_{\bar{I}\bar{J}} - \bar{B}^{p-1}_{ij})/2\}\bar{\mathbf{G}}^{\bar{I}} \otimes \bar{\mathbf{G}}^{\bar{J}} = \{(\delta^i_{\bar{I}}\delta^j_{\bar{J}}g_{ij} - \delta^I_{\bar{I}}\delta^J_{\bar{J}}G_{IJ})/2\}\bar{\mathbf{G}}^{\bar{I}} \otimes \bar{\mathbf{G}}^{\bar{J}} \\ \bar{\mathbf{E}}^e &\equiv {}^p\overrightarrow{\mathbf{E}^e}_{\bar{G}\bar{G}} = \mathbf{F}^{p-T}\mathbf{E}^e\mathbf{F}^{p-1} = (\bar{\mathbf{C}}^e - \bar{\mathbf{G}})/2 \\ &= \{(\bar{C}^e_{\bar{I}\bar{J}} - \bar{G}_{\bar{I}\bar{J}})/2\}\bar{\mathbf{G}}^{\bar{I}} \otimes \bar{\mathbf{G}}^{\bar{J}} = \{(\delta^i_{\bar{I}}\delta^j_{\bar{J}}g_{ij} - \bar{G}_{\bar{I}\bar{J}})/2\}\bar{\mathbf{G}}^{\bar{I}} \otimes \bar{\mathbf{G}}^{\bar{J}} \\ \bar{\mathbf{E}}^p &\equiv {}^p\overrightarrow{\mathbf{E}^p}_{\bar{G}\bar{G}} = \mathbf{F}^{p-T}\mathbf{E}^p\mathbf{F}^{p-1} = (\bar{\mathbf{G}} - \bar{\mathbf{B}}^{p-1})/2 \\ &= \{(\bar{G}_{\bar{I}\bar{J}} - \bar{B}^{p-1}_{ij})/2\}\bar{\mathbf{G}}^{\bar{I}} \otimes \bar{\mathbf{G}}^{\bar{J}} = \{(\bar{G}_{\bar{I}\bar{J}} - \delta^I_{\bar{I}}\delta^J_{\bar{J}}G_{IJ})/2\}\bar{\mathbf{G}}^{\bar{I}} \otimes \bar{\mathbf{G}}^{\bar{J}} \end{aligned}\right\}, \tag{8.61}$$

noting

$$\mathbf{F}^{p-T}\mathbf{E}\mathbf{F}^{p-1} = \mathbf{F}^{p-T}\{(\mathbf{F}^T\mathbf{F} - \mathbf{G})/2\}\mathbf{F}^{p-1} = (\mathbf{F}^{eT}\mathbf{F}^e - \bar{\mathbf{G}})/2 + \{\bar{\mathbf{G}} - (\mathbf{F}^p\mathbf{F}^{pT})^{-1}\}/2$$

with equations (8.25) and (8.30).

Further, the plastic strain tensor based in the kinematic hardening-intermediate configuration $\tilde{\mathcal{K}}$ is decomposed additively as follows:

$$\boxed{\tilde{\mathbf{E}}^p = \tilde{\mathbf{E}}^p_e + \tilde{\mathbf{E}}^p_d} \tag{8.62}$$

where

$$\left.\begin{aligned} \tilde{\mathbf{E}}^p &\equiv {}^p_d\overrightarrow{\mathbf{E}^p}_{\tilde{G}\tilde{G}} = \mathbf{F}^{p-T}_d\mathbf{E}^pF^{p-1}_d = (\tilde{\mathbf{C}}^p_e - \tilde{\mathbf{B}}^{p-1}_d)/2 \\ &= \{(\tilde{C}^p_{e\,\tilde{I}\tilde{J}} - \tilde{B}^{p-1}_{d\tilde{I}\tilde{J}})/2\}\tilde{\mathbf{G}}^{\tilde{I}} \otimes \tilde{\mathbf{G}}^{\tilde{J}} = \{(\delta^{\bar{I}}_{\tilde{I}}\delta^{\bar{J}}_{\tilde{J}}\bar{G}_{\bar{I}\bar{J}} - \tilde{B}^{p-1}_{d\tilde{I}\tilde{J}})/2\}\tilde{\mathbf{G}}^{\tilde{I}} \otimes \tilde{\mathbf{G}}^{\tilde{J}} \\ \tilde{\mathbf{E}}^p_e &\equiv (\mathbf{F}^{pT}_e\mathbf{F}^p_e - \tilde{\mathbf{G}})/2 = (\tilde{\mathbf{C}}^p_e - \tilde{\mathbf{G}})/2 \\ &= \{(\tilde{C}^p_{e\,\tilde{I}\tilde{J}} - \tilde{G}_{\tilde{I}\tilde{J}})/2\}\tilde{\mathbf{G}}^{\tilde{I}} \otimes \tilde{\mathbf{G}}^{\tilde{J}} = \delta^{\bar{I}}_{\tilde{I}}\delta^{\bar{J}}_{\tilde{J}}\{(\bar{G}_{\bar{I}\bar{J}} - \bar{G}_{\bar{I}\bar{J}})/2\}\tilde{\mathbf{G}}^{\tilde{I}} \otimes \tilde{\mathbf{G}}^{\tilde{J}} \\ \tilde{\mathbf{E}}^p_d &\equiv (\tilde{\mathbf{G}} - \tilde{\mathbf{B}}^{p-1}_d)/2 \\ &\quad + \{(\tilde{G}_{\tilde{I}\tilde{J}} - \tilde{B}^{p-1}_{d\tilde{I}\tilde{J}})/2\}\tilde{\mathbf{G}}^{\tilde{I}} \otimes \tilde{\mathbf{G}}^{\tilde{J}} = \{(\delta^{\bar{I}}_{\tilde{I}}\delta^{\bar{J}}_{\tilde{J}}\bar{G}_{\bar{I}\bar{J}} - \delta^I_{\tilde{I}}\delta^J_{\tilde{J}}G_{IJ})/2\}\tilde{\mathbf{G}}^{\tilde{I}} \otimes \tilde{\mathbf{G}}^{\tilde{J}} \end{aligned}\right\}, \tag{8.63}$$

noting

$$\begin{aligned} \tilde{\mathbf{E}}^p &\equiv \mathbf{F}^{p-T}_d\mathbf{E}^p\mathbf{F}^{p-1}_d = \mathbf{F}^{p-T}_d\{(\mathbf{F}^p_e\mathbf{F}^p_d)^T\mathbf{F}^p_e\mathbf{F}^p_d - \mathbf{G})/2\}\mathbf{F}^{p-1}_d \\ &= (\mathbf{F}^{pT}_e\mathbf{F}^p_e - \tilde{\mathbf{G}})/2 + \{\tilde{\mathbf{G}} - (\mathbf{F}^{p-T}_d\mathbf{F}^{p-1}_d)\}/2 \end{aligned}$$

with equations (8.44), (8.49) and (8.50). $\tilde{\mathbf{E}}^p$ can be also expressed by the covariant pull-back of $\bar{\mathbf{E}}^p$ by the kinematic hardening-elastic deformation gradient $\mathbf{F}_e^p$ as follows:

$$\tilde{\mathbf{E}}^p \equiv {}^p_e\overleftarrow{\bar{\mathbf{E}}^p}_{\tilde{G}\tilde{G}} = \mathbf{F}_e^{pT}\bar{\mathbf{E}}^p\mathbf{F}_e^p. \tag{8.64}$$

The generalized strain measures are described in Section 4.2.2. The generalized measures for the multiplicative decomposition are discussed in Menzel (2006).

8.6 Strain Rate and Spin Tensors

Various types of velocity gradient and strain rate tensors have been proposed for the finite elasto-plasticity theory, as will be described below.

8.6.1 Strain Rate and Spin Tensors in Current Configuration

Substituting equation (8.1) into equation (4.81), the velocity gradient is additively decomposed into elastic and plastic parts based in the current configuration as follows:

$$\boxed{\mathbf{l} = \mathbf{l}^e + \mathbf{l}^p} \tag{8.65}$$

where

$$\left.\begin{aligned}
\mathbf{l} &= \dot{\mathbf{F}}\mathbf{F}^{-1} = \dot{\mathbf{g}}_i \otimes \mathbf{g}^i = (\mathbf{g}^i \cdot \dot{\mathbf{g}}_j)\mathbf{g}_i \otimes \mathbf{g}^j = (\mathbf{g}_i \cdot \dot{\mathbf{g}}_j)\mathbf{g}^i \otimes \mathbf{g}^j\\
\mathbf{l}^e &\equiv \dot{\mathbf{F}}^e\mathbf{F}^{e-1} = \dot{\mathbf{g}}_i \otimes \mathbf{g}^i - \delta^i_{\bar{I}}\delta^{\bar{J}}_j(\bar{\mathbf{G}}^{\bar{I}} \cdot \dot{\bar{\mathbf{G}}}_{\bar{J}})\mathbf{g}_i \otimes \mathbf{g}^j\\
\bar{\mathbf{L}}^p &\equiv \dot{\mathbf{F}}^p\mathbf{F}^{p-1} = \dot{\bar{\mathbf{G}}}_{\bar{I}} \otimes \bar{\mathbf{G}}^{\bar{I}}\\
\mathbf{l}^p &\equiv \mathbf{F}^e\dot{\mathbf{F}}^p\mathbf{F}^{p-1}\mathbf{F}^{e-1} = \mathbf{F}^e\bar{\mathbf{L}}^p\mathbf{F}^{e-1}(= {}^e\overrightarrow{\bar{\mathbf{L}}^p}_{gg}) = \delta^i_{\bar{I}}\delta^{\bar{J}}_j(\bar{\mathbf{G}}^{\bar{I}} \cdot \dot{\bar{\mathbf{G}}}_{\bar{J}})\mathbf{g}_i \otimes \mathbf{g}^j
\end{aligned}\right\}, \tag{8.66}$$

noting

$$\begin{aligned}
\dot{\mathbf{F}}\mathbf{F}^{-1} &= (\mathbf{F}^e\mathbf{F}^p)^\bullet(\mathbf{F}^e\mathbf{F}^p)^{-1} = (\dot{\mathbf{F}}^e\mathbf{F}^p + \mathbf{F}^e\dot{\mathbf{F}}^p)\mathbf{F}^{p-1}\mathbf{F}^{e-1}\\
&= \dot{\mathbf{F}}^e\mathbf{F}^{e-1} + \mathbf{F}^e\dot{\mathbf{F}}^p\mathbf{F}^{p-1}\mathbf{F}^{e-1}
\end{aligned}$$

with

$$\begin{aligned}
\dot{\mathbf{F}}^e\mathbf{F}^{e-1} &= (\delta^i_{\bar{I}}\mathbf{g}_i \otimes \bar{\mathbf{G}}^{\bar{I}})^\bullet\delta^{\bar{J}}_j\bar{\mathbf{G}}_{\bar{J}} \otimes \mathbf{g}^j = (\delta^i_{\bar{I}}\dot{\mathbf{g}}_i \otimes \bar{\mathbf{G}}^{\bar{I}} + \delta^i_{\bar{I}}\mathbf{g}_i \otimes \dot{\bar{\mathbf{G}}}^{\bar{I}})\delta^{\bar{J}}_j\bar{\mathbf{G}}_{\bar{J}} \otimes \mathbf{g}^j\\
&= \delta^i_{\bar{I}}\delta^{\bar{J}}_j(\bar{\mathbf{G}}^{\bar{I}} \cdot \bar{\mathbf{G}}_{\bar{J}})\dot{\mathbf{g}}_i \otimes \mathbf{g}^j + \delta^i_{\bar{I}}\delta^{\bar{J}}_j(\dot{\bar{\mathbf{G}}}^{\bar{I}} \cdot \bar{\mathbf{G}}_{\bar{J}})\mathbf{g}_i \otimes \mathbf{g}^j,\\
&= \dot{\mathbf{g}}_i \otimes \mathbf{g}^i + \delta^i_{\bar{I}}\delta^{\bar{J}}_j(\dot{\bar{\mathbf{G}}}^{\bar{I}} \cdot \bar{\mathbf{G}}_{\bar{J}})\mathbf{g}_i \otimes \mathbf{g}^j = \dot{\mathbf{g}}_i \otimes \mathbf{g}^i - \delta^i_{\bar{I}}\delta^{\bar{J}}_j(\bar{\mathbf{G}}^{\bar{I}} \cdot \dot{\bar{\mathbf{G}}}_{\bar{J}})\mathbf{g}_i \otimes \mathbf{g}^j,\\
\dot{\mathbf{F}}^p\mathbf{F}^{p-1} &= (\delta^{\bar{I}}_I\dot{\bar{\mathbf{G}}}_{\bar{I}} \otimes \mathbf{G}^I)(\delta^J_{\bar{J}}\mathbf{G}_J \otimes \bar{\mathbf{G}}^{\bar{J}}) = \dot{\bar{\mathbf{G}}}_{\bar{J}} \otimes \bar{\mathbf{G}}^{\bar{J}},\\
\mathbf{F}^e(\dot{\mathbf{F}}^p\mathbf{F}^{p-1})\mathbf{F}^{e-1} &= (\delta^i_{\bar{I}}\mathbf{g}_i \otimes \bar{\mathbf{G}}^{\bar{I}})(\dot{\bar{\mathbf{G}}}_{\bar{R}} \otimes \bar{\mathbf{G}}^{\bar{R}})(\delta^{\bar{J}}_j\bar{\mathbf{G}}_{\bar{J}} \otimes \mathbf{g}^j) = \delta^i_{\bar{I}}\delta^{\bar{J}}_j\delta^{\bar{R}}_{\bar{J}}(\bar{\mathbf{G}}^{\bar{I}} \cdot \dot{\bar{\mathbf{G}}}_{\bar{R}})\mathbf{g}_i \otimes \mathbf{g}^j
\end{aligned}$$

with $(\bar{\mathbf{G}}^{\bar{I}} \cdot \bar{\mathbf{G}}_{\bar{J}})^\bullet = \bar{\mathbf{G}}^{\bar{I}} \cdot \dot{\bar{\mathbf{G}}}_{\bar{J}} + \dot{\bar{\mathbf{G}}}^{\bar{I}} \cdot \bar{\mathbf{G}}_{\bar{J}} = 0.$

Here, $\bar{\mathbf{L}}^p$ is referred to as the *plastic velocity gradient defined in the intermediate configuration* because of

$$(d\bar{\mathbf{X}})^\bullet = \bar{\mathbf{L}}^p d\bar{\mathbf{X}}, \tag{8.67}$$

noting $(d\bar{\mathbf{X}})^\bullet = \dot{\mathbf{F}}^p d\mathbf{X} = \dot{\mathbf{F}}^p \mathbf{F}^{p-1} d\bar{\mathbf{X}}$.

It follows from equation (8.65) that

$$\boxed{\mathbf{d} = \mathbf{d}^e + \mathbf{d}^p, \quad \mathbf{w} = \mathbf{w}^e + \mathbf{w}^p} \tag{8.68}$$

where

$$\mathbf{d} \equiv \mathrm{sym}[\boldsymbol{l}], \quad \mathbf{d}^e \equiv \mathrm{sym}[\boldsymbol{l}^e], \quad \mathbf{d}^p \equiv \mathrm{sym}[\boldsymbol{l}^p], \tag{8.69}$$

$$\mathbf{w} \equiv \mathrm{ant}[\boldsymbol{l}], \quad \mathbf{w}^e \equiv \mathrm{ant}[\boldsymbol{l}^e], \quad \mathbf{w}^p \equiv \mathrm{ant}[\boldsymbol{l}^p]. \tag{8.70}$$

$\mathbf{w}^e$ and $\mathbf{w}^p$ are the spin of substructure and the plastic spin, respectively. Noting equation $(4.97)_1$, it follows that

$$\left.\begin{aligned}
\mathbf{d} &= \frac{1}{2}\dot{g}_{ij}\mathbf{g}^i \otimes \mathbf{g}^j \\
\mathbf{d}^e &= \frac{1}{2}\{\dot{g}_{ij}\mathbf{g}^i \otimes \mathbf{g}^j - (\delta^r_{\bar{I}}\delta^{\bar{J}}_j g_{ir} + \delta^r_{\bar{I}}\delta^{\bar{J}}_i g_{jr})(\bar{\mathbf{G}}^{\bar{I}} \cdot \dot{\bar{\mathbf{G}}}_{\bar{J}})\mathbf{g}^i \otimes \mathbf{g}^j\} \\
\mathbf{d}^p &= \frac{1}{2}(\delta^r_{\bar{I}}\delta^{\bar{J}}_j g_{ir} + \delta^r_{\bar{I}}\delta^{\bar{J}}_i g_{jr})(\bar{\mathbf{G}}^{\bar{I}} \cdot \dot{\bar{\mathbf{G}}}_{\bar{J}})\mathbf{g}^i \otimes \mathbf{g}^j
\end{aligned}\right\} \tag{8.71}$$

because of

$$\begin{aligned}
2\mathrm{sym}[\boldsymbol{l}^e] &= \dot{g}_{ij}\mathbf{g}^i \otimes \mathbf{g}^j - \delta^i_{\bar{I}}\delta^{\bar{J}}_j(\bar{\mathbf{G}}^{\bar{I}} \cdot \dot{\bar{\mathbf{G}}}_{\bar{J}})\mathbf{g}_i \otimes \mathbf{g}^j - \delta^i_{\bar{I}}\delta^{\bar{J}}_j(\bar{\mathbf{G}}^{\bar{I}} \cdot \dot{\bar{\mathbf{G}}}_{\bar{J}})\mathbf{g}^j \otimes \mathbf{g}_i \\
&= \dot{g}_{ij}\mathbf{g}^i \otimes \mathbf{g}^j - \delta^i_{\bar{I}}\delta^{\bar{J}}_j(\bar{\mathbf{G}}^{\bar{I}} \cdot \dot{\bar{\mathbf{G}}}_{\bar{J}})(\mathbf{g}_r \cdot \mathbf{g}_i)\mathbf{g}^r \otimes \mathbf{g}^j - \delta^i_{\bar{I}}\delta^{\bar{J}}_j(\bar{\mathbf{G}}^{\bar{I}} \cdot \dot{\bar{\mathbf{G}}}_{\bar{J}})\mathbf{g}^j \otimes (\mathbf{g}_r \cdot \mathbf{g}_i)\mathbf{g}^r \\
&= \dot{g}_{ij}\mathbf{g}^i \otimes \mathbf{g}^j - \delta^r_{\bar{I}}\delta^{\bar{J}}_j(\bar{\mathbf{G}}^{\bar{I}} \cdot \dot{\bar{\mathbf{G}}}_{\bar{J}})(\mathbf{g}_i \cdot \mathbf{g}_r)\mathbf{g}^i \otimes \mathbf{g}^j - \delta^r_{\bar{I}}\delta^{\bar{J}}_i(\bar{\mathbf{G}}^{\bar{I}} \cdot \dot{\bar{\mathbf{G}}}_{\bar{J}})\mathbf{g}^i \otimes (\mathbf{g}_j \cdot \mathbf{g}_r)\mathbf{g}^j \\
&= \dot{g}_{ij}\mathbf{g}^i \otimes \mathbf{g}^j - (\delta^r_{\bar{I}}\delta^{\bar{J}}_j g_{ir} + \delta^r_{\bar{I}}\delta^{\bar{J}}_i g_{jr})(\bar{\mathbf{G}}^{\bar{I}} \cdot \dot{\bar{\mathbf{G}}}_{\bar{J}})\mathbf{g}^i \otimes \mathbf{g}^j, \\
2\mathrm{sym}[\boldsymbol{l}^p] &= \delta^i_{\bar{I}}\delta^{\bar{J}}_j(\bar{\mathbf{G}}^{\bar{I}} \cdot \dot{\bar{\mathbf{G}}}_{\bar{J}})\mathbf{g}_i \otimes \mathbf{g}^j + \delta^i_{\bar{I}}\delta^{\bar{J}}_j(\bar{\mathbf{G}}^{\bar{I}} \cdot \dot{\bar{\mathbf{G}}}_{\bar{J}})\mathbf{g}^j \otimes \mathbf{g}_i \\
&= \delta^r_{\bar{I}}\delta^{\bar{J}}_j(\bar{\mathbf{G}}^{\bar{I}} \cdot \dot{\bar{\mathbf{G}}}_{\bar{J}})\mathbf{g}_r \otimes \mathbf{g}^j + \delta^r_{\bar{I}}\delta^{\bar{J}}_i(\bar{\mathbf{G}}^{\bar{I}} \cdot \dot{\bar{\mathbf{G}}}_{\bar{J}})\mathbf{g}^i \otimes \mathbf{g}_r \\
&= \delta^r_{\bar{I}}\delta^{\bar{J}}_j(\bar{\mathbf{G}}^{\bar{I}} \cdot \dot{\bar{\mathbf{G}}}_{\bar{J}})(\mathbf{g}_i \cdot \mathbf{g}_r)\mathbf{g}^i \otimes \mathbf{g}^j + \delta^r_{\bar{I}}\delta^{\bar{J}}_i(\bar{\mathbf{G}}^{\bar{I}} \cdot \dot{\bar{\mathbf{G}}}_{\bar{J}})\mathbf{g}^i \otimes (\mathbf{g}_j \cdot \mathbf{g}_r)\mathbf{g}^j \\
&= (\delta^r_{\bar{I}}\delta^{\bar{J}}_j \mathbf{g}_{ir} + \delta^r_{\bar{I}}\delta^{\bar{J}}_i \mathbf{g}_{jr})(\bar{\mathbf{G}}^{\bar{I}} \cdot \dot{\bar{\mathbf{G}}}_{\bar{J}})\mathbf{g}^i \otimes \mathbf{g}^j.
\end{aligned}$$

As shown in equations (8.68) with (8.69), the strain rate $\mathbf{d}$ is not additively decomposed into purely elastic and purely plastic parts since $\mathbf{d}^p = \mathrm{sym}[\mathbf{F}^e\dot{\mathbf{F}}^p\mathbf{F}^{p-1}\mathbf{F}^{e-1}]$ depends on the elastic deformation gradient. However, an additive decomposition holds by assuming that the elastic deformation and rotation is infinitesimal, that is $\mathbf{F}^e = \delta^i_{\bar{I}}\mathbf{g}_i \otimes \bar{\mathbf{G}}^{\bar{I}} \cong \mathbf{g}_i \otimes \mathbf{g}^i = \mathbf{g}$ $(\bar{\mathbf{G}}^{\bar{I}} \cong \mathbf{g}^i)$, leading to $\mathbf{d}^p \cong \mathrm{sym}[\mathbf{g}\bar{\mathbf{L}}^p\mathbf{g}^{-1}] = \mathrm{sym}[\dot{\mathbf{F}}^p\mathbf{F}^{p-1}]$ (Dafalias, 1984).

8.6.2 Contravariant–Covariant Pulled-Back Strain Rate and Spin Tensors in Intermediate Configuration

The following velocity gradient tensor based on the contravariant–covariant pull-back of the velocity gradient tensor $\boldsymbol{l}$ by $\mathbf{F}^e$ to the intermediate configuration is additively decomposed into purely elastic and plastic parts, noting equations (8.65) and (8.66):

$$\boxed{\bar{\mathbf{L}} = \bar{\mathbf{L}}^e + \bar{\mathbf{L}}^p} \tag{8.72}$$

where

$$\left.\begin{aligned}
\bar{\mathbf{L}} &\equiv {}^{e}\overleftarrow{\boldsymbol{l}}^{\bar{G}}_{\bullet\bar{G}} = \mathbf{F}^{e-1}\boldsymbol{l}\mathbf{F}^e = \delta^{\bar{I}}_i\delta^j_{\bar{J}}(\mathbf{g}^i\cdot\dot{\mathbf{g}}_j)\bar{\mathbf{G}}_{\bar{I}}\otimes\bar{\mathbf{G}}^{\bar{J}}\\
\bar{\mathbf{L}}^e &\equiv {}^{e}\overleftarrow{\boldsymbol{l}}^{e\bar{G}}_{\bullet\bar{G}} = \mathbf{F}^{e-1}\boldsymbol{l}^e\mathbf{F}^e = \mathbf{F}^{e-1}\dot{\mathbf{F}}^e\\
&= \delta^{\bar{I}}_i\delta^j_{\bar{J}}(\mathbf{g}^i\cdot\dot{\mathbf{g}}_j)\bar{\mathbf{G}}_{\bar{I}}\otimes\bar{\mathbf{G}}^{\bar{J}} - \dot{\bar{\mathbf{G}}}_{\bar{I}}\otimes\bar{\mathbf{G}}^{\bar{I}}\\
\bar{\mathbf{L}}^p &\equiv {}^{e}\overleftarrow{\boldsymbol{l}}^{p\bar{G}}_{\bullet\bar{G}} = \mathbf{F}^{e-1}\boldsymbol{l}^p\mathbf{F}^e = \dot{\mathbf{F}}^p\mathbf{F}^{p-1} = \dot{\bar{\mathbf{G}}}_{\bar{I}}\otimes\bar{\mathbf{G}}^{\bar{I}}
\end{aligned}\right\}, \tag{8.73}$$

noting equation (4.84) and

$$\begin{aligned}
\mathbf{F}^{e-1}\boldsymbol{l}\mathbf{F}^e &= \delta^{\bar{I}}_i\bar{\mathbf{G}}_{\bar{I}}\otimes\mathbf{g}^i(\dot{\mathbf{g}}_r\otimes\mathbf{g}^r)\delta^j_{\bar{J}}\mathbf{g}_j\otimes\bar{\mathbf{G}}^{\bar{J}}\\
&= \delta^{\bar{I}}_i\delta^j_{\bar{J}}\delta^r_j\delta(\mathbf{g}^i\cdot\dot{\mathbf{g}}_r)\bar{\mathbf{G}}_{\bar{I}}\otimes\bar{\mathbf{G}}^{\bar{J}} = \delta^{\bar{I}}_i\delta^j_{\bar{J}}(\mathbf{g}^i\cdot\dot{\mathbf{g}}_j)\bar{\mathbf{G}}_{\bar{I}}\otimes\bar{\mathbf{G}}^{\bar{J}},\\
\mathbf{F}^{e-1}\boldsymbol{l}^e\mathbf{F}^e &= \delta^{\bar{I}}_i\bar{\mathbf{G}}_{\bar{I}}\otimes\mathbf{g}^i\{\dot{\mathbf{g}}_r\otimes\mathbf{g}^r - \delta^r_{\bar{R}}\delta^{\bar{S}}_s(\bar{\mathbf{G}}^{\bar{R}}\cdot\dot{\bar{\mathbf{G}}}_{\bar{S}})\mathbf{g}_r\otimes\mathbf{g}^s\}\delta^j_{\bar{J}}\mathbf{g}_j\otimes\bar{\mathbf{G}}^{\bar{J}}\\
&= \delta^{\bar{I}}_i\delta^j_{\bar{J}}\delta^r_j(\mathbf{g}^i\cdot\dot{\mathbf{g}}_r)\bar{\mathbf{G}}_{\bar{I}}\otimes\bar{\mathbf{G}}^{\bar{J}} - (\delta^{\bar{I}}_i\delta^r_{\bar{R}}\delta^i_r)(\delta^{\bar{S}}_s\delta^j_{\bar{J}}\delta^s_j)(\bar{\mathbf{G}}^{\bar{R}}\cdot\dot{\bar{\mathbf{G}}}_{\bar{S}})\bar{\mathbf{G}}_{\bar{I}}\otimes\bar{\mathbf{G}}^{\bar{J}}\\
&= \delta^{\bar{I}}_i\delta^j_{\bar{J}}(\mathbf{g}^i\cdot\dot{\mathbf{g}}_j)\bar{\mathbf{G}}_{\bar{I}}\otimes\bar{\mathbf{G}}^{\bar{J}} - (\bar{\mathbf{G}}^{\bar{I}}\cdot\dot{\bar{\mathbf{G}}}_{\bar{J}})\bar{\mathbf{G}}_{\bar{I}}\otimes\bar{\mathbf{G}}^{\bar{J}}\\
&= \delta^{\bar{I}}_i\delta^j_{\bar{J}}(\mathbf{g}^i\cdot\dot{\mathbf{g}}_j)\bar{\mathbf{G}}_{\bar{I}}\otimes\bar{\mathbf{G}}^{\bar{J}} - \dot{\bar{\mathbf{G}}}_{\bar{I}}\otimes\bar{\mathbf{G}}^{\bar{I}}.
\end{aligned}$$

The following strain rate and the spin tensors are defined based on equation (8.72):

$$\boxed{\bar{\mathbf{D}} = \bar{\mathbf{D}}^e + \bar{\mathbf{D}}^p,\quad \bar{\mathbf{W}} = \bar{\mathbf{W}}^e + \bar{\mathbf{W}}^p} \tag{8.74}$$

where

$$\bar{\mathbf{D}} \equiv \mathrm{sym}[\bar{\mathbf{L}}],\quad \bar{\mathbf{D}}^e \equiv \mathrm{sym}[\bar{\mathbf{L}}^e],\quad \bar{\mathbf{D}}^p \equiv \mathrm{sym}[\bar{\mathbf{L}}^p], \tag{8.75}$$

$$\bar{\mathbf{W}} \equiv \mathrm{ant}[\bar{\mathbf{L}}],\quad \bar{\mathbf{W}}^e \equiv \mathrm{ant}[\bar{\mathbf{L}}^e],\quad \bar{\mathbf{W}}^p \equiv \mathrm{ant}[\bar{\mathbf{L}}^p]. \tag{8.76}$$

Noting equation (2.28), it follows from equation (8.73) that

$$\left.\begin{aligned}
\bar{\mathbf{D}} &= \frac{1}{2}(\delta_i^{\bar{R}}\delta_{\bar{J}}^{j}G_{\bar{I}\bar{R}} + \delta_i^{\bar{R}}\delta_{\bar{I}}^{j}G_{\bar{J}\bar{R}})(\mathbf{g}^i\cdot\overset{\bullet}{\mathbf{g}}_j)\bar{\mathbf{G}}^{\bar{I}}\otimes\bar{\mathbf{G}}^{\bar{J}}\\
\bar{\mathbf{D}}^e &= \frac{1}{2}\{(\delta_i^{\bar{R}}\delta_{\bar{J}}^{j}G_{\bar{I}\bar{R}} + \delta_i^{\bar{R}}\delta_{\bar{I}}^{j}G_{\bar{J}\bar{R}})(\mathbf{g}^i\cdot\overset{\bullet}{\mathbf{g}}_j) - \overset{\bullet}{\bar{G}}_{\bar{I}\bar{J}}\}\bar{\mathbf{G}}^{\bar{I}}\otimes\bar{\mathbf{G}}^{\bar{J}}\\
\bar{\mathbf{D}}^p &= \frac{1}{2}\overset{\bullet}{\bar{G}}_{\bar{I}\bar{J}}\bar{\mathbf{G}}^{\bar{I}}\otimes\bar{\mathbf{G}}^{\bar{J}}
\end{aligned}\right\} \tag{8.77}$$

because of

$$\begin{aligned}
2\mathrm{sym}[\bar{\mathbf{L}}] &= 2\mathrm{sym}[\delta_i^{\bar{I}}\delta_{\bar{J}}^{j}(\mathbf{g}^i\cdot\overset{\bullet}{\mathbf{g}}_j)\bar{\mathbf{G}}_{\bar{I}}\otimes\bar{\mathbf{G}}^{\bar{J}}]\\
&= \delta_i^{\bar{I}}\delta_{\bar{J}}^{j}(\mathbf{g}^i\cdot\overset{\bullet}{\mathbf{g}}_j)\bar{\mathbf{G}}_{\bar{I}}\otimes\bar{\mathbf{G}}^{\bar{J}} + \delta_i^{\bar{I}}\delta_{\bar{J}}^{j}(\mathbf{g}^i\cdot\overset{\bullet}{\mathbf{g}}_j)\bar{\mathbf{G}}^{\bar{J}}\otimes\bar{\mathbf{G}}_{\bar{I}}\\
&= \delta_i^{\bar{I}}\delta_{\bar{J}}^{j}(\mathbf{g}^i\cdot\overset{\bullet}{\mathbf{g}}_j)\bar{\mathbf{G}}_{\bar{I}}\otimes\bar{\mathbf{G}}^{\bar{J}} + \delta_i^{\bar{J}}\delta_{\bar{I}}^{j}(\mathbf{g}^i\cdot\overset{\bullet}{\mathbf{g}}_j)\bar{\mathbf{G}}^{\bar{I}}\otimes\bar{\mathbf{G}}_{\bar{J}}\\
&= \delta_i^{\bar{I}}\delta_{\bar{J}}^{j}(\mathbf{g}^i\cdot\overset{\bullet}{\mathbf{g}}_j)(\bar{\mathbf{G}}_{\bar{I}}\cdot\bar{\mathbf{G}}_{\bar{R}})\bar{\mathbf{G}}^{\bar{R}}\otimes\bar{\mathbf{G}}^{\bar{J}} + \delta_i^{\bar{J}}\delta_{\bar{I}}^{j}(\mathbf{g}^i\cdot\overset{\bullet}{\mathbf{g}}_j)\bar{\mathbf{G}}^{\bar{I}}\otimes(\bar{\mathbf{G}}_{\bar{J}}\cdot\bar{G}_{\bar{R}})\bar{\mathbf{G}}^{\bar{R}}\\
&= \delta_i^{\bar{R}}\delta_{\bar{J}}^{j}(\mathbf{g}^i\cdot\overset{\bullet}{\mathbf{g}}_j)(\bar{\mathbf{G}}_{\bar{I}}\cdot\bar{\mathbf{G}}_{\bar{R}})\bar{\mathbf{G}}^{\bar{I}}\otimes\bar{\mathbf{G}}^{\bar{J}} + \delta_i^{\bar{R}}\delta_{\bar{I}}^{j}(\mathbf{g}^i\cdot\overset{\bullet}{\mathbf{g}}_j)\bar{\mathbf{G}}^{\bar{I}}\otimes(\bar{G}_{\bar{J}}\cdot\bar{G}_{\bar{R}})\bar{\mathbf{G}}^{\bar{J}}\\
&= (\delta_i^{\bar{R}}\delta_{\bar{J}}^{j}G_{\bar{I}\bar{R}} + \delta_i^{\bar{R}}\delta_{\bar{I}}^{j}G_{\bar{J}\bar{R}})(\mathbf{g}^i\cdot\overset{\bullet}{\mathbf{g}}_j)\bar{\mathbf{G}}^{\bar{I}}\otimes\bar{\mathbf{G}}^{\bar{J}},\\
2\mathrm{sym}[\bar{\mathbf{L}}^p] &= 2\mathrm{sym}[\overset{\bullet}{\bar{\mathbf{G}}}_{\bar{I}}\otimes\bar{\mathbf{G}}^{\bar{I}}] = \overset{\bullet}{\bar{G}}_{\bar{I}\bar{J}}\bar{\mathbf{G}}^{\bar{I}}\otimes\bar{\mathbf{G}}^{\bar{J}}.
\end{aligned}$$

Further, it follows from equations (2.29) and $(8.73)_3$ that

$$\bar{\mathbf{W}}^p = \mathrm{ant}[\overset{\bullet}{\bar{\mathbf{G}}}_{\bar{I}}\otimes\bar{\mathbf{G}}^{\bar{I}}] = \frac{1}{2}(\bar{\mathbf{G}}_{\bar{I}}\cdot\overset{\bullet}{\bar{\mathbf{G}}}_{\bar{J}} - \overset{\bullet}{\bar{\mathbf{G}}}_{\bar{I}}\cdot\bar{\mathbf{G}}_{\bar{J}})\bar{\mathbf{G}}^{\bar{I}}\otimes\bar{\mathbf{G}}^{\bar{J}}. \tag{8.78}$$

The plastic strain rate $\bar{\mathbf{D}}^p$ is related to the rate of the plastic right Cauchy–Green deformation tensor as follows:

$$\overset{\bullet}{\mathbf{C}}^p = 2\mathbf{F}^{pT}\bar{\mathbf{D}}^p\mathbf{F}^p = {}^p\overset{\leftarrow}{\bar{\mathbf{D}}}{}^p_{GG}, \tag{8.79}$$

noting that

$$(\mathbf{F}^{pT}\mathbf{F}^p)^\bullet = \overset{\bullet}{\mathbf{F}}^{pT}\mathbf{F}^p + \mathbf{F}^{pT}\overset{\bullet}{\mathbf{F}}^p = \mathbf{F}^{pT}(\mathbf{F}^{p-T}\overset{\bullet}{\mathbf{F}}^{pT} + \overset{\bullet}{\mathbf{F}}^p\mathbf{F}^{p-1})\mathbf{F}^p.$$

The rate of $\bar{\mathbf{C}}^e$ is given from equations (8.25) and (8.72) with equation $(8.73)_2$ as

$$\overset{\bullet}{\bar{\mathbf{C}}}^e = 2\mathrm{sym}[\bar{\mathbf{C}}^e\bar{\mathbf{L}}^e)] = 2\mathrm{sym}[\bar{\mathbf{C}}^e(\bar{\mathbf{L}} - \bar{\mathbf{L}}^p)], \tag{8.80}$$

noting that

$$\begin{aligned}
\overset{\bullet}{\bar{\mathbf{C}}}^e &= (\mathbf{F}^{eT}\mathbf{F}^e)^\bullet = \mathbf{F}^{eT}\overset{\bullet}{\mathbf{F}}^e + \overset{\bullet}{\mathbf{F}}^{eT}\mathbf{F}^e = \mathbf{F}^{eT}\mathbf{F}^e(\mathbf{F}^{e-1}\overset{\bullet}{\mathbf{F}}^e) + (\overset{\bullet}{\mathbf{F}}^{eT}\mathbf{F}^{e-T})\mathbf{F}^{eT}\mathbf{F}^e.\\
&= \bar{\mathbf{C}}^e\bar{\mathbf{L}}^e + \bar{\mathbf{L}}^{eT}\bar{\mathbf{C}}^e.
\end{aligned}$$

$\bar{\mathbf{W}}^p$ is written from equation $(8.76)_3$ with equation $(8.66)_3$ as

$$\boxed{\bar{\mathbf{W}}^p = \bar{\boldsymbol{\Omega}}^p + \bar{\boldsymbol{\Omega}}^p_R} \tag{8.81}$$

where

$$\boxed{\bar{\boldsymbol{\Omega}}^p \equiv \mathrm{ant}[\mathbf{R}^p\,\dot{\mathbf{U}}^p\,\mathbf{U}^{p-1}\mathbf{R}^{pT}]} \tag{8.82}$$

$$\boxed{\bar{\boldsymbol{\Omega}}^p_R \equiv \dot{\mathbf{R}}^p\mathbf{R}^{pT}} = \bar{\boldsymbol{\Omega}}_{\bar{N}} + \sum_{\alpha=1}^{3}\bar{\mathbf{N}}^{(\alpha)} \otimes \dot{\mathbf{N}}^{(\alpha)} \sum_{\beta=1}^{3}\mathbf{N}^{(\beta)} \otimes \bar{\mathbf{N}}^{(\beta)}, \tag{8.83}$$

$$\bar{\boldsymbol{\Omega}}_{\bar{N}} \equiv \sum_{\alpha=1}^{3}\dot{\bar{\mathbf{N}}}^{(\alpha)} \otimes \bar{\mathbf{N}}^{(\alpha)}, \tag{8.84}$$

noting

$$\bar{\mathbf{W}}^p = \mathrm{ant}[\dot{\mathbf{F}}^p\mathbf{F}^{p-1}] = \mathrm{ant}[(\mathbf{R}^p\,\mathbf{U}^p)^{\bullet}(\mathbf{R}^p\,\mathbf{U}^p)^{-1}] = \mathrm{ant}[\mathbf{R}^p\,\dot{\mathbf{U}}^p\,\mathbf{U}^{p-1}\mathbf{R}^{pT}] + \dot{\mathbf{R}}^p\mathbf{R}^{pT} \tag{8.85}$$

with equation $(8.17)_2$.

The three spins are incorporated above. $\bar{\boldsymbol{\Omega}}^p$ is the *plastic spin*, and $\bar{\mathbf{W}}^p$ and $\bar{\boldsymbol{\Omega}}^p_R$ are referred to as the *plastic material spin* and the *constitutive spin*, respectively by Dafalias (1998), which are all based in the intermediate configuration.

8.6.3 *Covariant Pulled-Back Strain Rate and Spin Tensors in Intermediate Configuration*

The following velocity gradient tensor with additive decomposition (Simo, 1998; Belytschko et al., 2000) is defined as the covariant pull-back of the strain rate tensor $\boldsymbol{l}$ by $\mathbf{F}^e$ to the intermediate configuration, noting equations (8.65) and (8.66):

$$\boxed{\bar{\mathsf{L}} = \bar{\mathsf{L}}^e + \bar{\mathsf{L}}^p} \tag{8.86}$$

where

$$\left.\begin{aligned}
\bar{\mathsf{L}} &\equiv {}^{e}\overleftarrow{\boldsymbol{l}}_{\bar{G}\bar{G}} = \mathbf{F}^{eT}\boldsymbol{l}\mathbf{F}^e = \bar{\mathbf{C}}^e\bar{\mathbf{L}} = \delta^i_{\bar{I}}\delta^j_{\bar{J}}(\mathbf{g}_i \cdot \dot{\mathbf{g}}_j)\bar{\mathbf{G}}^{\bar{I}} \otimes \bar{\mathbf{G}}^{\bar{J}} \\
&= {}^{p}\overrightarrow{\mathsf{L}}_{\bar{G}\bar{G}} \\
\bar{\mathsf{L}}^e &\equiv {}^{e}\overleftarrow{\boldsymbol{l}}^{\,e}_{\bar{G}\bar{G}} = \mathbf{F}^{eT}\boldsymbol{l}^e\mathbf{F}^e = \mathbf{F}^{eT}\dot{\mathbf{F}}^e = \bar{\mathbf{C}}^e\bar{\mathbf{L}}^e \\
&= \delta^i_{\bar{I}}\delta^j_{\bar{J}}(\mathbf{g}_i \cdot \dot{\mathbf{g}}_j)\bar{\mathbf{G}}^{\bar{I}} \otimes \bar{\mathbf{G}}^{\bar{J}} + \delta^i_{\bar{I}}\delta^r_{\bar{R}}\delta^{\bar{S}}_s\delta^s_{\bar{J}}g_{ir}(\bar{\mathbf{G}}^{\bar{R}} \cdot \dot{\bar{\mathbf{G}}}_{\bar{S}})\bar{\mathbf{G}}^{\bar{I}} \otimes \bar{\mathbf{G}}^{\bar{J}} \\
\bar{\mathsf{L}}^p &\equiv {}^{e}\overleftarrow{\boldsymbol{l}}^{\,p}_{\bar{G}\bar{G}} = \mathbf{F}^{eT}\boldsymbol{l}^p\mathbf{F}^e = \bar{\mathbf{C}}^e\bar{\mathbf{L}}^p = \delta^i_{\bar{I}}\delta^j_{\bar{R}}g_{ij}(\bar{\mathbf{G}}^{\bar{R}} \cdot \dot{\bar{\mathbf{G}}}_{\bar{J}})\bar{\mathbf{G}}^{\bar{I}} \otimes \bar{\mathbf{G}}^{\bar{J}}
\end{aligned}\right\}, \tag{8.87}$$

noting

$$\begin{aligned}
\mathbf{F}^{eT}\boldsymbol{l}\mathbf{F}^{e} &= \mathbf{F}^{eT}\dot{\mathbf{F}}\mathbf{F}^{-1}\mathbf{F}^{e} = \delta^{i}_{\bar{I}}\bar{\mathbf{G}}^{\bar{I}}\otimes\mathbf{g}_{i}\dot{\mathbf{g}}_{r}\otimes\mathbf{g}^{r}\delta^{j}_{\bar{J}}\mathbf{g}_{j}\otimes\bar{\mathbf{G}}^{\bar{J}}\\
&= \delta^{i}_{\bar{I}}\delta^{j}_{\bar{J}}\delta^{r}_{j}\delta(\mathbf{g}_{i}\cdot\dot{\mathbf{g}}_{r})\bar{\mathbf{G}}^{\bar{I}}\otimes\bar{\mathbf{G}}^{\bar{J}} = \delta^{i}_{\bar{I}}\delta^{j}_{\bar{J}}(\mathbf{g}_{i}\cdot\dot{\mathbf{g}}_{j})\bar{\mathbf{G}}^{\bar{I}}\otimes\bar{\mathbf{G}}^{\bar{J}},\\
\mathbf{F}^{eT}\boldsymbol{l}\mathbf{F}^{e} &= \mathbf{F}^{p-T}\mathbf{F}^{T}\boldsymbol{l}\mathbf{F}\mathbf{F}^{p-1} = \mathbf{F}^{p-T}\mathbf{L}\mathbf{F}^{p-1}\\
\mathbf{F}^{eT}\boldsymbol{l}^{e}\mathbf{F}^{e} &= \delta^{i}_{\bar{I}}\bar{\mathbf{G}}^{\bar{I}}\otimes\mathbf{g}_{i}\{\dot{\mathbf{g}}_{r}\otimes\mathbf{g}^{r}+\delta^{r}_{\bar{R}}\delta^{\bar{S}}_{s}\dot{\bar{\mathbf{G}}}^{\bar{R}}\cdot\bar{\mathbf{G}}_{\bar{S}}\mathbf{g}_{r}\otimes\mathbf{g}^{s}\}\delta^{j}_{\bar{J}}\mathbf{g}_{j}\otimes\bar{\mathbf{G}}^{\bar{J}}\\
&= \delta^{i}_{\bar{I}}\delta^{r}_{j}\delta^{j}_{\bar{J}}(\mathbf{g}_{i}\cdot\dot{\mathbf{g}}_{r})\bar{\mathbf{G}}^{\bar{I}}\otimes\bar{\mathbf{G}}^{\bar{J}}\\
&\quad+\delta^{i}_{\bar{I}}\delta^{r}_{\bar{R}}\delta^{\bar{S}}_{s}\delta^{j}_{\bar{J}}\delta^{s}_{j}\mathbf{g}_{ir}(\bar{\mathbf{G}}^{\bar{R}}\cdot\dot{\bar{\mathbf{G}}}_{\bar{S}})\bar{\mathbf{G}}^{\bar{I}}\otimes\bar{\mathbf{G}}^{\bar{J}}\\
&= \delta^{i}_{\bar{I}}\delta^{j}_{\bar{J}}(\mathbf{g}_{i}\cdot\dot{\mathbf{g}}_{j})\bar{\mathbf{G}}^{\bar{I}}\otimes\bar{\mathbf{G}}^{\bar{J}}+\delta^{i}_{\bar{I}}\delta^{r}_{\bar{R}}\mathbf{g}_{ir}(\bar{\mathbf{G}}^{\bar{R}}\cdot\dot{\bar{\mathbf{G}}}_{\bar{J}})\bar{\mathbf{G}}^{\bar{I}}\otimes\bar{\mathbf{G}}^{\bar{J}},\\
\mathbf{F}^{eT}\boldsymbol{l}^{p}\mathbf{F}^{e} &= \delta^{i}_{\bar{I}}\bar{\mathbf{G}}^{\bar{I}}\otimes\mathbf{g}_{i}\{\delta^{r}_{\bar{R}}\delta^{\bar{S}}_{s}(\bar{\mathbf{G}}^{\bar{R}}\cdot\dot{\bar{\mathbf{G}}}_{\bar{S}})\mathbf{g}_{r}\otimes\mathbf{g}^{s}\}\delta^{j}_{\bar{J}}\mathbf{g}_{j}\otimes\bar{\mathbf{G}}^{\bar{J}}\\
&= \delta^{i}_{\bar{I}}\delta^{r}_{\bar{R}}\delta^{\bar{S}}_{s}\delta^{j}_{\bar{J}}\delta^{s}_{j}g_{ir}(\bar{\mathbf{G}}^{\bar{R}}\cdot\dot{\bar{\mathbf{G}}}_{\bar{S}})\bar{\mathbf{G}}^{\bar{I}}\otimes\bar{\mathbf{G}}^{\bar{J}} = \delta^{i}_{\bar{I}}\delta^{j}_{\bar{R}}g_{ij}(\bar{\mathbf{G}}^{\bar{R}}\cdot\dot{\bar{\mathbf{G}}}_{\bar{J}})\bar{\mathbf{G}}^{\bar{I}}\otimes\bar{\mathbf{G}}^{\bar{J}}_{Hv},
\end{aligned}$$

or

$$\begin{aligned}
\bar{\mathbf{C}}^{e}\bar{\mathbf{L}}^{p} &= \delta^{i}_{\bar{I}}\delta^{j}_{\bar{J}}g_{ij}\bar{\mathbf{G}}^{\bar{I}}\otimes\bar{\mathbf{G}}^{\bar{J}}\dot{\bar{\mathbf{G}}}_{\bar{R}}\otimes\ \bar{\mathbf{G}}^{\bar{R}} = \delta^{i}_{\bar{I}}\delta^{j}_{\bar{J}}g_{ij}(\bar{\mathbf{G}}^{\bar{J}}\cdot\dot{\bar{\mathbf{G}}}_{\bar{R}})\bar{\mathbf{G}}^{\bar{I}}\otimes\bar{\mathbf{G}}^{\bar{R}}\\
&= \delta^{i}_{\bar{I}}\delta^{j}_{\bar{R}}g_{ij}(\bar{\mathbf{G}}^{\bar{J}}\cdot\dot{\bar{\mathbf{G}}}_{\bar{J}})\bar{\mathbf{G}}^{\bar{I}}\otimes\bar{\mathbf{G}}^{\bar{J}}
\end{aligned}$$

with equations (3.10), (3.11), (4.83), (4.92) and (8.66).

It follows from equation (8.86) with equation (8.87) that

$$\boxed{\bar{\mathbf{D}}=\bar{\mathbf{D}}^{e}+\bar{\mathbf{D}}^{p},\ \bar{\mathbf{W}}=\bar{\mathbf{W}}^{e}+\bar{\mathbf{W}}^{p}} \tag{8.88}$$

where

$$\left.\begin{aligned}
\bar{\mathbf{D}} &\equiv {}^{e}\overleftarrow{\mathbf{d}}_{\bar{G}\bar{G}} = \mathbf{F}^{eT}\mathbf{d}\mathbf{F}^{e} = \mathrm{sym}[\bar{\mathbf{L}}] = \mathrm{sym}[\bar{\mathbf{C}}^{e}\bar{\mathbf{L}}]\\
&= {}^{p}\overrightarrow{\mathbf{D}}_{\bar{G}\bar{G}}\\
\bar{\mathbf{D}}^{e} &\equiv {}^{e}\overleftarrow{\mathbf{d}}^{e}_{\bar{G}\bar{G}} = \mathbf{F}^{eT}\mathbf{d}^{e}\mathbf{F}^{e} = \mathrm{sym}[\bar{\mathbf{L}}^{e}] = \mathrm{sym}[\bar{\mathbf{C}}^{e}\bar{\mathbf{L}}^{e}]\\
\bar{\mathbf{D}}^{p} &\equiv {}^{e}\overleftarrow{\mathbf{d}}^{p}_{\bar{G}\bar{G}} = \mathbf{F}^{eT}\mathbf{d}^{p}\mathbf{F}^{e} = \mathrm{sym}[\bar{\mathbf{L}}^{p}] = \mathrm{sym}[\bar{\mathbf{C}}^{e}\bar{\mathbf{L}}^{p}]
\end{aligned}\right\}, \tag{8.89}$$

$$\left.\begin{aligned}
\bar{\mathbf{W}} &\equiv {}^{e}\overleftarrow{\mathbf{w}}_{\bar{G}\bar{G}} = \mathbf{F}^{eT}\mathbf{w}\mathbf{F}^{e} = \mathrm{ant}[\bar{\mathbf{L}}] = \mathrm{ant}[\bar{\mathbf{C}}^{e}\bar{\mathbf{L}}]\\
\bar{\mathbf{W}}^{e} &\equiv {}^{e}\overleftarrow{\mathbf{w}}^{e}_{\bar{G}\bar{G}} = \mathbf{F}^{eT}\mathbf{w}^{e}\mathbf{F}^{e} = \mathrm{ant}[\bar{\mathbf{L}}^{e}] = \mathrm{ant}[\bar{\mathbf{C}}^{e}\bar{\mathbf{L}}^{e}]\\
\bar{\mathbf{W}}^{p} &\equiv {}^{e}\overleftarrow{\mathbf{w}}^{p}_{\bar{G}\bar{G}} = \mathbf{F}^{eT}\mathbf{w}^{p}\mathbf{F}^{e} = \mathrm{ant}[\bar{\mathbf{L}}^{p}] = \mathrm{ant}[\bar{\mathbf{C}}^{e}\bar{\mathbf{L}}^{p}]
\end{aligned}\right\}. \tag{8.90}$$

$\bar{\mathbf{W}}^e$ and $\bar{\mathbf{W}}^p$ are the spin of substructure and the plastic spin based in the intermediate configuration, respectively. It follows from equations (8.87) and (8.89) that

$$\left.\begin{aligned}\bar{\mathbf{D}} &= \frac{1}{2}\delta^i_{\bar{I}}\delta^j_{\bar{J}}\overset{\bullet}{g}_{ij}\bar{\mathbf{G}}^{\bar{I}}\otimes\bar{\mathbf{G}}^{\bar{J}}\\ \bar{\mathbf{D}}^e &= \frac{1}{2}[\delta^i_{\bar{I}}\delta^j_{\bar{J}}\overset{\bullet}{g}_{ij} - \{\delta^i_{\bar{I}}(\bar{\mathbf{G}}^{\bar{R}}\cdot\overset{\bullet}{\bar{\mathbf{G}}}_{\bar{J}}) + \delta^i_{\bar{J}}(\bar{\mathbf{G}}^{\bar{R}}\cdot\overset{\bullet}{\bar{\mathbf{G}}}_{\bar{I}})\}\delta^j_{\bar{R}}g_{ij}]\bar{\mathbf{G}}^{\bar{I}}\otimes\bar{\mathbf{G}}^{\bar{J}}\\ \bar{\mathbf{D}}^p &= \frac{1}{2}\{\delta^i_{\bar{I}}(\bar{\mathbf{G}}^{\bar{R}}\cdot\overset{\bullet}{\bar{\mathbf{G}}}_{\bar{J}}) + \delta^i_{\bar{J}}(\bar{\mathbf{G}}^{\bar{R}}\cdot\overset{\bullet}{\bar{\mathbf{G}}}_{\bar{I}})\}\delta^j_{\bar{R}}g_{ij}\bar{\mathbf{G}}^{\bar{I}}\otimes\bar{\mathbf{G}}^{\bar{J}}\end{aligned}\right\}, \tag{8.91}$$

noting that

$$\begin{aligned}2\mathrm{sym}[\bar{\mathbf{L}}] &= \delta^i_{\bar{I}}\delta^j_{\bar{J}}(\mathbf{g}_i\cdot\overset{\bullet}{\mathbf{g}}_j)\bar{\mathbf{G}}^{\bar{I}}\otimes\bar{\mathbf{G}}^{\bar{J}} + \delta^i_{\bar{I}}\delta^j_{\bar{J}}(\mathbf{g}_i\cdot\overset{\bullet}{\mathbf{g}}_j)\bar{\mathbf{G}}^{\bar{J}}\otimes\bar{\mathbf{G}}^{\bar{I}}\\ &= \delta^i_{\bar{I}}\delta^j_{\bar{J}}(\mathbf{g}_i\cdot\overset{\bullet}{\mathbf{g}}_j)\bar{\mathbf{G}}^{\bar{I}}\otimes\bar{\mathbf{G}}^{\bar{J}} + \delta^j_{\bar{J}}\delta^i_{\bar{I}}(\mathbf{g}_j\cdot\overset{\bullet}{\mathbf{g}}_i)\bar{\mathbf{G}}^{\bar{I}}\otimes\bar{\mathbf{G}}^{\bar{J}}\\ &= \delta^i_{\bar{I}}\delta^j_{\bar{J}}(\mathbf{g}_i\cdot\overset{\bullet}{\mathbf{g}}_j + \overset{\bullet}{\mathbf{g}}_i\cdot g_j)\bar{\mathbf{G}}^{\bar{I}}\otimes\bar{\mathbf{G}}^{\bar{J}} = \delta^i_{\bar{I}}\delta^j_{\bar{J}}\overset{\bullet}{\mathbf{g}}_{ij}\bar{\mathbf{G}}^{\bar{I}}\otimes\bar{\mathbf{G}}^{\bar{J}},\end{aligned}$$

$$\begin{aligned}2\mathrm{sym}[\bar{\mathbf{L}}^p] &= \mathrm{sym}[\delta^i_{\bar{I}}\delta^j_{\bar{R}}g_{ij}(\bar{\mathbf{G}}^{\bar{R}}\cdot\overset{\bullet}{\bar{\mathbf{G}}}_{\bar{J}})\bar{\mathbf{G}}^{\bar{I}}\otimes\bar{\mathbf{G}}^{\bar{J}}]\\ &= \delta^i_{\bar{I}}\delta^j_{\bar{R}}g_{ij}(\bar{\mathbf{G}}^{\bar{R}}\cdot\overset{\bullet}{\bar{\mathbf{G}}}_{\bar{J}})\bar{\mathbf{G}}^{\bar{I}}\otimes\bar{\mathbf{G}}^{\bar{J}} + \delta^i_{\bar{I}}\delta^j_{\bar{R}}g_{ij}(\bar{\mathbf{G}}^{\bar{R}}\cdot\overset{\bullet}{\bar{\mathbf{G}}}_{\bar{J}})\bar{\mathbf{G}}^{\bar{J}}\otimes\bar{\mathbf{G}}^{\bar{I}}\\ &= \delta^i_{\bar{I}}\delta^j_{\bar{R}}g_{ij}(\bar{\mathbf{G}}^{\bar{R}}\cdot\overset{\bullet}{\bar{\mathbf{G}}}_{\bar{J}})\bar{\mathbf{G}}^{\bar{I}}\otimes\bar{\mathbf{G}}^{\bar{J}} + \delta^i_{\bar{J}}\delta^j_{\bar{R}}g_{ij}(\bar{\mathbf{G}}^{\bar{R}}\cdot\overset{\bullet}{\bar{\mathbf{G}}}_{\bar{I}})\bar{\mathbf{G}}^{\bar{I}}\otimes\bar{\mathbf{G}}^{\bar{J}}\\ &= \{\delta^i_{\bar{I}}(\bar{\mathbf{G}}^{\bar{R}}\cdot\overset{\bullet}{\bar{\mathbf{G}}}_{\bar{J}}) + \delta^i_{\bar{J}}(\bar{\mathbf{G}}^{\bar{R}}\cdot\overset{\bullet}{\bar{\mathbf{G}}}_{\bar{I}})\}\delta^j_{\bar{R}}g_{ij}\bar{\mathbf{G}}^{\bar{I}}\otimes\bar{\mathbf{G}}^{\bar{J}}.\end{aligned}$$

Substituting $\mathbf{d} = \mathbf{F}^{-T}\overset{\bullet}{\mathbf{E}}\mathbf{F}^{-1}$ in equation $(4.126)_1$ into equation $(8.89)_1$, $\bar{\mathbf{D}}$ is rewritten as

$$\bar{\mathbf{D}} = \mathbf{F}^{eT}\mathbf{F}^{-T}\overset{\bullet}{\mathbf{E}}\mathbf{F}^{-1}\mathbf{F}^{e} = \mathbf{F}^{p-T}\overset{\bullet}{\mathbf{E}}\mathbf{F}^{p-1} = {}^{p}\overset{\rightarrow}{\overset{\bullet}{\mathbf{E}}}_{\bar{G}\bar{G}} = \mathbf{F}^{p-T}(\overset{\bullet}{\mathbf{C}}/2)\mathbf{F}^{p-1} = {}^{p}\overset{\rightarrow}{\overset{\bullet}{\mathbf{C}}}_{\bar{G}\bar{G}}/2. \tag{8.92}$$

Further, it follows from equations (8.61), (8.80) and $(8.89)_2$ that

$$\boxed{\overset{\bullet}{\bar{\mathbf{C}}}{}^{e} = 2\overset{\bullet}{\bar{\mathbf{E}}}{}^{e} = 2\bar{\mathbf{D}}^{e}} \tag{8.93}$$

The additive decomposition holds for the rate of volume change as follows:

$$d_v = d_v^e + d_v^p, \tag{8.94}$$

noting equation (4.86), where

$$\left.\begin{aligned}d_v &\equiv \mathrm{tr}\mathbf{d} = \mathrm{tr}\bar{\mathbf{D}} = \mathrm{tr}\frac{\partial\overset{\bullet}{\mathbf{x}}}{\partial\mathbf{x}} = \frac{(dv)^{\bullet}}{dv} = \overset{\bullet}{\varepsilon}_v\\ d_v^e &\equiv \mathrm{tr}\mathbf{d}^e = \mathrm{tr}\mathbf{d} - \mathrm{tr}\mathbf{d}^p = \mathrm{tr}\bar{\mathbf{D}}^e = \mathrm{tr}\bar{\mathbf{D}} - \mathrm{tr}\bar{\mathbf{D}}^p = \frac{(dv)^{\bullet}}{dv} - \frac{(d\bar{V})^{\bullet}}{d\bar{V}} = \overset{\bullet}{\varepsilon}{}_v^e\\ d_v^p &\equiv \mathrm{tr}\mathbf{d}^p = \mathrm{tr}\bar{\mathbf{D}}^p = \mathrm{tr}\frac{\partial\overset{\bullet}{\bar{\mathbf{X}}}}{\partial\bar{\mathbf{X}}} = \frac{(d\bar{V})^{\bullet}}{d\bar{V}} = \overset{\bullet}{\varepsilon}{}_v^p\end{aligned}\right\}. \tag{8.95}$$

dV, dv and $d\bar{V}$ are the infinitesimal volumes in the initial, current and the intermediate configurations, respectively. ε_v, ε_v^e and ε_v^p are the total, elastic and plastic logarithmic volumetric strains, respectively, that is

$$\varepsilon_v \equiv \ln\frac{dv}{dV},\ \varepsilon_v^e \equiv \varepsilon_v - \varepsilon_v^p = \ln\frac{dv}{dV} - \ln\frac{d\bar{V}}{dV},\ \varepsilon_v^p \equiv \ln\frac{d\bar{V}}{dV}. \quad (8.96)$$

8.6.4 Strain Rate Tensors for Kinematic Hardening

From the relation

$$\begin{aligned}\bar{\mathbf{L}}^p &\equiv \dot{\mathbf{F}}^p\mathbf{F}^{p-1} = (\mathbf{F}_e^p\mathbf{F}_d^p)^\bullet(\mathbf{F}_e^p\mathbf{F}_d^p)^{-1} = (\dot{\mathbf{F}}_e^p\mathbf{F}_d^p + \mathbf{F}_e^p\dot{\mathbf{F}}_d^p)\mathbf{F}_d^{p-1}\mathbf{F}_e^{p-1}\\ &= \dot{\mathbf{F}}_e^p\mathbf{F}_e^{p-1} + \mathbf{F}_e^p\dot{\mathbf{F}}_d^p\mathbf{F}_d^{p-1}\mathbf{F}_e^{p-1}\end{aligned}$$

it follows that

$$\boxed{\bar{\mathbf{L}}^p = \bar{\mathbf{L}}_e^p + \bar{\mathbf{L}}_d^p} \quad (8.97)$$

where

$$\bar{\mathbf{L}}_e^p \equiv \dot{\mathbf{F}}_e^p\mathbf{F}_e^{p-1},\ \ \bar{\mathbf{L}}_d^p \equiv {}^{p}_{e}\overrightarrow{\tilde{\mathbf{L}}}{}^{p\bar{G}}_{d\bullet\tilde{G}} = \mathbf{F}_e^p\tilde{\mathbf{L}}_d^p\mathbf{F}_e^{p-1},\ \ \tilde{\mathbf{L}}_d^p \equiv \dot{\mathbf{F}}_d^p\mathbf{F}_d^{p-1}. \quad (8.98)$$

Thus, it must follow that

$$\bar{\mathbf{D}}^p = \bar{\mathbf{D}}_e^p + \bar{\mathbf{D}}_d^p, \quad (8.99)$$

$$\bar{\mathbf{D}}_e^p = \mathrm{sym}[\bar{\mathbf{L}}_e^p],\ \ \bar{\mathbf{D}}_d^p = \mathrm{sym}[\bar{\mathbf{L}}_d^p]. \quad (8.100)$$

Further, one obtains

$$\bar{\mathbf{D}}_e^p = \frac{1}{2}\bar{\dot{\mathbf{C}}}{}_e^p, \quad (8.101)$$

where $\bar{\dot{\mathbf{C}}}{}_e^p$ is defined by the push-forward of $\dot{\tilde{\mathbf{C}}}{}_e^p$ to the intermediate configuration by $\mathbf{F}_e^p$, that is

$$\bar{\dot{\mathbf{C}}}{}_e^p \equiv {}^{p}_{e}\overrightarrow{\dot{\tilde{\mathbf{C}}}}{}^{p}_{e\bar{G}\bar{G}} = \mathbf{F}_e^{p-T}\dot{\tilde{\mathbf{C}}}{}_e^p\mathbf{F}_e^{p-1}, \quad (8.102)$$

noting that

$$2\bar{\mathbf{D}}_e^p = \dot{\mathbf{F}}_e^p\mathbf{F}_e^{p-1} + \mathbf{F}_e^{p-T}\dot{\mathbf{F}}_e^{pT} = \mathbf{F}_e^{p-T}(\mathbf{F}_e^{pT}\dot{\mathbf{F}}_e^p + \dot{\mathbf{F}}_e^{pT}\mathbf{F}_e^p)\mathbf{F}_e^{p-1} = \mathbf{F}_e^{p-T}\dot{\tilde{\mathbf{C}}}{}_e^p\mathbf{F}_e^{p-1}. \quad (8.103)$$

Here, note that there does not exist the variable denoted by the symbol $\bar{\mathbf{C}}_e^p(\neq \mathbf{F}_e^{pT}\mathbf{F}_e^p)$, whilst $\tilde{\mathbf{C}}_e^p(=\mathbf{F}_e^{pT}\mathbf{F}_e^p)$ is not based in the intermediate configuration $\bar{\mathcal{K}}$ but in the kinematic hardening-intermediate configuration $\tilde{\mathcal{K}}$, and thus it is impertinent to use the symbol $\dot{\bar{\mathbf{C}}}{}_e^p$ (time-derivative of indefinable variable) in stead of $\bar{\dot{\mathbf{C}}}{}_e^p$ in equation (8.102).

The rate variable $\dot{\tilde{\mathbf{C}}}{}_e^p$ involved in equation (8.48) is expressed as

$$\dot{\tilde{\mathbf{C}}}{}_e^p = \mathbf{F}_d^{p-T}\dot{\mathbf{C}}^p\mathbf{F}_d^{p-1} - 2\mathrm{sym}[\tilde{\mathbf{C}}_e^p\tilde{\mathbf{L}}_d^p]. \quad (8.104)$$

from equation (8.48), noting that

$$\begin{aligned}
\dot{\tilde{\mathbf{C}}}_e^p &= (\mathbf{F}_d^{p-T}\,\mathbf{C}^p\,\mathbf{F}_d^{p-1})^{\bullet} \\
&= \dot{\mathbf{F}}_d^{p-T}\,\mathbf{C}^p\,\mathbf{F}_d^{p-1} + \mathbf{F}_d^{p-T}\,\dot{\mathbf{C}}^p\,\mathbf{F}_d^{p-1} + \mathbf{F}_d^{p-T}\,\mathbf{C}^p\dot{\mathbf{F}}_d^{p-1} \\
&= \dot{\mathbf{F}}_d^{p-T}\,(\mathbf{F}_e^p\mathbf{F}_d^p)^T\,(\mathbf{F}_e^p\mathbf{F}_d^p)\mathbf{F}_d^{p-1} + \mathbf{F}_d^{p-T}\,\dot{\mathbf{C}}^p\,\mathbf{F}_d^{p-1} + \mathbf{F}_d^{p-T}\,(\mathbf{F}_e^p\mathbf{F}_d^p)^T\,(\mathbf{F}_e^p\mathbf{F}_d^p)\dot{\mathbf{F}}_d^{p-1} \\
&= \dot{\mathbf{F}}_d^{p-T}\,\mathbf{F}_d^{pT}\,\mathbf{F}_e^{pT}\,\mathbf{F}_e^p\mathbf{F}_d^p\,\mathbf{F}_d^{p-1} + \mathbf{F}_d^{p-T}\,\dot{\mathbf{C}}^p\,\mathbf{F}_d^{p-1} + \mathbf{F}_d^{p-T}\,\mathbf{F}_d^{pT}\,\mathbf{F}_e^{pT}\,\mathbf{F}_e^p\mathbf{F}_d^p\dot{\mathbf{F}}_d^{p-1} \\
&= (\mathbf{F}_d^p\dot{\mathbf{F}}_d^{p-1})^T\,\tilde{\mathbf{C}}_e^p + \mathbf{F}_d^{p-T}\,\dot{\mathbf{C}}^p\,\mathbf{F}_d^{p-1} + \tilde{\mathbf{C}}_e^p\mathbf{F}_d^p\dot{\mathbf{F}}_d^{p-1} \\
&= (-\dot{\mathbf{F}}_d^p\,\mathbf{F}_d^{p-1})^T\,\tilde{\mathbf{C}}_e^p + \mathbf{F}_d^{p-T}\,\dot{\mathbf{C}}^p\,\mathbf{F}_d^{p-1} + \tilde{\mathbf{C}}_e^p(-\dot{\mathbf{F}}_d^p\,\mathbf{F}_d^{p-1}) \\
&= \mathbf{F}_d^{p-T}\,\dot{\mathbf{C}}^p\,\mathbf{F}_d^{p-1} - \{\tilde{\mathbf{C}}_e^p\dot{\mathbf{F}}_d^p\,\mathbf{F}_d^{p-1} + (\tilde{\mathbf{C}}_e^p\dot{\mathbf{F}}_d^p\,\mathbf{F}_d^{p-1})^T\}.
\end{aligned}$$

Further, substituting equation (8.79) into equation (8.104), we have

$$\boxed{\dot{\tilde{\mathbf{C}}}_e^p = 2(\mathbf{F}_e^{pT}\,\bar{\mathbf{D}}^p\,\mathbf{F}_e^p - \mathrm{sym}[\tilde{\mathbf{C}}_e^p\tilde{\mathbf{L}}_d^p])} \tag{8.105}$$

$\tilde{\mathbf{L}}_d^p$ in equation (8.98) is expressed as

$$\tilde{\mathbf{L}}_d^p = \dot{\tilde{\mathbf{G}}}_{\tilde{I}} \otimes \tilde{\mathbf{G}}^{\tilde{I}}, \tag{8.106}$$

noting

$$\dot{\mathbf{F}}_d^p\,\mathbf{F}_d^{p-1} = \delta_I^{\tilde{I}}\dot{\tilde{\mathbf{G}}}_{\tilde{I}} \otimes \mathbf{G}^I\delta_{\tilde{J}}^J\mathbf{G}_J \otimes \tilde{\mathbf{G}}^{\tilde{J}} = \delta_I^{\tilde{I}}\delta_{\tilde{J}}^J\delta_J^I\dot{\tilde{\mathbf{G}}}_{\tilde{I}} \otimes \tilde{\mathbf{G}}^{\tilde{J}}$$

with equation $(8.44)_3$. The symmetric part of $\tilde{\mathbf{L}}_d^p$ is expressed as

$$\tilde{\mathbf{D}}_d^p \equiv \mathrm{sym}[\tilde{\mathbf{L}}_d^p] = \dot{\tilde{G}}_{\tilde{I}\tilde{J}}\tilde{\mathbf{G}}^{\tilde{I}} \otimes \tilde{\mathbf{G}}^{\tilde{J}}, \tag{8.107}$$

noting equation (2.28).

The rate of $\mathbf{C}_d^p$ is given by the covariant pull-back of $\tilde{\mathbf{D}}_d^p$ by $\mathbf{F}_d^p$ to the reference configuration defined as follows:

$$\dot{\mathbf{C}}_d^p \equiv 2{}_d^p\,\overleftarrow{\tilde{\mathbf{D}}}{}_{d\,GG}^p = 2\mathbf{F}_d^{pT}\,\tilde{\mathbf{D}}_d^p\mathbf{F}_d^p, \tag{8.108}$$

noting that

$$\dot{\mathbf{C}}_d^p = (\mathbf{F}_d^{pT}\,\mathbf{F}_d^p)^{\bullet} = \dot{\mathbf{F}}_d^{pT}\,\mathbf{F}_d^p + \mathbf{F}_d^{pT}\,\dot{\mathbf{F}}_d^p = 2\mathbf{F}_d^{pT}\,[\{(\dot{\mathbf{F}}_d^p\,\mathbf{F}_d^{p-1} + (\dot{\mathbf{F}}_d^p\,\mathbf{F}_d^{p-1})^T\}/2]\mathbf{F}_d^p.$$

The dissipative velocity gradient for kinematic hardening $\tilde{\mathbf{L}}_d^p$ in equation $(8.98)_3$ can be decomposed into the symmetric and the anti-symmetric parts as follows:

$$\boxed{\tilde{\mathbf{L}}_d^p = \tilde{\mathbf{D}}_d^p + \tilde{\mathbf{W}}_d^p} \tag{8.109}$$

where

$$\tilde{\mathbf{D}}_d^p = \mathrm{sym}[\tilde{\mathbf{L}}_d^p], \quad \tilde{\mathbf{W}}_d^p = \mathrm{ant}[\tilde{\mathbf{L}}_d^p]. \tag{8.110}$$

8.7 Stress and Kinematic Hardening Variable Tensors

The total, elastic and plastic work rates are given by

$$\boxed{\left.\begin{array}{l} \boldsymbol{\tau}:\mathbf{d} = \boldsymbol{\tau}:\boldsymbol{l} = \bar{\mathbf{M}}:\bar{\mathbf{L}} = \bar{\mathbf{S}}:\bar{\mathbf{D}} = \bar{\mathbf{S}}:\bar{\mathbf{L}} \\ \boldsymbol{\tau}:\mathbf{d}^e = \boldsymbol{\tau}:\boldsymbol{l}^e = \bar{\mathbf{M}}:\bar{\mathbf{L}}^e = \bar{\mathbf{S}}:\bar{\mathbf{D}}^e = \bar{\mathbf{S}}:\bar{\mathbf{L}}^e \\ \boldsymbol{\tau}:\mathbf{d}^p = \boldsymbol{\tau}:\boldsymbol{l}^p = \bar{\mathbf{M}}:\bar{\mathbf{L}}^p = \bar{\mathbf{S}}:\bar{\mathbf{D}}^p = \bar{\mathbf{S}}:\bar{\mathbf{L}}^p \end{array}\right\}} \tag{8.111}$$

because of

$$\begin{aligned} \boldsymbol{\tau}:\boldsymbol{l} &= \mathrm{tr}\{(\mathbf{F}^e\bar{\mathbf{S}}\mathbf{F}^{eT})(\mathbf{F}^e\bar{\mathbf{L}}\mathbf{F}^{e-1})^T\} = \mathrm{tr}(\mathbf{F}^e\bar{\mathbf{S}}\mathbf{F}^{eT}\mathbf{F}^{e-T}\bar{\mathbf{L}}^T\mathbf{F}^{eT}))\} \\ &= \mathrm{tr}(\mathbf{F}^{eT}\mathbf{F}^e\bar{\mathbf{S}}\bar{\mathbf{L}}^T) = \mathrm{tr}(\bar{\mathbf{C}}^e\bar{\mathbf{S}}\bar{\mathbf{L}}^T) \\ &= \begin{cases} \bar{\mathbf{C}}^e\bar{\mathbf{S}}:\bar{\mathbf{L}} \\ \mathrm{tr}\{(\bar{\mathbf{C}}^e\bar{\mathbf{S}})^T\bar{\mathbf{L}}\} = \mathrm{tr}(\bar{\mathbf{S}}\bar{\mathbf{C}}^e\bar{\mathbf{L}}) = \bar{\mathbf{S}}:(\bar{\mathbf{C}}^e\bar{\mathbf{L}}) \end{cases} \end{aligned}$$

with equations (8.22), (8.66), (8.73) and (8.87); similar relations hold for the elastic and the plastic work rates by replacing the velocity gradient by the elastic and plastic velocity gradients. Here, $\bar{\mathbf{M}}$ is called the *Mandel stress* (Mandel, 1972b) and is obtained by applying the covariant–contravariant push-forward operation of $\mathbf{F}^p$ to the Eshelby-like stress tensor $\mathbf{M}$ in equation (6.49) or the covariant–contravariant pull-back operation of $\mathbf{F}^e$ from the Kirchhoff stress tensor $\boldsymbol{\tau}$ as follows:

$$\boxed{\bar{\mathbf{M}} = \bar{M}_{I}^{\bullet J}\bar{\mathbf{G}}^I \otimes \bar{\mathbf{G}}_J \equiv \bar{\mathbf{C}}^e\bar{\mathbf{S}}} = \begin{cases} \mathbf{F}^{p-T}\mathbf{M}\mathbf{F}^{pT} = {}^p\overrightarrow{\mathbf{M}}{}_{\bar{\mathbf{G}}}^{\bullet\bar{\mathbf{G}}} \\ \mathbf{F}^{eT}\boldsymbol{\tau}\mathbf{F}^{e-T} = {}^e\overleftarrow{\boldsymbol{\tau}}{}_{\bar{\mathbf{G}}}^{\bullet\bar{\mathbf{G}}} \end{cases} \quad (\neq \bar{\mathbf{M}}^T), \tag{8.112}$$

noting

$$\begin{aligned} \bar{\mathbf{C}}^e\bar{\mathbf{S}} &= (\mathbf{F}^{eT}\mathbf{F}^e)(\mathbf{F}^{e-1}\boldsymbol{\tau}\mathbf{F}^{e-T}) \\ &= \begin{cases} \mathbf{F}^{p-T}\mathbf{F}^{pT}\mathbf{F}^{eT}\boldsymbol{\tau}\mathbf{F}^{e-T}\mathbf{F}^{p-T}\mathbf{F}^{pT} = \mathbf{F}^{p-T}\mathbf{F}^T\boldsymbol{\tau}\mathbf{F}^{-T}\mathbf{F}^{pT} = \mathbf{F}^{p-T}\mathbf{M}\mathbf{F}^{pT} \\ \mathbf{F}^{eT}\boldsymbol{\tau}\mathbf{F}^{e-T} \end{cases}. \end{aligned}$$

The components of $\bar{\mathbf{M}}$ are given by

$$\bar{M}_{I}^{\bullet J} = \delta_I^i\delta_j^J\tau_i^{\bullet j} = \delta_I^i\delta_J^j\delta_R^J g_{ij}S^{RJ},$$

referring to equation (6.44) with equation $(2.79)_3$. Besides, it follows from equation $(8.112)_1$ that

$$\mathrm{tr}\mathbf{M} = \mathrm{tr}\bar{\mathbf{M}}. \tag{8.113}$$

As indicated by Lubliner (1986), the Mandel stress is subjected to the constraint on the six-dimensional space by the symmetry of the tensor

$$\bar{\mathbf{M}}\bar{\mathbf{C}}^e = (\bar{\mathbf{M}}\bar{\mathbf{C}}^e)^T \tag{8.114}$$

which imposes three restrictions on the nine original components.

Equation (8.111) shows that $\bar{\mathbf{M}}$ is work-conjugate with $\bar{\mathbf{L}}^p$. Here, $\bar{\mathbf{L}}^p$ depends only on the plastic deformation gradient as shown in equation $(8.66)_3$ or $(8.73)_3$. Further, the plastic deformation gradient $\mathbf{F}^p$ can be updated easily by time-integration of

$$\boxed{\dot{\mathbf{F}}^p = \bar{\mathbf{L}}^p\mathbf{F}^p} \tag{8.115}$$

based on equation $(8.66)_3$. On the other hand, the elastic deformation gradient $\mathbf{F}^e$ is calculated by excluding the plastic deformation gradient $\mathbf{F}^p$ from the deformation gradient $\mathbf{F}$. Consequently, $\bar{\mathbf{M}}$ and $\bar{\mathbf{L}}^p$ will be adopted for the formulation of the plastic constitutive equation later in this chapter. On the other hand, the work-conjugate pair of $\bar{\mathbf{S}}$ and $\bar{\mathbf{D}}^p$ is adopted for the plastic flow rule by Moran et al. (1990) and Simo (1998), but $\bar{\mathbf{D}}^p$ is influenced by the elastic deformation as shown in equation $(8.89)_3$. $\bar{\mathbf{S}}$ and $\bar{\mathbf{D}}^p$ were adopted by Dafalias (1998) and $\bar{\mathbf{S}}$ and $\bar{\mathbf{L}}^p$ by Belytschko et al. (2000), but these pairs do not satisfy work-conjugacy.

The constitutive equation is formulated primarily in the intermediate configuration, but the stress-update computation will be carried out in the reference configuration in Chapters 9 and 10. The pull-back of $\bar{\mathbf{S}}$ and $\bar{\mathbf{M}}$ to the reference state is given from equations (8.22) and (8.112) as follows:

$$\mathbf{S} = S^{IJ}\mathbf{G}_I \otimes \mathbf{G}_J = \begin{cases} \overleftarrow{\boldsymbol{\tau}}^{GG} = \mathbf{F}^{-1}\boldsymbol{\tau}\mathbf{F}^{-T} = \delta^I_i\delta^J_j\tau^{ij}\mathbf{G}_I \otimes \mathbf{G}_J \\ {}^p\overleftarrow{\bar{\mathbf{S}}}^{GG} = \mathbf{F}^{p-1}\bar{\mathbf{S}}\mathbf{F}^{p-T} = \delta^I_{\bar{I}}\delta^J_{\bar{J}}\bar{S}^{\bar{I}\bar{J}}\mathbf{G}_I \otimes \mathbf{G}_J \end{cases} (= \mathbf{S}^T), \tag{8.116}$$

$$\begin{aligned} \mathbf{M} &= M_I^{\bullet J}\mathbf{G}^I \otimes \mathbf{G}_J \\ &= \begin{cases} \mathbf{C}\mathbf{S} = {}^p\overleftarrow{\bar{\mathbf{C}}}{}^e_{GG}\,{}^p\overleftarrow{\bar{\mathbf{S}}}^{GG} = \delta^i_I\delta^j_R g_{ij}S^{RJ}\mathbf{G}^I \otimes \mathbf{G}_J \\ {}^p\overleftarrow{\bar{\mathbf{M}}}{}_G^{\bullet G} = \delta^{\bar{I}}_I\delta^J_{\bar{J}}\bar{M}_{\bar{I}}^{\bullet\bar{J}}\mathbf{G}^I \otimes \mathbf{G}_J \end{cases} (\neq \mathbf{M}^T), \end{aligned} \tag{8.117}$$

noting that

$$\begin{cases} {}^p\overleftarrow{\bar{\mathbf{C}}}{}^e_{GG}\,{}^p\overleftarrow{\bar{\mathbf{S}}}^{GG} = (\mathbf{F}^{pT}\bar{\mathbf{C}}^e\mathbf{F}^p)(\mathbf{F}^{p-1}\bar{\mathbf{S}}\mathbf{F}^{p-T}) = (\mathbf{F}^{pT}(\mathbf{F}^{eT}\mathbf{F}^e)\mathbf{F}^p)(\mathbf{F}^{p-1}(\mathbf{F}^p\mathbf{S}\mathbf{F}^{pT})\mathbf{F}^{p-T}) \\ {}^p\overleftarrow{\bar{\mathbf{M}}}{}_G^{\bullet G} = \mathbf{F}^{pT}\bar{\mathbf{M}}\mathbf{F}^{p-T} = \mathbf{F}^{pT}\bar{\mathbf{C}}^e\bar{\mathbf{S}}\mathbf{F}^{p-T} = \mathbf{F}^{pT}(\mathbf{F}^{eT}\mathbf{F}^e)(\mathbf{F}^p\mathbf{S}\mathbf{F}^{pT})\mathbf{F}^{p-T} \end{cases} \tag{8.118}$$

and

$$\begin{aligned} \mathbf{C}\mathbf{S} &= \delta^i_I\delta^j_J g_{ij}\mathbf{G}^I \otimes \mathbf{G}^J S^{RS}\mathbf{G}_R \otimes \mathbf{G}_S = \delta^i_I\delta^{jJ}_{\ JR}g_{ij}S^{RS}\mathbf{G}^I \otimes \mathbf{G}_S \\ &= \delta^i_I\delta^j_R g_{ij}S^{RS}\mathbf{G}^I \otimes \mathbf{G}_S = \delta^i_I\delta^j_R g_{ij}S^{RJ}\mathbf{G}^I \otimes \mathbf{G}_J \end{aligned}$$

with equations (4.1) and (6.44).

On the other hand, let the kinematic hardening variable, that is back stress $\tilde{\mathbf{S}}_e^p$, based in the kinematic hardening-intermediate configuration $\tilde{\mathcal{K}}$, be defined by

$$\tilde{\mathbf{S}}_e^p \equiv \tilde{S}_e^{p\tilde{I}\tilde{J}}\tilde{\mathbf{G}}_{\tilde{I}} \otimes \tilde{\mathbf{G}}_{\tilde{J}} \; (=\tilde{\mathbf{S}}_e^{pT}). \tag{8.119}$$

The contravariant push-forward of $\tilde{\mathbf{S}}_e^p$ by $\mathbf{F}_e^p$ to the intermediate configuration $\bar{\mathcal{K}}$ is given by

$$\bar{\mathbf{S}}_e^p = \bar{S}_e^{p\bar{I}\bar{J}}\bar{\mathbf{G}}_{\bar{I}} \otimes \bar{\mathbf{G}}_{\bar{J}} \equiv {}_e^p\overrightarrow{\tilde{\mathbf{S}}_e^p}{}^{\bar{G}\bar{G}} = \mathbf{F}_e^p\tilde{\mathbf{S}}_e^p\mathbf{F}_e^{pT} \; (=\bar{\mathbf{S}}_e^{pT}), \tag{8.120}$$

with the components

$$\bar{S}_e^{p\bar{I}\bar{J}} = \delta_{\tilde{R}}^{\bar{I}}\delta_{\tilde{S}}^{\bar{J}}\tilde{S}_e^{p\tilde{R}\tilde{S}}. \tag{8.121}$$

On the other hand, the contravariant pull-back of $\tilde{\mathbf{S}}_e^p$ by $\mathbf{F}_d^p$ to the reference configuration $\mathcal{K}_0$ is given by

$$\mathbf{S}_e^p = S_e^{pIJ}\mathbf{G}_I \otimes \mathbf{G}_J = {}_d^p\overleftarrow{\tilde{\mathbf{S}}_e^p}{}^{GG} = \mathbf{F}_d^{p-1}\tilde{\mathbf{S}}_e^p\mathbf{F}_d^{p-T} \; (=\mathbf{S}_e^{pT}), \tag{8.122}$$

with the components

$$S_e^{pIJ} = \delta_{\tilde{I}}^I\delta_{\tilde{J}}^J\tilde{S}_e^{p\tilde{I}\tilde{J}}. \tag{8.123}$$

Further, $\mathbf{S}_e^p$ is related to $\bar{\mathbf{S}}_e^p$ as follows:

$$\mathbf{S}_e^p = {}^p\overleftarrow{\bar{\mathbf{S}}_e^p}{}^{GG} = \mathbf{F}^{p-1}\bar{\mathbf{S}}_e^p\mathbf{F}^{p-T}. \tag{8.124}$$

The Mandel-type kinematic hardening variable in the kinematic hardening-intermediate configuration $\tilde{\mathcal{K}}$ is defined by

$$\tilde{\mathbf{M}}_e^p = \tilde{M}_{e\tilde{I}}^{p\bullet\tilde{J}}\tilde{\mathbf{G}}^{\tilde{I}} \otimes \tilde{\mathbf{G}}_{\tilde{J}} \equiv \tilde{\mathbf{C}}_e^p\tilde{\mathbf{S}}_e^p \; (\neq \tilde{\mathbf{M}}_e^{pT}) \tag{8.125}$$

with the components

$$\tilde{M}_{e\tilde{I}}^{p\bullet\tilde{J}} = \delta_{\tilde{I}}^{\bar{I}}\delta_{\tilde{R}}^{\bar{J}}\bar{G}_{\bar{I}\bar{J}}\tilde{S}_e^{p\tilde{R}\tilde{J}}, \tag{8.126}$$

noting that

$$\begin{aligned}\tilde{\mathbf{C}}_e^p\tilde{\mathbf{S}}_e^p &= \delta_{\tilde{I}}^{\bar{I}}\delta_{\tilde{R}}^{\bar{J}}\bar{G}_{\bar{I}\bar{J}}\tilde{\mathbf{G}}^{\tilde{R}} \otimes \tilde{\mathbf{G}}^{\tilde{J}}\tilde{S}_e^{p\tilde{S}\tilde{J}}\tilde{\mathbf{G}}_{\tilde{S}} \otimes \tilde{\mathbf{G}}_{\tilde{J}} \\ &= \delta_{\tilde{I}}^{\bar{I}}\delta_{\tilde{R}}^{\bar{J}}\delta_{\tilde{S}}^{\tilde{J}}\bar{G}_{\bar{I}\bar{J}}\tilde{S}_e^{p\tilde{R}\tilde{S}}\tilde{\mathbf{G}}^{\tilde{I}} \otimes \tilde{\mathbf{G}}_{\tilde{J}} = \delta_{\tilde{I}}^{\bar{I}}\delta_{\tilde{R}}^{\bar{J}}\bar{G}_{\bar{I}\bar{J}}\tilde{S}_e^{p\tilde{R}\tilde{J}}\tilde{\mathbf{G}}^{\tilde{I}} \otimes \tilde{\mathbf{G}}_{\tilde{J}}\end{aligned}$$

by virtue of equations (8.48) and (8.119).

The push-forward of $\tilde{\mathbf{M}}_e^p$ by $\mathbf{F}_e^p$ to the intermediate configuration $\bar{\mathcal{K}}$ is given as

$$\bar{\mathbf{M}}_e^p = \bar{M}_{e\bar{I}}^{p\bullet\bar{J}}\bar{\mathbf{G}}^{\bar{I}} \otimes \bar{\mathbf{G}}_{\bar{J}} = \bar{\mathbf{C}}_e^p\bar{\mathbf{S}}_e^p = \bar{\mathbf{G}}\bar{\mathbf{S}}_e^p = \bar{\mathbf{S}}_e^p \; (=\bar{\mathbf{M}}_e^{pT}), \tag{8.127}$$

noting that

$$\bar{\mathbf{M}}_e^p = \begin{cases} {}^p_e\overrightarrow{\tilde{\mathbf{C}}}{}^p_{e\,\bar{G}\bar{G}}\,{}^p_e\overrightarrow{\tilde{\mathbf{S}}}{}^{p\,\bar{G}\bar{G}}_e = (\mathbf{F}_e^{p-T}\tilde{\mathbf{C}}_e^p\mathbf{F}_e^{p-1})(\mathbf{F}_e^p\tilde{\mathbf{S}}_e^p\mathbf{F}_e^{pT}) = (\mathbf{F}_e^{p-T}(\mathbf{F}_e^{pT}\,\mathbf{F}_e^p)\mathbf{F}_e^{p-1})(\mathbf{F}_e^p(\mathbf{F}_e^{p-1}\bar{\mathbf{S}}_e^p\mathbf{F}_e^{p-T})\mathbf{F}_e^{pT}) \\ {}^p_e\overrightarrow{\tilde{\mathbf{M}}}{}^{p\,\bullet G}_{e\,G} = \mathbf{F}_e^{p-T}\tilde{\mathbf{M}}_e^p\mathbf{F}_e^{pT} = \mathbf{F}_e^{p-T}\tilde{\mathbf{C}}_e^p\tilde{\mathbf{S}}_e^p\mathbf{F}_e^{pT} = \mathbf{F}_e^{p-T}(\mathbf{F}_e^{pT}\,\mathbf{F}_e^p)(\mathbf{F}_e^{p-1}\bar{\mathbf{S}}_e^p\mathbf{F}_e^{p-T})\mathbf{F}_e^{pT} \end{cases} \tag{8.128}$$

by equations (8.54) and (8.120), with the components

$$\bar{M}_{e\,\bar{I}}^{p\bullet\bar{J}} = \delta_{\bar{R}}^{\tilde{R}}\delta_{\tilde{J}}^{\bar{J}}\delta_{\tilde{R}}^{\tilde{S}}\bar{G}_{\bar{I}\bar{S}}\tilde{S}_e^{p\tilde{R}\tilde{J}}, \tag{8.129}$$

noting

$$\bar{\mathbf{G}}\bar{\mathbf{S}}_e^p = \bar{G}_{\bar{I}\bar{S}}\bar{\mathbf{G}}^{\bar{I}} \otimes \bar{\mathbf{G}}^{\bar{S}}\delta_{\bar{R}}^{\tilde{R}}\delta_{\tilde{J}}^{\bar{J}}\tilde{S}_e^{p\tilde{R}\tilde{J}}\bar{\mathbf{G}}_{\bar{R}} \otimes \bar{\mathbf{G}}_{\bar{J}} = \delta_{\bar{R}}^{\tilde{R}}\delta_{\tilde{J}}^{\bar{J}}\delta_{\tilde{R}}^{\tilde{S}}\bar{G}_{\bar{I}\bar{S}}\tilde{S}_e^{p\tilde{R}\tilde{J}}\bar{\mathbf{G}}^{\bar{I}} \otimes \bar{\mathbf{G}}_{\bar{J}}.$$

The pull-back of $\tilde{\mathbf{M}}_e^p$ to the reference configuration is given by

$$\mathbf{M}_e^p = M_{e\,I}^{p\bullet J}\mathbf{G}^I \otimes \mathbf{G}_J = \mathbf{C}^p\mathbf{S}_e^p = \mathbf{F}_d^{pT}\tilde{\mathbf{S}}_e^p\mathbf{F}_d^{p-T}\,(\neq \mathbf{M}_e^{pT}), \tag{8.130}$$

noting that

$$\mathbf{M}_e^p = \begin{cases} {}^p_d\overleftarrow{\tilde{\mathbf{C}}}{}^p_{e\,GG}\,{}^p_d\overleftarrow{\tilde{\mathbf{S}}}{}^{p\,\bar{G}\bar{G}}_e = (\mathbf{F}_d^{pT}\tilde{\mathbf{C}}_e^p\mathbf{F}_d^p)(\mathbf{F}_d^{p-1}\tilde{\mathbf{S}}_e^p\mathbf{F}_d^{p-T}) = (\mathbf{F}_d^{pT}(\mathbf{F}_e^{pT}\,\mathbf{F}_e^p)\mathbf{F}_d^p)(\mathbf{F}_d^{p-1}(\mathbf{F}_d^p\mathbf{S}_e^p\mathbf{F}_d^{pT})\mathbf{F}_d^{p-T}) \\ {}^p_d\overleftarrow{\tilde{\mathbf{M}}}{}^{p\bullet G}_{e\,G} = \mathbf{F}_d^{pT}\tilde{\mathbf{M}}_e^p\mathbf{F}_d^{p-T} = \mathbf{F}_d^{pT}\tilde{\mathbf{C}}_e^p\tilde{\mathbf{S}}_e^p\mathbf{F}_d^{p-T} = \mathbf{F}_d^{pT}(\mathbf{F}_e^{pT}\,\mathbf{F}_e^p)(\mathbf{F}_d^p\mathbf{S}_e^p\mathbf{F}_d^{pT})\mathbf{F}_d^{p-T} \end{cases} \tag{8.131}$$

by equations $(8.52)_1$, (8.122), with the components

$$M_{e\,I}^{p\bullet J} = \delta_I^{\bar{I}}\delta_{\tilde{S}}^{\bar{R}}\delta_{\tilde{J}}^{J}\bar{G}_{\bar{I}\bar{R}}\tilde{S}_e^{p\tilde{S}\tilde{J}},$$

noting that

$$\begin{aligned} \mathbf{C}^p\mathbf{S}_e^p &= \delta_I^{\bar{I}}\delta_R^{\bar{R}}\bar{G}_{\bar{I}\bar{R}}\mathbf{G}^I \otimes \mathbf{G}^R\delta_{\tilde{S}}^{S}\delta_{\tilde{J}}^{J}\tilde{S}_e^{p\tilde{S}\tilde{J}}\mathbf{G}_S \otimes \mathbf{G}_J \\ &= \delta_I^{\bar{I}}\delta_R^{\bar{R}}\delta_S^R\delta_{\tilde{S}}^S\delta_{\tilde{J}}^J\bar{G}_{\bar{I}\bar{R}}\mathbf{G}^I \otimes \mathbf{G}^R\tilde{S}_e^{p\tilde{S}\tilde{J}}\mathbf{G}_S \otimes \mathbf{G}_J = \delta_I^{\bar{I}}\delta_{\tilde{S}}^{\bar{R}}\delta_{\tilde{J}}^J\bar{G}_{\bar{I}\bar{R}}\tilde{S}_e^{p\tilde{S}\tilde{J}}\mathbf{G}^I \otimes \mathbf{G}_J. \end{aligned}$$

Further, $\mathbf{M}_e^p$ is related to $\bar{\mathbf{M}}_e^p$ as follows:

$$\mathbf{M}_e^p = {}^p\overleftarrow{\bar{\mathbf{M}}}{}^{p\bullet G}_{e\,G} = \mathbf{F}^{pT}\bar{\mathbf{M}}_e^p\mathbf{F}^{p-T}. \tag{8.132}$$

Here, it follows from equations (8.131) and (8.132) that

$$\mathrm{tr}\mathbf{M}_e^p = \mathrm{tr}\bar{\mathbf{M}}_e^p = \mathrm{tr}\tilde{\mathbf{M}}_e^p. \tag{8.133}$$

The stress tensors and kinematic hardening variables in the reference, intermediate for kinematic hardening, intermediate and current configurations and their relations are illustrated in Table 8.2.

The Oldroyd rate, that is, contraviant convected rate of the second Piola–Kirchhoff stress $\bar{\mathbf{S}}$ in the intermediate configuration is given by

$$\overset{\circ}{\bar{\mathbf{S}}}{}^{ol} \equiv {}^p\overset{\overrightarrow{\bullet}}{\overleftarrow{\bar{\mathbf{S}}}} = \dot{\bar{\mathbf{S}}} - \mathrm{sym}[\bar{\mathbf{L}}^p\bar{\mathbf{S}}], \tag{8.134}$$

Table 8.2 Stress and kinematic hardening tensors in four configurations

Deformation gradient			$\mathbf{F}$						
Multiplicative decomposition			$\mathbf{F}^p$					$\mathbf{F}^e$	
Multiplicative decomposition for kinematic hardening			$\mathbf{F}_d^p$		$\mathbf{F}_e^p$				
Variants	Pull-back	Push-forward	$\mathcal{K}_0$		$\tilde{\mathcal{K}}$		$\bar{\mathcal{K}}$		$\mathcal{K}$
Contravariant	$\mathcal{F}^{-1}(\ )\mathcal{F}^{-T}$	$\mathcal{F}(\ )\mathcal{F}^{T}$	$\mathbf{S}$				$\bar{\mathbf{S}}$		$\boldsymbol{\tau}$
			$\mathbf{S}_e^p$		$\tilde{\mathbf{S}}_e^p$		$\bar{\mathbf{S}}_e^p$		
Covariant–contravariant	$\mathcal{F}^{T}(\ )\mathcal{F}^{-T}$	$\mathcal{F}^{-T}(\ )\mathcal{F}^{T}$	$\mathbf{M}$				$\bar{\mathbf{M}}$		$\boldsymbol{\tau}$
			$\mathbf{M}_e^p$		$\tilde{\mathbf{M}}_e^p$		$\bar{\mathbf{M}}_e^p$		

Note: $\mathcal{F} = \mathbf{F};\ \mathbf{F}^e, \mathbf{F}^p;\ \mathbf{F}_e^p, \mathbf{F}_d^p$

noting the following relation with equation $(8.66)_3$:

$$
\begin{aligned}
{}^{p}\overset{\rightarrow}{\overset{\bullet}{\overset{\leftarrow}{\bar{\mathbf{S}}}}}{}^{\bar{G}\bar{G}} &= \mathbf{F}^{p}(\mathbf{F}^{p-1}\bar{\mathbf{S}}\mathbf{F}^{p-T})^{\bullet}\mathbf{F}^{pT}\,(=\mathbf{F}^{p}\dot{\mathbf{S}}\mathbf{F}^{pT} = {}^{p}\overset{\rightarrow}{\dot{\bar{\mathbf{S}}}})\\
&= \mathbf{F}^{p}(\dot{\mathbf{F}}^{p-1}\bar{\mathbf{S}}\mathbf{F}^{p-T} + \mathbf{F}^{p-1}\dot{\bar{\mathbf{S}}}\mathbf{F}^{p-T} + \mathbf{F}^{p-1}\bar{\mathbf{S}}\dot{\mathbf{F}}^{p-T})\mathbf{F}^{pT}\\
&= \mathbf{F}^{p}\dot{\mathbf{F}}^{p-1}\bar{\mathbf{S}} + \dot{\bar{\mathbf{S}}} + \bar{\mathbf{S}}\dot{\mathbf{F}}^{p-T}\mathbf{F}^{pT} = (-\dot{\mathbf{F}}^{p}\mathbf{F}^{p-1})\bar{\mathbf{S}} + \dot{\bar{\mathbf{S}}} + \bar{\mathbf{S}}(-\mathbf{F}^{p-T}\dot{\mathbf{F}}^{pT})\\
&= \dot{\bar{\mathbf{S}}} - \dot{\mathbf{F}}^{p}\mathbf{F}^{p-1}\bar{\mathbf{S}} - \bar{\mathbf{S}}(\dot{\mathbf{F}}^{p}\mathbf{F}^{p-1})^{T} = \dot{\bar{\mathbf{S}}} - \bar{\mathbf{L}}^{p}\bar{\mathbf{S}} - \bar{\mathbf{S}}\bar{\mathbf{L}}^{pT}.
\end{aligned}
$$

The material-time derivative of the kinematic hardening variable $\bar{\mathbf{S}}_e^p$ in the intermediate configuration is given by

$$
\dot{\bar{\mathbf{S}}}_e^p = \mathbf{F}_e^p\dot{\tilde{\mathbf{S}}}_e^p\mathbf{F}_e^{pT} + 2\mathrm{sym}[\bar{\mathbf{L}}_e^p\bar{\mathbf{S}}_e^p], \tag{8.135}
$$

noting that

$$
\begin{aligned}
(\mathbf{F}_e^p\tilde{\mathbf{S}}_e^p\mathbf{F}_e^{pT})^{\bullet} &= \dot{\mathbf{F}}_e^p\tilde{\mathbf{S}}_e^p\mathbf{F}_e^{pT} + \mathbf{F}_e^p\dot{\tilde{\mathbf{S}}}_e^p\mathbf{F}_e^{pT} + \mathbf{F}_e^p\tilde{\mathbf{S}}_e^p\dot{\mathbf{F}}_e^{pT}\\
&= \dot{\mathbf{F}}_e^p\mathbf{F}_e^{p-1}\mathbf{F}_e^p\tilde{\mathbf{S}}_e^p\mathbf{F}_e^{pT} + \mathbf{F}_e^p\dot{\tilde{\mathbf{S}}}_e^p\mathbf{F}_e^{pT} + \mathbf{F}_e^p\tilde{\mathbf{S}}_e^p\mathbf{F}_e^{pT}\mathbf{F}_e^{p-T}\dot{\mathbf{F}}_e^{pT}\\
&= \mathbf{F}_e^p\dot{\tilde{\mathbf{S}}}_e^p\mathbf{F}_e^{pT} + \bar{\mathbf{L}}_e^p\bar{\mathbf{S}}_e^p + \bar{\mathbf{S}}_e^p\bar{\mathbf{L}}_e^{pT}.
\end{aligned}
$$

Further, the material-time derivative of $\bar{\mathbf{M}}_e^p$ is given from equation (8.127) and (8.135) by

$$
\dot{\bar{\mathbf{M}}}_e^p = \dot{\bar{\mathbf{S}}}_e^p = \mathbf{F}_e^p\dot{\tilde{\mathbf{S}}}_e^p\mathbf{F}_e^{pT} + \mathrm{sym}[\bar{\mathbf{L}}_e^p\bar{\mathbf{S}}_e^p]. \tag{8.136}
$$

8.8 Influences of Superposed Rotations: Objectivity

Consider the objectivities of the stress and deformation tensors described in the foregoing. Then, consider the superposition of the rotations designated by the orthogonal tensor $\mathbf{Q}$ in the current configuration and by the orthogonal tensor $\hat{\mathbf{Q}}$ in the intermediate configuration shown in equation (8.8) as follows:

$$\mathbf{F}^* = \mathbf{Q}\mathbf{F}, \quad \mathbf{F}^{*-1} = \mathbf{F}^{-1}\mathbf{Q}^T, \tag{8.137}$$

$$\mathbf{F}^{e*} = \mathbf{Q}\mathbf{F}^e\hat{\mathbf{Q}}^T, \quad \mathbf{F}^{e*-1} = \hat{\mathbf{Q}}\mathbf{F}^{e-1}\mathbf{Q}^T, \tag{8.138}$$

$$\mathbf{F}^{p*} = \hat{\mathbf{Q}}\mathbf{F}^p, \quad \mathbf{F}^{p*-1} = \mathbf{F}^{p-1}\hat{\mathbf{Q}}^T. \tag{8.139}$$

Denoting the current, the intermediate and the reference base vectors by $\mathbf{g}$, $\bar{\mathbf{G}}$ and $\mathbf{G}$, respectively, for both of the covariant and the cotravariant base vectors collectively eliminating the index, the following relations hold:

$$\left.\begin{array}{l}
\mathbf{g}^* \otimes \mathbf{g}^* = \mathbf{Q}\mathbf{g} \otimes \mathbf{Q}\mathbf{g} = \mathbf{Q}\mathbf{g} \otimes \mathbf{g}\mathbf{Q}^T \\
\mathbf{g}^* \otimes \bar{\mathbf{G}}^* = \mathbf{Q}\mathbf{g} \otimes \hat{\mathbf{Q}}\bar{\mathbf{G}} = \mathbf{Q}\mathbf{g} \otimes \bar{\mathbf{G}}\hat{\mathbf{Q}}^T, \quad \bar{\mathbf{G}}^* \otimes \mathbf{g}^* = \hat{\mathbf{Q}}\bar{\mathbf{G}} \otimes \mathbf{Q}\mathbf{g} = \hat{\mathbf{Q}}\bar{\mathbf{G}} \otimes \mathbf{g}\mathbf{Q}^T \\
\mathbf{g}^* \otimes \mathbf{G}^* = \mathbf{Q}\mathbf{g} \otimes \mathbf{G}, \quad \mathbf{G}^* \otimes \mathbf{g}^* = \mathbf{G} \otimes \mathbf{Q}\mathbf{g} = \mathbf{G} \otimes \mathbf{g}\mathbf{Q}^T \\
\bar{\mathbf{G}}^* \otimes \bar{\mathbf{G}}^* = \hat{\mathbf{Q}}\bar{\mathbf{G}} \otimes \hat{\mathbf{Q}}\bar{\mathbf{G}} = \hat{\mathbf{Q}}\bar{\mathbf{G}} \otimes \bar{\mathbf{G}}\hat{\mathbf{Q}}^T \\
\bar{\mathbf{G}}^* \otimes \mathbf{G}^* = \hat{\mathbf{Q}}\bar{\mathbf{G}} \otimes \mathbf{G}, \quad \mathbf{G}^* \otimes \bar{\mathbf{G}}^* = \mathbf{G}^* \otimes \hat{\mathbf{Q}}\bar{\mathbf{G}} = \mathbf{G} \otimes \bar{\mathbf{G}}\hat{\mathbf{Q}}^T \\
\mathbf{G}^* \otimes \mathbf{G}^* = \mathbf{G} \otimes \mathbf{G}
\end{array}\right\}. \tag{8.140}$$

Taking account of these relations, the influences of superposed rotations are shown below for tensors used in the finite elasto-plastic constitutive equation formulated in this chapter.

$$\boldsymbol{\tau}^* = \mathbf{Q}\boldsymbol{\tau}\mathbf{Q}^T, \tag{8.141}$$

$$\mathbf{S}^* = \mathbf{S}, \quad \mathbf{M}^* = \mathbf{M}, \tag{8.142}$$

$$\bar{\mathbf{S}}^* = \hat{\mathbf{Q}}\bar{\mathbf{S}}\hat{\mathbf{Q}}^T, \quad \bar{\mathbf{M}}^* = \hat{\mathbf{Q}}\bar{\mathbf{M}}\hat{\mathbf{Q}}^T, \tag{8.143}$$

$$\bar{\mathbf{C}}^{e*} = \hat{\mathbf{Q}}\bar{\mathbf{C}}^e\hat{\mathbf{Q}}^T, \quad \bar{\mathbf{U}}^{e*} = \hat{\mathbf{Q}}\bar{\mathbf{U}}^e\hat{\mathbf{Q}}^T, \quad \bar{\mathbf{E}}^{e*} = \hat{\mathbf{Q}}\bar{\mathbf{E}}^e\hat{\mathbf{Q}}^T, \tag{8.144}$$

$$\mathbf{C}^{p*} = \mathbf{C}^p, \quad \mathbf{U}^{p*} = \mathbf{U}^p, \tag{8.145}$$

$$\mathbf{b}^{e*} = \mathbf{Q}\mathbf{b}^e\mathbf{Q}^T, \quad \mathbf{V}^{e*} = \mathbf{Q}\mathbf{V}^e\mathbf{Q}^T, \tag{8.146}$$

$$\bar{\mathbf{B}}^{p*} = \hat{\mathbf{Q}}\bar{\mathbf{B}}^p\hat{\mathbf{Q}}^T, \quad \mathbf{V}^{p*} = \hat{\mathbf{Q}}\mathbf{V}^p\hat{\mathbf{Q}}^T, \tag{8.147}$$

$$\mathbf{R}^{e*} = \mathbf{Q}\mathbf{R}^e\hat{\mathbf{Q}}^T, \quad \mathbf{R}^{p*} = \hat{\mathbf{Q}}\mathbf{R}^p, \tag{8.148}$$

$$\bar{\mathsf{D}}^* = \hat{\mathbf{Q}}\bar{\mathsf{D}}\hat{\mathbf{Q}}^T, \tag{8.149}$$

$$\bar{\mathbf{L}}^{p*} = \hat{\mathbf{Q}}\bar{\mathbf{L}}^p\hat{\mathbf{Q}}^T + \hat{\boldsymbol{\Omega}} \tag{8.150}$$

$$(\dot{\mathbf{F}}^{p*}\mathbf{F}^{p*-1} = (\hat{\mathbf{Q}}\mathbf{F}^p)^\bullet(\hat{\mathbf{Q}}\mathbf{F}^p)^{-1} = (\dot{\hat{\mathbf{Q}}}\mathbf{F}^p + \hat{\mathbf{Q}}\dot{\mathbf{F}}^p)\mathbf{F}^{p-1}\hat{\mathbf{Q}}^T),$$

$$\bar{\mathbf{D}}^{p*} = \hat{\mathbf{Q}}\bar{\mathbf{D}}^p\hat{\mathbf{Q}}^T, \tag{8.151}$$

$$\bar{\mathbf{W}}^{p*} = \hat{\mathbf{Q}}\bar{\mathbf{W}}^{p}\hat{\mathbf{Q}}^{T} + \hat{\boldsymbol{\Omega}}, \tag{8.152}$$

$$\bar{\boldsymbol{\Omega}}^{p*} \equiv \hat{\mathbf{Q}}\bar{\boldsymbol{\Omega}}^{p}\hat{\mathbf{Q}}^{T}, \ \ \bar{\boldsymbol{\Omega}}_{R}^{p*} \equiv \hat{\mathbf{Q}}\bar{\boldsymbol{\Omega}}_{R}^{p}\hat{\mathbf{Q}}^{T} + \hat{\boldsymbol{\Omega}}, \tag{8.153}$$

$$\boldsymbol{l}^{p*} = \mathbf{Q}\boldsymbol{l}^{p}\mathbf{Q}^{T} + \mathbf{Q}\mathbf{F}^{e}\hat{\mathbf{Q}}^{T}\dot{\hat{\mathbf{Q}}}\mathbf{F}^{e-1}\mathbf{Q}^{T} = \mathbf{Q}\boldsymbol{l}^{p}\mathbf{Q}^{T} + \mathbf{F}^{e*}\hat{\boldsymbol{\Omega}}\mathbf{F}^{e*-1}, \tag{8.154}$$

$$(\mathbf{F}^{e*}\bar{\mathbf{L}}^{p*}\mathbf{F}^{e*-1} = \mathbf{Q}\mathbf{F}^{e}\hat{\mathbf{Q}}^{T}(\hat{\mathbf{Q}}\bar{\mathbf{L}}^{p}\hat{\mathbf{Q}}^{T} + \dot{\hat{\mathbf{Q}}}\hat{\mathbf{Q}}^{T})\hat{\mathbf{Q}}\mathbf{F}^{e-1}\mathbf{Q}^{T}$$

$$= \mathbf{Q}\mathbf{F}^{e}\bar{\mathbf{L}}^{p}\mathbf{F}^{e-1}\mathbf{Q}^{T} + \mathbf{Q}\mathbf{F}^{e}\hat{\mathbf{Q}}^{T}\dot{\hat{\mathbf{Q}}}\hat{\mathbf{Q}}^{T}\hat{\mathbf{Q}}\mathbf{F}^{e-1}\mathbf{Q}^{T}),$$

where

$$\hat{\boldsymbol{\Omega}} \equiv \dot{\hat{\mathbf{Q}}}\hat{\mathbf{Q}}^{T}. \tag{8.155}$$

The basic variables for describing the current thermomechanical state of material are: the purely elastic deformation measure $\bar{\mathbf{C}}^{e}(=\mathbf{U}^{e2})$ or $\bar{\mathbf{E}}^{e}$ (note that $\mathbf{b}^{e}$ is influenced by the rigid-body rotation); internal variables for describing the isotropic/anisotropic hardening; and temperature. On the other hand, the plastic deformation measures, that is the plastic deformation gradient $\mathbf{F}^{p}$, or the related variables, such as $\mathbf{C}^{p}$, are not state variables, as is known from the fact that they do not influence the mechanical state of material, while influencing slightly the mechanical state through the hardening variables induced by the plastic deformation (typically the mechanical state of material does not change by plastic deformation in perfectly plastic material so that the material structure is observed to be identical independent of the plastic deformation). Now, it is known from equations (8.143) and (8.144), first that tensors such as $\bar{\mathbf{C}}^{e}$ and $\bar{\mathbf{S}}$ or $\bar{\mathbf{M}}$ based in the intermediate base are not influenced by the superposition of rigid-body rotation in the current state but dependent on the choice of the intermediate configuration; and second, that Eulerian tensors such as $\mathbf{b}^{e}$ and $\boldsymbol{\tau}$ in the current base are influenced by the superposition of rigid-body rotation in the current state but independent of choice of intermediate configuration.

Various finite elasto-plastic constitutive equations based on $\bar{\mathbf{C}}^{e}$ for the elastic deformation along with $\bar{\mathbf{M}}$ and $\bar{\mathbf{L}}^{p}$ for the plastic flow rule and evolution rules of internal variables have been proposed. Among them, the formulations by Dettmer and Reese (2004) and Vladimirov et al. (2008, 2010) are useful, as they incorporate the further multiplicative decomposition of the deformation gradient for the nonlinear kinematic hardening advocated by Lion (2000) as elaborated in Section 8.4. However, these formulations are based on the *Clausius–Duhem inequality* and thus they are limited to von Mises materials with associated flow rule; this inequality must be satisfied but is not be helpful for the concrete formulation of constitutive relations as will be discussed in Section 8.14. The generalized formulation, in which the Clausius–Duhem inequality is not used, will be described below. Further, it will be given in the extended form for unconventional elasto-plasticity based on the subloading surface model (Hashiguchi, 1980, 1989) capable of describing the plastic strain rate caused by the rate of stress inside the yield surface, fulfilling the continuity and the smoothness conditions described in Section 1.11.

8.9 Hyperelastic Equations for Elastic Deformation and Kinematic Hardening

Hyperelastic constitutive equations for elastic deformation and kinematic hardening are described in this section. The component description of constitutive equations in the convected coordinate system is usually not simple and thus the direct notation and the component description of constitutive equations in the rectangular coordinate system are shown below as described in the beginning of Chapter 7.

8.9.1 Hyperelastic Constitutive Equation

The second Piola–Kirchhoff stress pushed forward to the intermediate configuration is given by the strain energy function of the right Cauchy–Green tensor as

$$\bar{\mathbf{S}} = \mathbf{F}^p\mathbf{S}\mathbf{F}^{pT} = 2\frac{\partial\psi^e(\bar{\mathbf{C}}^e)}{\partial\bar{\mathbf{C}}^e}, \tag{8.156}$$

and the Mandel stress is given by

$$\boxed{\bar{\mathbf{M}} \equiv \bar{\mathbf{C}}^e\bar{\mathbf{S}} = 2\bar{\mathbf{C}}^e\frac{\partial\psi^e(\bar{\mathbf{C}}^e)}{\partial\bar{\mathbf{C}}^e}} \tag{8.157}$$

If ψ^e is an isotropic function of $\bar{\mathbf{C}}^e$ only, the stress $\bar{\mathbf{S}}$ becomes coaxial to $\bar{\mathbf{C}}^e$ and thus the Mandel stress $\bar{\mathbf{M}}$ results in the symmetric tensor, that is

$$\bar{\mathbf{M}} = \bar{\mathbf{M}}^T. \tag{8.158}$$

The rate of $\bar{\mathbf{S}}$ is given by

$$\dot{\bar{\mathbf{S}}} = \bar{\mathbb{C}}^e : \frac{1}{2}\dot{\bar{\mathbf{C}}}^e = \bar{\mathbb{C}}^e : \dot{\bar{\mathbf{E}}}^e, \tag{8.159}$$

where

$$\boxed{\bar{\mathbb{C}}^e \equiv 2\frac{\partial\bar{\mathbf{S}}}{\partial\bar{\mathbf{C}}^e} = 4\frac{\partial^2\psi^e(\bar{\mathbf{C}}^e)}{\partial\bar{\mathbf{C}}^e\otimes\partial\bar{\mathbf{C}}^e}} \tag{8.160}$$

The hyperelastic tangent modulus tensor $\bar{\mathbb{C}}^e$ generally possesses not only the minor symmetry $\bar{\mathbb{C}}^e_{\bar{I}\bar{J}\bar{K}\bar{L}} = \bar{\mathbb{C}}^e_{\bar{J}\bar{I}\bar{K}\bar{L}} = \bar{\mathbb{C}}^e_{\bar{I}\bar{J}\bar{L}\bar{K}}$ by virtue of the symmetries of $\bar{\mathbf{S}}$ and $\bar{\mathbf{C}}^e$ but also the major symmetry $\bar{\mathbb{C}}^e_{\bar{I}\bar{J}\bar{K}\bar{L}} = \bar{\mathbb{C}}^e_{\bar{K}\bar{L}\bar{I}\bar{J}}$.

Substituting equations (8.88) and (8.93) into equation (8.159), it follows that

$$\dot{\bar{\mathbf{S}}} = \bar{\mathbb{C}}^e : \bar{\mathbf{D}}^e = \bar{\mathbb{C}}^e : (\bar{\mathbf{D}} - \bar{\mathbf{D}}^p). \tag{8.161}$$

Inserting equations $(8.89)_3$ and (8.161) into equation (8.134), we have

$$\overset{\circ}{\bar{\mathbf{S}}}{}^{Ol} = \bar{\mathbb{C}}^e : (\bar{\mathbf{D}} - \mathrm{sym}[\bar{\mathbf{C}}^e\bar{\mathbf{L}}^p]) - 2\mathrm{sym}[\bar{\mathbf{L}}^p\bar{\mathbf{S}}]. \tag{8.162}$$

Further, the rate of the Mandel stress is given as

$$\overset{\bullet}{\bar{\mathbf{M}}} = (\bar{\mathbf{C}}^e\bar{\mathbf{S}})^\bullet = \bar{\mathbb{L}}^e : \overset{\bullet}{\bar{\mathbf{C}}}{}^e, \tag{8.163}$$

where $\bar{\mathbb{L}}^e$ is the fourth-order hyperelastic tangent modulus tensor given by

$$\boxed{\bar{\mathbb{L}}^e \equiv \frac{\partial \bar{\mathbf{M}}}{\partial \bar{\mathbf{C}}^e} = \frac{1}{2}\bar{\mathbf{C}}^e\bar{\mathbb{C}}^e + \mathbf{G}\tilde{\otimes}\bar{\mathbf{S}}}, \; \bar{\mathcal{L}}^e_{\bar{I}\bar{J}\bar{K}\bar{L}} \equiv \frac{\partial \bar{M}_{\bar{I}\bar{J}}}{\partial \bar{C}^e_{\bar{K}\bar{L}}} = \frac{1}{2}\bar{C}^e_{\bar{I}\bar{P}}\bar{\mathcal{C}}^e_{\bar{P}\bar{J}\bar{K}\bar{L}} + \delta_{\bar{I}\bar{K}}\bar{S}_{\bar{L}\bar{J}}, \tag{8.164}$$

noting

$$(\bar{\mathbf{C}}^e\bar{\mathbf{S}})^\bullet = \bar{\mathbf{C}}^e\overset{\bullet}{\bar{\mathbf{S}}} + \overset{\bullet}{\bar{\mathbf{C}}}{}^e\bar{\mathbf{S}} = \bar{\mathbf{C}}^e\left(\bar{\mathbb{C}}^e : \frac{1}{2}\overset{\bullet}{\bar{\mathbf{C}}}{}^e\right) + \overset{\bullet}{\bar{\mathbf{C}}}{}^e\bar{\mathbf{S}} = \left(\frac{1}{2}\bar{\mathbf{C}}^e\bar{\mathbb{C}}^e + \mathbf{G}\tilde{\otimes}\bar{\mathbf{S}}\right) : \overset{\bullet}{\bar{\mathbf{C}}}{}^e,$$

$$(\overset{\bullet}{\bar{\mathbf{C}}}{}^e\bar{\mathbf{S}})_{\bar{I}\bar{J}} = \overset{\bullet}{\bar{C}}{}^e_{\bar{I}\bar{R}}\bar{S}^{\bar{R}\bar{J}} = G^{\bar{S}}_{\bar{I}}\overset{\bullet}{\bar{C}}{}^e_{\bar{S}\bar{R}}\bar{S}^{\bar{R}\bar{J}} = G^{\bar{S}}_{\bar{I}}\bar{S}^{\bar{R}\bar{J}}\overset{\bullet}{\bar{C}}{}^e_{\bar{S}\bar{R}},$$

while the symbol $\tilde{\otimes}$ is defined in $(1.208)_4$.

$\overset{\bullet}{\bar{\mathbf{M}}}$ is also written via equations $(8.89)_3$ and (8.93) as

$$\boxed{\overset{\bullet}{\bar{\mathbf{M}}} = (\bar{\mathbf{C}}^e\bar{\mathbf{S}})^\bullet = \bar{\mathbb{L}}^e : \overset{\bullet}{\bar{\mathbf{C}}}{}^e = 2\bar{\mathbb{L}}^e : (\bar{\mathbf{D}} - \bar{\mathbf{D}}^p) = 2\bar{\mathbb{L}}^e : (\bar{\mathbf{D}} - \mathrm{sym}[\bar{\mathbf{C}}^e\bar{\mathbf{L}}^p])} \tag{8.165}$$

If ψ^e is the isotropic function of only $\bar{\mathbf{C}}^e$ leading to isotropic elasticity, $\bar{\mathbb{L}}^e$ possesses a minor symmetry but it does not possess a major symmetry.

8.9.2 Hyperelastic Type Constitutive Equation for Kinematic Hardening

The back stress in the kinematic hardening-intermediate configuration is given by

$$\tilde{\mathbf{S}}^p_e = 2\frac{\partial \psi^p_e(\tilde{\mathbf{C}}^p_e)}{\partial \tilde{\mathbf{C}}^p_e}, \tag{8.166}$$

from which it follows that

$$\bar{\mathbf{S}}^p_e = {}^p_e\overset{\rightarrow}{\tilde{\mathbf{S}}}{}^{p\bar{G}\bar{G}}_e = 2\mathbf{F}^p_e\frac{\partial \psi^p_e(\tilde{\mathbf{C}}^p_e)}{\partial \tilde{\mathbf{C}}^p_e}\mathbf{F}^{pT}_e, \tag{8.167}$$

$$\mathbf{S}^p_e = {}^p_d\overset{\leftarrow}{\tilde{\mathbf{S}}}{}^{pGG}_e = 2\mathbf{F}^{p-1}_d\frac{\partial \psi^p_e(\tilde{\mathbf{C}}^p_e)}{\partial \tilde{\mathbf{C}}^p_e}\mathbf{F}^{p-T}_d, \tag{8.168}$$

and

$$\tilde{\mathbf{M}}^p_e = \tilde{\mathbf{C}}^p_e\tilde{\mathbf{S}}^p_e = 2\tilde{\mathbf{C}}^p_e\frac{\partial \psi^p_e(\tilde{\mathbf{C}}^p_e)}{\partial \tilde{\mathbf{C}}^p_e}, \tag{8.169}$$

$$\boxed{\bar{\mathbf{M}}^p_e = \bar{\mathbf{S}}^p_e = 2\mathbf{F}^p_e\frac{\partial \psi^p_e(\tilde{\mathbf{C}}^p_e)}{\partial \tilde{\mathbf{C}}^p_e}\mathbf{F}^{pT}_e} \tag{8.170}$$

$$\mathbf{M}_e^p = \mathbf{C}^p\mathbf{S}_e^p = 2\mathbf{F}^{pT}\mathbf{F}_e^p\frac{\partial\psi_e^p(\tilde{\mathbf{C}}_e^p)}{\partial\tilde{\mathbf{C}}_e^p}\mathbf{F}_d^{p-T}, \tag{8.171}$$

noting equations (8.120), (8.122), (8.125), (8.127) and (8.130). If ψ_e^p is an isotropic function of $\tilde{\mathbf{C}}_e^p$ only, the back stress $\tilde{\mathbf{S}}_e^p$ becomes coaxial to $\tilde{\mathbf{C}}_e^p$ and thus $\tilde{\mathbf{M}}_e^p$ results in the symmetric tensors

$$\tilde{\mathbf{M}}_e^p = \tilde{\mathbf{M}}_e^{pT}. \tag{8.172}$$

The rate of $\tilde{\mathbf{S}}_e^p$ is given from equation (8.166) as

$$\overset{\bullet}{\tilde{\mathbf{S}}}{}_e^p = \tilde{\mathbb{C}}_e^p : \frac{1}{2}\overset{\bullet}{\tilde{\mathbf{C}}}{}_e^p, \tag{8.173}$$

where

$$\boxed{\tilde{\mathbb{C}}_e^p \equiv 2\frac{\partial\tilde{\mathbf{S}}_e^p}{\partial\tilde{\mathbf{C}}_e^p} = 4\frac{\partial^2\psi_e^p(\tilde{\mathbf{C}}_e^p)}{\partial\tilde{\mathbf{C}}_e^p\otimes\partial\tilde{\mathbf{C}}_e^p}} \tag{8.174}$$

The hyperelastic modulus tensor $\tilde{\mathbb{C}}_e^p$ for the kinematic hardenig generally possesses not only the minor symmetry $\tilde{\mathbb{C}}^p_{e\tilde{I}\tilde{J}\tilde{K}\tilde{L}} = \tilde{\mathbb{C}}^p_{e\tilde{J}\tilde{I}\tilde{K}\tilde{L}} = \tilde{\mathbb{C}}^p_{e\tilde{I}\tilde{J}\tilde{L}\tilde{K}}$ by virtue of the symmetries of $\tilde{\mathbf{S}}_e^p$ and $\tilde{\mathbf{C}}_e^p$ but also the major symmetry $\tilde{\mathbb{C}}^p_{e\tilde{I}\tilde{J}\tilde{K}\tilde{L}} = \tilde{\mathbb{C}}^p_{e\tilde{K}\tilde{L}\tilde{I}\tilde{J}}$ also holds.

Substituting equation (8.173) into equation (8.135), $\overset{\bullet}{\bar{\mathbf{S}}}{}_e^p$ is expressed as

$$\overset{\bullet}{\bar{\mathbf{S}}}{}_e^p = \mathbf{F}_e^p\bar{\mathbb{C}}_e^p : (\mathbf{F}_e^{pT}\bar{\mathbf{D}}^p\mathbf{F}_e^p - \mathrm{sym}[\tilde{\mathbf{C}}_e^p\tilde{\mathbf{L}}_d^p])\mathbf{F}_e^{pT} + 2\mathrm{sym}[(\bar{\mathbf{L}}^p - \mathbf{F}_e^p\tilde{\mathbf{L}}_d^p\mathbf{F}_e^{p-1})\bar{\mathbf{S}}_e^p], \tag{8.175}$$

noting that

$$\begin{aligned}
&\mathbf{F}_e^p\overset{\bullet}{\tilde{\mathbf{S}}}{}_e^p\mathbf{F}_e^{pT} + 2\mathrm{sym}[\bar{\mathbf{L}}_e^p\bar{\mathbf{S}}_e^p]\\
&= \frac{1}{2}\mathbf{F}_e^p\bar{\mathbb{C}}_e^p : \overset{\bullet}{\tilde{\mathbf{C}}}{}_e^p\mathbf{F}_e^{pT} + 2\mathrm{sym}[(\bar{\mathbf{L}}^p - \bar{\mathbf{L}}_d^p)\bar{\mathbf{S}}_e^p]\\
&= \mathbf{F}_e^p\bar{\mathbb{C}}_e^p : (\mathbf{F}_e^{pT}\bar{\mathbf{D}}^p\mathbf{F}_e^p - \mathrm{sym}[\tilde{\mathbf{C}}_e^p\tilde{\mathbf{L}}_d^p])\mathbf{F}_e^{pT} + 2\mathrm{sym}[(\bar{\mathbf{L}}^p - \mathbf{F}_e^p\tilde{\mathbf{L}}_d^p\mathbf{F}_e^{p-1})\bar{\mathbf{S}}_e^p]
\end{aligned}$$

with the aid of equations (8.97), $(8.98)_2$ and (8.105). Then, $\overset{\bullet}{\bar{\mathbf{M}}}{}_e^p$ in equation (8.136) is described by

$$\boxed{\overset{\bullet}{\bar{\mathbf{M}}}{}_e^p = \mathbf{F}_e^p\bar{\mathbb{C}}_e^p : (\mathbf{F}_e^{pT}\bar{\mathbf{D}}^p\mathbf{F}_e^p - \mathrm{sym}[\tilde{\mathbf{C}}_e^p\tilde{\mathbf{L}}_d^p])\mathbf{F}_e^{pT} + 2\bar{\mathbf{D}}^p - 2\mathrm{sym}[\mathbf{F}_e^p\tilde{\mathbf{L}}_d^p\mathbf{F}_e^{p-1})\bar{\mathbf{S}}_e^p]} \tag{8.176}$$

8.10 Plastic Constitutive Equations

A basic formulation of plastic constitutive relations is given in this section, based on the concepts and basic variables described in the preceding sections.

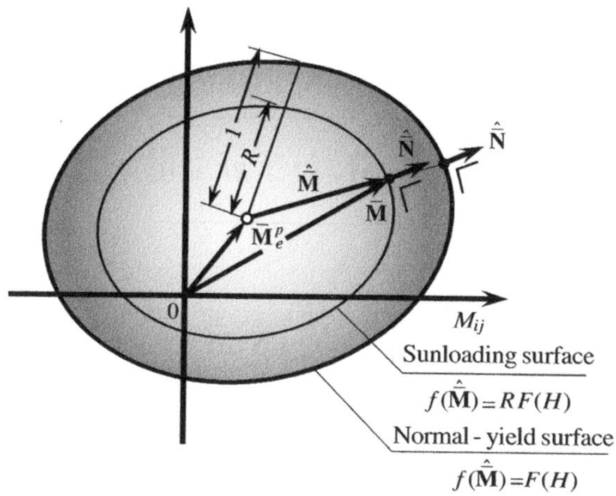

Figure 8.6 Normal-yield and subloading surfaces

8.10.1 Normal-Yield and Subloading Surfaces

Assume the yield condition

$$f(\hat{\bar{\mathbf{M}}}) = F(H), \tag{8.177}$$

where

$$\hat{\bar{\mathbf{M}}} \equiv \bar{\mathbf{M}} - \bar{\mathbf{M}}_e^p. \tag{8.178}$$

Here F is the isotropic hardening function of the isotropic hardening variable H, while the rate $\overset{\bullet}{H}$ is a homogeneous function of $\bar{\mathbf{D}}^p$ of degree 1 on account of the time-independent deformation behavior. $f(\hat{\bar{\mathbf{M}}})$ is assumed to be homogeneous function of $\hat{\bar{\mathbf{M}}}$ of degree 1.

The *conventional elasto-plasticity model* assumes that the interior of the yield surface is assumed to be a purely elastic domain. It has been extended to describe the plastic strain rate caused by the rate of stress inside the yield surface, and the extended models are called *unconventional elasto-plasticity models* (Drucker, 1998). The *subloading surface model* (Hashiguchi, 1980, 1989) as an unconventional elasto-plasticity model is based on a quite natural postulate that the plastic strain rate is induced progressively as the stress approaches the yield surface, fulfilling the smoothness condition described in Section 1.11. In this model, the surface, termed the *subloading surface*, is incorporated, which always passes through the current stress point and maintains a similar shape and the same orientation to the yield surface, renamed the *normal-yield surface*. Then, let the similarity ratio, that is the ratio of the size of the subloading surface to that of the normal-yield surface, termed the *normal-yield ratio* and designated by the symbol $R (0 \leq R \leq 1)$, be introduced as the general measure for approach of stress to the normal-yield surface. The subloading surface is represented by the equation (see Figure 8.6)

$$\boxed{f(\hat{\bar{\mathbf{M}}}) = RF(H)} \tag{8.179}$$

8.10.2 Consistency Condition

The time-differentiation of the subloading surface equation (8.179) reads

$$\boxed{\frac{\partial f(\hat{\bar{\mathbf{M}}})}{\partial \hat{\bar{\mathbf{M}}}} : \overset{\bullet}{\hat{\bar{\mathbf{M}}}} - \dot{R}F - R\dot{F} = 0} \tag{8.180}$$

which can be described as

$$\hat{\bar{\mathbf{N}}} : \overset{\bullet}{\hat{\bar{\mathbf{M}}}} - \left(\frac{\dot{F}}{F} + \frac{\dot{R}}{R}\right) \hat{\bar{\mathbf{N}}} : \hat{\bar{\mathbf{M}}} = 0, \tag{8.181}$$

where

$$\hat{\bar{\mathbf{N}}} \equiv \frac{\partial f(\hat{\bar{\mathbf{M}}})}{\partial \hat{\bar{\mathbf{M}}}} \bigg/ \left\| \frac{\partial f(\hat{\bar{\mathbf{M}}})}{\partial \hat{\bar{\mathbf{M}}}} \right\| \left(\left\|\hat{\bar{\mathbf{N}}}\right\| = 1 \right). \tag{8.182}$$

The following relation based on Euler's theorm of homogenous function is used for $\hat{\bar{\mathbf{M}}}$ is used in the derivation of equation (8.181) from equation (8.180), noting equation (8.179):

$$\frac{\partial f(\hat{\bar{\mathbf{M}}})}{\partial \hat{\bar{\mathbf{M}}}} = \frac{\frac{\partial f(\hat{\bar{\mathbf{M}}})}{\partial \hat{\bar{\mathbf{M}}}} : \hat{\bar{\mathbf{M}}}}{\frac{\partial f(\hat{\bar{\mathbf{M}}})}{\partial \hat{\bar{\mathbf{M}}}} \Big/ \left\|\frac{\partial f(\hat{\bar{\mathbf{M}}})}{\partial \hat{\bar{\mathbf{M}}}}\right\| : \hat{\bar{\mathbf{M}}}} \frac{\frac{\partial f(\hat{\bar{\mathbf{M}}})}{\partial \hat{\bar{\mathbf{M}}}}}{\left\|\frac{\partial f(\hat{\bar{\mathbf{M}}})}{\partial \hat{\bar{\mathbf{M}}}}\right\|} = \frac{\frac{\partial f(\hat{\bar{\mathbf{M}}})}{\partial \hat{\bar{\mathbf{M}}}} : \hat{\bar{\mathbf{M}}}}{\hat{\bar{\mathbf{N}}} : \hat{\bar{\mathbf{M}}}} \hat{\bar{\mathbf{N}}} = \frac{f(\hat{\bar{\mathbf{M}}})}{\hat{\bar{\mathbf{N}}} : \hat{\bar{\mathbf{M}}}} \hat{\bar{\mathbf{N}}} = \frac{RF}{\hat{\bar{\mathbf{N}}} : \hat{\bar{\mathbf{M}}}} \hat{\bar{\mathbf{N}}}. \tag{8.183}$$

Assume the following evolution rule for the normal-yield ratio (Hashiguchi, 1989):

$$\boxed{\dot{R} = U(R) \| \bar{\mathbf{D}}^p \| \quad \text{for} \quad \bar{\mathbf{D}}^p \neq \mathbf{0}} \tag{8.184}$$

where $U(R)$ is a monotonically decreasing function of normal-yield ratio fulfilling the conditions (see Figure 8.7)

$$U(R) \begin{cases} \rightarrow +\infty \text{ for } 0 \leq R \leq R_e \text{ (quasi-elastic state)} \\ > 0 \text{ for } R_e < R < 1 \text{ (sub-yield state)} \\ = 0 \text{ for } R = 1 \text{ (normal-yield state)} \\ < 0 \text{ for } R > 1 \text{ (supernormal-yield state)} \end{cases} . \tag{8.185}$$

$R_e (< 1)$ is the material constant denoting the value of R below which the elastic deformation is induced in practice (Tsutsumi et al., 2006). Here, we assume the explicit form of $U(R)$ as follows:

$$U(R) = u \cot\left(\frac{\pi}{2} \frac{\langle R - R_e \rangle}{1 - R_e}\right), \tag{8.186}$$

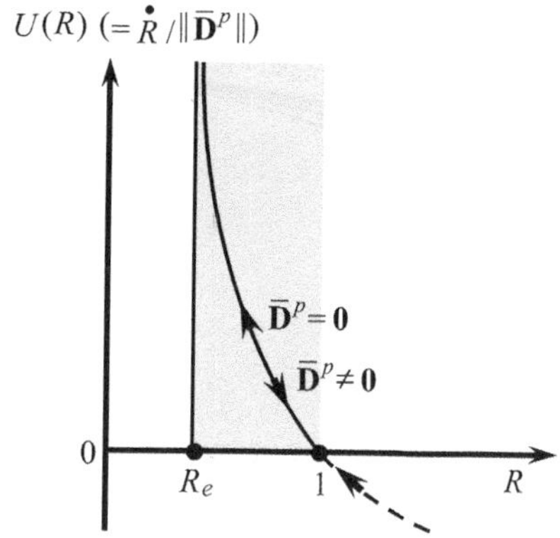

Figure 8.7 Function $U(R)$ in the evolution rule of normal-yield ratio

where u is material parameter. $\langle\ \rangle$ is Macaulay's bracket implying $\langle s\rangle = (s + |s|)/2$, that is $\langle s\rangle = s$ for $s \geq 0$ and $\langle s\rangle = 0$ for $s < 0$ for an arbitrary scalar variable s. Equation (8.184) can be analytically integrated in time as follows:

$$R = \frac{2}{\pi}(1 - R_e)\cos^{-1}\left\{\cos\left(\frac{\pi}{2}\frac{R_0 - R_e}{1 - R_e}\right)\exp\left(-u\frac{\pi}{2}\frac{\bar{D}^p - \bar{D}^p_0}{1 - R_e}\right)\right\} + R_e \tag{8.187}$$

under the initial condition $R = R_0$ for $\bar{D}^p = \bar{D}^p_0$, where $\bar{D}^p = \int ||\bar{\mathbf{D}}^p||dt$.

The subloading surface model possesses an automatic controlling function to attract the stress to the yield surface in the plastic loading process as illustrated in Figure 8.8.

The subloading surface model is drastically modified from the conventional elasto-plastic constitutive equation in the mathematical structures as follows:

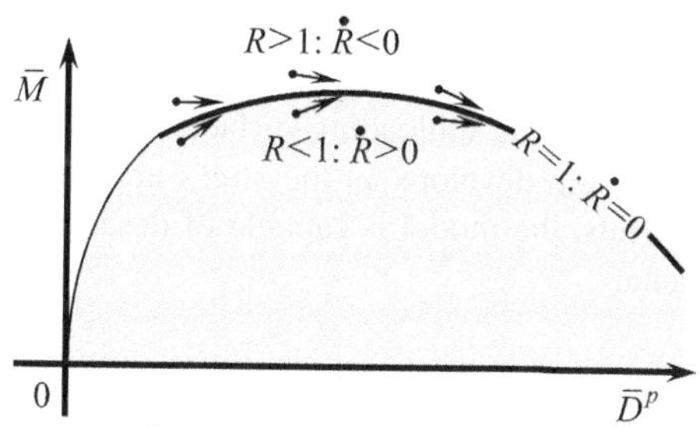

$$\dot{R} = U(R)\|\bar{\mathbf{D}}^p\| \quad \text{for } \bar{\mathbf{D}}^p \neq \mathbf{0}$$

$$U(R)\begin{cases} > 0 & \text{for } R < 1 \text{ (sub-yield state)}, \\ = 0 & \text{for } R = 1 \text{ (normal-yield state)}, \\ < 0 & \text{for } R > 1 \text{ (over normal-yield state)} \end{cases}$$

Figure 8.8 Stress-controlling function in subloading surface model: Stress is automatically attracted to yield surface in the plastic loading process

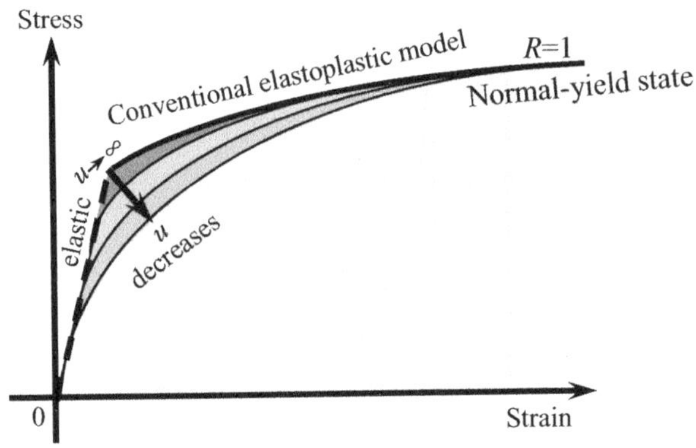

Figure 8.9 Influence of material parameter u on smoothness of elastic-plastic transition, that is curvature of stress-strain curve predicted by subloading surface model

1. No singularity of the tangent modulus is involved, so that the smoothness condition in equation (1.489) is always satisfied.
2. Judgment on the fulfillment of the yield condition is not required in the loading criterion.
3. The model includes a controlling function to attract the stress automatically to the yield surface, resulting in big advantages not only in analytical but also in numerical calculations.

In physical contexts:

4. A smoother transition from elastic to plastic state is described for a smaller value of the material parameter u in the evolution rule of the normal-yield ratio. The conventional elasto-plastic constitutive equation postulating the purely elastic domain inside the yield surface is reproduced by setting $R = 1$ and $U(R) = 0$ or choosing a large value of u. The influence of the material parameter u is depicted in Figure 8.9.
5. The tangent modulus always changes continuously.

The above-mentioned advantages of the subloading surface model are based on the natural assumption that the plastic strain rate develops as the stress approaches the yield surface, as is observed in any material. Thus, the model is capable of describing the wide classes of irreversible (inelastic) phenomena:

(1) the elasto-plastic deformation behavior of pressure-independent metals with high accuracy (Hashiguchi et al., 2012),
(2) the elasto-plastic deformation behavior of plastically compressible, pressure-dependent (frictional) materials, for example geomaterials (Hashiguchi and Chen, 1998; Hashiguchi et al., 2002, Hashiguchi and Mase, 2007);
(3) the rate-dependent plastic (viscoplastic) deformation behavior in the general rate of deformation including the impact loading behavior (Hashiguchi, 2009);
(4) the friction phenomenon caused by the sliding between solid bodies (Hashiguchi et al., 2005; Hashiguchi and Ozaki, 2008).

On the other hand, the other unconventional models, that is the multi-surface model (Mroz, 1967; Iwan, 1967), the two-surface model (Dafalias and Popov, 1975; Krieg, 1975), and the single-surface model with the superposition of nonlinear kinematic hardening rules (Chaboche et al., 1979), are based on the kinematic hardening concept which was proposed primarily for the description of induced anisotropy of metals by Prager (1956) in which the yield surface enclosing the purely elastic domain translates with the plastic deformation. It should be noted that the physical background for the development of the plastic strain rate and that for the anisotropy are substantially irrelevant to each other. They are limited only to the description of monotonic loading behavior of plastically incompressible and pressure independent metals, while they cannot describe the cyclic loading behavior pertinently (cf. Hashiguchi, 2009; Hashiguchi et al., 2012). Needless to say, they cannot describe the viscoplastic deformation behavior in the general rate of deformation and the friction behavior between solids pertinently. Besides, the purely elastic domain is assumed so that the singularity of tangent modulus is induced, changing abruptly from the elastic to the yield state and thus the smoothness condition in equation (1.486) is violated. In addition, they violate not only the smoothness condition in equation (1.486) but also the continuity condition in equation (1.485) when the tangential inelastic strain rate is incorporated as was described in Section 1.11.

Substituting equations (8.165), (8.176) and (8.184) into equation (8.181), the following *consistency condition* of the subloading surface is given:

$$
\begin{aligned}
&\hat{\bar{\mathbf{N}}} : [2\bar{\mathbb{L}}^e : (\bar{\mathbf{D}} - \mathrm{sym}[\bar{\mathbf{C}}^e \bar{\mathbf{L}}^p]) \\
&\quad - \mathbf{F}_e^p \bar{\mathbb{C}}_e^p : (\mathbf{F}_e^{pT} \bar{\mathbf{D}}^p \mathbf{F}_e^p - \mathrm{sym}[\tilde{\mathbf{C}}_e^p \tilde{\mathbf{L}}_d^p])\mathbf{F}_e^{pT} - 2\bar{\mathbf{D}}^p + 2\mathrm{sym}[\mathbf{F}_e^p \tilde{\mathbf{L}}_d^p \mathbf{F}_e^{p-1})\bar{\mathbf{S}}_e^p]] \\
&\qquad - \left(\frac{\dot{F}}{F} + \frac{U}{R} \|\bar{\mathbf{D}}^p\| \right) \hat{\bar{\mathbf{N}}} : \hat{\bar{\mathbf{M}}} = 0.
\end{aligned}
\tag{8.188}
$$

8.10.3 *Plastic and Kinematic Hardening Flow Rules*

Mandel (1972b) postulated the associated flow (normality) rule:

$$\bar{\mathbf{L}}^p = \dot{\lambda} \hat{\bar{\mathbf{N}}}, \tag{8.189}$$

noting the work conjugacy of $\bar{\mathbf{M}}$ and $\bar{\mathbf{L}}^p$ shown in equation $(8.111)_3$ and postulating the maximum plastic dissipation

$$(\bar{\mathbf{M}} - \bar{\mathbf{M}}^{\otimes}) : \bar{\mathbf{L}}^p \geq 0, \tag{8.190}$$

where $\dot{\lambda}(> 0)$ is the plastic multiplier and $\bar{\mathbf{M}}^{\otimes}$ is the admissible Mandel stress inside the convex yield surface. The plastic strain rate and the plastic material spin are given from equation (8.189) by

$$\bar{\mathbf{D}}^p = \dot{\lambda} \mathrm{sym}[\hat{\bar{\mathbf{N}}}], \tag{8.191}$$

$$\bar{\mathbf{W}}^p = \dot{\lambda} \mathrm{ant}[\hat{\bar{\mathbf{N}}}]. \tag{8.192}$$

If it is assumed that the elastic response is isotropic by imposing ψ^e as an isotropic function of $\bar{\mathbf{C}}^e$ only, the symmetries $\bar{\mathbf{M}} = \bar{\mathbf{M}}^T$ in equation (8.158), $\hat{\bar{\mathbf{N}}} = \hat{\bar{\mathbf{N}}}^T$ in equation (8.182) and thus

$\bar{\mathbf{W}}^p = \mathbf{O}$ hold, noting the symmetry of $\bar{\mathbf{M}}_e^p$ as shown in equation (8.127). Then, the plastic strain rate is given by the associated flow rule as

$$\bar{\mathbf{D}}^p = \dot{\lambda}\hat{\bar{\mathbf{N}}}. \tag{8.193}$$

On the other hand, let the flow rule for the plastic strain rate for the kinematic hardening be postulated as follows:

$$\tilde{\mathbf{L}}_d^p = b\frac{\tilde{\mathbf{M}}_e^{p\prime}}{c}||\bar{\mathbf{D}}^p|| = \frac{b}{c}\dot{\lambda}\tilde{\mathbf{M}}_e^{p\prime}, \tag{8.194}$$

where b and c are material constants with no dimension and stress dimension, respectively. Note that equation (8.194) possesses the same form as equation $(8.37)_2$, regarding $\dot{\varepsilon}_d^p \to \tilde{\mathbf{D}}_d^p$, $\alpha \to \tilde{\mathbf{M}}_e^p$ and $\dot{\varepsilon}^p \to \bar{\mathbf{D}}^p$ with $\tilde{\mathbf{L}}_d^p = \tilde{\mathbf{D}}_d^p$, while the back stresses α and $\tilde{\mathbf{M}}_e^p$ are derived from the strain energy function ψ_e^p of kinematic hardening as shown in equations (8.39) and (8.169). Therefore, equation (8.194) would result in the nonlinear kinematic hardening rule of Armstrong and Frederick (1966).

The anti-symmetric tensor $\bar{\mathbf{\Omega}}^p$ in equation (8.82) is the objective tensor induced by the plastic deformation as shown in equation $(8.153)_1$ so that it has to be formulated as a constitutive equation related to the plastic strain rate, which is called the *plastic spin* (Dafalias, 1985). On the other hand, $\bar{\mathbf{W}}^p$ in equation (8.85) and $\bar{\mathbf{\Omega}}_R^p$ in equation (8.83) are non-objective tensors as shown in equations (8.152) and $(8.153)_2$. $\bar{\mathbf{W}}^p$ and $\bar{\mathbf{\Omega}}_R^p$ are called the *plastic material spin* and the *constitutive spin*, respectively, as described in 8.6.2.

Mandel (1972a, b, 1973) introduced the *isoclinic concept* implying $\bar{\mathbf{\Omega}}_R^p = \mathbf{O}$, leading to $\bar{\mathbf{W}}^p = \bar{\mathbf{\Omega}}^p$ in equation (8.81) so that one need only consider the plastic spin $\bar{\mathbf{\Omega}}^p$. The arbitrariness of the intermediate configuration described in Sections 8.2 and 8.8 can now be neglected. Extending equation (5.65) proposed by Zbib and Aifantis (1988) to the intermediate configuration, the plastic spin $\bar{\mathbf{\Omega}}^p$ is formulated by the anti-symmetric tensor (cf. Wallin et al., 2003; Wallin and Ristinmaa, 2005)

$$\bar{\mathbf{\Omega}}^p = \eta^p(\bar{\mathbf{M}}\bar{\mathbf{D}}^p - \bar{\mathbf{D}}^p\bar{\mathbf{M}}), \tag{8.195}$$

where η^p is the material constant. Here, note that the plastic work rate is given by $\bar{\mathbf{M}}:\bar{\mathbf{L}}^p = \bar{\mathbf{M}}:\bar{\mathbf{D}}^p$ if it is assumed that the elastic response is isotropic by imposing the strain energy function ψ^e in equation (8.157) to be an isotropic function of $\bar{\mathbf{C}}^e$ only, leading to $\bar{\mathbf{M}} = \bar{\mathbf{M}}^T$ and thus $\bar{\mathbf{M}}:\bar{\mathbf{\Omega}}^p = 0$ (no work done by plastic spin). In a similar vein, the plastic spin of kinematic hardening in equation $(8.109)_2$ is formulated as

$$\tilde{\mathbf{W}}_d^p = \eta^k(\tilde{\mathbf{M}}_e^p\tilde{\mathbf{D}}_d^p - \tilde{\mathbf{D}}_d^p\tilde{\mathbf{M}}_e^p), \tag{8.196}$$

where η^k is the material constant (cf. Wallin et al., 2003; Wallin and Ristinmaa, 2005). Consequently, for the isotropy in the elastic and kinematic hardening energy functions, the plastic spin terms in equations (8.195) and (8.196) can be added to the flow rules of the plastic strain rate in equation (8.189) and that for the kinematic hardening in equation (8.194), respectively, satisfying the plastic work-conjugacy.

As was described in the end of Chapter 5, the hypoelastic-based plasticity based on the Zaremba-Jaumann rates of stress and back stress predicts a serious oscillatory stress–strain behavior if the plastic spin is not adopted (Nagtegaal and de Jong, 1981; Dafalias, 1985) since it is formulated in the current configuration which is directly influenced by the rigid-body rotation and the rotation of material substructure is overestimated by the continuum spin. On the other hand, there would be less necessity to incorporate the plastic spin in the hyperelastic-based plasticity formulation in the intermediate configuration since the intermediate configuration is not influenced by the rigid-body rotation and the stress and the back stress are calculated directly from the elastic deformation variable and the elastic-like deformation variable for back stress, respectively. Then, plastic spins have often been ignored since the formulation and analysis become far simpler than with plastic spin, a situation described as *spinless* by Dafalias (1998), *inelastically spin-free* by Gurtin and Anand (2005) and *irrotational* by Henann and Anand (2009). Thus, the spins in plastic deformation and kinematic hardening have often been ignored, that is

$$\bar{\boldsymbol{\Omega}}^p = \tilde{\mathbf{W}}_d^p = \mathbf{O}, \tag{8.197}$$

resulting in

$$\boxed{\bar{\mathbf{L}}^p = \bar{\mathbf{D}}^p = \dot{\lambda}\hat{\bar{\mathbf{N}}}} \tag{8.198}$$

$$\boxed{\tilde{\mathbf{L}}_d^p = \tilde{\mathbf{D}}_d^p = \frac{b}{c}\dot{\lambda}\tilde{\mathbf{M}}_e^{p\prime}} \tag{8.199}$$

The flow rules in equations (8.198) and (8.199) are adopted in the formulations below. Here, it is emphasized that the Clausius–Duhem inequality and the postulate of maximum plastic dissipation are not adopted.

8.10.4 Plastic Strain Rate

Substituting equations (8.198) and (8.199) into equation (8.188), we obtain

$$\begin{aligned}\hat{\bar{\mathbf{N}}}:\Bigg\{&2\bar{\mathbb{L}}^e:(\bar{\mathbf{D}}-\mathrm{sym}[\bar{\mathbf{C}}^e\dot{\lambda}\hat{\bar{\mathbf{N}}}])-\mathbf{F}_e^p\bar{\mathbb{C}}_e^p:\left(\mathbf{F}_e^{pT}\dot{\lambda}\hat{\bar{\mathbf{N}}}\mathbf{F}_e^p-\mathrm{sym}\left[\bar{\mathbf{C}}_e^p b\dot{\lambda}\frac{\tilde{\mathbf{M}}_e^p}{c}\right]\right)\mathbf{F}_e^{pT}\\ &-2\mathrm{sym}\left[\left(\dot{\lambda}\hat{\bar{\mathbf{N}}}-\mathbf{F}_e^p b\dot{\lambda}\frac{\tilde{\mathbf{M}}_e^p}{c}\mathbf{F}_e^{p-1}\right)\bar{\mathbf{S}}_e^p\right]\Bigg\}-\left(\frac{F'}{F}\dot{\lambda}h+\frac{U}{R}\dot{\lambda}\right)\hat{\bar{\mathbf{N}}}:\hat{\bar{\mathbf{M}}}=0,\end{aligned} \tag{8.200}$$

where

$$F'\equiv dF/dF,\quad h(\hat{\bar{\mathbf{N}}})=\dot{H}(\bar{\mathbf{D}}^p)/\dot{\lambda} \tag{8.201}$$

and h is the homogeneous function of $\hat{\bar{\mathbf{N}}}$ of degree 1.

The positive proportionality factor is derived from equation (8.200) as follows:

$$\dot{\lambda}=\frac{2\hat{\bar{\mathbf{N}}}:\bar{\mathbb{L}}^e:\bar{\mathbf{D}}}{M^p},\quad \boxed{\bar{\mathbf{L}}^p=\frac{2\hat{\bar{\mathbf{N}}}:\bar{\mathbb{L}}^e:\bar{\mathbf{D}}}{M^p}\hat{\bar{\mathbf{N}}}} \tag{8.202}$$

where the plastic modulus M^p is given by

$$\begin{aligned}M^p\equiv\hat{\bar{\mathbf{N}}}:\Bigg\{&2\bar{\mathbb{L}}^e:\mathrm{sym}[\bar{\mathbf{C}}^e\hat{\bar{\mathbf{N}}}]+\mathbf{F}_e^p\bar{\mathbb{C}}_e^p:(\mathbf{F}_e^{pT}\mathrm{sym}[\hat{\bar{\mathbf{N}}}]\mathbf{F}_e^p\mathbf{F}_e^{pT})+2\mathrm{sym}[\hat{\bar{\mathbf{N}}}\bar{\mathbf{S}}_e^p]\\ &-\mathbf{F}_e^p\bar{\mathbb{C}}_e^p:\mathrm{sym}\left[\bar{\mathbf{C}}_e^p b\frac{\tilde{\mathbf{M}}_e^p}{c}\right]\mathbf{F}_e^{pT}-2\mathrm{sym}\left[\mathbf{F}_e^p b\frac{\tilde{\mathbf{M}}_e^p}{c}\mathbf{F}_e^{p-1}\bar{\mathbf{S}}_e^p\right]+\left(\frac{F'}{F}h+2\frac{U}{R}\right)\hat{\bar{\mathbf{M}}}\Bigg\}.\end{aligned} \tag{8.203}$$

The subloading surface model is the natural extension of the conventional plastic constitutive equation with the restriction that the yield surface interior is a purely elastic domain. In fact, equation (8.203) reduces to the plastic modulus for the conventional plastic constitutive equation by choosing the material parameter as $u \to \infty$ in equation (8.186) or by directly setting $R = 1$ and $U = 0$ so that the description of the abrupt transition from the elastic to the plastic state is recovered.

The current stress always lies on the subloading surface, playing the role of loading surface, and thus no judgment is rqeuired as to whether the normal-yield condition is satisfied. The loading criterion is given simply as follows:

$$\left.\begin{array}{l} \bar{\mathbf{L}}^p \neq \mathbf{O} \ \text{ for } \ \dot{\lambda} > 0 \\ \bar{\mathbf{L}}^p = \mathbf{O} \ \text{ otherwise} \end{array}\right\} \tag{8.204}$$

which can also be stated as

$$\left.\begin{array}{l} \bar{\mathbf{L}}^p \neq \mathbf{O} \ \text{ for } \ \hat{\bar{\mathbf{N}}} : \bar{\mathbb{L}}^e : \bar{\mathbf{D}} > 0 \\ \bar{\mathbf{L}}^p = \mathbf{O} \ \text{ otherwise} \end{array}\right\} \tag{8.205}$$

except for the infinite softening state $M^p \to -\infty$. The loading criterion in equation (8.204) or (8.205) was given first by Hill (1967) and the physical background was clarified by Hashiguchi (2000, 2009).

8.11 Relation between Stress Rate and Strain Rate

The stress–strain relations based on the elastic and the plastic constitutive relations in the intermediate, reference, and current configurations are given in this section.

8.11.1 Description in Intermediate Configuration

Substituting equation (8.202) into equation (8.162), we derive

$$\overset{\circ}{\bar{\mathbf{S}}}{}^{Ol} = \bar{\mathbb{C}}^{ep} : \bar{\mathbf{D}}, \tag{8.206}$$

where

$$\bar{\mathbb{C}}^{ep} \equiv \bar{\mathbb{C}}^{e} - (\bar{\mathbb{C}}^{e} : \mathrm{sym}[\bar{\mathbf{C}}^{e}\hat{\bar{\mathbf{N}}}] + 2\mathrm{sym}[\hat{\bar{\mathbf{N}}}\,\bar{\mathbf{S}}])\frac{2\hat{\bar{\mathbf{N}}} : \bar{\mathbb{L}}^{e}}{M^p} \tag{8.207}$$

$$\left(\bar{\mathbb{C}}^{ep}_{\bar{I}\bar{J}\bar{K}\bar{L}} = \bar{\mathbb{C}}^{e}_{\bar{I}\bar{J}\bar{K}\bar{L}} - \left\{\bar{\mathbb{C}}^{e}_{\bar{I}\bar{J}\bar{P}\bar{Q}}\frac{1}{2}(\bar{C}^{e}_{\bar{P}\bar{T}}\hat{\bar{N}}_{\bar{T}\bar{Q}} + \hat{\bar{N}}_{\bar{P}\bar{T}}\bar{C}^{e}_{\bar{T}\bar{Q}}) + (\hat{\bar{N}}_{\bar{P}\bar{T}}\bar{S}_{\bar{T}\bar{Q}} + \bar{S}_{\bar{P}\bar{T}}\hat{\bar{N}}_{\bar{T}\bar{Q}})\right\}\frac{2\hat{\bar{N}}_{\bar{R}\bar{S}}\bar{\mathbb{L}}^{e}_{\bar{R}\bar{S}\bar{K}\bar{L}}}{M^p}\right).$$

8.11.2 Description in Reference Configuration

Equation (8.206) is given in the reference configuration by

$$\dot{\mathbf{S}} = \mathbf{F}^{p-1}\,\overset{\circ}{\bar{\mathbf{S}}}\,\mathbf{F}^{p-T} = \mathbf{F}^{p-1}(\bar{\mathbb{C}}^{ep} : \bar{\mathbf{D}})\mathbf{F}^{p-T} = \mathbf{F}^{p-1}\{\bar{\mathbb{C}}^{ep} : (\mathbf{F}^{p-T}\dot{\mathbf{E}}\,\mathbf{F}^{p-1})\}\,\mathbf{F}^{p-T} \tag{8.208}$$

$$\left(\dot{S}_{IJ} = F^{p-1}_{I\bar{I}}\{\bar{\mathbb{C}}^{ep}_{\bar{I}\bar{J}\bar{K}\bar{L}}(F^{p-1}_{K\bar{K}}\dot{E}_{KL}F^{p-1}_{L\bar{L}})\}F^{p-1}_{J\bar{J}} = F^{p-1}_{I\bar{I}}F^{p-1}_{J\bar{J}}F^{p-1}_{K\bar{K}}F^{p-1}_{L\bar{L}}\bar{\mathbb{C}}^{ep}_{\bar{I}\bar{J}\bar{K}\bar{L}}\dot{E}_{KL}\right),$$

resulting in

$$\overset{\bullet}{\mathbf{S}} = \mathbb{C}^{ep} : \overset{\bullet}{\mathbf{E}}, \tag{8.209}$$

where

$$\mathbb{C}^{ep}_{IJKL} \equiv F^{p-1}_{I\bar{I}} F^{p-1}_{J\bar{J}} F^{p-1}_{K\bar{K}} F^{p-1}_{L\bar{L}} \bar{\mathbb{C}}^{ep}_{\bar{I}\bar{J}\bar{K}\bar{L}}. \tag{8.210}$$

8.11.3 Description in Current Configuration

Substituting the Oldroyd rate of Kirchhoff stress $\overset{\circ}{\boldsymbol{\tau}}{}^{Ol} (= \mathbf{F}\overset{\bullet}{\mathbf{S}}\mathbf{F}^T)$ in equation (6.103) and the rate of the Green strain $\overset{\bullet}{\mathbf{E}}$ $(= \mathbf{F}^T \mathbf{dF})$ in equation (4.126) into equation (8.209), we obtain

$$\overset{\circ}{\boldsymbol{\tau}}{}^{Ol} = \mathbf{F}\{\mathbb{C}^{ep} : (\mathbf{F}^T \mathbf{dF})\}\,\mathbf{F}^T \tag{8.211}$$

$$\left(\overset{\circ}{\tau}{}^{Ol}_{ij} = F_{iI}(\mathbb{C}^{ep}_{IJKL} F_{kK} d_{kl} F_{lL}) F_{jJ} = F_{iI} F_{jJ} F_{kK} F_{lL} \mathbb{C}^{ep}_{IJKL} d_{kl} \right),$$

which results in

$$\overset{\circ}{\boldsymbol{\tau}}{}^{Ol} = \mathbb{c}^{ep} : \mathbf{d} \quad \left(\overset{\overleftarrow{\bullet}}{\overleftarrow{\bar{\boldsymbol{\tau}}}}{}^{gg} = \mathbb{c}^{ep} : \overset{\overleftarrow{\bullet}}{\overleftarrow{\bar{\mathbf{e}}}}_{gg} \right), \tag{8.212}$$

where

$$\mathbb{c}^{ep}_{ijkl} \equiv F_{iI} F_{jJ} F_{kK} F_{lL} \mathbb{C}^{ep}_{IJKL}. \tag{8.213}$$

It follows from equations (8.66) and (8.198) that

$$\boldsymbol{l}^p \equiv \mathbf{F}^e \bar{\mathbf{L}}^p \mathbf{F}^{e-1} = \mathbf{F}^e \dot{\lambda} \hat{\bar{\mathbf{N}}} \mathbf{F}^{e-1} = \dot{\lambda} \hat{\mathbf{n}}, \tag{8.214}$$

where

$$\left. \begin{aligned} \hat{\mathbf{n}} &\equiv \mathbf{F}^e \hat{\bar{\mathbf{N}}} \mathbf{F}^{e-1} = {}^e\overrightarrow{\hat{\bar{\mathbf{N}}}}{}^{g}_{\bullet g}, \quad \hat{n}_{ij} = F^e_{i\bar{R}} \hat{\bar{N}}_{\bar{R}\bar{S}} F^{e-1}_{\bar{S}j} \\ \hat{\bar{\mathbf{N}}} &= \mathbf{F}^{e-1} \hat{\mathbf{n}} \mathbf{F}^e = {}^e\overleftarrow{\hat{\mathbf{n}}}{}^{\bar{G}}_{\bullet \bar{G}}, \quad \hat{\bar{N}}_{\bar{I}\bar{J}} = F^{e-1}_{\bar{I}r} \hat{n}_{rs} F^e_{s\bar{J}} \end{aligned} \right\}. \tag{8.215}$$

Substituting equation (8.210) with equation (8.207) into equation (8.213) and noting equations (8.214) and (8.215), the tangent elasto-plastic modulus in the current configuration is represented explicitly as

$$\mathbb{c}^{ep} \equiv \mathbb{c}^e - \frac{1}{M^p} \left\{ \mathbb{c}^e : \mathrm{sym}[\hat{\mathbf{n}}] + 2\mathrm{sym}[\hat{\mathbf{n}}\boldsymbol{\tau}] \right\} \otimes (\hat{\mathbf{n}} : \mathbb{c}^e + 2\hat{\mathbf{n}}\boldsymbol{\tau}), \tag{8.216}$$

where

$$\hat{\mathbf{n}} \equiv \mathbf{F}^e \hat{\bar{\mathbf{N}}} \mathbf{F}^{e-1}. \tag{8.217}$$

If we assume the elastic and plastic isotropy as discussed for the symmetry of $\bar{\mathbb{C}}^{ep}$ in the intermediate configuration, $\hat{\mathbf{n}}$ reduces to $\mathbf{n}$ and thus $\boldsymbol{\tau}$ and $\mathbf{n}$ become coaxial and we have

$\boldsymbol{\tau}\mathbf{n} = \mathbf{n}\boldsymbol{\tau} = \mathrm{sym}[\mathbf{n}\boldsymbol{\tau}]$, so that $\mathbb{c}^{ep}$ in (8.216) is simply described as

$$\mathbb{c}^{ep} = \mathbb{c}^{e} - \frac{1}{M^{p}}(\mathbb{c}^{e} : \mathbf{n} + 2\boldsymbol{\tau}\mathbf{n}) \otimes (\mathbf{n} : \mathbb{c}^{e} + 2\mathbf{n}\boldsymbol{\tau}), \tag{8.218}$$

which possesses both the major and minor symmetries (Miehe, 1994), where

$$c^{e}_{ijkl} \equiv F^{e}_{i\bar{I}} F^{e}_{j\bar{J}} F^{e}_{k\bar{K}} F^{e}_{l\bar{L}} \bar{C}^{e}_{\bar{I}\bar{J}\bar{K}\bar{L}}, \tag{8.219}$$

$$c^{ep}_{ijkl} \equiv F^{e}_{i\bar{I}} F^{e}_{j\bar{J}} F^{e}_{k\bar{K}} F^{e}_{l\bar{L}} \bar{C}^{ep}_{\bar{I}\bar{J}\bar{K}\bar{L}}. \tag{8.220}$$

8.12 Material Functions of Metals

Material functions contained in the constitutive equations formulated in the preceding sections are formulated for metals in this section.

8.12.1 Strain Energy Function of Elastic Deformation

Let the following strain energy function of elastic deformation be adopted based on equation (7.58) in the modified neo-Hookean elasticity (2):

$$\boxed{\psi^{e}(\bar{\mathbf{C}}^{e}) = \frac{\Lambda}{4}\left\{\mathrm{det}\bar{\mathbf{C}}^{e} - 1 - 2\ln\left(\sqrt{\mathrm{det}\bar{\mathbf{C}}^{e}}\right)\right\} + \frac{\mu}{2}\left(\mathrm{tr}\bar{\mathbf{C}}^{e} - 3 - 2\ln\sqrt{\mathrm{det}\bar{\mathbf{C}}^{e}}\right)} \tag{8.221}$$

where μ and Λ are assumed to be material constants here also. It follows from equation (8.221) that

$$\bar{\mathbf{S}} = \frac{\Lambda}{2}(\mathrm{det}\bar{\mathbf{C}}^{e} - 1)\bar{\mathbf{C}}^{e-1} + \mu(\mathbf{G} - \bar{\mathbf{C}}^{e-1}), \tag{8.222}$$

noting equation (7.59), from which we have

$$\mathbf{S} = \mathbf{F}^{p-1}\bar{\mathbf{S}}\mathbf{F}^{p-T} = 2\mathbf{F}^{p-1}\left[\frac{\Lambda}{4}(\mathrm{det}\bar{\mathbf{C}}^{e} - 1)\bar{\mathbf{C}}^{e-1} + \frac{\mu}{2}(\mathbf{G} - \bar{\mathbf{C}}^{e-1})\right]\mathbf{F}^{p-T},$$

resulting in

$$\mathbf{S} = \frac{\Lambda}{2}\{\mathrm{det}\mathbf{C}(\mathrm{det}\mathbf{C}^{p})^{-1} - 1\}\mathbf{C}^{-1} + \mu(\mathbf{C}^{p-1} - \mathbf{C}^{-1}) \tag{8.223}$$

by use of

$$\begin{aligned}
\mathbf{F}^{p-1}\bar{\mathbf{C}}^{e-1}\mathbf{F}^{p-T} &= \mathbf{F}^{p-1}(\mathbf{F}^{eT}\mathbf{F}^{e})^{-1}\mathbf{F}^{p-T} = \mathbf{F}^{p-1}\mathbf{F}^{e-1}\mathbf{F}^{e-T}\mathbf{F}^{p-T}\\
&= (\mathbf{F}^{e}\mathbf{F}^{p})^{-1}(\mathbf{F}^{e}\mathbf{F}^{p})^{-T} = \mathbf{F}^{-1}\mathbf{F}^{-T} = (\mathbf{F}^{T}\mathbf{F})^{-1} = \mathbf{C}^{-1},\\
\mathrm{det}\bar{\mathbf{C}}^{e} &= \mathrm{det}(\mathbf{F}^{p-T}\mathbf{C}\mathbf{F}^{p-1}) = \mathrm{det}(\mathbf{C}\mathbf{C}^{p-1}) = \mathrm{det}\mathbf{C}(\mathrm{det}\mathbf{C}^{p})^{-1}
\end{aligned}$$

with the aid of equations (1.159) and (1.161).

Substituting equation (8.223) into (8.117), it follows that

$$\mathbf{M} = \frac{\Lambda}{2}\{\mathrm{det}\mathbf{C}(\mathrm{det}\mathbf{C}^{p})^{-1} - 1\}\mathbf{G} + \mu(\mathbf{C}\mathbf{C}^{p-1} - \mathbf{G}). \tag{8.224}$$

The hyperelastic tangent modulus tensor $\bar{\mathbb{C}}^e$ in equation (8.160) is given from equation (8.222) by

$$\bar{\mathbb{C}}^e = \Lambda[(\det\bar{\mathbf{C}}^e)\bar{\mathbf{C}}^{e-1} \otimes \bar{\mathbf{C}}^{e-1} + (\det\bar{\mathbf{C}}^e - 1)\bar{\mathfrak{C}}^e] - 2\mu\bar{\mathfrak{C}}^e, \tag{8.225}$$

where $\bar{\mathfrak{C}}^e$ is the fourth-order tensor defined as

$$\bar{\mathfrak{C}}^e \equiv \frac{\partial\bar{\mathbf{C}}^{e-1}}{\partial\bar{\mathbf{C}}^e} = -\bar{\mathbf{C}}^{e-1} \tilde{\otimes} \bar{\mathbf{C}}^{e-1} = -\bar{\mathbf{C}}^{e-1}\mathcal{I}\bar{\mathbf{C}}^{e-1},$$

$$\bar{\mathfrak{C}}^e_{\bar{I}\bar{J}\bar{K}\bar{L}} \equiv \frac{\partial\bar{C}^{e-1}_{\bar{I}\bar{J}}}{\partial\bar{C}^e_{\bar{K}\bar{L}}} = -\bar{C}^{e-1}_{\bar{I}\bar{K}}\bar{C}^{e-1}_{\bar{L}\bar{J}} = -(\bar{C}^{e-1}_{\bar{I}\bar{K}}\bar{C}^{e-1}_{\bar{J}\bar{L}} + \bar{C}^{e-1}_{\bar{I}\bar{L}}\bar{C}^{e-1}_{\bar{J}\bar{K}})/2, \tag{8.226}$$

noting equation (1.367), which possesses the minor symmetry $\bar{\mathfrak{C}}^e_{\bar{I}\bar{J}\bar{K}\bar{L}} = \bar{\mathfrak{C}}^e_{\bar{J}\bar{I}\bar{K}\bar{L}} = \bar{\mathfrak{C}}^e_{\bar{I}\bar{J}\bar{L}\bar{K}}$ and the major symmetry $\bar{\mathfrak{C}}^e_{\bar{I}\bar{J}\bar{K}\bar{L}} = \bar{\mathfrak{C}}^e_{\bar{K}\bar{L}\bar{I}\bar{J}}$.

8.12.2 Strain Energy Function for Kinematic Hardening

Assume the following strain energy function for kinematic hardening, which possesses the identical form to the shear part in equation (8.221).

$$\boxed{\psi^p_e(\tilde{\mathbf{C}}^p_e) = c\left\{\frac{1}{2}(\mathrm{tr}\tilde{\mathbf{C}}^p_e - 3) - \ln\sqrt{\det\tilde{\mathbf{C}}^p_e}\right\}} \tag{8.227}$$

where c is a material constant. From equation (8.227) we have

$$\tilde{\mathbf{S}}^p_e = 2\frac{\partial\psi^p_e(\tilde{\mathbf{C}}^p_e)}{\partial\tilde{\mathbf{C}}^p_e} = c(\mathbf{G} - \tilde{\mathbf{C}}^{p-1}_e). \tag{8.228}$$

It follows from equations (8.122) and (8.228) that

$$\mathbf{S}^p_e = c(\mathbf{C}^{p-1}_d - \mathbf{C}^{p-1}), \tag{8.229}$$

noting

$$\mathbf{F}^{p-1}_d\mathbf{G}\mathbf{F}^{p-T}_d = \mathbf{C}^{p-1}_d,$$

$$\begin{aligned}{}^p_d\overleftarrow{\tilde{\mathbf{C}}}{}^{p-1GG}_e &= \mathbf{F}^{p-1}_d\tilde{\mathbf{C}}^{p-1}_e\mathbf{F}^{p-T}_d = \mathbf{F}^{p-1}_d\mathbf{F}^{p-1}_e\mathbf{F}^{p-T}_e\mathbf{F}^{p-T}_d = (\mathbf{F}^p_e\mathbf{F}^p_d)^{-1}(\mathbf{F}^p_e\mathbf{F}^p_d)^{-T} = \mathbf{F}^{p-1}\mathbf{F}^{p-T}\\ &= \mathbf{C}^{p-1}.\end{aligned}$$

On the other hand, the variable $\mathbf{M}^p_e$ is given by

$$\mathbf{M}^p_e = c(\mathbf{C}^p\mathbf{C}^{p-1}_d - \mathbf{G}) \tag{8.230}$$

from equations (8.130) and (8.229).

The hyperelastic tangent modulus tensor $\tilde{\mathbb{C}}^p_e$ in equation (8.174) is given from equation (8.228) by

$$\tilde{\mathbb{C}}^p_e = -2c\tilde{\mathfrak{C}}^p_e, \tag{8.231}$$

where $\tilde{\mathfrak{C}}_e^p$ is the fourth-order tensor defined by

$$\tilde{\mathfrak{C}}_e^p \equiv \frac{\partial \tilde{\mathbf{C}}_e^{p-1}}{\partial \tilde{\mathbf{C}}_e^p} = -\tilde{\mathbf{C}}_e^{p-1} \bar{\otimes} \tilde{\mathbf{C}}_e^{p-1} = -\tilde{\mathbf{C}}_e^{p-1} \bar{\mathcal{I}} \tilde{\mathbf{C}}_e^{p-1},$$

$$\tilde{\mathfrak{C}}^p_{e\tilde{I}\tilde{J}\tilde{K}\tilde{L}} \equiv \frac{\partial \tilde{C}^{p-1}_{e\,\tilde{I}\tilde{J}}}{\partial \tilde{C}^p_{e\,\tilde{K}\tilde{L}}} = -\tilde{C}^{p-1}_{e\,\tilde{I}\tilde{K}} \tilde{C}^{p-1}_{e\,\tilde{J}\tilde{L}} = -(\tilde{C}^{p-1}_{e\,\tilde{I}\tilde{K}} \tilde{C}^{p-1}_{e\,\tilde{J}\tilde{L}} + \tilde{C}^{p-1}_{e\,\tilde{I}\tilde{L}} \tilde{C}^{p-1}_{e\,\tilde{J}\tilde{K}})/2, \tag{8.232}$$

which possesses the minor symmetry $\tilde{\mathfrak{C}}^p_{e\tilde{I}\tilde{J}\tilde{K}\tilde{L}} = \tilde{\mathfrak{C}}^p_{e\tilde{J}\tilde{I}\tilde{K}\tilde{L}} = \tilde{\mathfrak{C}}^p_{e\tilde{I}\tilde{J}\tilde{L}\tilde{K}}$ and the major symmetry $\tilde{\mathfrak{C}}^p_{e\tilde{I}\tilde{J}\tilde{K}\tilde{L}} = \tilde{\mathfrak{C}}^p_{e\tilde{K}\tilde{L}\tilde{I}\tilde{J}}$.

8.12.3 Yield Function

Assume the von Mises yield function and plastic equivalent hardening, that is

$$\boxed{f(\hat{\bar{\mathbf{M}}}') = \sqrt{\frac{3}{2}}||\hat{\bar{\mathbf{M}}}'||} \tag{8.233}$$

$$\boxed{F(H) = F_0[1 + h_1\{1 - \exp(-h_2 H)\}],\ \overset{\bullet}{H} = \sqrt{\tfrac{2}{3}}||\bar{\mathbf{D}}^p||} \tag{8.234}$$

for which we have

$$\overset{\bullet}{F} = F'\overset{\bullet}{H},\ F' \equiv F_0 h_1 h_2 \exp(-h_2 H),\ \overset{\bullet}{H} = \sqrt{\frac{2}{3}}\overset{\bullet}{\lambda}. \tag{8.235}$$

$F \to (1 + h_1)F_0$ holds for $H \to \infty$ in equation (8.234). It follows from equation (8.233) that

$$\hat{\bar{\mathbf{N}}} = \frac{\hat{\bar{\mathbf{M}}}'}{||\hat{\bar{\mathbf{M}}}'||}. \tag{8.236}$$

Note that in the general case, not limited to elastic isotropy, the Mandel stress is related to the Kirchhoff stress as follows:

$$\begin{aligned}\bar{\mathbf{M}} &= \bar{\mathbf{C}}^e\bar{\mathbf{S}}^e = \mathbf{F}^{eT}\mathbf{F}^e\mathbf{F}^p\mathbf{S}\mathbf{F}^{pT} = \mathbf{F}^{eT}\mathbf{F}^e\mathbf{F}^p\mathbf{F}^{-1}\boldsymbol{\tau}\mathbf{F}^{-T}\mathbf{F}^{pT} \\ &= \mathbf{F}^{eT}\boldsymbol{\tau}\mathbf{F}^{e-T}\mathbf{F}^{p-T}\mathbf{F}^{pT} = \mathbf{F}^{eT}\boldsymbol{\tau}\mathbf{F}^{e-T}.\end{aligned}$$

It follows that

$$\left.\begin{aligned}\bar{\mathbf{M}} : \bar{\mathbf{M}}^T &= \mathrm{tr}\{(\mathbf{F}^{eT}\boldsymbol{\tau}\mathbf{F}^{e-T})(\mathbf{F}^{eT}\boldsymbol{\tau}\mathbf{F}^{e-T})\} = \mathrm{tr}(\boldsymbol{\tau}\boldsymbol{\tau})\} = \boldsymbol{\tau} : \boldsymbol{\tau} \\ \bar{\mathbf{M}} : \bar{\mathbf{M}} &= \mathrm{tr}\{(\mathbf{F}^{eT}\boldsymbol{\tau}\mathbf{F}^{e-T})(\mathbf{F}^{eT}\boldsymbol{\tau}\mathbf{F}^{e-T})^T\} \neq \boldsymbol{\tau} : \boldsymbol{\tau}\end{aligned}\right\},$$

resulting in $||\hat{\bar{\mathbf{M}}}'|| \equiv \sqrt{\hat{\bar{\mathbf{M}}}' : \hat{\bar{\mathbf{M}}}'} \neq \sqrt{\hat{\bar{\mathbf{M}}}' : \hat{\bar{\mathbf{M}}}'^T} = \sqrt{\hat{\boldsymbol{\tau}}' : \hat{\boldsymbol{\tau}}'} = ||\hat{\boldsymbol{\tau}}'||$. Consequently, the von Mises yield condition for equation (8.233) is not equivalent to the magnitude of the deviatoric Kirchhoff stress in the current configuration in the elastic anisotropy. However, various constitutive equations making use of equation (8.233) have been proposed. In order that the yield condition described in terms of the Mandel stress is equivalent to the magnitude of the deviatoric

Kirchhoff stress in the current configuration, the yield stress function must be given by $f(\hat{\bar{\mathbf{M}}}') = (\hat{\bar{\mathbf{M}}}' : \hat{\bar{\mathbf{M}}}'^{T})^{1/2}$, for which $\partial f/\partial \bar{\mathbf{M}} = \hat{\bar{\mathbf{M}}}'^{T} / (\hat{\bar{\mathbf{M}}}' : \hat{\bar{\mathbf{M}}}'^{T})^{1/2}$.

Now, introduce the pull-back of $\hat{\bar{\mathbf{M}}}$ to the reference configuration given by

$$\hat{\mathbf{M}} \equiv {}^{p}\overleftarrow{\hat{\bar{\mathbf{M}}}}{}^{\bullet G}_{G} = \mathbf{F}^{pT}\hat{\bar{\mathbf{M}}}\mathbf{F}^{p-T} = \mathbf{C}\mathbf{S} - \mathbf{C}^{p}\mathbf{S}^{p}_{e}, \tag{8.237}$$

where

$$\hat{\bar{\mathbf{M}}} \equiv \bar{\mathbf{M}} - \bar{\mathbf{M}}^{p}_{e}, \tag{8.238}$$

noting equation (6.49) and (8.130) or

$$\begin{aligned}\mathbf{F}^{pT}(\bar{\mathbf{M}} - \bar{\mathbf{M}}^{p}_{e})\mathbf{F}^{p-T} &= \mathbf{F}^{pT}(\mathbf{F}^{eT}\mathbf{F}\mathbf{S}\mathbf{F}^{pT} - \mathbf{F}^{p}\mathbf{S}^{p}_{e}\mathbf{F}^{pT})\mathbf{F}^{p-T} \\ &= \mathbf{F}^{T}\mathbf{F}\mathbf{S} - \mathbf{F}^{pT}\mathbf{F}^{p}\mathbf{S}^{p}_{e} = \mathbf{C}\mathbf{S} - \mathbf{C}^{p}\mathbf{S}^{p}_{e}.\end{aligned}$$

Henceforth, we assume the isotropic functions for $\psi^{e}(\bar{\mathbf{C}}^{e})$ of $\bar{\mathbf{C}}^{e}$ and $\psi^{p}_{e}(\tilde{\mathbf{C}}^{p}_{e})$ of $\tilde{\mathbf{C}}^{p}_{e}$ leading to the symmetry of $\hat{\bar{\mathbf{M}}}$. Therefore, it follows that

$$\hat{\bar{\mathbf{M}}}' : \hat{\bar{\mathbf{M}}}' = \hat{\mathbf{M}}' : \hat{\mathbf{M}}'^{T}, \tag{8.239}$$

noting that

$$\hat{\bar{\mathbf{M}}}' : \hat{\bar{\mathbf{M}}}' = \hat{\bar{\mathbf{M}}}' : \hat{\bar{\mathbf{M}}}'^{T} = \mathrm{tr}(\hat{\bar{\mathbf{M}}}'\hat{\bar{\mathbf{M}}}') = \mathrm{tr}\{(\mathbf{F}^{pT}\hat{\bar{\mathbf{M}}}'\mathbf{F}^{p-T})(\mathbf{F}^{pT}\hat{\bar{\mathbf{M}}}'\mathbf{F}^{p-T})\} = \hat{\mathbf{M}}' : \hat{\mathbf{M}}'^{T}.$$

By virtue of equation (8.239), the yield function $f(\hat{\bar{\mathbf{M}}}')$ in equation (8.233) can be also described in terms of the variable in the reference configuration as follows:

$$f(\hat{\bar{\mathbf{M}}}') = \sqrt{\frac{3}{2}}\sqrt{\hat{\mathbf{M}}' : \hat{\mathbf{M}}'^{T}}. \tag{8.240}$$

8.12.4 Plastic Strain Rate and Kinematic Hardening Strain Rate

Substituting equation (8.233) into equation (8.198), the plastic strain rate is given by

$$\boxed{\bar{\mathbf{L}}^{p} = \bar{\mathbf{D}}^{p} = \dot{\lambda}\frac{\hat{\bar{\mathbf{M}}}'}{\|\hat{\bar{\mathbf{M}}}'\|}} \tag{8.241}$$

which is described in the reference configuration as follows:

$$\boxed{\dot{\mathbf{C}}^{p} = \mathbf{F}^{pT}2\bar{\mathbf{D}}^{p}\mathbf{F}^{p} = 2\dot{\lambda}\frac{\hat{\mathbf{M}}'\mathbf{C}^{p}}{\sqrt{\hat{\mathbf{M}}' : \hat{\mathbf{M}}'^{T}}}} \tag{8.242}$$

In addition, it follows from equations (8.108), (8.130) and (8.199) that

$$\boxed{\dot{\mathbf{C}}^{p}_{d} = 2\frac{b}{c}\dot{\lambda}\mathbf{M}^{p\prime}_{e}\mathbf{C}^{p}_{d},} \tag{8.243}$$

noting that

$$\overset{\bullet}{\mathbf{C}}{}_d^p = 2\mathbf{F}_d^{pT}\tilde{\mathbf{D}}_d^p\mathbf{F}_d^p = 2\mathbf{F}_d^{pT} b\overset{\bullet}{\lambda}\frac{\tilde{\mathbf{M}}_e^{p\prime}}{c}\mathbf{F}_d^p = 2\frac{b}{c}\overset{\bullet}{\lambda}(\mathbf{F}_d^{pT}\tilde{\mathbf{M}}_e^{p\prime}\mathbf{F}_d^{p-T})\mathbf{F}_d^{pT}\mathbf{F}_d^p.$$

8.13 On the Finite Elasto-Plastic Model in the Current Configuration by the Spectral Representation

The constitutive equation in terms of tensors based in the intermediate base has to be formulated so as not to be influenced by the non-uniqueness of the intermediate configuration. On the other hand, the Eulerian tensors $\mathbf{b}^e$ and $\boldsymbol{\tau}$ based in the current base are influenced by the superposition of rigid-body rotation in the current state but independent of selection of the intermediate configuration as shown in (5.39) in general and in equation (8.146) for $\mathbf{b}^e$. In that respect, the finite elastoplastic constitutive equation has been preferably formulated in the current configuration in terms of the Eulerian tensors $\mathbf{b}^e$ and $\boldsymbol{\tau}$ as represented by Simo (1992, 1998) (cf. also Bonet and Wood, 2008; de Souza et al., 2008) before the development of the formulation in the intermediate configuration described comprehensively in this chapter. Therein, noting that the work-conjugate plastic deformation rate measure with $\boldsymbol{\tau}$ is $\boldsymbol{l}^p$ as shown in equation $(8.111)_4$. Now, consider the variable, the following contravariant convected derivative of $\mathbf{b}^e$ is incorporated.

$$-\frac{1}{2}\overset{\leftarrow\bullet}{\mathbf{b}}{}^{egg} = \overset{\bullet}{\mathbf{b}}{}^e - \boldsymbol{l}\mathbf{b}^e - \mathbf{b}^e\boldsymbol{l}^T = \mathrm{sym}[\boldsymbol{l}^p\mathbf{b}^e], \tag{8.244}$$

noting that

$$\begin{aligned}
-\overset{\leftarrow\bullet}{\mathbf{b}}{}^{egg} &= -\mathbf{F}(\mathbf{F}^{-1}\boldsymbol{b}^e\mathbf{F}^{-T})^{\bullet}\mathbf{F}^T = -\mathbf{F}^e\mathbf{F}^p(\mathbf{F}^{p-1}\mathbf{F}^{e-1}\mathbf{F}^e\mathbf{F}^{eT}\mathbf{F}^{e-T}\mathbf{F}^{p-T})^{\bullet}\mathbf{F}^{pT}\mathbf{F}^{eT}\\
&= -\mathbf{F}^e\mathbf{F}^p\overset{\bullet}{\mathbf{F}}{}^{p-1}\mathbf{F}^{eT} - \mathbf{F}^e\overset{\bullet}{\mathbf{F}}{}^{p-T}\mathbf{F}^{pT}\mathbf{F}^{eT} = \mathbf{F}^e\overset{\bullet}{\mathbf{F}}{}^p\mathbf{F}^{p-1}\mathbf{F}^{eT} + \mathbf{F}^e\mathbf{F}^{p-T}\overset{\bullet}{\mathbf{F}}{}^{pT}\mathbf{F}^{eT}\\
&= \mathbf{F}^e\overset{\bullet}{\mathbf{F}}{}^p\mathbf{F}^{p-1}\mathbf{F}^{eT} + (\mathbf{F}^e\overset{\bullet}{\mathbf{F}}{}^p\mathbf{F}^{p-1}\mathbf{F}^{eT})^T = 2\mathrm{sym}[\mathbf{F}^e\overset{\bullet}{\mathbf{F}}{}^p\mathbf{F}^{p-1}\mathbf{F}^{eT}]\\
&= 2\mathrm{sym}[\mathbf{F}^e\overset{\bullet}{\mathbf{F}}{}^p\mathbf{F}^{p-1}\mathbf{F}^{e-1}\mathbf{b}^e] = 2\mathrm{sym}[\mathbf{F}^e\bar{\mathbf{L}}^{p-1}\mathbf{F}^{e-1}\mathbf{b}^e]
\end{aligned}$$

by equation (5.5) and (8.66). The associated flow rule is applied to the plastic velocity gradient $\boldsymbol{l}^p$. However, calculation of deformation is carried out in the principal direction based on the spectral decomposition of equation (8.244) under the assumption of the elastic and plastic isotropy leading to the coaxiality of $\mathbf{b}^e$ and $\boldsymbol{l}^p$. Here, note that the elastic isotropy leads to the coaxiality of stress and elastic strain and the plastic isotropy leads to the coaxiality of plastic strain rate and stress so that all of the stress, the elastic strain and the plastic strain rate become coaxial. This approach has been in vogue and applied widely to the finite elasto-plastic deformation analysis. However, it requires the spectral decomposition of deformation gradient and is limited to the isotropy. It was replaced recently by the finite elasto-plastic constitutive equation formulated in the intermediate configuration described in this chapter.

8.14 On the Clausius–Duhem Inequality and the Principle of Maximum Dissipation

Almost all finite elasto-plastic constitutive equations have been formulated based on the *Clausius–Duhem inequality* (Truesdell and Noll, 1965) and the *principle of maximum dissipation* (or *plastic work*) (Hill, 1950).

The Clausius–Duhem inequality must be fulfilled in any material. It is given in the present model by

$$\mathfrak{D} = \mathbf{S} : \overset{\bullet}{\mathbf{C}} - \overset{\bullet}{\psi} \geq 0, \tag{8.245}$$

where ψ is the *Helmholtz free energy* given by

$$\psi = \psi(\bar{\mathbf{C}}^e) + \psi(\bar{\mathbf{C}}^p_e) + \psi(H). \tag{8.246}$$

Inequality (8.245) can be rewritten as

$$\mathfrak{D} = \bar{\mathbf{M}} : \bar{\mathbf{D}}^p + \bar{\mathbf{B}} : \tilde{\mathbf{D}}^p_d - F(H)\overset{\bullet}{H} \geq 0, \tag{8.247}$$

where $\bar{\mathbf{M}}$, $\bar{\mathbf{B}}$ and $F(H)$ are regarded as thermodynamic forces.

On the other hand, by the principle of maximum dissipation the dissipation rate is maximum under the fulfillment of the yield condition so that the *Lagrange function* takes the maximum stationary value in the plastic loading process. This principle has been widely adopted in the derivation of the plastic flow rule (Mandel, 1972b; Ziegler, 1977; Collins and Houlsby, 1997; Simo, 1998; Simo and Hughes, 1998; Wallin et al., 2003; Tsakmakis, 2004; Wallin and Ristinmaa, 2005; de Souza Neto et al., 2008). The Lagrange function $\mathfrak{L}$ in the present yield condition $f(\bar{\mathbf{M}} - \bar{\mathbf{B}}) - F = 0$ is given by

$$\mathfrak{L} = \mathfrak{D} - \lambda\{f(\bar{\mathbf{M}} - \bar{\mathbf{B}}) - F\}. \tag{8.248}$$

Further, substituting equations (8.247) into equation (8.248), the Lagrange function $\mathfrak{L}$ is given by

$$\mathfrak{L} = \bar{\mathbf{M}} : \bar{\mathbf{D}}^p + \bar{\mathbf{B}} : \tilde{\mathbf{D}}^p_d - F\overset{\bullet}{H} - \lambda\{f(\bar{\mathbf{M}} - \bar{\mathbf{B}}) - F\}. \tag{8.249}$$

The variation of $\mathfrak{L}$ with respect to the thermodynamic forces $\bar{\mathbf{M}}$, $\bar{\mathbf{B}}$ and F is expressed as

$$\begin{aligned}
\delta\mathfrak{L} &= \delta\bar{\mathbf{M}} : \bar{\mathbf{D}}^p + \delta\bar{\mathbf{B}} : \tilde{\mathbf{D}}^p_d - \delta F\overset{\bullet}{H} - \delta\lambda\{f(\bar{\mathbf{M}} - \bar{\mathbf{B}}) - F\} \\
&\quad -\lambda\left\{\frac{\partial f}{\partial \bar{\mathbf{M}}} : \delta\bar{\mathbf{M}} + \left(-\frac{\partial f}{\partial \bar{\mathbf{B}}}\right) : (-\delta\bar{\mathbf{B}}) - \delta F\right\} \\
&= \left(\bar{\mathbf{D}}^p - \lambda\frac{\partial f}{\partial \bar{\mathbf{M}}}\right) : \delta\bar{\mathbf{M}} + \left(\tilde{\mathbf{D}}^p_d - \lambda\frac{\partial f}{\partial \bar{\mathbf{B}}}\right) : \delta\bar{\mathbf{B}} - \delta F(\overset{\bullet}{H} - \lambda).
\end{aligned} \tag{8.250}$$

The following flow rules are obtained because the Lagrange function $\mathfrak{L}$ must take the stationary value in the plastic loading process if the principle of maximum dissipation is accepted:

$$\bar{\mathbf{D}}^p = \lambda\frac{\partial f}{\partial \bar{\mathbf{M}}}, \quad \tilde{\mathbf{D}}^p_d = \lambda\frac{\partial f}{\partial \bar{\mathbf{B}}}, \quad \overset{\bullet}{H} = \lambda. \tag{8.251}$$

Therefore, only the associated flow rules are derived by this approach.

The postulate of maximum plastic work would not hold in general and thus this postulate would impose excessive limitations on the description of mechanical behavior of materials. In fact, the yield function in equation (8.249) was replaced by a convex potential function in order to obtain the non-associated flow rule (Wallin et al., 2003; Wallin and Ristinmaa, 2005). Further, particular material functions have been assumed, for which the fulfillment of the Clausius–Duhem inequality is easily recognized (Dettmer and Reese, 2004; Vladimirov et al., 2008, 2010). However, it cannot be verified easily that an assumed relation violates this inequality if a very particuliar constitutive relation is not assumed. Only the associated flow rule is derived based on the inequality. Besides, these approaches contradict the incorporation of tangential inelastic strain rate induced by the rate of stress tangential to the loading surface described in Section 1.11.2, which cannot be ignored in the plastic instability analyses.

The formulations based on the Clausius–Duhem inequality and the principle of maximum dissipation often result in thermodynamic formalism. This leads to the abandonment of physical background considerations in constitutive relations represented by that for associated flow rule, while the physical interpretation for the associated flow rule has been investigated by Prager (1949), Drucker (1950), Ilyushin (1961), Hill (1968), Hashiguchi (1993a) from the general point of view without falling into thermodynamic formalism.

The Clausius-Duhem inequality and the principle of maximum dissipation would have contributed to the formulation of the *variational principle* by which approximate solutions have been investigated for deformations of simple solids and structures as seen in the limit analysis and the slip-line method, etc. in the past. Here, recall that variational principle is useful for the selection of a solution closer to a correct solution from several approximate solutions and that the associated flow rule based on the maximum plastic dissipation is convenient for the formulation of the variational principle but does not possess generality. Thermodynamics currently represents an obstacle to the general formulation of concrete constitutive equations in isothermal processes (cf. Hashiguchi, 2001b). Nowadays, however, accurate solutions are investigated for complex solids and structures by the development of numerical methods, as represented by FEM. Then, we would have to formulate plastic constitutive equations with a generality free of thermodynamic laws including the Clausius–Duhem inequality and the principle of maximum plastic work.

9

Computational Methods for Finite Strain Elasto-Plasticity

The present chapter outlines the numerical methods for finite strain elasto-plasticity. We develop the stress-update algorithm based on the return-mapping scheme applicable to the finite strain hyperelasto-plasticity models with a combined isotropic and kinematic hardening. This constitutive model stems from the underlying concept of the multiplicative decomposition of the deformation gradient tensor into elastic and plastic parts and a further decomposition of the plastic part into an energetic and a dissipative part. In the course of this, an incremental format of the plastic evolution rules within the 'time-discrete framework' is derived and discussed in detail. The consistent (algorithmic) tangent modulus tensor (Simo and Taylor, 1985), which can be regarded as an algorithmic counterpart of the elasto-plastic tangent modulus tensor in the rate-form constitutive equations, is also introduced with its derivation.

The outline of this chapter is as follows. In Section 9.1, we briefly review the progress of research into the theory and numerical methods for finite strain elasto-plasticity. Section 9.2 briefly summarizes the elasto-plastic constitutive formulation for a specific material model within the 'time-continuum framework', that is the formulation in 'rate form', which was presented in Chapter 8, as a preliminary for deploying the constitutive formulation within the 'time-discrete framework', that is the incremental format of the constitutive equation. In Section 9.3 the constitutive equations are transformed to provide a description relative to the reference configuration $\mathcal{K}_0$ to unify the configuration on which the constitutive equations are represented. Section 9.4 presents the derivation of the incremental format of the plastic evolution rules suitable for the stress-update computations within the time-discrete setting by integrating the rate-form equations with respect to time. Section 9.5 addresses the procedure for updating the deformation gradient tensor, as the first process of the stress-update scheme. Section 9.6 describes the elastic predictor step and the judgment of the loading criterion on the basis of the elastic trial state. This is followed, in Section 9.7, by the plastic corrector step, in which the return-mapping procedure is carried out to determine the constitutive quantities at the updated state. Section 9.8 presents the Jacobian matrix to be used for the iterative solution

Introduction to Finite Strain Theory for Continuum Elasto-Plasticity, First Edition. Koichi Hashiguchi and Yuki Yamakawa.

scheme of the nonlinear equation for the return-mapping. Section 9.9 derives the consistent (algorithmic) tangent modulus tensor by means of two approaches: the analytical derivation by straightforward calculation of the derivatives, and the numerical computation by numerical differentiation using a perturbation technique. Finally, Section 9.10 shows several numerical examples, including finite element simulations of large elasto-plastic deformation problems, to demonstrate the capability of the constitutive model for large deformations as well as the numerical performance of the stress-update algorithm.

In order to enhance the reader's understanding of the numerical methods for finite strain elasto-plasticity, a Fortran program, named `returnmap`, for stress-update computation on the basis of the return-mapping algorithm is presented in Chapter 10, in which the source code is presented in full, together with a line-by-line explanation. The reader interested in the numerical implementation of the stress-update algorithm is referred to Chapter 10.

9.1 A Brief Review of Numerical Methods for Finite Strain Elasto-Plasticity

In the last three decades, there has been considerable research effort into the subject of computational plasticity, the aim of which has been to develop efficient and accurate stress-update algorithms, and remarkable progress has been made in this endeavor (Simo, 1998; Simo and Hughes, 1998). Since the pioneering work of Simo and Ortiz (1985), Simo and Taylor (1985), and Ortiz and Simo (1986), extensive research work on the development of the constitutive equations and stress-update algorithms on the basis of the return-mapping scheme has been done for metallic materials (Amirkhizi and Nemat-Nasser, 2007; Auricchio and Taylor, 1999; Dunne and Petrinic, 2005; Ekh and Runesson, 2001; Menzel et al., 2005; Miehe, 1996a; Papadopoulos and Lu, 1998; Weber and Anand, 1990) and for other engineering or natural materials, such as geomaterials (Borja and Tamagnini, 1998; Callari et al., 1998; Jeremić et al., 2001; Jeremić and Sture, 1997; Rouainia and Muir Wood, 2000; Runesson 1987).

In the early work on the constitutive formulation for finite deformations and its numerical algorithm, the hypoelastic-based formulation using the objective stress rates was adopted and the development of objective numerical time-integration algorithms for rate-form constitutive equations was a major focus of research (Fish and Shek, 1999; Flanagan and Taylor, 1987; Hughes and Winget, 1980; Pinsky et al., 1983; Rubinstein and Atluri, 1983). Later, the hyperelastic-based formulation for finite strain elasto-plasticity was devised by Simo (1986); Simo and Ortiz (1985), in which the multiplicative decomposition of the deformation gradient tensor into elastic and plastic parts is adopted as a fundamental kinematic assumption. Nowadays, the hyperelastic-based formulation of finite strain elasto-plasticity has gained widespread acceptance among researchers working on computational techniques as well as on constitutive models.

However, the development of constitutive models for large elasto-plastic deformations and effective stress-update algorithms is still an issue attracting intense research interest, especially for the incorporation of anisotropic plastic hardening, including kinematic hardening, and elastic and/or plastic anisotropy (Dettmer and Reese, 2004; Ekh and Runesson, 2001; Menzel et al., 2005; Menzel and Steinmann, 2003b; Shutov and Keißig, 2008a; Shutov et al., 2011; Vladimirov et al., 2008, 2010; Wallin et al., 2003; Wallin and Ristinmaa, 2005).

9.2 Brief Summary of Model Formulation

As a preliminary to the discussion of numerical methods for the stress-update computation based on the elastic predictor/return-mapping scheme, this section briefly summarizes the constitutive formulation. As a specific example of a material model for finite strains, we shall focus our attention on the von Mises plasticity model with a combined isotropic and kinematic hardening, which was presented and discussed in Chapter 8. Since the mechanical and theoretical background of this constitutive model as well as the detailed formulation and its derivation have already been outlined in that chapter, only the main constituents of the model necessary for developing the stress-update algorithm will be summarized in this section.

In this model the kinematic hardening is incorporated by introducing a dual multiplicative decomposition of the deformation gradient tensor, namely, the usual elastic–plastic decomposition $\mathbf{F} = \mathbf{F}^e\mathbf{F}^p$ and a further decomposition of the plastic part $\mathbf{F}^p = \mathbf{F}^p_e\mathbf{F}^p_d$ with $\mathbf{F}^p_e$ and $\mathbf{F}^p_d$ denoting, respectively, the energy storing plastic part and the energy dissipating plastic part; thereby a new intermediary configuration associated to the kinematic hardening, denoted by $\tilde{\mathcal{K}}$, is introduced. By this fundamental assumption on the kinematics, a constitutive model incorporating kinematic hardening can be accomplished as an extension of the classical one-dimensional rheological model of the Armstrong–Frederick kinematic hardening. In addition, we incorporate into the model the subloading surface concept, introduced in Sections 8.10.1 and 8.10.2 as a class of the unconventional plasticity theory in the sense that the model is capable of describing the evolution of plastic deformations even when a current stress point lies inside the yield surface.

In the following sections, we separately summarize the main constituents of the particular material model to be focused on in this chapter: the constitutive equations, the yield function, and the plastic evolution rules. Definitions of the constitutive quantities are omitted in this chapter to avoid repetition of Chapter 8.

9.2.1 *Constitutive Equations for Elastic Deformation and Isotropic and Kinematic Hardening*

The specific constitutive equations for hyperelasticity and isotropic and kinematic hardening adopted in the present model, as well as their tangent moduli, are as follows. As a matter of terminology, the term 'constitutive equation' here and henceforth refers to an equation which defines the relationship between the deformation and the stress, or between the deformation-like internal variable and the stress-like variable.

Hyperelastic constitutive equation and tangent modulus tensor for elastic deformation

By specifying the hyperelastic energy function ψ^e of neo-Hookean type given in equation (8.221), the hyperelastic constitutive equation is derived as a description in the intermediate configuration $\bar{\mathcal{K}}$, that is

$$\bar{\mathbf{S}} = 2\frac{\partial \psi^e}{\partial \bar{\mathbf{C}}^e} = \frac{\Lambda}{2}\{(J^e)^2 - 1\}\bar{\mathbf{C}}^{e-1} + \mu(\mathbf{I} - \bar{\mathbf{C}}^{e-1}). \tag{9.1}$$

The material constants Λ and μ for hyperelasticity correspond to the Lamé constants for linear elasticity in the infinitesimal deformation theory.

Differentiating equation (9.1) with respect to $\bar{\mathbf{C}}^e$ leads to the fourth-order hyperelastic tangent modulus tensor

$$\begin{aligned}\bar{\mathbb{C}}^e &:= 2\frac{\partial \bar{\mathbf{S}}}{\partial \bar{\mathbf{C}}^e} = 4\frac{\partial^2 \psi^e}{\partial \bar{\mathbf{C}}^e \otimes \partial \bar{\mathbf{C}}^e} \\ &= \Lambda\Big[(J^e)^2 \bar{\mathbf{C}}^{e-1} \otimes \bar{\mathbf{C}}^{e-1} + \big\{(J^e)^2 - 1\big\}\bar{\mathfrak{C}}^e\Big] - 2\mu\bar{\mathfrak{C}}^e,\end{aligned} \tag{9.2}$$

which is also relative to the intermediate configuration $\bar{\mathcal{K}}$. The explicit expression of the fourth-order tensor $\bar{\mathfrak{C}}^e := \partial\bar{\mathbf{C}}^{e-1}/\partial\bar{\mathbf{C}}^e$ relative to the intermediate configuration $\bar{\mathcal{K}}$ involved in equation (9.2) is given in equation (8.226).

Hyperelastic-type constitutive equation and tangent modulus tensor for kinematic hardening

We adopt the energy function ψ_e^p given in equation (8.227) for kinematic hardening. The corresponding constitutive equation described in the configuration $\tilde{\mathcal{K}}$ associated to kinematic hardening is given by

$$\tilde{\mathbf{S}}_e^p = 2\frac{\partial \psi_e^p}{\partial \tilde{\mathbf{C}}_e^p} = c(\mathbf{I} - \tilde{\mathbf{C}}_e^{p-1}). \tag{9.3}$$

The constant c is a shear modulus-like material parameter for kinematic hardening.

Taking the derivative of equation (9.3) with respect to $\tilde{\mathbf{C}}_e^p$, we obtain the fourth-order tangent modulus tensor for kinematic hardening,

$$\tilde{\mathbb{C}}_e^p := 2\frac{\partial \tilde{\mathbf{S}}_e^p}{\partial \tilde{\mathbf{C}}_e^p} = 4\frac{\partial^2 \psi_e^p}{\partial \tilde{\mathbf{C}}_e^p \otimes \partial \tilde{\mathbf{C}}_e^p} = -2c\tilde{\mathfrak{C}}_e^p \tag{9.4}$$

relative to the configuration $\tilde{\mathcal{K}}$, where the explicit expression of the fourth-order tensor $\tilde{\mathfrak{C}}_e^p := \partial\tilde{\mathbf{C}}_e^{p-1}/\partial\tilde{\mathbf{C}}_e^p$ relative to the configuration $\tilde{\mathcal{K}}$ involved in equation (9.4) is given in equation (8.232).

Constitutive equation and tangent modulus for isotropic hardening

We employ the isotropic hardening rule

$$F = F_0\big[1 + h_1\big\{1 - \exp(-h_2 H)\big\}\big], \tag{9.5}$$

where the material constant F_0 denotes the reference yield stress in uniaxial loading, h_1 is the parameter controlling the saturation limit stress, and h_2 is the exponential hardening parameter. A rule of this form exhibits an exponential hardening response which approaches a saturation limit stress $(1 + h_1)F_0$ as $H \to +\infty$.

By differentiating equation (9.5) with respect to H, the tangent modulus of isotropic hardening reads

$$F' := \frac{dF}{dH} = F_0 h_1 h_2 \exp(-h_2 H), \tag{9.6}$$

which, as expected, vanishes corresponding to the saturation of the isotropic hardening in the limit $H \to +\infty$.

9.2.2 Normal-Yield and Subloading Functions

The yield function is taken as a von Mises criterion expressed in terms of the reduced Mandel stress tensor $\hat{\bar{\mathbf{M}}} := \bar{\mathbf{M}} - \bar{\mathbf{M}}_e^p$ relative to the intermediate configuration $\bar{\mathcal{K}}$. Recalling equations (8.177), (8.178) and (8.233), we employ the yield function

$$f_{\text{yld}} = \sqrt{\frac{3}{2}}\|\hat{\bar{\mathbf{M}}}'\| - F. \tag{9.7}$$

It should be noted that, as was shown in equation (8.112), the Mandel stress tensor $\bar{\mathbf{M}}$ does not necessarily possess the symmetry property. In the case of elastic isotropy, however, by noting the coaxiality between $\bar{\mathbf{C}}^e$ and $\bar{\mathbf{S}}$, the symmetry property of $\bar{\mathbf{M}}$ follows from the identity $\bar{\mathbf{M}} = \bar{\mathbf{C}}^e\bar{\mathbf{S}}$. In contrast, the symmetry property of $\bar{\mathbf{M}}_e^p$ follows from equation (8.127), irrespective of the isotropy of the energy function ψ_e^p with respect to $\tilde{\mathbf{C}}_e^p$.

We further incorporate the subloading surface concept into the present model. The subloading surface concept was devised by Hashiguchi and Ueno (1977) and Hashiguchi (1980, 1989) as a class of unconventional elasto-plasticity model (Drucker, 1988) in order to express evolution of plastic deformations in the sub-yield state when a current stress point lies inside the yield surface. A detailed explanation of the subloading surface concept has already been given in Sections 8.10.1 and 8.10.2. Referring to the generic form (8.179) of the subloading surface function, we take a specific form of the von Mises plasticity as

$$f_{\text{sub}} = \sqrt{\frac{3}{2}}\|\hat{\bar{\mathbf{M}}}'\| - RF. \tag{9.8}$$

As described in Section 8.10, the scalar-valued state variable R is termed the 'normal-yield ratio', which means the ratio of the size of the subloading surface to the size of the yield surface, taking a value in the range $0 \le R \le 1$. To emphasize the distinction between the yield surface and the subloading surface, we henceforth refer to the yield surface as the 'normal-yield surface'. The evolution rule of R is presented in Section 9.2.4.

9.2.3 Plastic Evolution Rules

This section describes the plastic evolution rules. We particularize the plastic flow rule and the evolution rules of the internal variables for isotropic and kinematic hardening to the present material model.

Plastic flow rule

Recalling equations (8.198) and (8.241), we adopt the plastic flow rule of an associative type

$$\bar{\mathbf{L}}^p = \dot{\lambda}\hat{\bar{\mathbf{N}}} \tag{9.9}$$

with the (non-negative) plastic multiplier $\dot{\lambda}$ and the outward unit normal $\hat{\bar{\mathbf{N}}}$ to the subloading surface at the current stress point in the stress space given by

$$\hat{\bar{\mathbf{N}}} =: \frac{\partial f_{\text{sub}}}{\partial \bar{\mathbf{M}}} \Big/ \left\| \frac{\partial f_{\text{sub}}}{\partial \bar{\mathbf{M}}} \right\|. \tag{9.10}$$

Note that $\hat{\bar{\mathbf{N}}}$, which determines the evolutionary orientation of the plastic flow, is normalized so that the equality $\|\hat{\bar{\mathbf{N}}}\| = 1$ holds. From the subloading function (9.8), we have a specific form for the von Mises plasticity,

$$\hat{\bar{\mathbf{N}}} = \frac{\hat{\bar{\mathbf{M}}}'}{\|\hat{\bar{\mathbf{M}}}'\|}. \tag{9.11}$$

It should be noted here that the symmetry property of $\hat{\bar{\mathbf{M}}}$, and thus the symmetry property of $\hat{\bar{\mathbf{N}}}$ hold, and therefore $\bar{\mathbf{L}}^p = \bar{\mathbf{D}}^p$ follows, since we employ the isotropic hyperelastic energy functions (8.221) and the isotropic energy function associated to kinematic hardening (8.227).

In the present model, we adopt a particular choice $\bar{\mathbf{W}}^p = \mathbf{O}$ for the anti-symmetric part of $\bar{\mathbf{L}}^p$, that is an additional constitutive equation for spin is not introduced (see the detailed discussion in Section 8.10.3 and equations (8.193) and (8.198)). The reader is referred to work by Wallin et al. (2003) and Wallin and Ristinmaa (2005) in which the incorporation of the plastic spin is considered, and also to Ekh and Runesson (2001) for the case of elastic anisotropy.

Evolution rule for kinematic hardening

We particularize the evolution rule of the deformation variable for kinematic hardening to the present model by postulating a specific form of equation (8.194),

$$\tilde{\mathbf{L}}_d^p = \dot{\lambda}\frac{b}{c}\tilde{\mathbf{M}}_e^{p\prime}, \tag{9.12}$$

where b is the non-dimensional material constant which controls the evolution. Similar to the above-stated assumption of vanishing plastic spin, we assume that $\tilde{\mathbf{W}}_d^p = \mathbf{O}$, resulting in $\tilde{\mathbf{L}}_d^p = \tilde{\mathbf{D}}_d^p$ (see equations (8.197) and (8.199)).

Evolution rule for isotropic hardening

We employ the evolution rule of the deformation-like variable H for isotropic hardening of the form

$$\dot{H} = \sqrt{\frac{2}{3}}\dot{\lambda}. \tag{9.13}$$

9.2.4 *Evolution Rule of Normal-Yield Ratio for Subloading Surface*

The evolution rule of the normal-yield ratio R for the subloading surface is presented. The fundamentals of the subloading surface concept have already been described, and the evolution rule of the normal-yield ratio R was specified in Section 8.10.2. Recalling equations (8.184) and (8.186), we adopt the form

$$\dot{R} = U(R)||\bar{\mathbf{D}}^p|| = \dot{\lambda}U(R), \quad U(R) = u\cot\left(\frac{\pi}{2}\frac{\langle R - R_e\rangle}{1 - R_e}\right), \tag{9.14}$$

where $U(R)$ is the function controlling the evolution of R, whose functional property is shown in equation (8.185); u is the non-negative evolutionary coefficient; and R_e (< 1) is the material constant denoting the value of R below which the model exhibits a purely elastic response, and thus, no plastic deformation evolves.

The specific form $(9.14)_2$ of the evolutionary function $U(R)$ has a noteworthy feature; the particular choice of this functional form enables us to integrate the evolution rule with respect to time in an analytical way, and thus it is suitable for the stress-update algorithm with incremental steps. A result of the time-integration will be presented and further details will be discussed in Section 9.4.

9.3 Transformation to Description in Reference Configuration

The main constituents of the constitutive model were summarized in the previous section. It should be noted, however, that these are defined on the different configurations because of the mechanical basis or the convenience for constitutive formulation; for instance, the hyperelastic constitutive equation (9.1) is defined on the intermediate configuration $\bar{\mathcal{K}}$, whereas the constitutive equation (9.3) for kinematic hardening is defined on the configuration $\tilde{\mathcal{K}}$. Similarly, the plastic flow rule (9.9) and the evolution rule for kinematic hardening (9.12) are defined by the descriptions on the intermediate configuration $\bar{\mathcal{K}}$ and on the configuration $\tilde{\mathcal{K}}$, respectively, while the evolution rule for isotropic hardening (9.13) is the scalar equation.

In this section, we transform the tensor-valued constitutive functions to the description relative to the reference configuration $\mathcal{K}_0$ by means of the proper pull-back operations. This unification of the configuration on which the constitutive equations are described enables us to develop a simple and orderly stress-update algorithm without transformation of the configurations, and to derive a suitable form of the plastic evolution rule which is readily applicable to the tensor exponential-based effective time-integration scheme.

9.3.1 *Constitutive Equations for Elastic Deformation and Isotropic and Kinematic Hardening*

Hyperelastic constitutive equation

Transforming the hyperelastic constitutive equation (9.1) described in the intermediate configuration $\bar{\mathcal{K}}$ to the reference configuration $\mathcal{K}_0$ by means of the pull-back operation (8.116), we have an expression relative to the reference configuration $\mathcal{K}_0$:

$$\mathbf{S} = \frac{\Lambda}{2}\left(\frac{\det\mathbf{C}}{\det\mathbf{C}^p} - 1\right)\mathbf{C}^{-1} + \mu(\mathbf{C}^{p-1} - \mathbf{C}^{-1}). \tag{9.15}$$

Of course, in von Mises plasticity, $\det\mathbf{C}^p = (J^p)^2 = 1$ holds, implying plastic incompressibility, and thus we have $\det\mathbf{C} = J^2 = (J^e)^2 = \det\bar{\mathbf{C}}^e$ from $J = J^e J^p$.

Hyperelastic-type constitutive equation for kinematic hardening

The constitutive equation for kinematic hardening (9.3) is transformed from the description in the configuration $\bar{\mathcal{K}}$ to the reference configuration $\mathcal{K}_0$:

$$\mathbf{S}_e^p = c\big(\mathbf{C}_d^{p-1} - \mathbf{C}^{p-1}\big), \tag{9.16}$$

using the pull-back operation (8.122).

Constitutive equation for isotropic hardening

Since the constitutive equation for isotropic hardening (9.5) is a scalar-valued function of the scalar deformation variable H, there is no need to transform equation (9.5).

The constitutive equations (9.15), (9.16), and (9.5) transformed above are expressed in terms of the deformation variables $\mathbf{C}$, $\mathbf{C}^p$, $\mathbf{C}_d^p$, and H; therefore, with these deformation variables in hand, we can readily compute all the stress variables $\mathbf{S}$, $\mathbf{S}_e^p$, and F.

9.3.2 *Normal-Yield and Subloading Functions*

The normal-yield function (9.7) and the subloading function (9.8) presented in the previous section are expressed in terms of the stress variables relative to the intermediate configuration $\bar{\mathcal{K}}$. In this section, we transform these functions into descriptions relative to the reference configuration $\mathcal{K}_0$ by means of the proper pull-back operations.

For simplicity, we shall limit our discussion to the case of elastic isotropy. In this context, the symmetry of the Mandel stress tensor $\bar{\mathbf{M}}$ follows. Note also that, as shown in equation (8.127), the symmetry property of the back stress tensor $\bar{\mathbf{M}}_e^p$ always holds irrespective of the isotropy of the constitutive equation for kinematic hardening, although it possesses the Mandel stress-like structure, as can be observed from equation (8.127).

Using the pull-back operations (8.117) and (8.133) with the help of equation (1.81), and after some straightforward tensorial manipulations, we arrive at an equality

$$\|\hat{\mathbf{M}}'\| = \sqrt{\hat{\mathbf{M}}' : (\hat{\mathbf{M}}')^T}, \tag{9.17}$$

where we define the Eshelby-like stress tensor $\mathbf{M}$ and the Eshelby-like back stress tensor $\mathbf{M}_e^p$ relative to the reference configuration $\mathcal{K}_0$ as

$$\mathbf{M} := \mathbf{C}\mathbf{S} \quad \text{and} \quad \mathbf{M}_e^p := \mathbf{C}^p\mathbf{S}_e^p, \tag{9.18}$$

and the Eshelby-like reduced stress tensor

$$\hat{\mathbf{M}} := \mathbf{M} - \mathbf{M}_e^p. \tag{9.19}$$

It should be noted here that the symmetry properties of the Eshelby-like stress tensors $\mathbf{M}$, $\mathbf{M}^p_e$, and thus $\hat{\mathbf{M}}$ do not necessarily hold due to plastic anisotropy even in the case of isotropic energy functions ψ^e and ψ^p_e.

By using equality (9.17), the normal-yield function (9.7) and the subloading function (9.8) can then be rewritten, respectively, in the equivalent descriptions in the reference configuration $\mathcal{K}_0$ as

$$f_{\text{yld}} = \sqrt{\frac{3}{2}\hat{\mathbf{M}}' : (\hat{\mathbf{M}}')^T} - F \tag{9.20}$$

and

$$f_{\text{sub}} = \sqrt{\frac{3}{2}\hat{\mathbf{M}}' : (\hat{\mathbf{M}}')^T} - RF. \tag{9.21}$$

9.3.3 *Plastic Evolution Rules*

Similar to the descriptions of the constitutive equations, the plastic evolution rules are defined in the different configurations; that is the plastic flow rule (9.9) is expressed in terms of the quantities relative to the intermediate configurations $\bar{\mathcal{K}}$, and the evolution rule for kinematic hardening (9.12) in the configuration $\tilde{\mathcal{K}}$. We transform these two evolution rules into descriptions relative to the reference configuration $\mathcal{K}_0$.

Plastic flow rule

In the reference configuration $\mathcal{K}_0$, the plastic flow rule (9.9) is transformed to

$$\dot{\mathbf{C}}^p = 2\dot{\lambda}\frac{\hat{\mathbf{M}}'}{\sqrt{\hat{\mathbf{M}}' : (\hat{\mathbf{M}}')^T}}\mathbf{C}^p, \tag{9.22}$$

where we have utilized the pull-back operations (8.79), (8.117), and (8.133) and the equalities $\text{tr}\bar{\mathbf{M}} = \text{tr}\mathbf{M}$, $\text{tr}\bar{\mathbf{M}}^p_e = \text{tr}\mathbf{M}^p_e$, and (9.17). It should be noted that in the case of isotropic energy functions ψ^e and ψ^p_e, symmetry properties of the products $\mathbf{M}\mathbf{C}^p$ and $\mathbf{M}^p_e\mathbf{C}^p$ follow from the symmetry of $\bar{\mathbf{M}}$ and $\bar{\mathbf{M}}^p_e$, whereas $\mathbf{M}$, $\mathbf{M}^p_e$, and thus $\hat{\mathbf{M}}$ are non-symmetric even in this context of isotropy.

Evolution rule for kinematic hardening

The transformation of the evolution rule for kinematic hardening (9.12) from the configuration $\tilde{\mathcal{K}}$ to the reference configuration $\mathcal{K}_0$ leads to

$$\dot{\mathbf{C}}^p_d = 2\dot{\lambda}\frac{b}{c}\mathbf{M}^{p\prime}_e\mathbf{C}^p_d, \tag{9.23}$$

where the pull-back operations (8.108) and (8.133), and the equality $\text{tr}\tilde{\mathbf{M}}^p_e = \text{tr}\mathbf{M}^p_e$ have been utilized. By simple inspection, it is evident that $\mathbf{M}^p_e\mathbf{C}^p_d$ is a symmetric tensor in the case of the isotropic energy function ψ^p_e.

Evolution rule for isotropic hardening

The evolution rule (9.13) of the deformation-like scalar variable H for isotropic hardening does not need to be transformed, since it is the scalar function.

9.3.4 Evolution Rule of Normal-Yield Ratio for Subloading Surface

There is no need to transform the evolution rule of the normal-yield ratio R for the subloading surface defined by the scalar function (9.14).

9.4 Time-Integration of Plastic Evolution Rules

The solution of the nonlinear equilibrium equations for elasto-plastic boundary value problems is usually carried out by taking a finite number of load/displacement increments. In this context, the rate-form constitutive formulation within the 'time-continuum framework' (also referred to as the 'time-continuous framework') presented in the previous sections needs to be rewritten as an equivalent expression in an incremental setting at finite time steps. Thus, the establishment of the incremental format of the constitutive equations for the present model within the time-discrete framework will be our central objective in this section.

The derivation of the incremental format of the constitutive equations requires an integration of the rate-form plastic evolution rules defined in the time-continuum framework over the time interval $[t_n, t_{n+1}]$. However, the time-integration of the plastic evolution rules in an analytical way is quite complicated in most cases, and for sophisticated constitutive models practically impossible. In this context, it is of crucial importance to adopt a proper numerical time-integration scheme. An implicit time-integration scheme based on the backward Euler difference is often preferred because of its computationally sound characteristics of simplicity, robustness, and unconditional stability (Johansson et al., 2005). Alternative numerical integration schemes are of course available; for instance, an explicit time-integration scheme based on the forward Euler difference can be implemented by a simple numerical algorithm that is computationally less costly, since it requires only the solution data at the previous time step. Other alternative choices are a generalized mid-point Euler difference scheme as well as higher-order schemes which generally improve the accuracy of the approximate integration, although they may yield a complicated and computationally somewhat costly algorithm.

In general, however, an implicit scheme is preferred because of the above-mentioned advantages. Despite these advantages, it should be noted that this class of time-integration scheme has merely first-order accuracy, and thus some errors may be involved if a very large incremental deformation is prescribed within a time step. Moreover, it has been pointed out that the backward Euler integration scheme has a serious drawback in that, when applied to the computation of plastically incompressible constitutive models, the plastic incompressibility is not automatically fulfilled unless some special modifications to the scheme are incorporated. To resolve the above-mentioned disadvantages of the standard backward Euler scheme, the tensor exponential-based time-integration scheme is widely accepted as a suitable scheme for stress-update computations of finite strain plasticity models (Miehe, 1996a;Weber and Anand, 1990).

This section derives the incremental format of the plastic evolution rules suitable for numerical computations within the time-discrete setting by integrating the rate-form equations with respect to time. The tensor exponential-based time-integration scheme is presented and its relation to the standard backward Euler difference scheme is also discussed. A comparative numerical study on the accuracy of time-integration schemes is presented in Section 9.10.

In an attempt to derive the incremental format of the plastic evolution rules by time-integration, let us start by considering an initial-value problem of the differential equation with respect to time

$$\dot{\mathbf{A}}(t) = \mathbf{Z}\mathbf{A}(t), \tag{9.24}$$

where $\mathbf{A}(t)$ and $\mathbf{Z}$ are generally non-symmetric second-order tensors. $\mathbf{A}(t)$ has a functional dependence on time $t \in [t_n, t_{n+1}]$ with an initial value $\mathbf{A}_n := \mathbf{A}(t_n)$, while $\mathbf{Z}$ is constant. As seen in equations (8.115), (9.22), and (9.23), this class of initial-value problems is quite relevant to stress-update computations of finite strain plasticity models. Note here that, in the application to plasticity computations, equation (9.24) needs to be generalized by letting the tensor $\mathbf{Z}$ have a functional dependence on time t, that is $\mathbf{Z} = \mathbf{Z}(t)$. Then the problem is to find the updated quantity $\mathbf{A}_{n+1}$ at t_{n+1}. By integrating equation (9.24) over the time interval $[t_n, t_{n+1}]$ with an assumption that $\mathbf{Z}(t)$ is constant within this time interval, we have an update formula for $\mathbf{A}$ in an incremental format within the time interval $[t_n, t_{n+1}]$:

$$\mathbf{A}_{n+1} = \exp[\mathbf{Z}\Delta t]\mathbf{A}_n, \tag{9.25}$$

where $\exp[\mathbf{Z}\Delta t]$ denotes an exponential function of the tensor $\mathbf{Z}\Delta t$ (note here that the tensor $\mathbf{Z}$ is multiplied by the time increment $\Delta t := t_{n+1} - t_n$). In equation (9.25), $\mathbf{A}_{n+1} \to \mathbf{A}_n$ when $\Delta t \to 0$ since $\exp \mathbf{O} = \mathbf{I}$. Differentiating equation (9.25) with respect to t_{n+1} $(= t_n + \Delta t)$, with the help of equation (1.258), leads to

$$\begin{aligned}\frac{d\mathbf{A}_{n+1}}{dt_{n+1}} &= \left\{\mathbf{Z} + \mathbf{Z}^2\Delta t + \frac{1}{2!}\mathbf{Z}^3(\Delta t)^2 + \cdots\right\}\mathbf{A}_n = \mathbf{Z}\left\{\mathbf{I} + \mathbf{Z}\Delta t + \frac{1}{2!}\mathbf{Z}^2(\Delta t)^2 + \cdots\right\}\mathbf{A}_n \\ &= \mathbf{Z}\exp[\mathbf{Z}\Delta t]\mathbf{A}_n = \mathbf{Z}\mathbf{A}_{n+1},\end{aligned}$$

from which equation (9.24) is recovered.

The explicit expression for the exponential function of a generic tensor based on the infinite series representation was given in equation (1.258) and its numerical implementation is presented in Appendix D. Alternative approaches to evaluating the tensor exponential function by means of the spectral representation and the recursion formula based on the Cayley–Hamilton theorem have been proposed by Ortiz et al. (2001) and Miehe (1996a). Moreover, the derivative of the tensor exponential function $\partial(\exp[\mathbf{Z}\Delta t])/\partial(\mathbf{Z}\Delta t)$ is required in an iterative solution procedure for the return-mapping. The closed-form formula to evaluate the derivative of the tensor exponential function was given in equation (1.374), and its numerical implementation is presented in Appendix D.

While the tensor $\mathbf{Z}$ may indeed change even within the time interval on account of its time-dependence, we here employ an updated value $\mathbf{Z}_{n+1}$ at t_{n+1} in equation (9.25) to obtain an approximate solution to equation (9.24). We thereby have

$$\mathbf{A}_{n+1} = \exp[\mathbf{Z}_{n+1}\Delta t]\mathbf{A}_n, \tag{9.26}$$

which provides an incremental format for the implicit time-integration scheme over the time interval $[t_n, t_{n+1}]$. Alternatively, in a manner analogous to the generalized mid-point rule for the standard Euler difference scheme, a similar rule can be applied to equation (9.25) to obtain an approximate solution to equation (9.24), which yields

$$\mathbf{A}_{n+1} = \exp[\mathbf{Z}_{n+\theta}\Delta t]\mathbf{A}_n, \tag{9.27}$$

where $t_{n+\theta} := t_n + \theta \Delta t$ with the prescribed parameter θ satisfying $0 \leq \theta \leq 1$. The integration scheme based on the generalized mid-point rule contains the following schemes which are often used:

$$\begin{aligned} &\text{forward Euler-like explicit scheme for } \theta = 0,\\ &\text{backward Euler-like implicit scheme for } \theta = 1, \text{ and}\\ &\text{mid-point Euler-like scheme for } \theta = \tfrac{1}{2}. \end{aligned} \tag{9.28}$$

An alternative approach can be obtained by taking only the first-order term of the series representation of the tensor exponential function (D.1). This leads to an update formula

$$\mathbf{A}_{n+1} = \mathbf{A}_n + (\mathbf{Z}\Delta t)\mathbf{A}_n. \tag{9.29}$$

This form can also be derived by replacing the time-derivative $\dot{\mathbf{A}}(t)$ on the left-hand side of equation (9.24) with the difference quotient $(\mathbf{A}_{n+1} - \mathbf{A}_n)/\Delta t$, and taking $\mathbf{A}(t)$ on the right-hand side as $\mathbf{A}_n$. If we adopt $\mathbf{Z}_n$ in equation (9.29), we have the update formula based on the forward Euler explicit scheme, which was employed by Moran et al. (1990) to update the plastic deformation gradient tensor $\mathbf{F}^p$. Alternatively, by adopting $\mathbf{Z}_{n+1}$ and/or replacing $\mathbf{A}_n$ with $\mathbf{A}_{n+1}$ in the second term of the right-hand side of equation (9.29), we have the update formula based on the standard backward Euler implicit scheme, which was employed by Johansson et al. (2005), for example. Section 9.10.4 gives a numerical example comparing the accuracy of the tensor exponential-based scheme based on equation (9.26) and the standard backward Euler implicit scheme based on equation (9.29).

With these preliminaries for the general expression in hand, one should notice that the plastic flow rule (9.22) and the evolution rule for kinematic hardening (9.23) transformed into the reference configuration $\mathcal{K}_0$ in Section 9.3 can be written in the symbolic representation of the form (9.24). The correspondence between the two tensors $\mathbf{A}(t)$ and $\mathbf{Z}(t)$ in equation (9.24) and the variables in the evolution rules (9.22) and (9.23) is shown in Table 9.1. Similar to the above discussion for the general expression, we henceforth employ the backward Euler-type tensor exponential-based scheme, taking $\theta = 1$, while, as can be observed from Table 9.1, the evolutionary tensors in equations (9.22) and (9.23) which correspond to $\mathbf{Z}(t)$ may change within the time interval due to the change in stresses $\hat{\mathbf{M}}$ and $\mathbf{M}_e^p$. Then we have the incremental format of the plastic evolution rules for the implicit time-integration scheme:

$$\mathbf{C}^p_{n+1} = \mathbf{Q}^p_{n+1}\mathbf{C}^p_n \tag{9.30}$$

and

$$\mathbf{C}^p_{d,n+1} = \mathbf{Q}^p_{e,n+1}\mathbf{C}^p_{d,n}, \tag{9.31}$$

Table 9.1 Correspondence between equation (9.24) and equations (9.22) and (9.23)

Equation (9.24)	Plastic flow rule (9.22)	Evolution rule for kinematic hardening (9.23)
$\mathbf{A}(t)$	$\mathbf{C}^p$	$\mathbf{C}^p_d$
$\mathbf{Z}(t)$	$2\dot{\lambda}\dfrac{\hat{\mathbf{M}}'}{\sqrt{\hat{\mathbf{M}}' : (\hat{\mathbf{M}}')^T}}$	$2\dot{\lambda}\dfrac{b}{c}\mathbf{M}^{p\prime}_e$

where we have defined the tensor exponential functions

$$\mathbf{Q}^{p}_{n+1} := \exp\left[2\Delta\lambda \frac{\hat{\mathbf{M}}'_{n+1}}{\sqrt{\hat{\mathbf{M}}'_{n+1} : (\hat{\mathbf{M}}'_{n+1})^{T}}}\right] \tag{9.32}$$

and

$$\mathbf{Q}^{p}_{e,n+1} := \exp\left[2\Delta\lambda \frac{b}{c} \mathbf{M}^{p\prime}_{e,n+1}\right]. \tag{9.33}$$

Here $\Delta\lambda := \dot{\lambda}\Delta t$ denotes the (positive) incremental plastic multiplier within the time-discrete setting, which is related to the plastic deformation increment over the time interval. In the stress-update computation, the tensor exponential functions $\boldsymbol{Q}^{p}_{n+1}$ and $\boldsymbol{Q}^{p}_{e,n+1}$ serve as a 'constitutive integrator' giving the incremental format of the plastic evolution rules, and are sometimes referred to as an 'exponential integrator'; see, for example de Souza Neto et al. (2008) and Menzel et al. (2005).

The time-integration scheme based on the tensor exponential is widely accepted as a quite suitable tool for the stress-update computations of finite strain plasticity models, and plays a crucial role in the development of an accurate and efficient algorithm. As noted above, the tensor exponential-based scheme conserves the plastic incompressibility. In fact, for a generic deviatoric tensor $\mathbf{Z}$ and a generic unimodular tensor $\mathbf{A}_n$ (that is $\mathbf{Z} = \mathbf{Z}'$, $\mathrm{tr}\mathbf{Z} = 0$, and $\det\mathbf{A}_n = 1$; the tensorial quantities listed in Table 9.1 possess these properties), by taking the determinant of both sides of equation (9.25) with the aid of formulae $(1.159)_{4,6}$ and $(1.159)'_{6}$ for tensor exponential functions, we have

$$\begin{aligned} \det\mathbf{A}_{n+1} &= \det(\exp[\mathbf{Z}\Delta t])\det\mathbf{A}_n = \exp[(\mathrm{tr}\mathbf{Z})\Delta t]\det\mathbf{A}_n \\ &= (\exp 0)\det\mathbf{A}_n = 1, \end{aligned} \tag{9.34}$$

which implies that the time-integration scheme based on equation (9.25) conserves the plastic incompressibility. Dettmer and Reese (2004) and Vladimirov et al. (2008) have proposed a modified version of the tensor exponential-based time-integration scheme that preserves the symmetry property of the tensorial variables involved in the integration formula. By using this symmetry property, the number of simultaneous equations to be solved in the return-mapping scheme can be reduced.

The scalar evolution rule for isotropic hardening (9.13) can be integrated properly by means of the standard backward Euler difference approach, which yields

$$H_{n+1} = H_n + \sqrt{\frac{2}{3}}\Delta\lambda. \tag{9.35}$$

With the particular choice of the functional form $(9.14)_2$ for the evolutionary function $U(R)$, the evolution rule for the subloading surface $(9.14)_1$ can be integrated over the time interval

$[t_n, t_{n+1}]$ in an analytical way. Once $\Delta\lambda$ is determined, then, with the known value R_n, we have an updated value of the normal-yield ratio

$$R_{n+1} = \frac{2}{\pi}(1 - R_e)\cos^{-1}\left\{\cos\left(\frac{\pi}{2}\frac{\langle R_n - R_e\rangle}{1 - R_e}\right)\exp\left(-u\frac{\pi}{2}\frac{\Delta\lambda}{1 - R_e}\right)\right\} + R_e. \tag{9.36}$$

In a monotonic loading case, in fact, it is possible to integrate equation $(9.14)_1$ even over the entire time interval $[0, t_{n+1}]$. However, the incremental format (9.36) with respect to the time interval $[t_n, t_{n+1}]$ is preferred in actual computations for general loading process, since, as will be mentioned in Section 9.6, once unloading occurs, the normal-yield ratio R needs to be updated by equation (9.51) without regard to the evolution rule so that the subloading surface passes through the stress point after the unloading.

9.5 Update of Deformation Gradient Tensor

As the first process of the stress-update scheme, this section describes a procedure to update the deformation gradient tensor. This procedure is independent of the particular constitutive model, and is, in general, indispensable in the numerical analysis of boundary value problems for large deformations.

Let $\mathbf{X}$ denote a specific material point of interest in the body. In practice, in the finite element analysis, this corresponds to a quadrature point, or an integration point (for example the Gauss numerical integration point), in a typical finite element. Further, let $\boldsymbol{\varphi}_n(\mathbf{X})$ denote a motion which maps the initial position of the material point $\mathbf{X}$ to the current position of $\mathbf{x}$ at a time t_n whose spatial coordinate is $\mathbf{x}_n$, that is $\mathbf{x}_n = \boldsymbol{\varphi}_n(\mathbf{X})$. Here the subscript n indicates that the quantity is evaluated at the time t_n. We assume that the deformations and internal variables at the material point $\mathbf{X}$ at the time t_n are completely known, and thus the state of stresses at the time t_n can completely be determined via the constitutive equations. Assume further that the deformed configuration at t_n is equilibrated, that is the state of stresses and deformations in the body satisfies the equilibrium equation. For the specific constitutive model discussed in this chapter, of course, the total deformation gradient tensor $\mathbf{F}_n$ as well as its elastic and plastic parts, $\mathbf{F}^e_n$, $\mathbf{F}^p_{e,n}$, $\mathbf{F}^p_{d,n}$, and thus $\mathbf{F}^p_n$, and the scalar variables H_n and R_n are assumed to be known at t_n, and hence, all stress variables can readily be determined via the constitutive equations. Suppose that an incremental displacement field $\Delta\mathbf{u}$ is imposed on the configuration at t_n within a time interval $[t_n, t_{n+1}]$, where $t_{n+1} := t_n + \Delta t$ with a time increment Δt.

In this context, the problem we are concerned with is to determine and update the constitutive variables at the time t_{n+1} on the basis of the constitutive model summarized in the previous sections. This type of constitutive problem in which the total deformation is given as the primary variable is commonly referred to as a 'strain-driven problem', and is a relevant setting for the numerical solution of the elasto-plastic boundary-value problem by means of displacement-type finite element methods. It is not an obvious procedure to determine the variables in a manner consistent with the constitutive equations, the plastic evolution rules and the loading criterion. This procedure based on the time-discrete framework indeed requires the time-integration of the plastic evolution rules over the time interval $[t_n, t_{n+1}]$ and the return-mapping procedure to obtain an updated stress point lying on the subloading surface. These two procedures are crucial steps in the stress-update scheme, and will be addressed in detail in the subsequent sections.

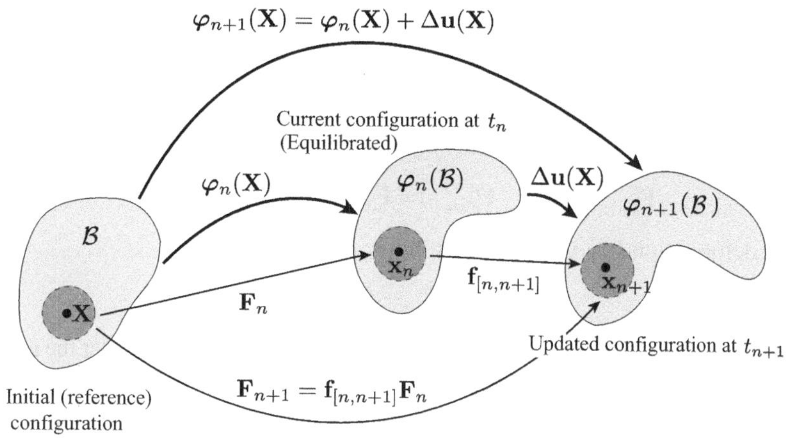

Figure 9.1 Schematic illustration of the update process of the deformation gradient tensor within the time-discrete framework

As a first step to advance the stress-update computation, we update the deformation gradient tensor $\mathbf{F}_n$ to the time t_{n+1} by

$$\mathbf{F}_{n+1} = \mathbf{f}_{[n,n+1]}\mathbf{F}_n, \tag{9.37}$$

noting that $\mathbf{F}_n = \varphi_n(\mathbf{X}) \otimes \nabla_{\mathbf{X}})$ and $\mathbf{F}_{n+1} = \varphi_{n+1}(\mathbf{X}) \otimes \nabla_{\mathbf{X}}$ by definition. Here we have introduced the relative deformation gradient tensor from the configuration at t_n into that at t_{n+1}:

$$\mathbf{f}_{[n,n+1]} := \mathbf{F}_{n+1}\mathbf{F}_n^{-1} = \mathbf{I} + \Delta\mathbf{u} \otimes \nabla_{\mathbf{x}_n}, \tag{9.38}$$

noting that the updated spatial coordinates of the material point $\mathbf{X}$ at the time t_{n+1} are given by $\mathbf{x}_{n+1} := \mathbf{x}_n + \Delta\mathbf{u}$. The nabla operator $(\cdot) \otimes \nabla_{\mathbf{x}_n}$ for a vector field means the Cartesian spatial derivative with respect to the coordinate $\mathbf{x}_n$ at the time t_n, and produces a gradient tensor. The relative deformation gradient tensor $\mathbf{f}_{[n,n+1]}$ in equation (9.38) represents the incremental deformation imposed on the configuration at time t_n by the incremental displacement $\Delta\mathbf{u}$ within the time interval $[t_n, t_{n+1}]$.

A schematic illustration on the update process of the deformation gradient tensor discussed above is shown in Figure 9.1.

9.6 Elastic Predictor Step and Loading Criterion

The stress-update algorithm presented in this chapter consists essentially of the following two steps: an elastic predictor step and a plastic corrector step. This section describes the elastic predictor step, in which an elastic trial state is computed under an assumption of a fictitious purely elastic response to the prescribed incremental deformation, and then the judgment of the loading criterion is carried out on the basis of the elastic trial state. The plastic corrector step, in which the return-mapping is performed in the plastic case after the loading judgment, is discussed in the next section.

The elastic predictor is obtained under an assumption that the response of the constitutive model to an incremental deformation imposed on the known and equilibrated state at t_n is purely elastic or, equivalently, no additional plastic deformation occurs within the time interval $[t_n, t_{n+1}]$. This assumption is also referred to as 'freezing of the plastic flow' in the literature on computational plasticity. This preliminary assumption implies

$$\mathbf{F}^{p(\mathrm{tr})}_{n+1} := \mathbf{F}^p_n, \quad \mathbf{F}^{p(\mathrm{tr})}_{e,n+1} := \mathbf{F}^p_{e,n}, \quad \mathbf{F}^{p(\mathrm{tr})}_{d,n+1} := \mathbf{F}^p_{d,n}, \tag{9.39}$$

and, for the deformation-like variable for isotropic hardening,

$$H^{(\mathrm{tr})}_{n+1} := H_n, \tag{9.40}$$

where the quantity with the superscript '(tr)' means a trial value obtained under the assumption of fictitious freezing of the plastic flow. In an analogous manner, the trial value of the normal-yield ratio for the subloading surface is assumed to be

$$R^{(\mathrm{tr})}_{n+1} := R_n. \tag{9.41}$$

With the use of the updated total deformation gradient tensor (9.37) and the trial plastic deformation gradient tensor $(9.39)_1$, we can compute the trial elastic deformation gradient tensor as

$$\mathbf{F}^{e(\mathrm{tr})}_{n+1} = \mathbf{F}_{n+1}\left(\mathbf{F}^{p(\mathrm{tr})}_{n+1}\right)^{-1} = \mathbf{F}_{n+1}\mathbf{F}^{p-1}_n = \mathbf{f}_{[n,n+1]}\mathbf{F}^e_n. \tag{9.42}$$

From equations (9.39) and (9.42), we have the trial right Cauchy–Green-type deformation tensors

$$\bar{\mathbf{C}}^{e(\mathrm{tr})}_{n+1} = \left(\mathbf{F}^{e(\mathrm{tr})}_{n+1}\right)^T \mathbf{F}^{e(\mathrm{tr})}_{n+1} = \mathbf{F}^{p-T}_n \mathbf{C}_{n+1}\mathbf{F}^{p-1}_n, \tag{9.43}$$

$$\mathbf{C}^{p(\mathrm{tr})}_{n+1} = \left(\mathbf{F}^{p(\mathrm{tr})}_{n+1}\right)^T \mathbf{F}^{p(\mathrm{tr})}_{n+1} = \mathbf{F}^{pT}_n \mathbf{F}^p_n = \mathbf{C}^p_n, \tag{9.44}$$

$$\tilde{\mathbf{C}}^{p(\mathrm{tr})}_{e,n+1} = \left(\mathbf{F}^{p(\mathrm{tr})}_{e,n+1}\right)^T \mathbf{F}^{p(\mathrm{tr})}_{e,n+1} = \mathbf{F}^{pT}_{e,n}\mathbf{F}^p_{e,n} = \tilde{\mathbf{C}}^p_{e,n}, \tag{9.45}$$

$$\mathbf{C}^{p(\mathrm{tr})}_{d,n+1} = \left(\mathbf{F}^{p(\mathrm{tr})}_{d,n+1}\right)^T \mathbf{F}^{p(\mathrm{tr})}_{d,n+1} = \mathbf{F}^{pT}_{d,n}\mathbf{F}^p_{d,n} = \mathbf{C}^p_{d,n}. \tag{9.46}$$

In equation (9.43) we have used the updated right Cauchy–Green deformation tensor given as $\mathbf{C}_{n+1} := \mathbf{F}^T_{n+1}\mathbf{F}_{n+1}$ with $\mathbf{F}_{n+1}$ in equation (9.37). Then, by substituting $\bar{\mathbf{C}}^{e(\mathrm{tr})}_{n+1}$, $\tilde{\mathbf{C}}^{p(\mathrm{tr})}_{e,n+1}$, and $H^{(\mathrm{tr})}_{n+1}$, respectively, into the constitutive equations (9.1), (9.3), and (9.5), we can readily compute the trial stresses $\bar{\mathbf{S}}^{(\mathrm{tr})}_{n+1}$, $\tilde{\mathbf{S}}^{p(\mathrm{tr})}_{e,n+1}$, and $F^{(\mathrm{tr})}_{n+1}$. Then, by using equations (8.112), (8.120), (8.127) and (8.178), the trial Mandel stress tensor $\bar{\mathbf{M}}^{(\mathrm{tr})}_{n+1}$, the trial Mandel-type back stress tensor $\bar{\mathbf{M}}^{p(\mathrm{tr})}_{e,n+1}$, and thus the trial reduced stress tensor $\hat{\bar{\mathbf{M}}}^{(\mathrm{tr})}_{n+1}$ can readily be computed. The deformation and stress variables computed above are collectively called the 'elastic predictor', and a state defined by the elastic predictor is referred to as the 'elastic trial state'. We can observe that the elastic trial state is determined only in terms of the known quantities at t_n and the updated total deformation gradient tensor $\mathbf{F}_{n+1}$; note here that $\mathbf{F}_{n+1}$ is computed from equation (9.37) via equation (9.38) using the incremental displacement $\Delta\mathbf{u}$ which is prescribed over the time interval $[t_n, t_{n+1}]$. With these trial quantities in hand, the normal-yield function

(9.7) as well as the subloading function (9.8) can be determined with respect to the elastic trial state.

Recall, however, that as discussed in Section 9.3, we intend to establish the stress-update algorithm based on the description relative to the reference configuration $\mathcal{K}_0$. Accordingly, we take the trial values $\mathbf{C}^{p(\text{tr})}_{n+1}$, $\mathbf{C}^{p(\text{tr})}_{d,n+1}$, and $H^{(\text{tr})}_{n+1}$, and upon substitution of these trial values into equations (9.15), (9.16), and (9.5), we get the trial stresses $\mathbf{S}^{(\text{tr})}_{n+1}$, $\mathbf{S}^{p}_{e,n+1}$, and $F^{(\text{tr})}_{n+1}$. Then, by using equations (9.18) and (9.19), the Eshelby-type trial stress tensor $\mathbf{M}^{(\text{tr})}_{n+1}$, the trial back stress tensor $\mathbf{M}^{p(\text{tr})}_{e,n+1}$, and the trial reduced stress tensor $\hat{\mathbf{M}}^{(\text{tr})}_{n+1}$ can be computed, and therefore, the normal-yield function (9.20) and the subloading function (9.21) are determined for the elastic trial state. It should be emphasized, of course, that the elastic trial state is not necessarily the actual solution for the updated state unless the response of the model within the increment is purely elastic, while, as discussed in the following, it is used in the judgment of the loading criterion and is used as an initial value for the iterative solution process of the return-mapping in a plastic case.

The next task is the judgment of the loading criterion, already presented in equations (8.204) and (8.205) in the time-continuum setting. The judgment of the loading for the subloading surface model in the time-discrete framework is concluded by means of the following criterion in terms of the elastic trial state:

$$\begin{aligned} &f^{(\text{tr})}_{\text{sub},n+1} \leq 0 \ \text{ or } \ f^{(\text{tr})}_{\text{elast},n+1} \leq 0 \quad \rightsquigarrow \quad \text{elastic case}, \\ &f^{(\text{tr})}_{\text{sub},n+1} > 0 \ \text{ and } \ f^{(\text{tr})}_{\text{elast},n+1} > 0 \quad \rightsquigarrow \quad \text{plastic case}. \end{aligned} \tag{9.47}$$

Here we have defined the value of the subloading function evaluated for the trial stresses as

$$f^{(\text{tr})}_{\text{sub},n+1} := \sqrt{\frac{3}{2}\hat{\mathbf{M}}^{(\text{tr})\prime}_{n+1} : (\hat{\mathbf{M}}^{(\text{tr})\prime}_{n+1})^T} - R^{(\text{tr})}_{n+1}F^{(\text{tr})}_{n+1}, \tag{9.48}$$

and the value of a function evaluated for the trial stresses as

$$f^{(\text{tr})}_{\text{elast},n+1} := \sqrt{\frac{3}{2}\hat{\mathbf{M}}^{(\text{tr})\prime}_{n+1} : (\hat{\mathbf{M}}^{(\text{tr})\prime}_{n+1})^T} - R_e F^{(\text{tr})}_{n+1}, \tag{9.49}$$

where the function f_{elast} in equation (9.49) is defined by replacing R in the subloading function (9.48) with the material constant R_e introduced in equation (9.14). The equation $f_{\text{elast}} = 0$ represents the elasticity limit surface in the stress space; if the stress point lies within this surface, the instantaneous response of the model is purely elastic. In line with textbooks on computational plasticity, we regard the loading criterion of the form (9.47) as the algorithmic counterpart of the loading criterion within the time-continuum framework.

The criterion (9.47) can be interpreted to mean that, if the stress point of the elastic trial state lies within or on either the subloading surface or the elasticity limit surface in the stress space, the elastic trial state is accepted as the actual solution of the updated state at t_{n+1}. In this case, we set $\Delta\lambda = 0$ and

$$(\cdot)_{n+1} := (\cdot)^{(\text{tr})}_{n+1} \tag{9.50}$$

except for the normal-yield ratio R. Recall that, as discussed in Section 8.10.1, the subloading surface always passes through the current stress point even after unloading. According to this intrinsic nature of the subloading surface model, the normal-yield ratio R needs to be updated so that the subloading function satisfies the equality $f_{\text{sub},n+1} = 0$ for the updated stresses. Straightforwardly, by solving equation (9.21) with respect to R, we have the updated normal-yield ratio

$$R_{n+1} = \frac{1}{F_{n+1}} \sqrt{\frac{3}{2} \hat{\mathbf{M}}'_{n+1} : (\hat{\mathbf{M}}'_{n+1})^T}, \tag{9.51}$$

and then the stress-update procedure for the elastic case is terminated.

Otherwise, if the stress point of the elastic trial state lies outside the subloading surface and the elasticity limit surface, the elastic trial state is not plastically admissible, and hence cannot be accepted as the solution. In this case, we have to proceed to the plastic corrector step, carrying out the return-mapping described in the next section in order to determine the plastically admissible solution.

Note that in the conventional plasticity models without the subloading surface, the interior of yield surface is a purely-elastic domain, so that the loading criterion is given solely in terms of the normal-yield function:

$$\begin{aligned} f^{(\text{tr})}_{\text{yld},n+1} \leq 0 \quad &\leadsto \quad \text{elastic case,} \\ f^{(\text{tr})}_{\text{yld},n+1} > 0 \quad &\leadsto \quad \text{plastic case,} \end{aligned} \tag{9.52}$$

where the value of $f^{(\text{tr})}_{\text{yld},n+1}$ is computed by substituting the trial stresses into the normal-yield function (9.20).

9.7 Plastic Corrector Step by Return-Mapping

If the incremental process within the time interval turns out to be 'plastic case' on the basis of the loading criterion (9.47) (the criterion (9.52) is adopted for conventional plasticity models), we proceed to the plastic corrector step in which the return-mapping is carried out. The objective in the plastic corrector step is to determine the updated quantities at t_{n+1} which simultaneously fulfill the incremental format of the plastic evolution rules (9.30), (9.31), and (9.35), as well as the subloading function (9.21). This task is accomplished by solving the following system of four simultaneous equations for the set of unknown quantities $\{\mathbf{C}^p_{n+1}, \mathbf{C}^p_{d,n+1}, H_{n+1}, \Delta\lambda\}$:

$$\left\{ \begin{aligned} &\mathbf{Y}_{\mathbf{C}^p} := \mathbf{C}^p_{n+1} - \mathbf{Q}^p_{n+1} \mathbf{C}^p_n = \mathbf{O}, \\ &\mathbf{Y}_{\mathbf{C}^p_d} := \mathbf{C}^p_{d,n+1} - \mathbf{Q}^p_{e,n+1} \mathbf{C}^p_{d,n} = \mathbf{O}, \\ &Y_H := H_{n+1} - H_n - \sqrt{\frac{2}{3}} \Delta\lambda = 0, \\ &Y_{f_{\text{sub}}} := \sqrt{\frac{3}{2} \hat{\mathbf{M}}'_{n+1} : (\hat{\mathbf{M}}'_{n+1})^T} - R_{n+1} F_{n+1} = 0, \end{aligned} \right. \tag{9.53}$$

in which the fourth equation, that is the subloading function $f_{\text{sub},n+1} = 0$ is included as a constraint to ensure that the updated stress point at t_{n+1} lies on the updated subloading surface in the stress space. The set of equations (9.53) needs to be complemented by the updating

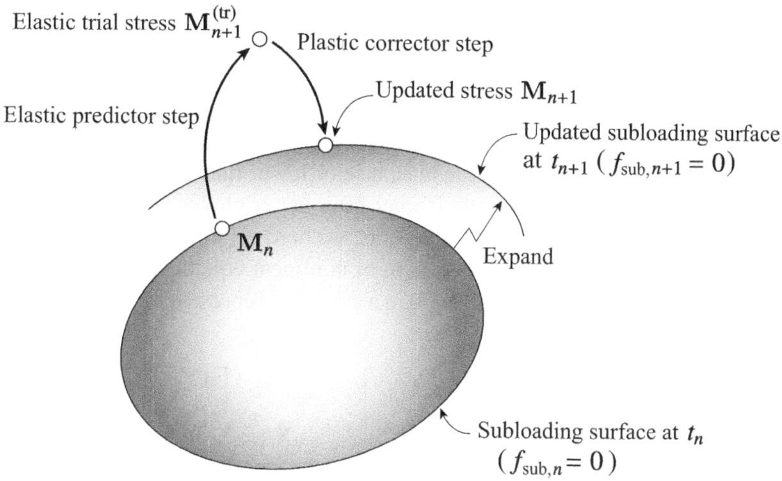

Figure 9.2 Geometric interpretation of the return-mapping procedure for isotropic and/or kinematic hardening plasticity

equation (9.36) for the normal-yield ratio R, because R involved in equation $(9.53)_4$ depends on one of the unknowns, $\Delta\lambda$.

The above-described procedure is generally referred to as the 'return-mapping' in accordance with the geometric interpretation in the stress space described below. In the plastic case, the elastic trial stress point computed as the elastic predictor lies outside the subloading surface at the elastic trial state, and thus it is undoubtedly plastically inadmissible. However, the updated stress point that is determined in the plastic corrector step as the solution of equation (9.53) via the constitutive equations lies on the subloading surface at the updated state, and accordingly, it is plastically admissible. To sum up, the stress-update process consisting of the elastic predictor step and the plastic corrector step as a whole implies that the stress point returns from the exterior of the subloading surface at the elastic trial state to the updated subloading surface. This geometric interpretation of the return-mapping process is schematically illustrated in Figure 9.2. This explanatory figure depicts the return-mapping process for general plasticity models with isotropic and/or kinematic hardening in which the size and position may change along with the plastic hardening induced by the evolution of plastic deformations within the time interval.

Let us now move on to the procedure for solving the system of simultaneous equations (9.53). It should be noted that equation (9.53) is generally nonlinear with respect to the unknowns $\{\mathbf{C}_{n+1}^p, \mathbf{C}_{d,n+1}^p, H_{n+1}, \Delta\lambda\}$. In fact, the tensor exponentials $\mathbf{Q}_{n+1}^p$ and $\mathbf{Q}_{e,n+1}^p$ as well as equation $(9.53)_4$ itself depend implicitly on the unknowns, because, as one can observe from equations (9.32) and (9.33), they depend on the stresses, and further, the stresses are nonlinear functions of the unknowns through the constitutive equations. Consequently, the solution procedure of equation (9.53) requires an iterative scheme. A simple choice and the one most commonly employed is the Newton–Raphson method, whereby, in principle, a quadratic asymptotic rate of convergence to the solution and, thus, a computationally efficient return-mapping procedure can be achieved, although it requires the exact Jacobian matrix in each iteration (see equation

(9.56) below), whose derivation and computational implementation are generally cumbersome even for relatively simple constitutive models, as will be described in Section 9.8. Of course, we can employ alternative options instead, such as the quasi-Newton method, the secant method, or the bisection method, which do not require the computation of the exact Jacobian matrix in contrast to the full Newton–Raphson scheme, although these schemes may result in less efficient return-mapping procedures because of lower rates of convergence. However, especially for complex material models, it may be a better choice at the expense of lower convergence rates. On the basis of these arguments, we here employ the full Newton–Raphson scheme in the solution procedure for the system of nonlinear equations (9.53).

To begin with, let us rewrite equation (9.53) in a symbolic representation using the following vector equation:

$$\mathbf{Y}(\mathbf{X}) = \mathbf{0}, \qquad \mathbf{Y} := \begin{bmatrix} \mathbf{Y}_{\mathbf{C}^p} \\ \mathbf{Y}_{\mathbf{C}^p_d} \\ Y_H \\ Y_{f_{\text{sub}}} \end{bmatrix}, \qquad \mathbf{X} := \begin{bmatrix} \mathbf{C}^p_{n+1} \\ \mathbf{C}^p_{d,n+1} \\ H_{n+1} \\ \Delta\lambda \end{bmatrix}. \tag{9.54}$$

Here we define the residual vector $\mathbf{Y}$ and the vector of the unknown quantities $\mathbf{X}$. Note that, in the present model, the third equation, $Y_H = 0$, does not necessarily need to be included in the above simultaneous equations, since it depends solely on H_{n+1} and $\Delta\lambda$, and therefore H_{n+1} can be computed by substituting the updated value of $\Delta\lambda$ after each iteration. In general, however, the evolution rule for H can be dependent not only on $\Delta\lambda$ but also on stresses in some complex constitutive model. Accordingly, we include the third equation so as to preserve the generality of the algorithm. Also note again that in the solution procedure, the evolution rule (9.36) for R is required as the complementary equation, whereas it is not included in equation $(9.54)_1$. This is because equation (9.36) is a function solely of $\Delta\lambda$, and thus R can be updated by using the updated value of $\Delta\lambda$ after each iteration.

The linearization of equation $(9.54)_1$ with the aid of equation (1.403) leads to

$$\mathbf{Y}(\mathbf{X}^{(k)}) + \mathbf{J}(\mathbf{X}^{(k)}) * \delta\mathbf{X} = \mathbf{0}, \tag{9.55}$$

where $\delta\mathbf{X}$ is the corrector vector, the superscript '(k)' denotes the number of iterations, and we define the Jacobian matrix for the return-mapping procedure as

$$\mathbf{J}(\mathbf{X}^{(k)}) := \left.\frac{\partial \mathbf{Y}}{\partial \mathbf{X}}\right|_{\mathbf{X}^{(k)}}. \tag{9.56}$$

More explicitly, it is written as

$$\mathbf{J} = \begin{bmatrix} \dfrac{\partial \mathbf{Y}_{\mathbf{C}^p}}{\partial \mathbf{C}^p} & \dfrac{\partial \mathbf{Y}_{\mathbf{C}^p}}{\partial \mathbf{C}^p_d} & \dfrac{\partial \mathbf{Y}_{\mathbf{C}^p}}{\partial H} & \dfrac{\partial \mathbf{Y}_{\mathbf{C}^p}}{\partial (\Delta\lambda)} \\ \dfrac{\partial \mathbf{Y}_{\mathbf{C}^p_d}}{\partial \mathbf{C}^p} & \dfrac{\partial \mathbf{Y}_{\mathbf{C}^p_d}}{\partial \mathbf{C}^p_d} & \dfrac{\partial \mathbf{Y}_{\mathbf{C}^p_d}}{\partial H} & \dfrac{\partial \mathbf{Y}_{\mathbf{C}^p_d}}{\partial (\Delta\lambda)} \\ \dfrac{\partial Y_H}{\partial \mathbf{C}^p} & \dfrac{\partial Y_H}{\partial \mathbf{C}^p_d} & \dfrac{\partial Y_H}{\partial H} & \dfrac{\partial Y_H}{\partial (\Delta\lambda)} \\ \dfrac{\partial Y_{f_{\text{sub}}}}{\partial \mathbf{C}^p} & \dfrac{\partial Y_{f_{\text{sub}}}}{\partial \mathbf{C}^p_d} & \dfrac{\partial Y_{f_{\text{sub}}}}{\partial H} & \dfrac{\partial Y_{f_{\text{sub}}}}{\partial (\Delta\lambda)} \end{bmatrix}. \tag{9.57}$$

From the standpoint of computer implementation, it should be noted that, as seen in equation $(9.54)_{2,3}$, the residual vector $\mathbf{Y}$ and the unknown vector $\mathbf{X}$ contain scalars and second-order tensors, and thus the Cartesian components of the tensors involved in equations (9.54) and (9.57) need to be rearranged properly into the vector–matrix expression merely for a computational convenience. Note also that the operator symbol '$*$' is introduced in equation (9.55) in order to denote collectively appropriate contractions between tensors in various orders. In the present case, specifically, we have the vectors $\mathbf{X}$ and $\mathbf{Y}$ with 14 components and the 14×14 square matrix $\mathbf{J}$ for three-dimensional problems, considering the symmetry property of $\mathbf{C}^p$ and $\mathbf{C}^p_d$. The derivation requires cumbersome differential operations and algebraic manipulations, and a detailed derivation and the resulting form of the Jacobian matrix $\mathbf{J}$ will be given in Section 9.8.

By solving equation (9.55), we readily compute the corrector vector

$$\delta\mathbf{X} = -\left[\mathbf{J}(\mathbf{X}^{(k)})\right]^{-1} * \mathbf{Y}(\mathbf{X}^{(k)}) \rightsquigarrow \mathbf{X}^{(k+1)} := \mathbf{X}^{(k)} + \delta\mathbf{X}, \qquad (9.58)$$

whereby the unknown vector $\mathbf{X}$ is updated. The initial value $\mathbf{X}^{(0)}$ of the unknown vector for beginning the iteration is usually taken as the elastic trial values with $\Delta\lambda^{(0)} := 0$, as implied by the assumption of the elastic trial state. Alternatively, we can specifically take $\Delta\lambda^{(0)} := f^{(\mathrm{tr})}_{n+1}/(\sqrt{6}\mu)$, which gives an exact solution of $\Delta\lambda$ for a model in the small strain framework with linear elasticity and von Mises plasticity without isotropic or kinematic hardening; see Wallin and Ristinmaa (2005) for a use of such an initial value within the finite strain setting.

The iteration of the Newton–Raphson scheme is continued until the residual is sufficiently small; that is to say, it is judged that the convergence of the solution is attained when the following criterion in terms of the residual norm is fulfilled:

$$\left\|\mathbf{Y}(\mathbf{X}^{(k+1)})\right\| < \mathrm{TOL}. \qquad (9.59)$$

Then, the updated quantities are accepted as the solution and the iteration scheme is terminated. In the numerical examples presented in Section 9.10, the tolerance for convergence is taken as $\mathrm{TOL} = 1.0 \times 10^{-12}$. From the standpoint of numerical computations, however, attention to the fact that the residual vector $\mathbf{Y}$ consists of components with different dimensions may be required in some cases, since the convergence may be impeded by larger component(s) of the residual vector. In fact, the component $Y_{f_{\mathrm{sub}}}$ in equation (9.54) has the dimension of stress, whereas the other components have the dimension of plastic strain.

For complex or highly nonlinear constitutive models, some numerical difficulty may be encountered during the iterative solution process of the return-mapping equation. A large curvature, or even a corner or vertex, of the yield surface will often cause a problem. This aspect is of importance with regard to a robust and stable numerical implementation of the constitutive models. In such cases, the use of treatments such as the line-search method to improve the initial guess and regularizations to avoid the singularity of the yield function will stabilize the iterative solution scheme (Arghavani et al., 2011; Dutko et al., 1993; Lambrecht and Miehe, 1999; Perić and de Souza Neto, 1999). In the computer program presented in Chapter 10, the reduced Newton–Raphson method is implemented to ensure convergence.

9.8 Derivation of Jacobian Matrix for Return-Mapping

In the iterative solution procedure by means of the Newton–Raphson scheme for the nonlinear simultaneous equations (9.53) involved in the return-mapping scheme, the Jacobian matrix $\mathbf{J}$ defined in equation (9.56) has a crucial role in ensuring the quadratic rate of convergence. The exact Jacobian matrix can be obtained by a straightforward analytical calculation of the derivative of the residual vector $\mathbf{Y}(\mathbf{X})$ with respect to the unknowns $\mathbf{X}$, although it requires cumbersome differentiations and lengthy algebraic manipulations. This section presents in detail the derivation and the resulting form of the Jacobian matrix for the specific constitutive model discussed in this chapter.

An alternative approach to obtain the Jacobian matrix is an approximate numerical computation of the derivatives by means of a first-order difference perturbation technique. This approach is computationally somewhat costly, but is quite useful especially for constitutive models with complex forms of evolution rules or yield function, and has been applied to various inelastic constitutive models (Johansson and Ekh, 2006; Menzel et al., 2005), since it does not require any cumbersome analytical differential calculation for the specific form of the incremental plastic evolution rules and the yield function. Furthermore, numerical experiment has shown that even the approximated Jacobian matrix computed by this approach is capable of preserving efficient asymptotic rates of convergence in the iterative solution procedure of the return-mapping equation. The detailed explanation of this numerical approximation approach is omitted here, and the interested reader is referred to the detailed description within the finite strain framework by Menzel and Steinmann (2003a,b), and also to the investigation based on a numerical assessment of the convergence property by Pérez-Foguet et al. (2000a,b) in a small strain setting.

9.8.1 Components of Jacobian Matrix

This section derives the components of the Jacobian matrix $\mathbf{J}$ in equation (9.57). For convenience in the computer implementation, we present the explicit expression for the Cartesian components in an index notation. For the sake of conciseness we use an abbreviated notation, omitting the subscript '$n+1$' while leaving the subscript 'n' to specify that the quantities at t_n are known and thus fixed during the return-mapping.

Firstly, by taking the partial derivatives of equation $(9.53)_1$ with respect to the unknowns $\mathbf{C}^p$, $\mathbf{C}^p_d$, H, and $\Delta\lambda$, we arrive at the Cartesian components of the Jacobian matrix

$$\frac{\partial Y_{\mathbf{C}^p,IJ}}{\partial C^p_{KL}} = \frac{1}{2}(\delta_{IK}\delta_{JL} + \delta_{IL}\delta_{JK}) - \frac{\partial Q^p_{IM}}{\partial C^p_{KL}} C^p_{n,MJ}, \tag{9.60}$$

$$\frac{\partial Y_{\mathbf{C}^p,IJ}}{\partial C^p_{d,KL}} = -\frac{\partial Q^p_{IM}}{\partial C^p_{d,KL}} C^p_{n,MJ}, \tag{9.61}$$

$$\frac{\partial Y_{\mathbf{C}^p,IJ}}{\partial H} = -\frac{\partial Q^p_{IM}}{\partial H} C^p_{n,MJ}, \tag{9.62}$$

$$\frac{\partial Y_{\mathbf{C}^p,IJ}}{\partial (\Delta\lambda)} = -\frac{\partial Q^p_{IM}}{\partial (\Delta\lambda)} C^p_{n,MJ} \tag{9.63}$$

for $I, J, K, L = 1, 2, 3$, with δ_{IJ} being the Kronecker delta. Here and henceforth, the summation convention over repeated indices is implied. Noting the symmetric property of $\mathbf{C}^p$, we used in the above derivation a trivial identity $\partial \mathbf{C}^p/\partial \mathbf{C}^p = (\mathbf{I}\bar{\otimes}\mathbf{I}+\mathbf{I}\underline{\otimes}\mathbf{I})/2 = \boldsymbol{\mathcal{S}}$ with the two dyadic multiplying operations denoted by the symbols '$\bar{\otimes}$' and '$\underline{\otimes}$' defined in equation (1.208) and the notation of the fourth-order symmetrizing tensor $\boldsymbol{\mathcal{S}}$ in Table 1.1. Equations (9.60)–(9.63) contain the derivatives of the tensor exponential functions $\partial \mathbf{Q}^p/\partial \mathbf{C}^p$, $\partial \mathbf{Q}^p/\partial \mathbf{C}^p_d$, $\partial \mathbf{Q}^p/\partial H$, and $\partial \mathbf{Q}^p/\partial(\Delta\lambda)$; these explicit expressions will be presented in Section 9.8.2.

The derivatives of equation $(9.53)_2$ can be obtained in an analogous manner; we have

$$\frac{\partial Y_{\mathbf{C}^p_d,IJ}}{\partial C^p_{KL}} = -\frac{\partial Q^p_{e,IM}}{\partial C^p_{KL}} C^p_{d,n,MJ}, \tag{9.64}$$

$$\frac{\partial Y_{\mathbf{C}^p_d,IJ}}{\partial C^p_{d,KL}} = \frac{1}{2}(\delta_{IK}\delta_{JL} + \delta_{IL}\delta_{JK}) - \frac{\partial Q^p_{e,IM}}{\partial C^p_{d,KL}} C^p_{d,n,MJ}, \tag{9.65}$$

$$\frac{\partial Y_{\mathbf{C}^p_d,IJ}}{\partial H} = -\frac{\partial Q^p_{e,IM}}{\partial H} C^p_{d,n,MJ} \tag{9.66}$$

$$\frac{\partial Y_{\mathbf{C}^p_d,IJ}}{\partial(\Delta\lambda)} = -\frac{\partial Q^p_{e,IM}}{\partial(\Delta\lambda)} C^p_{d,n,MJ} \tag{9.67}$$

for $I, J, K, L = 1, 2, 3$. Here we used the identity $\partial \mathbf{C}^p_d/\partial \mathbf{C}^p_d = (\mathbf{I}\bar{\otimes}\mathbf{I}+\mathbf{I}\underline{\otimes}\mathbf{I})/2 = \boldsymbol{\mathcal{S}}$. The explicit expressions for the derivatives of the tensor exponential functions $\partial \mathbf{Q}^p_e/\partial \mathbf{C}^p$, $\partial \mathbf{Q}^p_e/\partial \mathbf{C}^p_d$, $\partial \mathbf{Q}^p_e/\partial H$, and $\partial \mathbf{Q}^p_e/\partial(\Delta\lambda)$ will be given in Section 9.8.2.

The derivatives of the scalar equation $(9.53)_3$ can be easily calculated as

$$\frac{\partial Y_H}{\partial C^p_{KL}} = 0, \quad \frac{\partial Y_H}{\partial C^p_{d,KL}} = 0, \quad \frac{\partial Y_H}{\partial H} = 1, \quad \frac{\partial Y_H}{\partial(\Delta\lambda)} = -\sqrt{\frac{2}{3}}. \tag{9.68}$$

Finally, differentiation of equation $(9.53)_4$ along with the chain rule leads to

$$\frac{\partial Y_{f_{\text{sub}}}}{\partial C^p_{IJ}} = \frac{\partial f_{\text{sub}}}{\partial M_{MN}}\frac{\partial M_{MN}}{\partial C^p_{IJ}} + \frac{\partial f_{\text{sub}}}{\partial M^p_{e,MN}}\frac{\partial M^p_{e,MN}}{\partial C^p_{IJ}}, \tag{9.69}$$

$$\frac{\partial Y_{f_{\text{sub}}}}{\partial C^p_{d,IJ}} = \frac{\partial f_{\text{sub}}}{\partial M^p_{e,MN}}\frac{\partial M^p_{e,MN}}{\partial C^p_{d,IJ}}, \tag{9.70}$$

$$\frac{\partial Y_{f_{\text{sub}}}}{\partial H} = \frac{\partial f_{\text{sub}}}{\partial F} F', \tag{9.71}$$

$$\frac{\partial Y_{f_{\text{sub}}}}{\partial(\Delta\lambda)} = \frac{\partial f_{\text{sub}}}{\partial R}\frac{\partial R}{\partial(\Delta\lambda)} \tag{9.72}$$

for $I, J = 1, 2, 3$. The explicit form of the isotropic hardening modulus h has already been given in equation (9.6).

The derivative $\partial R/\partial(\Delta\lambda)$ involved in equation (9.72) is obtained by differentiating equation (9.36); we get

$$\frac{\partial R}{\partial(\Delta\lambda)} = u\frac{\cos\left(\dfrac{\pi}{2}\dfrac{\langle R_n - R_e\rangle}{1 - R_e}\right)}{\sin\left(\dfrac{\pi}{2}\dfrac{\langle R - R_e\rangle}{1 - R_e}\right)} \exp\left(-u\frac{\pi}{2}\frac{\Delta\lambda}{1 - R_e}\right). \tag{9.73}$$

Note that in actual numerical computations, division by zero, which may be encountered when the value of R approaches R_e, needs to be avoided.

Equations (9.69)–(9.72) involve the derivatives of the subloading function (9.21) with respect to the stresses and R, which read

$$\frac{\partial f_{\text{sub}}}{\partial \mathbf{M}} = \sqrt{\frac{3}{2}} \frac{(\hat{\mathbf{M}}')^T}{\sqrt{\hat{\mathbf{M}}' : (\hat{\mathbf{M}}')^T}}, \qquad \frac{\partial f_{\text{sub}}}{\partial \mathbf{M}_e^p} = -\frac{\partial f_{\text{sub}}}{\partial \mathbf{M}},$$
$$\frac{\partial f_{\text{sub}}}{\partial F} = -R, \qquad \frac{\partial f_{\text{sub}}}{\partial R} = -F. \tag{9.74}$$

Note the transpose in the numerator in the first equation. The derivatives $\partial \mathbf{M}/\partial \mathbf{C}^p$, $\partial \mathbf{M}_e^p/\partial \mathbf{C}^p$, and $\partial \mathbf{M}_e^p/\partial \mathbf{C}_d^p$ involved in equations (9.69)–(9.72) will be derived in Section 9.8.3.

9.8.2 Derivatives of Tensor Exponentials

This section describes the derivation of the derivatives of the tensor exponential functions $\mathbf{Q}^p$ and $\mathbf{Q}_e^p$ which are involved, respectively, in equations (9.60)–(9.63) and (9.64)–(9.67). We derive these derivatives in an analytical way by employing the closed-form formula (1.374), in which the exact derivative of the tensor exponential function is calculated in a quite straightforward manner by differentiating the series representation (1.258) term by term.

Alternative approaches to compute the tensor exponential derivative are also available. Ortiz et al. (2001) have devised two simple and straightforward methods based on the series expansion and the spectral function theorem, and have carried out a comparative numerical investigation of these two approaches, providing guidelines for the appropriate choice of the evaluation method for the tensor exponential derivative on the basis of the size of the argument. Another alternative is the approach which evaluates the tensor exponential derivative from the spectral representation by applying a perturbation formula to the eigenvalues and eigenvectors (Simo and Taylor, 1991), although it requires a cumbersome numerical scheme and may encounter difficulties especially in the presence of repeated eigenvalues.

To begin with, we rewrite equations (9.32) and (9.33) in the following form for the sake of brevity, omitting the subscript $n+1$:

$$\mathbf{Q}^p = \exp[\mathbf{Z}^p], \quad \mathbf{Z}^p := 2\Delta\lambda \frac{\hat{\mathbf{M}}'}{\sqrt{\hat{\mathbf{M}}' : (\hat{\mathbf{M}}')^T}}, \tag{9.75}$$

and

$$\mathbf{Q}_e^p = \exp[\mathbf{Z}_e^p], \quad \mathbf{Z}_e^p := 2\Delta\lambda \frac{b}{c} \mathbf{M}_e^{p\prime}. \tag{9.76}$$

Hereafter, we use the index notation, when needed, to provide the Cartesian components of the tensor exponential derivative for convenience in the computer implementation. By differentiating the tensor exponential $\mathbf{Q}^p$ with respect to the unknowns $\mathbf{C}^p$, $\mathbf{C}_d^p$, H, and $\Delta\lambda$ along with the chain rule, we arrive at

$$\frac{\partial Q_{IJ}^p}{\partial C_{KL}^p} = \frac{\partial (\exp[\mathbf{Z}^p])_{IJ}}{\partial Z_{MN}^p} \frac{\partial Z_{MN}^p}{\partial \hat{M}_{PQ}} \frac{\partial \hat{M}_{PQ}}{\partial C_{KL}^p}, \tag{9.77}$$

$$\frac{\partial Q^p_{IJ}}{\partial C^p_{d,KL}} = \frac{\partial(\exp[\mathbf{Z}^p])_{IJ}}{\partial Z^p_{MN}} \frac{\partial Z^p_{MN}}{\partial \hat{M}_{PQ}} \frac{\partial \hat{M}_{PQ}}{\partial C^p_{d,KL}}, \tag{9.78}$$

$$\frac{\partial Q^p_{IJ}}{\partial H} = 0, \tag{9.79}$$

$$\frac{\partial Q^p_{IJ}}{\partial(\Delta\lambda)} = \frac{\partial(\exp[\mathbf{Z}^p])_{IJ}}{\partial Z^p_{MN}} \cdot 2 \frac{\hat{M}'_{MN}}{\sqrt{\hat{\mathbf{M}}' : (\hat{\mathbf{M}}')^T}} \tag{9.80}$$

for $I, J, K, L = 1, 2, 3$. The closed-form representation of the derivative of the tensor exponential function $\partial \exp[\mathbf{Z}^p]/\partial \mathbf{Z}^p$ involved in the above equations and its numerical aspect will be described in Appendix D. Differentiation of equation $(9.75)_2$ leads to

$$\frac{\partial \mathbf{Z}^p}{\partial \hat{\mathbf{M}}} = 2\Delta\lambda \frac{1}{A}\left(\mathbf{I}\bar{\otimes}\mathbf{I} - \frac{1}{3}\mathbf{I}\otimes\mathbf{I} - \frac{1}{A^2}\hat{\mathbf{M}}' \otimes (\hat{\mathbf{M}}')^T\right) \tag{9.81}$$

with

$$A := \sqrt{\hat{\mathbf{M}}' : (\hat{\mathbf{M}}')^T}. \tag{9.82}$$

From equation (9.19) defining the Eshelby-like reduced stress tensor $\hat{\mathbf{M}}$ and the trivial identity $\partial \mathbf{M}/\partial \mathbf{C}^p_d = \mathbf{O}$, we have

$$\frac{\partial \hat{\mathbf{M}}}{\partial \mathbf{C}^p} = \frac{\partial \mathbf{M}}{\partial \mathbf{C}^p} - \frac{\partial \mathbf{M}^p_e}{\partial \mathbf{C}^p}, \quad \frac{\partial \hat{\mathbf{M}}}{\partial \mathbf{C}^p_d} = -\frac{\partial \mathbf{M}^p_e}{\partial \mathbf{C}^p_d}. \tag{9.83}$$

The explicit expressions for $\partial \mathbf{M}/\partial \mathbf{C}^p$, $\partial \mathbf{M}^p_e/\partial \mathbf{C}^p$, and $\partial \mathbf{M}^p_e/\partial \mathbf{C}^p_d$ will be given in Section 9.8.3.

Next, differentiation of the tensor exponential $\mathbf{Q}^p_e$ with respect to the unknowns $\mathbf{C}^p$, $\mathbf{C}^p_d$, H, and $\Delta\lambda$ using the chain rule gives the derivatives

$$\frac{\partial Q^p_{e,IJ}}{\partial C^p_{KL}} = \frac{\partial(\exp[\mathbf{Z}^p_e])_{IJ}}{\partial Z^p_{e,MN}} \cdot 2\Delta\lambda \frac{b}{c} \frac{\partial M^{p\prime}_{e,MN}}{\partial M^p_{e,PQ}} \frac{\partial M^p_{e,PQ}}{\partial C^p_{KL}}, \tag{9.84}$$

$$\frac{\partial Q^p_{e,IJ}}{\partial C^p_{d,KL}} = \frac{\partial(\exp[\mathbf{Z}^p_e])_{IJ}}{\partial Z^p_{e,MN}} \cdot 2\Delta\lambda \frac{b}{c} \frac{\partial M^{p\prime}_{e,MN}}{\partial M^p_{e,PQ}} \frac{\partial M^p_{e,PQ}}{\partial C^p_{d,KL}}, \tag{9.85}$$

$$\frac{\partial Q^p_{e,IJ}}{\partial H} = 0, \tag{9.86}$$

$$\frac{\partial Q^p_{e,IJ}}{\partial(\Delta\lambda)} = \frac{\partial(\exp[\mathbf{Z}^p_e])_{IJ}}{\partial Z^p_{e,MN}} \cdot 2\frac{b}{c} M^{p\prime}_{e,MN} \tag{9.87}$$

for $I, J, K, L = 1, 2, 3$, in which

$$\frac{\partial \mathbf{M}^{p\prime}_e}{\partial \mathbf{M}^p_e} = \mathbf{I}\bar{\otimes}\mathbf{I} - \frac{1}{3}\mathbf{I}\otimes\mathbf{I} = \mathcal{D}. \tag{9.88}$$

The closed-form representation of the tensor exponential derivative $\partial \exp[\mathbf{Z}^p_e]/\partial \mathbf{Z}^p_e$ involved in the above equations is given in Appendix D. The explicit expressions for $\partial \mathbf{M}^p_e/\partial \mathbf{C}^p$ and $\partial \mathbf{M}^p_e/\partial \mathbf{C}^p_d$ are given in Section 9.8.3.

9.8.3 Derivatives of Stresses

By using the definition (9.18) of the Eshelby-like stress tensor $\mathbf{M}$ and the Eshelby-like back stress tensor $\mathbf{M}_e^p$ relative to the reference configuration $\mathcal{K}_0$, the hyperelastic constitutive equation (9.15) and the hyperelastic-type constitutive equation for kinematic hardening (9.16) are transformed into expressions involving $\mathbf{M}$ and $\mathbf{M}_e^p$:

$$\mathbf{M} = \frac{\Lambda}{2}\left(\frac{\det\mathbf{C}}{\det\mathbf{C}^p} - 1\right)\mathbf{I} + \mu(\mathbf{C}\mathbf{C}^{p-1} - \mathbf{I}), \tag{9.89}$$

$$\mathbf{M}_e^p = c(\mathbf{C}^p\mathbf{C}_d^{p-1} - \mathbf{I}), \tag{9.90}$$

from which we obtain the derivatives

$$\frac{\partial M_{IJ}}{\partial C_{KL}^p} = -\frac{\Lambda}{2}\frac{\det\mathbf{C}}{\det\mathbf{C}^p}\delta_{IJ}C_{LK}^{p-1} - \mu C_{IM}C_{MK}^{p-1}C_{LJ}^{p-1}, \tag{9.91}$$

$$\frac{\partial M_{e,IJ}^p}{\partial C_{KL}^p} = c\delta_{IK}C_{d,LJ}^{p-1}, \tag{9.92}$$

$$\frac{\partial M_{e,IJ}^p}{\partial C_{d,KL}^p} = -cC_{IM}^pC_{MK}^{p-1}C_{LJ}^{p-1} \tag{9.93}$$

for $I, J, K, L = 1, 2, 3$, with δ_{IJ} being the Kronecker delta. Note that in the present model, $\det\mathbf{C}^p = (J^p)^2 = 1$ follows from the plastic incompressibility. In the above derivation we utilized the common formulae (1.367) and (1.371).

9.9 Consistent (Algorithmic) Tangent Modulus Tensor

In Section 8.11 we derived the rate-form elasto-plastic constitutive equation within the time-continuum setting, in which the incrementally linear relationship between the stress rate and the strain rate is represented via the elasto-plastic tangent modulus tensor (that is the so-called 'continuum tangent modulus tensor'). However, it is well known that in the iterative procedure for solving the nonlinear equilibrium equations for elasto-plastic boundary value problems (so-called global problems), the use of the continuum elasto-plastic tangent modulus tensor spoils the quadratic asymptotic rate of convergence to the solution which is intrinsic in the Newton–Raphson scheme, and consequently, results in lower convergence rates, leading to inefficiency of the analysis. The use of the so-called 'consistent tangent modulus tensor' (also referred to as the 'algorithmic tangent modulus tensor'), which can be regarded as an algorithmic or a time-discrete counterpart of the continuum tangent modulus tensor, is essential for efficient and numerically stable solution of nonlinear elasto-plastic problems by an iterative algorithm with quadratic convergence (Simo and Taylor, 1985). The phrase 'consistent' implies that the tangent modulus tensor is consistent with the stress-update algorithm for elasto-plastic constitutive models within the time-discrete setting.

There are two main approaches to obtaining the consistent tangent modulus tensor. One approach is a straightforward analytical calculation by differentiation. For simple constitutive models, the consistent tangent modulus tensor can be calculated analytically. However, for more sophisticated constitutive models, an analytical approach generally requires differentiation which may be cumbersome or in some cases infeasible, because of complicated couplings between stresses and deformation variables. Another alternative is a numerical approach in which the consistent tangent modulus tensor is computed by means of numerical differentiation using a perturbation technique. Unlike the analytical approach, this approach does not require any tedious differentiation, and is applicable regardless of the type of material models, although computationally somewhat costly. The adoption of numerical differentiation as a suitable technique to compute the consistent tangent modulus tensor within the finite strain framework was first proposed by Miehe (1996b). The reader is also referred to work by Pérez-Foguet et al. (2000a) and Pérez-Foguet et al. (2000b); Fellin and Ostermann (2002) for a formulation of this approach within the small strain setting and a numerical assessment of the convergence property.

This section describes the two approaches mentioned above to evaluating the consistent tangent modulus tensor. The analytical derivation by differentiation is presented in Section 9.9.1, and the numerical computation by the numerical differentiation based on the perturbation technique is described in Section 9.9.2. The expression relative to the reference configuration $\mathcal{K}_0$ in the so-called Lagrangian geometric setting is shown, and then it is transformed into the expression relative to the current configuration $\mathcal{K}$ in the so-called Eulerian geometric setting via the push-forward operation.

9.9.1 *Analytical Derivation of Consistent Tangent Modulus Tensor*

We derive the consistent tangent modulus tensor by differentiation in an analytical way. The consistent tangent modulus tensor in the description of the reference configuration $\mathcal{K}_0$ is defined by the generic expression

$$\mathbb{C}^{ep,\text{algo}} := 2\frac{d\mathbf{S}}{d\mathbf{C}}, \tag{9.94}$$

with which we have an incremental form of the elasto-plastic constitutive equation representing a relationship between the differentials of the second Piola–Kirchhoff stress tensor $\mathbf{S}$ and the right Cauchy–Green deformation tensor $\mathbf{C}$:

$$d\mathbf{S} = \mathbb{C}^{ep,\text{algo}} : \frac{1}{2}d\mathbf{C}. \tag{9.95}$$

Since, as can be observed from equation (9.15), the second Piola–Kirchhoff stress tensor $\mathbf{S}$ is the function of $\mathbf{C}$ and $\mathbf{C}^p$, the total differential $d\mathbf{S}$ is expressed as

$$d\mathbf{S} = 2\frac{\partial \mathbf{S}}{\partial \mathbf{C}} : \frac{1}{2}d\mathbf{C} + \frac{\partial \mathbf{S}}{\partial \mathbf{C}^p} : d\mathbf{C}^p = \left(2\frac{\partial \mathbf{S}}{\partial \mathbf{C}} + 2\frac{\partial \mathbf{S}}{\partial \mathbf{C}^p} : \frac{d\mathbf{C}^p}{d\mathbf{C}}\right) : \frac{1}{2}d\mathbf{C}, \tag{9.96}$$

using the chain rule. By comparing equations (9.95) and (9.96), we have

$$\mathbb{C}^{ep,\text{algo}} = 2\frac{\partial \mathbf{S}}{\partial \mathbf{C}} + 2\frac{\partial \mathbf{S}}{\partial \mathbf{C}^p} : \frac{d\mathbf{C}^p}{d\mathbf{C}}. \tag{9.97}$$

The derivatives $\partial\mathbf{S}/\partial\mathbf{C}$ and $\partial\mathbf{S}/\partial\mathbf{C}^p$ involved in equation (9.97) are obtained by differentiating equation (9.15):

$$\begin{aligned}\frac{\partial S_{IJ}}{\partial C_{KL}} &= \frac{\Lambda}{2}\frac{\det\mathbf{C}}{\det\mathbf{C}^p}C^{-1}_{IJ}C^{-1}_{LK}\\ &\quad-\frac{\Lambda}{2}\left(\frac{\det\mathbf{C}}{\det\mathbf{C}^p}-1\right)C^{-1}_{IK}C^{-1}_{LJ}+\mu C^{-1}_{IK}C^{-1}_{LJ},\end{aligned} \tag{9.98}$$

$$\frac{\partial S_{IJ}}{\partial C^p_{KL}} = -\frac{\Lambda}{2}\frac{\det\mathbf{C}}{\det\mathbf{C}^p}C^{-1}_{IJ}C^{-1}_{LK}-\mu C^{-1}_{IK}C^{-1}_{LJ}, \tag{9.99}$$

using the differential formulae (1.367) and (1.371) for the inverse and determinant of a tensor, respectively.

In the elastic case, the fourth-order tensor $d\mathbf{C}^p/d\mathbf{C}$ in equation (9.97) vanishes, and thus, the consistent tangent modulus tensor $\mathbb{C}^{ep,\text{algo}}$ coincides with the hyperelastic tangent modulus tensor $\mathbb{C}^e$ relative to the reference configuration $\mathcal{K}_0$, that is

$$\mathbb{C}^{ep,\text{algo}} = 2\frac{\partial\mathbf{S}}{\partial\mathbf{C}} =: \mathbb{C}^e, \tag{9.100}$$

where $\mathbb{C}^e$ can also be evaluated as a pull-back of the hyperelastic tangent modulus tensor $\bar{\mathbb{C}}^e$ in the intermediate configuration $\bar{\mathcal{K}}$ to the reference configuration $\mathcal{K}_0$:

$$\mathbb{C}^e_{IJKL} = F^{p-1}_{I\bar{I}}F^{p-1}_{J\bar{J}}F^{p-1}_{K\bar{K}}F^{p-1}_{L\bar{L}}\bar{\mathbb{C}}^e_{\bar{I}\bar{J}\bar{K}\bar{L}}. \tag{9.101}$$

In the elasto-plastic case, the fourth-order tensor of the total derivative $d\mathbf{C}^p/d\mathbf{C}$ involved in equation (9.97) can be obtained via the Jacobian matrix $\mathbf{J}$ defined in equations (9.56) and (9.57) (Ekh and Runesson, 2001; Johansson et al., 2005). To this end, we consider the relation $\mathbf{Y}=\mathbf{0}$ in equation $(9.54)_1$ and a total differentiation of $\mathbf{Y}$ with respect to $\mathbf{C}$:

$$d\mathbf{Y} = \left(\frac{\partial\mathbf{Y}}{\partial\mathbf{X}}*\frac{d\mathbf{X}}{d\mathbf{C}}+\frac{\partial\mathbf{Y}}{\partial\mathbf{C}}:\frac{d\mathbf{C}}{d\mathbf{C}}\right):d\mathbf{C} = \left(\mathbf{J}*\frac{d\mathbf{X}}{d\mathbf{C}}+\frac{\partial\mathbf{Y}}{\partial\mathbf{C}}\right):d\mathbf{C}=\mathbf{0},$$

noting that $\mathbf{Y}$ involves plural variables depending on $\mathbf{C}$ as shown in equation $(9.54)_3$, and thus performing the total differentiation of $\mathbf{Y}$ actually as $\mathbf{Y}(\mathbf{X}(\mathbf{C}),\mathbf{C})$, where we use $d\mathbf{Y}=\mathbf{0}$ which follows from $\mathbf{Y}=\mathbf{0}$. Note again that, as mentioned in Section 9.7, the appropriate contraction operation of tensors must be applied in the multiplication denoted by the symbol '$*$' for the vector–matrix expression. Then we have

$$\frac{d\mathbf{X}}{d\mathbf{C}} = -\mathbf{J}^{-1}*\frac{\partial\mathbf{Y}}{\partial\mathbf{C}}, \tag{9.102}$$

in which the derivative $\partial\mathbf{Y}/\partial\mathbf{C}$ can be obtained explicitly, while a cumbersome task of analytical differentiation is required. A glance at the definition of the vector $\mathbf{X}$ in equation$(9.54)_3$ reveals that $\mathbf{C}^p$ is a member of $\mathbf{X}$, and thus the fourth-order tensor $d\mathbf{C}^p/d\mathbf{C}$ is given as a part of $d\mathbf{X}/d\mathbf{C}$. Substitution of this result along with equations (9.98) and (9.99) into equation (9.97) yields the consistent tangent modulus tensor $\mathbb{C}^{ep,\text{algo}}$ relative to the reference configuration $\mathcal{K}_0$, that is in the material description.

Similar to the transformation of the continuum tangent modulus tensor, the push-forward operation from the reference configuration $\mathcal{K}_0$ to the current configuration $\mathcal{K}$ leads to the the consistent tangent modulus tensor relative to the current configuration $\mathcal{K}$, that is in the spatial description:

$$\mathbb{C}^{ep,\text{algo}}_{n+1,i,j,k,l} = F_{iI}F_{jJ}F_{kK}F_{lL}\mathbb{C}^{ep,\text{algo}}_{n+1,I,J,K,L} \tag{9.103}$$

for $i, j, k, l = 1, 2, 3$.

9.9.2 *Numerical Computation of Consistent Tangent Modulus Tensor*

The underlying concept in the numerical computation of the consistent tangent modulus tensor is numerical differentiation based on a forward difference approximation using a perturbation technique, which can be implemented by a quite simple numerical algorithm. In what follows, we describe the formulation of the perturbation technique developed by Miehe (1996b) within the finite strain framework.

Referring to equation (9.94), we can recast the consistent tangent modulus tensor in a component form,

$$\mathbb{C}^{ep,\text{algo}}_{n+1} = \sum_{I,J,K,L=1}^{3} 2\frac{\Delta S_{IJ}}{\Delta C_{KL}}\mathbf{E}_{(I)} \otimes \mathbf{E}_{(J)} \otimes \mathbf{E}_{(K)} \otimes \mathbf{E}_{(L)}, \tag{9.104}$$

where $\mathbf{E}_{(I)}$ $(I = 1, 2, 3)$ denote the Cartesian basis in the reference configuration $\mathcal{K}_0$, and we define the increment of the total right Cauchy–Green deformation tensor as

$$\Delta\mathbf{C} := \Delta(\mathbf{F}^T\mathbf{F}) = \mathbf{F}^T\Delta\mathbf{F} + (\Delta\mathbf{F})^T\mathbf{F}. \tag{9.105}$$

We can observe from equation (9.104) that the consistent tangent modulus tensor represents the sensitivity of the stresses to the change in total deformations with a finite step size. Thus, the perturbation process has a crucial importance in the numerical computation of the consistent tangent modulus tensor by numerical differentiation. We choose the following perturbation to the deformation gradient tensor with the perturbation parameter ϵ:

$$\Delta\mathbf{F}_{(KL)}(\epsilon) := \frac{\epsilon}{2}\left\{\left(\mathbf{F}^{-T}_{n+1}\mathbf{E}_{(K)}\right) \otimes \mathbf{E}_{(L)} + \left(\mathbf{F}^{-T}_{n+1}\mathbf{E}_{(L)}\right) \otimes \mathbf{E}_{(K)}\right\} \tag{9.106}$$

for $K, L = 1, 2, 3$, by which the total deformation gradient tensor at t_{n+1} is perturbed as

$$\mathbf{F}_{n+1,(KL)}(\epsilon) := \mathbf{F}_{n+1} + \Delta\mathbf{F}_{(KL)}(\epsilon), \tag{9.107}$$

which leads to

$$\begin{aligned}\Delta\mathbf{C}_{(KL)}(\epsilon) &= \epsilon\left(\mathbf{E}_{(K)} \otimes \mathbf{E}_{(L)} + \mathbf{E}_{(L)} \otimes \mathbf{E}_{(K)}\right)\\ &= 2\epsilon\,\text{sym}[\mathbf{E}_{(\text{K})} \otimes \mathbf{E}_{(\text{L})}].\end{aligned} \tag{9.108}$$

In elasto-plastic cases, careful consideration of the update procedure for the stresses and the plastic internal variables is required to ensure consistency between the tangent modulus tensor and the stress-update scheme within finite step sizes. Let $\mathbf{H}$ be a collective and symbolic representation of the plastic internal variables. Once the known state $\{\mathbf{F}_n, \mathbf{H}_n\}$ is given and the perturbed deformation gradient $\mathbf{F}_{n+1,(KL)}(\epsilon)$ is prescribed by equation (9.107), we can evaluate the stress corresponding to the perturbed state through the return-mapping, and thus the stress increment $\Delta\mathbf{S}$ can be evaluated as

$$\Delta\mathbf{S} \approx \mathbf{S}(\mathbf{F}_{n+1,(KL)}(\epsilon); \{\mathbf{F}_n, \mathbf{H}_n\}) - \mathbf{S}(\mathbf{F}_{n+1}; \{\mathbf{F}_n, \mathbf{H}_n\}). \tag{9.109}$$

Note that the updated stress at t_{n+1}, denoted here by $\mathbf{S}(\mathbf{F}_{n+1}; \{\mathbf{F}_n, \mathbf{H}_n\})$, is already evaluated. By substituting equations (9.108) and (9.109) into equation (9.104), the consistent tangent modulus tensor is evaluated as the forward difference approximation

$$\mathbb{C}^{ep,\text{algo}}_{n+1,IJKL} \approx \frac{1}{\epsilon}\Big[S_{IJ}(\mathbf{F}_{n+1,(KL)}(\epsilon); \{\mathbf{F}_n, \mathbf{H}_n\}) - S_{IJ}(\mathbf{F}_{n+1}; \{\mathbf{F}_n, \mathbf{H}_n\})\Big]. \tag{9.110}$$

In regard to the computation of the components $\mathbb{C}^{ep,\text{algo}}_{n+1,IJKL}$ of the fourth-order tensor, we need to compute the stresses for the perturbed deformation gradient tensor (9.107) via the return-mapping procedure for $(K, L) = (1, 1), (2, 2), (3, 3), (1, 2), (1, 3)$, and $(2, 3)$, and then evaluate the components of the stress increment ΔS_{IJ} $(I, J = 1, 2, 3)$ by means of equation (9.109).

The rate of convergence in the iterative solution procedure obtained using the numerically computed consistent tangent depends crucially on the setting of the perturbation parameter ϵ. Obviously, the appropriate choice of the value of ϵ depends on the precision of the computation as well as on the algorithm adopted in the numerical differentiation. The common choice for the double precision calculation is roughly $\epsilon = 1.0 \times 10^{-6} \sim 1.0 \times 10^{-8}$. Extensive numerical investigations on the effect of the perturbation parameter ϵ to the convergence property have been carried out by Miehe (1996b) and Pérez-Foguet et al. (2000a,b).

Numerical computation of the consistent tangent modulus tensor, as described above, has been applied to sophisticated inelastic models incorporating anisotropy and a damage effect within the finite strain framework by Eidel and Gruttmann (2003), Menzel and Steinmann (2003b), and Menzel et al. (2005). See also the applications to elasto-plastic models for cohesive–frictional materials by Pérez-Foguet et al. (2000a,b), and to the hypoplastic constitutive model by Fellin and Ostermann (2002) within the small strain framework.

9.10 Numerical Examples

In this section we present a set of numerical examples of the stress calculation based on the return-mapping algorithm described in the previous sections. Basic behavior of the constitutive model is examined under large shear deformation and cyclic deformation. Accuracy, robustness, and convergence of the stress calculations by means of the return-mapping algorithm are also assessed.

In the first four examples a simple shear problem is analyzed. This problem is often considered to examine the capability of constitutive models for finite strains, since, as is well known, some models predict a physically unrealistic oscillatory response of shear stress under large shear deformation. In the last two examples, fully three-dimensional finite element simulations of large elasto-plastic deformations are performed to demonstrate the applicability of the return-mapping algorithm to the numerical analysis of elasto-plastic boundary value problems. The finite element method for large deformations is beyond the scope of this book, since our attention is focused throughout upon the theory and numerics of finite strain elastoplatic constitutive equations. The reader interested in the nonlinear finite element method for elastoplatic deformations is referred to other textbooks (Belytschko et al., 2000; Bonet and

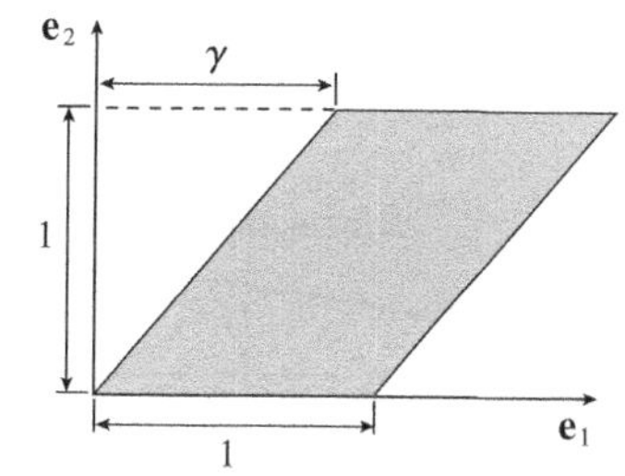

Figure 9.3 Simple shear deformation prescribed by equation (9.111)

Wood, 2008; Crisfield, 1991, 1997; de Souza Neto et al., 2008; Smith and Griffiths, 2004; Zienkiewicz and Taylor, 2005; Zienkiewicz et al., 2005) for further details.

To solve the set of nonlinear equations (9.54), we employ the full Newton–Raphson iterative algorithm, in which the values of the elastic trial state evaluated by equations (9.40), (9.41), (9.44), and (9.46) are used for the initial value $\mathbf{X}^{(0)}$ of the vector of unknown variables $\mathbf{X}$. For the normal-yield ratio R of the subloading surface, however, we set $R^{(\text{tr})}_{n+1} := R_e$ if, in the plastic case, the trial value $R^{(\text{tr})}_{n+1}$ $(= R_n)$ evaluated by equation (9.41) is smaller than the material constant R_e. The initial value of incremental plastic multiplier $\Delta\lambda^{(0)}$ is taken as zero, as usual, whereas an initial value $f^{(\text{tr})}_{n+1}/(\sqrt{6}\mu)$ gives an exact solution for the von Mises plasticity model without isotropic or kinematic hardening. The convergence tolerance for the return-mapping in equation (9.59) is set to 1.0×10^{-12}.

In all numerical examples in this section we carry out a set of analyses of homogeneous simple shear deformation. In a rectangular Cartesian coordinate system with an orthonormal basis $\{\mathbf{e}_1, \mathbf{e}_2, \mathbf{e}_3\}$, the simple shear deformation in the x_1–x_2 plane (see Figure 9.3) is described by the deformation gradient tensor

$$\mathbf{F} = \mathbf{I} + \gamma \mathbf{e}_1 \otimes \mathbf{e}_2, \tag{9.111}$$

where $\mathbf{I}$ is the second-order identity tensor and γ is a magnitude of shear. The fundamental kinematics of simple shear is explained in detail in Section 6.9.2. Let t_n denote a time instant at which all constitutive quantities are known. With N_{step} denoting the number of incremental steps, the simple shear deformation described by equation (9.111) can be divided into N_{step} equal increments, and the incremental deformation over a certain time interval $[t_n, t_{n+1}]$ $(n = 0, \ldots, N_{\text{step}} - 1)$ is then prescribed by the relative deformation gradient tensor

$$\mathbf{f}_{[n,n+1]} = \mathbf{F}_{n+1}\mathbf{F}_n^{-1} = \mathbf{I} + (\gamma/N_{\text{step}})\mathbf{e}_1 \otimes \mathbf{e}_2. \tag{9.112}$$

We solve the time-discrete constitutive problem over the time interval $[t_n, t_{n+1}]$ for a deformation gradient tensor at t_{n+1} given by

$$\mathbf{F}_{n+1} = \mathbf{f}_{[n,n+1]}\mathbf{F}_n = (\mathbf{f}_{[n,n+1]})^{n+1}, \quad n = 0, \ldots, N_{\text{step}} - 1. \tag{9.113}$$

The material constants adopted throughout this section, unless otherwise specified, are

$$\Lambda = 121.2\,\text{GPa}, \quad G = 80.7\,\text{GPa}, \quad F_0 = 500\,\text{MPa}, \quad R_e = 0.1. \tag{9.114}$$

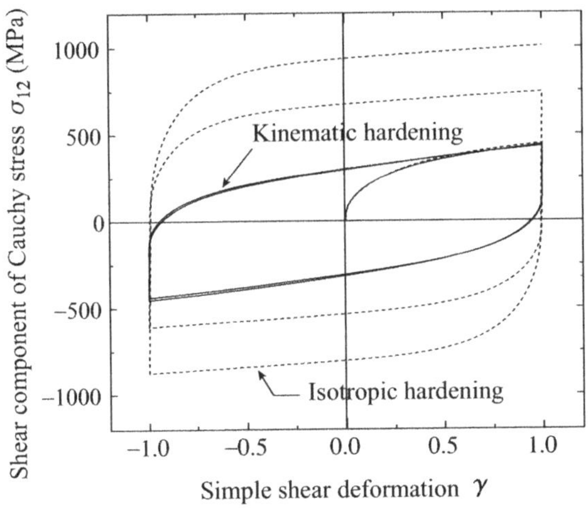

Figure 9.4 Cyclic strain-controlled simple shear analysis. Comparison of isotropic and kinematic hardening

9.10.1 Example 1: Strain-Controlled Cyclic Simple Shear Analysis

In order to demonstrate the constitutive behavior under cyclic deformation predicted by distinct types of plastic hardening rules, two analyses of cyclic strain-controlled simple shear deformation for $\gamma \in [-1, +1]$ are performed for:

(i) isotropic hardening only: $h_1 = 0.4286$ and $h_2 = 2.451$ for nonlinear isotropic hardening, and a linear isotropic hardening with a hardening coefficient of 200 MPa is added; and
(ii) kinematic hardening only: $c = 175$ MPa and $b = 0.18$.

In both cases the evolution coefficient of the subloading surface is taken as $u = 5.0$. The two analyses are performed with a sufficiently small increment $\Delta\gamma = +0.001$ per step for the imposition of shear and $\Delta\gamma = -0.001$ per step for the reversal of shear to analyze in detail the cyclic deformation accompanied by loading, unloading, and reloading responses.

Figure 9.4 depicts the responses obtained for cases (i) and (ii), showing the evolution of the Cauchy stress component σ_{12} with respect to the Cartesian coordinate system versus the amount of shear γ. In both cases, a smooth transition from elastic state to plastic state is predicted by virtue of incorporating the subloading surface concept. The solid curve predicted by the kinematic hardening model shows a steady shape for the hysteresis loop and a clear Bauschinger effect. With the present choice of the constant b, which controls the saturation of kinematic hardening, the curve does not reach a saturation state. On the other hand, as shown by a dotted line, the isotropic hardening model predicts a physically unrealistic continuous plastic hardening behavior without saturation.

9.10.2 Example 2: Elastic–Plastic Transition

The second numerical example demonstrates the effect of the subloading surface. A set of strain-controlled simple shear and reversal analyses by $\gamma \in [0 \to 1.0, 1.0 \to 0]$ is carried out for values 5.0, 20.0, 50.0, and 1000.0 of the evolutionary coefficient u in the evolution rule

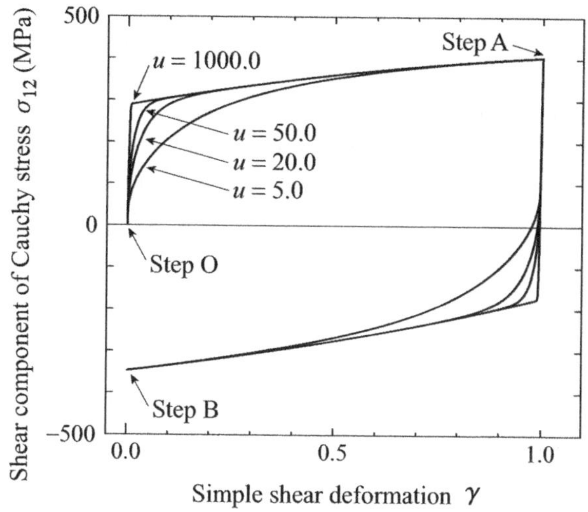

(a) Stress versus strain curves

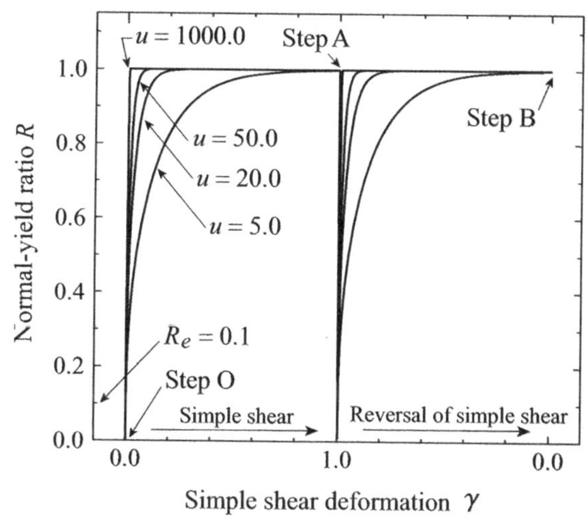

(b) The normal-yield ratio R versus the amount of simple shear γ relations

Figure 9.5 Influence of the evolutionary coefficient u in the evolution rule (9.14) of the normal-yield ratio R for the subloading surface. Results are plotted for the imposition and reversal of simple shear for $u = 5.0$, 20.0, 50.0, and 1000.0

of the subloading surface $(9.14)_2$. In all the analyses, kinematic hardening with $c = 175$ MPa and $b = 0.50$ is employed, while isotropic hardening is not used. The analyses are performed with a sufficiently small increment $\Delta\gamma = +0.0002$ per step for the imposition of shear and $\Delta\gamma = -0.0002$ for the reversed shear to examine in detail the variation of the normal-yield ratio R accompanying unloading and reloading responses during the reversed shear.

Figure 9.5(a) plots the Cauchy stress component σ_{12} versus the amount of shear γ computed for $u = 5.0$, 20.0, 50.0, and 1000.0. The smaller the value of u, the smoother the predicted transition from the elastic state to the plastic state. In contrast, larger values of u predict a response similar to the 'conventional' plasticity model, as observed in the curve of $u = 1000.0$, which exhibits an abrupt transition from the elastic state to the plastic state.

Figure 9.5(b) monitors the variation of the normal-yield ratio R versus the amount of shear γ. The curve obtained for smaller u shows a gentler approach of R to unity, that is the normal-yield state, with increasing shear. In contrast, the curve for larger u shows a rapid increase of R up to unity. It should be also observed that, during unloading accompanying the early stage of reversed shear, the value of R decreases almost to zero, which corresponds to a degeneration of the subloading surface to a null size. By a further reversed shear, subsequently, the evolution of R resumes promptly along with the evolution of plastic strain, and then the value of R again approaches unity.

9.10.3 Example 3: Large Monotonic Simple Shear Analysis with Kinematic Hardening Model

The behavior of the kinematic hardening model under large shear strain is examined and the influence of the kinematic hardening parameters is investigated by performing a parametric study for various values of the shear modulus c in the constitutive equation (9.3) and of the evolution parameter b in the evolution equation (9.12) associated to kinematic hardening. A large monotonic simple shear deformation γ is imposed up to 5.0 with $N_{\text{step}} = 100$. Isotropic hardening is not incorporated by setting $h_1 = h_2 = 0$, and the evolution coefficient of the subloading surface is taken as $u = 5.0$.

We first carry out a set of analyses for the values of evolution parameter $b = 0.00$, 0.10, 0.50, 1.00, 2.00, and 10.0, in which the other parameter c in the kinematic hardening constitutive equation is set equal to 175 MPa. The stress–strain curves obtained are shown in Figure 9.6(a). The case of $b = 0.00$ exhibits a similar behavior to the linear kinematic hardening model of Melan–Prager type, showing a linear increase of the shear stress σ_{12} with increasing shear deformation γ. This behavior may be expected from the fundamental structure of this kinematic hardening model. In the case of $b = 0.00$, the kinematic hardening continues to develop without saturation, since, in the multiplicative decomposition of the plastic part $\mathbf{F}^p = \mathbf{F}_e^p \mathbf{F}_d^p$, the equalities $\mathbf{F}_d^p = \mathbf{I}$ and thus $\mathbf{F}^p = \mathbf{F}_e^p$ hold irrespective of evolution of plastic deformation. However, for larger values of the evolution constant b, that is in the cases of $b = 1.00$, 2.00, and 10.0, the kinematic hardening saturates rapidly. One can observe that the curves for $b = 0.10$ and 0.50 show a transition from initial hardening to gradual decrease of shear stress with increasing shear deformation, followed by an eventual saturation of the kinematic hardening. Note also that all the cases exhibit a smooth transition from the elastic state to the plastic state as an effect of the subloading surface. Even the curve for $b = 0.00$ shows a gradual decrease of the slope from elastic state to plastic state with linear hardening.

The second series of analyses is performed by setting the parameter c equal to 0, 175, and 350 MPa, while the values of evolution parameter b are taken as 0.00 and 1.00. Computed stress–strain curves are plotted in Figure 9.6(b). For $b = 0.00$, the larger the value of c, the steeper the slope of linear kinematic hardening. For $b = 1.00$ we observe an initial hardening behavior, which gradually approaches a saturation state. Since extent of the initial hardening until the saturation is dependent on the values of parameter c, a larger value of the saturation shear stress is predicted for a larger values of c. Note also that, in the case of $b = 1.00$ and $c = 350$ MPa, a slight decrease of the shear stress is observed after the initial hardening. For $c = 0$, perfectly plastic behavior without hardening is predicted irrespective of the value of evolution parameter b.

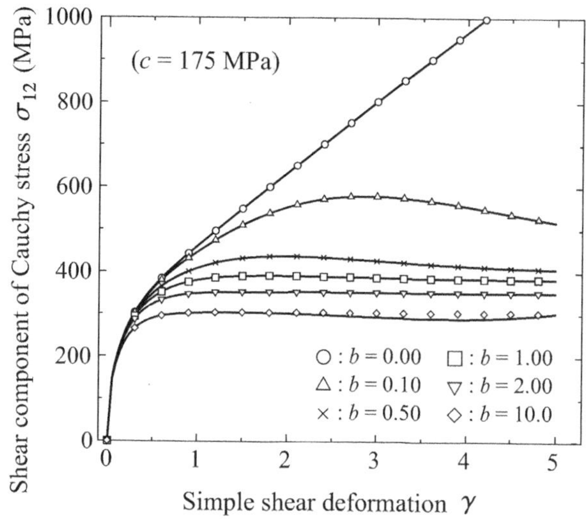

(a) Influence of the evolution parameter b associated with kinematic hardening. The value of the constitutive parameter c is chosen as 175 MPa

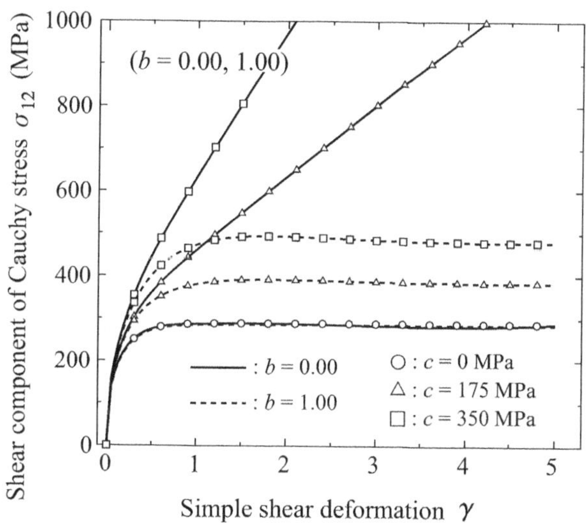

(b) Influence of the constitutive parameter c associated with kinematic hardening. The values of the evolution parameter b are chosen as 0.00 and 1.00

Figure 9.6 Influence of the material parameters associated with kinematic hardening

Throughout the above two series of analyses, the numerical results did not suffer from oscillations of shear stress even under large shear deformation up to $\gamma = 5.0$. The result of this numerical example therefore affirms the capability of the present constitutive model for large shear deformations.

9.10.4 Example 4: Accuracy and Convergence Assessment of Stress-Update Algorithm

As discussed in Section 9.7, in the numerical stress calculation by means of the return-mapping algorithm, the plastic corrector step is carried out by solving the system of nonlinear algebraic equations (9.53), which comprises the incremental format of the plastic evolution equations (9.30), (9.31), and (9.35), and the subloading function $f_{n+1} = 0$.

In order to ensure accuracy of the stress-update calculation of inelastic constitutive models, it is of crucial importance to select an appropriate time-integration scheme for the plastic evolution equations over the time interval $[t_n, t_{n+1}]$, as well as to compute the solution satisfying the convergence criterion (9.59). In most cases, however, the time-integration of the plastic evolution equations in an analytical way is practically impossible even in simple constitutive models.

To this end, an approximate numerical time-integration scheme is usually adopted in numerical stress-update computations. The backward Euler difference scheme is most commonly adopted because of its computationally sound characteristics of simplicity, robustness and unconditional stability. However, as mentioned in Section 9.4, the backward Euler difference-based time-integration scheme has a serious drawback in that, when applied to plastically incompressible finite strain constitutive models, the plastic incompressibility is not automatically retained unless some special modifications are made to the scheme. To resolve this disadvantage, the tensor exponential-based time-integration scheme is widely used in stress-update computations of finite strain plasticity models. This integration scheme allows larger incremental deformations than the standard backward Euler scheme, and retains the plastic incompressibility, when used in the von Mises plasticity model.

In order to examine the influence of the step size on the accuracy of the stress calculation, a set of monotonous simple shear analyses is carried out by imposing the same total shear deformation up to $\gamma = 5.0$ in a different number of steps, namely, $N_{\text{step}} = 10$, 20, 50, and 100. We also investigate the accuracy property of the time-integration schemes for the plastic evolution equations by performing the analyses using two types of scheme:

A. tensor exponential-based time-integration scheme using equations (9.30) and (9.31);
B. approximate evaluation of the tensor exponentials $\mathbf{Q}^p_{n+1}$ and $\mathbf{Q}^p_{e,n+1}$ by taking only up to the first-order term in the series representation of the tensor exponential function (D.1), that is a time-integration scheme based on equation (9.29) with $\mathbf{Z}_{n+1}$ for $\mathbf{Z}$, which can be regarded as a class of the standard backward Euler difference approximation.

A series of analyses are carried out for the kinematic hardening evolution parameter $b = 0.00$ and 0.50, while the parameter c is set equal to 175 MPa. The isotropic hardening is omitted by setting $h_1 = h_2 = 0$, and the evolution coefficient of the subloading surface is set to $u = 5.0$.

The results obtained for $b = 0.00$ and 0.50 are shown in Figures 9.7 and 9.8, respectively. In each figure, the results for cases A and B are shown in parts (a) and (b). respectively. In all cases the iterative computations in the return-mapping were found to be stable even for

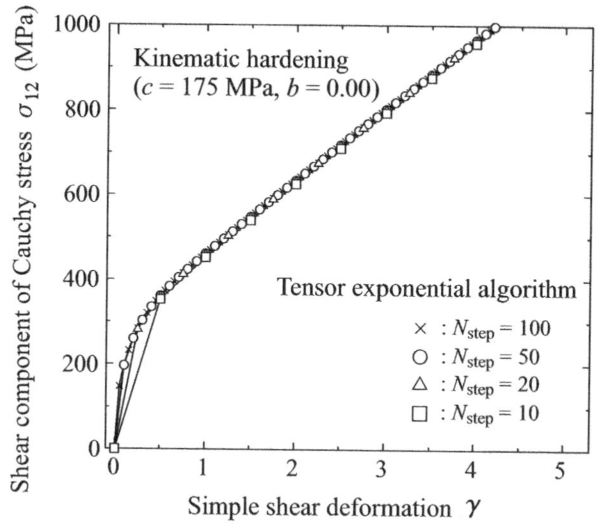

(a) Tensor exponential-based time-integration scheme

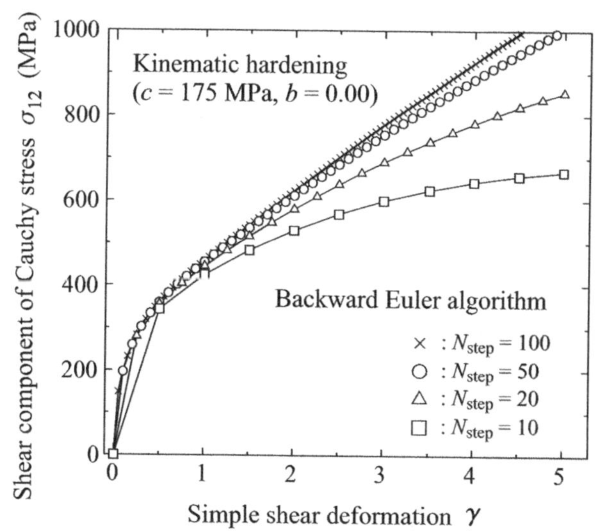

(b) Backward Euler time-integration scheme

Figure 9.7 Comparison of the time-integration schemes and the influence of the step size on the accuracy of stress calculation ($b = 0.00$)

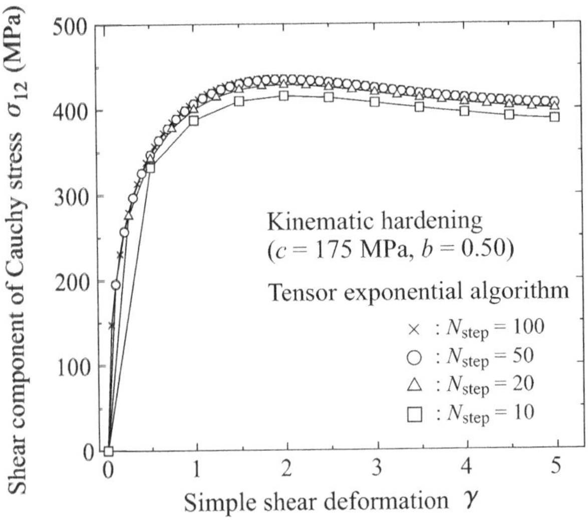

(a) Tensor exponential-based time-integration scheme

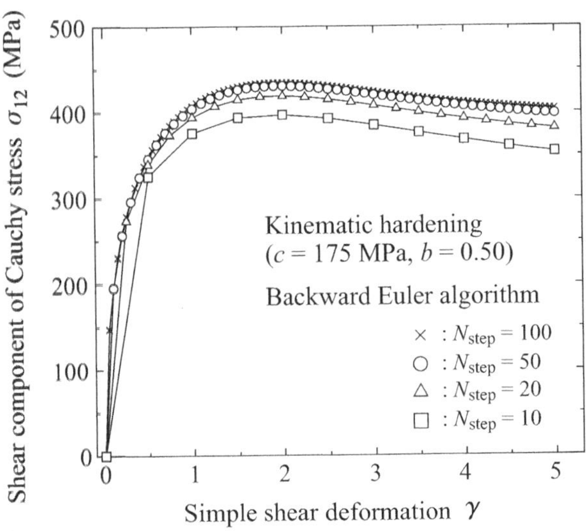

(b) Backward Euler time-integration scheme

Figure 9.8 Comparison of the time-integration schemes and the influence of the step size on the accuracy of stress calculation ($b = 0.50$)

large-deformation increments, as the solutions were convergent satisfying the convergence criterion (9.59). For both $b = 0.00$ and 0.50, the tensor exponential-based scheme gives more accurate results almost unaffected by step size than does the backward Euler scheme, while a relatively small error is observed in Figure 9.8(a) for $b = 0.50$. By contrast, the results for the backward Euler scheme involve significant error, which increases as the size of deformation increment increases. This may be attributed to the above-mentioned drawback of the standard backward Euler scheme that the plastic incompressibility is not automatically fulfilled. Modified versions of the backward Euler scheme that conserve the plastic incompressibility have been proposed and extensive comparative studies on these algorithms have been performed by Dettmer and Reese (2004), Helm (2006), Vladimirov et al. (2008), Shutov and Kreissig (2008a), and S-074. The analysis presented here confirms that in case A the tensor exponential-based scheme fulfills the plastic incompressibility (that is $J^p = 1$), whereas, for large step sizes, the backward Euler scheme may produce an incorrect result involving a spurious plastic volume change; the value $J^p = 0.3$ was obtained in case B for large step size in the case of $N_{\rm step} = 10$.

Next, we examine the convergence property of the iterative solution procedure for the nonlinear simultaneous equations involved in the return-mapping scheme. As a typical result, the convergence profile observed in the final loading step of case A for $b = 0.50$ is shown in Figure 9.9, in which the norm of the residual vector $\|\mathbf{Y}(\mathbf{X}^{(k)})\|$ in equation (9.59) is normalized by the initial residual norm $\|\mathbf{Y}(\mathbf{X}^{(0)})\|$ evaluated by the elastic trial values, and then is plotted versus the iteration number k. For smaller sizes of the deformation increments (that is $N_{\rm step} = 50$ and 100), a convergent profile with an asymptotic quadratic rate is observed, and hence the return-mapping requires only five or six iterations to obtain the converged solution. On the other hand, in the case of large-deformation increments, more than 10 iterations are required to attain convergence.

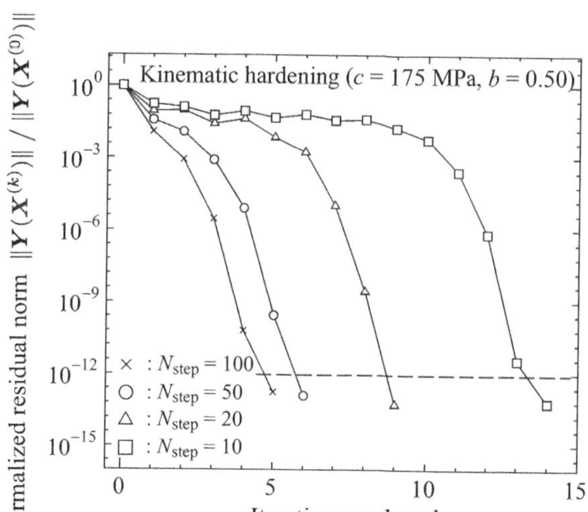

Figure 9.9 Convergence profile of the Newton–Raphson iterations in the return-mapping. The residual norm per iteration in the final loading step of case A for $b = 0.50$ is plotted

9.10.5 Example 5: Finite Element Simulation of Large Deflection of Cantilever

This example presents a finite element simulation of a large deflection of a cantilever beam in order to validate the capability of the elasto-plastic constitutive model for finite strains described in Chapter 8, and the applicability of the stress-update algorithm based on the return-mapping scheme to the numerical analysis of the elasto-plastic large-deformation boundary value problems. The problem of large bending of the cantilever is suitable for the validation of the constitutive model and the numerical algorithm, since it involves large deformation and large rotation.

To carry out the fully three-dimensional finite element analysis of the large elasto-plastic deformation, a Fortran program `returnmap` for the stress-update computation based on the return-mapping scheme, whose detailed explanation and source code are presented in Chapter 10, is implemented into a nonlinear finite element program developed for research by the authors (this finite element program is not presented since it is beyond the scope of this book), and then utilized for the analysis of the cantilever beam. The consistent tangent modulus tensor, computed by the numerical differentiation based on the perturbation approach (see the description in Section 9.9.2), is utilized in the iterative solution of the nonlinear equilibrium equation, that is the global iteration, to achieve a quadratic rate of convergence. In the analysis in this example, the perturbation parameter in equation (9.106) is taken as $\epsilon = 1.0 \times 10^{-8}$. A Fortran program `tgnum0` to evaluate the consistent tangent modulus tensor based on the numerical differentiation is also provided; see the description of the program in Section 10.3.6.

We consider a cantilever beam with length $L = 40\,\text{cm}$, width $W = 1\,\text{cm}$, and height $H = 4\,\text{cm}$, one end of which is fully fixed to a wall and the other end is free. The geometry of the cantilever is shown in Figure 9.10. The cantilever is discretized into a finite element mesh consisting of $20 \times 8 \times 2$ twenty-node quadratic brick elements, 1989 nodes, and 5759 effective degrees of freedom. The reduced Gauss integration with $2 \times 2 \times 2$ quadrature points is employed. The elastic cantilever and the elasto-plastic cantilever are considered for comparison. For the elasto-plastic cantilever, we used the material constants listed in equation (9.114), with the evolution coefficient of the subloading surface $u = 50.0$ and the parameters for kinematic hardening $c = 175$ MPa and $b = 0.50$, neglecting isotropic hardening by taking the relevant constants as $h_1 = h_2 = 0$. For the elastic cantilever, the same hyperelastic model as in the elasto-plastic cantilever is used with the material constants Λ and μ listed in equation (9.114). A monotonically increasing downward vertical concentrated load F is applied to a center point of the cross section at the free end and the vertical deflection u at this point is monitored. In addition, the stress–strain relations at the material points A, B, C, and D in Figure 9.10 (that is the Gauss integration points in the finite element context) are monitored and reported in the following to examine the local material behavior.

In the solution of the nonlinear equilibrium equation, the 'reduced' Newton–Raphson iteration is employed to improve the robustness of the solution of the nonlinear equilibrium equation even with large incremental steps. The convergence tolerance of the global iteration is taken as 1.0×10^{-10} for the norm of the global residual force vector. In both elastic and elasto-plastic analyses, the arc-length control method is utilized to track the equilibrium path, and a large deflection u almost equal to the cantilever length L is attained by 50 incremental steps.

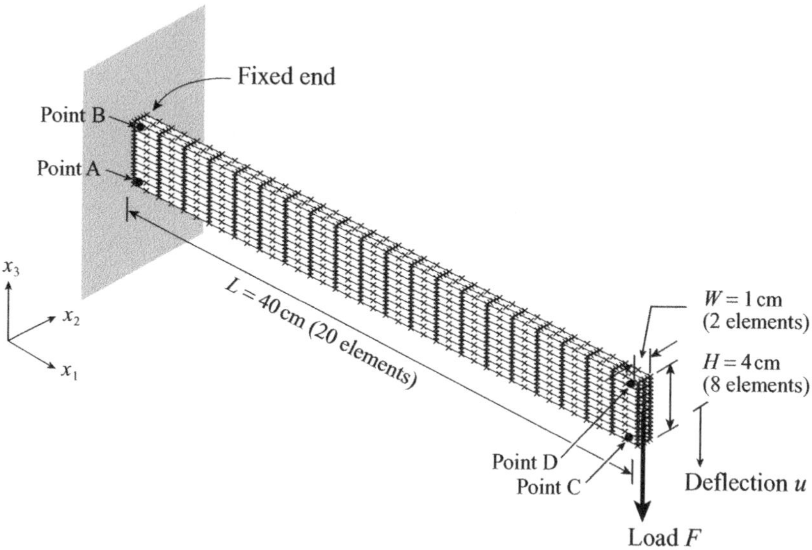

Figure 9.10 Finite element model of a cantilever discretized by 20 × 8 × 2 twenty-node quadratic elements

The overall view of the load versus deflection behaviors of the elastic and elasto-plastic cantilevers is plotted in Figure 9.11(a). The iterative solution of the equilibrium equation was stable as the solutions in all the incremental steps were convergent. The elastic cantilever exhibits a highly nonlinear behavior due to the change of geometry as a result of large deflection. The load of the elasto-plastic cantilever is very small compared with the elastic cantilever, so a detailed view is shown in Figure 9.11(b). For comparison, the solution of the Timoshenko beam theory

$$u = \frac{FL^3}{3EI} + \frac{FL}{AkG}$$

is also plotted by a thin dashed line, in which the Young modulus is $E = 209.8$ GPa, the moment of inertia of the cross section is $I = 5.33\,\mathrm{cm}^4$, the cross-sectional area is $A = 4.0\,\mathrm{cm}^2$, the shear correction factor $k = 10(1+\nu)/(12+11\nu) = 0.85$ for a rectangular cross section with the Poisson ratio $\nu = 0.30$ (Pilkey, 2002), and the elastic shear modulus is $G = 80.7$ GPa. The load–deflection curve of the elastic cantilever shows good agreement with the solution of the Timoshenko beam theory within a range of small deflection; however, as expected, it deviates from the theoretical solution with increasing deflection due to the effect of geometry change.

Figure 9.12 depicts the initial and deformed configurations of the elastic and elasto-plastic cantilevers, and the progress and distribution of the strain, in which the contours of the equivalent strain, defined as $\sqrt{2/3}\,\|\boldsymbol{\varepsilon}'\|$ in terms of the deviatoric part of the logarithmic strain tensor $\boldsymbol{\varepsilon}' := \boldsymbol{\varepsilon} - (\mathrm{tr}\boldsymbol{\varepsilon})\mathbf{I}/3$, are shown. The elasto-plastic cantilever exhibits a larger curvature than the elastic cantilever, and a larger amount of strain is observed in the elasto-plastic cantilever around the fixed end. Figure 9.13 shows the progress and distribution of the elastic

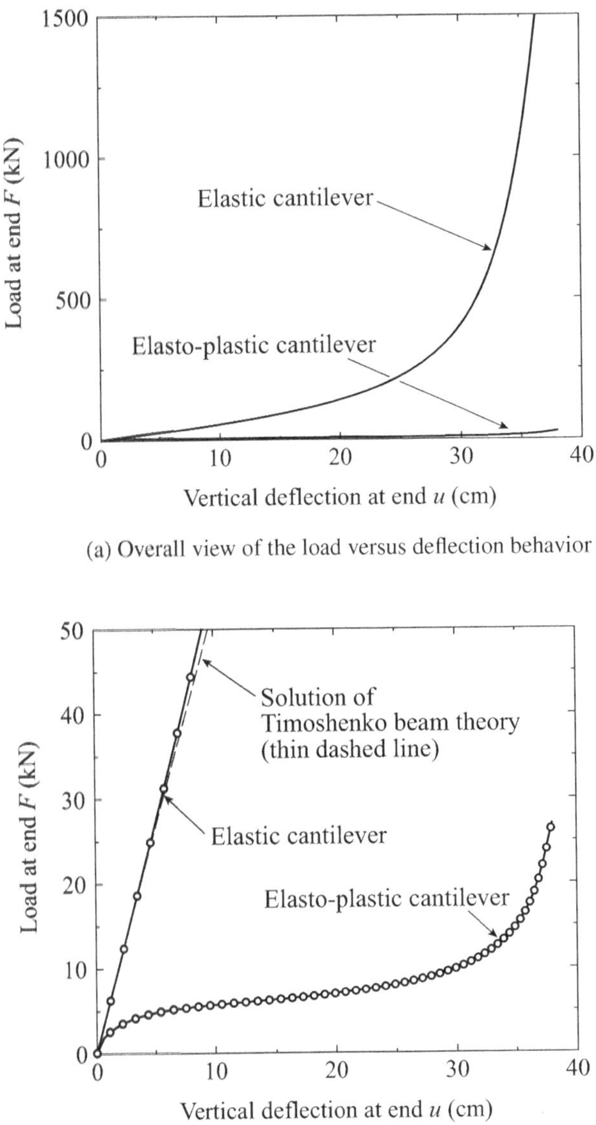

(a) Overall view of the load versus deflection behavior

(b) Detailed view of the load versus deflection behavior. The theoretical solution of the Timoshenko beam theory is also plotted by a thin dashed line for comparison

Figure 9.11 Load versus deflection behaviors of the elastic and elasto-plastic cantilevers

unloading and plastic loading regions in the elasto-plastic cantilever. The plastic loading region spreads out over almost the whole area of the cantilever in an earlier stage of deflection. However, by a further deflection, the elastic unloading region again spreads out.

In order to examine the local behavior at material points in the cantilever, the rotations in the x_1–x_3 plane at the points A, B, C, and D (the Gauss integration points used in

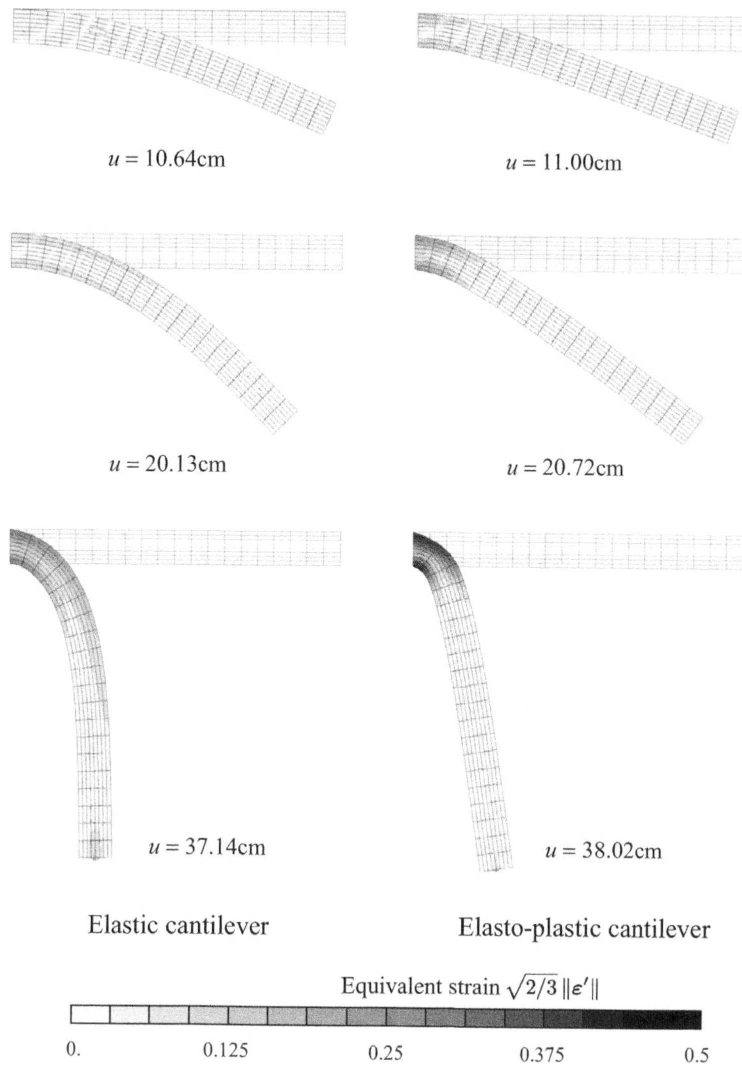

Figure 9.12 Initial and deformed configurations and contours of the equivalent strain. Comparison of the elastic and elasto-plastic cantilevers

the finite element analysis) designated in Figure 9.10 are plotted versus the deflection in Figure 9.14. Anti-clockwise rotation is defined as positive. In both of the elastic and the elasto-plastic cases, as can be expected from the deformed shape shown in Figure 9.12, the points C and D, close to the free end of the cantilever, undergo quite large rotations up to about 90 degrees. Figure 9.15 monitors the local stress–strain relations at the points A–D. The relations between the equivalent Kirchhoff stress $\sqrt{3/2}\,\|\boldsymbol{\tau}'\|$ and the equivalent strain $\sqrt{2/3}\,\|\boldsymbol{\varepsilon}'\|$ are plotted. Observe that, in both the elastic and the elasto-plastic cases, the amounts of strain at the points A and B are much larger than those at the points C and D. Similar to the previous

Figure 9.13 Distribution of elastic and plastic regions in the elasto-plastic cantilever

observation of Figure 9.12, larger amounts of strain are observed at the points A and B in the elasto-plastic cantilever than in the elastic cantilever.

Finally, we examine the convergence profile in the global iteration to validate the performance of the consistent tangent modulus tensor evaluated by numerical differentiation. Although the hyperelastic tangent modulus tensor can readily be evaluated by a straightforward differentiation as shown in equation (9.100), the numerically computed tangent modulus tensor is used even in the analysis of the elastic cantilever. In Figure 9.16, the norm of the residual force vector in the global iteration is plotted versus the iteration number k. The convergence profiles of the 1st and 50th (final) incremental steps in the analyses of the elastic and elastoplatic cantilevers are reported. In all the reported steps, less than 10 iterations are required to obtain the converged solution and strong convergent profiles are observed even in the analysis accompanied by large elasto-plastic deformation and rotation. This result confirms the effectiveness of the consistent tangent modulus tensor.

9.10.6 Example 6: Finite Element Simulation of Combined Tensile, Compressive, and Shear Deformation for Cubic Specimen

As the last numerical example in this section, a finite element simulation of a large elasto-plastic deformation of a cubic specimen is presented in order to demonstrate the applicability

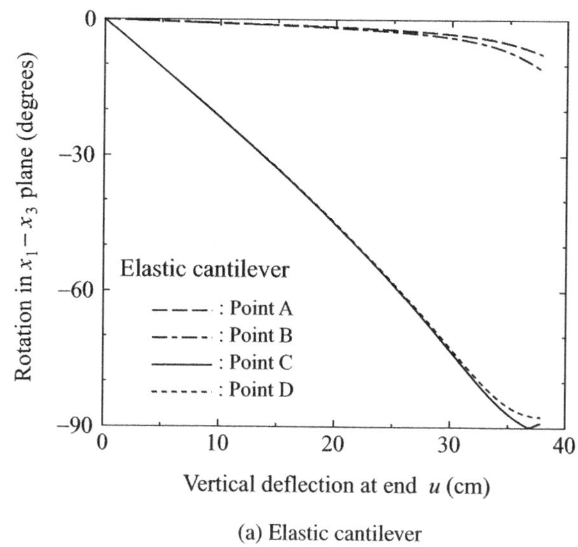

Figure 9.14 Rotation in x_1–x_3 plane at the points A, B, C, and D versus the deflection of the elastic and elasto-plastic cantilever

of the stress-update algorithm based on the return-mapping scheme to the elasto-plastic large-deformation analyses.

We consider a cubic specimen 1 m $\times$ 1 m $\times$ 1 m in size, discretized into a mesh comprised of $8 \times 8 \times 8$ twenty-node quadratic brick elements, as shown in Figure 9.17. The finite element mesh is composed of 2673 nodes, 512 elements, and 7344 effective degrees of freedom. The reduced Gauss integration with $2 \times 2 \times 2$ quadrature points is employed. As shown in

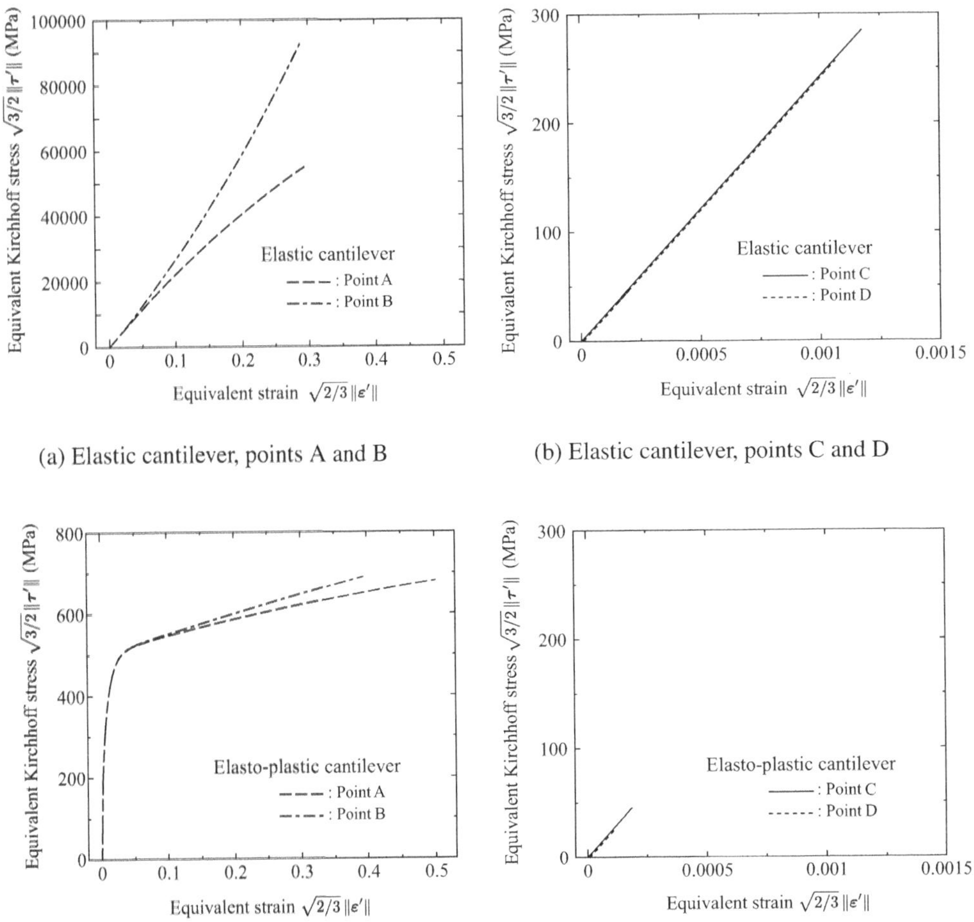

(a) Elastic cantilever, points A and B

(b) Elastic cantilever, points C and D

(c) Elasto-plastic cantilever,points A and B

(d) Elasto-plastic cantilever,points C and D

Figure 9.15 Local stress versus strain behaviors at the points A, B, C, and D

Figure 9.18, a combined tensile, compressive, and shear deformation whose track forms a closed loop is imposed on the specimen by prescribing a uniform displacement u_1 in x_1-direction and a uniform displacement u_2 in x_2-direction on the upper surface of the specimen, while all the nodes on the bottom surface are fully constrained. The x_1- and x_2-components of the reactive resultant force on the upper surface are denoted, respectively, by f_1 and f_2. No constraint on the x_3-direction is imposed except for the upper and bottom surfaces, so that a full three-dimensional analysis is carried out. We use the material constants listed in equation (9.114), with the evolution coefficient of the subloading surface $u = 50.0$ and the parameters for kinematic hardening $c = 175$ MPa and $b = 1.00$, neglecting isotropic hardening by taking the relevant constants as $h_1 = h_2 = 0$. The convergence tolerance of the global iteration is taken as 1.0×10^{-10} for the norm of the global residual force vector.

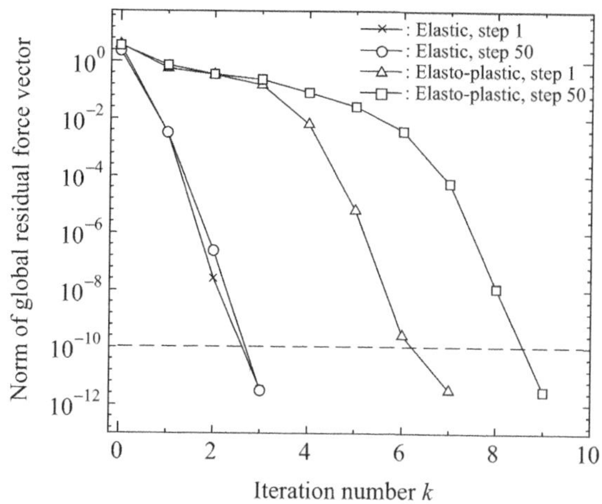

Figure 9.16 Convergence profile of the global iteration. The norm of the global residual force vector versus the iteration number is plotted

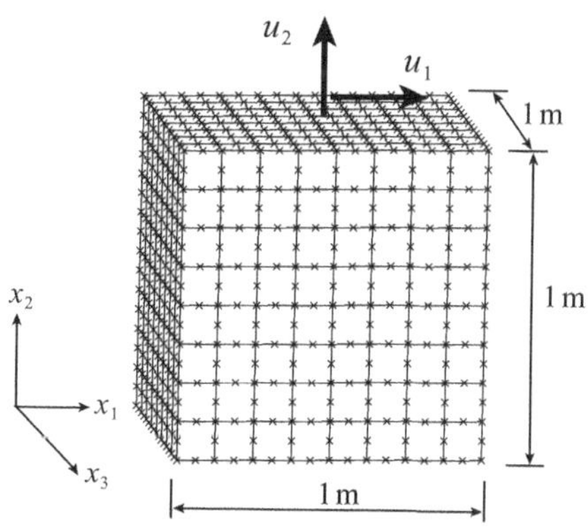

Figure 9.17 Finite element model of a cubic specimen discretized by 8 × 8 × 8 twenty-node quadratic elements

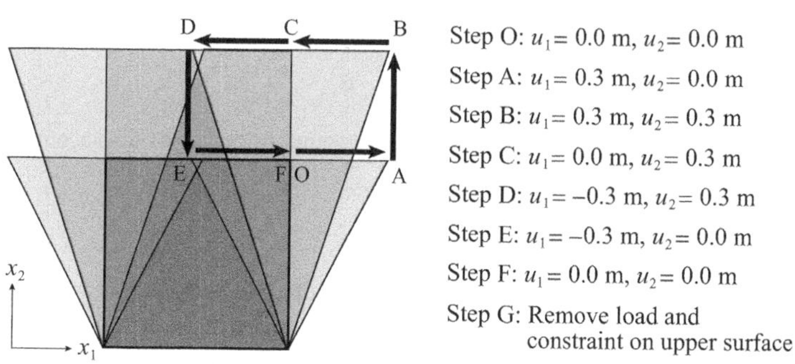

Figure 9.18 Combined tensile, compressive, and shear deformation imposed on the cubic specimen

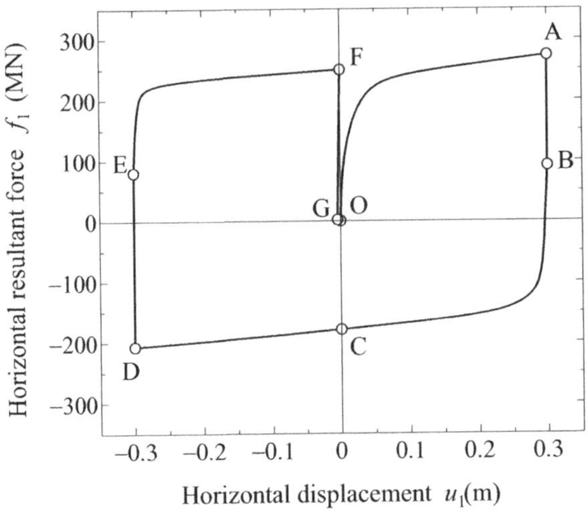

(a) f_1 versus u_1 curve

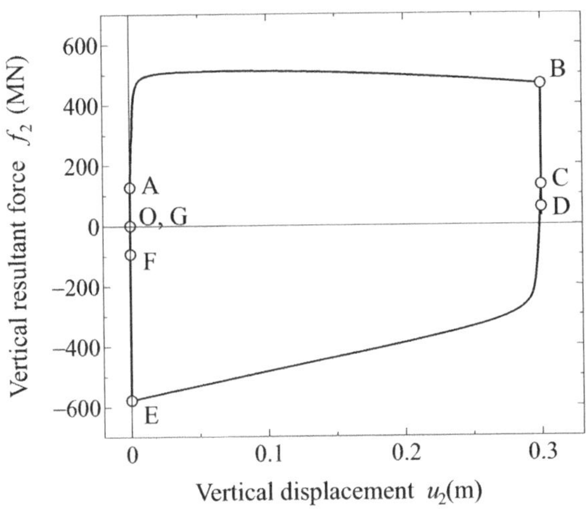

(b) f_2 versus u_2 curve

R

Figure 9.19 Global behavior of the cubic specimen. Reaction force versus displacement relations on the upper surface

The horizontal and vertical components of the reactive resultant force f_1 and f_2 on the upper surface of the specimen are plotted versus the displacements u_1 and u_2 in Figure 9.19 (a) and (b), respectively, in which the symbol '$\circ$' on the curve denotes the state at the end of each step O–G. Figure 9.20 depicts the deformed configurations of the specimen and the progress and distribution of the equivalent strain at the steps A–G, in which the contours of

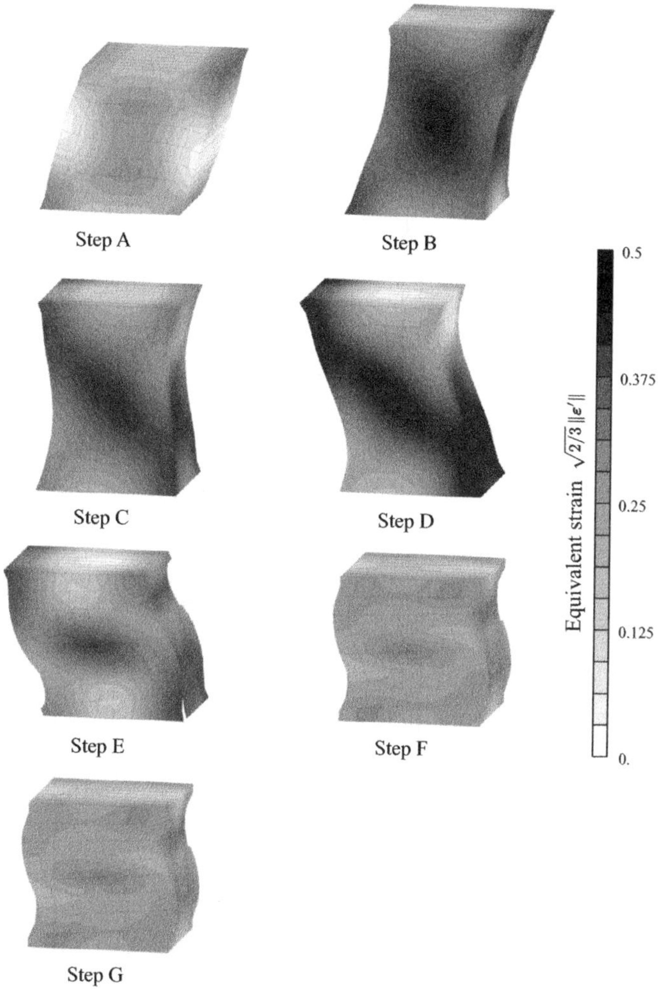

Figure 9.20 Deformed configurations and distribution of equivalent strain at steps O–G

the equivalent strain $\sqrt{2/3}\,\|\boldsymbol{\varepsilon}'\|$ are shown. The analysis verifies that the constitutive model is capable of representing the structural behavior under a large deformation. Moreover, the global as well as local iterative solution procedures were quite stable throughout the analysis even in a large-deformation regime, demonstrating the applicability and robustness of the developed stress-update algorithm and the evaluation of the consistent tangent modulus tensor based on numerical differentiation.

For a further investigation of the local constitutive behavior inside the specimen, we plot the shear and normal components of the Cauchy stress tensor, σ_{12} and σ_{22}, at the Gauss quadrature point nearest to the center of the specimen versus the corresponding components of the deformation gradient tensor, F_{12} and F_{22}, in Figure 9.21. It is observed from Figure 9.21 (a) that the value of F_{12} at Step A is about 0.3, which roughly corresponds to the deformation

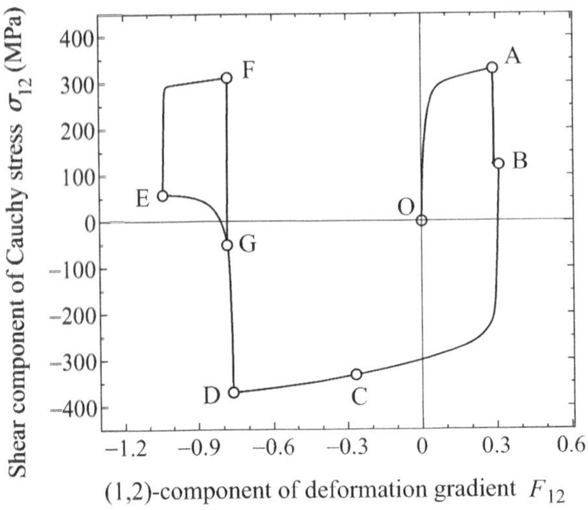

(a) σ_{12} versus F_{12} curve

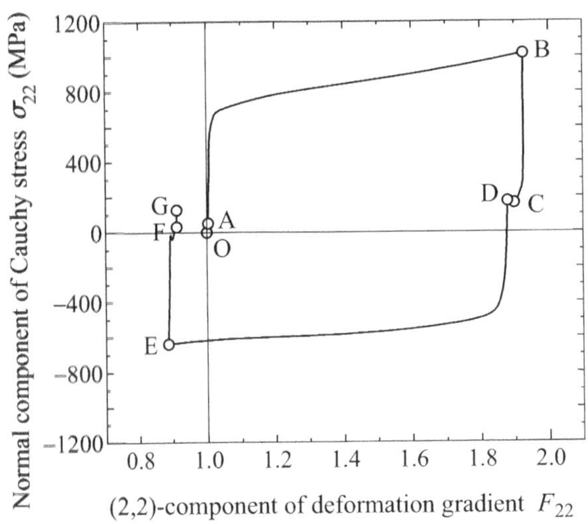

(b) σ_{22} versus F_{22} curve

Figure 9.21 Local material behavior of the inside of the specimen. Cauchy stress components are plotted versus the corresponding components of the deformation gradient tensor at the middle point of the specimen

expected, under an assumption of uniform deformation, to be induced by the imposed shear displacement ($u_1 = 0.3$ m), whereas F_{12} at Step E, that is, the state after the inverse shear, amounts to 0.9. Also observed from Figure 9.21 (b) is that the values of F_{22} at Steps B and C amount to about 1.9. These results imply that quite inhomogeneous and intense deformations evolve inside the specimen.

10

Computer Programs

This chapter presents a Fortran program, named `returnmap`, for the stress-update computation for the finite strain elasto-plastic constitutive model on the basis of the algorithm which comprises the elastic predictor step and the plastic corrector step presented in Chapter 9. The program `returnmap` deals with the material model based on the von Mises plasticity with combined isotropic and kinematic hardening, whose specific formulation was summarized and its counterpart within the time-discrete framework presented in Chapter 9. In addition, a Fortran subroutine, named `tgnum0`, to evaluate the consistent tangent modulus tensor by means of numerical differentiation (see Section 9.9.2) is presented.

The outline of this chapter is as follows. In Section 10.1 an introduction to the program `returnmap` is given, along with brief instructions on downloading, compilation and linking. In Section 10.2 the data structure of the output files generated by the program is described. In Section 10.3 the structure of the program `returnmap` is described, and the complete source codes of the routines, together with detailed line-by-line explanation of the codes as well as lists of the arguments and variables used in the program, are presented in order to enhance reader's understanding of the numerical implementation of the finite strain elasto-plastic constitutive models. The subroutine `tgnum0` is also explained and its source code is presented.

10.1 User Instructions and Input File Description

The Fortran program `returnmap` consists of a main routine and seven subroutines, as listed in Section 10.3. The complete source code is available on-line, together with sample data, from the author's website:

`http://msd.civil.tohoku.ac.jp/yamakawa/returnmap/`

The program `returnmap` is provided as a set of separate files, each of which contains a main routine or a subroutine (see Subsection 10.3.1 for the overview and structure of the program), so, by using the `make` command, together with a `makefile` that describes a rule for compilation and linking, these source codes are compiled all at once, and in turn, the object modules generated by the compilation are linked to generate the executable file, named `returnmap.exe`. The script file, named `makefile`, to be used for the compilation and linking

Introduction to Finite Strain Theory for Continuum Elasto-Plasticity, First Edition. Koichi Hashiguchi and Yuki Yamakawa.

Table 10.1 The contents and data structure of the input file `input.txt` (The variables `inel` and `ige` are provided for possible use in finite element codes to specify the element number and the Gaussian integration point number, respectively, but are left unused in the program `returnmap`)

Line	Format	Variable	Symbol	Description
1	a5	/RDGRD/	—	Headword
1	i5	mstep	—	The number of loading sets
1	i5,2i8	kstep, nbgn(kstep),nend(kstep)	—	Loading set number, step numbers (beginning and end of the kstepth loading set)
3	3es15.7	frx(kstep,1,1),...,	$\mathbf{f}_{[n,n+1]}$	Relative deformation gradient tensor, whose 3 × 3 Cartesian components are fully arranged.
	3es15.7	frx(kstep,2,1),...,		
	3es15.7	frx(kstep,3,1),...,		
		....		(mstep loading sets can be specified.)
1	a5	/MATER/	—	Headword
4	5es12.4	prope(1),...,prope(5),	—	Material constants (see Tables 10.2 and 10.20).
	5es12.4	prope(6),...,prope(10),		
	5es12.4	prope(11),...,prope(15),		
	5es12.4	prope(16),...,prope(20)		
1	a5	/STRES/	—	Headword
1	i8,i5,	inel,ige,	—	Dummy, dummy,
	10es15.7,	sg(1,1),sg(1,2),sg(1,3),...,	$\boldsymbol{\sigma}$	Cauchy stress tensor (its 3 × 3 Cartesian components are arranged in the line),
	i5,	idum,	—	dummy,
		idep,	—	elastic/plastic identifier (0, elastic; 1, plastic),
	i5	idum	—	dummy
1	a5	/PLAST/	—	Headword
1	i8,i5,	inel,ige,	—	Dummy, dummy,
	30es15.7	xp(1),...,xp(30)	—	plastic internal variables (see Tables 10.3 and 10.21).

is also available on-line from the author's website. Some modifications of `makefile` may be required for the compilation depending on the computer environment.

The program `returnmap` requires a single input file, named `input.txt`, which must contain the complete input data in the specified fixed format. The contents and structure of the required data are listed in Table 10.1. For a tensorial quantity, its Cartesian components should be given in this file. The program `returnmap` performs the deformation-controlled analysis in which an incremental deformation per step is prescribed by specifying the relative deformation gradient tensor in the entry `/RDGRD/`. Multiple loading sets can be prescribed by specifying `mstep` sets of the step numbers at the beginning and the end of the `kstep`th loading set, designated, respectively, by `nbgn(kstep)` and `nend(kstep)`, and the relative deformation gradient tensor corresponding to the loading set. Note that, by default, the maximum value of `mstep`, that

Table 10.2 Data description for the `/MATER/` entry in the input file `input.txt`

Variable	Symbol	Equation	Description
`prope( 1)`	Λ	(9.1)	Hyperelastic modulus corresponding to Lamé's first parameter in the linear elasticity
`prope( 2)`	μ	(9.1)	Hyperelastic shear modulus corresponding to Lamé's second parameter in the linear elasticity
`prope( 3)`	—	—	(not used)
`prope( 4)`	u	$(9.14)_2$	Evolutionary coefficient of the normal-yield ratio R
`prope( 5)`	R_e	$(9.14)_2$	Elasticity limit of the normal-yield ratio R
`prope( 6)`	F_0	(9.5)	Reference yield stress for isotropic hardening
`prope( 7)`	h_0	—	Linear isotropic hardening coefficient (unspecified in equation (9.5))
`prope( 8)`	h_1	(9.5)	Parameter controlling the saturation limit of isotropic hardening
`prope( 9)`	h_2	(9.5)	Exponential hardening parameter for isotropic hardening
`prope(10)`	—	—	(not used)
`prope(11)`	—	—	(not used)
`prope(12)`	c	(9.3)	Shear modulus-like kinematic hardening parameter
`prope(13)`	b	(9.12)	Evolutionary coefficient of kinematic hardening
`prope(14)-prope(20)`	—	—	(not used)

is the maximum number of loading sets designated by `lstep`, is specified as 10. Of course, the user can change the value of `lstep`, if needed; see the `parameter` statement of line 2 in Table 10.8.

The material constants and the plastic internal variables at the reference state need to be specified in the entries `/MATER/` and `/PLAST/`, respectively. A detailed explanation of the data description in the entries `/MATER/` and `/PLAST/` is given in Tables 10.2 and 10.3, respectively.

Table 10.3 Data description for the `/PLAST/` entry in the input file `input.txt`

Variable	Symbol	Description
`xp( 1)`	$\Delta\lambda$	Incremental plastic multiplier within the step
`xp( 2)`	H	Deformation-like variable for isotropic hardening
`xp( 3)`	R	Normal-yield ratio for the subloading surface
`xp( 4)-xp(10)`	—	(not used)
`xp(11)-xp(19)`	$\mathbf{C}^p$	Plastic right Cauchy–Green deformation tensor, whose Cartesian components are arranged in the order $C^p_{11}, C^p_{12}, C^p_{13}, C^p_{21}, C^p_{22}, C^p_{23}, C^p_{31}, C^p_{32}$, and C^p_{33}.
`xp(20)`	$\det\mathbf{C}^p$	Determinant of $\mathbf{C}^p$
`xp(21)-xp(29)`	$\mathbf{C}^p_d$	Right Cauchy–Green-type deformation tensor related to dissipative plastic deformation, whose Cartesian components are arranged in the order $C^p_{d,11}, C^p_{d,12}, C^p_{d,13}, C^p_{d,21}, C^p_{d,22}, C^p_{d,23}, C^p_{d,31}, C^p_{d,32}$, and $C^p_{d,33}$.
`xp(30)`	$\det\mathbf{C}^p_d$	Determinant of $\mathbf{C}^p_d$

Table 10.4 Sample data for the input file `input.txt`

Line	Data description
1	`/RDGRD/`
2	`2`
3	`1 1 100`
4	`1.0000000E+00 5.0000000E-02 0.0000000E+00`
5	`0.0000000E+00 1.0000000E+00 0.0000000E+00`
6	`0.0000000E+00 0.0000000E+00 1.0000000E+00`
7	`2 101 200`
8	`1.0000000E+00 -5.0000000E-02 0.0000000E+00`
9	`0.0000000E+00 1.0000000E+00 0.0000000E+00`
10	`0.0000000E+00 0.0000000E+00 1.0000000E+00`
11	`/MATER/`
12	`1.2120E+02 8.0700E+01 0.0000E+00 5.0000E+00 1.0000E-01`
13	`5.0000E-01 0.0000E+00 4.2860E-01 2.4510E+00 0.0000E+00`
14	`0.0000E+00 1.7500E-01 1.8000E-01 0.0000E+00 0.0000E+00`
15	`0.0000E+00 0.0000E+00 0.0000E+00 0.0000E+00 0.0000E+00`
16	`/STRES/`
17	`1 1 0.0000000E+00 0.0000000E+00 0.0000000E+00` *(continued)*
	`0.0000000E+00 0.0000000E+00 0.0000000E+00` *(continued)*
	`0.0000000E+00 0.0000000E+00 0.0000000E+00` *(continued)*
	`0.0000000E+00 0 0`
18	`/PLAST/`
19	`1 1 0.0000000E+00 0.0000000E+00 0.0000000E+00` *(continued)*
	`0.0000000E+00 0.0000000E+00 0.0000000E+00` *(continued)*
	`0.0000000E+00 0.0000000E+00 0.0000000E+00` *(continued)*
	`0.0000000E+00 1.0000000E+00 0.0000000E+00` *(continued)*
	`0.0000000E+00 0.0000000E+00 1.0000000E+00` *(continued)*
	`0.0000000E+00 0.0000000E+00 0.0000000E+00` *(continued)*
	`1.0000000E+00 1.0000000E+00 1.0000000E+00` *(continued)*
	`0.0000000E+00 0.0000000E+00 0.0000000E+00` *(continued)*
	`1.0000000E+00 0.0000000E+00 0.0000000E+00` *(continued)*
	`0.0000000E+00 1.0000000E+00 1.0000000E+00`

For the reader's convenience, sample data for a reversal simple shear analysis is also provided in Table 10.4. In this sample data, the first loading set (`kstep=1`) prescribes a simple shear deformation by 100 incremental steps (from the step number `istep=nbgn(1)=1` to `istep=nend(1)=100`), followed by the second loading set (`kstep=2`) which imposes a reversal simple shear deformation by 100 incremental steps (from the step number `istep=nbgn(2)=101` to `istep=nend(2)=200`).

10.2 Output File Description

The Fortran program `returnmap` produces two output files: `output1.txt` and `logfile.txt`. The file `output1.txt` contains the result of the deformation, the stresses, and the plastic

Table 10.5 Contents and data structure of `output1.txt`. For tensorial quantities, the Cartesian components are outputted to this file

Column	Format	Variable	Symbol	Description
1	i8	istep	n	Step number
2	i8	idepn	—	Elastic/plastic identifier (0, elastic; 1, plastic)
3	es15.7	djn	J	Total volume change
4	es15.7	djen	J^e	Elastic volume change
5	es15.7	djpn	J^p	Plastic volume change
6–14	9es15.7	ftn(1,1),ftn(1,2),...	$\mathbf{F}$	Deformation gradient tensor
15–23	9es15.7	sgn(1,1),sgn(1,2),...	σ	Cauchy stress tensor
24–53	30es15.7	xpn(1),...,xpn(30)	—	Plastic internal variables (see Table 10.21)

internal variables. The contents and data structure of `output1.txt` are explained in Table 10.5. The first line is the header, followed by the results computed for each incremental loading step. Note that for tensorial quantities, the Cartesian components are outputted in `output1.txt`. The convergence profile of the iterative solution procedure involved in the return-mapping described in Section 9.7 as well as other messages generated by the program during the computation are written into the file `logfile.txt`, whose data structure is shown in Table 10.6.

10.3 Computer Programs

10.3.1 Structure of Fortran Program `returnmap`

The Fortran program `returnmap` consists of a main routine and seven subroutines, as listed in this subsection. The subroutine `tgnum0` to evaluate the consistent tangent modulus tensor is also listed here, although it is not required in the program `returnmap`. The complete source code of the program `returnmap` is available, together with a sample data, from the author's website at the URL given in Section 10.1.

In order to provide an overview of the structure of the program `returnmap`, all the routines that make up the program are listed in Figure 10.1 as a tree diagram. The numbers of the tables

Table 10.6 Data structure of `logfile.txt`, into which the convergence profile of the iterative solution procedure in the return-mapping for each incremental step is outputted. Other messages generated by the program may be written into this file (if any)

Line	Format	Variable	Symbol	Description
1	'Load step:',*	istep	n	Step number
it	i5,5es12.4	it,	k,	Iteration number
		resy,	$\Vert\mathbf{Y}^{(k)}\Vert/\Vert\mathbf{Y}^{(0)}\Vert$,	Normalized residual norm
		res1,res2,	$\Vert\mathbf{Y}_{\mathbf{C}^{p(k)}}\Vert$, $\Vert\mathbf{Y}_{\mathbf{C}_d^{p(k)}}\Vert$,	Norms or absolute values of the residuals (see $(9.54)_2$ and (9.59))
		res3,res4	$\vert Y_{H^{(k)}}\vert$, $\vert Y_{f_{\rm sub}^{(k)}}\vert$	

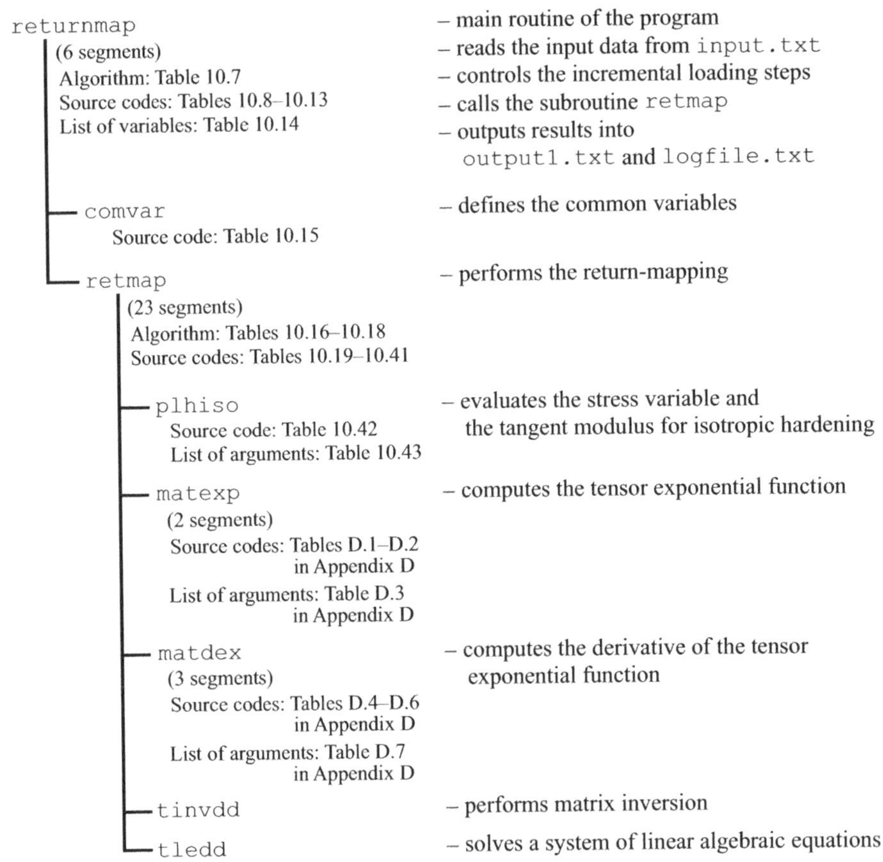

Figure 10.1 Structure of the program `returnmap`

in which the algorithm, the source codes, and the list of variables and arguments are shown are also listed in the tree diagram. Each routine is provided as a separate file with the same name as the name of the routine, with an added `.f` extension.

- Main routine `returnmap`: The main routine reads the input data from `input.txt`, controls the incremental loading steps, carries out the stress-update computation calling a subroutine for the return-mapping procedure, named `retmap`, and outputs the results into `output1.txt` and `logfile.txt`.
- Subroutine `comvar`: This subroutine defines the common variables used throughout the program.
- Subroutine `retmap`: This subroutine performs the stress-update computation based on the return-mapping scheme.
- Subroutine `plhiso`: This subroutine evaluates the stress variable and tangent modulus for isotropic hardening.

- Subroutine `matexp`: This subroutine computes the tensor exponential function required in the stress-update algorithm. The closed-form formula to evaluate the tensor exponential function is described in Appendix D.1, and the source code of the program `matexp` is presented in Appendix D.2.
- Subroutine `matdex`: This subroutine computes the derivative of the tensor exponential function required in the evaluation of the Jacobian matrix for the iterative procedure for solution of the nonlinear simultaneous equations involved in the return-mapping. The closed-form formula to evaluate the derivative of the tensor exponential function is described in Appendix D.3, and the source code of the program `matdex` is presented in Appendix D.4.
- Subroutine `tinvdd`: This subroutine, coded by Masato Sakata (former professor at Akita University), and provided as a FORTRAN77 library for ACOS SERIES 77 NEAC SYSTEM in the computer center of Tohoku University, carries out matrix inversion by means of the sweep-out method. The source code of `tinvdd` is not shown in this chapter, but it can be downloaded together with the source codes of all the other routines.
- Subroutine `tledd`: This subroutine, also coded by Masato Sakata, and provided as a FORTRAN77 library for ACOS SERIES 77 NEAC SYSTEM in the computer center of Tohoku University, solves a system of linear algebraic equations by means of the elimination method. The source code of `tledd` is not shown in this chapter, but it can be downloaded together with the source codes of all the other routines.
- Subroutine `tgnum0`: This subroutine evaluates the consistent (algorithmic) tangent modulus tensor in the description relative to the reference configuration by means of numerical differentiation as presented in Section 9.9.2. Although the consistent tangent modulus tensor is not actually required in the stress-update computation, and this subroutine is not required in the program `returnmap`, this subroutine is provided for the reader interested in attempting to implement the return-mapping scheme in analysis programs for elasto-plastic boundary value problems.

In the remainder of this section, the source codes of the programs `returnmap`, `comvar`, `retmap`, `plhiso`, and `tgnum0`, together with the lists of variables used in these routines, are presented. Furthermore, a line-by-line explanation of the source codes, showing the correspondence between the line numbers of the source code and equations in this book, is provided.

10.3.2 Main Routine of Program `returnmap`

The main routine of the program `returnmap` comprises six segments. The main role of this main routine is the control of the analysis according to the prescription of the incremental loading steps in `input.txt`. The algorithm and procedures in the main routine `returnmap` are summarized in Table 10.7, and the source codes of the segments are presented in Tables 10.8–10.13. The list of variables declared in the main routine and used throughout the program `returnmap` is shown in Table 10.14.

10.3.3 Subroutine to Define Common Variables: `comvar`

A simple subroutine `comvar` defines the variables that are commonly used throughout the program; see Table 10.14 for their definitions. The variables defined in the subroutine `comvar` are stored in the common block named `/etc/`.

Table 10.7 Algorithm and procedures in the main routine `returnmap` and the equations corresponding to the procedures

Segment Line	Table number of source code Description of procedure	Corresponding equation
Segment 1	**(Table 10.8)**	
1–17	Subroutine name, arguments, and variable declarations.	
20	Define the common variables by calling the subroutine `comvar`.	
23–25	Open the input and output files.	
Segment 2	**(Table 10.9)**	
28–62	Read the loading set number `mstep`, beginning step number `nbgn(kstep)`, end step number `nend(kstep)`, and the relative deformation gradient tensor $\mathbf{f}_{[n,n+1]}$ for the `kstep`th loading set.	
Segment 3	**(Table 10.10)**	
64–71	Read the material constants from the entry `/MATER/` of the file `input.txt`; see Table 10.2.	
74–82	Read the Cartesian components of the Cauchy stress tensor $\boldsymbol{\sigma}$ and the elastic/plastic identifier at the reference state from the entry `/STRES/` of the file `input.txt`.	
85–89	Read the plastic internal variables at the reference state from the entry `/PLAST/` of the file `input.txt`; see Table 10.3.	
92–98	Define $\mathbf{F}$, $\mathbf{F}^e$, $\mathbf{F}^p$, J, J^e, and J^p at the reference state ($\mathbf{F}^e$ and $\mathbf{F}^p$ are not used in the actual computation in this program).	
Segment 4	**(Table 10.11)**	
101–114	Output the header of the data to the file `output1.txt`.	
116–129	`format` statement for the header of the output data.	
131–139	Output the reference values to the file `output1.txt`.	
Segment 5	**(Table 10.12)**	
142	Begin the analysis of the loading step (`do 100`-loop, to line 209).	
145–158	Specify the loading set number `kstep` and the relative deformation gradient tensor $\mathbf{f}_{[n,n+1]}$.	
161	Update the total deformation gradient tensor by $\mathbf{F}_{n+1} = \mathbf{f}_{[n,n+1]}\mathbf{F}_n$.	(9.37)
164–169	Compute the total volume change $J_{n+1} = \det\mathbf{F}_{n+1}$.	
172	Initialize the incremental plastic multiplier as $\Delta\lambda = 0$.	
175	Output the step number `istep`.	
177–180	Perform the return-mapping by calling the subroutine `retmap`.	
Segment 6	**(Table 10.13)**	
183–192	Output the analysis result to the file `output1.txt`.	
195–206	Update $\mathbf{F}_{n+1}$, $\mathbf{F}^e_{n+1}$, $\mathbf{F}^p_{n+1}$, $\boldsymbol{\sigma}_{n+1}$, J_{n+1}, J^e_{n+1}, J^p_{n+1}, and the plastic internal variables.	
209	Go to the next loading step (`do 100`-loop, line 142).	
212–217	Close the input and output files.	
220–221	`stop` and `end` statements.	

Table 10.8 Main routine `returnmap`, segment 1: arguments and variable declaration

Program code

```
      program returnmap
      parameter ( lstep = 10 )
      implicit double precision (a-h,o-z)
      dimension nbgn(lstep),nend(lstep),frx(lstep,3,3)
      dimension prope(20)
      dimension fr(3,3)
      dimension ft(3,3),fe(3,3),fp(3,3),sg(3,3),xp(30)
      dimension ftn(3,3),fen(3,3),fpn(3,3),sgn(3,3),xpn(30)
      dimension cep(3,3,3,3)
      character*7 charin
      common /iofile/ior1,ior2,ior3,ior4,ior5,
     &               iow1,iow2,iow3,iow4,iow5,
     &               iow6,iow7,iow8,iow9,iow10
      data ior1,ior2,ior3,ior4,ior5 / 1, 2, 3, 4, 5/
     &     iow1,iow2,iow3,iow4,iow5 /11,12,13,14,15/
     &     iow6,iow7,iow8,iow9,iow10/16,17,18,19,20/
      common /etc/del(3),delta(3,3),p(3,3,3),dpai
c
c ----- define common block /etc/
      CALL comvar
c
c ----- open files
      open(ior1,file='input.txt',status='old')
      open(iow1,file='logfile.txt',status='unknown')
      open(iow2,file='output1.txt',status='unknown')
```

(continued)

10.3.4 Subroutine for Return-Mapping: `retmap`

The subroutine `retmap`, which consists of 23 segments, plays a crucial role in the program `returnmap`, performing the stress-update computation on the basis of the algorithm comprising the elastic predictor step and the plastic corrector step by means of the return-mapping scheme. A line-by-line explanation of the source code of the subroutine `retmap`, showing the correspondence between the line of the source code and the equations presented in Chapter 9, is given in Tables 10.16–10.18. The source codes of the segments are shown in Tables 10.19–10.41. A detailed explanation of the algorithm for the stress-update computation was given in Chapter 9. The subroutine `retmap` calls the subroutines `plhiso`, `matexp`, `matdex`, `tinvdd`, and `tledd`. Since all the arguments of the subroutine `retmap` have already been listed in Table 10.14, the list of arguments is omitted to avoid duplication.

Table 10.9 Main routine `returnmap`, segment 2: read input data from the file `input.txt` (1)

Line Program code

```
c
c ----- read relative deformation gradient for each loading step
         read(ior1,'(a)') charin
         if(charin.eq.'/RDGRD/') then
          read(ior1,540) mstep
   540  format(i5)
          if(mstep.gt.lstep) then
           write(*,*) '--- ERROR. Increase 'lstep'' parameter.'
           stop
          end if
          ichk1 = 0
          do kstep=1,mstep
           read(ior1,541) jstep,nbgn(kstep),nend(kstep)
   541   format(i5,2i8)
            if(jstep.eq.kstep) then
             ichk1 = ichk1+1
            end if
            read(ior1,542) frx(kstep,1,1),frx(kstep,1,2),frx(kstep,1,3)
            read(ior1,542) frx(kstep,2,1),frx(kstep,2,2),frx(kstep,2,3)
            read(ior1,542) frx(kstep,3,1),frx(kstep,3,2),frx(kstep,3,3)
   542   format(3es15.7)
           end do
           ichk2 = 0
           do kstep=1,mstep-1
            if((nend(kstep)+1).eq.nbgn(kstep+1)) then
             ichk2 = ichk2+1
            end if
           end do
           if(     (ichk1.eq.mstep).and.(ichk2.eq.(mstep-1))
     &     .and.(nbgn(1).eq.1) ) then
           else
            write(*,*) '--- invalid loading step setting.'
            stop
           end if
           nstep = nend(mstep)
          end if
c
```

(continued)

Table 10.10 Main routine `returnmap`, segment 3: read input data from the file `input.txt` (2)

Program code

```
c ----- material constants
        read(ior1,'(a)') charin
        if(charin.eq.'/MATER/') then
         read(ior1,550) (prope(k),k= 1, 5)
         read(ior1,550) (prope(k),k= 6,10)
         read(ior1,550) (prope(k),k=11,15)
         read(ior1,550) (prope(k),k=16,20)
  550   format(5es12.4)
        end if
c
c ----- Cauchy stress at reference state
        read(ior1,'(a)') charin
        if(charin.eq.'/STRES/') then
         read(ior1,551) idum,idum,
     &                  sg(1,1),sg(1,2),sg(1,3),
     &                  sg(2,1),sg(2,2),sg(2,3),
     &                  sg(3,1),sg(3,2),sg(3,3),
     &                  dum,idep,idum
  551   format(i8,i5,10es15.7,2i5)
        end if
c
c ----- plastic internal variables at reference state
        read(ior1,'(a)') charin
        if(charin.eq.'/PLAST/') then
         read(ior1,552) idum,idum,(xp(k),k=1,30)
  552   format(i8,i5,30es15.7)
        end if
c
c ----- initialize deformation gradient and volume change
        ft(1:3,1:3) = delta(1:3,1:3)
        fe(1:3,1:3) = delta(1:3,1:3)
        fp(1:3,1:3) = delta(1:3,1:3)
c
        dj  = 1.0d0
        dje = 1.0d0
        djp = 1.0d0
```

(continued)

Table 10.11 Main routine `returnmap`, segment 4: output header and reference values to the file `output1.txt`

Line Program code

```
c
c ----- output the reference state
         write(iow2,505) 'istep','idn',
     &                   'djn ','djen','djpn',
     &                   'ft11','ft12','ft13',
     &                   'ft21','ft22','ft23',
     &                   'ft31','ft32','ft33',
     &                   'sg11','sg12','sg13',
     &                   'sg21','sg22','sg23',
     &                   'sg31','sg32','sg33',
     &                   'xp01','xp02','xp03','xp04','xp05',
     &                   'xp06','xp07','xp08','xp09','xp10',
     &                   'xp11','xp12','xp13','xp14','xp15',
     &                   'xp16','xp17','xp18','xp19','xp20',
     &                   'xp21','xp22','xp23','xp24','xp25',
     &                   'xp26','xp27','xp28','xp29','xp30'
c
  505 format('( 1)',a5,1x,'( 2)',a3,1x,
     &'( 3)',a4,7x,'( 4)',a4,7x,'( 5)',a4,7x,
     &'( 6)',a4,7x,'( 7)',a4,7x,'( 8)',a4,7x,
     &'( 9)',a4,7x,'(10)',a4,7x,'(11)',a4,7x,
     &'(12)',a4,7x,'(13)',a4,7x,'(14)',a4,7x,
     &'(15)',a4,7x,'(16)',a4,7x,'(17)',a4,7x,
     &'(18)',a4,7x,'(19)',a4,7x,'(20)',a4,7x,
     &'(21)',a4,7x,'(22)',a4,7x,'(23)',a4,7x,
     &'(24)',a4,7x,'(25)',a4,7x,'(26)',a4,7x,'(27)',a4,7x,'(28)',a4,7x,
     &'(29)',a4,7x,'(30)',a4,7x,'(31)',a4,7x,'(32)',a4,7x,'(33)',a4,7x,
     &'(34)',a4,7x,'(35)',a4,7x,'(36)',a4,7x,'(37)',a4,7x,'(38)',a4,7x,
     &'(39)',a4,7x,'(40)',a4,7x,'(41)',a4,7x,'(42)',a4,7x,'(43)',a4,7x,
     &'(44)',a4,7x,'(45)',a4,7x,'(46)',a4,7x,'(47)',a4,7x,'(48)',a4,7x,
     &'(49)',a4,7x,'(50)',a4,7x,'(51)',a4,7x,'(52)',a4,7x,'(53)',a4,7x)
c
         write(iow2,512) 0,idep,
     &                   dj,dje,djp,
     &                   ft(1,1),ft(1,2),ft(1,3),
     &                   ft(2,1),ft(2,2),ft(2,3),
     &                   ft(3,1),ft(3,2),ft(3,3),
     &                   sg(1,1),sg(1,2),sg(1,3),
     &                   sg(2,1),sg(2,2),sg(2,3),
     &                   sg(3,1),sg(3,2),sg(3,3),
     &                   (xp(k),k=1,30)
```

(continued)

Table 10.12 Main routine `returnmap`, segment 5: carry out the return-mapping

Program code

```
c
c ----- incremental loading steps [do-loop: 100]
        do 100 istep=1,nstep
c
c ----- define relative deformation gradient
         ichk = 0
         do kstep=1,mstep
          kbgn = nbgn(kstep)
          kend = nend(kstep)
          if((kbgn.le.istep).and.(istep.le.kend)) then
           fr(1:3,1:3) = frx(kstep,1:3,1:3)
           ichk = ichk+1
          end if
         end do
c
         if(ichk.ne.1) then
          write(*,*) '--- invalid loading step setting.'
          stop
         end if
c
c ----- update total deformation gradient
         ftn(1:3,1:3) = matmul(fr(1:3,1:3),ft(1:3,1:3))
c
c ----- compute determinant of updated total deformation gradient
        djn =  ftn(1,1)*ftn(2,2)*ftn(3,3)
      &     + ftn(1,3)*ftn(2,1)*ftn(3,2)
      &     + ftn(3,1)*ftn(1,2)*ftn(2,3)
      &     - ftn(1,3)*ftn(2,2)*ftn(3,1)
      &     - ftn(1,1)*ftn(2,3)*ftn(3,2)
      &     - ftn(3,3)*ftn(1,2)*ftn(2,1)
c
c ----- initialize the plastic multiplier
        dp = 0.0d0
c
c ----- return-mapping algorithm
        write(iow1,*) 'Load step: ',istep
c
        CALL retmap(inel,ige,mtypee,prope,
      &            fr,djr,
      &            ft,dj,fe,dje,fp,djp,sg,idep,xp,
      &            ftn,djn,fen,djen,fpn,djpn,sgn,idepn,xpn)
```

(continued)

Table 10.13 Main routine `returnmap`, segment 6: output analysis result to the file `output1.txt` and go to the next step

```
c
c ----- output updated result
        write(iow2,512) istep,idepn,
     &                  djn,djen,djpn,
     &                  ftn(1,1),ftn(1,2),ftn(1,3),
     &                  ftn(2,1),ftn(2,2),ftn(2,3),
     &                  ftn(3,1),ftn(3,2),ftn(3,3),
     &                  sgn(1,1),sgn(1,2),sgn(1,3),
     &                  sgn(2,1),sgn(2,2),sgn(2,3),
     &                  sgn(3,1),sgn(3,2),sgn(3,3),
     &                  (xpn(k),k=1,30)
  512 format(2i8,3es15.7,9es15.7,9es15.7,30es15.7)
c
c ----- update variables
        ft(1:3,1:3) = ftn(1:3,1:3)
        fe(1:3,1:3) = fen(1:3,1:3)
        fp(1:3,1:3) = fpn(1:3,1:3)
        sg(1:3,1:3) = sgn(1:3,1:3)
c
        dj  = djn
        dje = djen
        djp = djpn
c
        xp(1:30) = xpn(1:30)
c
        idep = idepn
c
c ----- end of loading steps [loop = 100]
    100 continue
c
c ----- close files
        close(ior1)
        close(iow1)
        close(iow2)
        close(iow3)
        close(iow4)
        close(iow5)
c
c ----- end of program
        stop
        end program returnmap
```

Table 10.14 List of variables used throughout the program `returnmap`. Local variables used merely in the main routine are not listed in this table

Variable	Symbol	Description
`lstep`	—	Maximum number of loading sets
`mstep`	—	TNumber of loading sets
`kstep`	—	Loading set number
`nbgn(kstep)`	—	Step number at the beginning of `kstep`th loading set
`nend(kstep)`	—	Step number at the end of `kstep`th loading set
`frx(kstep,3,3)`	$\mathbf{f}_{[n,n+1]}$	Relative deformation gradient tensor within a step for `kstep`th loading set
`prope(20)`	—	Material constants (see Table 10.2)
`fr(3,3)`	$\mathbf{f}_{[n,n+1]}$	Relative deformation gradient tensor within a step in a loading set
`djr`	$j_{[n,n+1]}$	Relative volume change within a step ($j_{[n,n+1]} := \det\mathbf{f}_{[n,n+1]} = \det\mathbf{F}_{n+1}/\det\mathbf{F}_n = J_{n+1}/J_n$)
`ft(3,3)`	$\mathbf{F}_n$	Total deformation gradient tensor at t_n
`fe(3,3)`	$\mathbf{F}^e_n$	Elastic deformation gradient tensor at t_n (not used)
`fp(3,3)`	$\mathbf{F}^p_n$	Plastic deformation gradient tensor at t_n (not used)
`dj`	J_n	Total volume change at t_n
`dje`	J^e_n	Elastic volume change at t_n
`djp`	J^p_n	Plastic volume change at t_n
`sg(3,3)`	σ_n	Cauchy stress tensor at t_n
`idep`	—	Elastic/plastic identifier for the preceding step (0, elastic; 1, plastic)
`xp(30)`	—	Array of plastic internal variables at t_n (see Table 10.3)
`ftn(3,3)`	$\mathbf{F}_{n+1}$	Total deformation gradient tensor at t_{n+1}
`fen(3,3)`	$\mathbf{F}^e_{n+1}$	Elastic deformation gradient tensor at t_{n+1} (not used)
`fpn(3,3)`	$\mathbf{F}^p_{n+1}$	Plastic deformation gradient tensor at t_{n+1} (not used)
`djn`	J_{n+1}	Total volume change at t_{n+1}
`djen`	J^e_{n+1}	Elastic volume change at t_{n+1}
`djpn`	J^p_{n+1}	Plastic volume change at t_{n+1}
`sgn(3,3)`	σ_{n+1}	Cauchy stress tensor at t_{n+1}
`idepn`	—	Elastic/plastic identifier for the current step (0, elastic; 1, plastic)
`xpn(30)`	—	Array of plastic internal variables at t_{n+1} (see Table 10.3)
`cep(3,3,3,3)`	$\mathbb{C}^{ep,\text{algo}}_{n+1}$	Consistent tangent modulus tensor in the description relative to the reference configuration (evaluated by `tgnum0`, but not used in the stress-update computation)
`del(3)`	$\mathbf{1}$	Vector with its Cartesian components $(\mathbf{1})_i = 1$ for $i = 1, 2, 3$ (not used in the computation in `returnmap`)
`delta(3,3)`	$\mathbf{I}$	Cartesian components of the identity tensor $(\mathbf{I})_{ij} = \delta_{ij}$, that is the Kronecker delta
`p(3,3,3)`	ϵ_{ijk}	Permutation symbol (not used in the computation in `returnmap`)
`dpai`	π	The ratio of the circumference of a circle to its diameter
`inel`	—	Provided for possible use in finite element codes to specify the element number (not used and arbitrary in `returnmap`)
`ige`	—	Provided for possible use in finite element codes to specify the Gaussian integration point number (not used and arbitrary in `returnmap`)
`mtypee`	—	Identifier of material type (not used and arbitrary in `returnmap`)
`ior1,...,ior5`	—	Logical unit numbers for input files (only `ior1` is used for the input file `input.txt`)
`iow1,...,iow10`	—	Logical unit numbers for output files (only `iow1` and `iow2` are used for the output files `logfile.txt` and `output1.txt`, respectively)
`dum`	—	Dummy variable for double precision real numbers
`idum`	—	Dummy variable for integers

Table 10.15 Subroutine `comvar`: define the common variables

```
      subroutine comvar
      implicit double precision (a-h,o-z)
      common /iofile/ior1,ior2,ior3,ior4,ior5,
     &               iow1,iow2,iow3,iow4,iow5,
     &               iow6,iow7,iow8,iow9,iow10
      common /etc/del(3),delta(3,3),p(3,3,3),dpai
c
c ----- define del(3) and Kronecker delta(3,3) (common variables)
      del(1:3) = 1.0d0
c
      do i=1,3
       do j=1,3
        if(i.eq.j) then
         delta(i,j) = 1.0d0
        else if(i.ne.j) then
         delta(i,j) = 0.0d0
        end if
       end do
      end do
c
c ----- define permutation p(i,j,k) (common variables)
      p(1:3,1:3,1:3) = 0.0d0
c
      p(1,2,3) =  1.0d0
      p(3,1,2) =  1.0d0
      p(2,3,1) =  1.0d0
      p(3,2,1) = -1.0d0
      p(1,3,2) = -1.0d0
      p(2,1,3) = -1.0d0
c
c ----- define dpai=3.14159265... (common variables)
      dpai = 4.0d0*datan(1.0d0)
c
      return
      end subroutine comvar
```

Table 10.16 Algorithm of the subroutine `retmap` and the equations corresponding to the procedures (1 of 3)

Segment Line	Table number of source code Description of procedure	Corresponding equation
Segment 1	**(Table 10.19)**	
1–36	Subroutine name, arguments, and variable declarations.	
Segment 2	**(Table 10.20)**	
69–88	Set the material constants; see Table 10.2.	
Segment 3	**(Table 10.21)**	
128–157	Plastic internal variables at t_n, which are used as the trial values $H_{n+1}^{(\mathrm{tr})}$, $R_{n+1}^{(\mathrm{tr})}$, $\mathbf{C}_{n+1}^{p(\mathrm{tr})}$, and $\mathbf{C}_{d,n+1}^{p(\mathrm{tr})}$; see Table 10.3.	(9.40), (9.41), (9.44), and (9.46)
Segment 4	**(Table 10.22)**	
161–165	Compute $\mathbf{C}_{n+1}$ and its inverse $\mathbf{C}_{n+1}^{-1}$.	
168–173	Compute inverse tensors $(\mathbf{C}_{n+1}^{p(\mathrm{tr})})^{-1}$ and $(\mathbf{C}_{d,n+1}^{p(\mathrm{tr})})^{-1}$.	
177–183	Compute the trial stress tensors $\mathbf{S}_{n+1}^{(\mathrm{tr})}$ and $\mathbf{M}_{n+1}^{(\mathrm{tr})}$ by using the hyperelastic equation.	(9.15) and $(9.18)_1$
188–193	Compute the trial stress tensors $\mathbf{S}_{e,n+1}^{p(\mathrm{tr})}$ and $\mathbf{M}_{e,n+1}^{p(\mathrm{tr})}$ by using the hyperelastic-type constitutive equation for kinematic hardening.	(9.16) and $(9.18)_2$
197–199	Compute the trial reduced stress tensor $\hat{\mathbf{M}}_{n+1}^{(\mathrm{tr})}$, and its mean normal and deviatoric parts.	(9.19)
204–212	Compute $\sqrt{\hat{\mathbf{M}}_{n+1}^{(\mathrm{tr})\prime} : (\hat{\mathbf{M}}_{n+1}^{(\mathrm{tr})\prime})^T}$.	(9.17)
215	Evaluate the trial stress-like variable for isotropic hardening $F_{n+1}^{(\mathrm{tr})}$ by calling the subroutine `plhiso`.	(9.5)
Segment 5	**(Table 10.23)**	
218–219	Evaluate the subloading surface $f_{\mathrm{sub},n+1}^{(\mathrm{tr})}$ and the elasticity limit surface $f_{\mathrm{elast},n+1}^{(\mathrm{tr})}$ for the elastic trial state.	(9.48) and (9.49)
221–227	Judgment of loading criterion. (Elastic case: `go to 100`, line 230; Plastic case: `go to 110`, line 268)	(9.47)
Segment 6	**(Table 10.24)**	
230	Begin elastic case (`100 continue`)	
233–262	The elastic trial values are accepted as the updated values. We use $J_{n+1}^e = J_{n+1}/J_{n+1}^{p(\mathrm{tr})} = J_{n+1}/J_n^p$ and the push-forward operation $\sigma_{n+1} = (\mathbf{F}_{n+1}\mathbf{S}_{n+1}\mathbf{F}_{n+1}^T)/J_{n+1}$.	(9.50)
255	Update the normal-yield ratio R_{n+1} for the subloading surface.	(9.51)
265	End elastic case (`go to 199`, line 847).	
Segment 7	**(Table 10.25)**	
268	Begin plastic case (`110 continue`)	
271–298	Initial setting for the Newton–Raphson iteration. The initial values are set equal to the elastic trial values. The ‘reduced’ Newton–Raphson method is employed, if needed.	
Segment 8	**(Table 10.26)**	
301–302	Initialize counters for the full and reduced Newton–Raphson iterations.	
303	Begin the Newton–Raphson iteration (`900 continue`, from line 815).	
306–311	Compute $\mathbf{C}_{n+1}^{p-1}$ and $\mathbf{C}_{d,n+1}^{p-1}$.	
315–320	Compute the updated stresses $\mathbf{S}_{n+1}$ and $\mathbf{M}_{n+1}$. by using the hyperelastic equation.	(9.15) and $(9.18)_1$
325–337	Compute the derivative $\partial\mathbf{M}_{n+1}/\partial\mathbf{C}_{n+1}^p$.	(9.91)
	(continued)	

Table 10.17 Algorithm of the subroutine `retmap` and the equations corresponding to the procedures (2 of 3)

Segment Line	Table number of source code Description of procedure	Corresponding equation
Segment 9	**(Table 10.27)**	
342	Compute the updated stress for kinematic hardening $\mathbf{S}^p_{e,n+1}$.	(9.16)
348–350	Convert $\mathbf{S}^p_{e,n+1}$ to $\mathbf{M}^p_{e,n+1}$ and compute its mean normal and deviatoric parts.	$(9.18)_2$
356–368	Compute the derivatives $\partial\mathbf{M}^p_{e,n+1}/\partial\mathbf{C}^p_{n+1}$ and $\partial\mathbf{M}^p_{e,n+1}/\partial\mathbf{C}^p_{d,n+1}$.	(9.92) and (9.93)
Segment 10	**(Table 10.28)**	
372–374	Compute the updated reduced stress $\hat{\mathbf{M}}_{n+1}$ and its mean normal and deviatoric parts.	(9.19)
379–387	Compute $\sqrt{\hat{\mathbf{M}}'_{n+1} : (\hat{\mathbf{M}}'_{n+1})^T}$.	(9.17)
395	Evaluate the updated stress-like variable F_{n+1} and tangent modulus F'_{n+1} for isotropic hardening by calling the subroutine `plhiso`.	(9.5) and (9.6)
Segment 11	**(Table 10.29)**	
400–413	Update the normal-yield ratio R_{n+1} for the subloading surface.	(9.36)
417–427	Compute the tensor exponentials $\mathbf{Q}^p_{n+1}$ and $\mathbf{Q}^p_{e,n+1}$ by calling the subroutine `matexp`.	(9.75) and (9.76)
430–435	Evaluate the residuals $\mathbf{Y}_{\mathbf{C}^p}$, $\mathbf{Y}_{\mathbf{C}^p_d}$, Y_H, and $Y_{f_{\rm sub}}$ for the initial values.	(9.53)
Segment 12	**(Table 10.30)**	
437–450	Assemble the residual vector $\mathbf{Y}$, noting the symmetry property of $\mathbf{C}^p$ and $\mathbf{C}^p_d$.	$(9.54)_2$
453–465	Evaluate the residual norms.	
Segment 13	**(Table 10.31)**	
469–480	Check convergence rate. The reduced Newton–Raphson iteration with a reduction coefficient `dk` is employed, if needed (`go to 800`, line 789).	
482	Exit the reduced Newton–Raphson iteration (`899 continue`, from line 474).	
485–504	Check convergence and output the residual norm. If converged, exit (`go to 990`, line 818).	(9.59)
Segment 14	**(Table 10.32)**	
511–516	Compute the derivatives of the subloading surface $\partial f_{\rm sub}/\partial\mathbf{M}$, $\partial f_{\rm sub}/\partial\mathbf{M}^p_e$, $\partial f_{\rm sub}/\partial F$, and $\partial f_{\rm sub}/\partial R$.	(9.74)
523–537	Compute the derivatives of $\mathbf{Z}^p$ and $\mathbf{Z}^p_e$ with respect to $\hat{\mathbf{M}}$ and $\mathbf{M}^p_e$, respectively. Note that $\Delta\lambda$, assigned to `dp`, is not yet multiplied, though multiplied at lines 572 and 574.	(9.81) and (9.88)
543–553	Compute the derivative $\partial R_{n+1}/\partial(\Delta\lambda)$.	(9.73)
Segment 15	**(Table 10.33)**	
557–560	Compute the derivatives of the tensor exponentials $\exp[\mathbf{Z}^p]$ and $\exp[\mathbf{Z}^p_e]$ by calling the subroutine `matdex`.	(D.4)
563–602	Compute the derivatives of $\mathbf{Q}^p$ with respect to $\mathbf{C}^p$ and $\mathbf{C}^p_d$.	(9.77) and (9.78)
Segment 16	**(Table 10.34)**	
605–644	Compute the derivatives of $\mathbf{Q}^p_e$ with respect to $\mathbf{C}^p$ and $\mathbf{C}^p_d$.	(9.84) and (9.85)

(continued)

Table 10.18 Algorithm of the subroutine `retmap` and the equations corresponding to the procedures (3 of 3)

Segment Line	Table number of source code Description of procedure	Corresponding equation
Segment 17	**(Table 10.35)**	
649–665	Compute the derivatives of $\mathbf{Y}_{\mathbf{C}^p}$.	(9.60)–(9.63)
668–684	Compute the derivatives of $\mathbf{Y}_{\mathbf{C}^p_d}$.	(9.64)–(9.67)
Segment 18	**(Table 10.36)**	
687–690	Compute the derivatives of Y_H.	(9.68)
693–708	Compute the derivatives of $Y_{f_{\text{sub}}}$.	(9.69)–(9.72)
Segment 19	**(Table 10.37)**	
711–758	Assemble the Jacobian matrix $\mathbf{J}$, noting the symmetry property of $\mathbf{C}^p$ and $\mathbf{C}^p_d$.	(9.57)
Segment 20	**(Table 10.38)**	
761–766	Compute the corrector vector $\delta\mathbf{X}$ by solving the system of simultaneous equations for return-mapping.	$(9.58)_1$
770–786	The corrector vector $\delta\mathbf{X}$ is stored before the reduced Newton–Raphson iteration.	
Segment 21	**(Table 10.39)**	
789	Loop `800` (from line 479): reduced Newton–Raphson iteration.	
791–812	Update the unknown variables $\mathbf{C}^p_{n+1}$, $\mathbf{C}^p_{d,n+1}$, H_{n+1}, and $\Delta\lambda$.	$(9.54)_3$ and $(9.58)_2$
815	Go to the next iteration (`go to 900`, line 303).	
818	End the iteration (`990 continue`, from lines 495 and 501).	
820–824	Check $\Delta\lambda > 0$. If $\Delta\lambda \not> 0$, stop.	
Segment 22	**(Table 10.40)**	
833–834	Update the elastic and plastic volume changes. We use $J^e_{n+1} = J_{n+1}/J^p_{n+1}$.	
839–841	Compute the updated Cauchy stress tensor by the push-forward operation $\boldsymbol{\sigma}_{n+1} = (\mathbf{F}_{n+1}\mathbf{S}_{n+1}\mathbf{F}^T_{n+1})/J_{n+1}$.	
844	End plastic case (`go to 199`, line 847).	
879	End elastic and plastic cases (`199`, from lines 265 and 844).	
Segment 23	**(Table 10.41)**	
850–879	Update the plastic internal variables; see Table 10.3.	
881–882	`return` and `end` statements.	

Table 10.19 Subroutine `retmap`, segment 1: arguments and variable declaration

Program code

```
      subroutine retmap(inel,ige,mtypee,prope,
     &                  fr,djr,
     &                  ft,dj,fe,dje,fp,djp,sg,idep,xp,
     &                  ftn,djn,fen,djen,fpn,djpn,sgn,idepn,xpn)
      implicit double precision (a-h,o-z)
      dimension prope(20)
      dimension fr(3,3)
      dimension ft(3,3),fe(3,3),fp(3,3),sg(3,3),xp(30)
      dimension ftn(3,3),fen(3,3),fpn(3,3),sgn(3,3),xpn(30)
      dimension iwork(6)
      dimension ctn(3,3),ctninv(3,3)
      dimension cptr(3,3),cptrinv(3,3),cpdtr(3,3),cpdtrinv(3,3)
      dimension st20tr(3,3),st30tr(3,3)
      dimension sk20tr(3,3),sk30tr(3,3)
      dimension sx30tr(3,3),sx30trd(3,3)
      dimension cpn(3,3),cpninv(3,3),cpdn(3,3),cpdninv(3,3)
      dimension st20n(3,3),st30n(3,3)
      dimension sk20n(3,3),sk30n(3,3),sk30nd(3,3)
      dimension tge30cp(3,3,3,3),tgk30cp(3,3,3,3),tgk30cpd(3,3,3,3)
      dimension sx30n(3,3),sx30nd(3,3)
      dimension dfst(3,3),dfsk(3,3)
      dimension gp(3,3),gk(3,3),dgp(3,3,3,3),dgk(3,3,3,3)
      dimension q1tmp(3,3),q1exp(3,3),dq1exp(3,3,3,3)
      dimension q2tmp(3,3),q2exp(3,3),dq2exp(3,3,3,3)
      dimension dq11(3,3,3,3),dq12(3,3,3,3),dq13(3,3),dq14(3,3)
      dimension dq21(3,3,3,3),dq22(3,3,3,3),dq23(3,3),dq24(3,3)
      dimension tmp11(3,3,3,3),tmp12(3,3,3,3)
      dimension tmp21(3,3,3,3),tmp22(3,3,3,3)
      dimension y1(3,3),y2(3,3),y(14),dx(14),x1(14),aj(14,14)
      dimension aj11(3,3,3,3),aj12(3,3,3,3),aj13(3,3),aj14(3,3)
      dimension aj21(3,3,3,3),aj22(3,3,3,3),aj23(3,3),aj24(3,3)
      dimension aj31(3,3),aj32(3,3),aj41(3,3),aj42(3,3)
      common /iofile/ior1,ior2,ior3,ior4,ior5,
     &               iow1,iow2,iow3,iow4,iow5,
     &               iow6,iow7,iow8,iow9,iow10
      common /etc/del(3),delta(3,3),p(3,3,3),dpai
c
c ----- return-mapping algorithm for hyperelasto-plasticity model
c       with combined isotropic and kinematic hardening based on the
c       'dual' multiplicative decompositions
c       F = Fe * Fp and Fp = Fpe * Fpd.
c       The subloading surface concept is incorporated.
c       Tensor exponential-based algorithm is employed
c       for time integration of the plastic evolution equations.
```

(continued)

Table 10.20 Subroutine `retmap`, segment 2: material constants (comment lines 47–67 are omitted here for brevity)

Line	Program code
45	`c`
46	`c ----- material constants`
68	`c`
69	`      dlam   = prope( 1)`
70	`      dmyu   = prope( 2)`
71	`c     ------ = prope( 3)`
72	`      u0     = prope( 4)`
73	`      rse    = prope( 5)`
74	`      yld0   = prope( 6)`
75	`      plhi0  = prope( 7)`
76	`      plhi1  = prope( 8)`
77	`      plhi2  = prope( 9)`
78	`c     ------ = prope(10)`
79	`c     ------ = prope(11)`
80	`      dmyuk  = prope(12)`
81	`      plhk   = prope(13)`
82	`c     ------ = prope(14)`
83	`c     ------ = prope(15)`
84	`c     ------ = prope(16)`
85	`c     ------ = prope(17)`
86	`c     ------ = prope(18)`
87	`c     ------ = prope(19)`
88	`c     ------ = prope(20)`
89	`c`
90	`c --- to avoid division by zero`
91	`      if(dmyuk.eq.0.0d0) then`
92	`       dmyuk = 1.0d-14`
93	`      end if`

(continued)

Table 10.21 Subroutine `retmap`, segment 3: internal variables at the previous step (comment lines 96–126 are omitted here for brevity)

Line	Program code
94	`c`
95	`c ----- plastic internal variables`
127	`c`
128	`      dp          = xp( 1)`
129	`      xpitr       = xp( 2)`
130	`      rstr        = xp( 3)`
131	`c    ------      = xp( 4)`
132	`c    ------      = xp( 5)`
133	`c    ------      = xp( 6)`
134	`c    ------      = xp( 7)`
135	`c    ------      = xp( 8)`
136	`c    ------      = xp( 9)`
137	`c    ------      = xp(10)`
138	`      cptr(1,1)   = xp(11)`
139	`      cptr(1,2)   = xp(12)`
140	`      cptr(1,3)   = xp(13)`
141	`      cptr(2,1)   = xp(14)`
142	`      cptr(2,2)   = xp(15)`
143	`      cptr(2,3)   = xp(16)`
144	`      cptr(3,1)   = xp(17)`
145	`      cptr(3,2)   = xp(18)`
146	`      cptr(3,3)   = xp(19)`
147	`      djcptr      = xp(20)`
148	`      cpdtr(1,1)  = xp(21)`
149	`      cpdtr(1,2)  = xp(22)`
150	`      cpdtr(1,3)  = xp(23)`
151	`      cpdtr(2,1)  = xp(24)`
152	`      cpdtr(2,2)  = xp(25)`
153	`      cpdtr(2,3)  = xp(26)`
154	`      cpdtr(3,1)  = xp(27)`
155	`      cpdtr(3,2)  = xp(28)`
156	`      cpdtr(3,3)  = xp(29)`
157	`      djcpdtr     = xp(30)`

(continued)

Table 10.22 Subroutine `retmap`, segment 4: elastic predictor step

Program code

```
c
c ----- compute 'updated' total right Cauchy-Green deformation tensor
c       and its inverse
        ctn(1:3,1:3) = matmul(transpose(ftn(1:3,1:3)),ftn(1:3,1:3))
        ctninv(1:3,1:3) = ctn(1:3,1:3)
c
        eps = 1.0d-18
        CALL tinvdd(3,ctninv,djctn,eps,3,iwork,ill)
c
c ----- compute inverse of trial values cptr(3,3) and cpdtr(3,3)
        cptrinv(1:3,1:3)  = cptr(1:3,1:3)
        cpdtrinv(1:3,1:3) = cpdtr(1:3,1:3)
c
        eps = 1.0d-18
        CALL tinvdd(3,cptrinv,djcptr,eps,3,iwork,ill)
        CALL tinvdd(3,cpdtrinv,djcpdtr,eps,3,iwork,ill)
c
c ----- compute trial 2nd Piola-Kirchhoff stress on reference config.
c       by neo-Hookean hyperelastic model
        st20tr(1:3,1:3)
      &  = (dlam/2.0d0)*(djctn/djcptr-1.0d0)*ctninv(1:3,1:3)
      &                +dmyu*(cptrinv(1:3,1:3)-ctninv(1:3,1:3))
c
c ----- convert trial 2nd Piola-Kirchhoff stress on reference config.
c       to trial (Eshelby-like) stress on reference config.
        st30tr(1:3,1:3) = matmul(ctn(1:3,1:3),st20tr(1:3,1:3))
c
c ----- compute trial 2nd Piola-Kirchhoff-type stress variable
c       for kinematic hardening on reference config.
c       by neo-Hookean-type model (only deviatoric part)
        sk20tr(1:3,1:3) = dmyuk*(cpdtrinv(1:3,1:3)-cptrinv(1:3,1:3))
c
c ----- convert trial 2nd Piola-Kirchhoff-type stress variable
c       for kinematic hardening to trial (Eshelby-like) back stress
c       on reference config.
        sk30tr(1:3,1:3) = matmul(cptr(1:3,1:3),sk20tr(1:3,1:3))
c
c ----- compute 'reduced' (Eshelby-like) stress on reference config.
c       at trial state and its isotropic and deviatoric parts
        sx30tr(1:3,1:3) = st30tr(1:3,1:3)-sk30tr(1:3,1:3)
        sx30trv = (sx30tr(1,1)+sx30tr(2,2)+sx30tr(3,3))/3.0d0
        sx30trd(1:3,1:3) = sx30tr(1:3,1:3)-sx30trv*delta(1:3,1:3)
c
c --- NOTE: norm of deviatoric Mandel stress on intermediate config.
c           is equal to SQRT ( Sdev(i,j) * Sdev(j,i) ), where 'S'
```

(continued)

Table 10.22 (*Continued*)

Program code

```
c               denotes Eshelby-like stress on reference config.
          dnormxtr = dsqrt( sx30trd(1,1)*sx30trd(1,1)
     &                     +sx30trd(1,2)*sx30trd(2,1)
     &                     +sx30trd(1,3)*sx30trd(3,1)
     &                     +sx30trd(2,1)*sx30trd(1,2)
     &                     +sx30trd(2,2)*sx30trd(2,2)
     &                     +sx30trd(2,3)*sx30trd(3,2)
     &                     +sx30trd(3,1)*sx30trd(1,3)
     &                     +sx30trd(3,2)*sx30trd(2,3)
     &                     +sx30trd(3,3)*sx30trd(3,3) )
c
c ----- isotropic hardening stress at trial state
          CALL plhiso(xpitr,yld0,plhi0,plhi1,plhi2,sitr,tgitr)
```

(continued)

Table 10.23 Subroutine `retmap`, segment 5: judgment of loading criterion

Program code

```
c
c ----- check loading criterion
          ftrial = dsqrt(3.0d0/2.0d0)*dnormxtr-rstr*sitr
          felast = dsqrt(3.0d0/2.0d0)*dnormxtr-rse*sitr
c
          if((ftrial.le.0.0d0).or.(felast.le.0.0d0)) then
           idepn = 0
           goto 100
          else if((ftrial.gt.0.0d0).and.(felast.gt.0.0d0)) then
           idepn = 1
           goto 110
          end if
```

(continued)

Table 10.24 Subroutine `retmap`, segment 6: elastic case – accept the elastic trial values as the updated solution

Program code

```
c
c ------ elastic case
  100 continue
c
c  --- initialize incremental plastic multiplier
        dp = 0.0d0
c
c  --- update elastic/plastic/kinematic hardening
c      deformation gradient tensors
        fen(1:3,1:3) = fe(1:3,1:3)
        fpn(1:3,1:3) = fp(1:3,1:3)
c
c  --- update elastic and plastic volume changes
        djen = djn/dsqrt(djcptr)
        djpn = dsqrt(djcptr)
c
c  --- compute Cauchy stress on current config.
c      by push-forward of trial 2nd Piola-Kirchhoff stress on
c      reference config.
        sgn(1:3,1:3) = matmul(ftn(1:3,1:3),
     &                        matmul(st20tr(1:3,1:3),
     &                               transpose(ftn(1:3,1:3))))/djn
c
c  --- update strain variable for isotropic hardening
        xpin = xpitr
c
c  --- update normal-yield ratio for subloading surface
        rsn = dsqrt(3.0d0/2.0d0)*dnormxtr/sitr
c
c  --- update tensor variables for plastic deformation
c      and their determinants
        cpn(1:3,1:3)  = cptr(1:3,1:3)
        cpdn(1:3,1:3) = cpdtr(1:3,1:3)
        djcpn  = djcptr
        djcpdn = djcpdtr
c
c ----- end routine for elastic case
        goto 199
```

(continued)

Table 10.25 Subroutine `retmap`, segment 7: plastic case – initial setting of return-mapping

Program code

```
c
c ----- plastic case (execute return mapping algorithm)
  110 continue
c
c   --- initial settings for Newton's iteration
        itmax = 50
        tol = 1.0d-12
c
c   --- initial settings for 'reduced' Newton's iteration
        itsmax = 10
        res0s = 1.0d+10
        dk = 1.0d0
c
c   --- initial guess for Newton's iteration
        cpn(1:3,1:3)  = cptr(1:3,1:3)
        cpdn(1:3,1:3) = cpdtr(1:3,1:3)
c
        djcpn  = djcptr
        djcpdn = djcpdtr
c
        xpin = xpitr
c
c   NOTE: initial value for the plastic multiplier 'dp'
c         is taken as zero.
c
        if(rstr.gt.rse) then
         dp = 0.0d0
        else if(rstr.le.rse) then
         dp = 0.0d0
         rstr = rse+(1.0d-12)
        end if
c
        rsn = rstr
```

(continued)

Table 10.26 Subroutine `retmap`, segment 8: begin return-mapping iteration and compute updated stresses

Program code

```
c
c   --- begin iteration
         it  = 0
         its = 0
   900 continue
c
c ----- compute inverse of updated values cpn(3,3) and cpdn(3,3)
         cpninv(1:3,1:3)  = cpn(1:3,1:3)
         cpdninv(1:3,1:3) = cpdn(1:3,1:3)
c
         eps = 1.0d-18
         CALL tinvdd(3,cpninv,djcpn,eps,3,iwork,ill)
         CALL tinvdd(3,cpdninv,djcpdn,eps,3,iwork,ill)
c
c ----- compute updated 2nd Piola-Kirchhoff stress on reference config.
c       by neo-Hookean hyperelastic model
         st20n(1:3,1:3)
       & = (dlam/2.0d0)*(djctn/djcpn-1.0d0)*ctninv(1:3,1:3)
       &             +dmyu*(cpninv(1:3,1:3)-ctninv(1:3,1:3))
c
c ----- compute updated (Eshelby-like) stress on reference config.
         st30n(1:3,1:3) = matmul(ctn(1:3,1:3),st20n(1:3,1:3))
c
c ----- 4th-order tangent modulus tensor on reference config.
c       defined as the derivative of (Eshelby-like) stress
c       w.r.t. plastic right Cauchy-Green deformation tensor
         do i=1,3
          do j=1,3
           do k=1,3
            do l=1,3
             tge30cp(i,j,k,l) =
       &      -(dlam/2.0d0)*(djctn/djcpn)*delta(i,j)*cpninv(l,k)
       &      -dmyu*( ctn(i,1)*cpninv(1,k)*cpninv(l,j)
       &             +ctn(i,2)*cpninv(2,k)*cpninv(l,j)
       &             +ctn(i,3)*cpninv(3,k)*cpninv(l,j) )
            end do
           end do
          end do
         end do
```

(continued)

Table 10.27 Subroutine `retmap`, segment 9: compute updated stresses

Program code

```
c
c ----- compute updated 2nd Piola-Kirchhoff-type stress variable
c       for kinematic hardening on reference config.
c       by neo-Hookean-type model (only deviatoric part)
        sk20n(1:3,1:3) = dmyuk*(cpdninv(1:3,1:3)-cpninv(1:3,1:3))
c
c ----- convert updated 2nd Piola-Kirchhoff-type stress variable
c       for kinematic hardening to updated (Eshelby-like) back stress
c       on reference config. and compute its isotropic and deviatoric
c       components
        sk30n(1:3,1:3) = matmul(cpn(1:3,1:3),sk20n(1:3,1:3))
        sk30nv = (sk30n(1,1)+sk30n(2,2)+sk30n(3,3))/3.0d0
        sk30nd(1:3,1:3) = sk30n(1:3,1:3)-sk30nv*delta(1:3,1:3)
c
c ----- 4th-order tangent modulus tensors on reference config.
c       defined as the derivative of (Eshelby-like) back stress
c       w.r.t. plastic right Cauchy-Green deformation tensor and
c       deformation variable 'cpdn(3,3)' for kinematic hardening
        do i=1,3
         do j=1,3
          do k=1,3
           do l=1,3
            tgk30cp(i,j,k,l)  =  dmyuk*delta(i,k)*cpdninv(l,j)
            tgk30cpd(i,j,k,l) = -dmyuk
     &                   *( cpn(i,1)*cpdninv(1,k)*cpdninv(l,j)
     &                     +cpn(i,2)*cpdninv(2,k)*cpdninv(l,j)
     &                     +cpn(i,3)*cpdninv(3,k)*cpdninv(l,j) )
           end do
          end do
         end do
        end do
```

(continued)

Table 10.28 Subroutine `retmap`, segment 10: compute updated stresses

Program code

```
c
c ----- compute 'reduced' (Eshelby-like) stress on reference config.
c        at updated state and its isotropic and deviatoric parts
          sx30n(1:3,1:3) = st30n-sk30n(1:3,1:3)
          sx30nv = (sx30n(1,1)+sx30n(2,2)+sx30n(3,3))/3.0d0
          sx30nd(1:3,1:3) = sx30n(1:3,1:3)-sx30nv*delta(1:3,1:3)
c
c  --- NOTE: norm of deviatoric Mandel stress on intermediate config.
c            is equal to SQRT ( Sdev(i,j) * Sdev(j,i) ), where 'S'
c            denotes Eshelby-like stress on reference config.
          dnormx = dsqrt( sx30nd(1,1)*sx30nd(1,1)
     &                   +sx30nd(1,2)*sx30nd(2,1)
     &                   +sx30nd(1,3)*sx30nd(3,1)
     &                   +sx30nd(2,1)*sx30nd(1,2)
     &                   +sx30nd(2,2)*sx30nd(2,2)
     &                   +sx30nd(2,3)*sx30nd(3,2)
     &                   +sx30nd(3,1)*sx30nd(1,3)
     &                   +sx30nd(3,2)*sx30nd(2,3)
     &                   +sx30nd(3,3)*sx30nd(3,3) )
c
c  --- to avoid division by zero
          if(dnormx.lt.1.0d-14) then
           dnormx = 1.0d-14
          end if
c
c  --- isotropic hardening stress at updated state
          CALL plhiso(xpin,yld0,plhi0,plhi1,plhi2,si,tgi)
```

(continued)

Table 10.29 Subroutine `retmap`, segment 11: compute updated variables and residuals

Line Program code

```
c
c   --- compute updated normal-yield ratio for subloading surface
c       NOTE: function acos(x) in Fortran is defined
c             on the domain -1 <= x <= +1.
          if(rstr.gt.rse) then
           tmprs = dcos((dpai/2.0d0)*(rstr-rse)/(1.0d0-rse))
     &           *dexp(-u0*(dpai/2.0d0)*(dp/(1.0d0-rse)))
          else if(rstr.le.rse) then
           tmprs = dexp(-u0*(dpai/2.0d0)*(dp/(1.0d0-rse)))
          end if
c
          if(tmprs.gt.1.0d0) then
           tmprs =  1.0d0
          else if(tmprs.lt.-1.0d0) then
           tmprs = -1.0d0
          end if
c
          rsn = rse+(2.0d0/dpai)*(1.0d0-rse)*dacos(tmprs)
c
c   --- definition of directional tensors for plastic evolution eqs.,
c       that is plastic flow rule and evolution for kinematic hardening
          gp(1:3,1:3) = 2.0d0*sx30nd(1:3,1:3)/dnormx
          gk(1:3,1:3) = 2.0d0*(plhk/dmyuk)*sk30nd(1:3,1:3)
c
c   --- exponential maps 'Qp' and 'Qk'
          q1tmp(1:3,1:3) = dp*gp(1:3,1:3)
          q2tmp(1:3,1:3) = dp*gk(1:3,1:3)
c
          nmax = 20
          etol = 1.0d-18
          CALL matexp(nmax,etol,q1tmp,q1exp,ierr)
          CALL matexp(nmax,etol,q2tmp,q2exp,ierr)
c
c   --- residual vector of equations to be solved
          y1(1:3,1:3) = cpn(1:3,1:3)
     &                -matmul(q1exp(1:3,1:3),cptr(1:3,1:3))
          y2(1:3,1:3) = cpdn(1:3,1:3)
     &                -matmul(q2exp(1:3,1:3),cpdtr(1:3,1:3))
          y3 = xpin-xpitr-dsqrt(2.0d0/3.0d0)*dp
          y4 = dsqrt(3.0d0/2.0d0)*dnormx-rsn*si
```

(continued)

Table 10.30 Subroutine `retmap`, segment 12: assemble residual vector and compute residual norm

Program code

```
c
         y( 1) =  y1(1,1)
         y( 2) = (y1(1,2)+y1(2,1))/2.0d0
         y( 3) = (y1(1,3)+y1(3,1))/2.0d0
         y( 4) =  y1(2,2)
         y( 5) = (y1(2,3)+y1(3,2))/2.0d0
         y( 6) =  y1(3,3)
         y( 7) =  y2(1,1)
         y( 8) = (y2(1,2)+y2(2,1))/2.0d0
         y( 9) = (y2(1,3)+y2(3,1))/2.0d0
         y(10) =  y2(2,2)
         y(11) = (y2(2,3)+y2(3,2))/2.0d0
         y(12) =  y2(3,3)
         y(13) = y3
         y(14) = y4
c
c   --- compute norm of residual
         res1 = dsqrt( y( 1)**2+y( 2)**2+y( 3)**2
        &             +y( 4)**2+y( 5)**2+y( 6)**2 )
         res2 = dsqrt( y( 7)**2+y( 8)**2+y( 9)**2
        &             +y(10)**2+y(11)**2+y(12)**2 )
         res3 = dabs(y(13))
         res4 = dabs(y(14))
c
         res = dsqrt(res1**2+res2**2+res3**2+res4**2)
c
c   --- compute norm of residual at the beginning of iteration
         if(it.eq.0) then
          res0 = res
         end if
```

(continued)

Table 10.31 Subroutine `retmap`, segment 13: check convergence, and use 'reduced' Newton–Raphson scheme if necessary

Program code

```
c
c   --- check rate of convergence, and then,
c       'reduced' Newton's iteration is employed, if necessary.
         if(    ((res-(1.0d0-dk/4.0d0)*res0s).lt.0.0d0)
       &   .or.(its.ge.itsmax)) then
          its = 0
          res0s = 1.0d+10
          dk = 1.0d0
          goto 899
         else if(     ((res-(1.0d0-dk/4.0d0)*res0s).ge.0.0d0)
       &        .and.(its.lt.itsmax)) then
          its = its+1
          dk = dk/2.0d0
          goto 800
         end if
c
   899 continue
c
c   --- check convergence
         if(res0.gt.1.0d0) then
          resy = res/res0
         else if(res0.le.1.0d0) then
          resy = res
         end if
c
         write(iow1,555) it,resy,res1,res2,res3,res4
   555 format(i5,5es12.4)
c
         if(resy.lt.tol) then
          goto 990
         else if((it.ge.itmax).and.(resy.ge.tol)) then
          write(iow1,500) '--- return mapping failed.',
       &                  ' it =',it,' res =',resy,' at',
       &                  ' inel =',inel,' ige =',ige,'.'
   500 format(a26,a5,i3,a6,1pd12.4,a3,a7,i8,a6,i3,a1)
          goto 990
         else
          it = it+1
         end if
```

(continued)

Table 10.32 Subroutine `retmap`, segment 14: Jacobian matrix for return-mapping equations (1)

Program code

```
c
c     --- 1st derivative of the subloading function w.r.t.
c         (Eshelby-like) stress variables on reference config.
c         (note the transpose of the stress tensor in numerator),
c         scalar stress variable for isotropic hardening, and
c         normal-yield ratio
          dfst(1:3,1:3) =  dsqrt(3.0d0/2.0d0)
     &                   *transpose(sx30nd(1:3,1:3))/dnormx
          dfsk(1:3,1:3) = -dsqrt(3.0d0/2.0d0)
     &                   *transpose(sx30nd(1:3,1:3))/dnormx
          dfsi = -rsn
          dfrs = -si
c
c     --- 1st derivatives of directional tensors 'gp(3,3)' and 'gk(3,3)'
c         for plastic evolution w.r.t. 'reduced' (Eshelby-like) stress
c         and (Eshelby-like) back stress on reference config.
c         (note the transpose 'sx30nd(l,k)' of the latter tensor
c          in the last term)
          do i=1,3
           do j=1,3
            do k=1,3
             do l=1,3
              dgp(i,j,k,l) = 2.0d0*(1.0d0/dnormx)
     &                      *( delta(i,k)*delta(j,l)
     &                        -delta(i,j)*delta(k,l)/3.0d0
     &                        -(sx30nd(i,j)/dnormx)*(sx30nd(l,k)/dnormx) )
              dgk(i,j,k,l) = 2.0d0*(plhk/dmyuk)
     &                      *( delta(i,k)*delta(j,l)
     &                        -delta(i,j)*delta(k,l)/3.0d0)
             end do
            end do
           end do
          end do
c
c     --- derivative of normal-yield ratio w.r.t. incremental
c         plastic multiplier
c         (division by zero should be avoided and 'drs' must be non-negative
c          even when 'rsn' is less than or equal to 'rse')
          tmprs = (dpai/2.0d0)*(rsn-rse)/(1.0d0-rse)
c
          if(tmprs.ge.1.0d-08) then
           drs = u0*dexp(-u0*(dpai/2.0d0)*(dp/(1.0d0-rse)))
     &           *dcos((dpai/2.0d0)*(rstr-rse)/(1.0d0-rse))
     &           /dsin((dpai/2.0d0)*(rsn-rse)/(1.0d0-rse))
          else if(tmprs.lt.1.0d-08) then
           drs = u0*dexp(-u0*(dpai/2.0d0)*(dp/(1.0d0-rse)))
     &           *dcos((dpai/2.0d0)*(rstr-rse)/(1.0d0-rse))
     &           /(1.0d-08)
          end if
```

(continued)

Table 10.33 Subroutine `retmap`, segment 15: Jacobian matrix for return-mapping equations (2)

Program code

```
c
c   --- compute derivative of exponential maps
c       d exp[A(i,j)]/ d A(i,j)
         nmax = 20
         etol = 1.0d-18
         CALL matdex(nmax,etol,q1tmp,dq1exp,ierr)
         CALL matdex(nmax,etol,q2tmp,dq2exp,ierr)
c
c   --- derivative of exponential map 'Qp'
         do i=1,3
          do j=1,3
           do k=1,3
            do l=1,3
             tmp11(i,j,k,l) = 0.0d0
             tmp12(i,j,k,l) = 0.0d0
             do m=1,3
              do n=1,3
               tmp11(i,j,k,l) = tmp11(i,j,k,l)
        &       +dp*dgp(i,j,m,n)*(tge30cp(m,n,k,l)-tgk30cp(m,n,k,l))
               tmp12(i,j,k,l) = tmp12(i,j,k,l)
        &       -dp*dgp(i,j,m,n)*tgk30cpd(m,n,k,l)
              end do
             end do
            end do
           end do
          end do
         end do
c
         do i=1,3
          do j=1,3
           dq13(i,j) = 0.0d0
           dq14(i,j) = 0.0d0
           do k=1,3
            do l=1,3
             dq11(i,j,k,l) = 0.0d0
             dq12(i,j,k,l) = 0.0d0
             dq14(i,j) = dq14(i,j)+dq1exp(i,j,k,l)*gp(k,l)
             do m=1,3
              do n=1,3
               dq11(i,j,k,l) = dq11(i,j,k,l)
        &       +dq1exp(i,j,m,n)*tmp11(m,n,k,l)
               dq12(i,j,k,l) = dq12(i,j,k,l)
        &       +dq1exp(i,j,m,n)*tmp12(m,n,k,l)
              end do
             end do
            end do
           end do
          end do
         end do
```

(continued)

Table 10.34 Subroutine `retmap`, segment 16: Jacobian matrix for return-mapping equations (3)

Program code

```
c
c   --- derivative of exponential map 'Qk'
        do i=1,3
         do j=1,3
          do k=1,3
           do l=1,3
            tmp21(i,j,k,l) = 0.0d0
            tmp22(i,j,k,l) = 0.0d0
            do m=1,3
             do n=1,3
              tmp21(i,j,k,l) = tmp21(i,j,k,l)
     &          +dp*dgk(i,j,m,n)*tgk30cp(m,n,k,l)
              tmp22(i,j,k,l) = tmp22(i,j,k,l)
     &          +dp*dgk(i,j,m,n)*tgk30cpd(m,n,k,l)
             end do
            end do
           end do
          end do
         end do
        end do
c
        do i=1,3
         do j=1,3
          dq23(i,j) = 0.0d0
          dq24(i,j) = 0.0d0
          do k=1,3
           do l=1,3
            dq21(i,j,k,l) = 0.0d0
            dq22(i,j,k,l) = 0.0d0
            dq24(i,j) = dq24(i,j)+dq2exp(i,j,k,l)*gk(k,l)
            do m=1,3
             do n=1,3
              dq21(i,j,k,l) = dq21(i,j,k,l)
     &          +dq2exp(i,j,m,n)*tmp21(m,n,k,l)
              dq22(i,j,k,l) = dq22(i,j,k,l)
     &          +dq2exp(i,j,m,n)*tmp22(m,n,k,l)
             end do
            end do
           end do
          end do
         end do
        end do
```

(continued)

Table 10.35 Subroutine `retmap`, segment 17: Jacobian matrix for return-mapping equations (4)

Program code

```
c
c   --- Jacobian matrix of return-mapping equations
c
c     --- equation (1) of plastic flow
         do i=1,3
          do j=1,3
           do k=1,3
            do l=1,3
             aj11(i,j,k,l) = 0.50d0*( delta(i,k)*delta(j,l)
        &                             +delta(i,l)*delta(j,k) )
             aj12(i,j,k,l) = 0.0d0
             do m=1,3
              aj11(i,j,k,l) = aj11(i,j,k,l)-dq11(i,m,k,l)*cptr(m,j)
              aj12(i,j,k,l) = aj12(i,j,k,l)-dq12(i,m,k,l)*cptr(m,j)
             end do
            end do
           end do
          end do
         end do
         aj13(1:3,1:3) = -matmul(dq13(1:3,1:3),cptr(1:3,1:3))
         aj14(1:3,1:3) = -matmul(dq14(1:3,1:3),cptr(1:3,1:3))
c
c     --- equation (2) of kinematic hardening
         do i=1,3
          do j=1,3
           do k=1,3
            do l=1,3
             aj21(i,j,k,l) = 0.0d0
             aj22(i,j,k,l) = 0.50d0*( delta(i,k)*delta(j,l)
        &                             +delta(i,l)*delta(j,k) )
             do m =1,3
              aj21(i,j,k,l) = aj21(i,j,k,l)-dq21(i,m,k,l)*cpdtr(m,j)
              aj22(i,j,k,l) = aj22(i,j,k,l)-dq22(i,m,k,l)*cpdtr(m,j)
             end do
            end do
           end do
          end do
         end do
         aj23(1:3,1:3) = -matmul(dq23(1:3,1:3),cpdtr(1:3,1:3))
         aj24(1:3,1:3) = -matmul(dq24(1:3,1:3),cpdtr(1:3,1:3))
```

(continued)

Table 10.36 Subroutine `retmap`, segment 18: Jacobian matrix for return-mapping equations (5)

Line | Program code

```
c
c     --- equation (3) of isotropic hardening
         aj31(1:3,1:3) = 0.0d0
         aj32(1:3,1:3) = 0.0d0
         aj33 =  1.0d0
         aj34 = -dsqrt(2.0d0/3.0d0)
c
c     --- equation (4) of yield function
         do i=1,3
          do j=1,3
           aj41(i,j) = 0.0d0
           aj42(i,j) = 0.0d0
           do m=1,3
            do n=1,3
             aj41(i,j) = aj41(i,j)+dfst(m,n)*tge30cp(m,n,i,j)
       &                          +dfsk(m,n)*tgk30cp(m,n,i,j)
             aj42(i,j) = aj42(i,j)+dfsk(m,n)*tgk30cpd(m,n,i,j)
            end do
           end do
          end do
         end do
c
         aj43 = dfsi*tgi
         aj44 = dfrs*drs
```

(continued)

Table 10.37 Subroutine `retmap`, segment 1: assemble Jacobian matrix for return-mapping equations

Program code

```
c
c   --- assemble Jacobian matrix [14x14]
        m = 0
        do i=1,3
        do j=i,3
         m = m+1
         n = 0
         do k=1,3
         do l=k,3
          n = n+1
          aj(  m,  n) = ( aj11(i,j,k,l)
     &                   +aj11(j,i,k,l)
     &                   +aj11(i,j,l,k)
     &                   +aj11(j,i,l,k) )/4.0d0
          aj(  m,6+n) = ( aj12(i,j,k,l)
     &                   +aj12(j,i,k,l)
     &                   +aj12(i,j,l,k)
     &                   +aj12(j,i,l,k) )/4.0d0
          aj(6+m,  n) = ( aj21(i,j,k,l)
     &                   +aj21(j,i,k,l)
     &                   +aj21(i,j,l,k)
     &                   +aj21(j,i,l,k) )/4.0d0
          aj(6+m,6+n) = ( aj22(i,j,k,l)
     &                   +aj22(j,i,k,l)
     &                   +aj22(i,j,l,k)
     &                   +aj22(j,i,l,k) )/4.0d0
         end do
         end do
        end do
        end do
c
        m = 0
        do i=1,3
         do j=i,3
          m = m+1
          aj(  m,13) = (aj13(i,j)+aj13(j,i))/2.0d0
          aj(6+m,13) = (aj23(i,j)+aj23(j,i))/2.0d0
          aj(  m,14) = (aj14(i,j)+aj14(j,i))/2.0d0
          aj(6+m,14) = (aj24(i,j)+aj24(j,i))/2.0d0
          aj(13,  m) = (aj31(i,j)+aj31(j,i))/2.0d0
          aj(13,6+m) = (aj32(i,j)+aj32(j,i))/2.0d0
          aj(14,  m) = (aj41(i,j)+aj41(j,i))/2.0d0
          aj(14,6+m) = (aj42(i,j)+aj42(j,i))/2.0d0
         end do
        end do
c
        aj(13,13) = aj33
        aj(13,14) = aj34
        aj(14,13) = aj43
        aj(14,14) = aj44
```

(continued)

Table 10.38 Subroutine `retmap`, segment 20: solve return-mapping equations, correct unknown variables

Program code

```
c
c   --- solve simultaneous equations
          do k=1,14
           y(k) = -y(k)
          end do
c
          eps = 1.0d-20
          CALL tledd(14,aj,y,dx,14,eps,ill)
c
c     --- before conducting 'reduced' Newton's iteration,
c          variables before update are saved
          if(its.eq.0) then
           res0s = res
           x1( 1) =  cpn(1,1)
           x1( 2) = (cpn(1,2)+cpn(2,1))/2.0d0
           x1( 3) = (cpn(1,3)+cpn(3,1))/2.0d0
           x1( 4) =  cpn(2,2)
           x1( 5) = (cpn(2,3)+cpn(3,2))/2.0d0
           x1( 6) =  cpn(3,3)
           x1( 7) =  cpdn(1,1)
           x1( 8) = (cpdn(1,2)+cpdn(2,1))/2.0d0
           x1( 9) = (cpdn(1,3)+cpdn(3,1))/2.0d0
           x1(10) =  cpdn(2,2)
           x1(11) = (cpdn(2,3)+cpdn(3,2))/2.0d0
           x1(12) =  cpdn(3,3)
           x1(13) = xpin
           x1(14) = dp
          end if
```

(continued)

Table 10.39 Subroutine `retmap`, segment 21: solve return-mapping equations, correct unknown variables

Program code

```
c
c     --- update variables (loop 800: 'reduced' Newton's iteration)
  800 continue
c
         cpn(1,1)  = x1( 1)+dk*dx( 1)
         cpn(1,2)  = x1( 2)+dk*dx( 2)/2.0d0
         cpn(1,3)  = x1( 3)+dk*dx( 3)/2.0d0
         cpn(2,1)  = x1( 2)+dk*dx( 2)/2.0d0
         cpn(2,2)  = x1( 4)+dk*dx( 4)
         cpn(2,3)  = x1( 5)+dk*dx( 5)/2.0d0
         cpn(3,1)  = x1( 3)+dk*dx( 3)/2.0d0
         cpn(3,2)  = x1( 5)+dk*dx( 5)/2.0d0
         cpn(3,3)  = x1( 6)+dk*dx( 6)
c
         cpdn(1,1) = x1( 7)+dk*dx( 7)
         cpdn(1,2) = x1( 8)+dk*dx( 8)/2.0d0
         cpdn(1,3) = x1( 9)+dk*dx( 9)/2.0d0
         cpdn(2,1) = x1( 8)+dk*dx( 8)/2.0d0
         cpdn(2,2) = x1(10)+dk*dx(10)
         cpdn(2,3) = x1(11)+dk*dx(11)/2.0d0
         cpdn(3,1) = x1( 9)+dk*dx( 9)/2.0d0
         cpdn(3,2) = x1(11)+dk*dx(11)/2.0d0
         cpdn(3,3) = x1(12)+dk*dx(12)
c
         xpin      = x1(13)+dk*dx(13)
         dp        = x1(14)+dk*dx(14)
c
c    --- go to next iteration
         goto 900
c
c    --- end of iteration
  990 continue
c
         if(dp.lt.0.0d0) then
          write(iow1,*) '--- negative plastic multiplier ',dp,' at',
     &                 ' inel =',inel,' ige =',ige,'.'
          stop
         end if
```

(continued)

Table 10.40 Subroutine `retmap`, segment 22: update variables and end subroutine

```
c
c ----- update variables
c
c   --- update elastic and plastic deformation gradient tensors
          fen(1:3,1:3) = fe(1:3,1:3)
          fpn(1:3,1:3) = fp(1:3,1:3)
c
c   --- update elastic and plastic volume changes
          djen = djn/dsqrt(djcpn)
          djpn = dsqrt(djcpn)
c
c   --- compute updated Cauchy stress by push-forward operation of
c       2nd Piola-Kirchhoff stress from reference config. to current
c       config.
          sgn(1:3,1:3) = matmul(ftn(1:3,1:3),
     &                        matmul(st20n(1:3,1:3),
     &                               transpose(ftn(1:3,1:3))))/djn
c
c ----- end routine for plastic case
          goto 199
c
c ----- end of return mapping routine
  199 continue
```

(continued)

10.3.5 *Subroutine for Isotropic Hardening Rule:* `plhiso`

The subroutine `plhiso` evaluates the scalar-valued stress-like variable F and the tangent modulus F' for isotropic hardening. As can be seen in the source code shown in Table 10.42, the evaluation of the variables F and F' by equations (9.5) and (9.6) can be implemented in a quite straightforward and self-evident way, so that the explanation of the source code is omitted, while the list of the arguments in `plhiso` is shown in Table 10.43.

10.3.6 *Subroutine for Numerical Computation of Consistent Tangent Modulus Tensor:* `tgnum0`

The subroutine `tgnum0`, consisting of three segments, evaluates the consistent (algorithmic) tangent modulus tensor in the description relative to the reference configuration (that is the Lagrangian or material description) by means of numerical differentiation using a perturbation technique; see Section 9.9.2 for the detailed explanation of this approach. Table 10.44 gives a line-by-line explanation of the source code of the segments, with a description of the procedures within the algorithm, together with the corresponding equation involved in the formulation presented in Section 9.9.2. The source codes of the segments are shown in Tables 10.45– 10.47. Most of the arguments of `tgnum0` have been listed previously in Table 10.14,

Table 10.41 Subroutine `retmap`, segment 23: update variables and end subroutine

```
c
c ----- update plastic internal variables
        xpn( 1) = dp
        xpn( 2) = xpin
        xpn( 3) = rsn
        xpn( 4) = 0.0d0
        xpn( 5) = 0.0d0
        xpn( 6) = 0.0d0
        xpn( 7) = 0.0d0
        xpn( 8) = 0.0d0
        xpn( 9) = 0.0d0
        xpn(10) = 0.0d0
        xpn(11) = cpn(1,1)
        xpn(12) = cpn(1,2)
        xpn(13) = cpn(1,3)
        xpn(14) = cpn(2,1)
        xpn(15) = cpn(2,2)
        xpn(16) = cpn(2,3)
        xpn(17) = cpn(3,1)
        xpn(18) = cpn(3,2)
        xpn(19) = cpn(3,3)
        xpn(20) = djcpn
        xpn(21) = cpdn(1,1)
        xpn(22) = cpdn(1,2)
        xpn(23) = cpdn(1,3)
        xpn(24) = cpdn(2,1)
        xpn(25) = cpdn(2,2)
        xpn(26) = cpdn(2,3)
        xpn(27) = cpdn(3,1)
        xpn(28) = cpdn(3,2)
        xpn(29) = cpdn(3,3)
        xpn(30) = djcpdn
c
        return
        end subroutine retmap
```

Table 10.42 Subroutine `plhiso`: compute stress variable F and tangent modulus F' for isotropic hardening

```
      subroutine plhiso(x,yld0,plhi0,plhi1,plhi2,si,tgi)
      implicit double precision (a-h,o-z)
      common /iofile/ior1,ior2,ior3,ior4,ior5,
     &               iow1,iow2,iow3,iow4,iow5,
     &               iow6,iow7,iow8,iow9,iow10
c
c ----- nonlinear isotropic hardening rule
      si = plhi0*x+yld0*(1.0d0+plhi1*(1.0d0-dexp(-plhi2*x)))
c
c ----- tangent modulus of isotropic hardening
      tgi = plhi0+yld0*plhi1*plhi2*dexp(-plhi2*x)
c
      return
      end subroutine plhiso
```

Table 10.43 List of the arguments for the subroutine `plhiso`

Argument	Input	Output	Symbol	Remark
`x`	Deformation-like variable for isotropic hardening	Unchanged	H	
`yld0`	Reference yield stress for isotropic hardening	Unchanged	F_0	`prope( 6)` in Table 10.2
`plhi0`	Linear isotropic hardening coefficient	Unchanged	h_0	`prope( 7)` in Table 10.2
`plhi1`	Parameter controlling the saturation limit of isotropic hardening	Unchanged	h_1	`prope( 8)` in Table 10.2
`plhi2`	Exponential hardening parameter for isotropic	Unchanged	h_2	`prope( 9)` in Table 10.2
`si`	—	Stress-like variable for isotropic hardening	F	
`tgi`	—	Tangent modulus for isotropic hardening	F'	

Table 10.44 Algorithm of the subroutine `tgnum0` and the equations corresponding to the procedures

Segment Line	Table number of source code Description of procedure	Corresponding equation
Segment 1	**(Table 10.45)**	
1–15	Subroutine name, arguments, and variable declarations.	
23–27	Invert $\mathbf{F}_n$ and $\mathbf{F}_{n+1}$.	
Segment 2	**(Table 10.46)**	
31–32	`do 100` and `do 200`-loops: Compute for $(K, L) = (1, 1), (1, 2), (1, 3), (2, 1), (2, 2), (2, 3)$, and $(3, 3)$.	
35–41	Compute the perturbed deformation gradient tensor $\mathbf{F}_*$. (To simplify the notation, we define $\mathbf{F}_* := \mathbf{F}_{n+1,(K,L)}(\epsilon)$ with the symbol '*' meaning the perturbed state.)	(9.106) and (9.107)
44–49	Compute the determinant $\det\mathbf{F}_*$.	
52–54	Invert $\mathbf{F}_*$.	
58	Compute the relative deformation gradient tensor from t_* to t_n by $\mathbf{f}_{[n,*]} = \mathbf{F}_*\mathbf{F}_n^{-1}$.	
61–66	Compute the determinant $j_{[n,*]} := \det\mathbf{f}_{[n,*]}$.	
Segment 3	**(Table 10.47)**	
69–72	For the known state at t_n and the prescribed relative deformation gradient tensor $\mathbf{f}_{[n,*]}$, evaluate the updated Cauchy stress tensor $\boldsymbol{\sigma}_*$ at the perturbed state by calling the subroutine `retmap`.	
75–77	Pull-back operation $\mathbf{S}_{n+1} = \mathbf{F}_{n+1}^{-1}(J_{n+1}\boldsymbol{\sigma}_{n+1})\mathbf{F}_{n+1}^{-T}$.	
79–80	Pull-back operation $\mathbf{S}_* = \mathbf{F}_*^{-1}(J_*\boldsymbol{\sigma}_*)\mathbf{F}_*^{-T}$.	
83–88	Compute the consistent tangent, considering the symmetry $(K, L) \leftrightarrow (L, K)$.	(9.109) and (9.110)
90–91	End `do 100` and `do 200`-loops.	
93–94	`return` and `end` statements.	

Table 10.45 Subroutine `tgnum0`, segment 1

Program code

```
      subroutine tgnum0(inel,ige,mtypee,prope,
     &                  cep,epsx,
     &                  ft,dj,fe,dje,fp,djp,sg,idep,xp,
     &                  ftn,djn,sgn)
      implicit double precision (a-h,o-z)
      dimension prope(20),bv0x(3,3),cep(3,3,3,3)
      dimension ft(3,3),fe(3,3),fp(3,3),sg(3,3),xp(30)
      dimension ftn(3,3),sgn(3,3)
      dimension ftinv(3,3),ftninv(3,3),ftxinv(3,3),iwork(6),frx(3,3)
      dimension ftx(3,3),fex(3,3),fpx(3,3),sgx(3,3),xpx(30)
      dimension st2n(3,3),st2x(3,3)
      common /iofile/ior1,ior2,ior3,ior4,ior5,
     &               iow1,iow2,iow3,iow4,iow5,
     &               iow6,iow7,iow8,iow9,iow10
      common /etc/del(3),delta(3,3),p(3,3,3),dpai
c
c ----- computation of algorithmic (consistent) tangent modulus tensor
c       by numerical differentiation using a perturbation technique
c
c       *) type-0: Lagrangian geometric setting
c
c ----- inverse of total deformation gradient at 't_n' and 't_n+1'
      ftinv(1:3,1:3)  = ft(1:3,1:3)
      ftninv(1:3,1:3) = ftn(1:3,1:3)
      eps = 1.0d-18
      CALL tinvdd(3,ftinv,detft,eps,3,iwork,ill1)
      CALL tinvdd(3,ftninv,detftn,eps,3,iwork,ill2)
```

(continued)

Table 10.46 Subroutine tgnum0, segment 2

```
c
c ----- compute for row indices
c          (K,L) = (1,1),(1,2),(1,3),(2,2),(2,3),(3,3)
           do 100 k=1,3
           do 200 l=k,3
c
c ----- compute perturbed deformation gradient
           do i=1,3
            do j=1,3
             ftx(i,j) = ftn(i,j)
          &             +(epsx/2.0d0)*( ftninv(k,i)*delta(l,j)
          &                            +ftninv(l,i)*delta(k,j) )
            enddo
           enddo
c
c ----- compute Jacobian of the perturbed deformation gradient
           djx = ftx(1,1)*ftx(2,2)*ftx(3,3)
          &     +ftx(1,3)*ftx(2,1)*ftx(3,2)
          &     +ftx(3,1)*ftx(1,2)*ftx(2,3)
          &     -ftx(1,3)*ftx(2,2)*ftx(3,1)
          &     -ftx(1,1)*ftx(2,3)*ftx(3,2)
          &     -ftx(3,3)*ftx(1,2)*ftx(2,1)
c
c ----- inverse of perturbed deformation gradient
           ftxinv(1:3,1:3) = ftx(1:3,1:3)
           eps = 1.0d-18
           CALL tinvdd(3,ftxinv,detftx,eps,3,iwork,ill1)
c
c ----- compute relative deformation gradient between
c       the configurations at 't_n' and the perturbed state
           frx(1:3,1:3) = matmul(ftx(1:3,1:3),ftinv(1:3,1:3))
c
c ----- compute determinant of relative deformation gradient
           djrx = frx(1,1)*frx(2,2)*frx(3,3)
          &      +frx(1,3)*frx(2,1)*frx(3,2)
          &      +frx(3,1)*frx(1,2)*frx(2,3)
          &      -frx(1,3)*frx(2,2)*frx(3,1)
          &      -frx(1,1)*frx(2,3)*frx(3,2)
          &      -frx(3,3)*frx(1,2)*frx(2,1)
```

(continued)

Table 10.47 Subroutine `tgnum0`, segment 3

Line	Program code

```
67  c
68  c ----- return-mapping algorithm
69            CALL retmap(inel,ige,mtypee,prope,
70          &              frx,djrx,
71          &              ft,dj,fe,dje,fp,djp,sg,idep,xp,
72          &              ftx,djx,fex,djex,fpx,djpx,sgx,idepx,xpx)
73  c
74  c ----- compute 2nd Piola-Kirchhoff stress by pull-back operation
75           st2n(1:3,1:3) = matmul(ftninv(1:3,1:3),
76          &                         matmul(djn*sgn(1:3,1:3),
77          &                                transpose(ftninv(1:3,1:3))))
78           st2x(1:3,1:3) = matmul(ftxinv(1:3,1:3),
79          &                         matmul(djx*sgx(1:3,1:3),
80          &                                transpose(ftxinv(1:3,1:3))))
81  c
82  c ----- compute Lagrangian algorithmic tangent moduli
83           do i=1,3
84            do j=1,3
85             cep(i,j,k,l) = (st2x(i,j)-st2n(i,j))/epsx
86             cep(i,j,l,k) = cep(i,j,k,l)
87            end do
88           end do
89  c
90    200 continue
91    100 continue
92  c
93           return
94           end subroutine tgnum0
```

so that the list of the arguments is omitted to avoid duplication. The only quantity which has not yet been defined is `epsx`, the perturbation parameter in equation (9.106).

Note that the subroutine `tgnum0` calls the subroutines `retmap` and `tinvdd`, both of which are also required in the program `returnmap`. As mentioned in Section 9.9, it should be emphasized that the algorithm implemented in this subroutine is completely material-independent, and is not limited to the particular model of present interest. Therefore, this subroutine can be used to evaluate the consistent tangent modulus tensor of generic elasto-plastic models merely by modifying the `call` statement (lines 69–72 in segment 3 of the source code shown in Table 10.47) to call a subroutine for the return-mapping scheme of the relevant material model.

Appendix A

Projection of Area

Consider the projection of the area on the surface having the unit normal vector **n** onto the surface having the normal vector **m** in Figure A.1.

Now, suppose the plane abcd which contains the unit normal vectors **m** and **n**. Consider the line $\overline{\text{ef}}$ which is the intersecting line of the area having the unit normal vector **n** with this plane. Divide the area having the unit normal vector **n** into narrow bands perpendicular to this line and their projections onto the surface having the normal vector **m**. The lengths of the projected bands are same as the those of the original bands but the widths become those multiplied by the scalar product of the unit normal vectors, that is, $\mathbf{m} \cdot \mathbf{n}$. Consequently, the projected area $d\bar{a}$ is related to the original area da as follows:

$$d\bar{a} = \mathbf{m} \cdot \mathbf{n} da. \tag{A.1}$$

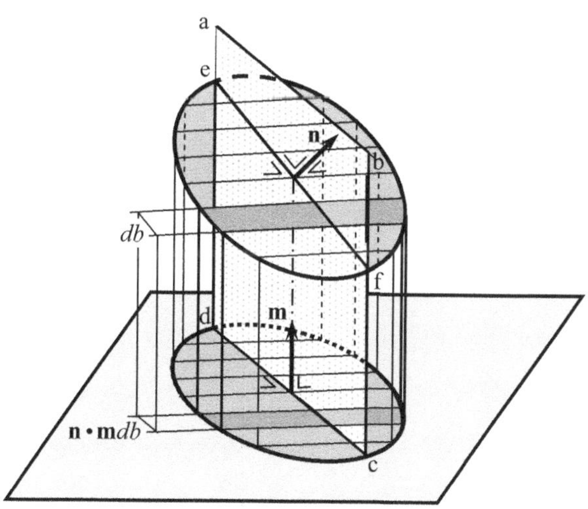

Figure A.1 Projection of area (Hashiguchi, 2009)

Appendix B

Geometrical Interpretation of Strain Rate and Spin Tensors

For the rate of the two infinitesimal line element vectors, from equation (4.80) we have

$$(d\mathbf{x}_i \cdot d\mathbf{x}_j)^\bullet = d\mathbf{x}_i \cdot (d\mathbf{x}_j)^\bullet + (d\mathbf{x}_i)^\bullet \cdot d\mathbf{x}_j = d\mathbf{x}_i \cdot \boldsymbol{l} d\mathbf{x}_j + \boldsymbol{l} d\mathbf{x}_i \cdot d\mathbf{x}_j = d\mathbf{x}_i \cdot (\boldsymbol{l} + \boldsymbol{l}^T) d\mathbf{x}_j$$
$$= d\mathbf{x}_i \cdot 2\mathbf{d} d\mathbf{x}_j \tag{B.1}$$

Denoting the angle between the two line elements by θ_{ij}, equation (B.1) can be rewritten as

$$(d\mathbf{x}_i \cdot d\mathbf{x}_j)^\bullet = (||d\mathbf{x}_i||\,||d\mathbf{x}_j|| \cos\theta_{ij})^\bullet$$
$$= \left\{ \left(\frac{(||d\mathbf{x}_i||)^\bullet}{||d\mathbf{x}_i||} + \frac{(||d\mathbf{x}_j||)^\bullet}{||d\mathbf{x}_j||} \right) \cos\theta_{ij} - (\sin\theta_{ij}) \overset{\bullet}{\theta}_{ij} \right\} ||d\mathbf{x}_i||\,||d\mathbf{x}_j|| \text{ (no sum).} \tag{B.2}$$

Equating the right hand sides in equations (B.1) and (B.2), we have

$$\frac{d\mathbf{x}_i}{||d\mathbf{x}_i||} \cdot \mathbf{d} \frac{d\mathbf{x}_j}{||d\mathbf{x}_j||} = \frac{1}{2} \left\{ \left(\frac{(||d\mathbf{x}_i||)^\bullet}{||d\mathbf{x}_i||} + \frac{(||d\mathbf{x}_j||)^\bullet}{||d\mathbf{x}_j||} \right) \cos\theta_{ij} - (\sin\theta_{ij}) \overset{\bullet}{\theta}_{ij} \right\} \text{ (no sum).} \tag{B.3}$$

For the case where the line elements $d\mathbf{x}_i$ and $d\mathbf{x}_j$ are parallel to each other, that is $\theta_{ij} = 0$, equation (B.3) reduces to

$$d_{ii} \equiv \frac{d\mathbf{x}_i}{||d\mathbf{x}_i||} \cdot \mathbf{d} \frac{d\mathbf{x}_i}{||d\mathbf{x}_i||} = \frac{(||d\mathbf{x}_i||)^\bullet}{||d\mathbf{x}_i||} \text{ (no sum),} \tag{B.4}$$

which denotes the rate of elongation.

For the case where the line elements $d\mathbf{x}_i$, $d\mathbf{x}_j$ are orthogonal to each other, that is $\theta_{ij} = \pi/2$, equation (B.3) becomes

$$d_{ij} \equiv \frac{d\mathbf{x}_i}{||d\mathbf{x}_i||} \cdot \mathbf{d} \frac{d\mathbf{x}_j}{||d\mathbf{x}_j||} = -\frac{1}{2} \overset{\bullet}{\theta}_{ij} \text{ (no sum),} \tag{B.5}$$

which denotes the rate of shearing.

Introduction to Finite Strain Theory for Continuum Elasto-Plasticity, First Edition. Koichi Hashiguchi and Yuki Yamakawa.

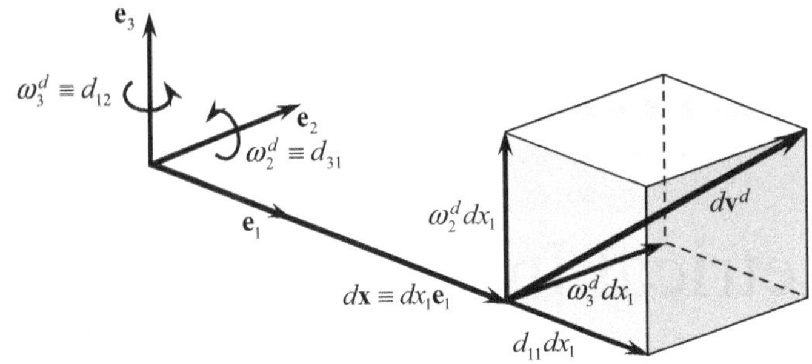

(a) Relative angular velocity caused by strain rate

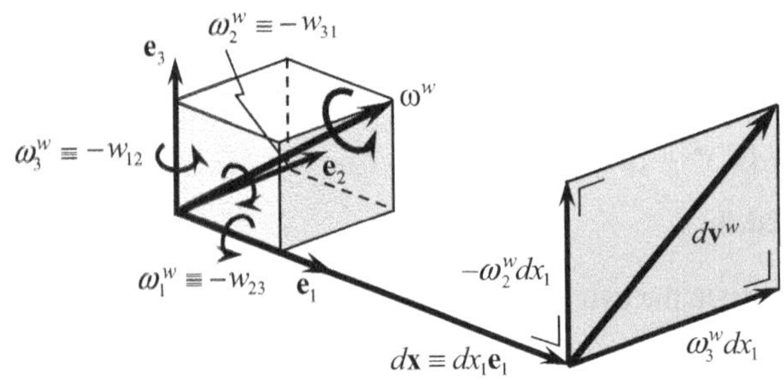

(b) Relative angular velocity caused by continuum spin

Figure B.1 Relative angular velovity of line element

The deformation induced by the strain rate **d** is described above. In what follows, consider the rate of rotation induced by the strain rate **d** and the continuum spin **w**, referring to Hisada (1992). The velocity increment $d\mathbf{v}$ can be decomposed based on equations (4.80) and (4.95) as

$$d\mathbf{v} = d\mathbf{v}^d + d\mathbf{v}^w, \tag{B.6}$$

where

$$d\mathbf{v}^d \equiv \mathbf{d}d\mathbf{x}, \quad d\mathbf{v}^w \equiv \mathbf{w}d\mathbf{x}. \tag{B.7}$$

Consider the infinitesimal line element $d\mathbf{x}$ parallel to the base vector $\mathbf{e}_1$ in the Cartesian base $\{\mathbf{e}_1, \mathbf{e}_2, \mathbf{e}_3\}$, that is, $d\mathbf{x} \equiv dx_1\mathbf{e}_1$. Then, we can write $d\mathbf{v}^d$ as follows:

$$d\mathbf{v}^d = \mathbf{d}d\mathbf{x} = (d_{rs}\mathbf{e}_r \otimes \mathbf{e}_s)(dx_1\mathbf{e}_1) = d_{r1}dx_1\mathbf{e}_r = d_{i1}dx_1\mathbf{e}_i \tag{B.8}$$

or

$$d\mathbf{v}^d = d_{11}dx_1\mathbf{e}_1 + \omega_3^d dx_1\mathbf{e}_2 + \omega_2^d dx_1\mathbf{e}_3, \tag{B.9}$$

where

$$d_{11} = \frac{\partial v_1}{\partial x_1}; \quad \omega_2^d \equiv d_{31} = \frac{1}{2}\left(\frac{\partial v_3}{\partial x_1} + \frac{\partial v_1}{\partial x_3}\right), \quad \omega_3^d \equiv d_{12} = \frac{1}{2}\left(\frac{\partial v_1}{\partial x_2} + \frac{\partial v_2}{\partial x_1}\right), \tag{B.10}$$

d_{11} denoting the rate of elongation, and ω_2^d and ω_3^d standing for the relative angular velocities around the base vectors $\mathbf{e}_2$ and $\mathbf{e}_3$, respectively, as illustrated in Figure B.1(a). Here, ω_2^d gives the minus angular velocity in the definition by the right-handed skew rule.

On the other hand, $d\mathbf{v}^w$ in equation $(\text{B.7})_2$ is written as

$$d\mathbf{v}^w = \mathbf{w}d\mathbf{x} = (w_{rs}\mathbf{e}_r \otimes \mathbf{e}_s)(dx_1\mathbf{e}_1) = w_{r1}dx_1\mathbf{e}_r = w_{i1}dx_1\mathbf{e}_i. \tag{B.11}$$

Here, denoting the axial vector caused by $\mathbf{w}$ as $\boldsymbol{\omega}^w$, it follows from equation (1.150) that

$$\boldsymbol{\omega}^w = -\frac{1}{2}\boldsymbol{\varepsilon}\mathbf{w}, \quad \omega_i^w = -\frac{1}{2}\varepsilon_{irs}w_{rs}, \tag{B.12}$$

which results in

$$\left.\begin{aligned} \omega_1^w &= -w_{23} = w_{32} = -\tfrac{1}{2}\left(\frac{\partial v_2}{\partial x_3} - \frac{\partial v_3}{\partial x_2}\right) \\ \omega_2^w &= -w_{31} = w_{13} = -\frac{1}{2}\left(\frac{\partial v_3}{\partial x_1} - \frac{\partial v_1}{\partial x_3}\right) \\ \omega_3^w &= w_{21} = -w_{12} = -\frac{1}{2}\left(\frac{\partial v_1}{\partial x_2} - \frac{\partial v_2}{\partial x_1}\right) \end{aligned}\right\} \tag{B.13}$$

Then, equation (B.11) is rewritten as

$$d\mathbf{v}^w = \omega_3^w dx_1\mathbf{e}_2 - \omega_2^w dx_1\mathbf{e}_3, \tag{B.14}$$

where ω_2^w and ω_3^w denote the relative angular velocities around the base vectors $\mathbf{e}_2$ and $\mathbf{e}_3$, respectively, as illustrated in Figure B.1(b).

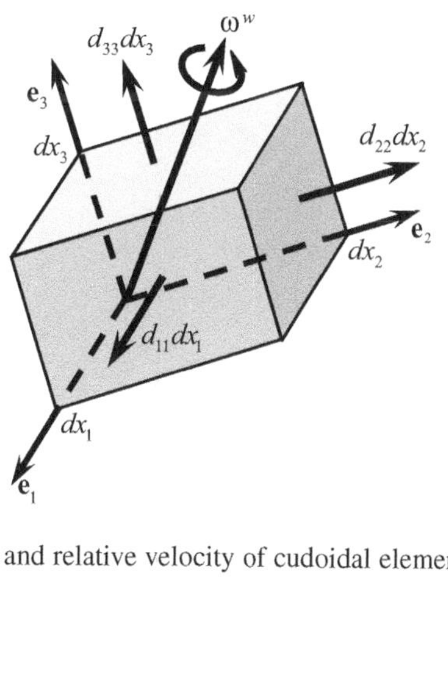

Figure B.2 Ddeformation rate and relative velocity of cudoidal element with sides in principle directions of strain rate

Having considered the rates of deformation and rotation with respect to the line element parallel to the base vector $\mathbf{e}_1$, similar results are obtained for the base vectors $\mathbf{e}_2$ and $\mathbf{e}_3$.

The resultant angular velocity of the infinitesimal line element $d\mathbf{x} \equiv dx_1\mathbf{e}_1$ caused by $\mathbf{d}$ and $\mathbf{w}$ are given from equations (B.10) and (B.13) by

$$\left.\begin{array}{ll} \omega_2 = -\omega_2^d + \omega_2^w = -d_{31} - w_{31} & \text{around } \mathbf{e}_2 \\ \omega_3 = \omega_3^d + \omega_3^w = d_{12} - w_{12} & \text{around } \mathbf{e}_3 \end{array}\right\} \tag{B.15}$$

If the base $\{\mathbf{e}_1, \mathbf{e}_2, \mathbf{e}_3\}$ is chosen in the principal direction of $\mathbf{d}$ leading to $d_{ij} = 0 (i \neq j)$, equation (B.15) becomes

$$\left.\begin{array}{ll} \omega_2 = \omega_2^w = -w_{31} & \text{around } \mathbf{e}_2 \\ \omega_3 = \omega_3^w = -w_{12} & \text{around } \mathbf{e}_3 \end{array}\right\} \tag{B.16}$$

and thus the rectangular parallelepiped element with sides parallel to $\{\mathbf{e}_1, \mathbf{e}_2, \mathbf{e}_3\}$ rotates by the relative angular velocity vector $\boldsymbol{\omega}^w$, deforming to keep the rectangular parallelepiped shape as shown in Figure B.2 (Leigh, 1968).

Appendix C

Proof for Derivative of Second Invariant of Logarithmic-Deviatoric Deformation Tensor

The following proof of equation $(7.47)_3$ is based on notes written by Prof. M. Wallin (2010) and supplemented by Prof. B. Raniecki (2010) in private correspondence with Hashiguchi.

The logarithm of the right stretch tensor, $\ln\mathbf{U}$, is the Hencky strain defined in equation (4.68), which is expressed by the Lagrangian triad $\mathbf{N}_\alpha$ and the principal values λ_α of $\mathbf{U}$ as follows:

$$\ln\mathbf{U} = \sum_{\alpha=1}^{3} \ln\lambda_\alpha \mathbf{N}_\alpha \otimes \mathbf{N}_\alpha, \tag{C.1}$$

the deviatoric tensor of which is given by

$$(\ln\mathbf{U})' = \sum_{\alpha=1}^{3} \{\ln\lambda_\alpha - (\ln J)/3\}\mathbf{N}_\alpha \otimes \mathbf{N}_\alpha = \sum_{\alpha=1}^{3} (\ln\underline{\lambda}_\alpha)\mathbf{N}_\alpha \otimes \mathbf{N}_\alpha, \tag{C.2}$$

where

$$\underline{\lambda}_\alpha = \lambda_\alpha J^{-1/3} = \lambda_\alpha (U_1U_2U_3)^{-1/3} = \sqrt{C_\alpha}(C_1C_2C_3)^{-1/6}, \tag{C.3}$$

which is the unimodular quantity defined in Section 4.6. The second invariant J_2 of deviatoric tensor $(\ln\mathbf{U})'$ (see equation (1.246)) is given by

$$J_2 = \frac{1}{2}\{(\ln\underline{\lambda}_1)^2 + (\ln\underline{\lambda}_2)^2 + (\ln\underline{\lambda}_3)^2\}, \tag{C.4}$$

from which we have

$$\frac{\partial J_2}{\partial \ln\underline{\lambda}_\alpha} = \ln\underline{\lambda}_\alpha. \tag{C.5}$$

Introduction to Finite Strain Theory for Continuum Elasto-Plasticity, First Edition. Koichi Hashiguchi and Yuki Yamakawa.

It follows from equation (C.3) that

$$\frac{\partial \ln \underline{\lambda}_\alpha}{\partial C_\beta} = \frac{\partial \ln\{\sqrt{C_\alpha}(C_1C_2C_3)^{-1/6}\}}{\partial C_\beta}$$

$$= \frac{1}{\sqrt{C_\alpha}(C_1C_2C_3)^{-1/6}}\left\{\frac{(C_1C_2C_3)^{-1/6}}{2\sqrt{C_\alpha}}\delta_{\alpha\beta} - \frac{1}{6}\sqrt{C_\alpha}(C_1C_2C_3)^{-7/6}\frac{C_1C_2C_3}{C_\beta}\right\},$$

from which

$$\frac{\partial \ln \underline{\lambda}_\alpha}{\partial C_\beta} = \frac{1}{2C_\alpha}\delta_{\alpha\beta} - \frac{1}{6}\frac{1}{C_\beta}. \tag{C.6}$$

Making use of equations (C.5) and (C.6), we further have

$$\frac{\partial J_2}{\partial C_\alpha} = \sum_{\beta=1}^{3}\frac{\partial J_2}{\partial \ln \underline{\lambda}_\beta}\frac{\partial \ln \underline{\lambda}_\beta}{\partial C_\alpha} = \sum_{\beta=1}^{3}\ln \underline{\lambda}_\beta\left(\frac{1}{2C_\beta}\delta_{\alpha\beta} - \frac{1}{6}\frac{1}{C_\alpha}\right) = \frac{1}{2}(\ln \underline{\lambda}_\alpha)C_\alpha^{-1} - \frac{1}{6C_\alpha}\sum_{\beta=1}^{3}\ln \underline{\lambda}_\beta,$$

from which it follows that

$$\frac{\partial J_2}{\partial \mathbf{C}} = \frac{1}{2}(\ln \mathbf{C}^{1/2})'\mathbf{C}^{-1} \tag{C.7}$$

because of $\sum_{\beta=1}^{3} \ln \underline{\lambda}_\beta = 0$.

Appendix D

Numerical Computation of Tensor Exponential Function and Its Derivative

The exponential function of a tensor along with its derivative plays a crucial role in computational plasticity, especially in the finite strain framework based on the multiplicative decomposition of the deformation gradient tensor. As discussed in Section 9.4, the time-integration scheme using the tensor exponential function is widely accepted as a suitable and efficient scheme to derive the incremental format of the finite strain elasto-plastic constitutive equation necessarily required in the stress-update algorithm within the time-discrete framework. The derivative of the tensor exponential function is required in the derivation of the Jacobian matrix and the consistent tangent modulus tensor, if the analytical approach using differentiation is employed; see Sections 9.8 and 9.9.

This appendix describes the explicit representation and the aspect of numerical computation of the tensor exponential function as well as its derivative. Moreover, Fortran computer programs for the tensor exponential function and its derivative are presented.

D.1 Numerical Computation of Tensor Exponential Function

The explicit expression for the exponential function of a generic (non-symmetric) second-order tensor $\mathbf{T}$ is given by the infinite series representation (1.258). The numerical computation of the tensor exponential function is done by truncating the infinite series with a prescribed accuracy; we compute taking a finite number of terms:

$$\exp[\mathbf{T}] \approx \sum_{n=0}^{N} \frac{1}{n!}\mathbf{T}^n = \mathbf{I} + \mathbf{T} + \frac{1}{2}\mathbf{T}^2 + \ldots + \frac{1}{N!}\mathbf{T}^N, \tag{D.1}$$

where the number of terms N required to fulfill the prescribed accuracy is determined by checking the convergence criterion

$$\frac{1}{N!}\left\|\mathbf{T}^N\right\| < \mathrm{TOL}, \tag{D.2}$$

where TOL is the prescribed tolerance. In the numerical analysis presented in Section 9.10, the tolerance was taken as $\mathrm{TOL} = 1.0 \times 10^{-18}$.

The tensor exponential function can also be expressed as formula (D.1) involves a costly computation of the power of the argument $\mathbf{T}^n$, though it can be reduced with the help of a recursion formula based on the Cayley–Hamilton theorem explained in Section 1.6.6 (Miehe, 1996a).

The evaluation algorithm for the tensor exponential function based on formula (D.1) is implemented in the Fortran subroutine `matexp` presented in the next section.

D.2 Fortran Subroutine for Tensor Exponential Function: `matexp`

The Fortran subroutine `matexp`, which consists of two segments shown in Tables D.1 and D.2, evaluates the exponential function of a 3×3 non-symmetric matrix on the basis of the series representation (D.1). The numerical implementation of the computation based on equation (D.1) is quite straightforward, and the source code includes detailed comment lines describing the procedures within the algorithm, so explanation of the source code is omitted here. The list of arguments for `matexp` is shown in Table D.3. If the convergence criterion (D.2) is fulfilled for $N < N_{\mathrm{max}}$, the error flag `ierr` is set equal to N; otherwise, `ierr` is set equal to -1 and then the calculation is terminated.

Table D.1 Subroutine `matexp`, segment 1

line program code

```
      subroutine matexp(nmax,etol,a,aexp,ierr)
      implicit double precision (a-h,o-z)
      dimension a(3,3),aexp(3,3),an(3,3),atmp(3,3)
      common /iofile/ior1,ior2,ior3,ior4,ior5,
     &               iow1,iow2,iow3,iow4,iow5,
     &               iow6,iow7,iow8,iow9,iow10
      common /etc/del(3),delta(3,3),p(3,3,3),dpai
c
c     compute exponential function of a matrix A(3,3)
c
c ----- initialize matrix (identity matrix as 0th-order approximation)
      aexp(1:3,1:3) = delta(1:3,1:3)
      an(1:3,1:3)   = delta(1:3,1:3)
      atmp(1:3,1:3) = delta(1:3,1:3)
c
c ----- initialize factorial of n and error flag
      dn = 1.0d0
      ierr = 0
c
c ----- 0th-order approximation
      if(nmax.eq.0) then
       goto 900
      end if
```

(continued)

Table D.2 Subroutine `matexp`, segment 2

```
c
        n = 1
  100 continue
c
c ----- compute A(3,3)^n and its norm
        an(1:3,1:3) = matmul(atmp(1:3,1:3),a(1:3,1:3))
        dd = 0.0d0
        do i=1,3
         do j=1,3
          dd = dd+an(i,j)**2
         end do
        end do
        dd = dsqrt(dd)
        atmp(1:3,1:3) = an(1:3,1:3)
c
c ----- compute factorial of n
        dn = dn*dble(n)
c
c ----- compute nth-order approximation of exp[A(3,3)]
        aexp(1:3,1:3) = aexp(1:3,1:3)+an(1:3,1:3)/dn
c
c ----- check convergence
        dx = dd/dn
        if(dx.lt.etol) then
         ierr = n
         goto 900
        else if((n.lt.nmax).and.(dx.ge.etol)) then
         n = n+1
         goto 100
        else if((n.ge.nmax).and.(dx.ge.etol)) then
         ierr = -1
         goto 900
        end if
  900 continue
c
        return
        end subroutine matexp
```

Table D.3 List of the arguments for the subroutine `matexp`

Argument	Input	Output	Symbol	Remark
`nmax`	Maximum number of terms of the series representation	Unchanged	N_{max}	
`etol`	Convergence tolerance	Unchanged	TOL	
`a(3,3)`	Matrix	Unchanged	$\mathbf{T}$	
`aexp(3,3)`	—	Tensor exponential function	$\exp[\mathbf{T}]$	
`ierr`	—	N or -1	—	N: converged, -1: not converged

D.3 Numerical Computation of Derivative of Tensor Exponential Function

The exact derivative of the tensor exponential function is obtained straightforwardly by differentiation of the series representation (1.258) term by term, which leads to the closed-form formula (1.374). This formula is applicable to a non-symmetric tensor. The detailed derivation of this formula is presented below equation (1.374).

The numerical implementation of the formula (1.374) based on the series representation has been described by de Souza Neto (2001), and is quite straightforward. Similar to equation (D.1), the numerical computation of the derivative of tensor exponential function is carried out by taking a finite number of terms, from which we have

$$\frac{\partial \exp[\mathbf{T}]}{\partial \mathbf{T}} \approx \sum_{n=1}^{N} \frac{1}{n!} \sum_{m=1}^{n} \mathbf{T}^{m-1} \widetilde{\otimes} \mathbf{T}^{n-m}, \tag{D.3}$$

in which we use the dyadic multiplication denoted by the symbol $\widetilde{\otimes}$ in equation $(1.208)_4$. For the convenience of numerical implementation, equation (D.3) is expressed in the index notation

$$\left(\frac{\partial \exp[\mathbf{T}]}{\partial \mathbf{T}}\right)_{ijkl} = \sum_{n=1}^{N} \frac{1}{n!} \sum_{m=1}^{n} (\mathbf{T}^{m-1})_{ik} (\mathbf{T}^{n-m})_{lj}. \tag{D.4}$$

The number of terms N is determined, for instance, by equation (D.2) or the following criterion so that the finite series representation (D.3) fulfills the prescribed accuracy:

$$\frac{1}{N!} \sum_{m=1}^{N} \|\mathbf{T}^{m-1} \widetilde{\otimes} \mathbf{T}^{N-m}\| < \mathrm{TOL}. \tag{D.5}$$

The formula (D.4) enables us to compute the derivative of the tensor exponential function to any prescribed degree of accuracy.

The algorithm to evaluate the derivative of tensor exponential function by means of the formula (D.4) is implemented in the Fortran subroutine `matdex` presented in the next section.

Table D.4 Subroutine matdex, segment 1

```
      subroutine matdex(nmax,etol,a,daexp,ierr)
      implicit double precision (a-h,o-z)
      dimension a(3,3),daexp(3,3,3,3)
      dimension an(20,3,3),tmpa(3,3,3,3),dn(20)
      common /iofile/ior1,ior2,ior3,ior4,ior5,
     &                iow1,iow2,iow3,iow4,iow5,
     &                iow6,iow7,iow8,iow9,iow10
      common /etc/del(3),delta(3,3),p(3,3,3),dpai
c
c     compute derivative of exponential function of a matrix A(i,j)
c
c ----- check array size limit
      if(nmax.gt.20) then
       write(iow1,*) '--- too large nmax is specified in matdex.'
       write(   *,*) '--- too large nmax is specified in matdex.'
       stop
      end if
c
c ----- initialize error flag
      ierr = 0
c
c ----- compute A(i,j)^n and n-factorial for n = 1, ..., nx,
c       which satisfies (1/nx!) norm{A^(nx)} < etol.
c   --- n = 1
      dd = 0.0d0
      do i=1,3
       do j=1,3
        an(1,i,j) = a(i,j)
        dd = dd+an(1,i,j)**2
       end do
      end do
      dd = dsqrt(dd)
      dn(1) = 1.0d0
      dx = dd/dn(1)
      if(dx.lt.etol) then
       ierr = 1
       nx = 1
       goto 110
      end if
```

(continued)

Table D.5 Subroutine `matdex`, segment 2

```
c   --- n = 2, 3, ..., nx
          ik = 2
  100 continue
          dd = 0.0d0
          do i=1,3
           do j=1,3
            an(ik,i,j) = an(ik-1,i,1)*a(1,j)
     &                  +an(ik-1,i,2)*a(2,j)
     &                  +an(ik-1,i,3)*a(3,j)
            dd = dd+an(ik,i,j)**2
           end do
          end do
          dd = dsqrt(dd)
          dn(ik) = dn(ik-1)*dble(ik)
          dx = dd/dn(ik)
          if(dx.lt.etol) then
           ierr = ik
           nx = ik
           goto 110
          else if((ik.lt.nmax).and.(dx.ge.etol)) then
           ik = ik+1
           goto 100
          else if((ik.ge.nmax).and.(dx.ge.etol)) then
           ierr = -1
           ik = nmax
           goto 110
          end if
  110 continue
c
c ----- daexp(i,j,k,l) = (1st term) + (2nd term)
          do i=1,3
           do j=1,3
            do k=1,3
             do l=1,3
              daexp(i,j,k,l) = delta(i,k)*delta(l,j)
              if(nx.ge.2) then
               daexp(i,j,k,l) = daexp(i,j,k,l)
     &                   +(delta(i,k)*a(l,j)+a(i,k)*delta(l,j))/dn(2)
              end if
             end do
            end do
           end do
          end do
c
          if(nx.le.2) then
           goto 900
          end if
```

(continued)

Table D.6 Subroutine `matdex`: segment 3

```
c
c ----- compute 3rd and higher order (pth) terms...
          do 200 ip=3,nx
           do i=1,3
            do j=1,3
             do k=1,3
              do l=1,3
c    ----- compute 1st term in pth term
                tmpa(i,j,k,l) = delta(i,k)*an(ip-1,l,j)/dn(ip)
                do 210 iq=2,ip-1
c    ----- compute qth term in pth term
                 tmpa(i,j,k,l) = tmpa(i,j,k,l)
     &                         +an(iq-1,i,k)*an(ip-iq,l,j)/dn(ip)
  210          continue
c    ----- compute pth (last) term in pth term
                tmpa(i,j,k,l) = tmpa(i,j,k,l)
     &                        +an(ip-1,i,k)*delta(l,j)/dn(ip)
c   ----- compute pth term
                daexp(i,j,k,l) = daexp(i,j,k,l)+tmpa(i,j,k,l)
               end do
              end do
             end do
            end do
  200 continue
c
  900 continue
          return
          end subroutine matdex
```

Table D.7 List of the arguments for the subroutine `matdex`

Argument	Input	Output	Symbol	Remark
`nmax`	Maximum number of terms of the series representation	Unchanged	$N_{\max}$	
`etol`	Convergence tolerance	Unchanged	TOL	
`a(3,3)`	Matrix	Unchanged	$\mathbf{T}$	
`daexp(3,3)`	—	Derivative of the tensor exponential function	$\partial \exp[\mathbf{T}]/\partial \mathbf{T}$	
`ierr`	—	N or -1	—	N: converged, -1: not converged

D.4 Fortran Subroutine for Derivative of Tensor Exponential Function: `matdex`

The Fortran subroutine `matdex`, which consists of the three segments shown in Tables D.4–D.6, evaluates the derivatives of the exponential function of a 3×3 non-symmetric matrix by means of the closed-form formula (D.4), which is derived by straightforward differentiation of the series representation. The numerical implementation of the algorithm for the calculation based on formula (D.4) is straightforward, and detailed comment lines are provided in the source code, so the explanation of the source code is omitted. The list of the arguments for `matdex` is shown in Table D.7. If the convergence criterion (D.5) is fulfilled for $N < N_{\max}$, the error flag `ierr` is set equal to N; otherwise, `ierr` is set equal to -1 and the calculation is terminated.

References

Amirkhizi, A.V. and Nemat-Nasser, S. (2007) A framework for numerical integration of crystal elasto-plastic constitutive equations compatible with explicit finite element codes, *Int. J. Plasticity*, **23**, 1918–1937.

Arghavani, J., Auricchio, F. Naghdabadi, R. and Reali, A. (2011) On the robustness and efficiency of integration algorithms for a 3D finite strain phenomenological SMA constitutive model, *Int. J. Numer. Meth. Eng.*, **85**, 107–134.

Aris, R. (1962) *Vectors, Tensors, and the Basic Equations of Fluid Mechanics*, New York: Dover Publishing, Inc.

Armstrong, P.J. and Frederick, C.O. (1966) *A mathematical representation of the multiaxial Bauschinger effect*, G.E.G.B. Report RD/B/N 731.

Asaro, R. and Lubarda, V.A. (2006) *Mechanics of Solids and Materials*, Cambridge: Cambridge Univ. Press.

Auricchio, F. and Taylor, R.L. (1999) A return-map algorithm for general associative isotropic elasto-plastic materials in large deformation regimes, *Int. J. Plasticity*, **15**, 1359–1378.

Başar, Y. and Weichert, D. (2000) *Nonlinear Continuum Mechanics: Fundamental Mathematical and Physical Concepts*, Berlin: Springer.

Becker, E. und Burger, W. (1975) *Kontinuumsmechanik*, Stattgart: B.G. Teubner.

Belytschko, T., Liu, W.K. and Moran, B. (2000) *Nonlinear Finite Elements for Continua and Structures*, Chichester: John Wiley and Sons, Ltd.

Besseling, J.F. and van der Gissen (1994) *Mathematical Modelling of Inelastic Deformation*, New York: Chapman & Hall.

Bertram, A. (2008) *Elasticity and Plasticity of Large Deformations*, Berlin: Springer.

Biot, M.A. (1965) *Mechanics of Incremental Deformations*, New York. John Wiley & Sons, Inc.

Bonet, J. and Wood, R.D. (2008) *Nonlinear Continuum Mechanics for Finite Element Analysis* (2nd edn), Cambridge: Cambridge University Press.

Borja, R.I. and Tamagnini, C. (1998) Cam-Clay plasticity, Part III: Extension of the infinitesimal model to include finite strains, *Comput. Meth. Appl. Mech. Eng.*, **155**, 73–95.

Borja, R.I., and Lin, C.H. and Montans, F.I. (2001) Cam-clay plasticity, Part IV: Implicit integration of anisotropic bounding surface model with nonlinear hyperplasticity and ellipsoidal loading function, *Comput. Meth. Appl. Mech. Eng.*, **190**, 3293–3323.

Bowen, R.M. and Wang, C.-C. (1976) *Introduction to Vectors and Tensors*, New York: Plenum Press.

Bowen, R.M. and Wang, C.-C. (2008) *Introduction to Vectors and Tensors*, Minneola, NY: Dover.

Bruhns, O.T. (2003) *Continuum Mechanics with Logarithmic Rate*, Lecture Notes, Kyushu University.

Brunig, M. (1998) Nonlinear finite element analysis based on a large strain deformation theory of plasticity, *Comput. Struct.*, **69**, 117–128.

Callari, C., Auricchio, F. and Sacco, E. (1998) A finite-strain Cam-clay model in the framework of multiplicative elasto-plasticity, *Int. J. Plasticity*, **14**, 1155–1187.

Carlson, D.E. and Hoger, A. (1986) The derivative of a tensor-valued function of a tensor, *Quart. Appl. Math.*, **406**, 409–423.

Chaboche, J.L. (2008) A review of some plasticity and viscoplasticity constitutive theories, *Int. J. Plasticity*, **24**, 1642–1693.

Introduction to Finite Strain Theory for Continuum Elasto-Plasticity, First Edition. Koichi Hashiguchi and Yuki Yamakawa.

Chaboche, J.L., Dang-Van, K. and Cordier, G. (1979) Modelization of the strain memory effect on the cyclic hardening of 316 stainless steel, *Trans. 5th Int. Conf. SMiRT*, Berlin, Division L., Paper No. L. 11/3.

Chadwick, P. (1976) *Continuum Mechanics*, London: George Allen & Unwin.

Chung, T.J. (2007) *General Continuum Mechanics*, Cambridge: Cambridge Univ. Press.

Crisfield, M.A. (1991) *Non-linear Finite Element Analysis of Solids and Structures, Vol. 1: Essentials*, Chichester: John Wiley & Sons, Ltd.

Ciarlet, P.G. (1988) *Mathematical Elasticity I: Three-Dimensional Elasticity*, Amsterdam: North-Holland.

Clifton, R.J. (1972) On the equivalence of $\mathbf{F} = \mathbf{F}^e\mathbf{F}^p$ and $\mathbf{F} = \mathbf{F}^p\mathbf{F}^e$, *J. Appl. Mech.* (ASMS), **14**, 703–717.

Collins, I.F. and Houlsby, G.T. (1997) Application of the thermomechanical principles to the modeling of geomechnical materials, *Proc. Royal Soc. London A*, **453**, 1975–2001.

Crisfield, M.A. (1997) *Non-linear Finite Element Analysis of Solids and Structures, Vol. 2: Advanced Topics*, Chichester: John Wiley & Sons, Ltd.

Dafalias, Y.F. (1983) Corotational rates for kinematic hardening at large plastic deformations. *J. Appl. Mech.* (ASME), **50**, 561–565.

Dafalias, Y.F. (1984) The plastic spin concept and a simple illustration of its role in finite plastic transformation, *Mech. Materials*, **3**, 223–233.

Dafalias, Y.F. (1985) The plastic spin, *J. Appl. Mech.* (ASME), **52**, 865–871.

Dafalias, Y.F. (1998) Plastic spin: Necessity or redundancy? *Int. J. Plasticity*, **14**, 909–931.

Dafalias, Y. F. and Popov, E. P. (1975) A model of nonlinearly hardening materials for complex loading. *Acta Mech.*, **23**, 173–192.

Davidson, L. (1995) Kinematics of finite elastoplastic deformation, *Mech. Materials*, **21**, 73–88.

de Souza Neto, E.A. (2001) The exact derivative of the exponential of an unsymmetric tensor, *Compt. Meh. Appl. Mech. Eng.*, **190**, 2377–2383.

de Souza Neto, E.A., Perić, D. and Owen, D.J.R. (2008) *Computational Methods for Plasticity: Theory and Applications*, Chichester: John Wiley & Sons, Ltd.

del Piero, G. (1979) Some properties of the set of fourth-order tensors, with application to elasticity, *J. Elasticity*, **9**, 245–261.

Dettmer, W. and Reese, S. (2004) On the theoretical and numerical modelling of Armstrong–Frederick kinematic hardening in the finite strain regime, *Compt. Meth. Appl. Mech. Eng.*, **193**, 87–116.

Dienes, J.K. (1979) On the analysis of rotation and stress rate in deforming bodies, *Acta Mech.*, **32**, 217–232.

Dimitrienko, Y.I. (2010) *Tensor Analysis and Nonlinear Tensor Functions*, Dordrecht: Kluwer Academic.

Dimitrienko, Y.I. (2011) *Nonlinear Continuum Mechanics and Large Inelastic Deformations*, Berlin: Springer.

Drucker, D.C. (1950) A more fundamental approach to plastic stress–strain relations, *Proc. 1st U.S. National Congr. Appl. Mech.* (ASME), 1, pp. 487–491.

Drucker, D.C. (1988) Conventional and unconventional plastic response and representation, *Appl. Mech. Rev.* (ASME), **41**, 151–167.

Dunne, F. and Petrinic, N. (2005) *Introduction to Computational Plasticity*, Oxford: Oxford University Press.

Dutko, M., Perić, D. and Owen, D.R.J. (1993) Universal anisotropic yield criterion based on superquadric functional representation: Part 1. Algorithmic issues and accuracy analysis, *Compt. Meth. Appl. Mech. Eng.*, **109**, 73–93.

Dvorkin, E.N. and Goldschmit, M.B. (2006) *Nonlinear Continua*, Berlin: Springer.

Eidel, B. and Gruttmann, F. (2003) Elastoplastic orthotropy at finite strains: multiplicative formulation and numerical implementation, *Compt. Materials Sci.*, **28**, 732–742.

Ekh, M. and Runesson, K. (2001) Modeling and numerical issues in hyperelasto-plasticity with anisotropy, *Int. J. Solids Struct.*, **38**, 9461–9478.

Eringen, A.C. (1962) *Nonlinear Theory of Continuous Media*, New York: McGraw-Hill.

Eringen, A.C. (1967) *Mechanics of Continua*, New York: Rebert E. Krieger Publishing Co.

Eringen, A.C. (1975) *Continuum Physics*, Part 2, New York: Academic Press.

Fellin, W. and Ostermann, A. (2002) Consistent tangent operators for constitutive rate equations, *Int. J. Numer. Anal. Meth. Geomech.*, **26**, 1213–1233.

Fish, J. and Shek, K. (1999) Computational aspects of incrementally objective algorithms for large deformation plasticity, *Int. J. Numer. Meth. Eng.*, **44**, 839–851.

Flanagan, D.P. and Taylor, L.M. (1987) An accurate numerical algorithm for stress integration with finite rotations, *Compt. Meth. Appl. Mech. Eng.*, **62**, 305–320.

Flügge, W. (1972) *Tensor Analysis and Continuum Mechanics*, Berlin: Springer-Verlag.

Fox, N. (1968) On the continuum theories of dislocations and plasticity, *Q. J. Mech. Appl. Math.*, **21**, 67–75.

Fung, Y.C. (1965) *Foundations of Solid Mechanics*, London: Prentice-Hall, Inc.

Fung, Y.C. (1969) *A First Course in Continuum Mechanics*, London: Prentice-Hall, Inc.

Green, A.E. and Naghdi, P.M. (1965) A general theory of an elastic–plastic continuum, *Arch. Ration. Mech. Anal.*, **18**, 251–281.

Gurtin, M.E. (1981) *An Introduction to Continuum Mechanics*, New York: Academic Press, Inc.

Gurtin, M.E. and Anand, L. (2005) The decomposition $\mathbf{F} = \mathbf{F}^e\mathbf{F}^p$, material symmetry, and plastic irrotationality for solids that are isotropic-viscoplastic or amorphous, *Int. J. Plasticity*, **21**, 1686–1719.

Han, C.-S., Chung, K., Wagoner, R.H. and Oh, S.-I. (2003) A multiplicative finite elasto–plastic formulation with anisotropic yield functions, *Int. J. Plasticity*, **19**, 197–211.

Hashiguchi, K. (1980) Constitutive equations of elastoplastic materials with elastic–plastic transition, *J. Appl. Mech.* (ASME), **47**, 266–272.

Hashiguchi, K. (1989) Subloading surface model in unconventional plasticity, *Int. J. Solids Struct.*, **25**, 917–945.

Hashiguchi, K. (1993a) Fundamental requirements and formulation of elastoplastic constitutive equations with tangential plasticity, *Int. J. Plasticity*, **9**, 525–549.

Hashiguchi, K. (1993b) Mechanical requirements and structures of cyclic plasticity models, *Int. J. Plasticity*, **9**, 721–748.

Hashiguchi, K. (1997) The extended flow rule in plasticity, *Int. J. Plasticity*, **13**, 37–58.

Hashiguchi, K. (2000) Fundamentals in constitutive equation: continuity and smoothness conditions and loading criterion, *Soils and Foundations*, **40**(3), 155–161.

Hashiguchi, K. (2001a) Description of inherent/induced anisotropy of soils: Rotational hardening rule with objectivity, *Soils and Foundations*, **41**(6), 139–145.

Hashiguchi, K. (2001b) On the thermomechanical approach to the formulation of plastic constitutive equations, *Soils and Foundations*, **41**(4), 89–94.

Hashiguchi, K. (2005) Subloading surface model with Tangential relaxation, *Proc. Int. Symp. Plasticity '05*, pp. 259–261.

Hashiguchi, K. (2009) *Elastoplasticity Theory*, Berlin: Springer: Second Edition, forthcoming, February 2013.

Hashiguchi, K. (2011) General interpretations and tensor symbols for pull-back, push-forward and convected derivative, *Proc. JSME 24th Comp. Mech. Conf.* (JSME), pp. 669–671.

Hashiguchi, K. and Chen, Z.-P. (1998) Elastoplastic constitutive equation of soils with the subloading surface and rotational hardening, *Int. J. Numer. Anal. Meth. Geomech.*, **22**, 197–227.

Hashiguchi, K. and Mase, T. (2007) Extended yield condition of soils with tensile strength and rotational hardening, *Int. J. Plasticity*, **23**, 1939–1956.

Hashiguchi, K. and Ozaki, S. (2008) Constitutive equation for friction with transition from static to kinetic friction and recovery of static friction, *Int. J. Plasticity*, **24**, 2102–2124.

Hashiguchi, K. and Protasov, A. (2004) Localized necking analysis by the subloading surface model with tangential-strain rate and anisotropy, *Int. J. Plasticity*, **20**, 1909–1930.

Hashiguchi, K. and Ueno, M. (1977) Elastoplastic constitutive laws of granular materials. In S. Murayama and A.N. Schofield (eds), *Constitutive Equations of Soils*, Proc. 9th Int. Conf. Soil Mech. Found. Eng., Special Series 9, Tokyo, JSSMFE, pp. 73–82.

Hashiguchi, K. and Tsutsumi, S. (2003) Shear band formation analysis in soils by the subloading surface model with tangential stress rate effect, *Int. J. Plasticity*, **19**, 1651–1677.

Hashiguchi, K. and Tsutsumi, S. (2007) Gradient plasticity with the tangential subloading surface model and the prediction of shear band thickness of granular materials, *Int. J. Plasticity*, **23**, 767–797.

Hashiguchi, K. and Yamakawa, Y. (2011) Unconventional finite strain theory based on the subloading surface model, *Proc. Plasticity 2011*, pp.106–108.

Hashiguchi, K. and Yamakawa, Y. (2012) Finite elastoplastic constitutive equation based on subloading surface model, *Proc. Symp. Theor. Appl. Mech., Japan.*

Hashiguchi, K., Saitoh, K., Okayasu, T. and Tsutsumi, S. (2002) Evaluation of typical conventional and unconventional plasticity models for prediction of softening behavior of soils, *Géotechnique*, **52**, 561–573.

Hashiguchi, K., Ozaki, S. and Okayasu, T. (2005) Unconventional friction theory based on the subloading surface concept, *Int. J. Solids Struct.*, **42**, 1705–1727.

Hashiguchi, K., Ueno, M. and Ozaki, T. (2012) Elastoplastic model of metals with smooth elastic–plastic transition, *Acta Mech.*, **23**, 985–1013.

Haupt, P. (2000) *Continuum Mechanics and Theory of Materials*, Berlin: Springer.

Heidari, M., Vafai, A. and Desai, C. (2009) An Eulerian multiplicative constitutive model of finite elastoplasticity, *Europ. J. Mech. A/Solids*, **28**, 1088–1097.

Helm, D. (2001) *Formgedächtnislegierungen: Experimentelle Untersuchung, phänomenologische Modellierung und numerische Simulation der thermomechanischen Materialeigenschaften*, Kassel: Universitätsbibliothek.

Helm, D. (2006) Stress computation in finite thermoviscoplasticity, *Int. J. Plasticity*, **22**, 1699–1727.

Henann, D.L. and Anand, L. (2009) A large deformation theory for rate-dependent elastic–plastic materials with combined isotropic and kinematic hardening, *Int. J. Plasticity*, **25**, 1833–1878.

Hill, R. (1950) *Mathematical Theory of Plasticity*, Oxford: Clarendon Press.

Hill, R. (1967) On the classical constitutive relations for elastic/plastic solids, *Recent Progress in Appl. Mech.*, pp. 241–249.

Hill, R. (1968) On constitutive inequality for simple materials–1, *J. Mech. Phys. Solids*, **16**, 229–242.

Hill, R. (1970) Constitutive inequalities for isotropic elastic solids under finite strain, *Proc. Roy. Soc., London A*, **314**, 457–472.

Hill, R. (1978) Aspects of invariance in solid mechanics, *Advances in Applied Mechanics*, **18**, 1–75.

Hirsch, N.W. and Small, S. (1974) *Differential Equations, Dynamic Systems and Linear Algebra*, Academic Press.

Hisada, T. (1992) *Tensor Analysis for Nonlinear Finite Element Method*, Maruzen (in Japanese).

Hughes, T.J.R. and Winget, J. (1980) Finite rotation effects in numerical integration of rate constitutive equations arising in large-deformation analysis, *Int. J. Numer. Meth. Eng.*, **15**, 1862–1867.

Ilyushin, A.A. (1961) On the postulate of plasticity. *Appl. Math, and Meek*, **25**, 746–752.

Irgens, F. (2008) *Continuum Mechanics*, Berlin: Springer.

Iwan, W.D. (1967) On a class of models for the yielding behavior of continuous and composite systems, *J. Appl. Mech.* (ASME), **34**, 612–617.

Jaumann, G. (1911) Geschlossenes System physikalischer und chemischer Differentialgesetze, *Sitzungsber. Akad. Wiss.*, Wien (IIa), **120**, 385–530.

Jeremić, B., Sture, S. (1997) Implicit integrations in elastoplastic geotechnics, *Mech. Cohes.-Frict. Mater.*, **2**, 165–183.

Jeremić, B., Runesson, K. and Sture, S. (2001) Finite deformation analysis of geomaterials, *Int. J. Numer. Anal. Methods Geomech.*, **25**, 809–840.

Johansson, G. and Ekh, M. (2006) On the modeling of evolving anisotropy and large strains in pearlitic steel, *Europ. J. Mech. A/Solids*, **25**, 1041–1060.

Johansson, G., Ekh, M. and Runesson, K. (2005) Computational modeling of inelastic large ratcheting strains, *Int. J. Plasticity*, **21**, 955–980.

Jeffreys, H. (1931) *Cartesian Tensors*, Cambridge: Cambridge University Press.

Kay, D.C. (1988) *Tensor Calculus*, New York: McGraw-Hill.

Khan, A.S. and Huang, S. (1995) *Continuum Theory of Plasticity*, New York: John Wiley & Sons, Inc.

Kintzel, O. and Başar, Y. (2006) Fourth-order tensors – tensor differentiation with applications to continuum mechanics, Part I: Classical tensor analysis, *ZAMM: Zeitschrift für Angewandte Mathematik und Mechanik*, **86**, 291–311.

Kratochvil, J. (1971) Finite-strain theory of crystalline elastic-inelastic materials. *J. Appl. Phys.*, **42**, 1104–1108.

Krieg, R.D. (1975) A practical two surface plasticity theory, *J. Appl. Mech.* (ASME), **42**, 641–646.

Kroner, E. (1959) Allgemeine Kontinuumstheoreie der Versetzungen und Eigenspannnungen, *Arch. Ration. Mech. Anal.*, **4**, 273–334.

Kyoya, T. (2008) *Continuum Mechanics Note*, Morikita (in Japanese).

Lambrecht, M. and Miehe, C. (1999) Two non-associated isotropic elastoplastic hardening models for frictional materials, *Acta Mech.*, **135**, 73–90.

Lee, E.H. (1969) Elastic–plastic deformation at finite strain, *J. Appl. Mech.* (ASME), **36**, 1–6.

Lee, E.H. and Liu, D.T. (1967) Finite-strain elastic–plastic theory with application to plane-wave analysis, *J. Appl. Phys.*, **38**, 19–27.

Leigh, D.C. (1968) *Nonlinear Continuum Mechanics: An Introduction to the Continuum Physics and Mechanical Theory of the Nonlinear Mechanical Behavior of Materials*, New York: McGraw-Hill.

Lion, A. (2000) Constitutive modeling in finite thermoviscoplasticity: a physical approach based on nonlinear rheological models, *Int. J. Plasticity*, **16**, 469–494.

Loret, B. (1983) On the effects of plastic rotation in the finite deformation of anisotropic elastoplastic materials. *Mech. Materials*, **2**, 287–304.

Lubarda, V.A. and Lee, E.H. (1981) A correct definition of elastic and plastic deformation and its computational significance, *J. Appl. Mech.* (ASME), **48**, 35–40.

Lubarda, V.A. (1999) Duality in constitutive formulation of finite-strain elastoplasticity based on $\mathbf{F} = \mathbf{F}^e\mathbf{F}^p$and $\mathbf{F} = \mathbf{F}^p\mathbf{F}^e$decompositions, *Int. J. Plasticity*, **15**, 1277–1290.

Lubarda, V.A. (2002) *Elastoplasticity Theory*, Boca Raton, FL: CRC Press.

Lubliner, J. (1986) Normality rules in large-deformation plasticity, *Mech. Materials*, **5**, 29–34.

Lubliner, J. (1990) *Plasticity Theory*, New York: Dover Publ.

Malvern, L.E. (1969) *Introduction to the Mechanics of a Continuous Medium*, London: Prentice-Hall.

Mandel, J. (1972a) Director vectors and constitutive equations for plastic and viscoplastic media, in A. Sawczuk (ed.), *Problems of Plasticity* (*Proc. Int. Symp. Foundation of Plasticity*), Leyden: Noordhoff, pp. 135–141.

Mandel, J. (1972b) *Plasticité Classique et Viscoplasticité*, Vienna : Springer-Verlag.

Mandel, J. (1973) Equations constitutives directeurs dans les milieux plastiques et viscoplastiques, *Int. J. Solids Struct.*, **9**, 725–740.

Marsden, J.E. and Hughes, T.J.R. (1983) *Mathematical Foundation of Elasticity*, Englewood Cliffs, NJ: Prentice Hall.

Maugin, G.A. (1994) Eshelby stress in elastoplasticity and ductile fracture, *Int. J. Plasticity*, **10**, 393–408.

Menzel, A. (2006) Relations between material, intermediate and spatial generalized strain measures for anisotropic multiplicative inelasticity, *Acta Mech.*, **182**, 231–252.

Menzel, A., Ekh, M., Runesson, K. and Steinmann, P. (2005) A framework for multiplicative elastoplasticity with kinematic hardening coupled to anisotropic damage, *Int. J. Plasticity*, **21**, 397–434.

Menzel, A., and Steinmann, P. (2003a) Geometrically non-linear anisotropic inelasticity based on fictitious configurations: Application to the coupling of continuum damage and multiplicative elasto-plasticity, *Int. J. Numer. Meth. Eng.*, **56**, 2233–2266.

Menzel, A., and Steinmann, P. (2003b) On the spatial formulation of anisotropic multiplicative elasto-plasticity, *Compt. Meth. Appl. Mech. Eng.*, **192**, 3431–3470.

Menzel, A., Ekh, M., Runesson, K. and Steinmann, P. (2005) A framework for multiplicative elastoplasticity with kinematic hardening coupled to anisotropic damage, *Int. J. Plasticity*, **21**, 397–434.

Miehe, C. (1994) On the representation of Prandtl-Reuss tensors within the framework of multiplicative elastoplasticity, *Int. J. Plasticity*, **10**, 609–621.

Miehe, C. (1996a) Exponential map algorithm for stress updates in anisotropic multiplicative elastoplasticity for single crystals, *Int. J. Numer. Meth. Eng.*, **39**, 3367–3390.

Miehe, C. (1996b) Numerical computation of algorithmic (consistent) tangent moduli in large computational inelasticity, *Compt. Meth. Appl. Mech. Eng.*, **134**, 223–240.

Moran, B., Ortiz, M. and Shih, C.F. (1990) Formulation of implicit finite element methods for multiplicative finite deformation plasticity, *Int. J. Numer. Meth. Eng.*, **29**, 483–514.

Mroz, Z. (1967) On the description of anisotropic workhardening, *J. Mech. Phys. Solids*, **15**, 163–175.

Nagtegaal, J.C. and de Jong, J.E. (1981) Some computational aspects of elastic–plastic large strain analysis, *Int. J. Numer. Meth. Eng.*, **17**, 15–41.

Negahban, M. (2012) *The Mathematical and Thermodynamical Theory of Plasticity*, New York, CRC Press.

Nemat-Nasser, S. (1979) Decomposition of strain measures and their rates in finite deformation elastoplasticity, *Int. J. Solids Struct.*, **15**, 155–166.

Nemat-Nasser, S. (2004) *Plasticity: A Treatise on Finite Deformation of Heterogeneous Inelastic Materials*, Cambridge: Cambridge Univ. Press.

Ogden, R.W. (1984) *Non-linear Elastic Deformations*, Chichester: Halsted.

Oldroyd, J.G. (1950) On the formulation of rheological equations of state. *Proc. Roy. Soc. London A*, **200**, 523–541.

Ortiz, M. and Simo, J.C. (1986) An analysis of a new class of integration algorithms for elastoplastic constitutive relations, *Int. J. Numer. Meth. Eng.*, **23**, 353–366.

Ortiz, M., Radovitzky, R.A. and Repetto, E.A. (2001) The computation of the exponential and logarithmic mappings and their first and second linearizations, *Int. J. Numer. Meth. Eng.*, **52**, 1431–1441.

Papadopoulos, P. and Lu, J. (1998) A general framework for the numerical solution of problems in finite elastoplasticity, *Comput. Meth. Appl. Mech. Eng.*, **159**, 1–18.

Pérez-Foguet, A., Rodríguez-Ferran, A. and Huerta, A. (2000a) Numerical differentiation for non-trivial consistent tangent matrices: an application to the MRS-Lade model, *Int. J. Numer. Meth. Eng.*, **48**, 159–184.

Pérez-Foguet, A., Rodríguez-Ferran, A. and Huerta, A. (2000b) Numerical differentiation for local and global tangent operators in computational plasticity, *Compt. Meth. Appl. Mech. Eng.*, **189**, 277–296.

Perić, D. and de Souza Neto, E.A. (1999) A new computational model for Tresca plasticity at finite strains with an optical parametrization in the principal space, *Compt. Meth. Appl. Mech. Eng.*, **171**, 463–489.

Perić, D., Owen, D.R.J. and Honnor, M.E. (1992) A model for finite strain elasto-plasticity based on logarithmic strains: Computational issues, *Compt. Meth. Appl. Mech. Eng.*, **94**, 35–61.

Pilkey, W.D. (2002) *Analysis and Design of Elastic Beams, Computational Methods*. New York: Wiley and Sons. Ltd.

Pinsky, P.M., Ortiz, M. and Pister, K.S. (1983) Numerical integration of rate constitutive equations in finite deformation analysis, *Compt. Meth. Appl. Mech. Eng.*, **40**, 137–158.

Prager, W. (1949) Recent development in the mathematical theory of plasticity, *J. Appl. Mech.* (ASME), **20**, 235–241.

Prager, W. (1956) A new method of analyzing stresses and strains in work hardening plastic solids, *J. Appl. Mech.* (ASME), **23**, 493–496.

Raniecki, B. (2004) *Selected Fragments of Hill's Course on Solid Mechanics*, Lecture Note in Kyushu University.

Raniecki, B. and Nguyen, H.V. (2005) Favorable theoretical framework for pressure-sensitive elastic–plastic damaging bodies under finite strain, *Proc. Symp. Elastoplasticity for Prof. K Hashiguchi Retirement Aniversary, Kyushu University*, pp.115–134.

Rosati, L. and Valoroso, N. (2004) A return map algorithm for general isotropic elasto/visco-plastic materials in principal space, *Int. J. Numer. Meth. Eng.*, **60**, 461–498.

Rouainia, M. and Muir Wood, D. (2000) An implicit constitutive algorithm for finite strain Cam-clay elasto-plastic model, *Mech. Cohes.–Frict. Mater.*, **5**, 469–489.

Rubinstein, R. and Atluri, S.N. (1983) Objectivity of incremental constitutive relations over finite time steps in computational finite deformation analyses, *Compt. Meth. Appl. Mech. Eng.*, **36**, 277–290.

Rudnicki, J.W. and Rice, J.R. (1975) Conditions for the localization of deformation in pressure-sensitive dilatant materials, *J. Mech. Phys. Solids*. **23**, 371–394.

Runesson, K., (1987) Implicit integration of elastoplastic relations with reference to soils, *Int. J. Numer. Anal. Meth. Geomech.*, **11**, 315–321.

Sansour, C., Karšaj, I. and Sorić, J. (2007) On anisotropic flow rules in multiplicative elastoplasticity at finite strains, *Compt. Meth. Appl. Mech. Eng.*, **196**, 1294–1309.

Seth, B.R. (1964) Generalized strain measure with applications to physical problems, in *Second-order Effects Inelasticity, Plasticity, and Fluid Dynamics*, Oxford: Pergamon.

Shen, L. (1998) On the decomposition of deformation into elastic–plastic parts, *J. Appl. Mech.* (ASME), **65**, 1059–1061.

Shutov, A.V. and Kreißig, R. (2008a) Finite strain viscoplasticity with nonlinear kinematic hardening: Phenomenological modeling and time integration, *Compt. Meth. Appl. Mech. Eng*, **197**, 2015–2029.

Shutov, A.V. and Kreißig, R. (2008b) Application of a coordinate-free tensor formalism to the numerical implementation of a material model, *ZAMM: Zeitschrift für Angewandte Mathematik und Mechanik*, **88**, 888–909.

Shutov, A.V. and Kreißig, R. (2010) Geometric integrators for multiplicative viscoplasticity: analysis of error accumulation, *Comput. Meth. Appl. Mech. Eng.*, **199**, 700–7011.

Shutov, A.V., Panhans, S. and Kreißig, R. (2011) A phenomenological model of finite strain viscoplasticity with distortional hardening, *ZAMM: Zeitschrift für Angewandte Mathematik und Mechanik*, **91**, 653–680.

Simo, J.C. (1986) On the computational significance of the intermediate configuration and hyperelatic relations in finite deformation elastoplasticity, *Mech. Materials*, **4**, 493–451.

Simo, J.C. (1992) Algorithms for static and dynamic multiplicative plasticity that preserve the classical return mapping schemes of the infinitesimal theory, *Comput. Meth. Appl. Mech. Eng.*, **99**, 61–112.

Simo, J.C. (1998) Numerical analysis and simulation of plasticity, in P.G. Ciarlet and J.L. Lions (eds.), *Handbook of Numerical Analysis*, Vol. IV, Amsterdam: North-Holland, pp. 183–499.

Simo, J.C. and Hughes, T.J.R. (1998) *Computational Inelasticity*, New York: Springer.

Simo, J.C. and Meschke, G. (1993) A new class of algorithms for classical plasticity extended to finite strains. Application to geomaterials, *Comput. Mech.*, **11**(4), 253–278.

Simo, J.C. and Ortiz, M. (1985) A unified approach to finite deformation elastoplastic analysis based on the use of hyperelastic constitutive equations, *Comput. Meth. Appl. Mech. Eng.*, **49**, 221–245.

Simo, J.C. and Pister, K.S. (1984) Remarks on rate constitutive equations for finite deformation problems: computational implications, *Compt. Meth. Appl. Mech. Eng*, **46**, 201–215.

Simo, J.C. and Taylor, R.L. (1985) Consistent tangent operators for rate independent elastoplasticity, *Comput. Meth. Appl. Mech. Eng.*, **48**, 101–118.

Simo, J.C. and Taylor, R.L. (1991) Quasi-incompressible finite elasticity in principal stretches. Continuum basis and numerical algorithms, *Compt. Meth. Appl. Mech. Eng.*, **85**, 273–310.

Smith, I.M. and Griffiths, D.V. (2004) *Programming the Finite Element Method* (4th edn), Chichester: John Wiley & Sons, Ltd.

Sokolnikoff, I.S. (1951) *Tensor Analysis: Theory and Applications*, New York: John Wiley & Sons, Inc.

Spencer, A.J.M. (1971) Theory of invariants. In A.C. Eringen (ed.), *Continuum Physics, Vol. 1, Mathematics*, New York: Academic Press.

Steinmann, P., Larsson, R. and Runesson, K. (1997) On the localization properties of multiplicative hyperelasto-pastic continua with strong discontinuities, *Int. J. Solids Struct.*, **8**, 969–990.

Storen, S. and Rice, J.R. (1975) Localized necking in thin sheet, *J. Mech. Phys. Solids*, **23**, 421–441.

Tamagnini, C., Castellanza, R. and Nova, R. (2002) A generalized backward Euler algorithm for the numerical integration of an isotropic hardening elastoplastic model for mechanical and chemical degradation of bonded geomaterials, *Int. J. Numer. Anal. Meth. Geomech.*, **26**, 963–1004.

Tanahashi, T. (1988) *Mechanics of Continua: (5) Vector Analysis and Physical Component*, Rikoh-Tosho (in Japanese)

Tanahashi, T. (2004) *Foundations of Deformation and Flow*, Sankei-sha (in Japanese)

Truedell, C. (1966) *The Elements of Continuum Mechanics*, Berlin: Springer-Verlag.

Truesdell, C. (1955) : Hypo-elasticity, *J. Rational Mech. Anal.*, **4**, 83–133.

Truesdell, C. and Toupin, R. (1960) *The Classical Field Theories*. In S. Flügge (ed.), *Encyclopedia of Physics*, Vol. III/1, Berlin: Springer-Verlag.

Truesdell, C. and Noll, W. (1965) *The Nonlinear Field Theories of Mechanics*. In S. Flügge (ed.), *Encyclopedia of Physics*, Vol. III/3, Berlin: Springer-Verlag.

Truesdell, C. (1977) *A First Course in Rational Continuum Mechanics, Vol. 1 General Concepts*, Academic Press.

Tsakmakis, Ch. (2004) Description of plastic anisotropy effects at large deformations – Part I: restrictions imposed by the second law and the postulate of Ilyushin, *Int. J. Plasticity*, **20**, 167–198.

Tsutsumi, S., Toyosada, M. and Hashiguchi, K. (2006) Extended subloading surface model incorporating elastic limit concept, *Proc. Plasticity '06*, Halifax, pp. 217–219.

Vladimirov, I.N., Pietryga, M.P. and Reese, S. (2008) On the modeling of nonlinear kinematic hardening at finite strains with application to springback—Comparison of time integration algorithm, *Int. J. Numer. Mth. Eng.*, **75**, 1–28.

Vladimirov, I.N., Pietryga, M.P. and Reese, S. (2010) Anisotropic finite elastoplasticity with nonlinear kinematic and isotropic hardening and application to shear metal forming, *Int. J. Plasticity*, **26**, 659–687.

Wallin, M., Ristinmaa, M. and Ottesen, N.S. (2003) Kinematic hardening in large strain plasticity, *European J. Mech. A/Solids*, **22**, 341–356.

Wallin, M. and Ristinmaa, M. (2005) Deformation gradient based kinematic hardening model, *Int. J. Plasticity*, **21**, 2025–2050.

Wang, Z.-Q. and Dui, G.-S. (2008) Two-point constitutive equations and integration algorithms for isotropic-hardening rate-independence elastoplastic materials in large deformation, *Int. J. Numer. Meth. Eng.*, **75**, 1435–1456.

Watanabe, O. (1990) Objective constitutive equation and numerical algorithm in finite strain, *Trans. Japan. Soc. Mech. Eng.*, 56–524, 893–902 (in Japanese).

Watanabe, O. (1991) Objective numerical integration algorithm for rate-type constitutive equation, *Comput. Mech. '91: Theory and Applications*, pp. 1170–1173.

Weber, G. and Anand, L. (1990) Finite deformation constitutive equations and a time integration procedure for isotropic, hyperelastic-viscoplastic solids, *Compt. Meth. Appl. Mech. Eng.*, **79**, 173–202.

Wu, H.-C. (2004) *Continuum Mechanics and Plasticity*, New York: Chapman & Hall/CRC.

Yamakawa, Y., Hashiguchi, K. and Ikeda, K. (2010) Implicit stress-update algorithm for isotropic Cam-clay model based on the subloading surface concept at finite strains, *Int. J. Plasticity*, **26**, 634–658.

Zaremba, S. (1903) Sur une forme perfectionnée de la théorie de la relaxation, *Bull. Int. Acad. Sci. Cracovie*, 594–614.

Zbib, H.M. (1993) On the mechanics of large inelastic deformations: kinematics and constitutive modeling, *Acta. Mech.*, **96**, 119–138.

Zbib, H.M. and Aifantis, E.C. (1988) On the concept of relative and plastic spins and its implications to large deformation theories. Part I: Hypoelasticity and vertex-type plasticity, *Acta Mech.*, **75**, 15–33.

Ziegler, H. (1977) *An Introduction to Thermodynamics*, Amsterdam: North-Holland.

Zienkiewicz, O.C. and Taylor, R.L. (2005) *The Finite Element Method for Solid and Structural Mechanics*, Amsterdam: Elsevier.

Zienkiewicz, O.C., Taylor, R.L. and Zhu, J.Z. (2005) *The Finite Element Method: Its Basis and Fundamentals* (6th edn), Amsterdam: Elsevier.

Index

Note: Page numbers with italicized *t*'s refer to tables, and numbers in **bold** designate pages on which the terms are mainly explained.

Introduction to Finite Strain Theory for Continuum Elasto-Plasticity, First Edition. Koichi Hashiguchi and Yuki Yamakawa.

Printed and bound by CPI Group (UK) Ltd, Croydon, CR0 4YY

Printed and bound by CPI Group (UK) Ltd, Croydon, CR0 4YY
21/09/2026
14965832-0001